Advances in Intelligent and Soft Computing 141

Editor-in-Chief: J. Kacprzyk

Advances in Intelligent and Soft Computing

Editor-in-Chief

Prof. Janusz Kacprzyk
Systems Research Institute
Polish Academy of Sciences
ul. Newelska 6
01-447 Warsaw
Poland
E-mail: kacprzyk@ibspan.waw.pl

Further volumes of this series can be found on our homepage: springer.com

Vol. 126. Khine Soe Thaung (Ed.)
Advanced Information Technology in Education, 2012
ISBN 978-3-642-25907-4

Vol. 127. Tianbiao Zhang (Ed.)
Instrumentation, Measurement, Circuits and Systems, 2012
ISBN 978-3-642-27333-9

Vol. 128. David Jin and Sally Lin (Eds.)
Advances in Multimedia, Software Engineering and Computing Vol.1, 2011
ISBN 978-3-642-25988-3

Vol. 129. David Jin and Sally Lin (Eds.)
Advances in Multimedia, Software Engineering and Computing Vol.2, 2011
ISBN 978-3-642-25985-2

Vol. 130. Kusum Deep, Atulya Nagar, Millie Pant, and Jagdish Chand Bansal (Eds.)
Proceedings of the International Conference on Soft Computing for Problem Solving (SOCPROS 2011) December 20–22, 2011, 2012
ISBN 978-81-322-0486-2

Vol. 131. Kusum Deep, Atulya Nagar, Millie Pant, and Jagdish Chand Bansal (Eds.)
Proceedings of the International Conference on Soft Computing for Problem Solving (SocProS 2011) December 20–22, 2011, 2012
ISBN 978-81-322-0490-9

Vol. 132. Suresh Chandra Satapathy, P.S. Avadhani, and Ajith Abraham (Eds.)
Proceedings of the International Conference on Information Systems Design and Intelligent Applications 2012 (India 2012) held in Visakhapatnam, India, January 2012, 2012
ISBN 978-3-642-27442-8

Vol. 133. Sabo Sambath and Egui Zhu (Eds.)
Frontiers in Computer Education, 2012
ISBN 978-3-642-27551-7

Vol. 134. Egui Zhu and Sabo Sambath (Eds.)
Information Technology and Agricultural Engineering, 2012
ISBN 978-3-642-27536-4

Vol. 135. Honghua Tan (Ed.)
Knowledge Discovery and Data Mining, 2012
ISBN 978-3-642-27707-8

Vol. 136. Honghua Tan (Ed.)
Technology for Education and Learning, 2012
ISBN 978-3-642-27710-8

Vol. 137. Jia Luo (Ed.)
Affective Computing and Intelligent Interaction, 2012
ISBN 978-3-642-27865-5

Vol. 138. Gary Lee (Ed.)
Advances in Intelligent Systems, 2012
ISBN 978-3-642-27868-6

Vol. 139. Anne Xie and Xiong Huang (Eds.)
Advances in Electrical Engineering and Automation, 2012
ISBN 978-3-642-27950-8

Vol. 140. Anne Xie and Xiong Huang (Eds.)
Advances in Computer Science and Education, 2012
ISBN 978-3-642-27950-8

Vol. 141. Dehuai Zeng (Ed.)
Advances in Computer Science and Engineering, 2012
ISBN 978-3-642-27947-8

Dehuai Zeng (Ed.)

Advances in Computer Science and Engineering

Springer

Editor
Dehuai Zeng
Shenzhen University
Guangdong
China, People's Republic

ISSN 1867-5662 e-ISSN 1867-5670
ISBN 978-3-642-27947-8 e-ISBN 978-3-642-27948-5
DOI 10.1007/978-3-642-27948-5
Springer Heidelberg New York Dordrecht London

Library of Congress Control Number: 2011945154

Printed on acid-free paper

Springer is part of Springer Science+Business Media (www.springer.com)

Preface

2012 the second International Conference on Advances in Computer Science and Engineering (CES 2012) will be held in Sanya, China from January 13 to 14, 2012. CES 2012 is sponsored by International Communication Sciences Association, Hong Kong. The aim of the conference is to provide an opportunity for experts, researchers and individuals who are interesting in Computer Science and Engineering to exchange their latest research progress and discuss the directions of future research of these fields, promoting the development of the Computer Science and Engineering. CES 2012 received hot response from participates, and they prepare extended contributions in order to allow for possibility comprehensive descriptions of the issues. Furthermore, at the especial invitation of the organizing committee of CES 2012, renowned scientists in the area participate the CES 2012 and give their keynote invited lecture.

It is well known that computer system can be divided into two major categories of software system and hardware system and each system includes many aspects. The main research topics CES 2012 include are as follows: (1) Software Engineering; (2) Intelligent Computing, (3) Computer Networks; (4) Artificial Intelligence Software, This conference proceedings contain selected the papers which discussing the most interesting and up-to-date topics mentioned above in Computer Science and Engineering.

Sanya, located in south of Hainan province, is the most southern seaside city in China. The city is renowned for its tropical climate and has emerged as a popular tourist destination, also serving as the training site of the Chinese national beach volleyball team. Sanya has a lot of attractions, such as Luhuitou Park, A park containing a large stone statue above Sanya harbor telling the story of a hunter pursuing a deer to an isolated headland near the sea.

The proceedings are the result of cooperation of experts, editors and publishers. We would like to express our sincerely acknowledgements to people who contribute to the proceedings and conference. Especially, our thanks are also due to the sponsors for providing much help for the conference. At last, we wish the attendees enjoy the conference and the proceedings are useful to the researchers in the fields Computer Science and Engineering.

CES 2012 Committee

CSE 2012 Organization

General Chairs

Jesscia Chang	International Communication Sciences Association (ICSA), Hong Kong
Jun Zhang	Huazhong University of Science and Technology, China

Program Chairs

Qingyuan Zhou	Jiangsu Teachers University of Technology, China
Jianwei Zhang	Suzhou University, China

Organizing Chair

Nian Zhang	Suzhou University, China

Publication Chairs

Jun Zhang	Huazhong University of Science and Technology, China
Dehuai Zeng	Shenzhen University, China

Program Committee

Liling Sun	Wuhan University, China
Man Cao	Hubei Normal University, China
Jiang Zhu	Beijing NormalUniversity, China
Wei Li	Asia Pacific Human-Computer Interaction Research Center, Hong Kong
Xiaoming Liu	Nankai University, China
Xiaoxiao Yu	Wuhan University
Chi Zhang	Nanchang University
Bo Zhang	Beijing University
Lei Zhang	Tianjin Institute of Urban Construction
Mark Zhou	Hong Kong Education Society, Hong Kong
Liangming Zhang	Hong Kong Education Society, Hong Kong
Alireza Yazdizadeh	International Science and Engineering Center

Wenjin Hu	Jiangxi Computer Study Institute, China
Qun Zeng	Nanchang University, China
Engineer. Yong Hu	Nanchang University, China
Yan Zhang	Nanchang University, China
Yaobin Liu	Nanchang University, China
Jianren Yang	Jingdezhen Ceramic Institute, China
Yong Wang	Jiangxi Industry and Information Committee, China
Lingyan Lu	Wuhan Institute of Technology, China
Lingmei Zeng	Wuhan Institute of Technology, China
Bin Pan	Wuhan Institute of Technology, China
Hongbin Sun	Wuhan Institute of Technology, China
Li Zhang	Wuhan Institute of Technology, China
Yuntao Wu	Wuhan Institute of Technology, China
Baozhong Liu	Wuhan Institute of Technology, China
Qiong Cai	Wuhan Institute of Technology, China
Huiqing Wang	Wuhan Institute of Technology, China
Airong Sun	Wuhan Institute of Technology, China
Jun Zou	Wuhan Institute of Technology, China
Guoqing Xu	Wuhan Institute of Technology, China
Anhua Ming	Wuhan Institute of Technology, China
Chengwan He	Wuhan Institute of Technology, China
Qingchun Wang	Wuhan Institute of Technology, China
Hongyin Hu	Wuhan Institute of Technology, China
Feng Yao	Wuhan Institute of Technology, China
Tao Lv	Wuhan Institute of Technology, China

Sponsored by

Huazhong University of Science and Technology, China
International Communication Sciences Association, Hong Kong

Contents

An Exploration on the Cultivating Mechanism for Economics and Management Innovative Talents

Shi Ronghui, Guo Xiaojing, and Yao Yang

School of Economics and Management
North China Electric Power University
Beijing, China
shirh@ncepu.edu.cn

Abstract. With the information era approaching and the boom of the knowledge economy, more and higher requirements in cultivating the professionals in the economic management aspect are placed by the whole society. This paper is aimed at establishing the suitable mechanism of cultivation innovative talents through setting up innovative experimental classes on account of current problems emerged in cultivation from the domestic and foreign situations including: the objectives in cultivating innovative talents, the selecting approach, cultivating strategy, cultivating approach, rolling mechanism, incentive mechanism, guarantee mechanism and so on.

Keywords: economic management, innovative talents, cultivation mechanism.

With the development of the global economy, the international competitions become more and more fierce, which embody the competition in the ability to innovate. Whether the country could face the challenge of the knowledge economy in the future and strengthen the com-petitiveness in innovative knowledge lies in the essential part that whether we can cultivate large numbers of talents with innovative knowledge and technology. Majors of business management are more and more treasured by the society as courses with pretty practical skills. It is an important program among universities on how to cultivate innovative talents in economic management and establish efficient cultivation mechanism.

1 A Summary of Development in Cultivating Innovative Talents Home and Abroad

1.1 Research in Development of Innovative Talents Abroad

United States is one of the earliest countries that focus on cultivating innovative talents. In 1973, American National Science Foundation assisted four colleges and universities such as MIT in establishing a "Innovation Center" to test the cultivation. In the report "Country is in danger" made by the American Education Quality Committee in 1983[1], it indicated that students' innovative ability should be cultivated. Since the middle 1980s, National Science Foundation invested in five immense projects

D. Zeng (Ed.): Advances in Computer Science and Engineering, AISC 141, pp. 1–8.
Springerlink.com

Including the plan of the increasing the undergraduates' research ability. American President Obama launched the campaign "Innovative Education" whose gist is improving the innovative ability of the students in America.

After the Second World War, advanced education boomed the swift economic development. German Congress passed the "General Compendium of the Universities" which stressed that the task of the colleges and universities was preparing for the scientific methods or the arts creation used by the students for careers[2]. These years, Germany reformed the inner organizations in universities to cultivate the innovative ability.

Japan also focuses on cultivating the innovative talents. In the 1980s, Japan established the "Temporary Education Deliberation Conference", putting the cultivation to priority ever. In 1995, Japanese Science department published "The White Book of Japanese Science", which stressed the necessity of create native new knowledge and technology. Meanwhile, the government proposed the concept "Science innovation supports establishing the country" and carry out the new measures to strengthen the self-innovative basis research.

1.2 The Current Cultivation Situation in the Colleges and Universities Home

The cultivation of innovative talents in colleges and universities started up late in China. In the 1980s, the thoughts were introduced from the West, however, it was not taken seriously by the society then. Since 1990s, with the promotion of "rejuvenate our country through science and education" and "sustainable development" strategies, the Education ministry proclaimed the " national rejuvenation action plan on education in the 21st century", and the "decision of deepening education reform under the comprehensive promotion of quality education" issued by the central committee of the Central Committee of the Communist Party of China about, highly explicated that higher education should cultivate specifically with innovative spirit and practice ability of senior specialized talents.

However, no matter home or abroad, the researches on the mechanism aiming at cultivating the innovative talents in economic management field are few.

2 The Existing Problems in the Cultivation of Our County

These years, colleges and universities in our country have explored a lot in the aspect of cultivating innovative talents in economic management field, however, there have been some shortcomings as a result of the late start, the aspects are specifically as follows:

2.1 Focus on the Knowledge Spread, However Neglect the Ability Cultivating. [3]

In most conventional economic management major trainings, the theoretical cultivation was focused especially, yet the practical operations were neglected. Examination-Oriented Education is the dominant trend in the education in which students were passively taught the knowledge teacher taught in a unilateral spoon-feed method that students cannot learn and think independently, all of these led to the unilateral pursuing the scores of the studies while ignored other abilities training.

2.2 Bounds between Majors Were So Limited, Giving Rise to the Limitation of the Ability

Though optional courses and minor courses were added in many colleges and universities, students were limited within the major range resulting from the heavy workloads in the economic management majors, ignored the cultural and social knowledge. These exert influence on the innovative ability cultivation.

2.3 The Deficiency in Scientific Research and Practice Blocked the Way to the Cultivating Innovative Abilities

Most colleges and universities pay little attention to the practical education resulting in the few opportunities to directly participate in the scientific projects and programs, the undergraduates do not have supervisors until the last school year dealing with the dissertations because of the unsound mechanism of supervisor distributions. Furthermore, few opportunities were given to the students to practice in the corporations. All these factors prevent the cultivation of the innovative talents.

3 Mechanism of Cultivating the Innovative Talents in Economic Man-Agement Field

It is crucial to establish a set of mechanism that cultivates the outstanding talents with all round qualities and society adapted abilities, the cultivation can be realized by setting up innovative experimental classes in the economic management filed.

It is inseparable between the economic and managerial majors on account of the broad application of economy as a basic course in social science while the management uses a lot of economic basic knowledge to do research on the organization management and enterprises management. Owing to the similarities between both courses that they are foundation courses with overlaps, it is very beneficial to set up the innovative experimental classes bringing the student majoring in economy and management together using the cross communications.

3.1 The Cultivation Goals of the Innovative Talents in Economy and management

Explicit the cultivation goals is to clearly answer what is innovative talents, chairman Hu Jintao addressed on the 13th academicians gathering of Chinese Academy of Sciences in June 16th ,2006, on which he explained comprehensively: first, have a noble life ideal ; second, possess aspiration to pursue truth and courage; third, qualified with rigorous scientific thinking ability and comprehensive scientific culture quality; fourth, have the solid foundation of professional; fifth, fulfilled with the strong teamwork spirits; sixth, form a dependable earnest work style. [4]

Combined with the education ministry of personnel training target, we can set the management of innovative talents training as a target of “generous scientific literacy, the strong the humanities, and excellent engineering quality and innovative talents" as the management of innovative talents training target.

3.2 The Selection Methods of the Experimental Classes

Select the students in each major to cultivate in terms of innovative talents cultivation mechanism in accordance with the major in the economy and management field. New students with the score exceeding the key score line in their province or city and at the same time rank top in certain proportion were provided with qualifications to register. After the volunteering enrollment and the first selection organized by the school, the leader group organized by the school interview the primarily selected students to finally select 30 to 40 from them.

3.3 The Cultivation Strategy of the Innovative Talents in Economy and management

We should follow the "solid foundation, broad caliber, strong ability" principle while combining the characters of these major to tailor the independent innovative talents cultivation mechanism.

1) Implement the tutorial system among the undergraduate students. Each student in the experimental class was allocated with tutor from the first school year, implementing the guidance of personal cultivation under the tutor. Each student could receive the tutor's help and guidance in the prospects such as selecting courses, choosing majors, reading the materials, plan for the study process, learning approach and managing the time. Tutors are the locomotive professors in each course and major with the recommendation of the college or the university. The broad range of the economic management major leads to the results of large amount of tutors' research projects and the broad range of the research field. It is a platform for the students to research scientifically and innovatively as well as a advantage which the colleges and the universities can utilize to encourage the students' participation in the tutors' projects as soon as possible and cultivate the innovative ability and the scientific research quality.

2) Carry out the cultivation pattern with the combination of in-class education and group education. In-class education leads the students to think and cultivate their exploring spirits through exploring teaching and learning methods with the main form of teachers' education. While the group education cultivate the teamwork spirits and the innovative ability by dividing the students into several groups carrying out the same task or cooperating with one another with an appropriate division of labor through distributing the projects.

3) Encourage and conduct the students to participate in all types of competitions. We should encourage the students to participate in all kinds of academic activities and scientific competitions besides arranging their free projects research or the tutor's projects' research, meanwhile encourage them to attend the foreign academic conferences by creating conditions positively for their short-term learning in the foreign high-leveled universities to broaden the horizon.

4) Arrange straight transaction for the inno-vative talents. The talents with the qualification of being sent on recommendation can choose the major during the postgraduate period according to their own desire. This also provides broader horizon to the experimental classes' students in the selecting of the major and the specific personality.

3.4 The Cultivation Pattern of Innovative Talents in Economic and Management Field

Cultivating the innovative talents is the core issue of all the educational activities and the setting up of the curriculum in accordance with the course in economy and management with the basic 4-year schooling. Educational activities are divided as foundational education section, professional major learning section and the comprehensive training section.

1) In the basic education pattern, enhance the students' solid foundation of the major. In the first two years of the students in the experimental classes, we should use the method of general education regardless of major. General education should cover foreign language, computer application, methodology, personality development, communication skills and economics, management etc., all of its subject categories according to cultivate broad-minded, through a series of basic courses to students in class and lay solid knowledge foundation for professional learning reserve. As for the students in experimental classes, executes tutorial system from grade one. For each student with tutor, implement personal cultivation under the tutor's guidance. Basic learning section, enhance the foundation of the students.

In the foundational education modules, we should design experiment sections in economic and management majors by organizing the students to visit the corporations, communicate with employees and managers, or invite the entrepreneurs to the lecture or discussion aimed at improving the knowledge of the different types of enterprises and institutions of the operation process and practical management majors of knowledge and arousing their interest in learning.

2) In the professional learning modules, cul-tivate students' professional qualities. In the classes of innovation based on a two-year general education, we could guide the students confirm the major into course according to their own course personal interest and specialty.

In professional learning modules, we can make full use of the training set up two-year broad caliber, encourage students to participate in the research with the tutor, exposed to the frontier area early. In addition, students will con-tinue to strengthen professional experimental learning, guaranteed the continuity of the study of foreign language, so that the students in this specialty area can learn the highest level of international latest technology and knowledge. Meanwhile, continue the research projects pro-posed by the tutor's research topic, students are required to list literature research progress and will regularly report within. the class and the grade.

In the practice module, direct experimental courses at each major in accordance with the necessity in each major. And management majors opened experiment courses include: financial management, marketing, marketing experiment, experiment of human resource management, management information system for experiment and I e-business experiment, with the approach in practicing professional courses to deepen the understanding of the students' knowledge.

3) In the comprehensive training modules, improve the students' comprehensive quality. After the foundational education of students in general and professional knowledge ,in the third stage, we should concentrate on the students in simulation experiment, the practice of enterprise and undergraduate graduation thesis of comprehensive training completion.

In the comprehensive training modules, we should use the approach of comprehensive to train the students in terms of the major know-ledge links in the courses , including roles and post collection of enterprise management experience in a simulated, ERP sand simulating, etc. Owing to the economic and management school covering large amount of majors, we can definitely satisfy the need in the simulative positions. In this kind of simulation experiment, some students form a team, rationally divide labor between team members and communicate effectively. Simulation experiments provided for students is under the condition of complete market competition of enterprise management environment, students must use the comprehensive knowledge to grasp the information of competitors, observation, analysis of the competition rivals competition strategy. This is helpful in cultivating students and management majors of enterprising spirits and sense of competition, improving its spirits of cooperation and coordination ability which can be used repeatedly decision-making of economic management of knowledge and methods in each round of simulation and test, and improve the management of students' comprehensive ability. [5]

In addition, establish innovation practice base, creating conditions for students in enterprise practice learning through cooperation with enterprises. Guide students in summer internship to combine the theoretical knowledge to solve practical problems. The graduation thesis writing is important at this stage of a comprehensive training innovative students of class, due to early intervention in project, students can select in accordance with the project, start the thesis in the fourth grade ahead of time. This allows the students to enter the real projects in the early times, strengthen the ability to combine theoretical knowledge to practice, cultivate the scientific quality and enhance the professional knowledge.

3.5 The Rolling Mechanism of the Innovative Experimental Classes

Using the rolling method of experimental innovation to students of class, main steps are as follows:

1) Each student of the class is graded each semester including the comprehensive assessment of morality, intellect and the physics, the results are determined in coexistence with cultural courses results, the results given by the tutor and the results in all kinds of competitions. In the first year and second year implement the mechanism of eliminating.

2) At the beginning of the second year, students with excellent results in study in the parallel classes of the same grade with good materials proving development potential can be selected through comprehensive tests according to respective situations.

4 Incentive Mechanism and Guarantee Mechanism in Cultivating the Innovative Talents

4.1 Good Incentive Mechanism Can Promote Innovative Talents into Full Play of Creation

The measures are mainly as follows: ①experimental classes' and normal students can attend school for evaluation of graduate student, scholarship, exchanges programs both at home and abroad, but the quota of students is higher;②can be absorbed in tutor's scientific research work, ③students in class is open to the laboratory, ④can enroll in the elective courses;⑤ can enjoy treatment of graduate student in the borrowing books aspect ⑥classes have certain special activity funds.

4.2 The Cultivation of Innovative Talents Need Effective Guarantee Mechanism of Support

Specific measures as follows: ① Clarify experimental teaching workload. For experimental teaching in class, the teacher can press allowance for post-graduate work in class; for as the guidance of teachers and students in class by guiding postgraduate courses (in) the computing workload. ② Divide the management responsibilities. The university is to be performed by the student affairs committee leading, and related departments of the university departments attended classes of leading group, "the training class management model of major issues and examination, and coordinating the work between different departments.

5 Conclusion

Talents cultivation is the essential foundation of modern development and progress of the society, while the talents in the economy and management are the key promotion to our country's economic boom. In other word, it exert great influence on our social economy whether the cultivation of the talents in the economy and management field succeeds. Only with the proper cultivation mechanism in talent development can we breed the talents in the economic and management field. This paper establishes a set of innovative economy and management talents oriented mechanism through setting up experimental classes to guarantee the talents quality and meet the demand of the country' development.

References

[1] Zhao, C.: Research on the College level Innoative Education at Home and Abroad. Higher Agricultural Education 7, 10–14 (2000)
[2] Zhao, C.: Research on the College level Innoative Education at Home and Abroad. Higher Agricultural Education 7, 10–14 (2000)

[3] Liu, S., Huang, J., Yang, Y., Ding, Z., Zhang, F.: Mechanism to bring up innovative talents at colleges and universities. Journal of Fujian Agriculture and Forestry University (Philosophy and Social Sciences) 11, 69–72 (2008)
[4] Hu, J.: China academy of 13th meeting and Chinese academy of the 8th meeting of speech, since the sixteenth congress important documents, June 5. The central literature press, Beijing (2008)
[5] Li, H.: Upon the Construction of Experimental Teaching System of Business Management. Journal of Ningbo University (Educational Science Edition) 2, 123–126 (2009)

Current Situations and Suggestions about Bilingual Education in Chinese High Education

Xin Luo

Department of Accounting
Zhengzhou Institute of Aeronautical Industry Management
Zhengzhou, Henan Province, China
luo_xin33@yahoo.com.cn

Abstract. With the economic globalization and education reform, the importance of bilingual education emerges step by step. This is just the focus of this paper: What are the present conditions in China and how to improve this new teaching method. Firstly, the growth of bilingual education in domestic and in foreign was discussed. Next, this paper illustrated some significant issues in Chinese high education, including lacking of professional teachers, inadequate appropriate text books and so on. Finally, some advices were presented, such as seeking high level lecturers. The major conclusion is that China education department should increase the input to strengthen bilingual education levels so that Chinese students can be better to serve for multinational companies, which is helpful to Chinese economic development.

Keywords: High Education, Bilingual Education, Professional Teachers of Bilingual Education.

1 Introduction

The foreign empirical testing about bilingual education started from the last century, in the end of the last century the relevant study reached maturity and brought out a series of principles which are helpful to students' growth. Bilingual education in Canada and Singapore has achieved great success, and has extensive international influence. In Canada, the government adopts official languages bilingualism (English and French) and multiculturalism policy. Also, the teaching mode is the French immersion teaching. In Singapore, the whole education system (from primary schools to universities) is just bilingual education including mother tongue and English. Since then, the "bilingual education" grows vigorously over the world.

In the early 1990s, China began to take bilingual education experiment. From primary schools to universities, the whole country started to adopt this type of teaching.

However, in many these institutions, the situations are fictitious and this phenomenon is more serious in universities. In many universities, teachers just only use these two kinds of languages (Chinese and English) to teach but cannot communicate these two languages. On the other hand, several universities indeed can understand the spirit of bilingual education and achieve some good results. Nevertheless, there are still several important issues in the whole education system of China.

D. Zeng (Ed.): Advances in Computer Science and Engineering, AISC 141, pp. 9–12.
springer link.com

2 The Situations of Universities' Bilingual Education in China and Major Problems

China Ministry of Education asks all universities should use bilingual education (Chinese and English) in public subjects and professional subjects in the bachelor-degree education. In addition, among three years the percentages of bilingual subjects should take 5% to 10% in all subjects. Moreover, the relevant department of government puts the level of bilingual education into the rating index system of the bachelor-degree education. However, since many reasons the results and levels of bilingual education are not very good in a lot of universities. In summary, vast majority of Chinese universities have the following issues about bilingual education in the different extent.

2.1 Lack of Professional Teachers of Bilingual Education

Bilingual Education has a high requirement to the professional teachers. They must have a series of professional abilities, such as profound professional knowledge, solid native language, outstanding English pronunciation, fluent oral English ability and good ability of using English language. However, according to the current situation of our country, the number of appropriate professional teachers is too little. Some teachers have the good oral English ability but have no the adequate relevant professional knowledge; Some people are good at the subjects but know a little about the relevant English. As a result, lack of professional teachers is one urgent problem and one approach must be found to solve this issue.

2.2 Lack of Appropriate Text Books

Bilingual education is asked to use foreign original edition books and relevant reference books. However, the different books have the different contents in levels and qualities, which lead to many difficulties when universities do the choice about appropriate text books. On the other hand, the issue of books' prices increases schools' and students' economic burdens.

2.3 Absence of Complete Management Mechanism of Bilingual Education

For one thing, the funds of using bilingual education are inadequate. Moreover, the attitudes of universities to bilingual education are different. In addition, many universities have no the relevant approaches to monitor and value the levels of bilingual education. Finally, the arrangement of relevant subjects is not reasonable in many universities.

3 Several Suggestions about Bilingual Education

3.1 Getting High-Level Professional Teachers

The teachers who use bilingual education method must have the following aspects of abilities. Firstly, they must have the ability of knowledge integration. The teachers should be able to do the organic integration of two kinds of cultures and to transform these two cultures. Also, they must understand the principle system about relevant subjects and have the ability to treat the differences between the two languages and cultures. Secondly, they should have the ability to apply modern educational technology. This aspect requires the relevant teachers can make use of multimedia technology, including courseware making. Thirdly, they should have the favorable communicable ability. The optimum interaction only can be gotten through the efficient communication between teachers and students, which can increase the quality of bilingual education. Also, the ability of thinking problems and solving problems using English can be improved by this point. Fourthly, they must have the good ability to control the teaching process. In general, the following four methods of bilingual education can be used. The first is lecture method. The second is game method. For example, such games as "to see who know more words" and "crossword" can be used to get a better classroom atmosphere. The third is self-studying and discussing method. One teaching programme can be used, that is questions, reading, discussing, presenting and summarizing. This method can train several students' abilities, including self-studying ability, analyzing ability and presenting ability. The last is writing method. This method requires students to write one essay using the relevant knowledge.

Nevertheless, not all universities have the appropriate teachers. Therefore, according to different situations, each school can choose one or all the following methods to improve the levels of professional teachers. Firstly, they can invite some foreign teachers with solid professional knowledge to teach bilingual courses. In addition, the PhDs and masters who have the oversea study experiments are good choice as well. Secondly, they can select outstanding teachers from the people who have the superior professional knowledge and English level. These teachers should be trained by several measures, such as studying the lecture of foreign experts, special training in domestic well-known universities, further special training in overseas and so on. Thirdly, they can construct source channels of professional teachers. The "double degree" method can be used to foster the inter-disciplinary talent who has the super ability in both professional knowledge and relevant foreign language.

3.2 Choosing the Reasonable Teaching Materials and Teaching Resources

Firstly, universities should know the range of choice of relevant text books. In addition, the professional teachers should pay more attention to the motivation of books' publishing and check the index of published books in some publishing house. Moreover, universities should ask the relevant teachers to analyze the contents of relevant books and study the special logical structures when they choose the original native text books. It is commonly thought that the good books should be written by famous writers, published by famous publishing house and used by well-known universities. On the other hand, if it was difficult to find the appropriate original native books, the professional teachers should be able to select the relevant contents from some original native books to construct some appropriate lecture notes. Secondly, universities should set up special funds to do bilingual education constructions, which can be used to buy some original native books and relevant references for the professional teachers. The last but not least, the relevant departments of government should support the importing and publishing of foreign text books and relevant references.

3.3 Establish and Perfect the Management Mechanism

Firstly, universities should establish incentive mechanism of bilingual education. Establish special funds of bilingual education has been put on the agenda in order to support the growth of bilingual education. In addition, universities should formulate corresponding encouraging policies and incentive mechanism. Secondly, standards of bilingual education should be set up as soon as possible. Universities should suggest explicit requirements to relevant text books, teaching frameworks, assignments and examinations. This approach can avoid the arbitrary and blindness of bilingual education. Thirdly, universities should monitor and evaluate the bilingual education effects. Universities can take several methods to value the teaching quality of relevant bilingual subjects, including constructing the valuing teams, periodic attending a lecture, and students' questionnaire. At last, universities should set up reasonable curriculum system of bilingual education. The bilingual subjects cannot be opened willfully and universities should choose the subjects which have the characteristics of rapid development of discipline and the international generality, such as Finance, Medicine, Information Technology and so on. At the mean time, universities should pay more attention to the coherence and the cohesion between bilingual teaching subjects and other courses. Prior to the bilingual subjects, the College English subjects should be strengthened. In addition, some other subjects should be opened, for example, Professional English and Professional Foreign Material Selections. Moreover, universities should choose the ideal opening time of bilingual subjects. In general, the best opening time of that should be the third grades. The College English should be opened in the first two grades and the Professional English should be opened in the fourth semester, which is helpful to the formal bilingual subjects teaching.

4 Conclusion

In summary, to improve the bilingual teaching quality is a gradual process and is a long-term and arduous task. As we all know, there is a considerable gap between the current situation of bilingual education and high levels. As a result, the relevant people must know the issues of bilingual education well and try their best to take measures to improve the bilingual teaching and make it to a new level.

References

1. Zhang, J.S.: The essence and characteristics of the case studying. Zhongguo Jiaoyu Xuekan (January 2004) (in Chinese)
2. Wang, J.J., Luo, L.: Some problems in the bilingual education. Beijing Daxue Xuebao (May 2007) (in Chinese)
3. Zeng, Y.B., Li, H.Q.: University's bilingual teaching: mode research and definition. Journal of Nangchang College (January 2005) (in Chinese)
4. Shi, Y.J.: The thinking of university's bilingual teaching. Liaoning Education Research (July 2005) (in Chinese)

Training Innovation Ability of Students in Practical Teaching of Regional Planning

Wei Yali[1,2], He Yuxiao[2], Gao Xuesong[1], and Gan Wenzhi[3]

[1] College of Resources and Environment, Sichuan Agricultural University, Chengdu, Sichuan ,China
weiyali@imde.ac.cn

[2] Institute of Mountain Hazards and Environment, Chinese Academy of Sciences and Ministry of Water Resources, Chengdu, Sichuan, China
heyuxiao@imde.ac.cn

[3] College of College of Political Science and Law, Sichuan Agricultural University, Yaan, Sichuan, China
329570927@qq.com

Abstract. In the background of economic globalization and regional integration, research and study on regional planning for the national suitably development of new concepts and theories is a major issue now. Regional planning is the important point and application field of discipline theory and practice. Based on the teaching experience in regional planning, the paper constructs practical teaching contents and practical teaching models, proposes to train students' innovation ability through the innovation of practical teaching model, and promotes students' cognitive ability to social reality and organization and coordination ability among students. The practice has proved that the practical teaching model of innovative regional planning can be achieved a win-win situation.

Keywords: Practical Teaching, Regional, Planning Innovation.

1 Introduction

With the development of economic globalization and regional integration, regional planning has become an important basis and means for a national macro-control of the regions. Experts and scholars are currently working on new ideas and theories of regional planning which are suitable for national development. Regional planning put the region and space as the main object of study, foe instance, human geography focus study on the regional spatial variation regulations and material elements distribution of regional space. Moreover, regional planning is the important point and application field of discipline theory and practice. Therefore, the study of innovative models for teaching practice of regional planning is particularly important. University practical teaching links are highlighting the dominant position of students and promoting the combination of theory and practice and making for training the cognitive, development and innovation abilities of students[1]. Training students'

D. Zeng (Ed.): Advances in Computer Science and Engineering, AISC 141, pp. 13–17.
Springerlink.com

creative thinking has become the trend of education around the world [2-3]. How to construct regional planning practical teaching system of technology application and innovative ability to train the innovative talents for knowledge and economic development is a new topic of deepening the practical teaching reform of colleges.

2 Innovation of Practical Teaching Models

2.1 Contents System Construction of Practical Teaching

Practical teaching is the essential part to improve the overall quality of students and practical ability, and the most effective way to train innovative spirit and practical ability of students[4]. Practical teaching contents of regional planning mainly are divided into two parts[5]:

1) The first part mainly focus on the system analysis of regional analysis for resources and environment bases, economic and social backgrounds, technology support conditions, the level of overall regional development, regional advantages and division of labor and regional industrial structure as the basis and foundation for regional planning.

2) The second part mainly focus on regional planning, illustrating the main theory, methodology and procedures and focusing on regional industrial development and distribution, land use and regional governance, regional infrastructure, urban systems and the ecological environment etc.

Regional analysis is the basis of regional planning which is the application result of regional analysis. The course content of two parts is complementary and indispensable. Based on the regional analysis and planning theory, the practical teaching draw up various regional planning to suit the different regional development.

2.2 Construction of Practical Teaching Models

After understanding of practical teaching contents for regional planning, the focus is to construct practical teaching model. Through years of teaching experience, practical teaching model of regional planning can be summarized as "two models", "three coordinations", "two bases." As shown in the figure 1. The teaching model of regional planning is divided into on-campus and off-campus "two models": the school should provide students with experimental teaching platform, including laboratory hardware devices (equipped with computers and measuring instruments); planning mapping software (hand-drawn, computer drawing of the CAD software, CPS Hongye planning software, the South CASS software, GIS software, etc.); experimental instruction (teachers and experimental instructions). The off-campus practice teaching platform mainly includes the establishment of practical teaching base to provide students with practical and effective workplace; the training of planning team where students are divided into different planning groups with clear division of labor to develop their planning, coordination and organization capacity; the test of planning results is mainly to provide information for decision-making to the local government with the planning results of the students.

Three co ordinations for practical teaching model of regional planning are reflected in the integration and unity of production, teaching, and research. The relationship within them is that teaching is the main part and research is the force and production is guarantee. The scientific research should be promoted by teaching and the scientific research can advance the teaching. The hemopoiesis function should be increased by their production to promote the comprehensive development of the whole regional planning practice.

The creation of innovative practice teaching base provide students with basic experimental conditions and site conditions which can stimulate the innovative spirit of students and develop their planning practical and creative abilities. In addition, combined with the government and the full use of academic research on discipline advantages and expertise teachers, colleges can set up research bases. By colleges and enterprises training, not only train the planning practical ability and innovation of students, but also improve the employment rate of graduates and defuse the employment pressure through the co-operation.

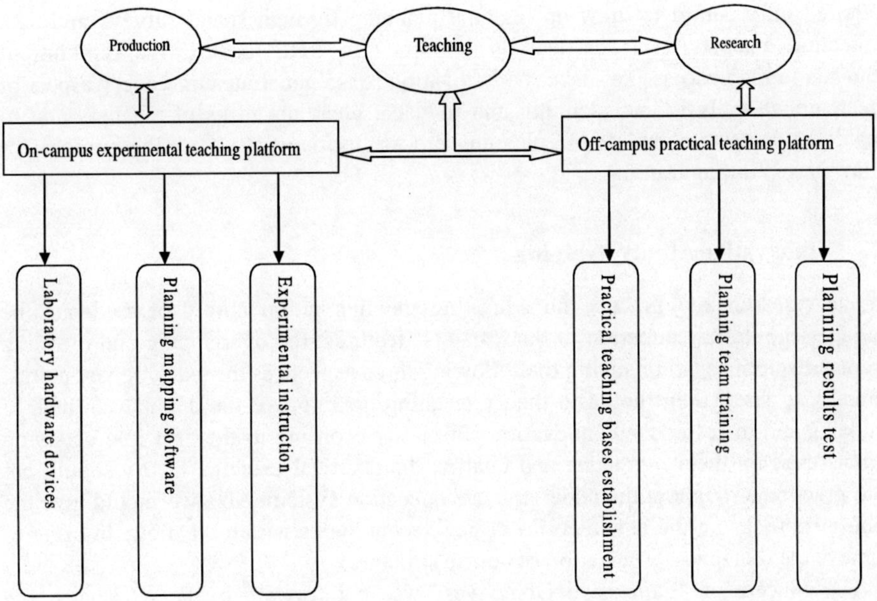

Fig. 1. Construction of Practical Teaching Model of Regional Planning

3 Innovation of Students' Synthetic Ability

3.1 Innovative Thinking Training

The key to innovation ability is the cultivation of innovative thinking through hands-on training. To help students master the creative way of thinking regional planning, colleges should offer some courses and seminars about creative thinking, creative skills and case analysis of creating success. Through the second classes, extra-curricular interest groups, research groups, technology collaboration groups and innovation training bases to carry out relevant innovative / creative knowledge learning. So that they can learn and master the self-breaking way of thinking, that is, creative techniques, for instance, diffusion discovery techniques, centralized integrated techniques, training techniques of creating awareness, etc. Moreover, innovative / creative thinking and capacity should be developed by analog simulation and practice training[6].

In the practical guidance of regional planning, the first is to train students' innovative thinking for overall regional analysis. Regional planning is a more integrated discipline that requires a clear understanding of the regional society, economy, natural conditions, regional development in the past, present and future prospects, and the location of large regions and the relation with surrounding areas. Therefore, it requires students to use various thinking methods and summarize the required information to draw up regional planning through social surveys and data collection analysis. It is particularly important to train the creative thinking of students in the process. To make the innovation sense penetrates into every aspect of the regional analysis and planning and to make clear the tasks of teachers and to enable students to complete the planning process by themselves are the purposes of innovative thinking training.

3.2 Innovative Ability Training

In the research and practice on educating the innovation ability of students, the colleges should pay attention to their majors' features, the overlapping and blending among disciplines, while taking the following three aspects as the teaching key points, namely, to strengthen the basic theory teaching, to improve the applied abilities of students and to foster their innovation abilities. According to the principle of laying equal stress on theory, practice and quality, the practical teaching theory should put into practice throughout the undergraduate education system. Also we should give the long-term focus on the new theories or new technologies within this field, in order to achieve the dynamic optimization of course structure.

The concept of "Expressive Objectives" was put forward by E. W. Eisner, the American curriculum theorist. An expressive objective describes an educational encounter: it identifies a situation in which students are to work, a problem with which they are to cope, a task they are to engage in; but it doesn't specify what from that encounter or task they are to learn[7-8]. Its characteristics are well illustrated by the practical teaching of regional planning, which is completed in all sorts of "encounters". During the case study of regional planning, students learn the whole process of drawing a planning and how to analyze its advantages and disadvantages. Often, with the aim of stimulating students' creativity, they are allowed to do their jobs with freedom and then make their works particular and unique. Rightly in this sense, it succeeds in circumventing the result of standardization actively, then tolerates the various but individual performance of students. Meanwhile, it also manifests the initiative and creativity of each learner by motivating his personalized performance.

4 Conclusions

By constructing an innovative teaching model, regional planning can impel the students to develop their innovative thinking and creativity in the process of practical teaching activity. Also, it can cultivate their cognitive competence towards the social reality and their abilities to cooperate or coordinate with others. Furthermore, by the application of innovative teaching model, a win-win situation can be achieved: firstly, to the students, it can enhance their comprehensive abilities, and lay a solid foundation for them to enter the society in future; secondly, to the schools, it can give the curriculum development a virtuous circle, and provide schools with opportunities to cooperate with local organizations.

References

1. Lingde, J.: On Social Synthesis Practice Teaching for Major of Urban Planning. Journal of Architectural Education in Institutions of Higher Learning 17(5), 114–116 (2008)
2. de Stavenga Jong, J.A., Wierstra, R.F.A., Hermanussen, J.: An exploration of the relationship between academic and experiential learning approaches in vocational education. British Journal of Educational Psychology 76, 155–169 (2006)
3. Shreeve, A.: A phenomenographic study of the relationship between professional practice and teaching your practice to others. Studies in Higher Education 35(6), 691–703 (2010)
4. Xianzhong, D.: The Content, Status and System of Anatomical Sciences and Technology Disciplines, pp. 78–84. High Education Press, Beijing (2003)
5. Gonghao, C., Qingquan, W., Kewei, L.: Regional Analysis and Planning, pp. 1–3. High Education Press, Beijing (2006)
6. Xiao, Y., Qian, J.P., Chen, X.: The Practical Teaching System of the Construction Technology Application and Innovative Ability Training. Chinese Art Education 1(20), 70–73 (2009)
7. Hong, D., Peipei, T.: Higher Vocational Courses Innovation and Practical Teaching Value Transition. China Agricultural Education (1), 62–64 (2010)
8. Eisner, E., Peshkin, A. (eds.): Qualitative Inquiry in Education, p. 7. Teachers college Press, New York (1990)

Simulation System of Examination Score Analysis Based on an Improved Apriori Algorithm*

Li Xiang

School of Computer Engineering
Huaiyin Institute of Technology
Huai'an 223003, China
LargePearStudio@gmail.com

Abstract. This paper introduces an improved algorithm based on Apriori algorithm to analysis examination score. The new algorithm is implemented with vertical data layout, breadth first searching, and intersecting. It takes advantage of the efficiency of vertical data layout and intersecting, and prunes candidate frequent item sets like Apriori. Finally, the new algorithm is applied in simulation system of examination score analysis. The result shows that the relations will be affected by the students' grades, and it can be applied in guiding students' study and teachers' teaching practice.

Keywords: apriori algorithm, score analysis, association rules.

1 Introduction

Apriori algorithm [1-2] is data mining association rules algorithm, put forward by Imielinski, Agrawal and Swami. Apriori algorithm is a kind of data mining algorithm which is based on the horizontal data representation and breadth first search .This paper goes into Apriori algorithm through the analysis of student score, puts forward a new algorithm that adopts intersecting count to implement breadth first searching on database of vertical data layout. The new algorithm takes advantage of the efficiency of vertical data layout and intersecting, and prunes candidate frequent item sets like Apriori. It is applied in emulation analysis of student examination score, the result can show the relations that affect the students' grades among related subjects. It can be applied in guiding teachers' teaching practice and students' study.Ease of Use.

2 Basic Conceptions

2.1 Association Rules[3-4]

Definition 1: Let $I=\{ i_1 , i_2 ,\cdots, i_m \}$ be a collection. D is a transactional database. D is the collection of transaction T, and T can be seen as the collection of item, and

* This work was supported by the Natural Science Foundation of Huai'an Technology under Grant HAG2010068.

D. Zeng (Ed.): Advances in Computer Science and Engineering, AISC 141, pp. 19–25.
Springerlink.com

$T \subseteq I$; TD is the only identifier of transaction T in transactional database. A is a collection, if $A \subseteq T$, and then we say T includes A. An association rule is a formula as $A \Rightarrow B$, $A \subseteq I$, $B \subseteq I$, and $A \cap B = \Phi$.

Conditions fill the association rule $A \Rightarrow B$:

- With Support $S\%$. $S\%$=support $(A \Rightarrow B) = P(A \cup B)$, namely the percentage of transactions in D that contain $A \cup B$ at least is $S\%$.
- With Confidence $C\%$. $C\%$ = confidence $(A \Rightarrow B) = P(B|A) = P(A \cup B) / P(A)$, namely the percentage of transactions in D contain A that also contain B at least is $C\%$.

Support is the percentage of transactions in database. Confidence is the intensity of the rule. Itemsets that satisfy minimum support (min_sup) is called frequent itemsets. Rules that satisfy both minimum support (min_sup) and minimum confidence (min_conf) are called strong association rules, usually noted $A \Rightarrow B$ $(S\%, C\%)$.

2.2 Apriori Algorithm

Apriori Algorithm is a basic algorithm that mining generates Boolean association rules needs frequent itemsets. Apriori algorithm mainly uses a circular system to rake though one gradation in turn to complete frequent itemsets mining. The main idea of this circular system is to generate $(k+1)$- itemsets from the k- the frequent itemsets. Detailed procedure: First, find frequent 1- itemsets, called L_1; then, use L_1 to mine L_2, namely frequent 2- itemsets; the circular system terminates when it can't find more frequent k- itemsets. Mine one gradation L_k needs traversing the whole database once.

Apriori Algorithm can be described as follows:
Input: data set D, minimum support min_sup
Output: frequent itemsets L in D
Steps:

```
L1 = find_frequent_1 - itemsets (D) ;
For ( k = 2; Lk-1 ! =Φ; k + + )
{Ck = apriori_gen (Lk-1 ) ;
For each transaction t ∈D{
Ct = subset (Ck , t)
   For each candidates c ∈Ct
c. count + + ; }
Lk = { c ∈C | c. count≥min_sup }}
L =L ∪ Lk;
```

Function apriori_gen
Input: L_{k-1} frequent(k - 1)_itemsets
Output: candidate itemsets C_k
Steps:

```
for each itemset  l1 ∈Lk-1
```

```
for each itemset  l2 ∈L(k-1)
if (l1 [ 1 ] = l2 [ 1 ] ) ∧l1 [ 2 ] = l2 [ 2 ] ) ∧··· ∧ (l1 [ k -1 ]≠l2 [ k - 1 ] ) {
c = l1∞l2 // generate candidate itemsets
if  has infrequent subset ( c,L(k-1) ) then
delete c //delete un-frequent itemsets
else add  c  to  Ck }
return Ck
```

3 Improved Apriori Algorithm

Through research we can discover that Apriori algorithm has important features as follows:

a) It needn't to judge all k *(k-1)* dimensional subsets when judging if the *(k-1)*-dimensional subsets of *k*-dimensional candidate itemsets R are included in L_{k-1}. It just needs to judge the other k-2 subsets because the two subsets, $\{ R [l] \mid 0 \le l \le k -2\}$ and $\{ R [l] \mid 0 \le l \le k - 1, l \ne k - 2\}$, which connect into candidate itemsets R are included in L_{k-1}.

b) If every transaction *T* and itemsets in transaction are sequenced in dictionary order, once two *(k-1)*-dimensional frequent itemsets I_x and I_y can't be connected, then all the itemsets after I_x and I_y not satisfy connection conditions, do not have to test the connection to determine.

By comprehensive application of these two important features, we can propose improved Apriori algorithm implemented with vertical data sets to perform breadth first searching and intersecting. Store itemset and item according to the dictionary order. That is, first dig out all the frequent k-itemset and record them, then (k-1)-same as the previous item of itemset connect to a candidate (k+1)-itemsets, when connecting use the important property 2 to reduce the number of judgments; use the important property 1 to pruning judge on candidate itemsets, if the candidate itemset does not meet the properties of Apriori ,then cut; if meets, then directly count support with intersecting.

```
// use breadth first searching
for ( k = 2; L(k-1) ≠Φ; k + + ) {
for all Xi ∈L(k-1) {
  for all Xj∈L(k-1) ( j > i) {
  if (Link Item (Xi , Xj, k - 1) ) {
R =Xi∪Xj ;
  if ( not Trimed (R, L(k-1) ) ) {
Tidset (R) = Tidset (Xi ) ∩Tidset (Xj ) ;
if ( Tidset (R) ≥minsupp) {
L =L∪R;
Lk =Lk∪R; }
              }
```

```
        } else
          break;
      }
   }
}
```

//use Link Item ( ) function to determine if X_i and X_j can be connected to a candidate (k+1)-dimensional itemset. As long as their forward (k-1)- items are all the same, k-th-item of the former is less than the latter, then X_i and X_j can be connected to a (k+1)-dimensional candidate itemset. Specific function as follows:

```
bool Link Item (Xi , Xj, k) {
for (i= 0; l < k - 1; i+ + )
  if (Xi [l] ≠Xj[ l] )
       return false;
if (Xi k - 1]< Xj [k - 1])
       return true;
else
       return false;
}
```

//use Trimed ( ) function to determine if *k-(k-1)*-dimensional subsets of k-dimensional itemset R are all in L_{k-1}. If one *(k-1)*-dimensional subset is not in L_{k-1}, the candidate itemset R can't be frequent, then return true. Reduce number of counts k to *(k-2)*, specific function as follows:

```
 bool Trimed (R, Lk-1 ) {
for each itemset r1 ∈R
for each itemset l1 ∈Lk-1{
  if (r1 ∈ Lk-1 )
        return true;
}
return false;
}
```

Experiment comparison of Apriori algorithm and improved Apriori algorithm. Hardware environment of experiment is Intel Core2 Duo T7350 , 3G DDR3 memory, software environment is Visual C++6.0, the experiment data acquisition uses T10 I4D100k, from Frequent ItemsetMining Implementations Repository (http://fimi. cs. helsinki. fidata). Time of the two algorithms are showed in Fig.1.

As we can see from the Figure, improved algorithm operating time is less than the original algorithm, and with the reduction of minimum support and the need for the number of frequent itemsets mining increased advantage of the improved Apriori algorithm is very significant.

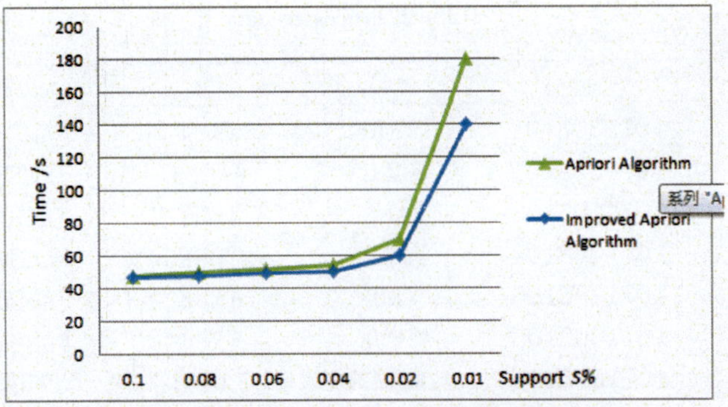

Fig. 1. Time of the two algorithms

4 Simulation

The simulation results use database technology to collect school students' educational system test results and forms a multi-door basis of this analysis database. Take randomly the following a session of several courses of students as examples: C++ programming, C# programming, JAVA programming, results published in Table 1.

Table 1. Student test scores

StdNum	C++	C#	JAVA
1071303101	82	89	72
1071303102	72	70	65
1071303103	74	83	71
1071303104	69	76	63
1071303105	77	91	70
1071303106	75	89	63
1071303107	62	63	75
1071303108	86	69	88
1071303109	74	78	63
1071303110	68	65	60
...	...	...	...

Experiments carried out association analysis of students' score when mining required a logical type data. Therefore turn the data of Table 1 into Boolean. Due to finding good relations between several courses, so the filed up to 75 points is set "TRUE", it shows this item is in the transaction. The others set "FALSE", it shows this item isn't in the transaction. Turn Table 1 into the style which is easy to be handled with association rules, in Table2.

Table 2. Processed data

StdNum	C++	C#	JAVA
1071303101	TRUE	TRUE	FALSE
1071303102	FALSE	FALSE	FALSE
1071303103	FALSE	TRUE	FALSE
1071303104	FALSE	TRUE	FALSE
1071303105	TRUE	TRUE	FALSE
1071303106	TRUE	TRUE	FALSE
1071303107	FALSE	FALSE	TRUE
1071303108	TRUE	FALSE	TRUE
1071303109	FALSE	TRUE	FALSE
1071303110	FALSE	FALSE	FALSE
...	...	...	...

To mine association rules in good courses in Table 2, this experiment in Table 2 set the minimum support is 20%, the confidence is 50%. The 180 records in Table 2 as Apriori mining association rules data, using the improved Apriori algorithm, obtain results in Table 3.

Table 3. Frequent Itemsets

Frequent Itemsets	Support Count
{C++, C#}	62
{C#, JAVA}	35
{C++}	72
{C#}	95
{JAVA}	40

Count the confidence of non-empty subset in final frequent itemsets and delete the records which are less than the minimum confidence. Eventually produce association rules as follows:

a) C++ $\geq$ C#, confidence=62/72 $\approx$ 86.11% ;
b) C# $\geq$ C++, confidence=62/95 $\approx$ 65.26% ;
c) JAVA $\geq$ C#, confidence=35/40 $\approx$ 87.50% .

From the above run results can obtain the following potential association:

- When C++ programming is good, C# programming has 86.11% of the good possible;

- When C# programming is good, JAVA programming has 65.26% of the good possible;
- When JAVA programming is good, C# programming has 87.50% of the good possible.

5 Conclusion

This paper aims for improving the efficiency of mining frequent itemsets when using mining association rules algorithm, puts forward an improved Apriori algorithm. And the algorithm is applied to the analysis of paper grades; simulation results show the affective relationships among students' examination paper grades of every subject, to provide decision support for school's guidance on students' studies and future work.

References

1. Agrawal, R., Imielinski, T., Swami, A.: Mining association rules between sets of item s in large databases. In: Proc of ACM SIGMOD Conference on Management of Data, pp. 207–216. ACM, New York (1993)
2. Agrawal, R., Srikant, R.: Fast algorithm form ining association rules in large databases. In: Proceedings of the 20 th International Conference on Very Large Data Bases, pp. 487–499. Morgan Kaufmann Publishers Inc., San Francisco (1994)
3. Zaki, M.J., Parthasarathy, S., Ogihara, M., Li, W.: New algorithms for fast discovery of association rules. In: Proc. of the 3 rd Intpl Conf. on KDD and DataMining (KDD 1997), NewportBeach, California (August 1997)
4. Han, J.W., Micheline, K.: Data Mining Concepts and Techniques. Machinery Industry Press, Beijing (2002); Fan Ming, Meng Xiaofeng,Translated

Practice of Flexible Talents Training Mode "2+0.5+0.5" for Surveying and Mapping Major for Higher Vocational Education

Xiaoning Liu

Department of Architectural Engineering
Henan Polytechnic Institute
Nanyang, China
Nylxn@yahoo.com.cn

Abstract. The characteristics of surveying and mapping industry determines that the cultivation of occupational ability and training of professional accomplishment shall be of equal importance for students in surveying and mapping major of higher vocational education. Based on college-enterprise cooperation, the flexible talents training mode "2+0.5+0.5" was established to realize the self-construction of occupational ability and development of professional accomplishment of students and explore a talents training mode that combines professional education and social practice through "Working-Schooling Alternation" with two-phase studies and practices at school or in enterprises respectively, especially "Work-study Combination" during the two-phase practices in enterprises.

Keywords: School-Enterprise Cooperation, Working-Schooling Alteration, Flexible Training, Accomplishment Training, Surveying and Mapping major.

1 Introduction

The surveying and mapping major of higher vocational education is specialized in application and operation. Students are required not only to master the specialized theories of surveying and mapping major in modern time, acquire professional operational skills through participation in social practice as an important task, and develop good teamwork spirit and hard-working tradition. After several years of practice and exploration, the flexible talents training mode "2+0.5+0.5" has been established for the engineering surveying and mapping major of Henan Polytechnic Institute (hereinafter referred to as "the Institute"), and it put emphasis on the training of specialized skills and professional accomplishment for students and primarily explores the talents training mode combing professional education with social practice, and finally good effects are achieved in training talents.

2 Exploring a Talents Training Mode of "Working-Schooling Alteration", Actively Implementing Flexible Teaching

The surveying and mapping major of higher vocational education is specialized in training highly qualified and skilled talents with excellent professional accomplishment

D. Zeng (Ed.): Advances in Computer Science and Engineering, AISC 141, pp. 27–32.
Springerlink.com

and occupational skills to master professional theories and skills of modern surveying and mapping, and engage in the front line surveying and mapping operation indoors or outdoors[1]. The characteristics of surveying and mapping industry, which requires field operation with instruments, determine the training of occupational ability and professional accomplishment shall be of equal importance. Based on the best practice experiences of other colleges and universities, a reform was intensified by the engineering surveying and mapping major of the Institute to implement the flexible talents training mode "2+0.5+0.5" in the form of "Working-Schooling Alteration", in which, "2" refers to two-phase professional study at school (the first academic year and the fourth and fifth semesters, or the first academic year and the third and fifth semester); the first "0.5" refers to the practices in enterprises sponsored by school; and the second "0.5" refers to the internship program sponsored by enterprises in the sixth semester. The flexible talents training mode "2+0.5+0.5" is characterized by the alteration of two-phase school study and two-phase practices in enterprises, especially targeted at realizing the goal of strengthening the occupational ability and professional accomplishment of students through the "0.5+0.5", namely participation in social practices in enterprises and "Work-study Combination", and finally the goal of "Training by practice" by way of integrating practice and professional education would be achieved.

As considering enterprise and school, technology and skill, accomplishment and ability, a multi-dimensional curriculum system has been established for the major to integrate and develop modular program courses based on common surveying and mapping instruments with focus on operation & maintenance of surveying instruments, processing of surveying and mapping data and analysis & application. A curriculum system with theory and practice integration corresponding to modular courses has been established based on the three phases of basic skill training, construction of position competence and promotion of comprehensive abilities. Finally, talent training is characterized by co-training between school and enterprise, co-development of quality education and occupational ability, as well as flexible combination of school education and enterprise practice.

The flexibility of flexible talents training mode "2 +0.5 +0.5" is reflected as follows: on one hand, the first "0.5" refers to the third semester or fourth or even fifth semester, as well as the holidays. The talents training program has been improved in rigidity, inflexibility and inadaptation to the business requirements, and the conflicts between student internship and production schedule or task allocation of enterprises are also solved. Before the summer holiday of 2009, due to the insufficient staff and instruments, the cadastral survey project in Anyuan, Ganzhou, Jiangxi Province contracted by Nanyang Three-dimensional Surveying and Mapping Co., Ltd. would not be completed as scheduled, and the Institute was requested to assign students to Jiangxi Ganzhou to assist in completing the 1:500 cadastral map. Upon request, a timely adjustment of teaching plan was made, and four teachers and a class of students were assigned to assist the Company in completing the project mapping and enabled the enterprise to deliver the project on schedule smoothly. Through this task, the cooperation between Three-dimensional Survey and Mapping Co., Ltd. and the Institute was raised to a new level. On the other hand, during talents training process, the setting and sequences of curriculum could be adjusted timely according to development requirements and specific circumstances of projects in surveying and mapping industry. During the practical cadastral survey in Xinyang Huangchuan, the courses "Digital Mapping Technology" and "Total Station Measurement Technology" were taught on site with practices by students immediately while listening. In the meantime, the investment in training and laboratories was also increased by the Institute based on the principle of "Satisfying teaching requirements, adapting to production, giving attention to researches", and more than RMB one million Yuan had been invested in updating and purchasing the latest mapping instruments. An effective management system had been developed and implemented and a management mechanism consistent with working-schooling alteration and flexible teaching had also been established.

3 Focusing on Deep Collage-Enterprise Cooperation, Building a Platform of "Teaching, Learning and Practicing Integration", Strengthening the Training of Students' Professional Skills

As an inevitable development trend of higher vocational colleges, college-enterprise cooperation is not only an objective requirement raised based on economic development, but also the internal demand for existence and development of higher vocational colleges. However needless to say, the way of cooperation between higher vocational colleges and enterprises is still in the early stages or the initial stage of middle level cooperation so far with the depth of cooperation far lagging behind the advanced stage of deep cooperation. The reason lies on that the deep economic relationship and long-term mechanism for college-enterprise cooperation have not been established.

Flexible talents training program "2+0.5+0.5" is established based on the deep college-enterprise cooperation. Normally, surveying and mapping enterprises hold some technicians, but when a project is undertaken, a large number of labors knowing surveying and mapping or even some advanced surveying and mapping instruments will be required especially by those small scale enterprises. Accordingly, a great importance had been attached to the establishment of deep cooperation with enterprises by the surveying and mapping major of the Institute at the beginning. The emphasis is placed on signing long-term agreements with respect to instrument procurement, maintenance, operation, undertaking production tasks from those enterprises with more businesses and strong technology, and establishing economic interest community, in order to guarantee a smooth development of college-enterprise cooperation. For example, cooperation agreements on undertaking part of production tasks had been signed between the surveying and mapping major of the Institute and Nanyang Three-dimensional Surveying and Mapping Co., Ltd., and Nanyang Chunyang Surveying and Mapping Co., Ltd. By such agreements, a large number of labors knowing surveying and mapping required by enterprises were solved. More importantly, a platform of teaching, learning and practicing integration was also established, providing students with opportunities to enhance their professional skills through practices.

In 2008, a cadastral survey in Huangchuan, Xinyang assigned by Nanyang Chunyang Surveying and Mapping Co., Ltd. was completed by students majored in surveying and mapping of the Institute as well as another cadastral survey in Zhenping, Nanyang assigned by Nanyang Three-dimensional Surveying and Mapping Co., Ltd. in 2009. Through these two practices in several months, students were enabled to learn the production process and business standards of enterprises, and their practical operational abilities had also been greatly improved. More importantly, students were provided with opportunities to conduct independent operations, explore and learn to solve problems, and train students' abilities to identify, analyze and solve problems. Some of the tasks have not been learned by the students before, but through the process of "do first - then know - redo" and implementation of the cycle of "practice - learning - re-practice", "learning in practice" and "practice in learning" were truly realized, and students were trained to take the initiative to study rather than to passively study. Consequently, the self-construction of occupational ability and professional accomplishment of students were achieved and improved gradually [2].

4 Strengthening Training of Professional Accomplishment, Realizing Self-developing of Professional Morality

4.1 Implementation of Accomplishment-Oriented Education in Talents Training Program

Surveying and mapping industry is engaged in long-term operations outdoors, where employees are required to possess good professional accomplishments. Importance has always been attached to the training of professional accomplishment by the Institute, and the accomplishment-oriented education has been listed in professional talents training programs. The students in the Institute are trained to form the moral character of "Loyalty, Perseverance", working style of "Strict, Careful" and quality concept of "Perfect, Excellent" upon the request with military characteristics brought forward by the Institute. These requirements are exactly consistent with the teamwork spirit, meticulous attitude to work and hardworking work style of surveying and mapping industry. Thus, targeting at the talents training program, a curriculum system of professional accomplishment was established with the courses such as career planning, quality expanding, etc. Detailed curriculum standards and assessment methods were also established in accordance with the curriculum system. While beyond the curriculum system, quality development & code of conduct, technological innovation quality education, vocational skills competitions, community activities and social practice courses were also carried out. The curricular and extracurricular courses were combined to promote multi-dimensional, entire-process and all-day education. The examination results of accomplishment-oriented education were set as a prerequisite for graduation. Meanwhile, the satisfaction index of enterprises was also taken as an important indicator for examining the quality of talents training. During participation in production internship and in-post internship, the assessment on students was carried out by the enterprises, so that enterprises and the society may participate in the assessment on quality of talents training, especially cultivation and assessment on professional accomplishments.

4.2 Self-construction of Professional Morality by Practice

The students of surveying and mapping major enrolled in the Institute in 2008 participated in the 1:500 cadastral mapping in Anyuan, Ganzhou, Jiangxi Province. A lot of difficulties were encountered, for example, Henan students unaccustomed to the hot weather in July and August, the food and poor living conditions in Ganzhou, Jiangxi Province, as well as the tight schedule, heavy task and strict requirements, while no one had ever attempted to quit or complained but remained working at the forefront. During the practical measurement, students were assigned to various towns with self-controlling and self-management and the measurement results were directly checked and accepted by enterprises. Unqualified results would be subject to reworking. A large proportion of task had been completed by the students for enterprises through two-month working, when the teamwork spirit, meticulous attitude to work and hardworking work style of students were trained, and the self-developing of professional morality was also realized. In addition, during the cadastral mapping in Zhenping County, an approach of "skilled students guiding new students" was adopted, namely skilled senior students guided junior students with few experiences, and each group was consisted of both skilled and new students. Through this approach, senior students became more responsible, new students became more initiative, and finally double ultimate effects were gained.

4.3 Actively Developing Skills Competitions and Knowledge-Oriented Seminars, Enriching Academic Atmosphere

Students' comprehensive abilities can be shown intensively in vocational skills competitions, having positive effects on enhancing students' initiative in learning and the communications between college and enterprises. In June, 2010, the First Surveying and Mapping Skill Competition "South Kolida" Cup was held by Henan Polytechnic Institute, where four teams from Surveying and Mapping Institute of Nanyang Planning Bureau and other enterprises were invited in this competition. In such competition, students were enabled to find their shortages and learn experiences, and enterprises were also enabled to know and understand the students. Thus, the exchange and communications between college and enterprises were promoted, and a solid foundation was also laid for further deep college-enterprise cooperation.

In August 2010, four students was assigned by the Institute to participate in the first Surveying and Mapping Skill Competition "Topcon" Cup sponsored by Education Advisory Committee on Surveying and Mapping Major in higher vocational colleges of the Ministry of Education, and State Bureau of Surveying and Mapping. Based on the knowledge and skill requirements specified in "National

Occupational Standard on Senior Engineering Surveyers" and more related new knowledge, technologies and skills, the competition was carried out by coordinating both theory examination and assessment on operation skills. After two-day keen competitions, the team of the Institute stood out from 40 colleges and universities and won a first prize in 1:500 Digital Mapping, and a second prize in First Level Electromagnetic Distance Measurement (EDM) Traverse, a third prize in Fourth-grade-leveling measurement and a second prize in Teamwork.

Meanwhile, related experts within the industry had been actively invited by the Institute to give various academic lectures and reports to enrich the academic atmosphere. In the past one year, Professor Tao Benzao, Professor Li Hanwu from Wuhan University, Researcher Liu Shaochuang from the Institute of Remote Sensing Applications of Chinese Academy of Science and various other experts and scholars had also been invited to give lectures in the Institute. Chairmen of Nanyang Three-Dimensional Surveying and Mapping Co., Ltd. and Nanyang Chunyang Surveying and Mapping Co., Ltd. had also made speeches for students. Through these academic activities, the interest and passion of students to their major were greatly motivated with more awareness of the development prospects of the industry, so as to expand their mind and knowledge.

5 Conclusions

The flexible talents training mode "2+0.5+0.5" was established based on the school-enterprise cooperation in the form of "Working-Schooling Alteration" with focus on "Work-study Combination", "Studying-Practicing Integration" and emphasis on practices in enterprises. Through the two practices in enterprises, the self-construction of students' occupational abilities and enhancement of professional morality were realized gradually. After several-year practices, this talents training mode was proved to be suitable for training qualified professional talents in surveying and mapping major of higher vocational education.

References

1. Ding, L.: 'Research on Talents Training by Mode of Work-study Combination in Surveying and Mapping Major of Higher Vocational Education. Education and Vocation, 95–97 (June 2010)
2. Zhu, J., Fan, G.: Practice and Exploration on the Talents Training Mode of "Major + Enterprise, Employment + Entrepreneurship. Chinese Vocational and Technical Education, 84–85 (August 2010)
3. Zhang, H.: The Research of School-Enterprise Cooperation Institutionalization for Higher Vocational Education. Education and Vocation, 11–13 (June 2010)
4. Zhang, D., Ming, Li, M.: Study and Practice of Talent Foster Mode on Cooperation of Learning with Working in Surveying and Mapping Specialty for High Vocational Education. Surveying and Mapping of Geology and Mineral Resources, 44–46 (June 2010)

The Construction of Entrepreneurial Practice Platform System for College Students

Yang Huanchun, Liu Kailian, Xu Shu, Zhang Xiutin, Zhou Fei, and Huang Xingli

School of Business
Wenzhou University
Wenzhou, China
yhchyhch@126.com

Abstract. The key point and difficult point of entrepreneurial education is to construct a complete entrepreneurial practice platform. Based on the objective understanding and analysis of entrepreneurial practice platform, this paper puts forward the construction form and path of the entrepreneurial practice platform system; combines the real example of one local university, and proposes concrete building contents, namely, campus entrepreneurial analog and simulation studio, campus entrepreneurial studio, off-campus enterprise incubator, entrepreneurial data center and other multi-level practice platform system.

Keywords: Terms – entrepreneurial education, entrepreneurial practice, entrepreneurial practice platform system, construction.

1 Introduction

Entrepreneurial education has attracted great attention of many colleges and universities in China. "21st Century Action Plan for Education Invigoration" announced by the Ministry of Education in January, 1999 pointed out that: "Strengthening the entrepreneurial education to teachers and students, and encouraging them to start high-tech enterprises." This is the first time to propose that universities and colleges should carry out entrepreneurial education. In April, 2002, the Department of Higher Education, the Ministry of Education, chose 9 universities and colleges such as Tsinghua University, Beijing University of Aviation Industry and etc. as pilot units to carry out entrepreneurial education, which opened the prelude to the Chinese entrepreneurial education. In recent years, through different entrepreneurial education methods, many universities and colleges have achieved some achievements.

However, the entrepreneurial education in China is still at the starting or bud stage, the entrepreneurial activities carried out by universities and colleges are nearly mere formality. For example, although some departments and universities organize various forms of entrepreneurial plan competitions for university students, most students just compete for the competition itself, rather than putting entrepreneurial plans into practice. Schools also lack some links of entrepreneurial education system, such as corresponding policies and measures, promotion mechanism, sound entrepreneurial education theory and practice teaching system, entrepreneurial education teachers and

D. Zeng (Ed.): Advances in Computer Science and Engineering, AISC 141, pp. 33–39.
Springerlink.com

teaching materials, etc., leading to the unsatisfactory development and effect of entrepreneurial education. So, how to effectively foster the entrepreneurial sense of students and improve their real entrepreneurial capabilities has always been a major difficult problem bothering university entrepreneurial education. This paper argues that universities and colleges should strengthen the building of entrepreneurial practice platform, which is one of the effective methods to improve entrepreneurial education system and ensure successful entrepreneurship.

2 Objective Cognition of the Construction of Entrepreneurial Practice Platform

2.1 Entrepreneurial Practice Platform Is an Effective Component of Entrepreneurial Education System

The entrepreneurial education of university students is a systematic project and an important component of university education, including not only curriculum system, but also non-curriculum system. The entrepreneurial education system not only requires the participation of schools, but also the involvement of the society; it is not only theoretical, but also practical. With the establishment and development of entrepreneurial education concept, more and more Chinese universities have recognized that as a general skill entrepreneurship has become one type of basic knowledge, one kind of skill impartation to win success in the unstable and changing work environment. To realize this skill impartation, we need hierarchical cultivation and anneal. Abilities like market forecasting, financial analysis, industry communication, laws and regulations interpretation and etc. must keep learning and improving on the practice platform. In addition, the edification of those non-intellectual factors should penetrate reasonably in the practice environment. At present, universities generally emphasize the education of entrepreneurial courses. When universities invest a lot in infrastructures, some fail to regard the building of entrepreneurial practice platform for university students as an effective component of university entrepreneurial education system, and fail to strengthen the construction of entrepreneurial practice platform for university students.

2.2 Entrepreneurial Practice Platform Is an Effective Carrier for Students' Entrepreneurial Activities

Our nation and universities encourage students to start undertakings of their own after graduation, and the entrepreneurial practice at school is an important link for students to achieve this goal. The entrepreneurial activity of students has strong practicality and complexity, which is a bridge of communication between students preparing to start undertakings and "society" as well as "market", and a form for students to get to know and learn entrepreneurship. Schools build entrepreneurial practice platform, improve the platform market factors, form real entrepreneurial environment, provide basic conditions and venture capital support for the entrepreneurship of students, provide guiding teachers to correct entrepreneurial practice, so that students can promote the healthy development of entrepreneurial projects in accordance with market rules, reduce entrepreneurial risks and boost success ratio of entrepreneurship in the process of entrepreneurial trial. Schools build entrepreneurial practice platform, and then students use the platform to conceive entrepreneurial models and routes, receive entrepreneurial training, learn entrepreneurial experience from successful people, assess the relation between individuals, enterprises and market, transform projects and technology, and incubate small enterprises. In this way, quasi-entrepreneurs and entrepreneurs carry out the operation and management of enterprises, look for entrepreneurial feelings, accumulate entrepreneurial experience and gain entrepreneurial experience. Therefore, university entrepreneurial practice platform become the public arena and effective carrier for university students to practice entrepreneurship.

2.3 Production-Education-Research Base Is an Important Platform for Students to Start Entrepreneurial Practice

The three basic functions of university are teaching, scientific research and social service, and teaching, scientific research and industry complement each other, with a natural link. The scientific research provides internal driving force for the industry, while industry provides material guarantee for scientific research. The combination of production, education and research expand the functions of the university, emphasize the real purpose and practicality of personnel training and scientific research, and improve the conversion rate of research fruits. Production, study and research base become the scientific technology innovation platform in real sense, the place for students to carry out entrepreneurial training and the incubator of entities at various levels. Students can conduct entrepreneurial exercise in enterprise incubator, with more complete entrepreneurial factors and more mature entrepreneurial environment.

3 The Construction Form and Path of the Entrepreneurial Practice Platform System

Today, universities attach great importance to the training and counseling of entrepreneurial theory, while this paper argues that the improvement of the success rate of students' entrepreneurship must base on the education of entrepreneurial theory, build complete entrepreneurial practice platform system, let students carry out entrepreneurial activities on the entrepreneurial practice platform system, accumulate entrepreneurial experience, and contact enterprises and market at zero distance. University entrepreneurial practice platform system should include three levels at least, namely, analog and simulation, campus entrepreneurship and enterprise incubator, thus students can participate systematic training at three levels according to their own entrepreneurial foundations and conditions, and can also choose one level of the platform system to practice entrepreneurship.

3.1 Building Analog and Simulation Platform

Analog and simulation is mainly based on analog and simulation. Students can foster entrepreneurial sense through various activities such as analog design of entrepreneurial plans and entrepreneurial seminars, campus simulation operation and etc, comprehend key points of entrepreneurship, understand factors of entrepreneurial market, establish entrepreneurial directions and lay the foundations for real entrepreneurship.

3.2 Establishing Campus Entrepreneurial Studio and Entrepreneurial Park

Entrepreneurial studio and entrepreneurial park directly provide students with entrepreneurial practice places, and allow students to start undertakings without restraint. Entrepreneurial studio and entrepreneurial park provide students with entrepreneurial spaces and related services, set up entrepreneurial venture capital, provide small loan, and create market environment for incubating enterprises. Students interested in entrepreneurship can apply for entering the studio and park, and registering in accordance with national industry and commerce, tax administration and other relevant provisions. The started enterprises will operate fully in accordance with the market, the school can offer entrepreneurial students with favors and cares in terms of places and capital, and give some incubation supporting period to create good entrepreneurial environment for students. The entrepreneurial backbone takes up "boss" of the enterprise or project, while other students become employees, and they exercise and grow together.

3.3 Establishing and Improving Campus and Off-Campus Research Base (Enterprise Incubator)

The platform provides students with all-true entrepreneurial environment. Enterprise managers or teachers with entrepreneurial experience lead students to make entrepreneurial design and market-oriented practice, master production management, cost accounting, marketing and social communication skills. When possible, let students operate independently, launch market campaign separately, carry out entrepreneurial practice, market exercise or transit to start undertakings directly.

4 The Specific Content of the Construction of Entrepreneurial Practice Platform System – A Case Study of the Economy and Management Experimental Teaching Center of Wenzhou University

Economy & Management Experimental Teaching Center of Wenzhou University relies on the Business School of Wenzhou University, making full use of the entrepreneurial resource of Wenzhou businessmen and resources of many small and medium sized Wenzhou enterprises, and providing services for university students to carry out entrepreneurial education and entrepreneurial practice. In recent years, the center has actively explored the entrepreneurial practice forms of university students, and gradually formed two interactive practice carriers, that is, campus and off-campus. Besides, the center fosters entrepreneurial entities and projects by campus entrepreneurial teaching, entrepreneurial circumstance and practice simulation, with related resources of Wenzhou Merchant Development & Research Association, and by using off-campus corporate venture capital and enterprise incubator. The process of entrepreneurial incubation is shown in Fig. 1.

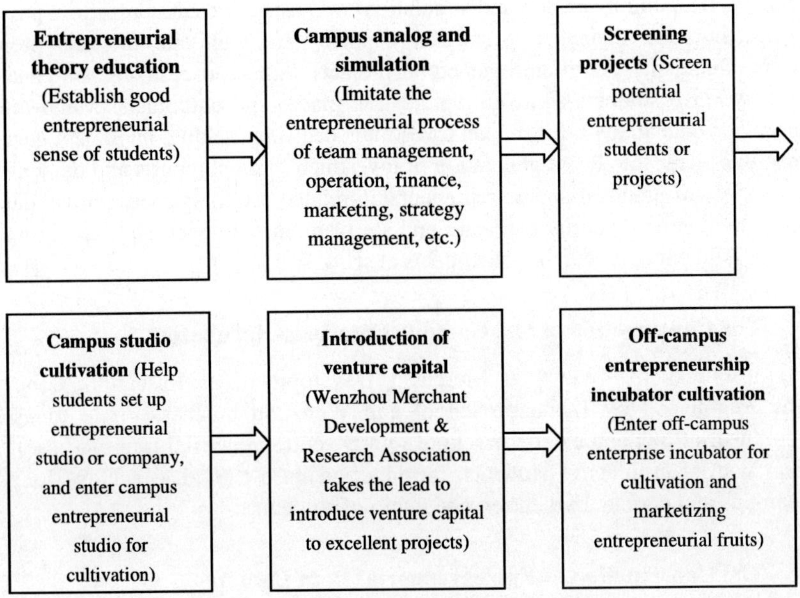

Fig. 1. Process of entrepreneurship cultivation

Meanwhile, in order to achieve the successful cultivation of entrepreneurial entities and projects, the center strengthens the construction of campus and off-campus interactive and multi-level entrepreneurial practice platform system.

4.1 The Construction of Campus Entrepreneurial Analog and Simulation Studio

1) Analog and simulation platform of independent entrepreneurship and experience: On this platform, each student is both the consumer and producer (investor), and they are given a certain amount of entrepreneurial capital. As a consumer, he has to survive, so he has to buy consumer goods; while as a producer (investor), he has to create wealth. A student gains earnings through his own production, management and investment, after excluding his own consumption and expense, we get the extent of the surplus accumulation which is used to evaluate the effect of his independent entrepreneurial practice, enabling students to experience and enjoy the individual entrepreneurial process and achievements.

2) University students' entrepreneurial psychology and quality evaluation system: The system adopts mathematical model to test the entrepreneurial potential, entrepreneurial capability, entrepreneurial psychology, entrepreneurial intelligence assessment, entrepreneurial personality, entrepreneurial success indicators and etc. of university students, plus corresponding report of the test result, in order to find out the entrepreneurial potential of students and establish the career development.

4.2 Construction of Campus Entrepreneurial Studio

Through corresponding analog and simulation, we inspire the entrepreneurial passion of students, improve their entrepreneurial capacity and cultivate the entrepreneurial projects of students. As to entrepreneurial projects with short realizing time and quick effects, we give students some entrepreneurial places and equipment. Meanwhile, we also provide data analysis for project investment decision-making, intelligent generation of business plans, intelligent generation of investment project reports and other systems, and a variety of practical and entrepreneurship-related economy management database, creating real entrepreneurial hardware and software environment, and rapidly realizing the successful entrepreneurship of students at school.

4.3 The Construction of Off-Campus Enterprise Incubator

Using resources of "Wenzhou Merchant Development & Research Association" jointly established by Business School and Wenzhou business circle to establish entrepreneurial incubator outside school jointly (entrepreneurial training base). As to entrepreneurial projects of students, introduce venture capital to achieve long-term, off-campus and cooperative entrepreneurship of students.

4.4 The Construction of Entrepreneurial Data Center

Data center is the application integration center and data sharing center of hardware system, system software, database management system, application database and etc. in the entrepreneurial practice platform system, providing hardware, software and data services for campus entrepreneurial analog and simulation studio, campus entrepreneurial studio and off-campus enterprise incubator.

5 Conclusion

The practice has proved that through entrepreneurial practice platform, the innovation and entrepreneurial ability, social responsibility and team spirit of students are well exercised, and fairly good and substantial entrepreneurial effects have been achieved. Cultivating highly qualified talents with innovative, creative and entrepreneurial abilities is an important task for the current colleges and universities, so university teachers and related personnel should always focus on this concept in their work, continuing to explore and practice, through the construction of various forms of entrepreneurial practice platform systems, in order to foster more and better entrepreneurial talents suiting the social development.

Acknowledgment. This is one of the achievements of "Construction and Management of the Experimental Teaching System of Economics and Management at Business School " project (key of education reform project of Wenzhou University 2007), the author appreciates the Wenzhou University funding. The authors also would like to thank the reviewers for their comments and suggestions that have greatly improved the content and presentation of this paper.

References

1. Tian, L.: Problems of the Entrepreneurial Education of University Students and Solution Exploration. Literature on Party Building 12, 71–73 (2009)
2. Zhou, J., Zhao, Y.: Research and Practice of the Teaching System of Innovative and Entrepreneurial Education Practice. Beihua University (Social Science Edition) 10(6), 104–106 (2009)
3. Shan, A., Gao, Y., Shu, Z., Xi, X.: Innovative Teaching Model Fosters Entrepreneurial Talents-Entrepreneurial Talent Cultivating Model and Innovation Experimental Area Practice of Animal Science Majors. Northeast Agricultural University (Social Science Edition) 7(1), 17–20 (2009)
4. Wu, W., Jin, J., Wang, W.: The Exploration and Practice of University Students' Innovative and Entrepreneurial Education. Journal of Zhejiang Sci-Tech University 23(4), 502–504 (2006)
5. Fan, X., Lin, W., Shen, J., Zi, W.: Exploration on the Operation and Management of the Entrepreneurial Base of University Students – a Case Study of the Innovative Entrepreneurial Practice Base of City College of Zhejiang University. Research Management (Suppl.) 29(12), 45–49 (2008)
6. Guo, H.: The Related Exploration on the Practice Link in the Entrepreneurial Education of University Students. Contemporary Education Forum 6, 84–85 (2009)
7. He, W., Chen, G., Shao, X.: Exploration and Practice of "Trinity" Entrepreneurial Education Model. Vocational Education Research 7, 8–10 (2008)
8. Zhao, T.: Exploration and Practice of Entrepreneurial Education in Local Universities. Hubei Normal University Journal (Social Science Edition) 29(1), 116–118 (2009)
9. Lei, F.: Study on the Establishment of Entrepreneurial Practice Platform System of University Students. Vocational Technology (6), 19–20 (2009)
10. Chen, H., Xing, S.: Study on the Construction of Trinity Entrepreneurial Practice Platform of University Students. China Adult Education 21, 22–23 (2009)
11. Guo, X.: The Engineering Construction of Entrepreneurial Practice Mechanism of University Students. The Employment of China University Students (3), 61–62 (2008)
12. Jinlin, G.: Optimizing Campus Entrepreneurial Practice and Enhancing the Effects of Entrepreneurial Education. Transportation Vocational Education (2), 55–57 (2008)

Construction of Practical Teaching System for Network Engineering Specialty in Local Undergraduate University—A Case Study of Chuzhou University*

Guilin Chen, Guangshun Yao, Chuanjian Yang, Huibin Wang, and Shenghui Zhao

Department of Computer Science and Technology
Chuzhou University
Chuzhou, Anhui Province, 239000, China
{glchen,gshyao,wanghuibin}@chzu.edu.cn

Abstract. Practical teaching plays an important role in the process of cultivating talents of network engineering; it has a great significance on developing the innovation spirit and the practice ability of students. A great number of universities started research on the practical teaching system according to their characteristics currently. On the basis of analyzing the problems existed in the practical teaching of network engineering of our university, a practical teaching system was constructed and implemented in accordance with the actual conditions of our university and experiences of practical teaching reform in other universities, which aimed at cultivating "applied talents with strong competitive power and innovation spirit". The practice proves that the system has achieved a good result in cultivating practical ability and innovative ability of the students.

Keywords: Network Engineering, Practical Teaching, Innovation spirit.

1 Introduction

Along with the rapid development of information technology in China, network has become an infrastructure of human society and a basic approach of working and living, demands of the society for professional talents of network engineering has risen sharply due to the extensive construction of network infrastructure and a large number of businesses on the network, a great crisis of network talents has emerged [1]. Meanwhile, "Structure Unemployment" has appeared, there are many graduates majored in network engineering are unemployed, the main cause of this phenomenon is the defects existed in the practical teaching system of universities. Universities should attach importance to the reform of practical teaching system for the purpose of ensuring the quality and level of practical teaching, promoting the coordinated development of knowledge, ability and quality, and cultivating network talents to meet the demands of society.

---

* This work is supported by Anhui Province Quality Engineering Projects of Colleges and Universities (20101042).

D. Zeng (Ed.): Advances in Computer Science and Engineering, AISC 141, pp. 41–49.
Springerlink.com

Practical teaching is a significant aspect in the talents cultivation of general universities, and it's an important part of cultivating innovative talents, students' innovation spirit and individuality development could be inspired with it. At present, there are many shortcomings in the practical teaching system of network engineering, such as paying less attention to the practice with too much importance attached to the theory teaching; insufficient practical teaching resources and outdated practical courses, because many teachers, administrative staff and students get used to class teaching of imparting book knowledge; universities didn't realize the importance of practical teaching and neglected the cultivation of practical teacher staff, which leads to shortage of practical teachers; flawed teaching arrangements and practice become a mere formality[2]; lacking needed practical base and support fund. Reference [3] also stresses the importance of practical teaching for engineering students. Currently, the competitions among the enterprises embody the competitions of innovative applied talents with strong competitive power, as a university which specialize in applied talents cultivation, Chuzhou University also needs to train talents with strong competitive power and innovation spirit to meet the market demand. So computer science and technology department adjusted practical teaching system timely and obtained a good achievement, it has a great significance on ensuring the quality of applied talents.

2 Proposal of Practical Teaching System

In order to meet the needs of new times, Department of Computer Science and Technology of Chuzhou University, for short CSCI, innovated and improved the training objective of network engineering specialty, proposed "cultivation of applied network engineers with strong competitive power and innovation spirit"; we also constructed practical teaching system fits for cultivating network engineers based on the training objective. As shown in Fig.1, practical teaching system includes practical teaching content system, practical teaching support system and practical teaching administrative system. The system developed student's ability to find and solve problems in some degree, equipped students with innovation spirit and practical ability, and achieved the aim of cultivating advanced talents with innovation spirit and practical ability.

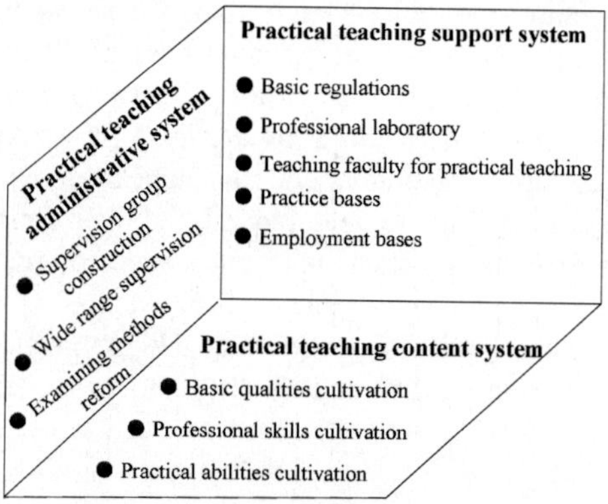

Fig. 1. Practical teaching system

3 Practical Teaching Content System

Practical teaching content system of network engineering is an important aspect of specialty construction; as Fig.2 shows, it is made up of three modules, includes basic qualities cultivation, professional skills cultivation and practical abilities cultivation.

Practical abilities cultivation	Graduation project
	Graduation practice
	Practical and innovative items, professional competitions
	Competitions of carve-out and employment
Professional skills cultivation	Academic yearly design and practice
	Curriculum-design
	Experiment curriculum
Basic qualities cultivation	Social and cognitive practice
	Training of professionalism cultivation of IT practitioner, intercommunication and social etiquette
	Moral and cultural cultivation

Fig. 2. Practical teaching content system

3.1 In Respect of Practical Abilities Cultivation

The students' practical abilities are improved by graduation project, graduation practice, practical and innovative items, professional skills competitions or competitions of carve-out and employment.

Graduation project is the most comprehensive practical teaching in the four years of college, and it is an important process of making students predominate basic theory and applied skills, and improving students' comprehensive qualities such as abilities of scientific research and management. Most topics of graduation projects come from the problems that students would meet in the process of exercitation or practical problems existing in teachers' research, it seems like students touching enterprises no distance.

We established a monitoring system of the practice quality; instructors of graduation practice are undertaken by experienced teachers. From graduation project plans, choosing topic and instructor of students, opening speech, arrangement of graduation practice experiment, to the end of oral defense, accomplishment evaluation and so on, and each step is standardized and refined; it leads to a marked advancement in the quality of graduation thesis.

Besides, we encourage and guild students to participate in practical activities, to apply for intramural scientific research projects and publish academic papers. Students are also encouraged to take an active part in contests, such as mathematical modelling contest, computer skills contest, programming contest, design contest of website. Students get the opportunities to participate in a multi-channelled and multi-level practice by taking part in these contests and their hands-on ability are greatly improved.

3.2 In Respect of Professional Skills Cultivation

Professional skills cultivation mainly depends on academic yearly design, academic yearly practice, curriculum-design, and experiment curriculum. As to the content of experiment curriculum, a number of replication experiments are replaced with experiments of integrity, designing and innovation, which help students to improve abilities. A transformation is achieved from isolated exercitation to comprehensive exercitation and from mastery of knowledge to cultivation of ability.

3.3 In Respect of Basic Qualities Cultivation

We train students' basic qualities mainly on social practice, cognitive practice, professionalism cultivation of IT practitioner, training of intercommunication and social etiquette, moral and cultural cultivation. The so-called basic qualities are related to knowledge and competence, a person without the basic qualities is worthless to the society even though he is very talented. Students access to the society and foster a sense of willing to help by social useful activities, such as organizing students to be volunteers in old folks' home, making money by selling papers and donating it to a leukaemia student. While studying foundation courses and professional courses, students take elective courses about intercommunication and social etiquette, moral and cultural cultivation and so on, and it wins the approval of students.

The focus of future construction of practical teaching content system is according to the concept of "intuition" → "theory" → "practice" → "method and application", on the basis of the trends of network technique development, by means of ways such as cooperation between colleges and enterprises, integration of production, teaching and research, improve the practical teaching contents like experiment curriculums, academic yearly practices, trainings of employment and innovation, practical and innovative projects, professionalism cultivation of IT practitioner, basic qualities education and so on. The main contents of construction are as follows:

1) Network and application integration lab and cloud computing platform of our university had been completed and are in operation, they are built in cooperation with Chuzhou telecom corporation which invested ￥600,000. We are doing researches in advanced network technology and tracking the developmental process of it based on the platform, and extracting experimental projects and various topics from academic research projects carried out in recent years. These projects mainly focused on network architecture, network security, parallel programming, Web Services and network technology of next generation.

2) Extracting design topics from projects in cooperation with enterprises which integrated with production, teaching and research in recent years (planning to increase the number of projects, and improving the level of cooperation).

3) Making students become the leading role in practical teaching; bringing their initiative into full play. Guiding students to participate in researches and projects of production, teaching and research, encouraging them to select topics in the process of development based on observation and analysis; formulate related measures of administration, and encouraging teachers to assume the mission of guiding students actively, stimulating students' interests to take part in projects.

4) Designing the syllabus, contents and items of experiment curriculums according to the application needs and technologic development tendency, ensure the advancement and applicability of experiment contents, and the proportion of comprehensive experiments and innovative experiments is over 50%.

5) Cooperating with practical education and employment bases; improving the program of graduation project and graduation practice; work out a new scheme of academic year design and academic year practice.

6) Organizing students to take part in contests like network design contest, programming contest, robot contest, design contest of single chip, business contest; setting up elective courses of communication and social etiquette in IT workplace.

4 Practical Teaching Support System

As Fig.3 shows, practical teaching support system of network engineering specialty is made up of basic regulations, teaching faculty for practical teaching, professional laboratory, and practice and employment bases and so on.

4.1 In Respect of Basic Regulations

Every department is organized to revise and improve regulations about experiment operation and practice works, a serious of administrative documents about examination of experiment's and practice's results, evaluation for the quality of practice and management of graduation thesis are issued. Explicit requirements and quality standards have been stipulated for practice and training in many aspects, which include purposes, methods, types, organization and management, working procedure and requirements, evaluation of accomplishment, use and management of practice fund, institutionally ensure that the work of practice and training to go on orderly and normatively.

<table>
<tr><td rowspan="3">Practice and employment bases</td><td colspan="2">Employment bases</td></tr>
<tr><td colspan="2">Practice bases outside of school</td></tr>
<tr><td colspan="2">Practice bases in school（Center of school network）</td></tr>
<tr><td rowspan="2">Laboratory</td><td>Professional laboratory</td><td rowspan="2">Open and innovative laboratories</td></tr>
<tr><td>College-enterprise jointly Constructing laboratory</td></tr>
<tr><td rowspan="3">Teaching faculty for practical teaching</td><td colspan="2">Enterprise's expert</td></tr>
<tr><td colspan="2">Theory and practice combined teachers</td></tr>
<tr><td colspan="2">Full-time practical teaching or administer-teachers</td></tr>
<tr><td rowspan="3">Basic regulations</td><td colspan="2">Administration regulations for practical teaching's quality supervision</td></tr>
<tr><td colspan="2">Administration regulations for kinds of practical action</td></tr>
<tr><td colspan="2">Administration regulations for Lab</td></tr>
</table>

Fig. 3. Practical teaching support system

4.2 In the Respect of Teaching Faculty for Practical Teaching

Teachers are the subject of practical system; the core and foundation of laboratory construction are brining up a well structured practical teaching faculty with high ability. Construction of teaching faculty are planned according to the laboratory development, attach importance to both recruiting and training, and combine professional and foundation, for the purpose of building up double-quality practical teachers.

We have built up a team about 20 people of experiment management and teaching, who are responsible full-time or part-time. Training teachers regularly to enrich their practical experiences, encouraging teachers to take part in various practical activities to improve their practical abilities, periodic evaluations are also conducted, and teachers who don't meet the requirements of practical instructor are disqualified, these measures are provided to ensure the reliability and the quality of practical teaching faculty, and to support the practical teaching system.

4.3 In Respect of Professional Laboratory

The scale of higher education institutions has been continuously enlarged, that makes the number of students increasing rapidly. Meanwhile, laboratory area and devices are insufficient relatively, the cultivation of students' practical abilities and improvement of practical teaching are greatly influenced and limited. It requires the universities to intensify construction of open experiment platforms, and to exert the vital function of laboratory in the process of higher education [4]. According to the characteristics of network engineering and network technology, the basic guidelines of laboratory construction are defined as "open, innovation, integration, management". Laboratories under network engineering practical teaching center are constructed to open and innovative laboratories by ways of institution innovation, management innovation and teaching innovation. At recently, laboratories are full opened to all of the college students, students can apply for using any device of laboratories at any time, that reflect the concept of "student-centered" . Simultaneity, tutors are designated to improve self learning and innovation of students, and incentives are made to encourage students to find solutions for the problems which met in the process of learning or practice.

Hardware facilities of network engineering practical teaching center have had a certain scale, which includes 9 professional labs of different kinds. Experiments of PDS, LAN, wireless networks, WAN, Internet, IPv6, Internet security and management, nerve of a covering, services computing and network application can be conducted. The labs own a number of advanced devices, which include series equipments of CISCO, Juniper and H3C (such as high-end router, layer 3 switch, radio network controller, wireless AP, firewall, network management system), Fluke network test equipments, servers and disk array of IBM. These facilities provide a practical teaching environment which is highly consistent with the practical applications for talents cultivation, so the students are able to learn and apply network technology in a real application environment.

Virtual learning systems are being used in computer information education since 2002. That system can reduce maintenance costs of hardware and accelerate the development of new technologies. Research indicates that students conduct experiments in virtual labs has the same effect as conducting in real labs [5]. We are also exploring the use of virtual teaching technique, for instance, PacketTracer of CISCO is used as virtual teaching environment, as shown in Fig.4, students of network engineering specialty can learn switch's configuration in that environment, which has achieved a better results.

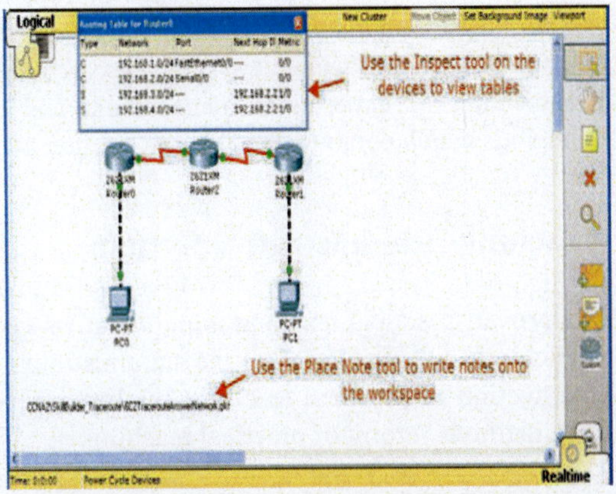

Fig. 4. Operator interface of PacketTracer

4.4 In Respect of Bases of Practice and Employment

Attaching great importance to the construction of practice bases and choosing appropriate extramural practice bases is the prerequisite to ensure the qualities of practical activities. Setting up bases of production, teaching and research is an effective way to broaden students' horizons and train their abilities of innovation and practice. The network center of our university is attached to computer science and technology department, which is intramural practical training base of network engineering specialty, and all students of the major participate in cognitive practice here. Besides, about 30 students go to network center for their graduation practice every year, and about 30 students work there to provide technical supports in technical posts at their spare time, about 10 students take part in intramural information system projects generally.

The construction of extramural practical teaching bases mainly concentrated in cooperation with enterprises. The development of cooperate projects and exchanges of teachers and technical personnel improve the cooperation with enterprises of practical teaching bases. Students are able to participate in the production and research of enterprises directly; help the enterprises resolve technical challenges. Besides, practice, graduation project and thesis of students are related to the production, both of enterprises and students benefit from it.

We focus on the improvement and enhancement in the future construction. The construction contents, goals and measures are mainly focus on the following aspects: upgrade and complete hardware devices of professional labs, the university also plans to increase investments for the development of network and engineering specialty, 500,000 Yuan are invested to upgrade experiment devices every year; improve the construction of soft environment by system innovation, transform professional labs into open and innovative labs; intensify development of practical teaching bases, increase the number of practical teaching bases, further discuss the approach of construction and management standardization of practical teaching bases; improve the construction of practical teaching faculty; on the basis of perfecting general management system, further improve practical teaching management system, research and formulate a new system to promote innovation of experiment management and practical teaching, research and explore methods which combine labs, practice center and research center.

5 Practical Teaching Administrative System

Our department has improved practical teaching administrative system to ensure the quality of practical teaching. As Fig.5 shows, the administrative system includes: supervision group construction of practical teaching, quality supervision of practical teaching and examining methods reform of practical teaching.

5.1 Control Group Construction of Practical Teaching

We have established a teaching quality supervision group, which responsible for checking the labs, all labs and practice bases are spot checked periodically, contents of the check includes rationality of teachers' teaching plans, arrangements of teaching contents, students' attendance, availability and utilization of equipments, security and sanitary conditions of the teaching spaces, results of the check are informed in time.

Supervision group construction of practical teaching	Teaching quality supervision group
Wide range supervision	Experiment
	Practice
	Practical training
	Curriculum design
	Graduation project
Examining methods reform	Combination process examination with result examination

Fig. 5. Practical teaching administrative systems

5.2 Wide Range Supervision of Practical Teaching

Practical teaching supervision of our department involves the key links of practical teaching, such as experiment, practical training, practice, curriculum design, and graduation project; it lasts from entrance to graduation.

5.3 Construction of Rational Examining Methods for Practical Teaching

There are differences between practical teaching and theory teaching on contents and methods, thus their examining methods are different too. We are exploring and trying to reform the examining methods of practical teaching, which includes combine process examination with result examination, pay attention to the process and give more concern to the results, increase the weights of the results. Setting a strict demand on the students, those students who fail the exam of practice will have to take a make-up exam, and students who fail the make-up exam will need to take the course again. To those courses which include experiment or other practical teaching contents, increase the proportion of practical teaching contents in the accomplishment evaluation.

6 Conclusion

The construction of a matured practical teaching system and advanced experiment platforms is very important in the cultivation of network engineering talents with innovation spirit and high quality. We constructed a practical teaching system which is guided by the aim of training applied talents with strong competitive power and innovation spirit. Practice proves that it has a great significance on developing the innovation spirit and the practice ability of the students, and improving the comprehensive quality of students. But, practical teaching reformation is a long-term and arduous task that calls for persevering endeavour, constantly adjustments and exploration in the process of teaching are required; only in this way can we cultivate qualified and innovative talents to meet the need of era.

References

1. Jiang, N., Yi, J., Chen, Q., Wu, J.: Research and Practice of Training Project for Network Engineering Specialty. Journal of Higher Education Research 28(3), 67–69 (2005) (Chinese)
2. Lin, H., Huang, H., He, G.: Construction of Practical Teaching System in University in Favor of Innovative Talent Cultivation. Journal of Changchun University of Science and Technology(Social Sciences Edition) 22(3), 457–458 (2009) (Chinese)
3. Lin, J.: On the Orientation of Engineering Talent Training in Higher Education Institutions. Researches in Higher Education of Engineering (5), 11–17 (2009) (Chinese)
4. Zhangf, J., Fang, H.: Strengthening the Open Experimental Platform Building, Promoting Student Innovation Ability Training. Modern Educational Technology 19(7), 132–134 (2009) (Chinese)
5. Stackpole, B.: The evolution of a virtualized laboratory environment. In: Pro. of the 9th ACM SIGITE Conference on information Technology Education, Cincinnati, OH, USA, October 16 - 18, pp. 243–248 (2008)

Construction of Ecological Curriculum Resource System for "Biology Pedagogy"*

Tu Min, Song Zhiyong, and Cui Hong

School of Life Science
HuaZhong Normal University
Wuhan, Hubei Province, China
tuerya111@163.com

Abstract. This thesis took HuaZhong Normal University (CCNU) as an example to introduce the existing states of the biology pedagogy course in universities, put forward the concepts for construction of the ecological curriculum resource system of "biology pedagogy" and presented the ideas for developing three resource environments: autonomous learning, interactive learning and dynamic circulation.

Keywords: biology pedagogy, ecological, curriculum resource.

"Biology Pedagogy" is the core curriculum of biology teacher education and has been gradually developing from a course to be a curriculum system in recent years. Taking HuaZhong Normal University (CCNU) as an example, currently it has already preliminarily established the curriculum system with the "Biology Pedagogy" as the core and with the characteristics of biology teacher education, including biology pedagogy, teaching skills training, history of life science, biological experiment teaching design in middle school, instructional technology and experiments of biology, biology teaching reform and research in middle school, sustainable development education, design and research of biology activity lessons and practice teaching etc., in total 305 credit hours of 24 academic credits.

With the construction and implementation of the curriculum system, the development of curriculum resources will surely be positioned and dug again. The relevant research introducing the ecological view into the curriculum is a hot issue in late 1980s in the research field of curriculum. The ecological view of the curriculum resource means that, directed by the fundamental ideas of the ecology, the three-dimensional teaching resource and its surrounding environment system are regarded as an ecological system, and the elements within the three-dimensional curriculum resource are mutually dependent and in mutual competition. These elements

* This work is supported by CCNU A1 curriculum resource construction project of Teacher Education, and this thesis is one of the achievements made in the innovative platform of CCNU country Teacher Education—the research on theoretical innovation (Grant No.985ZB0404).

D. Zeng (Ed.): Advances in Computer Science and Engineering, AISC 141, pp. 51–58.
Springerlink.com

incessantly exchange of materials, energy and information and thus making the system in a dynamic balance [1].The learning environment is the place for learners to explore and to autonomously learn, where there are abundant information resources and learning tools. It can be said that the learning environment is the important source of curriculum resources.

This concept of ecological construction of curriculum resource satisfy the very requirements for building up multimedia, three-dimensional and ecological high quality curriculum resource for tuition-free normal college students. Based on the ecological view of curriculum resource, the authors gave serious thinking on the construction of "biology pedagogy" curriculum resource and presented the idea of constructing the ecological curriculum resource system of "biology pedagogy".

1 Concepts of Constructing Ecological Curriculum System of "Biology Pedagogy"

The ecological resources are a kind of dynamic resources of self-growing and self-update and with the forms of life. The situational theory emphasizes that, effective teaching should give overall consideration to the factors such as educator, learner, teaching content, subject environment and social environment etc, and put the learning and development of students in an open ecological system which is dynamic interaction with the outside world [2]. The construction of ecological resource should be considered in this kind of ecological system. In the authors' opinion, the ecological curriculum resource should have the characteristics of integrity, situation, openness and generative etc.

Based on the concept of "environment is resource" emphasized by the ecological resources and the characteristics of biology pedagogy course, the three types of learning resource environments supported by information technology is constructed, i.e. autonomous learning resource environment, interactive resource environment and dynamically-circulating resource environment, which make up of the multimedia, three-dimensional and ecological resource system. The three types are independent resource environments and also may compose of a mutually-influential, mutually-utilizing and mutually-generative resource system.

2 Construction of Ecological "Biology Pedagogy" Curriculum Resource System

2.1 Construction of Autonomous-Learning Resource Environment Based on Digital Teaching Resource Bank

The construction of digital teaching resource bank refers to building up a unified teaching platform or utilizing the existing teaching platform (e.g. web-based courses and excellent course etc) in order to meet the need of the time, and integrating a variety of excellent teaching resources into the resource bank to present and deliver these curriculum resources to teachers and students. The construction of digital teaching resource bank of biology pedagogy is not only helpful to provide more means of obtaining resource for the teachers, expand their teaching vision of and reduce the time of selection and filtering from mass storage of information from Internet, but also conducive to the high degree sharing of teaching resources for the convenience of teachers to extract proper resources for preparing lesions and improve the efficiency, and for students to conduct their learning activities by obtaining the corresponding resources according to their needs and to break the teacher-centered, textbook-centered and classroom-centered traditional teaching model, and thus forming the student-centered new autonomous learning. The framework of the "biology pedagogy" curriculum resource platform is shown as follows:

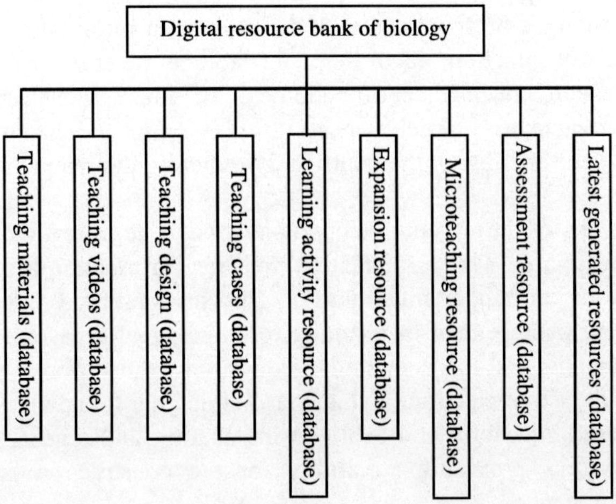

Fig. 1. Digital teaching resource bank chart of biology pedagogy

The construction of digital resource ban can simulate the construction of mature web-based course platform. In addition to traditional teaching materials such as text, pictures and animations, the resource bank also includes the real video of teachers' classroom teaching of this course, as well as the teaching design prepared by teachers in accordance with the teaching content. Also, in every unit, the teaching designs for different types of classes, e.g. instructional class, exploratory class and information class can be constructed to lay a theoretical foundation for isomeric classrooms between schools and classes. The resource bank of teaching cases and learning activities provide a large quantity of excellent case resources of biology classroom teaching and learning activities in middle school. The expansion resources are mainly web-based resources and research-oriented studying resources, supporting the extracurricular expansion research of students. The microteaching resources gather the excellent written microteaching plans, videos of splendid cases of normal university students' microteaching trainings, as well as the use of microteaching classroom and operation tutorial videos etc. The assessment resources include teachers' teaching introspection, students' gains from learning and classroom diagnosis resources etc. The latest generated resources cover the newest generated texts or videos of students' activities, schoolwork and assessment and teachers' introspection and self-diagnosis. The construction of the digital resource bank bases on the curriculum content to classify and store resources that might be involved in the learning process of students and meanwhile realizes the resource update and teacher-student interaction. By virtue of this resource bank, students can conduct better autonomous learning and research-oriented learning.

2.2 Construction of Interactive Learning Resource Environment Supported by Interaction Technology

The interactive electronic whiteboard inherits the functions of ordinary blackboard and the education equipments like traditional multimedia, but it also innovates and develops its own specific functions with the support of information technology, e.g.: interaction control function, recording and storage function, network connection function and resource management function etc. The interactive electronic whiteboard has obvious advantages in such aspects as displaying and handling the teaching content, storing the writing on the board and promoting the interaction of teacher and students [3].

As the carrier of information technology and a teaching aid, the interactive electronic whiteboard lays a foundation for the information-based teaching and realizes a series of important links of information-based education, such as organization and management of teaching resources, real-time recording and use of generative resources and the cooperative education from different places. It can be said that the interactive electronic whiteboard builds the foundation platform for the information-based education and achieves the sharing and regeneration of teaching resources, and thus providing possibility for the construction of the interactive resource environment.

The interactive electronic whiteboard itself has the resource database for storing various types of resources needed by classroom teaching and learning, e.g. videos, flash animations, texts, audio files, web-based materials and traditional teaching materials for direct display and use in classroom teaching, and it may also integrate and manage the resources like the instruction videos of expert teachers of this curriculum, teaching design resources based on students' activities, biology teaching videos of middle school teachers, audio and videos of expert assessment of teaching, used for students' classroom learning, autonomous learning and cooperated learning.

In the classroom teaching, dynamic generative resources will be produced continuously. The interactive electronic whiteboard has the function of recording writings on the board, and with the course information collecting system, the electronic whiteboard may also record the whole process of the classroom teaching as video files and form the vivid and abundant renewable resources. Through watching these teaching videos, students can learn outside the classroom, teachers may carry out the teaching retrospection and instructional diagnosis and experts can assess and comment the course.

Under the network environment, the interactive electronic whiteboard can achieve the synchronous classroom teaching from different places and complete the collaborative instruction of students. In the class of biology pedagogy, students can watch the real-time classroom teaching of biology by middle school teachers, and experts from different universities can give real-time supervision of biology pedagogy class by this technology. Therefore, the students can receive instructions from course teacher, expert and middle school teacher. The realizing process of collaborative education can generate new teaching resources ceaselessly, for students to learn and to think.

2.3 Construction of a Complete Dynamically-Circulating Resource Environment Supported by the Network Platform and Giving Prominence to Generative Resources

The digital learning platform and interactive resources are integrated via the network platform to construct a complete dynamically-circulating resource environment. Through this platform, students can obtain existing teaching resources and generative resources and know their own learning situations and processes. As a matter of fact, for the ecological "biology pedagogy" curriculum resource system, this "dynamically-circulating resource environment" is the major display form and core. Through this resource environment, the resources can be circulating in the ecological resource system, renewed and reused in the three environments.

At the time of using and consuming resources, only the active exploitation and utilization of renewable resources can prolong the life cycle of resources and make them cycling in the "ecological system". In the ecological system of resources, the educator and the educated are not only the consumer, feedback persons and beneficiary of, but also the producer of resources. In other words, on precondition that the educator maximizes the resource utilization with the assistance of information technology, the whole process of learning can regenerate abundant usable resources, i.e. generative resources. As shown in Fig. 2, the three resource environments are equivalent to various environmental systems in the ecological system and the arrows represent the material cycle, energy flow and information exchange between the systems. In the ecological curriculum resource system, the generative resources are the flowing materials and energy and maintain the dynamic balance of the whole ecological resource system.

The so-called "generative resources" refer to various kinds of teaching conditions and factors which are produced during the teaching process or after-class and able to progress the teaching and learning, but which are different from the existing teaching resources before the classroom teaching. For example: learning situations of students in the classroom activities, including interest in learning, activeness, attention, learning methods and ways of thinking, cooperation ability and quality, opinions, suggestions, ideas, questions asked, arguments and wrong answers, learning reflection, and diagnosis assessment data etc[4].

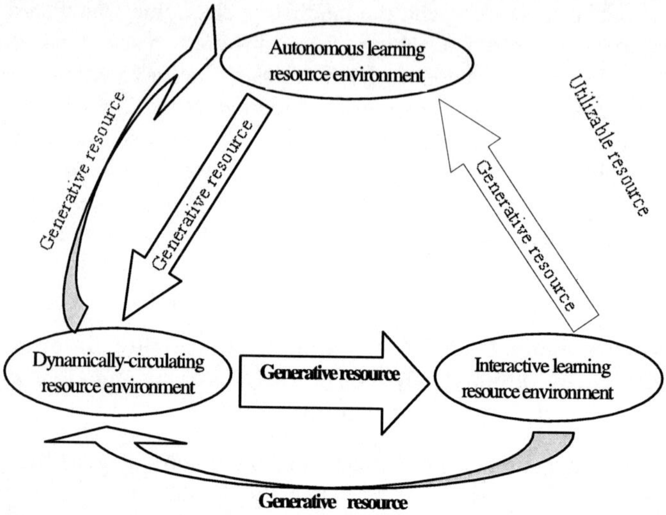

Fig. 2. Ecological resource system

The generative resources mainly comprise of process resource, retrospective resource and diagnosis assessment resource. The process resource is derived from thinking and confusions of students in the learning process and the randomly-generated non-prearranged questions asked by teachers in the teaching process, as well as the differential resources produced in teaching and learning by students with different cognition basis and styles. The retrospective resource is generated during the high-level thinking activities in the assessment, control and adjustment of the cognitive process and the retrospections of teachers and students are the major source of the retrospective resources. The diagnosis resource indicates the analysis of teaching and learning activities by use of professional teaching and learning diagnosis software, e.g. OLAP technology[5], to obtain the visual analysis results, which can give teachers and students the most visualized information about the teaching and learning activities. In the resource system of biology pedagogy, there are three sources of the diagnosis resources:

1) The classroom records: the diagnosis and analysis data for the biology pedagogy classroom records are helpful for students to optimize the teaching design and provide a reference case for students.

2) Learning activities under interactive environment: the diagnosis and analysis data for the learning activities under interactive environment can give prominence to the effect and feedback of individual learning activities, assist teachers and students to more profoundly know the differences between classroom learning activities and facilitate the better learning and use in the future.

3) Excellent middle-school classroom teaching cases: the diagnosis and analysis data for the excellent teaching cases in middle school classrooms are an important resource for the normal university students. For the teachers-to-be, it provides concrete and detailed data support for them to know about the classroom teaching. It is different from abstract concepts, so it is a very practical learning resource for teachers-to-be who can't do practices temporarily.

The resource cycling must be in an open system because the openness of the system provides a basis for the abundance of resources and promote the system's self-organization. Therefore, we must take open website as the platform of providing the resource exchange, integration and reutilization for "autonomous learning resource environment" and "interactive resource environment" and the platform for sharing and processing of "generative resources", thus forming a complete dynamically-circulating ecological curriculum resource system.

3 Conclusions

In the ecological resource system, the resources can incessantly update by itself as the teaching activities progress forward and accumulate spirally in the process of recycling and regeneration, so that the resource system becomes increasingly large. The effective resource utilization and the dynamic balance are the important conditions for the sustainable development of the resource system. The authors hold the opinion that, these measures must be taken for promoting the sustainable development of ecological curriculum resource system: firstly, to strengthen the infrastructure construction and improve the carrying capacity of resources, because the information technology is the basis supporting the construction of the system; secondly, to change the teaching ideas of resource constructors and teachers, improve teachers' ability in information technology via training and enable them to master the use and maintenance technologies of the equipment, and the learners must also have the ability of using resources; thirdly, to make good preliminary preparations for the resource construction, including the teaching design and course teaching record according to the construction conditions, and the design should adapt to students' characteristics and the resource types in need; fourthly, to normalize the construction standard and policies, avoid the redundant resource construction by clarifying the directions and content of construction, and make the construction smooth through utilizing related policies to guarantee the construction conditions.

Today when the tuition-free normal university education policy has been executed for three years, six normal universities directly under the Ministry of Education, while responding to the policy, are actively developing and exploring the means to improve the quality of instructional courses of tuition-free normal university students. In HuaZhong Normal University, during the construction of "Innovation Platform of Teaching for Tuition-free Normal University Students", the excellent curriculum resources are important to guarantee the quality of this platform. The "excellent curriculum resources" indicate the curriculum resources with high degree of display and as the benchmark in the field, which generally have the main features of ecological learning content, multimedia display and three-dimensional learning space[6]. The construction of the ecological "biology pedagogy" curriculum resource system is considered under the principle of excellent resources. The curriculum resource of biology pedagogy is seen as a natural ecological system, reviewing the exchange of material, energy and information between resource environments from the perspective of ecological view, and therefore constructing the curriculum resource system which enables the resource recycling. Meanwhile, the dynamic balance and sustainable development of the resource system can be achieved through various guarantee measures. Of course, the construction of this resource system needs the powerful support of information technology and the high concern of the university administration, which provides the construction conditions. In the further teaching and learning practices, the resource system will be perfected and adjusted according to the basic qualities of tuition-free normal university students, teaching environment and teaching and learning feedback from teachers and students. It attempts to be a curriculum resource system as the benchmark of biology pedagogy and finds out a new approach for the construction and development of China's biology pedagogy curriculum system.

Acknowledgment. I'm mostly grateful to my supervisors professor Cui Hong and associate professor Li Juan without whose support and patience this project would not even have got off the ground. I'm also grateful to my colleagues DengYanmin and Yu Xiao. For their time spent on brain storming and panel discussions with me .

References

1. Zhang, H., Li, J., Ding, S.: Construction of the student-oriented three-dimensional ecological system of teaching resources. Journal of Teaching and Management 1, 54 (2009)
2. Zhang, W., Guo, Y.: Exploration of ecological physics teaching based on the situational learning theory. Curriculum, Teaching Material and Method 5, 59 (2006)
3. Zhang, M., Wangxia.: The platform for constructing the information-based education by the interactive electronic whiteboard——basic idea description of the manuscript Interactive Whiteboard and Teaching and Learning Innovation. Modern Distance Education Research 1, 26–27 (2010)
4. Dong, Z.: With an eye to development, based on generative——on the characteristics of generative resources and its implementing strategies in information technology teaching. E-education Research 4, 67 (2006)
5. Wang, L., Li, Y.: Teaching diagnosis and evaluation model based on OLAP technology. Computer Engineering 5, 49 (2003)
6. Teaching Information Resource Construction Team of HuaZhong Normal University, Code for the High Quality Teaching Information Resource Construction for Tuition-free Normal College Students (2010)

Deepening "4E" Talents Cultivation Model and Exploring "4331" Systematic Operation Mechanism

Jiding Zhu and Rongrong Sun

Department of Architectural Engineering
Henan Polytechnic Institute
Nanyang, China
Zhujd777@sina.com, sunrr2006@126.com

Abstract. The strategy of "expert + enterprise", "employment + entrepreneurship" is the important connotation of College-Enterprise Cooperation and Work-Study Combination for higher vocational educations. Taking '4E' talents cultivation model as breakthrough point, the Construction Decoration Engineering Technology Major is established with a 3-in-1 course system of work-study combination, quality development and business incubation. It aims to realize the teaching organization mode based on mixed segmentation of three semesters, improve the workshop teaching model characterized as "integration of science, practice and employment" , and establish a set of "4331" systematic operation mechanism through constant exploration and gradually improve it.

Keywords: Higher Vocational Education, College-Enterprise Cooperation, Talents Cultivation Model, Operation Mechanism.

1 Introduction

The innovation on higher vocational education has developed from connotation construction into mechanism construction. Through innovation and practice for many years, diversified talents cultivation models in the higher vocational education have individualcharacteristics, but the operation mechanism of supporting talents cultivation model remains to be further developed. In the process of cultivating talents of higher vocational education, such aspects involved in the curricular system, teaching organization, curricular teaching mode and others need to be systematically rationalized so as to form a complete operation mechanism and guarantee the effective enforcement of talents cultivation model.

2 "4e" Talents Cultivation Model

"4E" talents cultivation model is formed based on talents cultivation model with both industry features and college features by our practice and exploration in the Construction Decoration Technology Engineering Major in nearly ten years. "4E"

D. Zeng (Ed.): Advances in Computer Science and Engineering, AISC 141, pp. 59–65.
Springerlink.com

refers to “expert+enterprise, employment+entrepreneurship”; “4E” talents cultivation model is focused on the deep integration and interaction of “expert, enterprise, employment and entrepreneurship” in the process of cultivating talents. Generally it can be summarized as follows: establishing expert by enterprise, starting an undertaking by expert, encouraging employment by developing experts and starting an undertaking by joining enterprises[1].

2.1 Developing Experts by Enterprise, Implementing Order-Based Education

In order to jointly cultivate talents, the order-based “harbor design class” has been set up in cooperation with Beijing Dongyi Risheng • Nanyang Harbor Decoration Company for the purpose of cultivating badly-needed talents for enterprises. Based on the demands and advantages of enterprises, the deep college-enterprise combination is actually achieved through cooperation in deeper level and establishment of the interest community for talents training plan, curricular construction, experiment and practice construction as well as the construction of teaching faculty. In order to make full use of talents and resources advantages in enterprises, the experts and technicians have been invited to act as part-time teachers, take part in curriculum construction and undertake teaching tasks. In this way, a good effect has been achieved, and a mutually beneficial talent sharing model has been formed between school and enterprise. School teachers are also assigned to practice in production and regularly take part in production practice so as to cultivate double competent teach faculty. The enterprises play a important role in the whole process of talent cultivation to continuously develop order-based education. While guaranteeing the stable development of “Harbor design class”, “Harbor supervisor assistant class” can also be incubated to implement the talent cultivation by way of Working-Schooling Alteration and strengthen the advantages of order-based education.

2.2 Starting an Undertaking by Experts, Establishing Entrepreneurial Platform

Zhongyuan Decoration Factory attached to the Institute has been reconstruced. Based on the technical advantages of college’s majors, and the requirements of modern enterprise mechanism, Nanyang Carnival Decoration Engineering Ltd. Co., which is focused on technology development and technical services, has been created and co-financed jointly by the society, school and professional teachers. Parts of the technical works of the enteprise can be undertaken by professional teachers, while experimental and practical rooms in the school may also be used for supporting enterprise technology research and development. In cooperation with enterprise, a series of teaching forms can be carried out based on work-study combination, such as practical examination, project driving, on-the-spot teaching and so on. Practice training, in-post practice, employment and others for students can be realized by relying on enterprises. Excellent students can start undertakings relying on enterprise’s brand, technology, funds and other resources. A good platform for students in entrepreneurship has been created by Carnival Decoration Company.

2.3 Establishing a Long Term College-Enterprise Cooperation Mechanism to Jointly Run School, Cultivate Talents, Obtain Employment and Make Progress

Cooperations have been achieved in school-running, talents cultivation, employment and development. Close cooperations in the aspects of talents, facilities, information and others have also been built to realize resources sharing, enrich technical talents for enterprises, improve productivity, train teachers' practical abilities, enhance the quality of professional talents cultivation, and guarantee students' employment so as to achieve three-win situation among school, enterprises and students. Based on mutual interlocking and employment, high-level interaction and regular symposiums, both school and enterprise.actively explore and improve the cooperation management system through constant cooperation and mutual adaptation, make clear rights and interests for both parties, establish a cooperation mechanism corresponding with '4E' talents cultivation model in the long term and ensure a sustainable development of cooperation between college and enterprise.

3 3-in-1 Curricular System

Focusing on the training of professional quality, a 3-in-1 curricular system of business/school partnership, quality expanding, enterprise incubation has been established jointly by college and enterprise; the work-study combination curriculum is developed on the basis of deeply analyzing job requirements and concluding typical work task; as being institutionalized, standardized and normalized, the quality development education is also implemented in the overall process of talent cultivation; Finally, the enterprise incubation education is systematized and blended with the process of talents cultivation to strengthen the continuity and durability of entrepreneurship education, and improve the enterpreneurial ability. This curricular system is consisted of the three systems as follows:

3.1 Business/School Partnership Curricular System

Project curricula are developed for courses of general education and,basic professional competencies; curricula in learning areas are also developd for courses of career development and professional core competencies In this way, theoretical teaching and practical teaching are integrated in the process of the project to realize the combination between theory and practices. Based on social resources, tasks are undertaken from enterprises, and students are encouraged to take part-time jobs in enterprises; Taking skill competitions as an important means for inspecting teaching effects, independent skill competition weeks can also be set up to promote study and assessment through competitions.

3.2 Quality Development Education System

It is required to enhance the enforcement of quality development education, increase enforcement effect. The courses like career planning, skill competition, quality development training, etc. are set up as required courses in college, and detailed curriculum standards are made and put into practice; Courses as quality cultivation and behavior code, technological innovation quality education, club activities and social practice, etc. are also opened as required courses after class, and counted towards graduate credits to improve the status of the courses and enhance the evaluation effect.

Enterprise satisfaction is taken as an important indicator for evaluating the quality of talent training. In the production practice, assessment on students is conducted by enterprises; vocational skill competitions are held jointly by college and enterprise to encourage students to actively take part in the competitions of industrial skills with aim to promote study and assessment through competitions. The enterprises and society are involved in the assessment on talents cultivation quality, especially cultivation and assessment on professional quality, and thus it is easy to find any deficiencies in time and actually improve talent cultivation quality. It is necessary to change the teaching concept, reform diversified academic assessment system and emphasize cultivation and assessment on professional quality.Students are enabled to find out problems, analyze and solve the problems through practical experiences of working tasks, and also trained to take the initiative to study rather than to passively study,so as to realize self-construction and improvement of professional accomplishment.

3.3 Business Incubation System

Business incubation system for students are constructed through college-enterprise cooperation to strengthen professional quality education and achieve sustainable development. A business incubation center for college students is jointly invested and built by both college and enterprise to create a platform of entrepreneurial practice and business incubation. Incubation center is managed and operated totally in line with the production mode of enterprises. Students are enabled to experience the whole process of enterprise operation. Relevant courses like Architectural Economy and Enterprise Management,Entrepreneurship education and Guidance, Entrepreneurship Planning and Experience, Business Incubation and Entrepreneurship supervision are set up covering the overall process of talents cultivation to form a complete entrepreneurship education system and implement "Three Steps" strategy of entrepreneurship education, entrepreneurship experiences and business incubation[2].

Step 1: Conducting entrepreneurship education, cultivating objects with potentials for entrepreneurship. Through the analysis of enterprepreneurship cases, Entrepreneurship education courses are opened to develop students' enterpreneurship consciousness so as to lay down the foundation of entrepreneurship knowledge and cultivate entrepreneurial abilities. Entrepreneurship contest can also be carried out to let students to organize teams, conduct market investigation and study, and write business plans so as to find and select students with potentials as training objects.

Step 2: Experiencing startup. The entrepreneurship education is required to fully reflect the practicality, implement entrepreneurship experience plan and improve the practical ability of starting an undertaking for students. Business incubation center for college students has been established in the decoration supermarket of 'I love my family'. Served as the platform of experiencing real business operations, students are assigned to to do in-post practice, experience the role of each post at the department, develop their abilities of organization management and social communication in the practice, accumulate the experience of enterprise operation and improve practical ability of starting an undertaking.

Step 3: Business incubation. As an entrepreneurial platform, Nanyang Carnival Decoration Company provide students with startup capital, small-sum loan, providing technology support and management mode, and the students are allowed to start an undertaking, set up branch offices, operate independently and take full responsibility of their own profits and losses relying on enterprise's brand, technology, funds and others. Meanwhile, the support on technology and management mode for students' start-ups is also strengthened and promote towards a healthy development. In order to finally realize the sound development of entrepreneurial system, it is necessary to improve the supporting system for initial companies, set up funds for entrepreneurship development, promote systematization, institutionalization and standardization and strengthen management and coordination.

4 Teaching Organization Mode Based on "Mixed Segmentation of Three Semesters"

Three-semester system is implemented by dividing the existing two-semester system into three semesters with two long and one short. The two long semesters are planed for theory teaching and practice teaching and last for 19 weeks respectively; the short semester (summer vacation between the two long semesters) lasts for 2-4 weeks, and an arrangement is made as below: social investigation of decoration company (in the first year), summer social practice outside school (in the second year) and professional practice outside school (in the third year).

Concentrated and mixed segment-based teaching for major professional courses is implemented in the long semesters. Segment-based mode refers to that parts of courses are arranged in segment and the other parts of courses are arranged under traditional mode in a semester. The mode is characterized by relative flexiblity, strong operability and convenience to make partial adjustment under the existing curricular mode, so the teaching arrangement of quality education courses and elective courses will never be affected. For example, if the following 5 courses: A. military training; B. cognition practice; C. art foundation and application; D. building decoration materials; E. CAD and building decoration mapping are offered in one semester, and on a weekly basis, the teaching of A-B-C-D-E courses will be completed by stages. As for courses like sports and politics, the entire semester teaching is maintained and implemented in parrallel with segment-based courses[3].

5 Workshop Teaching Model of "Integration of Theory, Practice and Employment"

Taking workshop as carrier, workshop teaching is a model integrating curricula, classrooms (workshop) and production practice, and transforming traditional closed teaching into open teaching. It is required to combine production with teaching closely based on the curriculum knowledge, as focusing on professional technology application, leading by professional teachers, and targeting at undertaking technology projects. The students are guided by teachers to finish the comprehensive and professional technology training in the process of undertaking and completing production technology projects[4].

Workshop teaching model is promoted to reveal the characteristcs of work-study combination with the integration of theory, practice and employment. As for workshop, a operation mode of combining school supervision with students' self-management are adopted to build a learning environment and atmosphere consistent with the actual post, and establish a platform of combining classroom teaching with enterprise objectives and achieve the concept of "learning in practicing, and practicing in learning". Based on the platform of workshop and engineering projects as carrier, the project teaching can be carried out by introducing productive tasks of enterprises at the right time and learning from operational process of enterprises. So it is required to change the traditional practices of "lecturing, and listening", set up "work group" as organizational form, implement teacher-to-student and student–to-student interactive teaching, and adopt the method of "self-assessment by students, assessment among groups and comments from teachers" to carry out teaching evaluation.

The teaching shall be well-organized through workshop teaching model and recycling process of ""practicing, learning and repracticing" and implementation of the concept of "learning in practicing" and "practicing in learning". Various teaching methods like project orientation and task driving are adopted to prove professionality, practicality and openness, and achieve the effects of "learning in practicing, practicing in learning" on the basis of students knowing "how to do it", and through the process of "doing first, understanding and re-doing". In the process of discussing, questioning and exploring, students will realize the way to think, master methods of work and achieve the conversion from passive learning to active learning.

6 Conclusions

"4E" talents cultivation model is a general guidance framework and "4331" systematic operation mechanism is built for the specific implementation. It is only a part of the mechanism construction of higher vocational education. The process from guidance framework to specific implementation involves many factors such as faculty, internship and practice training, quality assessment, and etc. The mechanism construction can be improved through implementation, and the development of higher vocational education reforms can be promoted further.

References

1. Fan, G., Zhu, J.: "4E" Talents Cultivation Model for Construction Decoration Major. Vocational Education of Machinery, 29–39 (March 2010)
2. Wang, C., Muo, G.: Practice and Exploration of the Economics and Management College Students for Their Abilities to Start an Undertaking. Education &Voacation, 84–85 (January 2010)
3. Zhu, J., Fan, G.: Practice and Exploration on the Talents Training Mode of "Major + Enterprise, Employment + Entrepreneurship. Chinese Vocational and Technical Education, 84–85 (August 2010)
4. Department of Higher Education, Ministry of Education of P.R. China. The CAHE's Affiliation of Combination of Production, Teaching and Research, "The Only Way". Higher Education Press (2004)

A Perspective on Technical and Vocational Education and Training

Sen Cong and Xiaomei Wang

College of Applied Science and Technology
Beijing Union University
Beijing, 102200, China
congsen@buu.edu.cn, bhxmem@163.com

Abstract. The technical and vocational education and training (TVET) is particularly important for promoting economic development, expanding employment size and improving the quality of employment. This paper presents a perspective on the development of TVET. To fit present social and economic developments, vocational school and industry enterprises cooperation is one of the most effective methods for vocational education. Training the skills-based teachers is the important guarantee for the development of vocational education. The information and communication technology (ICT) has provided a wide variety of activities and experiences that support vocational education. The reform will make the TVET more and more sensitive to social and economic changes and demands.

Keywords: technical and vocational education and training, cooperation education, teacher training, e-learning.

1 Introduction

In the current globalizing world, economic markets are shifting from a local to a global arena. Competition in economic markets has become more and more intense. The economic development drives the demand for expertise and high quality workforce in various fields of the whole society, leading to a new round of development of vocational education. In the global market place, skilled workforce is the key to competitive prosperity. The development of technical and vocational education and training (TVET) [1] has become one of the most important strategies in both developing and developed countries. Vocational Education is particularly important for promoting economic development, expanding employment size and improving the quality of employment. Governments around the world are currently undertaking reforms on their education and training systems to meet the demand for an appropriately skilled workforce in an evolving global economy.

Countries which want to survive in the global competition are facing tremendous pressures to improve workforce quality. China has made great efforts to develop Vocational Education. The State Council issued the State Council's decision on vigorously developing vocational education on October 28, 2005. Vocational education is positioned as an important foundation of social and economic development and

D. Zeng (Ed.): Advances in Computer Science and Engineering, AISC 141, pp. 67–75.
Springerlink.com

strategic focus of education. In recent years, Chinese vocation education has continued to grow and develop and it has supplied a large number of skilled workforces for the country. There are a lot of middle and higher vocational schools providing technical and vocational education and training which aim to educate well-qualified workforces. The significance of the cooperation between vocational school and industry enterprises has begun to obtain more attention in China. The vocational educational and training reforms have to be made based on the market needs.

The paper is organized as follows. Section 2 introduces the concepts of TVET and its role in economic development. Section 3 gives an overview of cooperation between vocational schools and industry enterprises. Section 4 presents the perspective on development of TVET using ICT. Section 5 describes qualifications of vocational education. Section 6 provides the conclusions.

2 Technical and Vocational Education

2.1 The Concepts of Technical and Vocational Education

The TVET is part of higher education and a special type of higher education. It is defined as a comprehensive term referring to those aspects of the education process involving the study of technologies and related sciences, and acquisition of practical skills, attitudes, understanding and knowledge relating to occupants in various sectors of economic and social life. The objective of technical and vocational education is to bring up application-oriented talents that have related technical theoretical knowledge and practical ability [2]. Its requirements of theoretical study are low, more emphasis on hands-on capacity and demanding actual working capacity. Vocational education was considered appropriate for professions like automobile mechanics, plumbers, electricians and other similar professions. It is geared towards the practical needs of a specific occupation. Vocation education is seen as strategy that contributes to increase efficiency in education investment and can provide young people with the skills needed for employment in industry.

2.2 TVET and Economic Development

The advent of globalization has demanded more specialized labor markets, higher levels of skills, and diversified vocational education. Improving the skills and knowledge of workforce is crucial for achieving and maintaining economic competitiveness, especially in a context of progressing globalization [3]. The sectors affected by globalization include [4]: (1) trade development; (2) technological innovation; (3) global social network; (4) opportunities for economic growth; (5) advancements in information and communication. Globalization is one of the most important factors causing the fall in the demand for unskilled labor and the rise in wage inequalities between unskilled and skilled labor. These changes created new demands for more adaptable, multi-skilled and creative labor. Vocational education is considered to be the most effective instrument of meeting globalization demands. The economy becomes more knowledge intensive. In the United States, it is realized that the new information-based economy requires a different kind of preparation than the old economy. For years, there has been a growing awareness that general education is often too academic and does not prepare young people adequately for the world of work. Undoubtedly, as compared to general education, the TVET has a closer and more direct link with economic and professional development. It is important element of economic development strategies.

3 Cooperation between School and Industry

3.1 The Teaching of Student

To fit present technological developments, the cooperation between vocational schools and industry enterprises is one of the most effective methods for vocational education [5]. Practices and studies on workplace learning have received increased attention as a result of the increasingly significant role of professional skills and expertise in vocational education. Cooperation makes them strengthen in scientific, technological, and economical ways by bonding school and industry present possibilities [6].

The need of the enterprise is the core of the vocational education system. Mass customization-oriented education is a low-cost and high speed style of cultivating talents with a certain amount of teaching to meet employers' random and uneven demands. On the one hand, it enables schools to personalize education for students, on the other hand, the cost of education is under control and education resources are fully utilized [7]. Education customization enables education resources and knowledge resources fully used, help to solve out of touch between teaching and practice, supply and demand do not match as well as the construction of teachers cannot keep up the pace of development [8].

The flow chart of vocational education is shown in figure 1. The suggestions are given as following: (1) Vocational teaching syllabus and course contents should be created with vocational school-industry cooperation, and should be updated according to technologic developments; (2) Inviting representatives of the industry to teach professional courses; (3) Vocational school should supply students an education opportunity in industry enterprises.

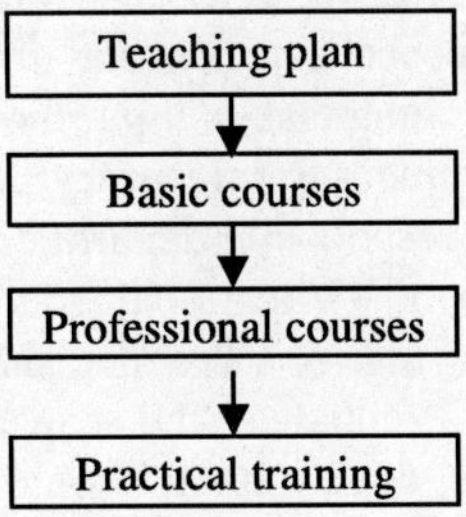

Fig. 1. The flow chart of vocational education

The cooperation between school and industry which is a tool that helps to convert the basic and theoretical knowledge which has been obtained in schools, into a practice is also being used in various countries. With cooperation, the technological information needed by the industry enterprises was transferred from enterprises to school and in return the industry has been providing to schools the opportunity of practical application. This mutual cooperation has been speeding the country's economic development. It is widely used in the United States, the United Kingdom, German, Japan and so on. In the United States, students spend half time learning in vocational school, and half time working in enterprises. In this way, theory can be applied in actual operation [9]. In Japan, in order to improve the efficiency of vocational education, enterprises and vocational school run education jointly. It makes the direction of vocational education more clearly and to be up to the mustard.

3.2 The Training of Teacher

A high level of professional teaching force is essential to ensure the quality of vocational education, to ensure training qualified skilled workforce for enterprises. So the construction of teaching staff is so important, and each teacher should always update his professional knowledge to fit the professional development.

Vocational teachers are disproportionate with the growing number of students in China. According to monitoring surveys, the student-teacher ratio is 26:1 [10]. In addition, skills and capabilities of the teachers are also urgent need to strengthen. Practicality is an important feature of vocational education, and the cultivation of teachers' practical skills is critical guarantee for the vocational education. From the experience of the foreign well-developed vocational education teachers' training, vocational teachers must accept practice training into the industry if they want to obtain rich practical experience. Vocational teacher training should take the initiative in cooperation with the industry [11]. It is easy to find the combination point of vocational education and the industrial production. The professional training can make the teachers perfect their own knowledge and skills. The introduction and cultivation of double qualified teacher has been valued in vocational school of teacher improvement. It requires and encourages vocational teachers to make more cooperation with industries.

Compared with foreign developed vocational education, there is a big gap on the aspect of Chinese vocational education in teacher training. Firstly, vocational teachers cannot get enough practice, especially practice in industries. Secondly, they are understaffed and their usual teaching task is tight, the practice time is limited. Thirdly, they want to obtain the improvement training very much, but they cannot get enough training funds and government responsible departments are in charge of limited funds which cannot well meet demand. Vocational schools need to be reformed. The suggestions are given as following: (1) Teacher should establish self-awareness of professional development, and enhance the capacity of individual professional practice; (2) The school and government should raise funds through multiple channels, and carry out various trainings of vocational teacher; (3) The appreciate training method will be selected according training content to improve pertinence and practicality of training mode.

4 Vet Reformation Using ICT

The transfer of knowledge is achieved mostly by lecturing in traditional education. The key shortcoming is that the students are not motivated enough to acquire knowledge actively [12].To respond to the resulting updated requirements of economic trends and goals, vocational educational planning has to be shifted to new forms of instructional content and delivery. Educational technology optimizes teaching process and improves educational quality through the design, development, utilization, management and evaluation of teaching process and related resources.

With the advent of information society, modern information and communication technology (ICT) has had a profound impact on the structure, content, resources and implementation of practice. The ICT has provided a wide variety of activities and experiences that support vocational education. Informational technology provides a wealth of resources and solid physical security for vocational curriculum integration. The e-learning method provides the base for efficient integration of the vocational education into global information society [13]. Educational process can be significantly improved by creating new educational resources with visually appealing multimedia contents that interactive elements and up-to-date reusable information.

The development of e-learning offers new possibilities for learning and leads to drastic changes in vocational education due to its flexibility to access, just-in-time delivery, and cost-effectiveness. E-learning is defined as the use of computer multimedia technologies and the Internet to deliver information and instruction to learners. Multimedia technology is increasingly being used to create instructional environments for vocational education. With multiple means such as diagrams, animations, textual explanations, photos, video and audio available for presenting the same information, teacher can create different environments to meet the different learning style of students. Multimedia fits well into moves to make students more active, autonomous, and able to take control over and responsibility for their own learning. E-learning breaks the restriction on the time and space of traditional classroom teaching, and makes it possible that students can learn through the network whenever, wherever and whatever they want. It enables vocational education to adjust learning requirement and update knowledge resources in a more efficient way. The architecture of ICT-based vocational teaching platform is shown in figure 2.

In the platform, three interfaces are provided for students, teachers, and experts coming from industries. The student interface enables the student to maintain personal information, access learning material, assess learning performance, and ask questions. The teacher interface enables the teacher to maintain learning material and assessment rules, and process students' inquiry. The expert interface enables the expert to process and facilitate discussion, maintain the expert knowledge, and make the assessment rules together with teacher.

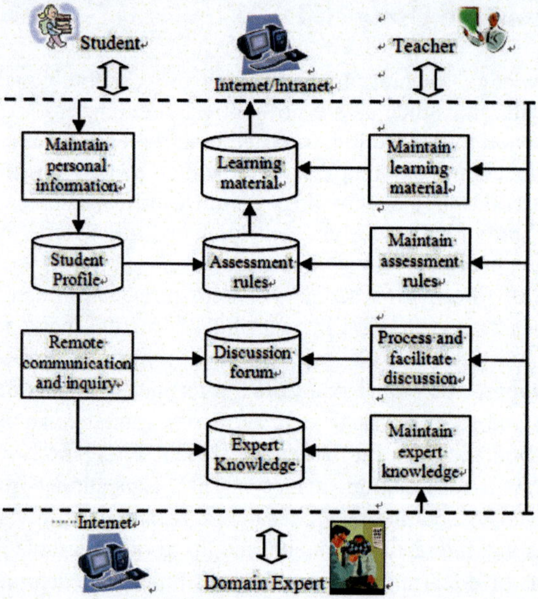

Fig. 2. The architecture of ICT-based vocational teaching platform

In the ICT-based vocational teaching platform, the discussion forum is an important communication tool for the students and teacher. It can provide the diversified interpersonal interaction such as between teacher and student, between student and student, between domain expert and student, and communication in a group. Students also can ask questions and submit their assignment by email.

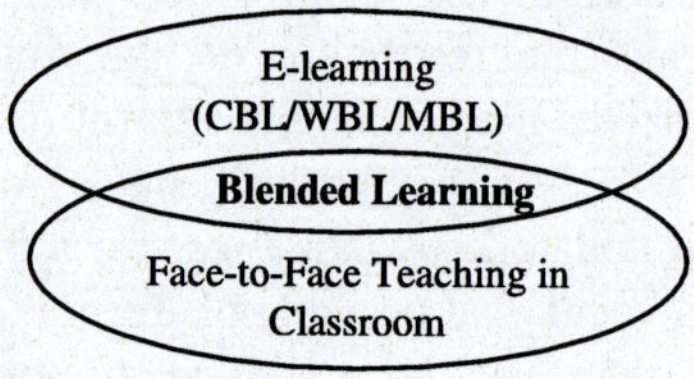

Fig. 3. Concept of blended learning

Blended learning is a connection point of traditional classroom teaching and E-learning and integrates the advantages of them. It is realized as a combination of a face-to-face environment and E-learning [14]. It aims at creating the most efficient learning environment [15]. The concept of blended learning is shown in figure 3.

5 Qualifications of Vocational Education

Achieving vocational qualification is one of the main objectives of education that are provided in vocational schools. In schools, theoretical and practical training is given to the trained staffs that are needed by the related industries. It has been expected for the graduates of vocational schools to contribute dynamic impacts to the development of the economy process. Therefore, every vocational school must be in accordance with professional competence with the training of its graduates had received and the expectations of contemporary human resources.

More and more graduates of vocational programs enter higher education. In the UK, schooling is compulsory from age 5 to 16. After completion of compulsory education in secondary schools, students remaining in education at a school may choose between general (academic) and vocational subjects or take a mixture of the two. School-based vocational qualifications and pathways are developing. The national vocational qualifications (NVQ) are designed for people to gain recognized qualifications for specific occupations. The general national vocational qualifications (GNVQ) have been introduced in the 1990s, and vocational general certificate of secondary education (GCSE) and vocational A-level have been introduced since 2000. The national qualification framework in UK is shown in table 1 [16].

Table 1. The National Qualification Framework in UK

Level of qualification	**General qualifications**	**Vocationally-related**	**Occupational qualifications**
5	Higher-level qualifications		Level 5 NVQ
4			Level 4 NVQ
3 Advanced	A-Level	Vocational A-level	Level 3 NVQ
2 Intermediate	GCSE Grades A - C	Vocational GCSEs	Level 2 NVQ
1 Entry Level	GCSE Grades D - G	Foundation GNVQ	Level 1 NVQ

A vocational GCSE is equivalent to two academic (general) GCSEs and enables progression to further education, training or employment. The government in the UK wants to promote the parity of esteem between vocational and more traditional academic subjects through the introduction of GCSEs in vocational subjects. It is of certain reference value for vocational education in China. The coordination will be made of general education together with theoretical and applied vocational education. Updating the contents of vocational education, recognition and provision of equivalence of titles, certificates and continuous education programs within the context of national vocational qualities are necessary.

6 Conclusions

There are improvements and changes in vocational education in concordance with continuous technological and social changes. To fit present social and economic developments, cooperation between vocational schools and industry enterprises cooperation is one of the most effective methods for vocational education. Training the skills-based teachers is the important guarantee for the development of vocational education. Information and communication technology (ICT) has provided a wide variety of activities and experiences that support vocational education. There is no doubt that TVET will become more and more sensitive to social and economic changes and demands.

Acknowledgment. This research is supported by Funding Project for Talents Scheme under the Jurisdiction of Beijing Municipality, Project No. 2010D005022000010.

References

1. Wang, M.D.: Speech on Chinese Vocational Education Conference. Education and Vocation 7, 15–21 (1996)
2. Deissinger: The evolution of the modern vocational training systems in England and Germany: A comparative view. Journal of Comparative Education 24, 17–36 (1994)
3. Mouzakitis, G.S.: The role of vocational education and training curricula in economic development. Procedia Social and Behavioral Sciences 2, 3914–3920 (2010)
4. Siochru, S.O.: Social consequences of the globalization of the media and communication. Labour Office, Policy Integration Department, Switzerland (2006)
5. Ivanco, V., Kostolny, K., Kubin, K.: Co-operation between the technical University of Kosice and Industry in its region. Global Journal of Engineering Education 2, 157–160 (1998)
6. Isgoren, N.C., Cinar, A., Tektas, N.: The importance of cooperation between vocational schools and industry. Procedia Social and Behavioral Sciences 1, 1313–1317 (2009)
7. Li, N.Q.: The design of teaching plan of higher vocational education under the circumstance of mass customization. Vocational and Technical Education 34, 39–41 (2002)
8. Chen, J.X., Gu, F.: Study on the method of mass customization in the higher education. Researches in Higher Education of Engineering 5, 20–23 (2007)
9. Leonard, C.: Vocational education and training in the developed world. The Vocational Aspect of Education 115, 173–182 (1991)
10. Chen, J.D.: Thinking about construction of teachers training system in Chongqing vocational shools. Journal of Chongqing Vocational & Technical Institute 17(2), 5–7 (2008)
11. Han, H.Z.: Comparation of the way to cultivate teachers for professional education in China and foreign countries. Journal of Changchun University 16(5), 107–108 (2006)
12. Bates, T.: Managing technological change: strategies for college and University leaders. Jossey-Bass, San Francisco (2000)
13. European Commission, Directorate General for Education and Culture. E-learning in continuing vocational training, Final Report, EAC-REP-003 (2005)

14. Bozic, N.H., Mornar, V., Boticki, I.: A blended learning approach to course design and implementation. IEEE Transactions on Education 52(1), 19–30 (2009)
15. Bersin, J.: The blended learning handbook. Wiley, New York (2004)
16. Cuddy, N., Leney, T.: Vocational education and training in the United Kingdom. Office for Official Publications of the European Communities, Luxembourg (2005)

Development of Mineral Processing Engineering Education in China University of Mining and Technology*

Haifeng Wang, Yaqun He, Chenlong Duan, Yuemin Zhao, Youjun Tao, and Cuiling Ye

School of Chemical Engineering and Technology, China University of Mining and Technology
Key Laboratory of Coal Processing and Efficient Utilization, Ministry of Education
Xuzhou, China
whfcumt@126.com

Abstract. Mineral processing plays a key role for preparing fuel or materials in coal industry, steel industry, and non-ferrous metals industry. Lack of energy resource and pollution has baffled the economy development of China. Research on mineral processing is giving more and more attention and Mineral processing engineering education has been given high priority in China. Taking the personnel training of coal processing and utilization as a background, the development of mineral processing engineering education in China University of Mining and Technology (CUMT) is introduced in this paper. The faculty group, laboratory and scientific research conditions of the discipline are presented. The personnel training scheme of the subject, the courses system, student criterion are also discussed, and how to cultivate excellent graduates in mineral processing engineering field to meet the demand of high education internationalization are discussed.

Keywords: engineering education, mineral processing, personal training, teaching reform.

1 Introduction

More than 3000 years ago, the ancient Chinese had mastered and practiced distinguishing and sorting diverse minerals and hence developed technology for the alloy's composition of bronze casting [1]. Now, Lack of energy resource and pollution has baffled the economy development of China. Research on mineral processing is giving more and more attention and Mineral processing engineering education has been given high priority in China. Now, more than 30 universities set up the discipline of mineral processing engineering. Accordingly, a large number of technical personnel including bachelor, master and Ph.D. in the field of mineral processing engineering graduated from many universities and research institutes, and

* This work is support by the project of the construction point of the national special discipline, the project of the national teaching group and the China Education Ministry Key Lab Open Project (CPEUKF08-05).

D. Zeng (Ed.): Advances in Computer Science and Engineering, AISC 141, pp. 77–83.
Springerlink.com

contribute to the sustainable development of coal industry, steel industry, and non-ferrous metals industry. In terms of the number of universities set up mineral processing major is 30, and in terms of number of graduates is about 1800 undergraduates, 500 masters and 100 doctors in 2009 [2].These universities and research institutes cultivate lots of mineral processing engineering persons to meet the demand of rapid economic development.

CUMT is a Chinese national key university being directed under the Ministry of Education, and it is one of university firstly setting up the discipline of mineral processing engineering. It has been officially listed as one of the selected university for the national ‘211 Project’. With the development of over 100 years, CUMT has formed a discipline major system with a priority to engineering, with mining feature, and combined with science, engineering, literature, management, law, economy and other multiple subjects developed harmoniously. CUMT currently has 2953 staffs, which includes more than 1600 teaching stuff, 7 academicians of the Chinese Academy of Engineering, 197 supervisors for doctoral graduates, 270 professors, and 385 associate professors. There are more than 44 000 students in CUMT Xuzhou campuses [3].

2 Establishment of Mineral Processing Engineering Discipline in China

There was not a separate discipline of mineral processing engineering before 1952 in China, which was a part of the mining engineering. In 1952, with the innovation of Chinese high education, the discipline of mineral processing engineering, then called concentration of useful minerals, was founded in the Beijing College of Mining and Technology, the former of CUMT, and associated with related disciplines of Tangshan Jiaotong University, Beiyang University, Tsinghua University, and Jiaozuo College of Technology. There is a new beginning of China’s mineral processing engineering education, and after then the discipline was going to develop and grow in strength on a never straight road. Now, it takes a leading role in the field of mineral processing engineering education. It was authorized to confer master’s degree in 1981, doctoral degree in 1984 and postdoctoral scientific research station in 1985. It entered into the key disciplines of Jiangsu province and the former ministry of coal industry in 1994, and was qualified to pass the inspection of the outstanding brand discipline of Jiangsu province in 2006, as well as was selected as the national key discipline again in 2007. Nowadays, the discipline of mineral processing engineering makes great contributions to national coal industry with the predominance of coal preparation, coal chemical engineering, coal combustion, and pollution control, etc.

3 Development of Mineral Processing Engineering Education in CUMT

3.1 Teaching Staff

There are about 30 teaching faculties in the discipline of mineral processing engineering of CUMT, which form a high-level scientific and researching faculty with two members of the Chinese Academy of Engineering as the leader of the staffs. It has some noted experts and professors, including two academician of the Chinese Academy of Engineering, 14 professors, 6 associate professors, 11 supervisors for doctoral graduates, 20 supervisors of master graduates, 2 'Yangtze River Scholar Prize Program Project' Scholars, 2 owners of 'National Outstanding Young Scholar Fund', and 1 'Cross-century talented person' [3, 4].

The faculty of the discipline obtained many kinds of honors, such as 'the national level youth expert with prominent contribution' owner, 'Chinese youth of science and technology prize' owner, ' National distinguished teacher' owner, 'Jiangsu Province top ten outstanding youths' owner, 'Jiangsu Province outstanding educator', etc. Over thirty times awards had been gained in various honor titles over provincial or national level. The discipline of mineral processing engineering attained the outstanding discipline team of 'Lan-Qing project' of Jiangsu Province in 2004. In 2007, a research echelon of this discipline (high efficient fluidization clean coal theory and technology) was selected as one of the Science and technology innovation groups of Jiangsu Province, and the research of the "theory and appliance for high efficient clean processing of coal resource" had gained the sustentation fund for the innovation research group of the national natural science fund. Taking the processing and utilization of coal resource and clean coal technology as the feature, the discipline of mineral processing engineering was approved as a national key discipline and a key construction unit of the national '211 project' of CUMT. Recently, it gets the national "985 innovation platform constructing project" for the discipline development and evolution. It has made a great contribution to the education and scientific research of mineral processing engineering in China and it also has got a high academic position and influence in this field of the world.

3.2 Experimental and Practical Conditions

This discipline has a strong back ground of engineering. Experiments and practical teaching are very important for the cultivation of the spirit of innovation and practical engineering capacity of the undergraduates. Accordingly, the top-ranking experimental and practical engineering cultivation system for the education of mineral processing engineering has been formed in CUMT. The laboratory possesses a lot of advanced instruments including high-speed video camera system for motion particle analysis, scan electrical mirror, SEM, gas chromatography/ infrared spectrum apparatus, the determinative instrument of molecular weight, etc. The laboratory of mineral processing engineering is the key lab of the ministry of education.

There are many traditional experimental lessons setting for the undergraduates, such as "experimental technology of mineral processing", "metal working practice", "experiment technology of electrician and electron". At the same time, some of the experiments were extended. The innovative experiments, especially combining with the advanced scientific and research projects, are enrolling the undergraduates. A number of experimental research platforms of new technology for mineral processing were established, and then the depth and extent of experiment teaching were also extended. In recent years, many kinds of pilot experimental research systems were set up with the support of the national “211 project” and the national “985 innovation platform constructing project”. The distinctive experimental researching systems were built based on researching projects, which including coal dry beneficiation and screening pilot scale system, coal slurry pilot-scale system, high sulphur coal desulphurization and ultra pure coal preparation pilot scale system, simulation and control system of jigging separation, etc.. Most students can take part in the practice research projects and to develop technique training for achieving scientific exercitation, and many kinds of practical training were developed to cultivate the students’ engineering capability.

The discipline of mineral processing engineering is the traditional advanced one in CUMT, and well cooperated with coal preparation plants, mines, and so on. This set up a firm foundation for the construction of practice base. So far, many coal mine groups such as Huaibei Coal Mine Corporation, Yanzhou Coal Mine Corporation, Datun Coal Mine Corporation and so on, had selected as the prior practice bases [5]. In the plants, the students can combine the knowledge from books with the practical work and confirm the theoretical knowledge via work experience or communion with the engineers.

3.3 Personnel Training

The compound high-level engineering capability persons and talented people of the exploitation, design and management of new techniques were trained in the discipline of mineral processing engineering. The undergraduate students enroll according to the specialty of mineral processing engineering, and the training model was '5+3' subsection education method, which based general education on the extensive and innovative personnel training. More attention attach to innovative ability and engineering accomplishment of the students training with the specialty. For the undergraduate students, required courses, such as Advanced mathematics, Physics, Descriptive geometry and Engineering drawing, Computer technology basis, Philosophy principle, English, etc have been opened. The specialty-related basic courses and some specialized courses are selected by the undergraduate students, such as Organic chemistry, Inorganic chemistry and analytical chemistry, Physical chemistry, Chemical principles, Mineral processing, Engineering mechanics, Hydrodynamics and mechanism, Electrical and electronic technology, Mathematical model of mineral processing, Design of mineral processing plant, Electrical engineering, Modern enterprise management, and so on [3, 6]. Meanwhile, three times exercitation in different coal preparation and mineral processing plants are arranged for the students to learn practical skills in the plants. The students' specialized training is strengthened at the 6th semester and the employment selections are free while students have graduated.

Since 1953, more than 3000 undergraduate students, about 390 graduate students, and approximately100 doctoral students have given the mineral processing engineering education in CUMT. Now, we have 567 undergraduate students, 89 graduate students, 20 doctoral students and 54 oversea students in this discipline.

4 Tearching and Scientific Research Achievements

4.1 Tearching Achievements

Many honors and achievements on course construction and teaching have obtained in the discipline of mineral processing engineering. Courses of 'Mineral processing' and 'Mineral processing plant design' were appraised the national excellence courses. There are 3 provincial excellence courses, 4 ministerial excellence teaching materials, and so on. Furthermore, the discipline has awarded the first prize of national teaching awards with modern teaching artifice. In 2006, 'Coal chemistry' won the provincial excellence course, and 'Design and management of coal preparation' won the key excellence teaching material of construction, 'Mineral processing mechanism' was ranked on the national planning teaching material of higher education in the '11th five plan'. In 2007 the teaching group was selected as a 'national outstanding teaching unit constructing project'. In 2008 the subject of mineral processing was approved as a 'national featured subject constructing project'.

4.2 Scientific Research Achievements

In recent years, the faculties of mineral processing engineering have undertaken many projects, such as '863' National High Tech Projects, '973' National Significant Basic Research Projects, National Outstanding Young Scholar Fund Projects, National Natural Scientific Funds Projects, National Key Industry Experiment Projects, National Innovation Engineering Projects, National Key Science and Technology Promotion Projects. Between year 2003 and 2006, 413 papers on mineral processing were published on open issued publications at home as well as abroad and 9 monographs or textbooks were published. More than 40 patents were obtained in this discipline, and we also attained the national and the province or the ministerial level rewards 38 items [3].

In the researching fields of coal dry screening and separation, advanced coal-based fuel, coal desulphurization and ultra pure coal preparation, simulation and control of mineral processing, comprehensive utilization of mineral resources and environment protection, we had got the independent intellectual property rights of research results and have been applying to the industry. The total outlay of scientific research is up to 54 600 000 Yuan (RMB).The theory of the dense phase and high density gas-solids fluidized bed, the particle screen-penetrating probability screening theory and elastic screen surface screening theory were founded. The method and separator of air dense medium fluidized bed were invented, and firstly realized the industrial application in the world. The theory of non-pressure center-feeding of heavy medium cyclone was developed and the flow sheet is commercial application. The theory of micro-bubble flotation column was put forward and the micro-bubble flotation column separator is well employed in coal industry and mineral processing industry. Three demonstrating plants of desulphurization and de-ash of high sulphur coal were built. More than 400 micro-bubble flotation column separators are in commercial application. These technologies have made great contribution to improve the efficiency of coal utilization and environment protection.

At the same way, the undergraduates were involved in and did many experimental or research works under the direction of supervisors. And they can use the equipments and systems in the laboratory or in the experiment classes skillfully. In this way, the students could learn and master the necessary engineering skills to do experiments or research by themselves. For the undergraduates, 'Scientific research training program' is set up to encourage the students to take part in the research works, and experienced professors and researchers were selected as supervisors strictly.

4.3 Academic Communication

With the internationalization of higher engineering education, the discipline of mineral processing engineering has established cooperative relationship with more than 20 abroad universities and institutes of United States, German, Australia, Canada, Japan, Korea, Italy, Russian, Poland, etc. More than 30 undergraduate and graduate students were associated training with these universities and institutes. Over 10 international or national academic conferences have sponsored by the department of mineral processing engineering. More than 100 teachers lectured abroad and attended international academic conferences. More and more international projects were completed by the international cooperation, and a large number of engineers were needed, especially in the field of mineral processing. International engineering education for mineral processing is being our new object in cultivation of undergraduates. And now, we also have been accepted international or national scholar in advanced studies by many abroad universities.

5 Epilogue

The discipline of mineral processing engineering in CUMT is building into an international top-ranking discipline with the effort of the faculties. With the trend of higher engineering education internationalization, Internationalization strategy for mineral processing engineering education is being our new object in cultivation of undergraduates. To make the distinctive fields, hold national researching projects and strengthen primary research, cultivate excellent undergraduates, it will attain the international top-ranking level in mineral processing engineering education.

Acknowledgment. This work is support by the project of the construction point of the national special discipline, the project of the national teaching group and the China Education Ministry Key Lab Open Project (CPEUKF08-05).

References

1. Wang, D., Qiu, G., Hu, Y., Yao, G.: Evolution and Innovation of Mineral Processing in China. In: Proc. XXIV International Mineral Processing Congress (IMPC 2008), Beijing, China, September 24-28, pp. 3–14 (2008)

2. Zhang, X., Zhao, Y., Liu, J., Xie, J., Wang, J., Li, H.: Mineral Processing Education in China. In: Proc. XXV International Mineral Processing Congress (IMPC 2010), Brisbane, QLD, Australia, September 6-10, pp. 3095–3098 (2010)
3. He, Y., Zhao, Y., Liu, J., Xie, J., Tao, Y., Duan, C., Wen, B.: Exerting the Advantages of National Key Discipline, Outstanding the Training of Engineering Capability, Cultivating Talent to be with Innovation Ability. In: Proc. XXV International Mineral Processing Congress (IMPC 2010), Brisbane, QLD, Australia, September 6-10, pp. 3063–3068 (2010)
4. Zhao, Y.M., Shen, L.J., He, Y.Q.: The Discipline of Mineral Processing Engineering and the Personnel Training in China University of Mining and Technology. In: Coal Processing & Comprehensive Utilization, May 18-20, vol. 5 (2006) (in Chinese)
5. Duan, C., He, Y., Zhao, Y., Tao, Y., Chen, Z., Ye, C.: The experimental and practical teaching system of the discipline of mineral processing engineering in China University of Mining and Technology. In: Proc. 2010 International Conference on Optics Photonics and Energy Engineering (OPEE), Wuhan, China, May 10-11, pp. 133–136 (2010)
6. Kuang, Y., Zhao, Y., Liu, J., He, Y., Tao, Y.: Mineral Processing Engineering Education in China – Bringing up the Engineering Diathesis and Design Competence of the College Students. In: Proc. XXV International Mineral Processing Congress (IMPC 2010), Brisbane, QLD, Australia, September 6-10, pp. 3077–3084 (2010)

The Application of Multimedia Technology in Foreign Language Teaching

Zhang Zhi-feng

Department of Business English arbin University of Commerce,
Harbin, Heilongjiang, China
zhangzhifengzzf@126.com

Abstract. The whole world is undergoing great changes and twenty first century is an internationalized and high-tech era which is transforming from industrialized society to informatization ones . With the rapid development of science and technology, the upsurge of the new and interdisciplinary course, along with the merge and combination of literal arts and scientific arts, all these factors have contributed to the grim challenge that we faced during the process of cultivating foreign language talents. In order to face the challenge, the foreign major must gear to the demands of the system of market economy and cultivate the compound talents. Therefore, it is essential to improve the teaching devices and adopt advanced teaching facilities. The development of science and technology have made it possible for us make good use of teaching facilities and provide abundant resources. Multimedia technology is becoming more and more popular with college students, nowadays, it plays an important role in enhancing the teaching efficiency.

Keywords: multimedia, foreign language teaching, web-based instruction.

1 Introduction

The teaching process under the network environment refers to classroom teaching which is performed in the computer classroom in the local area network on a small scale. The technological feature of web-based class is featured by the following aspects: one person operates one computer and can operate the software independently; one can surf the internet or campus network and consult some material; one can also send posts.

The development of information technology and multimedia technology which is based on the internet technology not only provide pictures , words, and sound and video but also provide the learners with vivid and interactive learning environment , these factors have contributed to the emergence of new teaching mode in the web-based instruction .

Literally, web-based instruction refers to the teaching activities performed by means of computer network. In essence, with the practice and the development of web-based instruction, it has motivated a deep reform in the field of teaching thought, teaching theory and learning theory.

D. Zeng (Ed.): Advances in Computer Science and Engineering, AISC 141, pp. 85–90.
Springerlink.com

The teaching model of web-based instruction is a brand-new teaching model which is based on the network technique, the core is to adopt the advantages of both teacher-centered and student-centered teaching model and combine the two with each other, to make the best of their advantages. The teaching model of web education requires that the teachers are well equipped with information technology and adopt brand-new educational thought, education approach and design.

2 The Advantages of Multimedia Teaching

2.1 The Usage of Multimedia Teaching Can Expand the Classroom Capacity and Provide a Wide Rang of Teaching Resources

While the teachers are preparing for the lessons, the teachers need a large amount of material and information. The traditional teaching in confined to books in terms of putting in the information to students, this kind of information is very limited and lack of flexibility, convenience and interactivity. The combination of the feature of multimedia and storage technology of web can provide a large amount of information and the advantages of large information capacity and fast speed make it possible for teachers to transmit the information, all kinds of forms of training approaches and non-word information can be provided, all these factors contribute to the improvement of teaching environment. Abundant supply of educational software such as disc, video and cassette are useful teaching resources, which make the contents of teaching more colorful and the teaching methods more diverse and flexible.

2.2 The Classroom Can Be a Vigorous and Creative Learning Space and Make the Student an Active Learner

In the case of multimedia teaching, we are concerned about two problems: the first one is the technological problems of classroom teaching and the second is the leading role that students play in the learning process. The popularity of multimedia technology and network technique has provided the classroom situation teaching reform with opportunities and made it more vigorous.

Let the students be immersed in the language environment and take in the language under a certain language environment, it can achieve better results by practicing and using the language. the situation design can be vivid and creative by means of information technology, and the combination of disc, picture, sound and photograph are able to stimulate the senses of students, in this way, the students will take an active part in the classroom activities and exert their leading role in the process of learning foreign languages, their interest of learning English will be greatly enhanced. It is said that interest is the best teacher and it is the infinite initiative to learn and to acquire the emotional orientation of knowledge, it is the students act.

During the process of teaching , we take advantage of information technology devices and create a relaxed and delightful learning atmosphere in which students can enjoy the process of learning and positive attitude towards learning can be formed and their interest can be aroused , which is beneficial to the enhancement of their language proficiency. We can also make use of computer animation demonstration and practical operation by the students to cultivate their probing study and to expand their, the students can determine individualistic learning process voluntarily and choose learning strategies freely. The process of learning English is no longer a passive one but an active participation process. The strong learning interest and active participation are helpful to the optimization of students' internal psychology; this kind of psychological process will interact with the optimized external stimulation, in this way the students can obtain different achievements in the process of learning English.

2.3 Prearranged Situation Will Make it Easier for Students to Assimilate Knowledge

Prearranged situation is a kind of situation that fits into the specific set and atmosphere of the teaching contents and the teacher designed it beforehand so that they can reach the teaching goal, the students can experience the situation emotionally and it is easy to understand the contents of the book and enhance teaching efficiency. multimedia has provided convenience for the creation of this kind of teaching situation , the following methods are frequently used : pause , slow , circular play and so on , among which the function of slow down the play can arouse the student's curiosity and replay the picture and make it alive in students' mind , to help them understand the difficult point .

2.4 To Cultivate Students' Competence in Processing the Information Efficiently

The usage of multimedia computer assisted instruction in the classroom can not only help the students with English learning , most important , let the students have a good command of how to use information technology , especially , to strengthen their power to probing , selecting , managing and processing information , to keep its essence and discard its dross , to broaden their horizon and tell the good from the bad so that they can live freely in the globe village . Nowadays, those who can not master information technology can never catch up with the development of society. From another perspective, the one who can obtain and process the information effectively from the internet will all along hold a leading role in the process of learning English.

2.5 The Use of Multimedia Can Help Students to Raise Cultural Awareness

As we know, language and culture are closely related with each other, and culture contains a wide range of fields, the use of multimedia-assisted instruction enables us to make use of the recourses on the network and acquaint the students with the western culture, which is good for successful communication. The demonstration of the multimedia video system let students experience. Teachers can use the courseware to cultivate their culture awareness, that is, to know more about the cultural differences between the East and the West so that they can communicate more efficiently with foreigners.

3 The Effect of Multimedia Teaching

3.1 The Usage of Integrated Educational Devices

Integrated educational devices refer to the combination of multimedia computer disc, video, television and computer, which can not only make good use of the advantages of languages and but also have the feature of interactive computer assisted instruction. Traditional teaching and the previous CAI can not achieve this.

3.2 Multimedia Teaching Is More Vivid, Visual and Lovely in Terms of Teaching Mode and Expressive Methods

The software of multimedia teaching puts more emphasis on the students' initiative and active participation in the process of learning. The organizing form of knowledge and information are helpful to the student's exploring and discovering study, thus, it can overcome the weaknesses of passiveness in the traditional teaching software. Meanwhile, interactivity makes it possible for teachers to keep grip of teaching contents and rhythm and to gear to the student's requirements, which can not be achieved by traditional course broadcasts on television.

3.3 Multimedia Teaching Can Greatly Enhance Teaching Efficiency

It is a kind of assisted instruction which can greatly reduce the time spent in writing on the blackboard and comparatively increase the time invested in the information capacity of teaching in each time unit. The rapid development of science and technology and the coming of the information age are challenging the foreign language teachers with higher requirements. We should not be content with enriching our professional knowledge but should adapt ourselves to the trend of the age, along with enhancing our own practice skill try to explore and invent new teaching devices. the teachers should be capable of mastering new technological fruits ---computer internet technology so that we can operate these devices with ease in the process of teaching , there is no doubt that it will play an indispensible role in improving our teaching standards , efficiency and results. In the case of students, it can improve students oral English and provide a large amount of material for imitation, imitation can cultivate students' sense of language and practice their listening. Traditional English teaching is that teachers instruct the spoken language of students directly. It is difficult for students to remember and to repeat the long sentences. But it is much simpler to use the multimedia courseware to solve this problem.

4 The Existing Problems of Multimedia Teaching

4.1 Some of the Teachers Can Not Make the Best of the Courseware

Although we have talked about the advantages of multimedia teaching, there still remain some disadvantages.

Traditional teaching is mainly teacher-centered and teachers play a leading role in the classroom. In order to achieve better teaching results due attention should be paid to student's initiative, multimedia teaching has deepened the knowledge understanding of students to some extent, while the students have fewer opportunities to operate the computer.

4.2 Lack of Interactivity

If the students sit in front of the computer all along and it is likely that the students do not know how to communicate with students or teachers, therefore, in the process of using multimedia, while we enhance students' initiative it is worthwhile to pay attention to the teachers' leading role which is an indispensible element in this case.

4.3 The Frequent Use of Internet Will Prevent Us from Thinking Thoroughly

Consulting the material on the internet is fast in nature and can enable us to think faster, but if you operate it incorrectly it will remain on a superficial level.

For example, we click different website link in a few seconds that is from one webpage to another or from one point to another, because we click it too fast it will prevent us from thinking thoroughly and limit competence of forming our opinion, hint and analyzing information and deferring the meaning between the lines.

5 Conclusion

From above it can be seen that the multimedia teaching are characteristic of visual , scientific and vivid, teachers and students are more liable to use multimedia teaching . It breaks the mould of traditional teaching which consists of teacher, blackboard and chalk. in the meantime, it makes it possible for students to take a fruitful and vivid course, students can enjoy the whole process of learning and they become more and more interested in English, students can practice listening, speaking, translating and reading more effectively on the computer, and the teachers can consult a large amount of information from the internet and present it to students very efficiently instead of writing it on the blackboard. The usage of multimedia teaching can help teacher to achieve their teaching goal and it is a breakthrough in modern education. The students as well as teachers profit a lot from it. yet we have to admit that there are some problems exists in it , for example , some of the teachers can not operate the computer correctly and can not familiarize themselves with modern devices , as a result , we must make good use of its advantages and make it serve us better to realize informatization in modern reduction.

References

1. Ping, L.: Discussion on the English multimedia teaching model and the cultivation of students' original cognitive competence. Foreign Language and Foreign Language Teaching (5) (2005)
2. Yan, H.: Multimedia and English talented education. People's Education Publishing House (2006)
3. Ma, C.: Optimization of English classroom teaching by using information technological device. Information Technology Education (3) (2007)
4. Cheng, W.: Discussion on the reasonable application of multimedia
5. assisted instruction. China Distant Education (15) (2003)
6. Chang, H.: Discussion on the advantages of multimedia teaching model. Liberal Education Material (21) (2006)
7. Wu, S.: The current situation the analysis on English multimedia teaching. Education and Profession (17) (2007)

The Research of Cultivation of Innovatory Spirit and Engineering Ability for the Students of the Specialty of Engineering Management

Ruiling Wang[1] and Jianhui Wang[2]

[1] School of Civil Engineering and Architecture
Chongqing University of Science and Technology
Chonqing, China
[2] LongFor Properties Co.Ltd.
Chonqing, China
{cqwrl73,geniushawk}@163.com

Abstract. For the students of the specialty of Engineering Management, innovatory spirit and engineering ability are most important and the two abilities are helpful in their all day lives. But, how to cultivate theses abilities of the students and attain the aim of the talents' cultivation? This is the purpose of this article. The article follows the mental characteristics and rules of cultivation, combined with its own characteristics of the specialty of Engineering Management, the article gives some suggestions about the cultivation of innovatory spirit and engineering ability of the students of the specialty of Engineering Management, it discusses from the aspects of the education of the students' ideology, from the study of special courses, from organizing the specialized activities, from taking part in all kinds of specialized competition and every practice links for the students of the specialty of Engineering Management. If follow with the ways above, the students will have rich and broad basic knowledge, reasonable knowledge structure and strong motive of research, their abilities of using their knowledge, analyzing and resolving the problem will be exercised well, creative thoughts and operating ability will be exercised too. Furthermore, the abilities of expressing and organizing of the students will affect their attainment and development of innovatory spirit and engineering ability.

Keywords: the specialty of Engineering Management, innovatory spirit, engineering ability, cultivation, research.

1 Introduction

For the students of the specialty of Engineering Management, innovatory ability and engineering practice ability are most important and they are helpful in the students' all day liives. So, the cultivation and education of innovatory spirit and engineering ability are very necessary. University stage is the key phase of cultivating the students' all kinds of abilities, so, we should catch the valuable opportunity and offer

D. Zeng (Ed.): Advances in Computer Science and Engineering, AISC 141, pp. 91–97.
Springerlink.com

more opportunity of cultivating the ability of innovatory and engineering of the students, so as to cultivate more students of the specialty of Engineering Management who meet the need of the society and have latent capacity.

But because of the effection of education which meets for the need of exam and because of the factors such as the scarcity of the skill training of specialty, the society and university do not pay more attention to the cultivation of innovatory thinking, and this result in lacking innovatory consciousness and work practice of the students, their abilities of innovatory and engineering are not strong. The article will give some suggestions from the view of how to cultivate the ability of innovatory and engineering of the students of the specialty of Engineering Management.

2 Guide the Students to Cultivate Their Innovatory Spirit and Engineering Abilities through Ideological Education

The teacher's thought and idea will affect his student mostly, so, for the students of the specialty of Engineering Management, the teacher's education will lay emphasis on imperceptibly influence and correct knowledge accumulation. Do attention, to cultivate the innovatory character of the students is not "don't manage them and the students can do what they want to do". Because the students are in puberty, they have the character of open, they like new and different things, and they are rebellious, self-important and emotional too, if they don't get good indication and education, their rebellion and emotion will corrode their reasonable thinkings. So, the teachers should act as good guiders, and grasp the degree of tightness. The teachers should educate their students to be ready to accept new theories, new things, and make them have good habits of thinking more and getting to work more. The teachers should adroitly guide action according to circumstances and teach their students in accordance with their aptitude and make their characteristic of every student is brought into play. Be sure not to constrain the development of their personalities. Here ,we suggest the teachers may use the way of "5W1H" to cultivate the students' abilities of full thought when they educate the students, that is why(why do so),what(do what),who(who are the correct person),when(when to start and when to end),where(where we should do this) and how(how to do well and how to avoid failure). The education practice shows that only the individual character is developed fully, the latent capacity of the students will be given full play, their thirst for knowledge will be aroused and met, and they can have true creativeness.

Through education the students' creative longing and creative intention can be aroused. The teachers should guide their students to overcome some unhealthy psychology in their study and life with consious, the unhealthy psychology includes satisfaction with things as they are, stubbornly adhere to his own opinions, etc. The teacher should help his students overcome the psychology of following the others, because the psychology of following the others will limit his development of innovatory idea, and make himself go with the other's way if he agrees with or yeilds to someone's thought, though sometimes, the psychology of following the others is helpful to following the old routaine and resolving the ordinary problem.

The teachers should pay attention to the students' future education, to tell them about competitive pressure and the pressure of abtaining employment from the present society, etc., to cultivate their crisis consciousness, and make them feel pressure, because where there is pressure, there is motive power, so that the students can be eager to make progress. The teachers may tell them some struggle stories about some successful persons with conscious when at class or in daily education work, so as to arouse the students' desire of attainment success.

The teachers should guide their students make clear and definite aim. Usually, when the aim is defined, the people will follow its rule to work,and the person will become active and initiative and be highly creative. So, the teachers should help their students to establish their life goal, erect spiritual prop. But they should insure that the aim established is specific and feasible. If the aim is too high, it will make the undergraduate lose their confidence, if it is too low, it will not be helpful to cultivate the undergraduate's strong will.

About the cultivation of the innovotory spirit, in order to assure the students to make reasonable aim, the teachers may lead their students to make the analysis of SWOT, that is strength; weakness; opportunity and threat, and lead them to find out that how to take full advantage of their strength to overcome their weakness, how to catch the opportunity to show their strength and avoid the threat. By the analysis of SWOT, the students will make great progress on recognizing themselves correctly, on orienting their own position, and they will not be limited by themselves and find their own position.

The teachers must cultivate their students to bear and deal with all kinds of setbacks arised suddenly and cultivate their abilities of defeating crisis. The teacher should also cultivate their spirit condition of activeness, the spirit of rousing themselves and the spirit of "never say failure" when they face with setback, the students can exercise their psychology support ability from practice, the teacher should also lead them to form their own ideas and thoughts when they follow the example of the others' successful experiences and doings.

3 To Cultivate the Students' Innovatory Ability and Engineering Ability during the Study of Their Specialized Courses

The teachers of the specialty of Engineering Management should lay stress on passing on positive information which is helpful to cultivating the students' innovatory ability and engineering practice ability combined with the reality of the specialized courses. The teachers should guide their students to break away from disturbance of thought during ideology education and teaching process, and the teacher should develop the students' creativity thinking, to cultivate their life attitudes which are vigorous, promising, diligent and full of vigor, to cultivate them and make them have responsible spirit and have thoughts. The teacher should guide his student to develop his interest, love and strong point combined with every teaching link. For example, some students are interested in budget, then the teacher may encourage them to take part in budget competition. If someone is interested in project management, the teacher may arrange him to attend project management sand table imitation. For the student who is lack of interest, love and strong point, the teacher may guide him to develop and cultivate his interest step by step according to his own condition and reality.

The teacher should pay more attention to cultivate his students' innovatory desire in order to cultivate their innovatory ability, to encourage them enlarge their innovatory idea of theory point, to bring their innovatory idea into fruit, for examle, in the course of real estate plan for the specialty of real estate, the teacher may combine the teaching and plan practice, let them complete a creative plan by surveying so as to cultivate the students' innovatory spirit with consicousness and encourage. The teacher should guide his students to survey and study the real estate market, take use of their special knowledge and his own innovatory idea to write a plan about some place and building. About this, in order to produce good effect, the university may carry out inspire mechanism, to reward the students with innovatory plan, this can admit the student's innovatory ability and it is a kind of guide about this practice of the university, and this is helpful to arouse the student's creativity enthusiasm, is also helpful to build up good innovatory atmosphere. when the students have strong innovatory desire and spirit, they can give play to their intelligence and wisdom to produce creative fruits. The teachers should help the students to erect the consciousness of resolving the problem of more way, more plans, to exercise the students' ability of seeking difference when they are thinking about the question and resolving the problem, so, the teacher s may teach with example of cases, let the students take part in case discussion, guide them to find out the way of resolving problem. Often, there are several solutions about a case, so, by this chance, the teacher may let the students think more and innovate more.

Resonable knowledge structure is important to cultivateing student to think alone, exploring boldly and innovating positively. If a student has solid basic knowledge, reasonable knowledge structure, his creative thinking will be brought into play easily, if not, he can hardly to innovate. The teacher should guide his students to build up reasonable knowledge structures. The students should grasp not only the basic knowledge and special skill of its own specialty, but also new knowledge of other deciplines such as the decipline in the neiborhood and crisscross. The teacher should ask his students put emphasis on professionall knowledge, at the same time, they should attach more importance to the humanities, they should enlarge their knowledge scope, open their thinking, arouse their explore spirit, enhance their adaptability. The teacher should guide the students to choose the courses across their specialty and to read more books out of their specialty. The teacher should offer more opportunity for their students to improve their innovatory spirit in order to cultivate know-all talents, the universities should lay stress on cultivating compound talents with good adaptability to future social change and career shift. When they accumulate enough knowledge, they will have great progree on analyzing questions, resolving problems, the talents of differentiating between right and wrong, work ability and creativity ability. Taylor ever said, someone who has rich knowledge and experiences will be easier to produce new connection in the mind and original view than those who have only one kind of knowledge and experience. In a word, the teacher should let their students know: only on the basis of grasping their own professional knowledge and have broad knowledge, can they use their knowledge to resolve problems correctly, can they find out the gap and to innovate.

4 Taking Part in All Kinds of Special Activities and Competition, Let the Student to Cultivate and Exercise His Innovatory Spirit and Engineering Ability

Only by practice can abilities be formed and developed, so, the students should take part in more practice drills. According to the demand of cultivating the students' innovatory spirit, the teacher may let his students to organize and conduct all sorts of special activities, for example, to set up engineering management association, to hold science and technology festivals, to put on a series of lecture or reports or exchange meetings, to hold professional knowledge competitions, to conduct technology competitions, etc., so that the students can grow up under the nature of innovatory spirit and enhance their innovatory ability when they take part in these activities. The teacher may organize his students to take part in "challenge cup" competition of science and technology works, do pioneering work competition, real estate plot competition, project management plot competition, etc. so as to cultivate their abilities of original thinking, their abilities of analyzing and resolving problem and abilities of practicing. The teacher should encourge his students to take part in teacher's question for study and improve their abilities of study and research. The teacher should lead his students to practice in enterprises in holidays and know about what is needed in practice, in his development, by this, the students will probably become to be bear hardships and stand hard work and responsible.

5 To Enhance the Cultivation of Innovatory Spirit and Engineering Ability in Every Practice Link of the Specialty

According to the demand of the specialty of Engineering Management, the specialty should cultivate the management talents who have more strong knowledge of construction technology, economics,management, law and certain special basic knowledge, at the same time, they should have certain practice ability, innovatory ability, and they must be compound, application and practice innovatory and they can do the whole-process work in engineering construction domain.

The abilities needed of the specialty of Engineering Management include: the ability of study and master the knowledge and practice which needed in engineering, that is, the ability of observation, resolving the practice problem, cooperation, exploration, using of professional knowledge [1].

In order to cultivate the students' above abilities, the teacher should guide his students to cultivate his own sharp observation ability, and lead his students to be good at imagination in the links of recognizaion field work, production field work, course experiment or graduation project. Because the specialty of Engineering Management is a specialty which attaches great importance to practice, the university should offer practice opportunities as much as possible in teaching in order to cultivate the students' engineering ability.

In order to cultivate the students'engineering ability of the specialty of Engineering Management effectively, the author's suggestion is : besides setting more practice links, the teacher has to arrange correct time and place for his students to practice. If the time is too short, the students can not learn more, it will become a mere formality. So, no matter how long the practice time is, it is not wrong. In field work link, the students can simulate the whole management, supervisory and control process of a project. And they can be arranged into the constrution field to practice. If possible, the university may engage some persons with experienced engineering technology and management as guiders, let them guide the students in all sorts of field works, courses design, graduate practices, etc., and by this way, field work, course design, graduate practice will be connected with production process of enterprises. The uninversity may arrange practice link for the students by organizing pioneering work or course simulation. In order to insure that the practice link is the same as the practice in enterprises, the university should invite the experienced engineering management persons to give lectures.

The university should also build engineering management labs for the students, in order to arouse the students and enhance their study abilities, the university should consider to open the labs and let the students to use freely at any time, thus, the abilities of resolving problems independently, creativity, abstract thought and synthetical judgement of the students will be improved .

When the students are doing their graduation project, the university may let the students go boldly into practical posts, and invite some field workers to guide them how to complete their graduation projects, and let the guiders assess the students, thus, the ability of engineering practice of the students will be strengthened without question, and this can be helpful to their adaptabilities for their work after graduation. Another way is to let the students who have signed a contract with enterprise to go to work in advance, at the same time, to complete their graduation projects, by this way, the ability of adaptabilities for the work circumstance and resolving problems independently will be enhanced too.

In a word, the cultivation of the abilities of innovatory and engineering is a systematic engineering for the students of the specialty of Engineering Management, we should study and explore the ways earnestly, the teacher should cultivate the abilities of innovatory and engineering of the students at any time when he educates and manages his students, and keep on perfecting them in practice. We believe, by the ways above, the students will have rich and broad basic knowledge, reasonable knowledge structure and strong motive of research, their abilities of using their knowledge, analyzing and resolving the problem, creative thoughts and operating will be exercised well. Futhermore, the abilities of expressing and organizing of the students will affect the attainment and development of innovatory spirit and engineering ability.

Acknowledgment. Thanks very much for my father and my husband, they gave me great help and support during the writing of this paper, without their support, there is no this paper.

References

1. Gai, Y., et al.: The study of practice teaching of cultivating the ability of Engineering Management undergraduate. Science Study of Teacher's Education (March 2009)
2. Gai, Y., et al.: The Study of Practice Teaching of Cultivating the Ability of Engineering Management Undergraduate. Science Study of Teacher's Education (March 2009)
3. Wang, R., et al.: Reformation and Study of the Practice Teaching of Specialty of Engineering and Management
4. Li, Y.: The Cultivation of Innovatory of the Undergraduates. Journal of Henan Business College 16(3) (May 2003)
5. Zheng, J.: Looking at Innovation Education:Some Questions about the Cultivation of Innovatory Ability. Education Research (3), 13–16 (2003)
6. Huang, X.: The Exploration of Innovatory Thinking and Ability of the Undergradutates. Journal of GuangDong Vocational College 2(5) (October 2003)

The Tentative Ideas for Solving the Difficult Problem about Residents to Buy Housing

Wang Xiaoyan and Zhang Rui

The school of civil engineering
Chang'an University
Xi'an, Shaanxi Province, China
{249373419,271149698}@qq.com

Abstract. The article analyzed the phenomenon that it is very difficult to common people to buy a commercial house because of the high price leaded by unusual behaviors. It believes that it is the behavior of speculation of real estate that causes the commercial housing price rising unusually. As to the behavior of speculation of real estate, it gave some corresponding policy scenarios and suggestions to resolve it. Simultaneously, it analyzed the results if the policies are carried out and the benefits to the related side.

Keywords: housing price, speculation of real estate, differential measures.

1 Introduction

In 2009, China real estate staged the craziest performance, according to the 2009 nationwide real estate market data by National Bureau of Statistics, annual housing price rose 24% and average 813 Yuan per square meter, hitting a 11-year high [1]. April 17, 2010, the State Council promulgated the "guo shi tiao" which produced only short-term effectiveness, but the September statistics show that the regulation and control once again become "useless regulation and control": October 15 the national bureau of statistics released a report show that the 70 cities' chain index of housing price rose 0.5% in September [2].

China's high housing prices not only can not meet the residents' demand for self-living, but also jeopardize the healthy development of the entire national economy and social harmony. And the again defeat of regulation and control make the residents' demand for self-living begun from disappointment to despair. September 29, the Ministry of Housing and Urban proposed the new five requirements of regulation and control which is only further deepening of "guo shi tiao", so predictable results will still be repeating cycle of "guo shi tiao". The new regulation increases the down-payment and interest rate for the first suite house, this "one size fits all" approach make the own housing dream of the lower-income residents and young people disillusioned. So how to control the housing price of abnormal rising and safeguard the residents' demand for self-living have become China's current social problems urgently to solve.

D. Zeng (Ed.): Advances in Computer Science and Engineering, AISC 141, pp. 99–104.
Springerlink.com

2 Status Analysis

China's urban housing prices rise abnormally, by both supply and demand factors caused. The reason of course include the government, banks, developers who formed interest groups and jointed rally pushing up prices and sharing the tremendous benefits by rising real estate prices. But we all know demand determines supply, that the most important factors of price rising must be exuberant demand. The demand for real estate can be divided into demand for self-living and demand for speculating and investing. The needs of ordinary citizen is self-living, while others are investing and speculating, in a certain sense these people can be collectively referred to as real estate speculators whose self-living demand to be satisfied began to shift the speculating and investing demand. As interest rates of bank deposits dropping in recent years, the money supply increasing and price of commodities rising, the stock market for a long declining, fund losing, but the hosing price climbing, the capital can not look for funding more than real estate to increase the value. So it flowed to real estate inevitably.

As real estate speculators raise money more easily than the self-living residents to buy a house, its speculating and investing demand can be met easily. After they a large number of buying, the housing shortage will be appeared and the demand for self-living will appear more urgent.In the case of in short supply situation, the supply-side can rise prices to reap higher profits, then they can scramble new land to develop new buildings; As land is scarce resources, local governments can raise land price which will result in rising housing price; real estate speculators who can sell houses for profit, and can continue to invest in new housing speculation. This will form a vicious circle which lead housing price rising constantly and residents of self-living demand buying a house more difficult. According to State Grid Corporation survey, in 660 cities of China, the electric meter reads zero for 6 months up to 65.4 million residential suites that can accommodate 2 billion people to live [3]. It can be seen that the phenomenon of the domestic real estate as a speculating and investing instrument is so serious and the object of regulation and control by Government to Opposite and suppress are mainly real estate speculators and investors. Therefore, speculating and investing in housing is the result of housing price rising abnormally.

3 Relevant Policy Tentative Ideas and Suggestions

September 26, 2010 to October 16, 2010, China Index Research Institute Conducted a questionnaire survey of the more than 1,500 homebuyers, dozens of real estate development enterprises and dozens of real estate policy experts that reached the following conclusions: 70% of homebuyers supported differentiated credit policy, but didn't think these policies to curb speculation and investment is very useful; the vast majority of experts supported the differentiated credit policy and suspending the mortgage of third suite house, and believed that these policies have had significant impact to curb speculation and investment; the support of differentiated credit policy in real estate development enterprises had low proportion [4]. To maintain the stability of the real estate market and inhibit housing price rising abnormally, it must make corresponding policy adjustments for real estate speculation and investment. Since most home buyers and experts to support differentiated policy, regulation may wish to introduce differential measures. Policies envisaged as follows:

3.1 Building a Nationwide Commercial Housing Network Information Platform

Build a nationwide commercial housing network information platform by the State Council, the Ministry of Housing and Urban, departments of statistics and so on, which provides real-name system to buy a house [5], and collected the country, all cities' new housing information in nearly five years. The entire build process will be supervised by central government: first built around the city and then brought together to their respective provinces, final summarized to the central authorities. Since it only counts the new housing in nearly five years, the statistical process is not difficult. Listed real estate information platform (including real estate location, planning units, and copy number, floor area ratio, virescence, developers, construction units, etc.) will be built on the Internet appear to accept public supervision.

This platform can effectively prevent the hoarding land behavior, if the hoarding land behavior of real estate development enterprises is found, the State can recover overdue land for the construction of affordable housing to supplement the shortage of affordable housing land supply situation. Additionally the most important advantage of this platform is that after the buildings completion, the information of the transaction price and buyers' name of all dwelling will be publicized on the platform which can completely eliminate the acts of developers covering plate reluctant sellers and strongly suppress real estate speculators to buy more houses to speculate. Moreover, according to housing transactions, it can set the program to automatically generate changes in prices which can be obtained more realistic and intuitive price trends. Thus Builders do not take the trouble to declare their own housing price, and statistical offices can also save data collecting links to save human resources and administrative costs.

3.2 Establishing a National Information System of Individual Housing

The information system of individual housing will be established by the State Council together with the relevant departments. It will record each person's housing purchasing information and occupied area. For example, if a single person has a set of housing with 90 square meters, its personal housing occupancy area is 90 square meters; if the person has two sets of housing with 90 square meters, its personal housing occupancy area is 180 square meters; if two persons have a set of housing with 90 square meters, their personal housing occupancy area is 45 square meters... The information system of individual housing network can build based on the country's second-generation ID card system, the nationwide commercial housing network information platform, and the residents account information and marriage information. The system's build process will also be supervised by central government: first built around the city and then brought together to their respective provinces, final summarized to the central authorities. It still only counts the information in nearly five years, so it can be done with the coordination of various departments.

After the system established, the person who buy the commercial housing and its personal housing occupancy area is clear. Occupying an area far beyond the use of housing is likely to be real estate speculators who can be paid more attention. In addition this system built in individual but not family as a unit, so it can effectively prevent the fake divorce phenomenon in currently which purpose was to access preferential policies for the first suite housing. The main purpose of creating this system lies in its service that the following differential regulation and control measures must be supported by this system to implement.

3.3 Implementing Differential Housing Price Subidies or Tax in Commercial Housing

The purpose of real estate speculators to buy housing is not to live, but to reap the profits of appreciation. Therefore, it must reduce their speculative profits, so that the residents can own their own house and solve the housing difficulties for ordinary residents. Implementing differential housing price can completed these two objectives: Estimate the reasonable occupied area of per person by the relevant state departments that the part of less than the area will get government subsidy but the excess part will impose luxury tax of real estate. For example, if the reasonable occupied area of per person is 40 square meters, when residents buy a house, the part below 40 square meters can be get housing subsidy: 0 to 10 square meters will get 30% subsidy of the housing price; 10 to 20 square meters will get 20% subsidy; 20 to 30 square meters will get 10% subsidy; 30 to 40 square meters have no subsidy and luxury tax; 40 to 50 square meters will levy 10% luxury tax of the housing price; 50 to 60 square meters will levy 20% luxury tax, and so on. The less area is occupied, the more subsidies are got; the more area is occupied, the higher luxury tax of real estate is imposed. If residents would like to redeem or transact housing, he must refund the housing subsidies and pay corresponding interest. The information system of individual housing will adjust the personal housing occupancy area after the transaction, thus the housing after redemption can still enjoy the corresponding subsidies.

Real estate speculators get profits through the cheap buying but expensive selling in the past, so the more they buy, the more profit will be get. However, this policy will make that buy the more they buy, the more luxury tax will be levied which means the cost became higher, so they must consider the limit price they can afford. The area to be bought by this limit price will be much less than area which the real estate speculators occupied now, so it will give a major blow to real estate speculators. On the other hand the residents of the demand for self-living get the government subsidies in their reasonable occupied area, so the buying housing cost will much lower than in the past, while following the withdrawal of a large number of real estate speculators, housing is no longer a shortage and the prices will be further decline, in this way it's much easier to achieve that the residents can own their own house. Thus the main demand will appear structural change and after the demand for speculating and investing exit the market, the demand for self-living will be much easier to enter the market than before. This can also weaken the rich-poor divide, conducive to social harmony and stability and strengthen the party's ruling.

3.4 Implementing Differential Down Payment and Loan Rate in Commercial Housing

From lots of data, we can see that four-fifths mobilizing funds of real estate speculators came from bank loans, so they can get through the leveraged loan get several times the excess profits. The bank loan which became the main funds' source of real estate speculators and contributed to the housing speculation behavior is an indisputable fact. Therefore, bank credit policy should be adjusted, on the one hand to inhibit the speculation of real estate and on the other hand to help residents buy their own house. Implementing differential down payment and loan interest rates can completed these two objectives: also estimate the reasonable occupied area of per person by the relevant state departments that the part of less than the area will apply low down payment and loan interest rates but the excess part will apply high down payment and loan rate. For example, if the reasonable occupied area of per person is 40 square meters, when residents buy a house, the part below 40 square meters can apply low down payment and loan rate: 0 to 10 square meters will be applied 5% down payment and 0.7 times loan rate; 10 to 20 square meters will be applied 10% down payment and 0.8 times loan rate; 20 to 30 square meters will be applied 15% down payment and 0.9 times loan rate; 30 to 40 square meters will be applied 20% down payment and 1 times loan rate; 40 to 50 square meters will be applied 25% down payment and 1.1 times loan rate; 50 to 60 square meters will be applied 30% down payment and 1.2 times loan rate, and so on, until 100% down payment. The less area is occupied, the lower down payment and loan rate is applied; the more area is occupied, the higher down payment and loan rate is applied.

In addition, the mortgage payment standard by banks is provided no longer according to the individual proof of income but to the personal income tax payment certificate in last year. Under the networked personal tax payment certificate, the bank can calculate the income of borrowers which is real income of borrowers. This Measure can curb the real estate speculators who can borrow another citizen identity card whose can not afford to buy a house that defraud preferential loans to real estate speculation.

After the implementation of this policy, in one hand It can maximize the inhibition of real estate speculators who use leverage to obtain bank loans to earn excess profits, and can reduce the risk of bank loans to guarantee credit security. On the other hand when the residents of demand for self-living buy housing, the low down payment reduces its barriers to entry and easier to raise funds; low loan rate reduce the repayment costs and easier to implement the dream of owning a house.

3.5 Implementing Differential Transaction Tax for Second-Hand Housing

The purpose of real estate speculators to buy a house is not for self-living, but to seek housing profit after appreciation: when they purchase a house, they will surely sell for profit after the housing price appreciation to their satisfied price. Implementing differential transaction tax for second-hand housing can reduce the real estate speculation profits and inhibit the real estate speculation. This policy should be based on the time of buying the house to levy second-hand housing transaction tax: Housing buying time is shorter, the higher transaction tax will be levied; Housing buying time is longer, the lower transaction tax will be levied. For example, when the commercial housing of buyer want to be transacted within one year , the transaction tax will be levied up to 20%; within 1-2 years transaction, 16% of transaction tax will be levied;

within 2-3 years transaction, 12% of transaction tax will be levied within 3-4 years transaction, 8% of transaction tax will be levied within 4-5 years transaction, 4% of transaction tax will be levied; five years or longer, the transaction tax can be reduced to base rate. The differential transaction tax rate for second-hand housing should be developed reasonably after in-depth study by related departments.

The purpose of this policy is not only to reduce the expect of real estate speculation, but also to curb real estate speculators who borrow another ID card to fraud preferential policies to speculate by their free funds. Even short-term its purchasing house has a lot of appreciation, but the high transaction greatly reduced its transaction tax profit; If the buyer afford the transaction tax that means the suite price rose 20%, a Sensible buyer will not buy this house.

4 Conclusion

The five policies complement each other: Implementing differential transaction tax for second-hand housing and establishing a national information system of individual housing need to rely on nationwide commercial housing network information platform, while implementing differential housing price subsidy or tax and Implementing differential down payment and loan rate need to rely on information system of individual housing. The five policies can be believed that can maximize inhibit real estate speculation and make a lot of residents of demand for self-living live in affordable housing. With the withdrawal of real estate speculators, the number of housing expansion and a large number of residents buying their house, the Contradictions of supply and demand can be resolved, thus price will return to reasonable and the purpose of the regulation and control can be achieved.

The problem that housing price increasing abnormally leads to residents buying housing difficultly is the most sensitive social hot spots, which relates to the people' living and build a harmonious society. Using a variety of measures to curb housing prices increasing fast is the Government's priorities and responsibilities. Using the above multi-pronged approach and solution will effectively solve the problem that housing price increasing abnormally leads to residents buying housing difficultly , thus real estate market even the whole national economy can maintain a healthy development.

References

1. Statistics: 2009 National housing prices rose an average of 813 yuan (January 2010) (in Chinese), http://finance.qq.com/a/20100120/003426.htm
2. The housing price of China's 70 large and medium cities rose 0.5% qoq in September (October 2010) (in Chinese), http://news.hexun.com/2010-10-15/125141470.html
3. The meaning of 6540 million vacant housing (July 2010) (in Chinese), http://www.chinavalue.net/Blog/418128.aspx
4. China's real estate policy review in the third quarter (October 2010) (in Chinese), http://fdc.soufun.com/report/ReportDetail.aspx?id=954&pageIndex=3
5. Hanbing, S.: three measures make housing price returned to reasonable price (March 2010) (in Chinese), http://news.sina.com.cn/pl/2010-03-04/133219788332.shtml

The Cost Control in the Exploitation of Real Estate

Wang Xiao-yan and Zhang Rui

The school of civil engineering
Chang'an University
Xi'an, Shaanxi Province, China
249373419@qq.com, 271149698@qq.com

Abstract. Based on the summary of housing prices and it's affection, the article supposes that the question of high housing price at present might be resolved from the cost aspect in the exploitation of real estate. It elaborated the significance of cost control. It also believed that the cost control should be manifested in the overall control and key control which mainly manifested in the earlier period investment decision stage and the plan design stag. Finally it gave corresponding suggestions of cost control in these two stages.

Keywords: real estate, cost control, overall control, key control.

1 Introduction

Housing price has become the hottest topic in China currently. The high housing price have a large number of macroeconomic reasons, such as the national real estate industry policy, local government finance dependence on land revenue, real estate speculation in real estate corporations, real estate developers seeking extra profit and other factors. But the simple formula of "commodity price = product cost + profit" in economics is not broken bumpy truth. Therefore, high housing price is not only controlled through the country's macroeconomic measures, but also to lead real estate developers to control in the cost aspects that the implementation of two-pronged approach would achieve the most reasonable means to control the high housing price.

2 The Significance of the Cost Control in the Exploitation of Real Estate

The cost control is a effective way to control cost and increase reasonable profit in the exploitation of real estate. Cost control has a great significance whether on the whole national economy or on the real estate exploitation enterprises.

2.1 Promoting the Healthy Development of National Economy

Reducing Cost has a positive meaning which can control housing price effectively and promote a healthy development of the national economy. Many experts and simple

D. Zeng (Ed.): Advances in Computer Science and Engineering, AISC 141, pp. 105–110.
springerlink.com

residents believe that housing prices can not be reduced for many years because of high housing cost, so only using effectively way to control costs, housing prices can have a reducing space. Further more only controlling the price rising housing price, the real estate market can be facilitated re-adjustment and land resources can be used rationally in order to avoid the real estate market bubble economy, thus the national economy can have a healthy development. Now, real estate is the pillar industry of national economy, the development of real estate industry can also lead to develop for other industries such as cement, steel, aluminum and so on. According to the mathematical model by World Bank, every 100 Yuan of real estate investment products will promote sales of 130-150 Yuan of other commodities. Therefore, the healthy development of the real estate industry and real estate market stability will lead the development of other related industries and promote the healthy development of the national economy.

2.2 Enhancing Core Competitiveness of Enterprises in Real Estate Development

After nearly two decades development of China's real estate, the industry itself and market development have a considerable progress that consumers became increasingly rational and mature, as well as competition was increasingly fierce. Real estate companies in order to be able to long-term survival and development must strive to reduce costs and maintaining the cost in a reasonable level. Only the real estate companies which reduce the cost leave under the industry average can be reflected their all the different management levels and its core competitiveness. Therefore, the large distance of cost between the real estate company and the industry average level has, the stronger core competitiveness can be reflected clearly.

2.3 Improving the Economic Benefits of Development Projects

In a market economy, the real estate business the fundamental purpose is to gain profit, and profit is the revenue obtained by the compensation cost to get. Profit is the difference between sales revenue and cost. It is sure that increasing the profit by raising housing price will stir all angry in this period of abnormally high housing price, so cost control to improve economic efficiency become the most direct and effective means.

2.4 Improving the Management of Development Projects

Most premises of the development enterprise project management and technical level is not high, especially in the low level of cost control. The project cost control as a management activity throughout the whole process will improve the management of the entire project management levels. Over time the management level of developer even the whole industry will be improved.

3 The Measures of the Cost Control in the Exploitation of Real Estate

Cost control in the exploitation of real estate development should adhere to the principle of combining with the overall control and the key control.

Overall control means the real estate development business should control all elements in the whole real estate exploitation process. Real estate development costs include the whole process from the investment decision-making to housing sale, including : (1) Land costs (land acquisition costs, demolition of ground attachments compensation, resettlement demolition costs, etc.); (2) construction cost (fee Feasibility Study, survey planning and design costs, construction and installation costs, infrastructure and public facilities, etc.); (3) the development of indirect costs (the development of wages and salaries, benefits, office expenses, etc.); (4) administrative costs; (5) Selling expenses (6) Finance costs [1]. The each cost Occurrence process of above listed should have monitored to avoid regulatory dead ends, because when an element occur a loophole, this loophole will inevitably be discovered and used by some people to seek illegal interests.

Although the overall control throughout the entire process of real estate development, the potential cost savings has a huge difference at each stage. Land costs and construction costs accounted for the largest proportion of the entire development cost, but theirs potential cost savings are small, such as the study shows that the potential cost savings in the construction only accounted for 5% to 15% of the total investment. The greatest impacts on the stages of the project cost are the early stage of the investment decision-making and planning design. Therefore, it should be implemented the key control on the stage of the investment decision-making and planning design.

4 The Key Control of the Cost Control in the Exploitation of Real Estate

The cost control in the exploitation of real estate should focus on the stage of investment decision-making and planning design.

4.1 The Cost Control in the Stage of Investment Decision-Making

According to statistics, the investment decision-making stage of the project cost has the highest impact level in various stages of construction, reaching 80% -90% [2]. Phase of the project investment decision can be seen on the project's significant impact on project cost, including the determination of construction standards, building site selection, technology selection, design selection, etc., which can directly related to the level of project cost. The correctness of Investment decision is directly related to the construction project cost and its social and economic benefits. So for real estate development projects need to establish a scientific decision-making mechanism. The main basis of decision-making is the scientific analysis and forecasting of project - the feasibility of the project analysis and research. However, seldom real estate companies are willing to invest in feasibility study, if the investments made decision only by their feeling, it will be very dangerous. Feasibility study is based on the accurate analysis and full estimated of psychological needs of consumers and the market capacity, cost, etc. to decide the feasible of the project. It must be completed the flow of funds and distribution plan at the feasibility stage, so that developers can try to reduce the cash investment, reducing capital costs, in order to facilitate the smooth progress of the project. Feasibility study is a prerequisite for success of the project, so the author believes that this stage should do the following:

1) Doing market research : Survey the recent market demand conditions at home and abroad, including forecasts of future market trends; Forecast the product sales, analysis the price and determine the product's market competitiveness and market prospects; have a economic and technical feasibility compared with Programs and development direction of product, on the base of full consideration of market, technology, environment and other factors determined the reasonable scale of the proposed construction to achieve economies of scale. The level of construction standards should be reasonably determined by current level of economic development, to distinguish between different regions, different sizes, different grades, different capabilities, different consumer groups, full of construction standards, quality requirements, the performance of building materials prices, etc.

2) Technology and economic integration, optimization of good planning : To be truly effective control of project cost, it must be integrated with actual situation of the project on the basis of completion of market research. With the premise of satisfying the functional and production requirements, combining with the technological advance and economic rationality to design multi-program comparison, strive to technology advanced under the conditions of economic rationality and the economic rationality on the basis of advanced technology.

Selection of real estate projects is necessary measure to find a reasonable mean to develop real estate programs. The various options for development and management programs in the plan of real estate projects should be carried out economic analysis and calculations to screen the minimum acceptable rate of return programs, and compare these programs. Comparing and Selecting the available program for comparison should follow the principle of corresponding diameter in costs and benefits calculation and select the appropriate indicators of economic evaluation as a comparison index by the actual situation of the project.

3) Careful preparation of investment estimates : Accurate estimation of construction investment can provide a reliable basis of raising capital for financing investment projects, engineering design bidding and construction limit designing. Preparation should be responsible, realistic : avoid overestimate the risk calculation that avoid overstocking and waste of funds ; avoid fewer estimates deliberately that keep the prices down and then invest in additional which would disrupt the investment plans. It should be noted that adjust the price of different regions and different time as well as pay attention to overall balance of the total project investment.

Cost control should make the accumulation of Estimating investment, establish the information network construction cost, collect the relevant indicators and information timely and combine with the actual situation to the amend, supplement and improve the indicators of the project. This will provide a reliable scientific decision-making for leaders and lay a solid foundation for the follow-up completion of the preliminary design estimates, construction budget and final accounts.

4.2 The Cost Control in the Stage of Designing

The design phase is the key and focus to real estate project cost management. While design fees in the whole process of construction costs is not significant which account for only 1.5% -2%, but can impact the project cost up to 75%[3].。 The current designers generally emphasis on technical design but ignore its economy. They fear the accident and take responsibility for any safer, so they Increase the design standards, but have less rational consideration of economy which fundamentally affect the effective control of project costs. Once the shape is designed, the investment is set on the basic. the economical, effective cost control of design is the key to the whole process of the project and the focus of cost control, so the development side should strengthen the control of the design institute in the following areas at the design stage:

1) implementing the design tender system : In order to effectively prevent irregularities in the field of design, to ensure fair competition in a good institutional environment for the qualified and strength of the design units, the tender system design should be implemented. The design bidding should be the whole process of bidding which not only pay attention to plans of design, but also give adequate attention to the construction design stage. This requires the preparation of high quality building design mission statement which should have a detailed description for the function of the proposed building, style, budget, schedule and general building materials, equipment selection and other aspects, and it also requires determining the depth of construction plan design, in order to anticipate and control future engineering change. After bidding a contract with the designer, the developers should restrict the design behavior of the designing side.

2) Accelerating the promotion of Design Supervision: Designing is a professional work, so the developers appeared to be inadequate in the design phase only relying on the internal management of design and theirs design contract. In this situation, the implementation of design supervision system can solve the above problems, and achieve the following results:

a) supervisory role: Designing supervision can strengthen the awareness of designer to control cost and can prevent to casual strengthen the design strength but disregard to its economy;

b) facilitating role: Designing supervision can also help to develop design ideas to optimize designing;

c) checking role: It can avoid the possible mistakes and defects timely in the design process.

3) implementing the method of value engineering : Project design phase should use value engineering method to optimiz the program and implement the quotas designing in order to make investment well controlled in the designing phase. Construction projects should be designed to take control standards, cost, manpower, and the use of new technologies, new processes, new materials, construction process and consider the degree of difficulty in the construction method to create favorable conditions for the construction, thus design changes can be avoided effectively which could reduce the project's investment cost.

4) strengthening the design review before drawing : Strengthening the audit before drawing should include the following three aspects:

a) engineering Rationality review: *The occurrence of the engineering change should be controled as possible before the change in the pre-construction because the costs in the pre-construction are the smallest but can be achieved the best effect;*

b) security Audit : *In a safe situation, it should minimize the project Quantity, thereby reducing the project cost;*

c) budget review : *The designing budget is an important part of the design which is submitted by the designing department which should reduce or avoid missing items, omissions, thus it can lay a solid foundation of success for the next phase of construction cost control.*

5 Summary

The cost control in the exploitation of real estate is one aspect for reasonable reducing housing price, and additionally it still needs to rely on the government's macro control. If two-pronged approach, I believe that the problem of unreasonably high housing price can be cured. The government's macro-control will inevitably harm the interests of certain interest groups, but cost control is the means to achieve a win-win for all the parties which not only benefits the country and people, but more conducive to real estate development enterprises. Real estate development companies should pay enough attention to cost control and do well in overall control and key control which would make due contributions for the housing price decline a reasonable level.

References

1. Li, Y.: Brief discussion about real estate project's cost. Sichuan Building Materials 36, 177–178 (2010) (in Chinese)
2. Cheng, R.: Real estate development project's cost control". Beijing University of Posts and Telecommunications Engineering Master Papers, p.16 (June 2008) (in Chinese)
3. Yao, Z.J.: Brief discussion about the cost management of real estate development enterprises. Real Estate Development, 46–178 (January 2010) (in Chinese)

Bibliometric Analysis on E-Sports in China

Zhang Guorui

Department of Sport Training
Shenyang Sport University
Shenyang, Liaoning 110102, China
Z_g_r_007@163.com

Abstract. Taking journal full-text database of CNKI(China National Knowledge Infrastructure) as data sources, the paper carried on statistical analysis on 151 papers on E-sports by September, 2010 from publishing years, journal distribution, author distribution, citation frequency, download frequency ,etc. to find questions of the research of E-sports and provide references for the research of E-sports in China. Through statistics, the author found research characteristics of E-sports in China: the number of papers creases rapidly since 2003; 50.30 percent of papers are published on journals of sports, but only 13.25 percent of papers are published on core journals; the research basis is weak, the research is still at the starting stage, emphasizes on theory but neglects practice. So the author puts forward suggestions including deepening the theoretical research, reducing the repeated research, emphasizing the practice research and introducing the foreign research on E-sports to promote the development of E-sports in China.

Keywords: E-sports, Bibliometric, Statistic, Analyze, China.

Electronic Sports is abbreviated e-sports, and the first e-sports approach is originated in the online competitive games. E-sports uses high-tech hardware and software devices as sports equipment, and is an intellectual confrontation sport between people. E-sports can exercise and improve participants' thinking ability, responsiveness, coordination of heart, eyes, and limb, and determination, and foster the team spirit. As a new athletic sport, e-sports became the number 99 sport event approved by General Administration of Sport in November 2003, but in 2008 with the integration of existing sport events in China, E-sports was re-defined the No.78 sport event.

1 Data Source and Analysis Method

Data were derived from China Academic Journal Network Publishing Database of CNKI, using the search terms 'e-sports' and limited field 'title'. After collating the data, 151 papers on e-sports was calculated out; using the method of bibliometric, this paper carried on analyzing obtained papers from the change tendency of the total number of papers, journal distribution, subject distribution, fund source, citation frequency, download frequency, core authors and so on. But as the journal published by the network exists the hysteresis nature, the statistical data from 2009-2010 is not comprehensive, and is only for reference.

D. Zeng (Ed.): Advances in Computer Science and Engineering, AISC 141, pp. 111–118.
Springerlink.com

Table 1. Time distribution of papers on e-sports

Year	2010	2009	2008	2007	2006	2005	2004	2003	2000
Number of papers	12	23	34	30	20	25	11	2	1
Percentage (%)	6.62	15.23	22.52	17.88	11.92	16.56	7.28	1.32	0.67

2 Statistical Analysis of Papers

2.1 Time Distribution of Papers

Since 2000, the growth of papers on e-sports fluctuates, overall shows rising trend, and the statistical number of papers on e-sports is shown in table 1.

The statistics show that the first paper on e-sports in China is 'new hot spot of international athletics——e-sports' written by An Xiaoqiang, published in 2000 No.4 of ' Tianjin Science & Technology'. Data from Table 1 show that the amount of papers on e-sports since 2003 showed an increasing trend, and indicates the industry's deepening understanding and study of e-sports for the year.

2.2 Journal Distribution of Papers

According to statistics, 151 papers were approximately distributed in four categories journals, shown in table 2. Of which 82 papers on e-sports were published in sports journals, occupied 54.30% in all papers, we can see sports journals mostly pay attention to the study of e-sports and provide a broad academic space for the study of e-sports.

Table 2. Discipline distribution of journal publishing papers on e-sports

category	sports	education	economics	others
Number of papers	82	16	15	38
percentage (%)	54.30	10.60	9.93	25.17

According to Bradford's law, papers on some subject are relatively concentrated in a number of journals, while other papers are scattered in other related journals. The statistics show that 151 papers on e-sports from 2000 to 2010 were published in 86 journals, average each journal published 1.76 papers, of which 32 journals published 97 papers over the above average, and we can see table 3. According to the determination of the core journals, the top 15 journals published 63 papers, occupying 41.72% and above 1/3 of total papers, that 15 journals is the core journals on the study of e-sports, and are important information sources on e-sports.

Table 3. Journal distribution published 2 and above papers on e-sports

Serial number	journal	Number of papers publish-ed in journal	percentage (%)
1	Bulletin of Sport Science & Technology	12	7.95
2	Journal of Jilin Institute of Physical Ecucation	5	3.31
3	Market Modernization	5	3.31
4	Inner Mongolia Sports Technology	4	2.65
5	Journal of Wuhan Institute of Physical Education*	4	2.65
6	Journal of Pla Institute of Physical Education	4	2.65
7	Journal of Pla Institute of Physical Education	4	2.65
8	Zhejiang Sport Science	4	2.65
9	Journal of Beijing Sport University*	3	1.99
10	Journal of Hubei Sports Science	3	1.99
11	Exam Weekly	3	1.99
12	Journal of Shandong Institute of Physical Education and Sports*	3	1.99
13	Sports Research and Education	3	1.99
14	Journal of Sports Adult Education	3	1.99
15	Journal of Tianjin University of Sport*	3	1.99
16	Science & Technology Information Scholarly	2	1.32
17	Science & Technology Information Scholarly	2	1.32
18	Journal of Hebei Institute of Physical Education	2	1.32
19	Heilongjiang Science and Technology Information	2	1.32
20	Cultural Life of the Barracks	2	1.32
21	Science and Gulture	2	1.32
22	Science & Technology Information	2	1.32
23	Science & Technology Information	2	1.32
24	Liaoning Sport Science and Technology	2	1.32
25	Neijiang Technology	2	1.32
26	Journal of Shanghai University of Sport*	2	1.32
27	Journal of Capital Institute of Physical Education*	2	1.32
28	Digital Users(Internet World)	2	1.32
29	Sports Expo	2	1.32
30	Sports Mentorship	2	1.32
31	Sports World Scholarly	2	1.32
32	Sports Culture Guide*	2	1.32
Total		97	64.24

Note: Journals plus * from the compilation of A Guide to the Core Journal of China(2008 version).

20 papers on e-sports published in core journals accounts for 13.25% of total papers, they were all published on 8 core journals, and core journals is all about sports. It can be seen that now papers on e-sports are mostly published in general journals, only a few is published in core journals, and the quality of papers need to be further improved.

2.3 Subject Distribution of Papers

According to subjects papers on e-sports are classified, that involves four subjects, and shown in table 4. As can be seen from table 4, the number of papers on the development of e-sports accounts for 39.07%, and is the largest proportion of papers, indicating that researchers generally concerned about the development of e-sports; the number of papers on other subjects accounts for 31.13%, indicating that subjects of e-sports are widely distributed. The number of papers on the development of e-sports industry is also many, reaching 23.18%, but the number of papers on the e-sports impact on the education is a few, only 6.62%, and researchers need to pay more attention in future in this subject.

Table 4. Subject distribution of papers on e-sports

subject	development	Industry development	Impact on education	other
Number of papers	59	35	10	47
percentage (%)	39.07	23.18	6.62	31.13

In statistics, the author found that the e-sports study of our country is still in the stage of the introduction, the substantive content of e-sports has not been involved, such as the study of some specific e-sports, the empirical study of some e-sports, the choosing and training of athletes in e-sports and so on, that did not involve, and to some extent that reflects the study of e-sports has yet to be expanding.

2.4 Fund Source of Papers

The statistics show that 8 papers on e-sports were supported by the fund, involving 10 funds, including 6 provincial or above funds, 1 municipal fund and 3 university's funds. Papers supported by funds accounts for 5.06% of all papers, so thus the current study on e-sports has not been sufficient attention.

2.5 Citation Frequency of Papers

Based on the citation frequency sorting of CNKI, there are 14 papers that the citation frequency is more than 10, shown in table 5. There are 11 papers of them published in core journals, it can be seen that currently papers published in core journals relatively have higher qualities, and can be better recognized.

Table 5. Citation frequency of papers on e-sports

Title	Author	Literature source	Publishing time	Citation frequency
Preliminary study about the development of E-sport	Feng Yuchao	Zhejiang Sport Science	2003/05	31
Study on Conception,Category and Development Process of the E-sports	Li Zonghao, Wang Jian, Li Bai	Journal of Tianjin University of Sport	2004/01	28
Make Digital Interpretation of Unlimited Exciting of Sports——E-sports and Development of It in China	He Huixian	Sports Culture Guide	2004/08	25
The Current Situation and Tendency Of E-SportsCompetition Under The Background Of Digital Sports	Hu Jian	Journal of Chendu Sport University	2004/03	21
Related Concepts and Types of E-sports	He Wei	Sports Culture Guide	2004/05	18
Feasibility Analysis of E-sports Industry Business in China	Dai Yunpeng, Yang Yugong	Sports Culture Guide	2004/05	17
Present situation and strategies of esports games in China	Lei Xi	Zhejiang Sport Science	2005/01	15
Research on the Current Situation and Strategy Consideration on the Development Way of E-sports Games Industry in China	Lei Xi, Xia Siyong	Journal of Beijing Sport University	2005/08	13
Development of campus electronic games	Ling Gang, Wang Feng-xian	Journal of Wuhan Institute of Physical Education	2005/08	13
Background of e-sports industry in China	Cui Kailing	Journal of Wuhan Institute of Physical Education	2006/08	11
Problems and Countermeasures about the Development of Chinese E-sport Industry	Du Yu, Zheng Zhiqiang	Journal of Beijing Sport University	2006/12	11
E-sports:a cognitive game based on virtual reality	Jia Peng, Yao Jiaxin	Journal of Wuhan Institute of Physical Education	2005/01	11
Electronic Sports Should be Catalogued to Game	Yang Fang	Journal of Shandong Institute of Physical Education and Sports	2005/01	11
On development mode of E-sports industry in China	Zhao Zijian, Ren Zhong	Journal of Wuhan Institute of Physical Education	2006/09	10

2.6 Download Frequency of Papers

Based on the download frequency sorting of CNKI, there are 11 papers that the download frequency is more than 200, shown in table 6, but statistics did not include data of mirror image station of colleges.

As can be seen from table 6, 9 papers of 11 papers were published in core journals, and have relatively high degree of concern. Of which 9 papers are about the concept and development of e-sports, that shows the study of e-sports is still at the initial stage and the demand for the basic study of e-sports is relatively large in China;2 papers are about e-sports industry, and that shows the combination of the management and e-sports has drawn increasing attention.

Table 6. Download frequency of papers on e-sports

Title	Author	Literature Source	Publishing Time	Download Frequency
Study on Conception,Category and Development Process of the E-sports	Li Zonghao, Wang Jian, Li Bai	Journal of Tianjin University of Sport	2004/01	367
Research on the Current Situation and Strategy Consideration on the Development Way of E-sports Games Industry in China	Lei Xi, Xia Siyong	Journal of Beijing Sport University	2005/08	350
The Current Situation and Tendency Of E-SportsCompetition Under The Background Of Digital Sports	Hu Jian	Journal of Chendu Sport University	2004/03	259
Development of campus electronic games	Ling Gang, Wang Fengxian	Journal of Wuhan Institute of Physical Education	2005/08	245
Problems and Countermeasures about the Development of Chinese E-sport Industry	Du Yu, Zheng Zhiqiang	Journal of Beijing Sport University	2006/12	238
Make Digital Interpretation of Unlimited Exciting of Sports——E-sports and Development of It in China	He Huixian	Sports Culture Guide	2004/08	228
E-sports:a cognitive game based on virtual reality	Jia Peng, Yao Jiaxin	Journal of Wuhan Institute of Physical Education	2005/01	227
Status Quo of Carrying out Electron-Sport Competitions in Shanghai Universities and Colleges and Some Suggestions	Yang Qian, Liu Bing	Journal of Shanghai University of Sport	2007/03	215
Present situation and strategies of E-sports games in China	Lei Xi	Zhejiang Sport Science	2005/01	206
Study on the Current Situation of Chinese E-sport	Li Li, Li Xiaolan, Lv Yanli	Journal of Physical Education Institute of Shanxi Teachers University	2005/02	202
Related Concepts and Types of E-sports	He Wei	Sports Culture Guide	2004/05	201

3 Statistical Analysis on Authors of Papers

3.1 Statistical Analysis on Author Affiliations

The statistics show that the first authors of papers on e-sports are 137, distributed in colleges, research institutes, middle schools and other affiliations. Which authors of 138 papers are from colleges, account for 91.39%, and colleges are dominated in all affiliations; authors of 13 papers are from research institutes and middle schools, account for 8.61%.It can be seen that colleges are the main body of study on e-sports.

3.2 Statistical Analysis on Co-authors

In the statistics of 151 papers, single authors account for 47.02%, two co-authors account for 39.74%, three co-authors account for 9.27%, four co-authors account for 2.65% and five co-authors account for 1.32%. It can be seen that the study on e-sports mainly relies on single authors and two co-authors.

4 Characteristics of the Study and Recommendations

Through statistical analysis of domestic papers on e-sports and overviews of statistical analysis results, the following characteristics of studies on e-sports can be found:

Number of papers has grown rapidly since 2003, showed an increasing trend;

50.30% of papers were published in sports journals, but only 13.25% of papers were published in core journals, indicating that the current domestic study on e-sports is still in its infancy, and it is not deep enough;

The study on e-sports emphasizes on the theoretical study, the domestic study base is relatively weak, the introduction abroad is not much, the study with practice is few, and the theory and practice are not synchronized.

Therefore, in the study on e-sports in the future, the author proposes deepening the theory study of e-sports, reducing the duplication of the study, strengthening the practice study of e-sports, paying attention on the abroad related study, and carrying on the practice study on the different e-sports events.

References

1. Electronic sports (September 30, 2010), `http://en.wikipedia.org/wiki/Electronic_sports`
2. Qiuyu, H.: In Addition to Entertainment, What We Can Also Do Using E-sports? Beijing Morning Post (January 16, 2004)
3. Li, Z., Wang, J., Li, B.: Study on Conception, Category and Development Process of the E-sports. Journal of Tianjin University of Sport 19, 1–3 (2004)
4. He, H.: Make Digital Interpretation of Unlimited Exciting of Sports——E-sports and Development of It in China. Sports Culture Guide, 3–7 (2004)

5. Hu, J.: The Current Situation and Tendency Of E-Sports Competition Under The Back-ground Of Digital Sports. Journal of Chendu Sport University 30, 18–21 (2004)
6. He, W.: Related Concepts and Types of E-sports. Sports Culture Guide, 11–13 (2004)
7. Dai, Y., Yang, Y.: Feasibility Analysis of E-sports Industry Business in China. Sports Culture Guide, 14–15 (2004)
8. Lei, X., Xia, S.: Research on the Current Situation and Strategy Consideration on the Development Way of E-sports Games Industry in China. Journal of Beijing Sport University 8, 1033–1035 (2005)
9. Ling, G., Wang, F.: Development of campus electronic games. Journal of Wuhan Institute of Physical Education 39, 110–112 (2005)
10. Cui, K.: Background of e-sports industry in China. Journal of Wuhan Institute of Physical Education 40, 34–36 (2008)

Experimental Investigations on the Performance of a Dual Fuel Diesel Engine with Hydrogen and LPG as Secondary Fuels

D.B. Lata* and Ashok Misra

Department of Mechanical Engineering, Birla Institute of Technology, Mesra, Ranchi, 835215, India
devdaslata@yahoo.com

Abstract. This paper presents experimental investigations on dual fuel operation of a 4 cylinder 39 kW diesel engine. To carry out detailed investigations hydrogen, liquefied petroleum gas (LPG) and mixture of LPG and hydrogen have been used as secondary fuels. The paper includes details on brake thermal efficiency and on emissions of un-burnt hydrocarbon (HC), carbon monoxide (CO), and NO_x. When only hydrogen is used as secondary fuel, maximum enhancement in the brake thermal efficiency is 14% which is obtained with 30% of secondary fuel. When only LPG is used as secondary fuel, maximum enhancement in the brake thermal efficiency (of 5%) is obtained with 40% of secondary fuel. Compared to the pure diesel operation, proportion of un-burnt HC and CO increases, while, emission of NO_x reduces in both Tests. There is significant enhancement in performances of dual fuel engine when hydrogen-LPG mixture is used as the secondary fuel. The highlight of this Test is that when 40% of mixture of hydrogen and LPG (30:70) is used as secondary fuel, enhancement in the brake thermal efficiency becomes 24% and HC emission is reduced by 64%.

Keywords: Dual fuel engine, diesel engine, alternative fuels, hydrogen, LPG.

1 Introduction

One of the main purposes for improving the combustion process of conventional internal combustion engine is to find out useful ways to reduce exhaust emissions without making major alteration on their mechanical configuration [1]. With the increasing need to conserve fossil fuels and minimize toxic emission, much effort is being focused on the advancement of current combustion technology [2]. This provoked for developing and testing of several alternative fuels. The main exhaust emissions from diesel engine are smoke, NO_x, particulate matter. The only option to reduce these pollutants is to use alternative fuels which does not have sulphur dioxide,

* Corresponding author.

D. Zeng (Ed.): Advances in Computer Science and Engineering, AISC 141, pp. 119–128.
Springerlink.com

aldehydes and ketones [3]. Various solution have been proposed like alcohols, vegetable oils, liquefied petroleum gas (LPG), compressed natural gas (CNG), hydrogen, biogas, producer gas and liquefied natural gas (LNG). Among these, the uses of gaseous fuels have a dominant position [4]. Hydrogen and LPG have auto-ignition temperatures of 858 K and 788 K respectively and require an ignition source to be burnt in an I.C engine [4]. The diesel fuel which has an auto-ignition temperature of 525 K can be used as a pilot fuel to ignite hydrogen and LPG. To improve the performance and reduce emission, Saravanan et al. [3] used hydrogen as an air-enrichment medium while diesel as an ignition source in a stationary diesel engine system.

Lambe and Watson [5] conducted an experimental study by optimizing C.I. engine for hydrogen combustion with diesel pilot fuel. The hydrogen contained 65-95% of the fuel energy. It was reported that smoke was reduced between 20% and 82% with dual fuel combustion depending on load condition. While nitrogen oxides were reduced up to 70% in some Tests.

The combustion analysis on a direct injection diesel engine by using hydrogen with diesel and hydrogen with diethyl ether (DEE) as ignition source was done by Saravanan et al. [3]. It increases brake thermal efficiency by 20% and oxides of nitrogen increased by 13% than diesel operation. Hydrogen assisted diesel combustion was investigated by Lilik et al. [6] on a DDC/VM Motori 2.5L, 4-cylinder, turbocharged, common rail, direct injection light-duty diesel engine, with a focus on exhaust emissions. Hydrogen was substituted for diesel fuel on an energy basis of 0%, 2.5%, 5%, 7.5%,10% and 15% by aspiration of hydrogen into the engine's intake air.

Due to availability and emission concern, LPG has also been tried as the secondary fuel. Poonia et al. [7] experimentally investigated the effects of intake charge temperature, pilot fuel quantity, exhaust gas recirculation and throttling of the intake charge can improve the performance of LPG-diesel dual fuel engine. Vijayabalan et al. [8] modified single cylinder vertical air-cooled diesel engine to use LPG in dual fuel mode to study the performance, emission, and combustion characteristics. The brake thermal efficiency was improved by 3% in the glow plug assisted dual fuel mode with reduction in hydrocarbon, carbon monoxide, and smoke emissions by 69%, 50% and 9% respectively at lower load.

In all above mentioned works, neither hydrogen nor LPG enhances all desired features of a dual fuel engine and therefore, it becomes sensible to try a mixture of both as a secondary fuel.

The understanding of brake thermal efficiency and emission of unburned hydrocarbon (HC), carbon monoxide (CO), oxides of nitrogen (NO_x) and smoke is indispensable to determine the over-all performance of a dual fuel diesel engine. In view of this, the present paper investigates the performances of a dual fuel diesel engine with hydrogen, LPG and mixture of LPG and hydrogen as secondary fuels at different load conditions and gaseous fuel substitutions.

2 Experimentation

A diesel engine setup was developed to carry out the study on dual fuel engines. A four stroke, compression ignition engine, model Mahindra was used for the experimental investigation. The diesel engine was modified to work on dual fuel mode by attaching hydrogen and LPG gas cylinder in connection with the intake manifold through flame traps, mass flow meters, followed by a one way non-return valve and common flame arrestor. The engine was coupled to a hydraulic dynamometer. The engine was run at constant speed of 1500 RPM. The amount of pilot diesel fuel was automatically controlled with the help of governor while the flow of gaseous fuels was controlled manually. The predetermined amount of gaseous fuel as percentage of diesel at each load percentage was inducted into the intake manifold through the gaseous fuel supply system. High precision optical crank angle encoder (Kistler make) was used to correlate pressure data on crank angle basis.

For gaseous fuels two separate flame traps and mass flow meters were used. Exhaust gas emissions namely CO, NO_x and unburnt hydrocarbons (UHC) were measured by an AVL 5000 DI Gas-analyzer. The experiments were performed on the test engine under the following four conditions.

Test I: Engine runs on diesel only.
Test II: Engine runs on diesel as pilot fuel and hydrogen as secondary fuel.
Test III: Engine runs on diesel as pilot fuel and LPG as secondary fuel.
Test IV: Engine runs on diesel as pilot fuel and LPG plus hydrogen as secondary fuel.

3 Results and Discussion

The experimental results at rated speed of 1500 rpm, injection pressure 220 bar and injection timing 18^0 BTDC are presented for Tests II, III, and IV at different load conditions. The gaseous fuel substitution with diesel is mainly presented at 10% and 80% of full load condition. The 10% of full load was selected to represent engine performance at light load condition whereas 80% load represents high load operation. Full load condition has not been taken into consideration due to knocking problem and limited to 50% and 70% respectively. The mixture of LPG and hydrogen were varied in the following proportions in each combination (M) (LPG-90% + H_2-10%), (LPG-80% + H_2-20%), (LPG-70% + H_2-30%), (LPG-60% + H_2-40%).

3.1 Brake Thermal Efficiency

The variation of brake thermal efficiency at different load condition for hydrogen (Test II), LPG (Test III) and mixture of LPG and hydrogen (Test IV) are shown in Figs. 1 (a, b). It is observed that at lower load conditions, the brake thermal efficiencies of the Tests II, III and IV are lower than pure diesel operation (Test I). At lower load condition, the lesser ignition centre is formed and it further reduces at higher substitution of gaseous fuel. Hence, poor ignition of gaseous fuel results in lower brake thermal efficiency. Further, at these conditions the inducted gaseous-air forms too lean mixture to burn well. This leads to low combustion rate and flame quenching. The brake thermal efficiencies at 10% load conditions for 30% substitution of gaseous fuel (hydrogen and LPG) as in Test II and III are 14.4% and 16.8% respectively as compared to 19.57% of Test I operation.

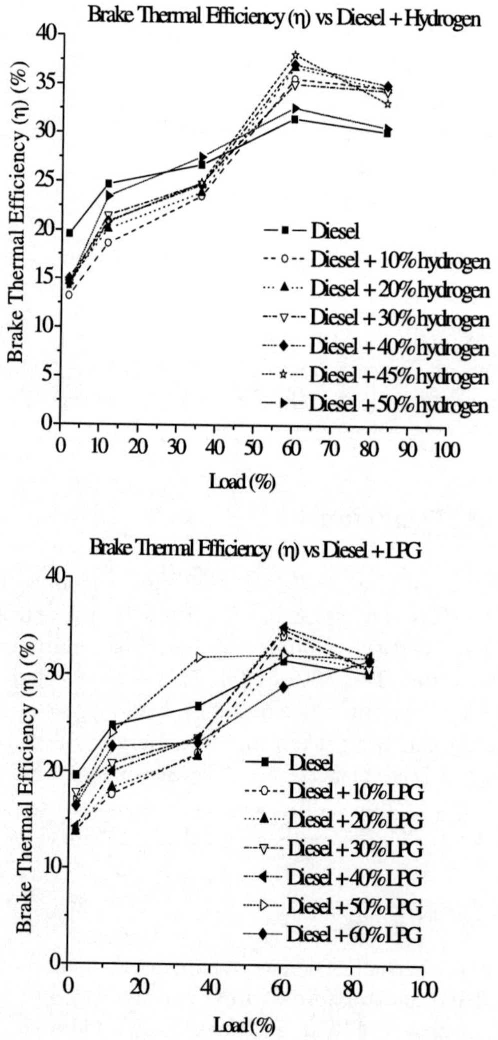

Fig. 1. Brake Thermal Efficiency) (%) vs. Load (%) a) for Diesel + Hydrogen. b) f or Diesel + LPG.

The brake thermal efficiency of Test II is less among other Tests due to lower quenching distance and higher thermal conductivity [5] resulting in higher cooling losses than LPG, while LPG has higher pre-ignition energy release rate as compared to hydrogen [2]. The lower brake thermal efficiency of Test III is due to sluggish combustion [7].

The brake thermal efficiency at 40% load conditions with the same gaseous fuels substitution for the Tests II and III are found to be 24.5% and 23.3% respectively against 26.75% of Test I operation. At 80% load condition, the brake thermal efficiency with the same gaseous fuels substitution for the Tests II and III are found to be 35.12% and 31.38% respectively as compared to 30.16% of Test I operation. This is because of increase in combustion rate due to larger pilot diesel quantities (as an average 5.44 mg/cycle) which leads to stronger ignition sources and hence, more complete and better combustion of gaseous fuel. It is worth to be mentioned here that beyond 45% of hydrogen substitution and 40% of LPG substitution, the thermal efficiency starts decreasing. This happens due to the fact that higher concentration of hydrogen or LPG leads to higher flame velocity and burning rate resulting in rapid consumption of oxygen, therefore, lesser oxygen becomes available in the diffusion combustion phase.

Figure 1 (b) shows variations in brake thermal efficiency at 10%, 40% and 80% of load condition respectively at each mixture combination of LPG and hydrogen (Test IV). The brake thermal efficiency at 10% load condition by substitution of 30% mixture (LPG and hydrogen) with hydrogen percentage varying from 10% to 40% are 15.1%, 14.7%, 14.3% and 13.9%, respectively as compared to 18.57% of diesel operation (Test I). It shows 7.45% drop in brake thermal efficiency when hydrogen percentage increased from 10% to 40% in the mixture of LPG and hydrogen. Further, 23% drop in brake thermal efficiency of the Test IV (60:40) was observed as compared to Test I operation. In pure diesel operation, the spray does not penetrate up to the cylinder walls due to less pilot quantity and the combustion is confined to piston chamber (bowl) only [5]. Moreover, this combustion zone is surrounded by air which acts as semi-insulator between the burned gases and cylinder walls.

Hence pure diesel operation has higher brake thermal efficiency than Tests II, III and IV at lower load conditions.

At lower load conditions, the propane (LPG)-air flame tends to be stable and hydrogen-air flame becomes unstable and by increasing hydrogen fraction flame destabilization increases due to reduction in Markstein length (Markstein length measures the effect of curvature on a flame; larger the Markstein length, greater the effect of curvature on burning velocity).

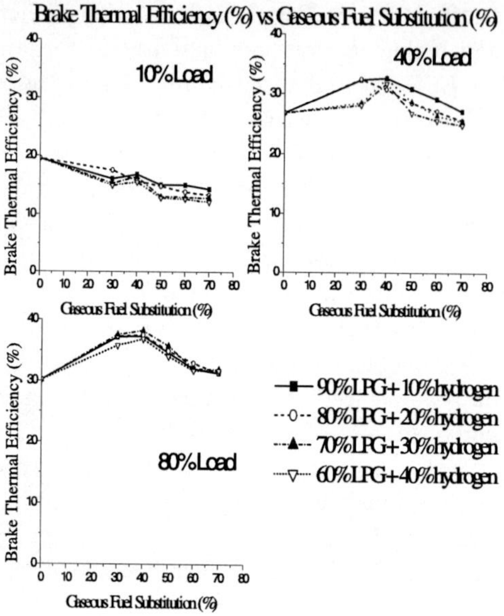

(a)

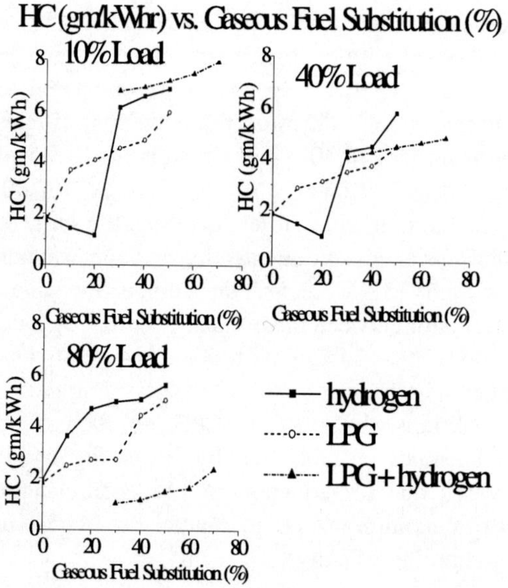

(b)

Fig. 2. a) Brake Thermal Efficiency (%), b) HC emissions vs Gaseous Fuel Substitution (%) for Diesel + Hydrogen + LPG Substitution.

At 80% load condition Test IV (70:30) shows 25%, 9% and 22% rise in brake thermal efficiency as compared to Test I, II and III, respectively (Fig. 2-a). At higher load conditions LPG flame tends to become unstable and hydrogen-air flames tend to be stable and increasing hydrogen fraction would lead to stabilization of flame [7] which is due to increase in Markstein length. This nature is because of opposite diffusion behavior of propane (LPG) and hydrogen, since diffusivity of hydrogen (0.61 cm^2/sec) is more than LPG (0.12 cm^2/sec) in air. Further, it was observed that 40% mixture of LPG and hydrogen (70:30) gives rise in brake thermal efficiency by 19% and 23% as compared to Test I at 40% and 80% load conditions respectively. However, 12% drop in efficiency was observed at 10% load condition.

4 Emissions

4.1 Un-Burnt Hydrocarbons (UHC)

Figure 2 (a) shows variation in unburned hydrocarbon at 10%, 40% and 80% load conditions for the Tests II, III and IV. The over leaning and under mixing is responsible for HC emission in diesel engine. At 10% load condition, Test II, III and IV shows HC emissions of 6.86 gm/kWh, 5.9 gm/kWh and 6.94 gm/kWh respectively as compared to 1.72 gm/kWh of Test I.

This is because of reduction in pilot quantity, which causes poor ignition of gaseous fuel and inducted mixture is too lean to burn. While, at 40% load condition, Test II, III and IV shows HC emissions of 4.45 gm/kWh, 4.65 gm/kWh and 4.37 gm/kWh respectively as compared to 2.13 gm/kWh of Test I.

At 80% load condition Test II, III and IV shows maximum HC emission of 5.64 gm/kWh, 4.57 gm/kWh and 1.07 gm/kWh respectively as compared to 1.8 gm/kWh of Test I operation. At higher load condition Test IV shows lower HC then Test I due to better and fast combustion rate leading to more complete combustion and hence low HC emissions.

4.2 Carbon Monoxide

The formation of carbon monoxide depends on post oxidation reaction. In normal diesel engine due to presence of more excess air, the carbon oxidation reaction almost completed [8]. The variation in CO is shown in Fig. 3 (a). At 10% load condition Test II, III and IV shows 89%, 77% and 75% raise in CO emission respectively as compared to Test I operation. At light load condition gaseous fuel-air mixture near the pilot is burned. Thus some partial oxidation product like carbon monoxide may come out in the exhaust. At higher concentration of gaseous fuel, the concentration of this partial oxidation product could increase [8]. Moreover, inducted mixture becomes rich due to more displacement of air. This is thought to be reason for the raise in CO emissions. While, at 40% load condition, Test II, III and IV shows CO emissions of 0.38 gm/kWh, 0.28 gm/kWh and 0.38 gm/kWh respectively as compared to 0.8 gm/kWh of Test I. At 80% load condition maximum raise in CO emission for the Test II, III and IV are 76%, 84% and 80% respectively as compared to Test I due to rich mixture.

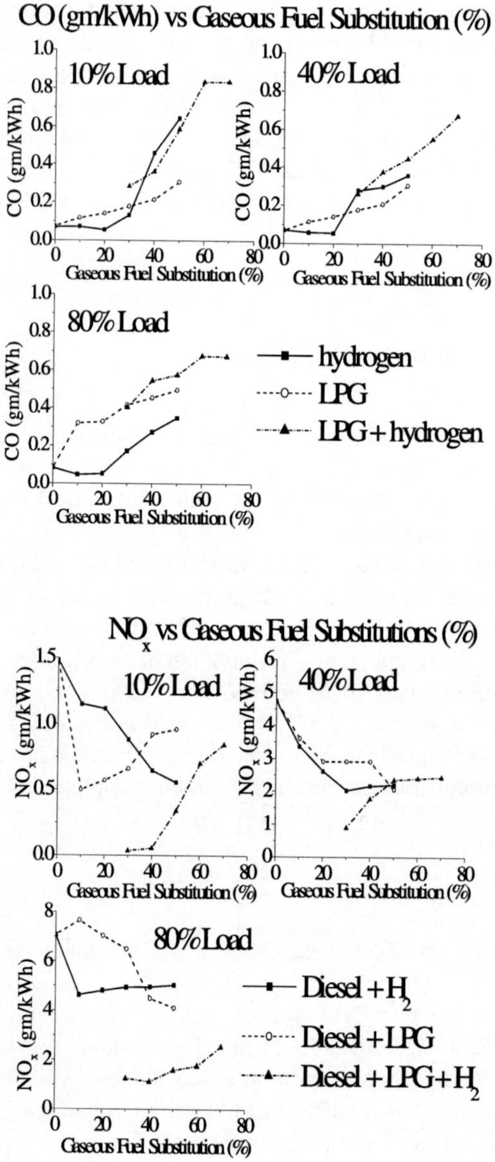

Fig. 3. a) Un-burnt HC (gm/kWh), b) NO_x vs Diesel + Gaseous Fuels Substitution (%) as function of Load (%)

4.3 NO_x

Figure 3 (b) shows variation of NO_x for the Test II, III and IV at 10% and 80% load condition. At 10% load condition 60%, 33% and 93% drop in NO_x emission were observed for the Test II, III and IV respectively as compared to Test I. Similarly, at 80% load condition 20%, 41% and 84% reduction in NO_x for the Test II, III and IV were observed respectively as compared to Test I. This is due to more uniform temperature distribution obtained with the gaseous fuel-air mixture results in high temperature region reduces around the diesel flame. While, at 40% load condition, Test II, III and IV shows NO_x emissions of 2.13 gm/kWh, 3.12 gm/kWh and 1.88 gm/kWh respectively as compared to 4.88 gm/kWh of Test I.

5 Conclusions

On the basis of the results and discussions presented above, the following conclusions may be drawn:

1) Substitution of diesel by 30% of hydrogen and 40% of LPG as in Test II and III enhances brake thermal efficiency by 16% and 5% respectively than Test I operation.
2) A 40% diesel substitution by the mixture of 70% LPG and 30% hydrogen in Test IV increases brake thermal efficiency by 26% compared to pure diesel operation.
3) At low load conditions for the Tests II to IV, the brake thermal efficiency is always lower than diesel operation (Test I). However, it has better efficiency at higher load condition.
4) A severe knock was noticed during the dual fuel diesel engine operation with 50% of hydrogen and 70% of LPG substitution as in Test II and III respectively.
5) Test II and III increases unburnt HC by 67% and 60% respectively while Test IV reduces by 67% and similarly carbon monoxide increases by 75%, 83% and 79% respectively.

The operation of the engine becomes most efficient and eco-friendly performances when a mixture of hydrogen and LPG is used as secondary fuel. Brake thermal efficiency may be increased and emission of NO_x, smoke and HC may be reduced.

References

1. Papagiannakis, R.G., Hountalas, D.T., Rakopoulos, C.D.: Theoretical study of the effects of pilot fuel quantity and its injection timing on the performance and emissions of a dual fuel diesel engine. Energy Conversion & Management 48, 2951–2961 (2007)
2. Choi, G.H., Chung, J.Y., Han, S.B.: Performance and emissions characteristics of a hydrogen enriched LPG internal combustion engine at 1400 rpm. Int. J. Hydrogen Energy 30, 77–82 (2005)

3. Saravanan, N., Nagarajan, G.: An experimental investigation of hydrogen-enriched air induc-tion in a diesel engine system. Int. J. Hydrogen Energy 33, 1769–1775 (2008)
4. Mansour, C., Abdelhamid, B., Abdelkader, A., Francoise, G.: Gas-diesel (dual-fuel) modeling in diesel engine environment. Int. J. Therm. Sci. 40, 409–424 (2001)
5. Lambe, S.M., Watson, H.C.: Low polluting, energy efficient C.I. hydrogen engine. Int. J. Hydrogen Energy 17, 513–525 (1992)
6. Lilik, G.K., Zhang, H., Herreros, J.M., Haworth, D.C.: Hydrogen assisted diesel combustion. Int. J. Hydrogen Energy 35, 4382–4398 (2010)
7. Poonia, M.P., Ramesh, A., Gaur, R.: Effect of intake air temperature and pilot fuel quantity on the combustion characteristics of a LPG diesel dual fuel Engine, 97–105 (1998) SAE paper 982455
8. Vijayabalan, P., Nagarajan, G.: Performance, emission and combustion of LPG diesel dual fuel engine using glow plug. JJMIE 2, 105–110 (2009)

A Graph Design for Nine Graphs with Six Vertices and Nine Edges

Caihua Wang

Basic Courses Teaching Department, Chinese people' Armed Police Force Academy, Lang Fang, 065000, HeBei, China
ahdxsxw@126.com

Abstract. In this paper, we investigate the existence problem of a graph design for a nine-edge six-vertex graph. Using standard recursions and combining with the necessary direct constructions, we give a near solution for the problem.

Keywords: G-design, holey G-design, pair wise balanced design.

1 Introduction

Let G be a set of graphs and H a graph. A G-decomposition of H is a partition of H into sub graphs (blocks) so that each sub graph is isomorphic to a graph of G. When H is a complete graph K_v of order v, the graphs in a G-decomposition of H form a G-design of order v, denoted by $G-GD(v)$. When G contains a single graph G, the design is a G-design. A $\{K_{k1},\cdots,K_{km}\}$-design of order v is called a pair wise balanced design and it is denoted by $PBD(v,\{k_1,\cdots,k_m\})$.

First, we define a complete multipartite graph to be of type $g_1^{u_1}\cdots g_s^{u_s}$ if it has exactly $\sum_{1\le i\le s}u_i$ classes (groups) in the multipartite, and there are u_i groups of size g_i for $i=1,2,\cdots,s$. A $G-$ design of the complete multipartite graph of type $g_1^{u_1}\cdots g_s^{u_s}$ is termed a G-group divisible design of type $g_1^{u_1}\cdots g_s^{u_s}$, and it is often called a GDD.A $\{K_{k1},\cdots,K_{km}\}-GDD$ is often denoted by $\{k_1,\cdots,k_m\}-GDD$. A $\{k_1,\cdots,k_m\}-GDD$ of type 1^v is actually a $PBD(v,\{k_1,\cdots,k_m\})$.

Graph design is an important research course, which puts its focus on the decomposition with exact balance. For the study of graph design, it is not only important for itself, but also important for transportation, website and DNA data choosing. The existence of graph design has been studied.

Numerous articles have been written on the existence of G-designs. The spectrum problem has been considered for all graphs with up to five vertices, and for all graphs with six vertices and up to eight edges. Particularly, [2] is the latest survey.

In this paper, we will discuss graph designs of one graph with six vertices and nine edges, which is denoted by G_8 (the notation in [1] Appendix I, see the figure below),

D. Zeng (Ed.): Advances in Computer Science and Engineering, AISC 141, pp. 129–134.
Springerlink.com

and prove that there exists a $G_8 - GD(v)$ when $v \equiv 0 (\bmod 9)$ and $v \geq 9$, except for $v = 9$ and except possibly for $v \in \{18,54,72,90\}$. For the graph G_8, as a block in graph designs, it will be denoted by $\{a,b,c,d,e,f\}$ according to the following vertex –labels.

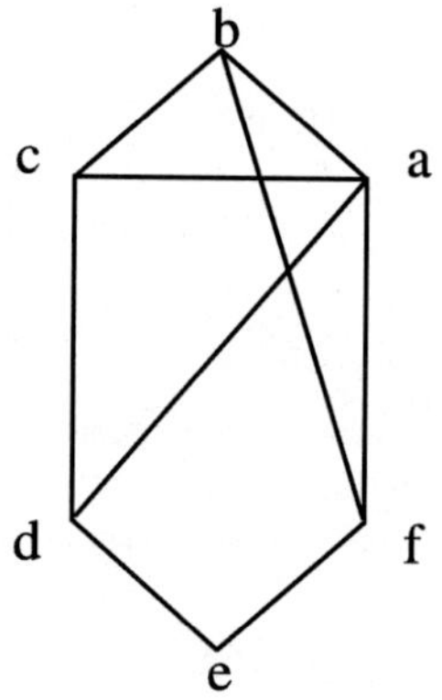

$\{ a,b,c,d,e,f \}$

2 Direct Constructions

First, we present direct constructions for small orders, which will be used in the last section to obtain the final theorem.

Lemma 2.1. The necessary conditions for the existence of $G_8 - GD(v)$ is $v \equiv 0,1(\bmod 9)$ and $v \geq 9$.

Lemma 2.2. For every $v \in \{10,19,27,36,45,55,63,73,99\}$, there exist a $G_8 - GD(v)$.

Proof: For $v \in \{10,19,55,73,\}$, let the point set Z_v, the desired design is obtained by adding $2(\bmod 10)$ for $v = 10$ and by adding $1(\bmod v)$ to the rest point to the following base blocks.

$v = 10: \{0,1,3,4,2,7\}$.

$v = 19: \{0,1,3,7,15,6\}$.

$v = 55: \{11,44,26,7,15,39\}, \{21,19,7,37,27,18\}, \{0,6,32,11,28,48\}$.

$v = 73: \{26,10,4,29,18,59\}, \{34,11,48,22,43,24\}, \{0,1,5,7,15,35\}, \{0,15,46,55,26,43\}$.

For $v \in \{27,36,45,63,99\}$, let the point set be $Z_{v-1} \cup \{\infty\}$. The desired design is obtained by adding $2(\bmod v-1)$ for $v \in \{27,63,99\}$, by adding $5(\bmod 35)$ for $v = 36$, by adding $4(\bmod 44)$ for $v = 45$ to the following base blocks. Here, ∞ is the fixed point.

$v = 27: \{6,7,21,1,\infty,0\}, \{0,2,5,8,1,16\}, \{1,10,14,23,7,9\}$.

$v=36:\{34,2,1,13,3,33\},\{30,1,31,17,0,32\},\{24,9,21,0,10,19\},\{7,28,15,27,11,29\},\{30,33,26,\infty,28,2\},$
$\{1,9,16,27,19,7\},\{0,30,11,8,25,29\},\{32,7,13,2,\infty,14\},\{3,13,19,5,31,21\},\{4,13,26,28,5,20\}$.

$v=45:\{42,32,35,10,6,22\},\{25,30,24,28,27,12\},\{3,40,8,43,\infty,16\},$
$\{4,20,18,23,21,41\},\{1,12,42,13,14,20\},\{2,$
$25,33,24,29,19\},\{12,29,38,39,17,35\},\{11,29,39,31,2,0\},\{15,27,18,26,\infty,29\},\{18,29,25,39,26,1\}$.

$v=63:\{27,11,13,58,43,46\},\{53,59,34,48,28,17\},\{54,48,49,27,17,26\},$
$\{40,30,38,47,51,7\},\{25,4,42,26,44,37\},$
$\{17,55,6,56,\infty,47\},\{1,22,26,29,16,48\}$.

$v=99:\{90,69,21,95,49,10\},\{43,27,46,18,\infty,15\},$
$\{84,97,17,74,38,60\},43,33,22,7,15,13\},$
$\{64,33,30,26,89,55\},\{74,72,16,89,29,58\},$
$\{22,14,57,31,36,68\},\{19,25,66,36,24,18\},$
$\{5,16,71,38,65,9\},\{0,19,72,23,68,20\},\{1,15,28,45,85,38\}$

Lemma 2.3. For $t\in\{3,4,5,6\}$, there exist a $G_8-HD(9^t)$.

Proof. Let the point set be Z_{9t} and let the group set be $\{\{0,t,\cdots,8t\}+i:0\le i\le t-1\}$. the desired design is obtained by adding $t(\bmod 9t)$ for $t\in\{3,4\}$, by adding $1(\bmod 45)$ for $t=5$,by adding $2(\bmod 54)$ for $t=6$ to the following base blocks.

$t=3:\{21,2,5,20,19,24\},\{20,10,25,16,18,3\},\{0,1,4,2,6,20\}$.

$t=4:\{28,18,14,0,20,15\},\{17,0,25,34,5,3\},\{12,19,3,25,27,8\},$
$\{14,30,7,3,13,29\},\{18,8,23,13,12,5\},\{0,5,9,34,19,2\}$

$t=5:\{4,23,6,39,26,1\},\{0,1,5,14,29,8\}$.

$t=6:\{11,31,27,10,17,38\},\{9,44,0,4,2,1\},$
$\{17,30,15,26,4,50\},\{6,22,37,34,3,35\},\{0,14,51,25,35,49\}$

Lemma2.4. There exist a $G_8-HD(27^6)$.

Proof. Let the point set be $Z_{135}\cup\{\infty_0,\infty_1,\cdots,\infty_{26}\}$, and let the group set be $\{\{0,5,10,\cdots,130\}+i:0\le i\le 4\}$ .The desired design is obtained by adding $1(\bmod 135)$ to the following base blocks. Here, $\infty_i+1=\infty_{(i+1)}(\bmod 27)$ for $i=0,1,\cdots,26$.

$\{116,69,17,\infty_{15},99,\infty_{19}\},\{2,\infty_5,50,59,75,44\},\{124,32,18,\infty_5 6,60\},$
$\{50,24,43,101,\infty_{25},106\},\{\infty,44,93,85,13,40\}$
$\{\infty_0,61,58,21,53,59\},\{77,110,64,88,105,36\},$
$\{0,1,22,123,\infty_6,97\},\{0,44,62,\infty,126,67\}$.

Lemma2.5. There exist a $G_8-HD(27^3 36^1)$.

Proof. Let the point set be $Z_{81}\cup\{\infty_0,\infty_1,\cdots,\infty_{35}\}$, and let the group set be $\{\{0,3,6,\cdots,78\}+i:0\le i\le 2\}$.The desired design is obtained by adding $1(\bmod 81)$ to the following base blocks. Here, $\infty_i+1=\infty_{(i+4)}(\bmod 36)$ for $i=0,1,\cdots,35$.

$\{28,35,\infty_{33},14,\infty_{30},24\},\{\infty_2,32,49,36,\infty_{11},6\},\{\infty_{31},44,36,67,77,25\},$
$\{46,\infty_0,3,\infty_{31},33,5\},\{71,\infty_{30},27,22,68,10\},$
$\{\infty_{13},42,64,62,\infty_1,58\},\{0,25,85,1,48,\infty_{20}\}$.

Lemma 2.6. There exist a $G_8-HD(27^3 45^1)$.

Proof. Let the point set be $Z_{81}\cup\{\infty_0,\infty_1,\cdots,\infty_{44}\}$, and let the group set be $\{\{0,3,6,\cdots,78\}+i:0\le i\le 2\}$.The desired design is obtained by adding $1(\bmod 81)$ to the following base blocks. Here, $\infty_i+1=\infty_{(i+5)}(\bmod 45)$ for $i=0,1,\cdots,44$.

$\{\infty_{25},7,68,63,85,78\},\{\infty_{29},9,43,39,\infty_{20},44\},\{45,\infty_6,67,\infty_{38},59,28\},$
$\{7,\infty_{43},63,93,4,53\},\{12,13,5,\infty_{14},28,\infty_{22}\},$
$\{23,75,\infty_{36},36,50,34\},\{70,\infty_6,32,\infty_0,3,47\},\{0,19,\infty_{32},79,26,\infty_{28}\}$.

Lemma 2.7. There exist a $G_8-HD(45^3 63^1)$.

Proof. Let the point set be $Z_{135}\cup\{\infty_0,\infty_1,\cdots,\infty_{62}\}$, and let the group set be $\{\{0,3,6,\cdots,132\}+i:0\le i\le 2\}$.The desired design is obtained by adding $1(\bmod 135)$ to the following base blocks. Here, $\infty_i+1=\infty_{(i+7)}(\bmod 63)$ for $i=0,1,\cdots,62$.

$\{56,48,\infty_{54},106,6,112\},\{27,47,130,\infty_8,19,\infty_{24}\},\{\infty_{40},30,85,98,\infty_{60},34\},$
$\{12,22,\infty_{15},52,14,\infty_7\},\{108,\infty_2,110,$
$\infty_{37},132,133\},\{49,35,\infty_{56},27,118,\infty_{37}\},\{111,\infty_{20},64,\infty_{46},54,80\},$
$\{80,114,\infty_{34,},85,17,91\},\{20,39,\infty_{36},27,125,$
$109\},\{114,56,7,\infty_{56},23,\infty_{17}\},\{0,76,\infty_{59},73,20,\infty_{48}\},\{\infty_4,1,95,52,\infty_{12},119\}$.

We will use the recurrence construction below later.

Construction 1. If there exists a $K-GDD$ of type $m_1^{r_1}\cdots m_t^{r_t}$ and a $G-HD$ of type n^k for each $k\in K$ .then there exists a $G-HD$ of type $(nm_1)^{r_1}\cdots(nm_t)^{r_t}$..

Construction 2. Suppose there exists a $G-HD$ of type $m_1^{r_1}\cdots m_t^{r_t}$. If there exists a $G-ID(m_i+w,w)$ for each $i=1,2,\cdots,t$ .Then there exists a $G-ID(u+w,w)$ for $u=m_1r_1+\cdots m_tr_t$. If there also exists a $G-ID(u+w,w)$, then there exists a $G-GD(u+w)$.

Lemma 2.8([3]) (1) : For any $v\ge 3$ and $v\notin(6,8)$, there exists a $B(\{3,4,5\},1;v)$.

(2) For any $v\ge 4$ and $v\notin\{7,8,9,10,11,12,14,15,18,19,23\}$, there exist a $B(\{4,5,6\},1;v)$.

Lemma 2.9. For any $n \geq 3$ and $n \notin \{6,8\}$ .there exist a $G_8 - GD(9n+1)$.

Proof: By lemma 2.8 we know, for any $n \geq 3$ and $n \notin \{6,8\}$, there exist a $B(\{3,4,5\},1;v)$. To this n using a weight of 9, then for $t \in \{3,4,5\}$, construct a $G_8 - HD(9^t)$ on every t length block. By construction 1 we get a $G_8 - HD(9^n)$ .adding a infinite point, applying Construction 2 and using a $G_8 - GD(10)$ to fill in the holes gives a $G_8 - GD(9n+1)$ .Here we need $G_8 - HD(9^t)$ for $t \in \{3,4,5\}$ as input designs, which all come form lemma 2.3. The proof is complete.

Lemma 2.10. For any $n \geq 4$ and $n \notin \{7,8,9,10,11,12,14,15,18,19,23\}$, there exist a $G_8 - GD(9n-9)$.

Proof: By lemma 2.8 we know, for any $n \geq 4$ and $n \notin \{7,8,9,10,11,12,14,15,18,19,23\}$, there exist a $B(\{4,5,6\},1;n\}$, delete one point, we get a $\{4,5,6\} - GDD(3^a 4^b 5^6)$ of $(n-1)$ order, where a,b,c is nonnegative integer. To this GDD, using a weight of 9 .for $t \in \{4,5,6\}$, construct a $G_8 - HD(9^t)$ on every t length block. By Construction 1 we get $G_8 - HD(27^a 36^b 45^c)$. By construction 2 and by lemma 2.2, using a $G_8 - GD(v)$ for $v \in \{27,36,45\}$ to fill in the holes gives $G_8 - GD(9n-9)$. Here, we need a $G_8 - HD(9^6)$.

The existence is obtained by lemma 2.3. In the following, we continue to discuss for $n \in \{9,13,14,17,18,22\}$, there exist a $G_8 - GD(9n)$.

Lemma 2.11. For $n \in \{9,13,14,17,18,22\}$, there exist a $G_8 - GD(9n)$.

Proof: For $n = 9$, start with a $3 - GDD(3^3)$, to this GDD using a weight of 9 and input a $G_8 - HD(9^3)$, by construction 1 we know ,we can obtain a $G_8 - HD(27^3)$, using a $G_8 - GD(27)$ to fill in the holes, by construction 2, we get a $G_8 - GD(81)$. Similarly, for $n = 17$, start with a $3 - GDD(3^4 5^1)$, to this GDD using a weight of 9 and input a $G_8 - HD(9^3)$, we have a $G_8 - HD(27^4 45^1)$, using a $G_8 - HD(27\ )$ and a $G_8 - GD(45)$ to fill in the holes gives a $G_8 - GD(153)$. For $n = 13$, by lemma 2.5 we know, there exist a $G_8 - HD(27^3 36^1)$, using a $G_8 - GD(27\ )$ and a $G_8 - GD(36)$ to fill in the holes gives a $G_8 - GD(117)$. For $n = 14$, by lemma 2.6 we know, there exist a $G_8 - HD(27^3 45^1)$, using a $G_8 - GD(27)$ and a $G_8 - GD(45)$ to fill in the holes gives a $G_8 - GD(126)$. For $n = 18$, by lemma 2.4 we know, there exist a $G_8 - HD(27^6)$, using a $G_8 - GD(27)$ to fill in the holes gives a $G_8 - GD(162)$. For $n = 22$, by lemma 2.7 we know, there exist a $G_8 - HD(45^3 63^1)$, using a $G_8 - GD(45)$ and a $G_8 - GD(63)$ to fill in the holes gives a $G_8 - GD(198)$. Here we need a $G_8 - GD(v)$ for $v \in \{27,36,45,63\}$ as input designs, which all come from lemma 2.2. The proof is complete.

3 Main Results

Theorem 3.1. Besides the inexistance of $G_8-GD(9)$ and $G_8-GD(v)$, the unknown existance of $v\in\{18,54,72,90\}$, the necessary condition of the existance of $G_8-GD(v)$, $v\equiv 0,1(\mathrm{mod}\,9)$ and $v\geq 9$ is sufficient.

Proof: First, computer search shows the nonexistence of $G_8-GD(9)$.Then combining the results in lemma 2.2 and lemma 2.9-2.11, the proof of theorem 3.1 is completed.

References

1. Colbourn, C.J., Ge, G.N., Ling, A.C.H.: Optical grooming with grooming ratio eight. Discrete Applied Math. 157, 2763–2772 (2009)
2. Adams, P., Bryant, D., Buchanan, M.: A survey on existence of G-Designs. J. Combin. Des. 16, 373–410 (2008)
3. Colbourn, C.J., Dinitz, J.H. (eds.): CRC Handbook of Combinatorial Designs, 2nd edn. Chapman and Hall/CRC, Boca Raton, FL (2007)

Positive Solutions for Singular Systems of Three-Point Boundary Value Problems

Caihua Wang

Basic Courses Teaching Department, Chinese people' Armed Police Force Academy, Langfang, 065000, Hebei, China
(ahdxsxw@126.com)

Abstract. This paper investigates the following singular systems of nonlinear second-order three-point boundary value problems

$$\begin{cases} -u'' = f(t,v), & t \in (0,1) \\ -v'' = g(t,u), & t \in (0,1) \\ u'(0) = v'(0) = 0, & u(1) = \alpha u'(\eta), v(1) = \alpha v'(\eta) \end{cases}$$

Where $\eta \in (0,1)$, $\alpha < 0$, Under some weaker conditions, the existence of positive solutions is obtained by applying the fixed point theorem of cone expansion and compression.

Keywords: positive solution, singular, Three-point boundary value problems.

1 Introduction

The study of three-point boundary value problems for nonlinear ordinary differential equations was initiated by Gupta [1]. Since then, many authors have studied three-point boundary value problems by applying the Leray-Schauder continuation theorem and its nonlinear alternatives, the coincidence degree theory of Krasnoselskii's fixed point theorem(see[1]-[4]).recently ,many authors investigated systems of boundary value problems(see[5][6]),However, to our knowledge, the existence of positive solutions of singular systems of three-point boundary value problems has not yet been studied only(see[7]).

In this paper, we shall study the following singular systems of nonlinear second-order three-point boundary value problems(BVP)

$$\begin{cases} -u'' = f(t,v), & t \in (0,1) \\ -v'' = g(t,u), & t \in (0,1) \\ u'(0) = v'(0) = 0, & u(1) = \alpha u'(\eta), v(1) = \alpha v'(\eta) \end{cases} \tag{1.1}$$

Where $\eta \in (0,1)$, $\alpha < 0$, $f, g \in C(0,1) \times C(0,1) \times R^+, R^+)$. The existence of positive solutions to BVP(1.1) is obtained under some weaker assumptions.

The vector $(u,v) \in C^2(0,1) \times C^2(0,1)$ is said to be a positive solution of BVP (1.1) if (u,v) satisfies (1.1) and $u(t) > 0, v(t) > 0$ for $t \in (0,1)$.

D. Zeng (Ed.): Advances in Computer Science and Engineering, AISC 141, pp. 135–141.
Springerlink.com

To derive the existence theorems for positive solutions, we make the following assumptions:

$$(H_1): f \in C((0,1)\times R^+, R^+), \quad g \in C((0,1)\times R^+, R^+),$$

And there exist $p_i \in C((0,1), R^+)$, $q_i \in C(R^+, R^+)$, $i=1,2$

$q_2(0)=0$, such that

$f(t,x) \le p_1(t)q_1(x), \quad g(t,y) \le p_2(t)q_2(y), \quad t\in(0,1), \quad x,y \in R^+$

And

$$a=\int_0^1 (1-s-\alpha)p_1(s)ds < +\infty \qquad b=\int_0^1 (1-s-\alpha)p_2(s)ds < +\infty .$$

(H_2) : There exist $\alpha_1,\alpha_2 \in (0,+\infty)$ with $\alpha_1\alpha_2 \le 1$ such that

$$\limsup_{x\to+\infty} \frac{q_1(x)}{x^{\alpha_1}} < +\infty, \qquad \limsup_{y\to+\infty} \frac{q_2(y)}{y^{\alpha_2}} = 0$$

(H_3) : There exist $\beta_1,\beta_2 \in (0,+\infty)$ with $\beta_1\beta_2 \le 1$ such that

$$\liminf_{x\to 0^+} \min_{t\in[0,\eta]} \frac{f(t,x)}{x^{\beta_1}} > 0, \qquad \liminf_{y\to 0^+} \min_{t\in[0,\eta]} \frac{g(t,y)}{y^{\beta_2}} = +\infty .$$

The main results obtained are as follows:

Theorem 1.1. Assume that (H_1)-(H_3) hold, Then BVP(1.1) has at least one positive solution.

The rest of paper is organized as follows: a number of Lemmas useful to the derivation of the main results are given in section 2, then the proof of the theorems is given in section 3.

2 Preliminaries and Lemmas

Lemma 2.1 ([8]). Let K be a cone of a real Banach space E, Ω_1 and Ω_2 be bounded open sets of E, $0\in \overline{\Omega}_1 \subset \Omega_2$, Suppose that $A: K\cap(\overline{\Omega}_2 \setminus \Omega_1) \to K$ is completely continuous such that one of the following conditions is satisfied:

(i) : $\|Au\| \le \|u\|, \; u\in \partial\Omega_1 \cap K; \; \|Au\| \ge \|u\|, \; u \in \partial\Omega_2 \cap K$.

(ii) : $\|Au\| \ge \|u\|, \; u\in \partial\Omega_1 \cap K; \; \|Au\| \le \|u\|, \; u \in \partial\Omega_2 \cap K$.

Then, A has a fixed point in $K\cap(\overline{\Omega}_2 \setminus \Omega_1)$.

Lemma 2.2 ([7],[9],[10]) . Let $0<\eta<1$, $\alpha<0$; then for any $y\in[0,1]$, the following boundary value problem :

$$\begin{cases} u''(t)+y(t)=0, & 0<t<1, \\ u'(0)=0, & u(1)=\alpha u'(\eta) \end{cases} \tag{2.1}$$

Has an unique solution.

$u(t) = \int_0^1 k(t,s)y(s)ds$, where $k(t,s)$ $[0,1]\times[0,1] \to R^+$ is defined by

$$k(t,s) = \begin{cases} (1-s)-(t-s)-\alpha, & 0 \le s \le t \le 1 \text{且} s \le \eta \\ (1-s)-\alpha, & 0 \le t \le s \le \eta \\ (1-s), & 0 \le t \le s \le 1 \text{且} s \ge \eta \\ (1-s)-(t-s), & \eta \le s \le t \le 1 \end{cases}$$

By Lemma 2.2, we easily obtain:

Lemma 2.3 ([10]): For any $(t,s) \in [0,1]\times[0,1]$, $1-\eta \le k(t,s) \le 1-s-\alpha$.

Let $E = C[0,1]$ be a *Banach* space endowed with norm $\|u\| = \max_{t\in J}|u(t)|$, $J = [0,1]$,

Let

$P = \{u \in E \,|\, u(t) \ge 0, t \in J\}$, Then P is a cone of E .For any $r > 0$,Let $B_r = \{u \in C[0,1] \,|\, \|u\| < r\}$, $\partial B_r = \{u \in C[0,1] \,|\, \|u\| = r\}$.

It is easy to prove that $(u,v) \in C^2(0,1)\times C^2(0,1)$ is a solution of BVP (1.1) if and only if $(u,v) \in C[0,1]\times C[0,1]$ is a solution of the following system of nonlinear integral equations:

$$\begin{cases} u(t) = \int_0^1 k(t,s) f(s, v(s)) ds \\ v(t) = \int_0^1 k(t,s) g(s, u(s)) ds, \quad t \in [0,1] \end{cases}$$

Obviously, the above system of nonlinear integral equations can be regarded as the following nonlinear integral equation:

$$u(t) = \int_0^1 k(t,s) f(s, \int_0^1 k(s,\tau) g(\tau, u(\tau)) d\tau) ds, \qquad t \in [0,1]$$

Define an operator $A : P \to P$, by

$$(Au)(t) = \int_0^1 k(t,s) f(s, \int_0^1 k(s,\tau) g(\tau, u(\tau)) d\tau) ds, \qquad t \in [0,1] \qquad (2.2)$$

It is easy to verify that $x(t)$ is a fixed point of A in $C[0,1]$; then BVP (1.1) has one solution (u,v),

$$\begin{cases} u(t) = x(t) \\ v(t) = \int_0^1 k(t,s) g(s, x(s)) ds, \quad t \in [0,1] \end{cases}$$

Lemma 2.4: Assume that (H_1) holds, Then $A : P \to P$ is completely continuous.

Proof. First it is easy to see that A map p into p.Then we prove that A maps bounded sets into bounded sets. Suppose $D \subset P$ is an arbitrary bounded set. Then there exist $M_1 > 0$, such that $\|u\| \le M_1$, for all $u \in D$. By the continuity of q_2, there is $M_2 > 0$ such that $M_2 = \sup_{x\in[0,M_1]} q_2(x)$. So for any $u \in D, s \in [0,1]$, by Lemma 2.3,we have

$$\int_0^1 k(s,\tau)g(\tau,u(\tau))d\tau \le \int_0^1 k(s,\tau)p_2(\tau)q_2(u(\tau))d\tau \le M_2\int_0^1(1-\tau-\alpha)p_2(\tau)d\tau = M_2 b \quad (2.3)$$

By the continuity of q_1, there is $M_3 > 0$ such that $M_3 = \sup_{x\in[0,M_2 b]} q_1(x)$. It follows from (2.3), (H_1) and Lemma 2.3 that

$$\begin{aligned}(Au)(t) &= \int_0^1 k(t,s)f(s,\int_0^1 k(s,\tau)g(\tau,u(\tau))d\tau)ds\\ &\le \int_0^1 k(t,s)p_1(s)q_1(\int_0^1 k(s,\tau)g(\tau,u(\tau))d\tau)ds\\ &\le M_3\int_0^1(1-s-\alpha)p_1(s)ds < +\infty \end{aligned} \quad (2.4)$$

Therefore, $A(D)$ is uniformly bounded. In the following we show that $A(D)$ is equicontinuous. According to Lemma 2.2, we have

$$(Au)'(t) = -\int_0^t f(s,\int_0^1 k(s,\tau)g(\tau,u(\tau))d\tau)ds$$

Therefore, by (H_1) ,we have

$$|(Au)'(t)| \le \int_0^t p_1(s)q_1(\int_0^1 k(s,\tau)g(\tau,u(\tau))d\tau)ds \le M_3\int_0^t p_1(s)ds < +\infty \quad (2.5)$$

Let $u(t) = \int_0^t p_1(s)ds$

Therefore, $\int_0^1 u(t)dt = \int_0^1\int_0^t p_1(s)dsdt < +\infty$

by (H_1) and (2.5), we have for any given $0 \le t_1 \le t_2 \le 1$ and $u \in D$,we have

$$|(Au)(t_1)-(Au)(t_2)| = \left|\int_{t_1}^{t_2}(Au)'(t)dt\right| \le M_3\int_{t_1}^{t_2}u(t)dt \quad (2.6)$$

We know from (2.5) and (2.6) , the absolute continuity of the integral function that $A(D)$ is equicontinuous. This together with (2.4) and the Ascoli--Arzela theorem guarantee that $A(D)$ is relatively compact. Therefore, A is a compact operator.

Now, we show that A is continuous. Suppose $u_m, u \in D$ and $\|u_m - u\| \to 0$ $(m \to +\infty)$ .Then there exist $M_4 > 0$ such that $\|u_m\| < M_4$ and $\|u\| < M_4$,from the above proof we know $\{Au_m\}$ is relatively compact .In the following we prove $\|Au_{mk} - Au\| \to 0$ $(m \to \infty)$ .In fact, if this is not true, then there exists $\varepsilon_0 > 0$ and $\{u_{mk}\} \subset \{u_m\}$ such that $\|Au_{mk} - Au\| \ge \varepsilon_0 \quad (k = 1, 2, \ldots)$. Since $\{Au_{mk}\}$ is relatively compact, there exists a sequence of $\{Au_{mk}\}$ which converges in p to some $u^* \in p$, without loss of generality, we assume that $\{Au_{mk}\}$ itself converges to u^* ,that is

$$\lim_{k \to +\infty} \|Au_{mk} - u^*\| = 0$$

Obviously, $(Au_{mk})(t) \to u^*(t)$, as $k \to +\infty$, for $t \in [0,1]$ .By (H_1) and Lemma 2.3,we obtain

$$k(s,\tau)g(\tau,u_{mk}(\tau)) \le (1-\tau-\alpha)p_2(\tau)q_2(u_{mk}(\tau)) \le M_5(1-\tau-\alpha)p_2(\tau) \qquad s \in [0,1]$$

Where $M_5 = \sup_{x \in [0, M_4]} q_2(x) < +\infty$, hence

$$k(t,s)f(s,\textstyle\int_0^1 k(s,\tau)g(\tau,x_{mk}(\tau))d\tau) \le (1-s-\alpha)p_1(s)q_1(\textstyle\int_0^1 k(s,\tau)g(\tau,x_{mk}(\tau))d\tau)$$

$$\le M_6(1-s-\alpha)p_1(s) \tag{2.7}$$

Where $M_6 = \sup_{x \in [0, M_5 b]} q_1(x)$ Then (H_1) , (2.7) and *Lebesgue's control Theorem* imply that

$$u^*(t) = \lim_{k \to +\infty} (Au_{mk})(t) = (Au)(t) \qquad t \in [0,1]$$

That is $u^* = Au$. This is a contradiction with $\|Au_{mk} - Au\| \ge \varepsilon_0 \quad (k = 1, 2, \ldots)$. Consequently, A is continuous on P . To sum up, Lemma 2.4 is proved.

Let $K = \left\{u \in P \,\middle|\, u \ge 0 \, and \, \min_{t \in [0,\eta]} u(t) \ge \gamma\|u\|\right\}$, where $0 < \gamma = \min\left\{\frac{1-\eta}{1-\alpha-\eta}, 1-\eta\right\} < 1$ .It is obvious that K is a sub-cone of P .

Lemma 2.5: $AK \subset K$

Proof: for any $u \in K$, we prove $Au \in K$. According to Lemma 2.3, we, know

$$\|Au\| = \max_{t \in [0,1]} (Au)(t) \le \int_0^1 (1-s-\alpha) f(s, \textstyle\int_0^1 k(s,\tau)g(\tau,u(\tau))d\tau)ds$$

$$\min_{t \in [0,\eta]} (Au)(t) \ge (1-\eta)\int_0^1 f(s, \textstyle\int_0^1 k(s,\tau)g(\tau,u(\tau))d\tau)ds$$

$$\ge \frac{1-\eta}{1-\eta-\alpha}\int_0^1 (1-s-\alpha) f(s, \textstyle\int_0^1 k(s,\tau)g(\tau,u(\tau))d\tau)ds$$

$$\ge \gamma\|Au\|$$

So $Au \subset K$.

3 Main Results

Proof of Theorem 1.1: By (H_2), there exist $c_1 > 0, \varepsilon_1 > 0, N_1 > 0, N_2 > 0$ and

$$(2\varepsilon_1)^{\alpha_1} c_1 ab^{\alpha_1} < 1 \tag{3.1}$$

Such that

$$q_1(x) \le c_1 x^{\alpha_1} + N_1, \qquad q_2(y) \le \varepsilon_1 y^{\alpha_2} + N_2, \qquad x, y \in R^+ \tag{3.2}$$

Therefore, by (H_1), (3.1), (3.2) and Lemma 2.3, we have

$$\begin{aligned}(Au)(t) &\le \int_0^1 k(t,s)p_1(s)q_1\left(\int_0^1 k(s,\tau)g(\tau,u(\tau))d\tau\right)ds \\ &\le \int_0^1 k(t,s)p_1(s)\left[c_3\left(\int_0^1 k(s,\tau)g(\tau,u(\tau))d\tau\right)^{\alpha_1} + N_1\right]ds \\ &\le N_1\int_0^1 k(t,s)p_1(s)ds + c_1\int_0^1 k(t,s)p_1(s)\left(\int_0^1 k(s,\tau)g(\tau,u(\tau))d\tau\right)^{\alpha_1} ds \\ &\le aN_1 + c_1\int_0^1(1-s-\alpha)p_1(s)\left[\int_0^1 k(s,\tau)p_2(\tau)(\varepsilon_1 u^{\alpha_2}(\tau) + N_2)d\tau\right]^{\alpha_1} ds \\ &\le aN_1 + c_1 b^{\alpha_1}\int_0^1(1-s-\alpha)p_1(s)(\varepsilon_1\|u\|^{\alpha_2} + N_2)^{\alpha_1} ds \\ &= aN_1 + c_1 ab^{\alpha_1}[2^{\alpha_1}(\varepsilon_1^{\alpha_1}\|u\|^{\alpha_1\alpha_2} + N_2^{\alpha_1})] \\ &= aN_1 + (2N_2)^{\alpha_1}c_1 ab^{\alpha_1} + (2\varepsilon_1)^{\alpha_1}c_1 ab^{\alpha_1}\|u\|^{\alpha_1\alpha_2}\end{aligned}$$

By (3.1), we can choose sufficiently large $R_1 > 0$, such that

$$\|Au\| \le \|u\|, \quad \text{for any} \quad u \in \partial B_{R_2} \cap K \tag{3.3}$$

On the other hand, by (H_3), we know that there exist $\varepsilon_2 > 0, c_2 > 0$ and $\rho \in (0,1)$ such that

$$f(t,x) \ge \varepsilon_2 x^{\beta_1}, \quad g(t,y) \ge c_2 y^{\beta_2}, \quad x, y \in [0,\rho], \quad t \in [0,\eta] \tag{3.4}$$

And ε_2, c_2, ρ satisfy

$$\varepsilon_2 c_2^{\beta_1} \gamma^{\beta_1\beta_2}\left(\int_0^\eta (1-s-\alpha)ds\right)^{\beta_1+1} \ge 1 \tag{3.5}$$

It follows from $q_2(0) = 0$ and the continuity of q_2 that there exists sufficiently small $\varepsilon \in (0,\rho)$, such that

$$q_2(u) \le b^{-1}\rho, \quad \text{for } u \in [0,\varepsilon] \tag{3.6}$$

Thus, for any $u \in \partial B_\varepsilon \cap K$ and $s \in [0,1]$, we have

$$\begin{aligned}\int_0^1 k(s,\tau)g(\tau,u(\tau))d\tau &\le \int_0^1 k(s,\tau)p_2(\tau)q_2(u(\tau))d\tau \\ &\le \int_0^1(1-\tau-\alpha)p_2(\tau)b^{-1}\rho d\tau \le \rho\end{aligned} \tag{3.7}$$

By (3.4) and (3.7), we know that for $t \in [0,\eta]$

$$(Au)(t) \geq \int_0^\eta k(t,s)f(s,\int_0^1 k(s,\tau)g(\tau,u(\tau))d\tau)ds$$
$$\geq \varepsilon_2 \int_0^\eta k(t,s)(\int_0^1 k(s,\tau)g(\tau,u(\tau))d\tau)^{\beta_1} ds$$
$$\geq \varepsilon_2 \int_0^\eta k(t,s)(c_2 \int_0^1 k(s,\tau)u^{\beta_2}(\tau)d\tau)^{\beta_1} ds$$
$$\geq \varepsilon_2 \int_0^\eta (1-s-\alpha)[c_2^{\beta_1}(\int_0^\eta (1-\tau-\alpha)u^{\beta_2}(\tau)d\tau)^{\beta_1}]ds$$
$$\geq \varepsilon_2 c_2^{\beta_1}\gamma^{\beta_1\beta_2}(\int_0^\eta (1-s-\alpha)ds)^{\beta_1+1}\|u\|^{\beta_1\beta_2} \geq \|u\|^{\beta_1\beta_2} \geq \|u\|$$

Consequently,

$$\|Au\| \geq \|u\|, \text{ 对 } \forall u \in \partial B_\varepsilon \cap K \qquad (3.8)$$

By Lemma2.2, (3.3) and (3.8), A has at least one fixed point $\omega_2 \in (\overline{B}_{R_1} \setminus B_\varepsilon) \cap K$, Then BVP (1.1) has at least one positive solution $(\overline{u},\overline{v}) \in (K,K)$ and

$$\begin{cases} \overline{u}(t) = \omega_2(t) \\ \overline{v}(t) = \int_0^1 k(t,s)g(s,\omega_2(s))ds \end{cases}$$

It is clear that $\overline{u}(t) > 0, \overline{v}(t) > 0$ for $t \in (0,1)$.

The proof of Theorem 1.1 is completed.

References

1. Gupta, C.P.: Solvability of a three-point nonlinear boundary value problem for a second order differential equations. Math. Anal. Appl. 168, 540–551 (1988)
2. Ma, R.: Positive solutions of a nonlinear three-point boundary value problem. Electron. J. Differential Equations 34, 1–8 (1999)
3. Timoshenko, T.: Theory of Elastic Theory. McGraw-Hill, New York (1971)
4. Webb, J.R.L.: Positive solutions of some three-point boundary value problems via fixed point index theory. Nonlinear Anal. 47, 4319–4332 (2001)
5. Dalmasso, R.: Existence and uniqueness of positive solutions of semilinear elliptic systems. Nonlinear Anal. 57, 341–348 (2004)
6. Ma, R.: Muitiple nonnegative solutions of second –order systems of boundary value problems. Nonlinear Anal. 42, 1003–1010 (2000)
7. Liu, B.: Positive solutions for singular systems of three-point boundary value problems. Computers and Mathematics with Applications 53, 1429–1438 (2007)
8. Guo, D., Lakshmikanham, V.: Nonlinear problems in Abstract Cone. Academic Press, New York (1988)
9. Gupta, C.P.: A sharper conditions for the solvability of a three-point second order boundary value problem. Math. Anal. Appl. 205, 586–597 (1997)
10. Li, S.: Existence of positive solutions for a three-point second order boundary value problem. Journal of Lishui College 26(5) (2004)

6.3 Sustained Optimization of the Value Chain

The value chain appraisal is a permanent process, in which the enterprise has to make long-term dynamic appraisal as well as short-term appraisal. It has to pay close attention to changes in the internal value chain, the horizontal value chain and the vertical value chain and feedback problems in the merger and integration in time to realize sustained optimization of the value chain and form a dynamic and benign cycle.

Acknowledgment. Supported by the MOE Project of Research Institute of Humanities and Social Sciences "Research on Enterprise Mergers and Acquisitions from the temporal and spatial dimensions" （09YJA790195）；

Supported by the Development Research Project of Hebei Social Sciences (201003089)"Research on Enterprise Mergers and Acquisitions in the Light of the Value Chain Optimization in Hebei Province".

References

1. Yin, G.: A Study on the Theory of theValue Chain and the Integrative Competence of the Enterprise in the Merger and Acquisition. A Master's Thesis of Northern Transportation University (2006)
2. Porter, M.: Competitive Advantages. Huaxia Publish House, Beijing (1997) (Tran. Xiaoyue Chen)
3. Liu, H.: The Merger and Acquisition Integration Based on the Value Chain Theory. Finance and Accounting Monthly, Compact Edition (May 2008)
4. Xiao, X.: The Value Chain Integration of the Enterprise Mergers and Acquisitions. China Social Press, Beijing (2008)
5. Xiao, Z.: A Study on the Selection of the Target Enterprise Based on the Synergic Effect. A doctoral dissertation of Harbin Institute of Technology (2007)

Study on the Integration of College Sports Network Education Resource under Modern Distance Education Background*

Liu Huaijin[1] and Zheng Weitao[2]

[1] Faculty of P.E Hunan Institute of Science and Technology
Yueyang,Hunan Province, 414006, China
[2] Key Sports Engin. Lab General Administration of Sport of China
Wuhan, Hubei Province, 430079, China
tyxlhj@126.com, zhengweitao@sina.com

Abstract. These instructions The development and utilization of colleague physical network curriculums is the complement and evolution of traditional educational methods, playing a positive role in promoting physical education reform in China. Modern distance education can utilize and optimize the state educational resources, improve the education quality, form the open education network and build lifelong learning system. The integration of modern distance education and college physical network curriculums can not only expand the teaching time and space to solve the "learning and training contradictions" and " learning and working contradictions" for the sports learners, but also contribute to the sustainable development of the existing elaborate physical courses and network curriculums, conducive to physical education resource sharing, which is an important way for school physical education and lifelong physical education in the future.

Keywords: Modern distance education, College sports network Education Resource, Integration, Study.

1 Introduction

With the development of computer technology and network communication technology, networks have become the main channel for the majority of people to learn and acquire information in life, and network teaching has become the modern means of education to keep up with the development of times and society. On account of the powerful information dissemination capacity of the computer network, the excellent performance of resource sharing, advanced two-way interactivity and multimedia

* This work is partially supported by National Natural Science Foundation of China (Grant No. 51009113),the research project of general administration of sport of the People`s Republic of China(Grant No. 2011A025、2011B023), the College Education Reform Project of Hunan Provincial (Grant No. 2011-315-282、521 ; 2010-243-285), Hunan Province Social Science Research Excellent Young Project(Grant No. Kt11-075).

D. Zeng (Ed.): Advances in Computer Science and Engineering, AISC 141, pp. 143–148.
springerlink.com

technologies, modern distance education can provide abundant teaching resources and flexible forms of teaching, and improve students' learning initiative and flexibility, conducive to individualized teaching and learning. Aiming at the space conflicts, learning and training conflicts and other issues in physical course teaching, the reform of college physical education should take full advantage of modern distance education, integrate the course resources, expand the teaching time and space, and enrich teaching ways to adjust physical education to the requirements of social development.

2 The Characteristics of Modern Distance Education

In China, the traditional teaching forms of one-way distance education have emerged, such as TV video teaching, TV university, correspondence and so on, which exist the study time, space and other limitations. Besides, as a demonstration tool instead of a cognitive tool, teaching medium can not achieve the interactive and independent study. Modern distance education is formed as a new two-way interactive teaching means associated with the development of multimedia technology, networks and communications technology, becoming the primary means of lifelong learning system at the times of knowledge economy construction. With the help of the advanced communication network to transmit information, it can achieve the across-time-and-space, interactive, and independent learning. Modern distance education is a significant project to fully utilize and optimize the state education resources, popularize and improve education quality, form the open education network and build lifelong learning system.

Modern distance education has the advantages of openness, interactivity, sharing, collaboration and independence, etc., it can conduct a wide range of information dissemination with high capacity and high speed. With the help of technical multimedia means to build rich educational resources information, it breaks the students' learning time and space to improve students' learning independence and initiative. For PE teaching, the learners can learn a variety of theoretical knowledge and physical skills from the network at any time and any place free from space, time or climate impact. Meanwhile, the difficulties of explanation, understanding and repetitiveness for the techniques of various body movements involved in physical skill teaching may be solved by multi-media video, animation or collecting the standard movements of the sports stars in major sports events from websites.

3 Current Situation and Existing Problems of Physical Network Teaching

Network course teaching is specific demonstration of computer network technology application to teaching, and a new teaching mode based on multimedia, computer networks and the Internet, which extends the classroom, to the campus network and the Internet, realizing the teaching resources sharing around the school or even the country and the world[1]. Physical network courses include all levels (national, provincial and university level) of the websites of elaborate courses and physical courses.

Guided by "College Teaching quality and Reform Project" issued by the Ministry of Education along with Ministry of Finance, the whole country has set off an upsurge in elaborate courses construction. After a decade, the colleges in China have built a surprising number of elaborate courses. Elaborate courses and network courses can be seen on the college home pages. However, the survey found that many network courses websites had become an "unfinished project "nobody concerned with, with old resources, lagged site updates, and even some existing in name only. Compared to other subjects, physical network course is in a more worrying situation. The most important factor for this issue is lack of financial support for maintenance and the policy to encourage students to use while during its construction, there is financial support and encouragement policy. Network courses should follow the development trend of educational informationization and lifelong learning needs, providing more convenient and powerful support for people to acquire new knowledge at any time [2]. Launching the physical distance education on the basis of the existing physical network courses resources can solve the learning problems for a great number of learners who can not learn in school (PE teachers in rural middle and primary school, social PE workers, athletes and coaches), and meanwhile, the tuition fee for distance education course may be used for follow-up course development, thereby the network courses resources achieving sustainable development.

4 Advantages of Physical Network Teaching

4.1 Physical Network Teaching Space Can Expand Time and Space to Address the Shortage of Teachers, Inadequate Facilities and Other Issues

The biggest difference of PE from other subjects is that many courses require a specific time and place, and there are theory and skill subjects. In the school, the traditional skill teaching is often subject to short supply of sports venues, weather and other factors, and the theory courses are also often affected by students training and competition. Outside the school, PE teachers, athletes, coaches, social sports instructors and other sports workers have to use their vacation or interrupt their work to take the refresher courses, which costs a lot of time as well as some learning and living expenses to achieve the desire to learn. The construction of physical network courses adopts the following means: teachers post the course content to the website, and conduct teacher-student interaction by posing and answering questions and other forms for extra-curricular learning and after-class tutorial to expand teaching space and time, making up for the school hours and venues inadequacy and overcoming many difficulties for the external learners.

4.2 The Network Teaching of Physical Education Courses May Update Teaching Resources in Time to Help Students and Teachers to Master the Latest and Most Cutting-Edge Knowledge

With the rapid development of world sports and the constant update and improvement of sports competition rules (such as soccer, volleyball, table tennis, etc.), teachers need to keep continuous learning and pay attention to sports hot issues, apply the most advanced teaching methods and means in teaching , and bestow the latest sports knowledge to students. Physical courses website may timely collect the media information on the latest events to enrich teaching resources, integrating teaching theory with practice to resolve teaching difficulties. Since the publishing cycle of traditional textbooks is long, the knowledge update is inevitably lagged off. Through network teaching of physical education courses, teachers can reprint or download the latest sports knowledge directly from the Internet to pass on the students to enrich the teaching content. Students may engage in the independent "ordered" study according to their needs and characteristics, which increases their interest in learning and help their fast access to the cutting-edge knowledge of sports, thereby enhancing the level and quality of teaching.

4.3 The Network Teaching of Physical Education Courses May Overcome the Teaching Obstacles Caused by Teachers' Age and Health Factors

In the traditional physical education, movement demonstration is the most intuitive and most important teaching method. However, old age can not follow sports. Due to the physical education teachers' own quality, age and other reasons, some skill movement demonstration could be done inaccurately, incoherently and out of standard, so that students in the beginners action correct action is difficult to grasp the concept, which make it hard for students to master the correct movement concept at the beginning stage. By means of the development and utilization of network resources, such as videos, students can view the standard movement of world-class elite athletes or PE teachers performed in youth through course website, establishing the correct movement concept. Moreover, the physical education course website information is accumulated over time, which can record and preserve the valuable teaching resources of physical education teachers.

4.4 The Network Teaching of Physical Education Courses Can Cultivate Students's Ability of Independent Learning

Network resources have the features of openness, sharing, interactivity, collaboration and independence, etc.. Network data access is the process of data mining, requiring information extractors to possess the necessary retrieval skills and memorizing capacity. The network teaching of physical education courses helps students to play the self-awareness and self-education function in the learning process; helps to cultivate the students' practical ability to think independently, pose questions and solve problems; help students to arrange for learning plans and learning schedule according to their own knowledge structure, time and appropriate place to achieve individualized teaching. Acquiring the knowledge outside of the teaching materials through websites, they can master the latest sports skilled movement and sports trends. It expands their horizons, promotes the interactive discussions with teachers and students through course websites, and also develops students self-learning ability.

5 The Meaning of the Integration of Modern Distance Education and Physical Network Courses

As an important part of internet-based teaching, modern distance education is a critical link of the implementation the modern distance education project, the formation of open education network, and the establishment of teaching resources of education for all people. "National Long-term Education Reform and Development Planning Guidelines" proposes "to strengthen the system establishment of network teaching resources, develop network learning courses, and carry out high-quality and high-degree in distance education." "National College Physical Education Teaching Guidelines "clearly states:" To attach importance to the integration of theory and practice, to permeate related theoretical knowledge into practical teaching, and to use various forms of modern teaching methods." Modern distance education has the characteristics of openness, interactivity, sharing, economical efficiency, collaboration and independence, etc.. It can promote students to form the lifelong sports awareness, and it is an effective way for college PE courses content expansion and courses content development. Resources are essential, and teaching resources construction is the foundation for the healthy and sustainable development of modern distance education[3]. Taking advantage of physical education network teaching resources (such as elaborate courses, network courses, etc.) to carry out distance education is the essential way to adjust the school physical education to quality-oriented education, and the primary means of future school physical education and lifelong sports construction. In the meanwhile, integrating the existing physical elaborate courses and network course resources may ensure sustainable development of the construction results of national quality projects. Modern distance education and physical education network teaching are conducive to building a learning-oriented society, facilitating learners' independent and lifelong learning.

6 Summary and Outlook

Modern distance education can fully utilize and optimize national educational resources, improve the quality of education, form the open education network and build a life-long learning system. The integration of modern distance education and college physical education courses enables to expand the teaching time and space to address the sports learners' "learning and training contradictions" and "working and learning contradictions". On the other hand, it contributes to the sustainable development of the existing elaborate courses and network course of physical education, conducive to physical education resource sharing, which is a critical way for the future school physical education and lifelong physical education.

Acknowledgment. This study is partially supported by the National Natural Science Foundation of China (Grant No. 51009113),the research project of general administration of sport of the People`s Republic of China(Grant No. 2011A025、2011B023), the College Education Reform Project of Hunan Provincial (Grant No. 2011-315-282, 521；2010-243-285)，Hunan Province Social Science Research Excellent Young Project(Grant No. Kt11-075).

References

1. Zuo-yong, W.: Study of Construction of Network Teaching of Physical Education. Journal of Qiqihar University (Philosophy & Social Science Edition) 2, 171–172 (2010)
2. A.Y.Y.S.P.: Present Situation and Characteristics of Construction of Network Courses of China s Physical Education Specialty. Journal of Shanghai Physical Education Institute 32(4), 68–72 (2008)
3. Lin, H.: Construction, Reorganization and Utilization on Teaching Resources in Distance Education. Journal of University of Electronic Science and Technology of China(Social Sciences Edition) 7(3), 88–91 (2005)
4. M.J.M.A.M.S.: Enhancing Social Presence in Online Learning: Mediation Strategies Applied to Social Networking Tools. Online Journal of Distance Learning Administration 12(4) (2009)

The Effect of Production-Study-Research Cooperation Mode to the Employment of Social Sports Major Graduates

Zhou Ci-bao and Liu Huaijin

Faculty of P.E
Hunan Institute of Science and Technology
Yueyang, Hunan Province, 414006, China
tyxlhj@163.com, tyxlhj@126.com

Abstract. Analysis of the positive effects of the model industry, learning and research for the improvement of the intellects' qualities and the development of the social sports major. the mode manufacture -learning-research cooperation to improve the quality of social sports, the positive role of professional development, social sports professional production has been widely applied mode, adapt to the needs of society, has the characteristics of universities, through the combination of the feasibility of the model to business social sports professionals, will help to solve social sports professional graduates' problem. the social sports major developped under the mode of will help to solve social sports professional graduates' employment problem.

Keywords: Manufacture-learning-research cooperation, Education modem Social sports major, Employment.

1 Introduction

Social sports professional sports education to adapt to the society is higher for PE talents of the new requirements and portfolio of a new major, 1998 the professional education department will be listed in the ordinary university undergraduate course catalogue. Since the social sports professional since its establishment, there had been some hinder the development of professional school confusion was not solved, serious obstacle to the development of the professional and the improvement of the quality of talents and the problem of employment, etc. Therefore, it is essential for the education teaching methods and social aspects such as employment, to the professional development of education teaching system and operating mechanism and the important problems are discussed. In order to find the right social sports professional development way, thus to promote social education teaching reform of professional sports development, in order to improve social sports professional students employment.

D. Zeng (Ed.): Advances in Computer Science and Engineering, AISC 141, pp. 149–155.
springer link.com

2 China's Social Sports Professional Students, the Employment Status

Social sports professional development so far only 10 years, in the training process in many of the problems that can not be neglected in, an obvious example is the professional students "entrance" and "export" not quite fluently. Generally speaking, the current social sports professional talent market supply and demand imbalances, serious contradictions, universities open society to professional sports, seem to be the supply, but the fact is not so. All provinces and cities of the average number of social sports professional point of 3.61 a. From professional absolute number see, mainly distributed in east China, north China, central China and northeast, the four regional social sports professional accounted for the national 70%; From professional distribution density look, central China, east China and northeast, south China social sports professional point averages more than the national average, were 5.33, 5.00, 4.67, 4.33; From the provinces social sports professional quantity look, in guangdong 10, jiangxi, jilin, hubei all seven become social sports professional point distribution of the most densely populated area, the emergence of a professional cluster phenomenon. Professional stick increased student's employment virtually pressure.

2.1 Social Factors

2.1.1 Social Demand Is Big and University Training Talents of "Relative Surplus"

Society for social sports specialized talented person's demand for greater shown in two aspects: one is the fitness of the social sports instructor's urgent needs. According to the "national fitness program" regulation for every 400 people with two-five of the instructor of social sports standards, our country social sports instructor has need at least 650000 people. However, at present, China's social sports instructor very lack, like hunan such a population 6.5 million of the province, the province social sports instructor only 5000, average every 12000 people have a social sports staff; Tianjin in the 124 street agency, only two full-time sports cadres; And like guangzhou this economy and sports are developed place, 82 street countries. combining community there is not at present a full-time sports cadres. In recent years, the sports industry in China is get considerable development, but from the quality and the number and the developed countries in the world is still in the lower level, in order to promote the rapid development of the sports industry in China, also need a large number of sports organization management and business development personnel involved, bear responsibility.

2.1.2 Demand Is High and the Employing Unit While University Talents Training Ability Is Low

21 century is a society full of competition, talent in social competition is extremely important. The social civilization, and the transformation of the notion of health for people, and make the society for social sports specialized talented person's demand multi-layered and diversity. Social sports professional graduates difficult to meet its requirements. Social sports professionals must have the special sports skills so that practical guiding need. So it requires universities social sports professional must offer

corresponding professional skills course and social actual demand match, but at present our country except a few professional of social sports colleges and universities offer professional skills extracurricular, most college and no special skills course opened, the training of professional talents can't meet the demand of fitness guidance skills, along with the recruitment of students scale are on the increase, and the insufficient funds, the serious shortage of equipment, the practice instruction work relative lack of ability to collapse, far from meeting the needs of the employing unit.

2.2 Universities Is a Factor

2.2.1 Cultivation Objective Orientation Deviation

China's higher JiaoYuSi issued "the ordinary university undergraduate course major directory and professional introduction to clearly define the social sports professional talent training goal:" to be in the field of social sports in mass sports activities of the organization management, consulting guidance, business development and the teaching and scientific research, and other aspects of the senior specialized talents However, through the different types of social sports colleges training aim that survey, most colleges and universities social sports professional training objectives common orientation in the organization management, business development, fitness guidance, sports teaching and so on, so that the training target is accord with the requirement, but not according to their own school-running characteristics, the region's cultural differences, and the local market economy level factors such as the training goal orientation.

2.2.2 Curriculum Not Clear

Social sports professional is a newly emerged cross major, its course main basis sociology, management science, sports study Settings. The influence of the traditional thought, social sports professional curriculum in there are still too many categories, main skills not outstanding, practice course proportion too little. Social sports professional has applied and practical characteristics, strengthen the practice guidance ability training is the prominent social sports professional "one only" the key to be. At present the recruitment of students scale expansion, fund shortage, some necessary practice facility lack, this has made social sports professional people in the actual practice ability in the above his business, go against professional talents of employment.

2.3 Students' Factors

2.3.1 Matriculate Quality Foundation Is Bad, Weak Economic Foundations

In recent years, the university as the rapid social development and with high-level talent demand, each major implementation "enrollment expansion", the society of professional sports is no exception. ChenYongJun basis for social sports professional growth in the investigation of that social sports professional 1999, only 14, 2001 31, 2003, 68, 2004, 106. Predictably, social sports professional is also in only after the expansion of higher education developing rapidly. Due to expansion, certainly will will appear on the recruit students of cultural quality low, sports skills poor phenomenon, go against real personnel selection and use.

2.3.2 Employment Idea Awareness Deviation

At present, our country college graduates employment mechanism has by original communication under the planning economic system that integrates transition to the market economy system of graduates and unit of choose and employ persons "two-way choice" the employment system. The influence of the traditional employment concept, most graduate employment expectation on the high side, in employment, often too much attention to work area, working environment, working remuneration, development foreground factor, makes them in employment loss when again and again of employment opportunities. Without fully understanding some graduate students of their actual ability, always think you in some ways than anyone else, when in employment should also obtain better than others superior work, this kind of undesirable 'psychological and led them to compare the growth of psychology to when choosing profession "GaoBuCheng, low not".

3 The Concept and the Function Combination

3.1 The Concept of the Combination

The combination is the production enterprise (produce), colleges and universities (learning), scientific research institutions (research), according to "share the benefits and risks, mutual complementarity and common development" the principle, to the technology contract as the foundation, in accordance with their respective advantages, sharing technology innovation work required to different stages of the resources, the cooperation of technology innovation activities. The combination of the joint technical innovation is platform, is the acceleration science and technology and the effective way of the rapid economic development, it has become the consensus of all countries. The cooperation of the basic connotation of education can be summarized as: it to cultivate students' comprehensive quality and comprehensive ability and employment competition ability as the key point, is universities and social departments close cooperation, through some organization form, the study and work in the combination of teaching process through a talent's education mode. UNESCO recently declared intensifying the joint as a key work to promote, the development of the world has had a huge influence.

3.2 The Function of the Combination of Analysis

3.2.1 The Power-and-Power Union, to Improve the Strength, and Promote the Development of Specialized Subject

The combination of education is to improve the combined strength, promote the development of the discipline of important ways. Practice shows that the model can help promote education combining study professional education the deepening of the reform of the practical education link construction, and the healthy development of the scientific research work; To help teachers team training; To improve the practice of college students' innovation ability.

The cooperation of education is the enterprise and research institutions involved in school, industry and scientific research institutions than the school education system more directly facing knowledge, technology innovation of the actual, especially some high-tech industry and high-tech development department is based on the frontier science and technology, not only in the innovation of information, innovation environment so as to rich advantage, and containing the enormous potential of education. To cultivate students' ability, improve students' quality.

3.2.2 Pay Attention to Link Out Cohesion, Perfect Model Links, Improve Mode Effect

Education mode is to adapt the unity of knowledge economy era is the inevitable requirement of talent training. Professional colleges and universities should training high quality, practical and innovative talents, and such talents in school to narrow range training out. Must develop new training mode. Therefore, in addition to further reform education thought, education method and way, strengthen the construction of the teaching staff and reform the management model outside, still should broaden the outlook, and aiming at the professional development Front, seize the progress of science and technology, facing the market for the society. Through the development of various forms of combining study education. Become a closed for running a school open, develop talent of education high quality new way. Form the university as the core, combined with the combination, face the national economic construction of higher education could new pattern. Realize the teaching, scientific research and between enterprise's benign circulation.

4 The Combination of Social Sports Professional Graduates' Influence

4.1 Enhancing University-Enterprise Contact, Practice and Theory, and the Combination of Increasing Employment Competitive

For social sports professional students in the investigation of the employment intention, found in the school students of the proposal, mentioned most is to increase the frequency practice opportunities for theory and practice union, many people think that it should be a freshman, I have been arranged more practice. But, as the current investigation to see, most of the social and professional body without a stable or fixed ShiXiDian, some existing social practice opportunities are also very stable. Such as social sports career now, health club management and management, all kinds of sports venues run with the students learn in the school sports management and management vary greatly. Therefore, but the school with the club or sports venues for cooperation. So that some of the senior managers to school sports business for students of special training. After the training can make the students to relevant sports place to practice, such, student's practice is very specific, and make students learn really useful knowledge and skills. Therefore, the combination of social sports is professional has been seeking a new mode of running. This will be fully promote students' knowledge in the combination of theory and practice, enough preparation for employment, to increase the competitiveness of the employment, improving the social sports professional students' employment.

4.2 Improve Teachers Team Construction, Improve the Students' Learning Passion

The cooperation of education is not only beneficial to the teachers' professional practice ability training and helps teach division team theory and the improvement of teaching quality. Because the teacher through production practice, accumulated the massive work experience. Improve the professional practice ability and promote the double-meaning teachers training. In the actual production of problem again become teachers' scientific research, through the achievement transfer, technology transfer, and expand the school-enterprise cooperation areas, helps to sustainable and stable development. The teachers can be held for students and lecture, write notes and so on, enrich the teaching content, so as to promote the teaching content and teaching method of the reform, to provide the power for the development of the subject construction.

And the student is by taking part in some scientific research, the production project, make learning rich content, intuitive, and students' motivation to learn greatly strengthened, the purpose of the study is more explicit, and at the same time, to realize the study knowledge significance and value. Aeriform in, students learn knowledge will for later obtain employment to lay a solid foundation, so as to improve the employment rate.

4.3 Improve Learning Style, to Improve the Students' Creative Ability

UNESCO in 1998 world conference on higher education published "world higher education declaration pointed out:" for the convenience of higher education should be graduates employment, develop their skills and active spirit, graduates are more and more is no longer just job seekers, and for the first time will be the job creator." China's higher schools accept this idea, and conduct business education, but the effect is not obvious. Surveys show that college students to China's current business education effect evaluation is not high, college students of business skill also has the disparity, the main reason is the lack of education also venture targeted.

Enterprise and school in training person have their own characteristics and advantages, school education and scientific research strength theory, strong, and the enterprise in the production practice, fund has an advantage. Through the cooperation, will be able to play out their advantages. The student to participate in the technology development practice, be helpful for application knowledge work site of technical problems to solve. Through the technology development, technology promotion and technical transformation of the industry practice experience, students gain a real open future technology innovation and start their own businesses the gate "golden key".

5 Conclusion

The cooperation of education is the unity of the concrete embodiment in personnel training, teaching content and the system of the deepening the reform of the power, is cooperative education training comprehensive application talents of effective way. Social sports professional go running to the form of combining study knowledge economy and knowledge sports fusion; Is beneficial to the improvement of the quality of talents of social sports, the professional development has a positive role in promoting; For the construction of the teaching staff and students' learning interest stimulating, improve employment competitive. China's social sports professional should insist on the cooperation of education the roads to make China's social sports professional can better and faster development.

References

1. Tian, Y.P.: The confusion and choice of the education of social sports major nanjing normal university. Sports & Science 6(3), 68–70 (2009)
2. Rong, J.J., Zhao, X.Q., Zhan, R.Q.: Some Thoughts on the Employment Status of the Professional Personnel in University Social Physical Education. Journal of Huaibei Coal Industry Teachers College(Natural Science edition) 29(4), 74–78 (2008)
3. T.L.T.Y.: A Feasibility Study on the Application of the Model "Industry,Learning and Research" for the Major of Social Sports in Institutions of Higher Learning. Journal of Capital College of Physical Education 17(4), 117–118 (2005)

Research and Design Based on Distributed Firewall Log Server

Lu Chang and Xu GuangMing

Public Experiment Center, Zhejiang Ocean University, ZhouShan, Zhejiang Province, China
robotlu@126.com

Abstract. Analyzed and proposed the model of distributed firewall log server, elaborated the characteristics of the log system, on the basis, gived its functional modules and design elements.

Keywords: Distributed Firewall, log server, log initialization, Log storage, IDS process.

Distributed Firewall Log Server is receiving, processing, storage, centralized management and auditing log information about the internal network, including all the protected host, the border firewall, policy server etc. It also monitored the internal network status in real time, and realized the function of statistics detection based on log information.

1 Log Server Task Description

Log server's features include three aspects, one is collection, classification and storage about a variety of information in system, and the other is processing, recording and display, the third is statistical detection based on log information. So, when log message generated in each module, the information should be written in database by the log server, statistic and analyze through the analysis engine. When the manager asked for statistical information, it will take out and analyze the information, extactc out the valuable information, provide audit interface, and requires the user interface intuitive and easy to query[1].

1.1 Log Server Design Features

Based on log system functions and analyzed how to achieve, we bring out desion elements below:

- Log Server is a c/s system, client makes a request to the log server as information collection engine, and the log server as the server side, all requests for specific treatment.
- Database connection pool technology used to avoid speed bottleneck by frequent database operations.

D. Zeng (Ed.): Advances in Computer Science and Engineering, AISC 141, pp. 157–163.
springer link.com

- Transmission of the log server should be encrypted to prevent internal network attacks such as DDOS.
- Because a server connect multiple clients through socket, the server processing should adopt multi-threaded way.
- In order to achieve the automatic response alarm in the internal network, intrusion detection linkage process should be established.
- The log system needed facilitate to query, graphical and intuitive user interface, which as record and For future reference.

1.2 The Overall Structure of the Log Server

Based on the analysis and consideration above, the system basic operation mode using Client / Server structure, the overall framework in Figure 1.

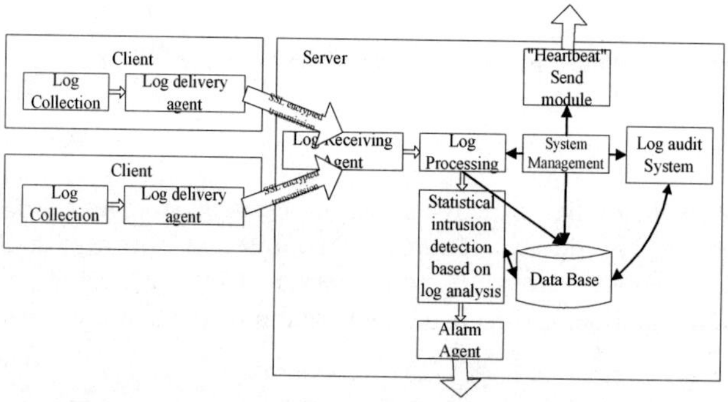

Fig. 1. Overall structure Log Server

Client refers to the hosts which need to audit in local network, send agents installed on the client collect raw logs, which completed by the corresponding host firewall, perimeter firewalls, policy center, and transport to server throuth conversion and SSL encrypted.

Server is mainly composed of three parts: data collection, audit log analysis system and statistical-based intrusion detection.

Data acquisition for receiving data from the client, and then, store into the database; Audit system coordinate the other part of work and receive the information which configurated and adjusted by administrator; Statistic intrusion detection based on log analysis achieved the function of intrusion detection and linkage to other modules.

2 Log Initialization Procedure

Log initialization extract relevant configuration information, open the log receiver module and establish interactive information connection with the Strategy Center when it started, consists of three parts:

1) Log server upload the IP address, port, host name and other related information by communicating with the Strategy Center, which is prerequisite for connection with other host, border host and log erver. Information will be stored in local server which Strategy Center obtained, when initialize log server, information will be transmitted to hosts, then, the host will connect to log server through obtained information, upload log information[2][6].

2) Obtaining initial heart rate serial number and other original information from Strategic Center, then, log server can keep in touch with Strategic Center notmally and timely, which prove itself active.

3) Initializing information related to log server startup, ensure normal start, such as the establishment of the list, the consistency of the system time and so on. Log initialization program running by linear manner, as Figure 2.

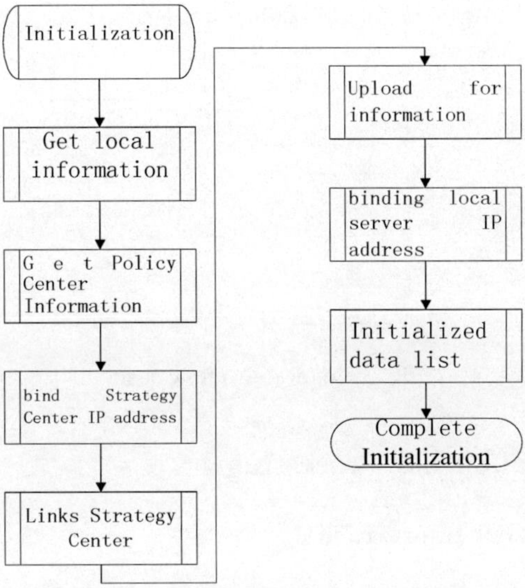

Fig. 2. Flow chart of the log initialization

3 Log Receiver Module

First, server address relevant binding work should completed by application initialization, then, the listener background step into the state of listening, waiting for connection requests from outside, when a new connection request, new thread will be allocated to perform after task, Diagram in Figure 3.

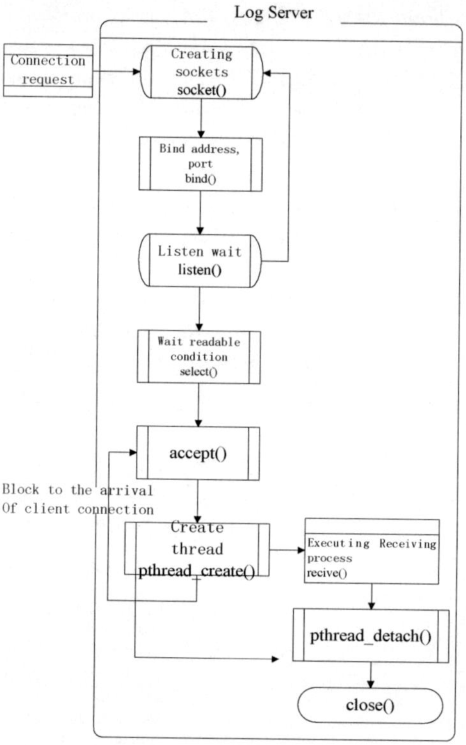

Fig. 3. Log Receive flow chart

4 Host Connection Log Processing

4.1 Receiving Header Information

First, call "IdTCPServerConnect (AThread: TIdPeerThread)" functions to receive packet header, in the packet header through "send_comm_type" to determine whether it is uploaded, if it was, continue, or, handled by other modules, and then, through "send_run_method" function to determine whether it was host firewall log information, if not, handled by other modules[3].

The log message size can't be affirmed advancely in one TCP connection, so, it can't not assign a block of memory to receive the uploaded log information. in the program, we should create variable quantity of memory stream firstly, then, reveiving data packets through the length of log information in head to allocate a block of memory.

4.2 Log Data Processing (trans_hc () function)

After receiving complete log data, it will use function of "trans_hc()" to deal with, and form a unified format to store into database, at the same time, extract the relevant information to start leaning profile of intrusion detection based on log[4].

Comparison of timestamp of upload in log information and time information in log server to comfirm realtime error, making IP address into characters to satisfy displaying of audit system.

4.3 Log Storage

In project of the data storage, we realize database interoperability by using of Borland DBExpress technology.

First, calling "sqlconnection()" function to exchange connection information with the MySQL database: user name, password, database name, connection style, and so on. Function contains "TSQLConnection" components. "TSQLConnect" component establishes a connection with the database. In connection, "TSQLConnect" always interact with DBEXpress drive and dbxdrivers/dbxconnections two files. Dbxdrivers file contains library which driver needed (dynamic link library or shared object) and the default settings of connection parameters. Dbxconnections file listed collection of connections, each name represented a group of settings of TSQLConnection and described a particular database connection.Addition to describing and establishing a database connection, TSQLConnection also provides properties, methods, and events of the server login, transaction management and remove data from the server[5][7].

After exchanging information in consultation, MySQL database will assign a thread to handle the connection, then create one-way data set of TSQLDataSet to operate the database through TSQLConnection. TSQLDataSet one-way data set provides a mechanism, through which the application can read data from a database table. One-way data set is designed for fast, portable access to database information, the overhead is very small. One-way data set sent SQL commands to the database server, if the command returns a record set, then get a one-way pointer to access these records. One-way data set is not in the memory buffer, which makes them more quickly and save resources compared to other types of data sets. But also because of not buffer records, it is not flexible than other. Finally, after writing the log information into the database, call the "Close()" function to Close the connection, complete the process of log storage.

4.4 IDS Process Based on the Log Statistics

Program calls "innerscrip()" function to handle the process of log-based IDS statistics, Receiving process as figue 4.

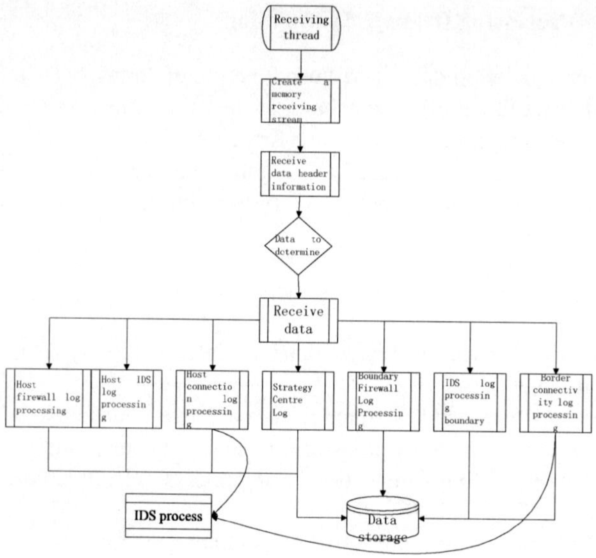

Fig. 4. Log Receive flow chart

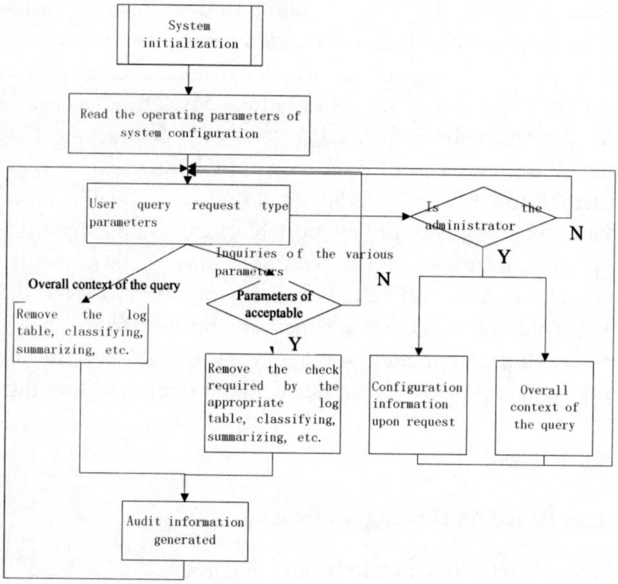

Fig. 5. Schematic diagram of the user audit

5 Other Log Processing

Similarly, the processing of host firewall logs, host IDS logs, policy center logs, firewall logs border, boundary connection logs, border IDS logs and the host connection log is similar.

6 Log User Audit Module

Audit system is a window of system's running for users, it will tell the user system, it will tell user what happened and what did not happen, workflow is shown in Figure 5.

Acknowledgment. This paper described a distributed firewall log server design, based on Linux operating system platform, as Kylix programming language and MySQL database tools to realize, Kylix language as a graphical interface tools and realize the connection with MySQL database through DBExpress technology.

References

1. Dulay, N., Lupu, E., Sloman, M., Damianou, N.: A Policy Deployment Model for the Ponder Language. In: Proc. IEEE/FIP International Symposium on Integrated Network Management (IM 2001), IEEE Press, Seattle (2001)
2. Scuba, C.L., Spafford, E.H.: Reference model for firewall technology, Source. In: Annual Computer Security Applications Conference, pp. 133–145 (1997)
3. Lyu, M.R., Lau, L.K.Y.: Firewall security, Policies, testing and performance evaluation, Source. In: Proceedings-IEEE Computer Society's International Computer Software and Applications Conference, pp. 116–121 (2000)
4. Duflos, S.: Architecture for policy-based security management for distributed multimedia services, Source. In: Proceedings of the ACM International Multimedia Conference and Exhibition, pp. 653–655 (2002)
5. Cheswick, W.R., Bellovin, S.M., Rubin, A.D.: Firewalls and Internet Security, Repelling the Wily Hacker, 2nd edn. AddisonWesley (2003)
6. Xing, L.M., Ping, H., Peiwu, D.: Research on artificial neural network method for credit application. Research Information Ltd(RIL) (2), 127–131 (2004)
7. Fung, G.: A comprehensive overview of basic clustering algorithms, pp. 1–36. University of Winsconsin, Madison (2001)

Research and Implementation of File Encryption and Decryption

Lu Chang and Xu GuangMing

Republic Experiment Center, Computer Center, ZheJiang Ocean University,
Zhou Shan, China
{robotlu,xuguangming}@126.com

Abstract. Analyzed and studied on public key and private key cryptography, studied and compared the RSA and DES algorithm, achieved a basic process of file encryption and decryption, provided effectivly protection to document security.

Keywords: Encryption, decryption, DES algorithm, MD5 algorithm, module.

In the process of design about file encryption and decryption, the system of password is an important guarantee for file security. In the paper, using DES encryption algorithm to encrypt and decrypt the contents of the file operations, and using MD5 algorithm to encrypt and decrypt the keys used in the process of transmission and preservation. It can effectively protect the security of the content of the document, but also has a faster processing and response speed[1].

1 Processing

File encryption process as show in Figure 1.

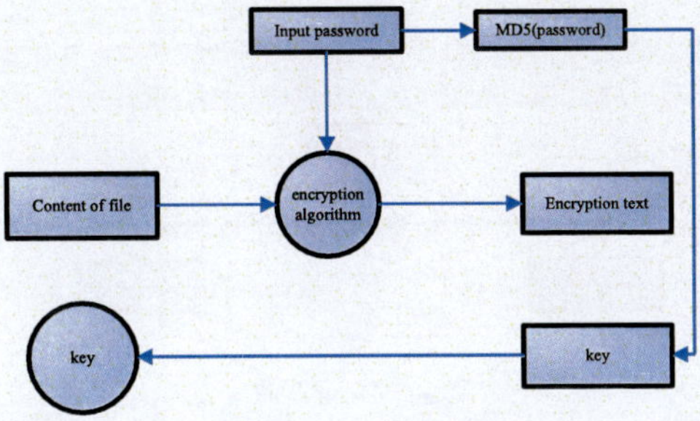

Fig. 1. File encryption process

System obtained detailed information such as the passwords of user, contents of the files, converted the passwords of user into keys with MD5 algorithm and stored into a

D. Zeng (Ed.): Advances in Computer Science and Engineering, AISC 141, pp. 165–170.
springer link.com

dictionary, in the same time, operated the contents of file according to the encryption algorithm and password, converted the plaintext into ciphertext. Then, we can verify the legitimacy of the user during the decryption.

File decryption process, as shown in Figure 2

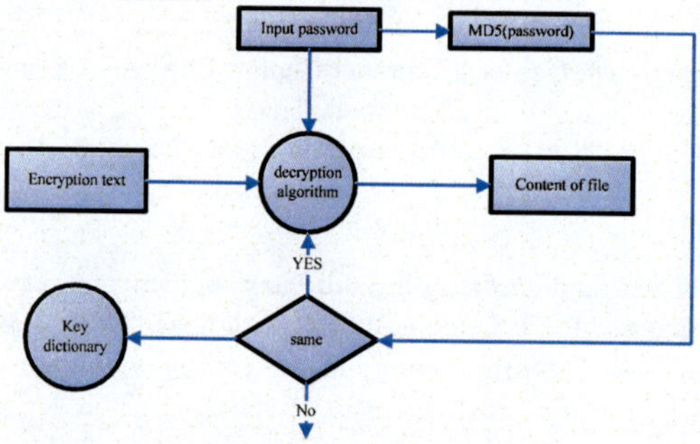

Fig. 2. File decryption process

In the process, according to the decrypted password user input, system compared with the key stored in dictionary after converted by MD5 algorithm, if the same, then the user is legitimate, and has the right to decrypt the corresponding ciphertext operation; if dnot the same, then the user is illegal, and has no right to decrypt the encrypted file, gives the wrong password prompt[2].

In the process, provided verification of the legality of the user, and extracted encrypted information, restored the encrypted file with corresponding decryption algorithm, to ensure effective security of the encrypted file.

2 Function Module

System mainly consists of two modules: Algorithm module and user module, as shown in Figure 3:

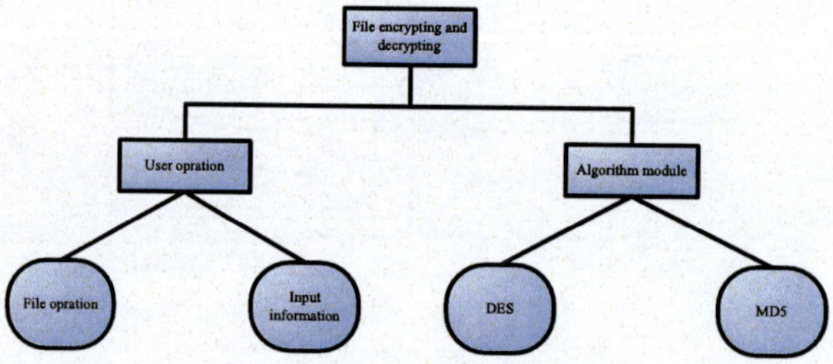

Fig. 3. Main modules

2.1 Algorithm Module

The template is used to format your paper and style the text. All margins, column widths, line spaces, and text fonts are prescribed; please do not alter them. You may note peculiarities. For example, the head margin in this template measures proportionately more than is customary. This measurement and others are deliberate, using specifications that anticipate your paper as one part of the entire proceedings, and not as an independent document. Please do not revise any of the current designations[3].

2.2 User Module

Providing friendly user interface, and receiving and processing inputed information, such as a password, file path and the operation mode, offered corresponding treatment tips. Provided functions of file-related operations, including file creation, deletion and obtation of the file path, and including reading and writing to file content.

3 DES Algorithm Module

3.1 Implementation Process

DES algorithm convert 64-bit plain text into 64-bit cipher text, the flow chart shown in Figure 4:

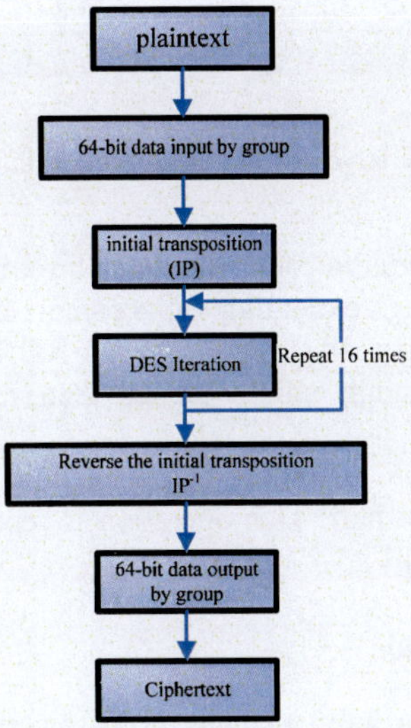

Fig. 4. DES algorithm flow chart

DES operate 64-bit plaintext block, though initial replacement, such as shown in Table 1,assembling the 64-bit plaintext blockby bit, split it into left part and right part by 32-bit length. Then, make 16 rounds of the same iteration, computing process shown in Figure 5. In the process, the plaintext data and the key combined. After 16 rounds, left and right through the end replacement (replacement of the inverse of the initial replacement), that is, the ciphertext.

Table 1. Initial replacement rules

58	50	42	34	26	18	10	2	60	52	44	36	28	20	12	4
62	54	46	38	20	22	14	6	64	56	48	40	32	24	16	8
57	49	41	33	25	17	9	1	59	51	43	35	27	19	11	3
61	53	45	37	29	21	13	5	63	55	47	39	31	23	15	7

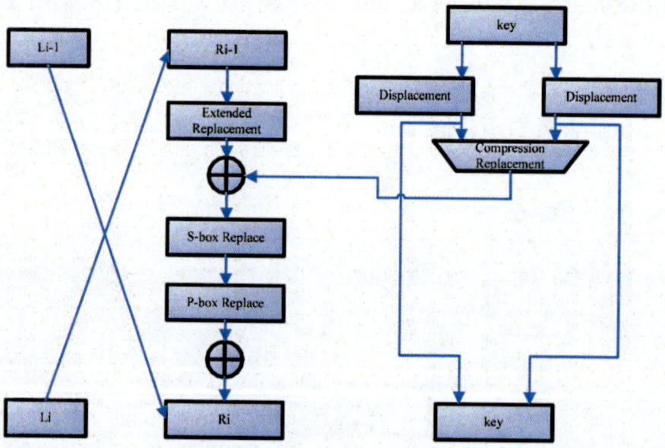

Fig. 5. One iterative process of DES algorithm Flow Chart

As shown above, the process in each round, each round key bit shift, and then selected 48b bits from 56, through extending replacement, extending the right side of the data to 48-bit, and combinding 48-bit key through XOR operation, replacing 48-bit data with new 32-bit data through eight S- boxes, then, replace one more time, called P- box replacement.

Then, through another XOR operation, output of function F and the left part combined, the results is new right half. This operation is repeated 16 times, reaching the purpose of DES encryption[4][7].

3.2 MD5 Algorithm Module

When input password, system will calculate the password into MD5 value, and then compare the MD5 value saved in the file system, and to determine the password is correct. Through it, system can determine the legality of user without know the plain code. It can prevent disclosure of user's password, and also increase the difficulty of beeing cracked to some degree[5][8].

3.3 Implementation Process

After initialization, MD5 handle the input text by 512- bit block, and each divided into 16 32- seat child group. Output of algorithm consists of four 32- bit sub-groups and cascade them to a 128 - bit hash value.

MD5 algorithm flow chart shown in Figure 6:

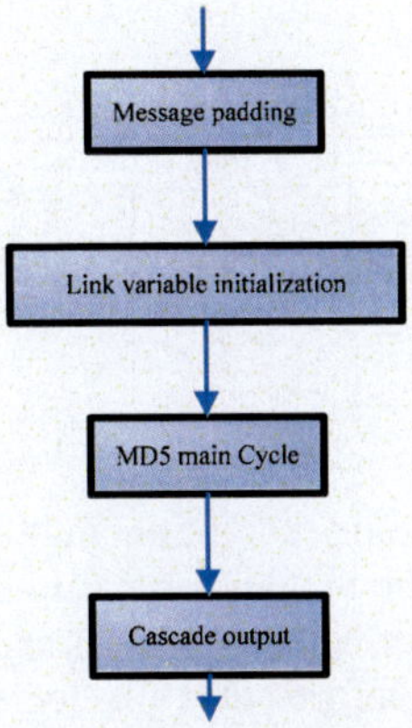

Fig. 6. MD5 flow chart

First fill the message, make the length small than multiple of 512-bit 64 bits. The method is attaching 1 behind the message, padding the number with 0, and then subsequently attached to a 64-bit message length (before padding). The role of the steps is to make the message length is exactly an integer multiple of 512 bits, and ensuring that the messages are different after padding.

Four 32-bit link variables (chaining variable) is initialized to:

A=0x01234567
B=0x89abcdef
C=0xfedcba98
D=0x76543210

Then make the algorithm loop, the frequency of loop is the number of 512 bit message groups.

copy four variables above to other variables: A to a, B to b, C to c, D to d[6].

The main loop has four rounds, each round is very similar, as shown in Figure 7. The first round operated 16 times. In each operation, nonlinear function operating on the three of a, b, c and d, then the result plus a fourth variable, the text of a sub-group and a constant. The results then move to the right of a variable number of rings, plus one of a, b, c or d. Finally, replacing one of a, b, c or d.

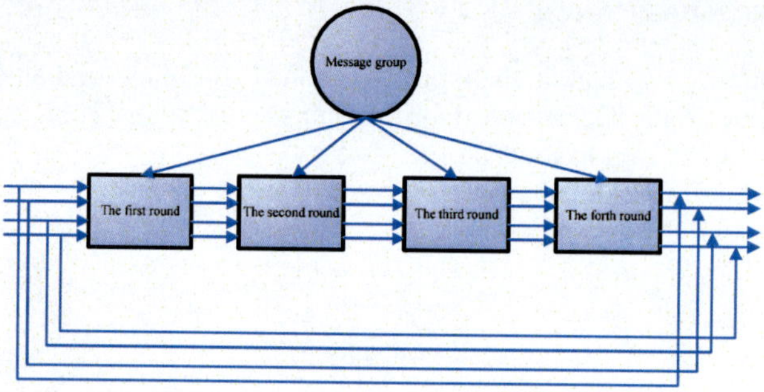

Fig. 7. MD5 main cycle

4 Conclusion

Simple small application software of encryption and decryption to file with the utility of DES, MD5, RSA algorithm, the system can easily encrypt and decrypt a variety of file (such as. TXT,. DOC, etc.), effectively protect and precisly restore the content of encrypted file, protected personal information to a degree.

References

1. Lu, D.: Computer Cryptography, Computer network data privacy and security. Tsinghua University Press (1998)
2. Li, K., Wang, D., Dong, X.: Practical Cryptography and computer data security. Northeastern University Press (1997)
3. Yang, Y., Sun, W.: Modon cryptography new theory. Science Press, Beijing (2001)
4. Schneier, B.: Application Cryptography - Protocols, Algorithms and C source code. Machinery Industry Press (2000)
5. Huang, Y., Chen, L., Tang, S.: Information security and encryption decryption core technology. Electronic Press (2001)
6. Li, S., Wang, D.: Modern Cryptography Theory, Method and frontier. Science Press, Beijing (2009)
7. Manzanares, A.I., Sierra, C.J.M., Marquez, J.T.: On the imp lementation of security policies with adaptative encryption. Computer Communications 29(2), 2750–2758 (2006)
8. Vaudenay, S.: On the Weak Keys of Blowfish. In: Gollmann, D. (ed.) FSE 1996. LNCS, vol. 1039, pp. 27–32. Springer, Heidelberg (1996)

Performance and Application of Computer Technology in Interior Design

Chunmi Zhang and Jiaying Hu

Hebei Normal University of Science &Technology, Qinhuangdao, Hebei, China
470234853@qq.com

Abstract. Excellent interior designers usually have their own unique idea, by which they could retain their own design style, as well as let the consignor get their intention and the final interior decoration effect in a very intuitive way. The initial manifestation mode of interior decoration is usually the free-hand drawing of a designer. This free-hand drawing requires a designer to have a solid art foundation and a preferable aesthetic flavor. They are able to endow actual interior space with a deeper sense of beauty through these free-hand drawings from the perspective of design aesthetics. Although free-hand drawing has a strong nature of expressiveness, it's not good enough in simulating the actual space and inconvenient for local adjustment and modification of the program. Design softwares, such as 3D Studio Max, Photoshop, have freed designers from complicated and redundant work of free-hand drawing. These softwares have a fairly strong function in simulating the reality, and the design rendering almost can reproduce a real scene. Thus, a designer's idea can be presented with a visual effect closed to the reality.

Keywords: Computer technology, interior design, design software, performance and application.

1 Introduction

Design rendering is of vital importance when a designer is conceiving how to decorate the interior space. By rendering, not only a designer's idea can be presented, but also the designer could communicate with the consignor. In the performing area of interior design, application of Computer-aided Design developed rapidly, with emergings of many design softwares used exclusively to drawing renderings of interior design. The extensive use of these design softwares not only improves the work efficiency of designers tremendously, but also presents a more real, more standard and more accurate effect, through which the art atmosphere of interior space can be more easily reflected and the designer is more likely to have resonance with the consignor.

In the early 1980s, computer graphics art design, as the most cutting-edge visual expressing means, started to be introduced from abroad and extensively applied in the mass media and visual art design fields, such as environmental art design, visual communication design, public art design, film and television animation and so on.

D. Zeng (Ed.): Advances in Computer Science and Engineering, AISC 141, pp. 171–176.
springer link.com

By giving an analysis of the application of computer design softwares, such as 3D Studio Max and photoshop, in renderings of interior design, and giving a comparison of differences and relations between traditional hand-drawing skills and computer graphics in technological age, this article further implores how should computer technology play a more important role in interior design performance.

2 Software for Drawing Rendering: 3D Studio Max

3D Studio Max, usually short for 3DS Max or MAX, is a 3D animation rendering and producing software developed by Autodesk Company, on the basis of PC system. It is now being widely used in areas of film and television advertisement, industrial design, architecture design, auxiliary teaching and engineering visualization, etc. The utilization of 3DS MAX occupies a position of absolute advantage in renderings of interior design and architecture design, which are relatively developed domestically.

2.1 Powerful Modeling Function of 3D Studio Max

First, confirm that you have the correct template for your paper size. This template has been tailored for output on the US-letter paper size. If you are using A4-sized paper, please close this template and download the file for A4 paper format called "CPS_A4_format".

Manufacturing of 3D model plays a very important role in rendering of interior design. The rendering is the visualized reappearance of designer's intention and idea. Only after a 3D virtual space model being created, can the further simulation of real material, illumination and other effects be realized.

Modeling tools provided by 3DS MAX include basic modeling tools and advanced modeling tools. Basic modeling tools provide cube, sphere, hemispheroid, cylinder, tube, donut, cone and polygon, while advanced modeling tools create mountain, water, ripple, wave, granule and irregular shape and object, such as human body and plant, etc. 3D graphics can be distorted, bended, zoomed, swung, transformed in different angles, sculpted and drilled into hole.

The basic process of producing small-sized model by 3DS MAX is: Building models. It's inseparable of "creating" and "modifying" in 3D MAX, yet a 3D model can be formed from a 2D graphic by rotating, setting out, pinching and other modifying commands. Geometry, being extracted directly, should also be modified before a modeling work being finished; Setting the material and adding it to the model; Installing lights and camera; Rendering the camera view after transforming the perspective view into the camera view.

Through these modeling tools and modifying tools, small-sized models can be created, as well as large-sized architectures. The function of 3D MAX in graphical expression of environmental art is that it can create random models of real world quite easily.

3D modeling function is also very powerful. The following home design fully display the powerful 3D modeling function. (Figure 1)

Fig. 1.

In this sitting room, all these objects that we can see belong to 3D models, including the wall and all kinds of small accessories. The creating of indoor space model is a relatively simple part of 3D modeling functions. Firstly, creating the wall of the room. There are mainly two ways of creating the wall: the first way is to create the solid box of each wall by Box, one of solid-creating tools. Then combined each box closely by moving and capturing tools. Hollow places, such as the pendulum frame on the background wall behind the sofa, should be cut out by Boolean operations. It goes the same with entries and windows. This is one of the methods of creating wall. The second way is relatively easier, in which AutoCAD needs to be used together. Introducing a CAD plan into a 3D space when lines of the CAD plan become existed in forms of 3D lines. Then pinching, one of modifying commands, so as to create many walls at a time. This method is simple and practical, so that many designers adopt it in wall modeling.

Then focusing on these small models in the space: sofa, tea table, dinning set, lamps and small accessories. These models can be created by designers themselves, but lots of modifying and editing commands are required. In most cases, a simple model can be created by commands, which are adaptable to the pinching and editing of 2D or 3D objects. There is also another way being widely used by designers. That is introducing a ready-made model into the space to conduct collocation and design. The popularization of 3D softwares enabled more and more enthusiasts to join in the world of 3D. They divided 3D modeling work into segments. Models of furnitures and decorations frequently used by designers are drafted by professional personnel, and then uploaded on the Internet so that everybody can share. Hence, a 3D furniture model base took shape. Models in the base are various, accurately drafted and frequently renewed, thus, the drawing work of interior designer becomes easier and more vivid.

2.2 Material and Rendering of 3D Studio Max

3DS MAX provides abundant materials and textures, and can be used for editing the whole entity or a certain part of the entity in terms of colour, shading, reflection, bump and transparency, etc. Objects can be transformed through displacing, rotating, zooming and stretching. Illuminant can be ambient light, floodlight, or spotlight. Multiple cameras not only can be fixed in different positions, directions and angles, but also can be adjusted flexibly, with the function of real-time preview of animation, so as to observe the effect of objects timely.

3DS MAX integrates functions of modeling, light material setting and rendering, so, in terms of the whole manufacturing process of rendering, it is very convenient in the procedure of setting materials that 3DS MAX can set the corresponding material right after modeling, thereby observe the whole effect of the model.

Synchronized material and nonsynchronized materials are two terms of material in 3D MAX, in particular, synchronized material connects material in the editor and material on the surface the an object in the scene, which means when we choose an object in the scene, there would be a suggestive sign of this material in the editor. This kind of connection improved the intuitiveness in material setting.

In 3D MAX, mapping is the most charming part of material. It not only provides the surface of an object with abundant colours, but also endows the material with a nature of transparency, self-illumination or the bump effect. Different versions of 3D MAX all provide various types of maps, by the using of which a lifelike material effect can be acquired.

The use of composite materials is a feature of material setting by 3D MAX. Composite materials combine two or more than two different materials together in a certain way to get an effect of complex materials.

3 Analysis of the Application of Photoshop in Post Processing of Interior Design Rendering

Photoshop is a professional image processing software developed by Adobe company. It has strong functions in image processing and is widely used in image design, typography and print, web design, visual identity design, news media and many other fields. In the environment design rendering, Photoshop is mainly used in the post processing of drawings, including adjusting colours, adding displays in the space and constructing artistic atmosphere, etc.

In 3D Studio Max and Lightscape, rendering still belongs to a "rough" phase which needs to be finished. 3D Studio Max and Lightscape have their incomparable advantages in modeling, light rendering and material mapping, yet obviously insufficient in terms of adding decorations, such as displays and plants, and building artistic atmosphere. However, Photoshop is able to accomplish this kind of tasks with high proficiency. The only work needed to be done is choosing pictures that suit the scene and merging them together with the final rendering in an artistic way. Sky, for instance, green plants, characters, transportations, interior displays and so on. All these things can be added and processed artistically by Photoshop.

3.1 The Function of "Cutout" in Marquee Tool Group of Photoshop

"Cutout" is a basic function of Photoshop in artistic processing of image. The system offers different kinds of marquee tools, such as the rectangular tool, the elliptical tool, the lasso tool, the single column/line pixel tool and the magic wand tool, etc. Besides, Photoshop provides many commands of Selection Editing, which are able to produce, zoom, calculate selections, as well as save and install selections. In the procedure of post processing of environmental art design rendering, the marquee tool group will give a full play to its advantages in terms of image compositing and colour adjustment.

3.2 Layer Function of Photoshop

Layer, or image layer, is an indispensable carrier in the artistic creation of graphics, as well as the most effective approach in image processing. In order to understand it more easily, one can imagine that a layer is formed by overlaying pieces of transparent colored papers together. Images in other layers will not be affected by the editing of a certain layer. Thus, new layers and a layer set can be used to edit zones which need adjustment during the post processing of rendering. It's easy to modify and distinguish editing procedures without confusions so that the design task can be accomplished in an orderly way.

3.3 Drawing and Modifying Functions of Tool Groups of Pen and Path-Selection in Photoshop

Photoshop provides powerful pen tool group and path-selection tool group which are used in creating, editing, path setting and graphic drawing. These tools can draw all kinds of plans and texts arranged in accordance with the path quickly. Through the "path" palette, one can get the thumbnail of all paths in the image and immediately fill the path, stroke, and transform selections, as well as flourish the artistic effect of rendering by using layer and commands of colour adjustment.

3.4 Colour Adjustment of Image and Special Decorative Function of Filter in Photoshop

There is a large quantity of commands for colour adjustment in the image menu in Photoshop. Designers are able to adjust the transparency, purity and contrast freely during the artistic processing of rendering, so as to acquire the effect of brightness and clearness. They can also use selection commands and layer tab to correct the colour in parts of a drawing so that the whole picture will present harmoniously, elegantly and clearly.

Filter also plays a very important role in Photoshop. The system offers numerous kinds of filters, including pixelate, distort, noise, blur, artistic brush, stylize, texture, render and etc. In addition, exclusive filters with special effects, developed by the third party, are also allowable in Photoshop, in order to create different kinds of special artistic effects.

In the process of designing the environmental effect of space, existing 3D graphics and all kinds of planar materials are frequently composited, generally in a way that compounding 2D and 3D graphics based on a united perspective angle by Photoshop, or drawing a planar rendering by hand according to the planar layout, and then importing the rendering into 3D MAX and Photoshop through scanning or digital device, so as to create a 3D scene. Large quantities of post processings of renderings need artistic blending of many planar materials. Thus, designers must be quite familiar with commands in all menus so that the rendering can fully meet the design requirements in terms of colour, light effect or artistic atmosphere. (Figure 2)

Fig. 2.

The widespread use of computer design software in environmental art design showed us a brand new visual world. Artistic effects, which are hardly acquired by free-hand drawing before, can now be achieved by getting twofold results with half the effort, through the aids of computer. Various speedy means and manners for designing, being offered by softwares like 3D Studio Max and Photoshop, have emancipated hands of designers from the shackle of heavy, slow and redundant work. The organic combination of 2D drafting techniques of Photoshop, scientific division of work in 3D softwares (3D Studio Max and Lightscape) and designer's free-hand drawing have tremendously promoted the efficiency of expressing means of environmental artistic design, and have given fully expression to the superiority of combining science and technology with art in modern environmental design. This organic combination represents the development trend of visual cultures and expressing means of interior environmental art design in the information times.

References

1. Yan, H.: Introduction to environmental art design, pp. 1–5. Chinese Youth Press (2007) (in Chinese)
2. Yu, X.: Rendering of interior, pp. 21–22. Chinese Art Academy Press (2001) (in Chinese)
3. Yuan, G., Ben, D.: Computer-Aided rapid expression of design, p. 11. Ocean Press (2009) (in Chinese)
4. Zheng, S.: Interior design intellection and means, pp. 10–15. China Architecture and Building Press (1997) (in Chinese)
5. Liu, Y.: Hand-drawing illustration for interior and exterior, p. 4. LiaoNing Art Press (2008) (in Chinese)

Embedded System Design for Speed Measurement and Timing of Vehicles

Cao Yong[1], Zhang Yu[1], Shao Si Jie[2], and Xiong Wei[3]

[1] Department of Control Engineering, Academy of Armored Force Engineering
Beijing, China
[2] Department of Weapon Engineering, Academy of Armored Force Engineering
Beijing, China
[3] Department of Radar Measurement Station, Troop No.63620 of PLA, Jiuquan, China
cycycy157@sohu.com,
{zhangyu3984412,8617143sj,xw7490}@163.com

Abstract. Embedded system for speed measurement and time is designed by adopting sensor and computer control technology to monitor operation condition of vehicles in real time, measuring and recording the speed while running in the examining locale. According to the function target, the design scheme is put forward to carry out the design of photoelectric receiving, single chip system, wireless communications circuit, and software realization. The mechanism is designed rationally, achieving the aim that unchanging vehicles and installing expediently.

Keywords: Speed measurement, timing, embedded system.

1 Introduction

In examination system of vehicle driving and training, a driver's assessment is determined by speed and total travel time during driving situation. Sensor and PC control technic is applied in embedded vehicle velocity detection and timing system, for real-time monitoring running condition of vehicle. Regarding vehicle bearing up for assessment area as relative initialized time, vehicle bearing off assessment area as relative finishing time, measurement and recoding instantaneous speed total travel time and average speed during driving situation, moreover, transmissing data after having been disposed to rear commanding center with method of wireless communication and real-time display. Achievement assessing and managing is dealed with software of achievement assess.

2 System Major Function Indices

- Automatism real-time and nicety noting assessing administrative time and speed;
- Equipping with function data wirelessly transmissing and transmissing longer distance more than 2 km;

D. Zeng (Ed.): Advances in Computer Science and Engineering, AISC 141, pp. 177–183.
springer link.com

- No need for reequipping;
- Acclimatizing itself to field travelling crane circumstance.

3 System Design Scheme

As is shown in figure 1, the system design scheme block diagrams are composed of three parts such as surveyor's pole on ground surface, embedded system of vehicle and terminal disposal.

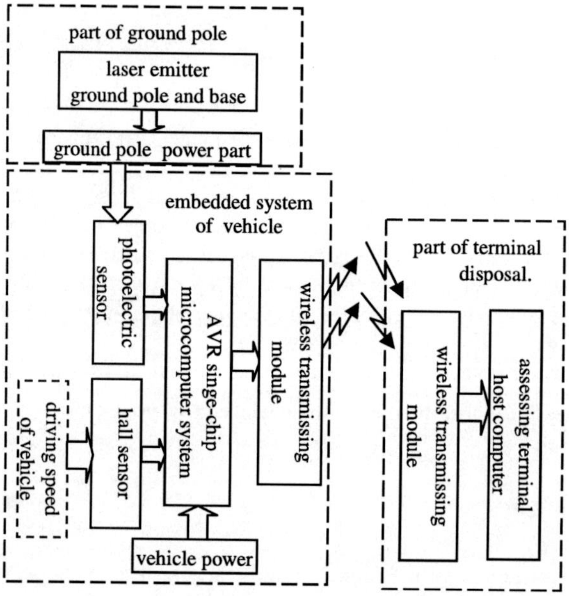

Fig. 1. Block diagram of system scheme

Separately locating two terminals of raceway, part of ground pole is setted on base. Based on ground pole, linear laser signal produced by linear laser emitter is received by photoelectric sensor locating testing part of vehicle, regarding which as initiation and finishing signal of vehicle driving total time. Setted on side board of vechile, hall sensor probe against alnico equally distributing around the circle. When wheel is turnning, each alnico passing probe makes hall sensor exporting signals. Output signal of photoelectric sensor and hall sensor is put into AVR singe-chip microcomputer system, after processing, gaining the speed and time of moving vechile. Then via wireless model transmissing to assessing terminal host computer located at terminal disposal. part, data is dealed with and analyzed. Working principle and working flow chart of system is shown in figure 2.

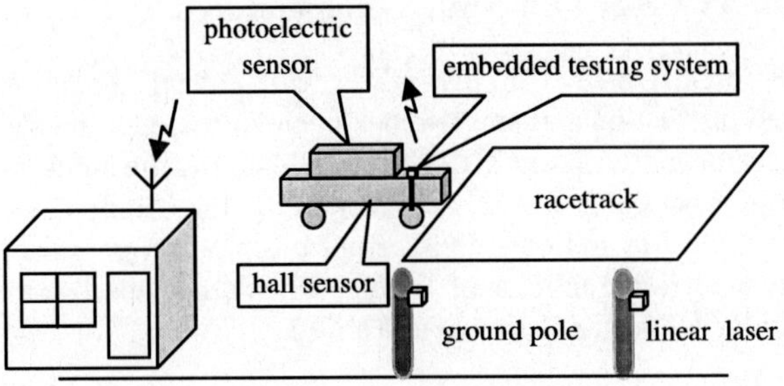

Fig. 2. Working principle and working flow chart of system

4 System Hardware Design

Circuit design of embedded system maily includes design of photoelectric received circuit, design of AVR singe-chip microcomputer system and circuit design of wireless communications.

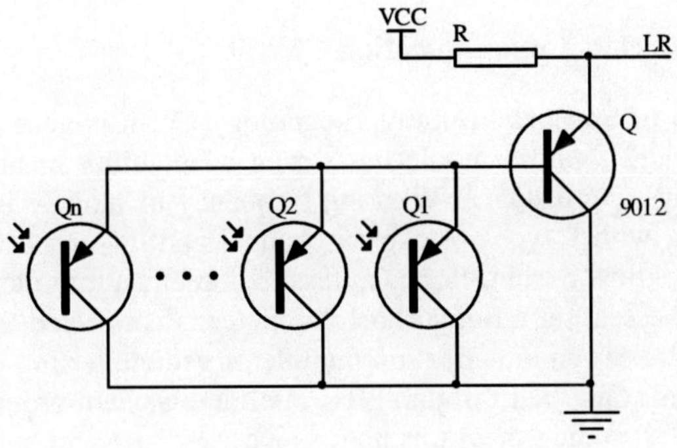

Fig. 3. Photoelectric receiving circuit

4.1 Design of Photoelectric Received Circuit

Using phototransistor as photoelectric sensor, with way of parallel connection, realizing function of wired or distributing on PCB board linearly. Using PNP photosensitive triode 9012 increases sensitivity of optical receiving greatly. Photoelectric receiving circuit is shown in figure 3.

When vehicle passing linear laser based on ground pole locating starting-point of racetrack, as long as existing photosensitive triode receiving light, so photosensitive triode Q is on state. Low level from LR is put into singe-chip microcomputer, as initial timing signal. The same principle, when vehicle passing linear laser based on ground pole locating finishing-point of racetrack, low level from LR is put into singe-chip microcomputer twicely, as terminational timing signal. Thus, through output signal of LR, singe-chip microcomputer system calculates total running time accurately, realizing timing function.

4.2 Design of AVR Singe-Chip Microcomputer System

Atmega8 is a 8 bit CMOS MCU in low-power, with enhanced AVR RISC structure[1]. For its advanced instruction and single-clock cycle of instruction executing time, high-throughput of ATmega8 is up to 1MIPS/MHz, alleviating the antinomy between power loss and speed of processing. Besides, ATmega8 has three agile models for comparing timer and counter, inside and outside interrupt of chip, programmable serial interface USART, easily realizing funtions of timing, calculating speed and communication. Pins distribution of ATmega8 is shown as Table 1.

Table 1. Pins distribution of ATmega8

ATmega8 pins	distribution	funtion
PD0(RXD)	wireless module TXD	wireless communication receiving
PD1(TXD)	wireless module RXD	wireless communication transmissing
PD2(INT0)	photoelectric receiving output LR	timing funtion
PD3(INT1)	output of hall sensor	speed measurement funtion
PD5(T1)		

4.3 Design of Wireless Communication Circuit

Using SRWF-1028 of ShangHai Sunray Technology Co, it is generalized transparent transmissing module[2], whose modle has strong adaptability of anti-interference as well as transmissing long distance. Working frequency of module includes 403, 433, 470, 868, 915MHz, with 8 types of chosen channel and three interface way of TTL, RS232, RS485. In this design, for wireless communication module of terminal processing portion, assessing terminal host computer is connected through RS232 interface way. For wireless communication module of vehicle testing portion, TXD and RXD of Atmega8 is connected through TTL interface way. In order to avoid interference among vehicles, adding measures photoelectric isolation, circuit design is shown in figure 4.

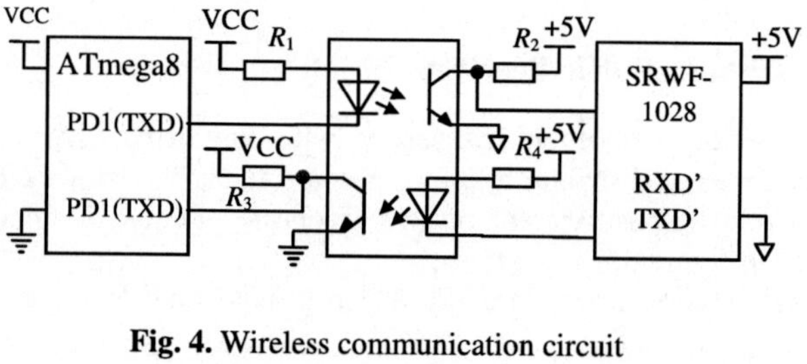

Fig. 4. Wireless communication circuit

5 Design of Mechanical Structure

In order to satisfy the needs of no reequipping and field travelling crane easily, using design of mechanical structure showing in figure 5. Part of broken line is structure of embedded system. Singe-chip microcomputer controlling circuit PCB and wireless communication model is installed in controlling box. Magnet is installed in hollow connecting rod of below controlling box, with adsorbing above tyre of vehicle. In order to avoid influence from sunlight to photoelectric receiving circuit, photoelectric receiving circuit PCB is setted inside hollow cylinder. Grating structure has been shaped when surface of cylinder collimation against photosensitive triode. Hall sensor receiving terminal against alnico equally distributing around the tyre, realizing the function of speed measurement. Output of hall sensor and photosensitive receiving circuit is put into controlling box by cable. Power supply of embedded system is from of vehicle.

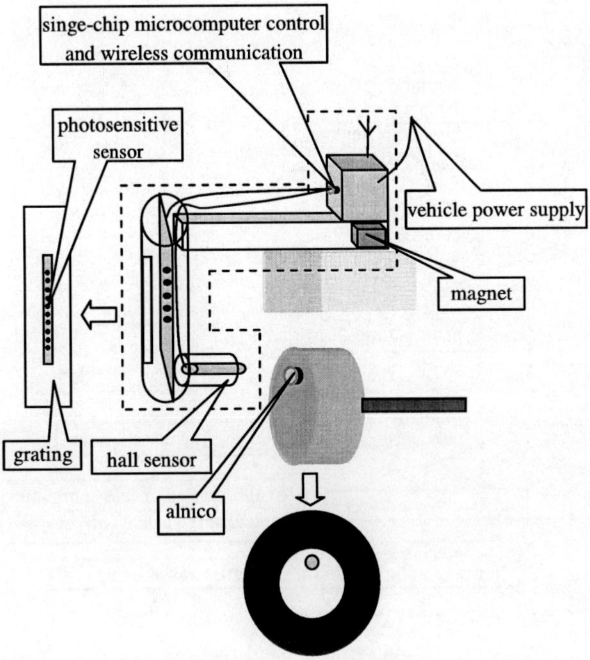

Fig. 5. Mechanical structure

6 Design of System Software

6.1 Speed Measurement and Timing Software

1) Design of speed measurement software: When using hall sensor carring through speed measurement[3-5], adoption method of pulse counter, if wheel turn around a circle, output of sensor is a pulse or fixed many pulses(decided by alnico of vechile equally distributing around the circle). Then AVR singe-chip microcomputer system takes count of pulses put in, by processing, gaining instantaneous speed of vehicle, through serial module transmissing host computer to process and display in real-time. Speed measurement software flow is shown in figure 6.

2) Design of timing software: Laser signal of initiative ground pole and finishing ground pole when vehicle running, received by photoelectric sensor, touching off counter of AVR singe-chip microcomputer to begin timing or stop timing, realizing the function of timing. Frondose timing software flow is shown in figure 7.

6.2 System Uplinking Communication Agreement

In order to reflect running status of vehicle and supply data for assessing, embedded system needs to transmit information to PC terminal real-time, system uplinking communication agreement is shown in figure 8.

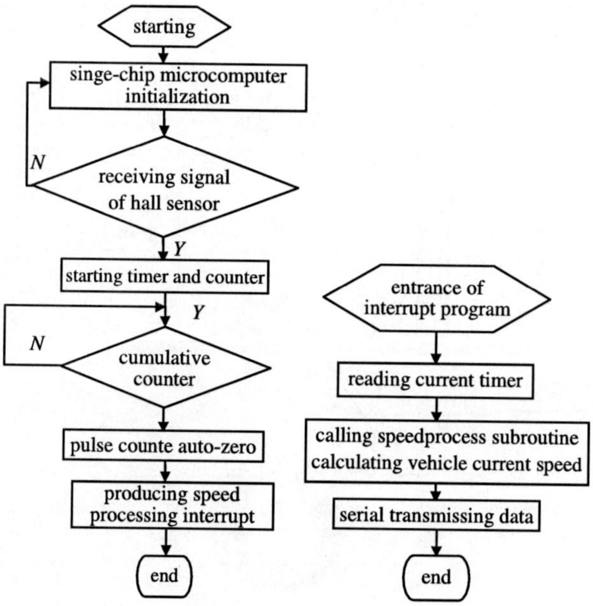

Fig. 6. Speed measurement software flow

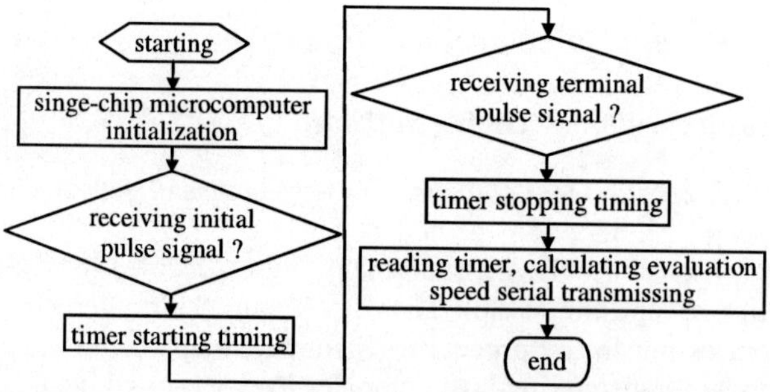

Fig. 7. Flow of timing software

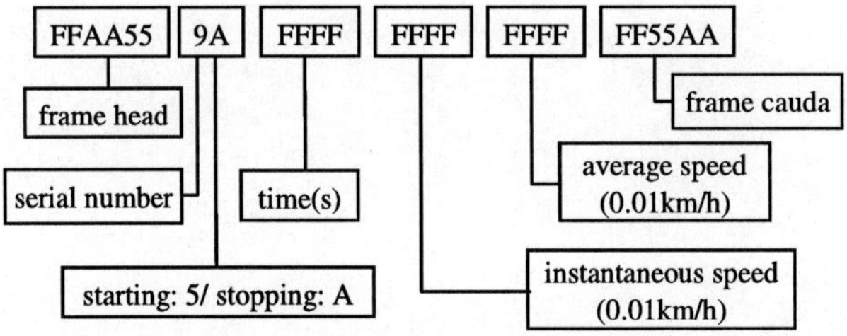

Fig. 8. Embedded system uplinking communication agreement

Frame head of communications data package begins with 0xFF, behind which following 0xAA and 0x55. Frame cauda of communications data package begins with 0xFF, behind which following 0xAA and 0x55. That type of data is not easily shaped by noise, so receiving agreement only receives data package with the type of frame head and frame cauda. Information such as serial number of vehicle, starting and stopping status of timer, average speed, instantaneous speed and so on. It is transmissed by embedded system agreement, which is transmitted to PC terminal, referrence for assession.

7 Conclusions

Depicted from passage of vehicle speed and timing of embedded system, real-time monitors running speed and total driving time of vehicle, transmissing data after having been disposed to assession terminal with method of wireless communication, referrence for assession. System is brief, agile and highly dependable, moreover, no need for reequipping, easy disassembly, especially acclimatizing itself to field travelling crane circumstance.

References

1. ATMEL. AVR singlechip ATmega8 Datesheet
2. ShangHai: Sunray Technology Co.,Ltd. SRWF-1028(V1.5). Using ofwireless model explaining book, `http://www.51sunray.com`
3. Yang, J., Liu, F.: Application study of hall sensor A44E in cartwheelps resurver. Electronic Measure Technology 32, 100–102 (2009)
4. Lu, G.Q.: Application of hall sensor in intelligent odometer based on pulse detection. Machinery Design AND Manufacture 1, 87–88 (2009)
5. Ding, Z.Q.: Design of motor speed measurement device based on hall effect sensor. Journal of Agricultural Mechanization Research 5, 81–83 (2010)
6. Chien, C.L., Westgate, C.R.: The Hall effect and its applications, pp. 523–545. Plenum Press, New York (1999)
7. Hafeman, D.G., Parce, J.W., McConnell, H.M.: Light-addressable potentiometric sensor for biochemical systems. Science 240, 1182–1185 (2008)

Enterprise Mergers and Acquisitions in the Light of the Value Chain Optimization

Congzhi Di[1,*] and Fusheng Yu[2]

[1] Hebei University of Economics and Business, Renmin University of China
Shijiazhuang, China; Beijing, China
dicongzhi@126.com
[2] Renmin University of China RUC, Beijing, China

Abstract. This paper is focused on the study of the application of the value chain theory in the enterprise merger and acquisition (M&A) from the temporal (the pre-event, mid-event and after-event) and spatial (the horizontal value chain, the vertical value chain and the internal value chain) dimensions. Then this paper constructs a theoretical frame of M&A in terms of the value chain optimization.

Keywords: Merger, Acquisition, Value Chain, Value Chain Optimization.

1 Introduction

Merger and Acquisition is an effective way for the enterprise to acquire core strategic resources rapidly. In the recent years, the economic globalization has brought a new trend of enterprise merger and acquisition which becomes an effective means for the enterprise to realize rapid expansion and better distribution of resources, to enhance it competitiveness, and resist various risks to compete for shares in the foreign markets, hence the fact that the 6th trend of mergers and acquisitions since 2003 has not yet lost its momentum until now. The global financial tsunami in 2008, with the subprime crisis as its starting point, had severally impacted the global economy, resulting in the breakup and M&A of a large amount of companies. But at the same time, it had also provided golden opportunities for the optimization and integration of the enterprise value chain and the industrial chain. Under such a background, the M&A have presented new characteristics: the strengthening of the enterprise's core competitiveness and the adding of its value have become the main objective of the M&A, and cross-border M&A and mergers between giants have become a mainstream which are to realize the optimization and integration of the value chain. This paper is intended to make a thorough study on the M&A in terms of the value chain optimization and integration and construct a theoretical frame of the M&A for the enterprise's practice, contributing to the study on the M&A and the value chain.

* Corresponding author.

D. Zeng (Ed.): Advances in Computer Science and Engineering, AISC 141, pp. 185–191.
springer link.com

2 The Value Chain Management Theory and Its Impacts on M&A

Michael E. Porter first proposed the Value Chain theory in 1985. He took the value chain as the basic means in analyzing the enterprise's competitive advantages and claimed that the value system consisted of the supplier value chain, the enterprise value chain, the channel value chain and the buyer value chain. Then appeared the value chain consisted of raw material and customers (Hines, Peter, 1993), and then the virtual value chain (Rayport and Sviokla, 1995) and the network value chain(Richard and Rafael, 2000).

The essence of the value chain management lies in the conception of viewing all the practical and potential value adding activities in the light of "chains". On one hand, it considers the suppliers, the customers and the competitors as the constituents of the external value chain, place emphasis on the cooperation with them and build the value adding of the enterprise on the basis of that of the entire value chain union, thus embodying an all-win conception; on the other hand, it regards all the value adding activities within the enterprise as organic, and interacting components of the internal value chain, and emphasize value maximization through coordinating and eliminating non-value adding activities.

The successful realization of M&A depends on the strengthening of the competitiveness of the enterprise and the increasing of resources and value after the M&A. The value chain theory has provided a new perspective for the M&A, and helps to analyze the competitiveness of the enterprise at the value chain level. If M&A can realize the optimization of the value chain, then they can enhance the competitiveness of the enterprise.

3 The Basic Frame of the M&A in the Light of Value Chain Optimization

As for the enterprise's motive to resort to mergers and acquisitions there are various versions of explanations, of which Porter's competitive advantage theory is one, which believes that the mergers and acquisitions of the enterprise are aiming at enhancing its competitive advantages. However, Porter's theory does not explain how the merger and acquisition can enhance the enterprise's competitive advantages and how to achieve successful mergers and acquisitions. Based on the competitive advantage theory, this paper analyzed how to take the optimization and integration of the value chain into consideration through the whole process of merger and acquisition, so as to increase the enterprise's competitive advantages. It constructed a frame of the enterprise mergers and acquisition in the light of the value chain optimization from two dimensions, that is, the temporal dimension and the spatial dimension.

Temporally, the merger and acquisition activities can be divided into pre-event activities, mid-event activities and after-event activities in accordance with the pre-event, mid-event and after-event stages of the value chain management. Each of the three parts will be studied in detail from a spatial perspective in terms of the horizontal value chain, the vertical value chain and the internal value chain.

The merger and acquisition of the enterprise concern the merger and acquisition enterprise and the target enterprise. This paper studied mainly from the merger and acquisition enterprise's point of view.

4 The Pre-event Activities of M&A

Merger and acquisition is a strategic investment on the part of the enterprise, which have profound influence on the enterprise's management and development. While the pre-event activities are critical to the success of the merger and acquisition, therefore, should be placed adequate emphasis on. In general, activities before merger and acquisition include determining the strategy of the merger and acquisition, deciding the target enterprise, making responsible investigation, working out the merger and acquisition budget and making merger and acquisition plan.

Enterprise merger and acquisition in the light of the value chain optimization is aiming at the optimization of the value chain, therefore, whether the merger and acquisition can bring value adding and optimization of the value chain must be taken into serious consideration before the merger and acquisition happen.

The pre-event activities of M&A include five steps.

4.1 Value Chain Analysis of the Acquirer

Whether it is necessary for the enterprise to resort to mergers and acquisitions, why the enterprise has to do this and how to realize successful merger and acquisition are major problems for the enterprise in deciding its merger and acquisition strategy. To solve these problems, the merger and acquisition enterprise should first make a value chain analysis on itself, which is the basis for formulating the merger and acquisition strategy. The value chain analysis before the merger and acquisition can help the merger and acquisition enterprise decide whether to resort to merger and acquisition and predict the possible effect of the merger and acquisition on its value chain.

The value chain analysis of the Acquirer includes internal value chain analysis, horizontal value chain analysis, and vertical value chain analysis.

a. Internal value chain analysis

Internal value chain analysis can be divided into three steps. First, analyze the basic activities and supplementary activities of the enterprise and determine the value chain. Second, make a further analysis of the operation chain. Thirdly, evaluate the enterprise value chain.

b. Horizontal value chain analysis

Horizontal value chain analysis can be divided in three parts. First, make a detailed investigation and simulated measurement of the competitors and identify the competitors' value chains. Second, compare the value chains of the two parties. Make a comparison between the competitor's value chain and its own value chain to find out the differences of the two and analyze the advantages and disadvantages of each of them. Finally, predict the possibility of the M&A with its competitors.

c. Vertical value chain analysis

Vertical value chain analysis mainly includes: First, identify the nodes on the industrial value chain. Second, identify the industrial value chain. Third, evaluate the position and role of the Acquirer in the industrial value chain, seek for opportunity to expand its value chain to its upstream and downstream enterprises and then predict the possibility of M&A.

4.2 The Formulation of the M&A Strategy

The formulation of the M&A strategy includes the definition of the M&A motive, the determination of the type of M&A, preliminary selection of the target enterprise and the selection of the M&A opportunity.

a. The definition of the M&A motive

The M&A should be in accordance with the enterprise's strategy in order to create value. The competitive advantage theory holds that the enterprise's motive of M&A rises from the competitive pressures. Through M&A, the enterprise can gain competitive advantages from outside and generate new competitive advantages.

b. The determination of the type of M&A

There are three types of M&A of the enterprise: horizontal M&A, vertical M&A and diversified M&A. The enterprise can either resort to mergers with its upstream or downstream enterprises to enhance its competitiveness, control and eliminate its competitors by merging with them, or expand its business scope by penetrating into relevant industries. The ultimate purpose of any type of M&A is to strengthen the competitiveness of the enterprise through the optimization and integration of the enterprise's value chain.

c. Preliminary selection of the target enterprise

The enterprise then can make a preliminary selection of its target enterprise, which can bring the optimization of the value chain to the enterprise and enhance it competitiveness.

d. Selection of the opportunity for M&A

The enterprise should choose a proper opportunity for the M&A, based on the analysis of the characters of the industry development and that of value chains.

4.3 The Analysis and Evaluation of the Value Chains of the Target Enterprises

The enterprise has to make a further value chain analysis of the preliminarily selected target enterprises so as to decide the final target enterprise and make an evaluation for it. The analysis and evaluation of the target enterprises' value chains include: the value chain analysis of the target enterprises; the comparative analysis of the value chains of the Acquirer and the target enterprise; the determination of the target enterprise; the evaluation of the target enterprise

4.4 The Responsible Investigation and the M&A Budgeting

The enterprise should make a responsible investigation on the target enterprise in terms of its background and history, the industry it belongs to, its marketing mode, manufacture method, financial information and system, and its research and development plan, etc. During the investigation, the Acquirer should make a further adjustment of the previous analysis results according to the value chain analysis of the target enterprise and work out the M&A budget.

4.5 Designing the M&A Plan

The Acquirer should work out a complete M&A plan in accordance with the M&A strategy, and make a detailed arrangement of the activities and steps in the M&A. Besides, it has also to choose a suitable way of payment, and formulate accounting and legal procedures of pricing, financing and taxation concerning the M&A.

5 The Mid-Event Activities of the M&A

The mid-event activities of the M&A are in a stage in which the M&A is practically accomplished, including the accomplishment of various legal and procedural activities and the integration after the M&A, which are critical to a successful M&A. The two parties in the M&A have to make a M&A plan based on adequate preparation, and carry out the integration after the M&A. The mid-event activities, especially the integration after the M&A, are decisive for the realization of the expected results of the M&A. Therefore, the enterprise should place much emphasis on the integration after the M&A to realize the optimization and integration of the value chains on the basis of the value chain analysis.

5.1 The Accomplishment of Legal and Procedural Activities

The enterprise should make a reasonable timetable for the accomplishment of various legal and procedural activities according to the M&A plan. These activities include signing the M&A agreement, the enterprise over registration, the capital transfer registration and issuing the M&A proclamation.

5.2 The Integration after the M&A

Integration is a critical factor for the success of the M&A. In the cases of M&A failures, a large proportion resulted from the failure in the integrating process. Porter's research shows that most enterprises cannot successfully integrate after the M&A. Haspeslagh and Jemison (1991) believed that the integration after the M&A is the source of value creation and important guarantee for a successful M&A. Corneal Consultants Corporation analyzed the reasons of the M&A failures, and believed that the integration after the M&A is where the greatest risk lied in the whole M&A process. Integration is a process in which the enterprise realizes value adding and the strengthening of its competitiveness. The M&A of the enterprise in the light of value chain optimization is to enhance the enterprise's competitiveness by optimizing its value chain in the M&A process. The integration of the value chains of the M&A includes the integration of the internal value chains, the integration of the horizontal value chains and the integration of the vertical value chains.

a. the integration of the internal value chains

The internal value chain integration can be preceded in three steps. First, the acquirer should develop the integration of the development strategies. Second, the acquirer should do the reconstruction and reengineering of the value chain. Third, the acquirer should do the comprehensive integration including culture, organization, system, human resources, etc.

b. The integration of the horizontal value chains

As the enterprise's competitiveness changes after the M&A, so does the relation between the enterprise and its competitors. The enterprise now have to recognize new competitors and make value chain analysis on them, know their advantages and disadvantages in the completion, grasp the new competitive situation, formulate new competitive strategies against its competitors and make corresponding arrangements of production, sales and raw material supply.

c. the integration of the vertical value chains

The enterprise after the M&A should make reanalysis of the vertical value chains to discern changes in the industrial value chain, redefine the role and position of the enterprise in the industrial value chain and coordinate its relation with its upstream and downstream enterprises and reconstruct a strategically cooperative relation with them, so as to generate effect that are favorable to the enterprise value chains and the industrial value chain.

6 The After-Event Activities of the M&A

The integration after the M&A is a long-term process, and which has profound influence on the enterprise. Therefore, the enterprise should place equal emphasis on the activities after the M&A in the light of the enterprise value chain optimization.

The after-event activities include performance appraisal and feedback and the sustained optimization of the value chains after the M&A.

6.1 The Value Chain Appraisal

The major current M&A appraisals are financial index appraisal and the stock market index appraisal, which, with their superficiality and one-sidedness, cannot reflect whether the core competitiveness has enhanced. The M&A appraisals in the light of the value chain optimization begin with the value chain appraisals, which include the internal value chain appraisal and the external value chain appraisal.

The internal value chain appraisal mainly analyzes whether the enterprise value chain has been optimized after the M&A, has met the demand of the strategy of the M&A and is in accordance with enterprise's development strategy, and thus bring more value-added to the enterprise.

The external value chain appraisal is to analyze the change of the enterprise's position in the entire industry, make comparison between the changes of the enterprise's competitiveness and that of its competitors' and evaluate whether the M&A has enhanced the competitiveness of the enterprise by analyzing the changes of the horizontal value chain and the vertical value chain after the M&A.

6.2 The Adjustment of the Enterprise Strategy

Through the value chain appraisal, the enterprise can have a better understanding of the effect of the merger an acquisition on it. Changes in the value chain can result in changes of the enterprise competitiveness, and thus affect the enterprise strategy. Therefore, the enterprise should make strategic adjustment in time to adapt to the changes in the enterprise's internal and external value chains.

Research on Fire Distance Education and Training System Based on Multi-Agent

Wang Wei

Department of fire fighting engineering, The Chinese People's Armed Police Forces Academy, Langfang, Hebei Province, China
wjww119@163.com

Abstract. Along with the continuous development of artificial intelligence, multi-Agent technology had been applied to the modern distance education unceasingly. This paper analyzed the function of the multi-Agent technology in the fire distance education system, and proposed one kind of architecture, which was based on the multi- Agent technology, and discussed the function of main Agents as the important aspects. Compared with the traditional distance education system, this system could reflect the distance education teaching's cross-time and cross- space personality characteristics more fully, really achieved teaching students individually in accordance with their aptitudes and improving teaching quality.

Keywords: Multi-Agent, fire, distance education.

1 Introduction

In 2008 the main points of the Ministry of Education requested that developing distance education and continuing education positively, based on experiment site, and making comprehensive deployment and every effort to promote the construction work of learning society that was learning for all and lifelong learning [1]. As the only institution of higher institute that serving the Ministry of Public Security directly, The Chinese People's Armed Police Forces Academy responded to the call of nation positively, explored new ideas, and took the best, it founded the distance education system that was suitable for public security, relying on the internal office network of the public security, which had contributed to the information technology construction of present public security [2]. And the application of fire distance education break the limit of the space and time, and made fire forces at the grassroots level learners follow their own learning program online, making full use of the internal network resource of public security. However, many of the current education and training systems mainly presented the teaching materials, lacking the necessary interactive tools. Because of using the same teaching strategies for the students who had the different cognitive levels and for the different teaching content, it was difficult to achieve individualized teaching. The same time with the usage of hypertext ways to organize teaching materials, the students would prone Trek phenomenon in the learning process and easy to deviate from the learning goals when they grasped the study initiative [3]. Therefore, the

D. Zeng (Ed.): Advances in Computer Science and Engineering, AISC 141, pp. 193–198.
springer link.com

design and development were based on Multi-Agent intelligent multimedia distance learning fire safety system, according to the teaching strategies used in the distance learning course, which was determined by the fire brigade cadres and soldiers that had different basis of the learning and different knowledge types, monitoring the behavior of dynamic adjustment of teaching strategies through the interaction of the Human-computer in the specific learning process by using intelligent Agent technology, which could achieve adaptive learning and intelligent teaching.

2 Agent Technology

2.1 The Synopsis of the Agent Technology

Agent technology is a very practical computer software technology which emerges with the development of Internet technology, and it used to describe hardware that have adaptive and autonomous capabilities, software, any other natural objects and man-made. There is no more precise definition at present. Smith and others defined it as " a software entity that was used to complete a specific task, but moved continually "; Selker thought that Agent was "a computer program through acting you to complete some task to simulate the human relations"[4]; Janca defined an Agent as " a software entity that could be entrusted to a task ", summed up, a relatively loose understanding of Agent was: a self-contained program that could control their own decision and behavior according to its perception of a certain environment [5].

2.2 The Function of the Agent in the Fire Distance Education System

There was no teaching strategy could solve all the problems in the process of teaching and learning, and excellent teaching system must choose teaching strategies intelligently and adjust the micro-teaching strategies dynamically in the process of teaching and learning according to the teaching objectives, teaching content and the specific details of the students [3]. At present, the application of Agent to the internet was increasing rapidly, that was because the Intelligent Multi-Agent distance learning system showed great potential for improving the quality of the client, improving the usability, improving the discovery and filtering of the information and the automatic implementation. The function of the Agent in the fire distance education system was as follows [6]:

(1) Information retrieval. The learners often did not know where could find the needed information when they were facing the rich and varied and array of information resources. Agent could solve the problem through finding automatically and filtering information. Comparing with traditional manual searching, its speed was quick and timeliness was good.

Regardless of the soldier student or the cadre students were no longer limited by the teacher professional field and Knowledge level for the needs of learning business fast by using of public security net's precious resources fully on the basis of fire distance education.

(2) Information organization. Along with the increase of information, sorting and organizing information was onerous and time-consuming, and by using of Agent to

complete the model analysis, filtering, classification, associating for a large amount of data and so on was a good way. It could not only bring people from the tedious Information classification, induction, and organizations working and improve efficiency, but also could select the appropriate content for students automatically according to the learner's personality and learning, which could provide support for the learner's individual guidance and the instruction in the process of learning.

(3) Environment response. Aware of surrounding environment's changes, and take the appropriate action immediately, such as the emergence of new information that related to the teaching resources within the net of The Ministry of Public Security. Such as the changes of the teaching content and the analysis of the Fire case, Agent could replace teacher or Students to complete some operations, such as update or download new teaching resources.

(4) Social exchange. Each Agent was not isolated; it had some kinds of connections with its periphery Agent inevitably. By using of the Agent's automatic consultative function to resolve issues for mutual exchange of learning that a single Agent could not do.

3 The System Design Based on the Multi- agent

By using the thought of the multi-Agent to analyze the overall demand and design of the distance fire education system could reflect the intelligence and initiative of the teaching fully. The system was using three-tier architectures of the Browser / Agent / DBMS Server model, the system was divided into the client's presentation layer, the Middle agent layer, and the data layer. This was a "thin client" model, it took few resources of the client and the requirement of the client was low that could help with accelerating the access speed, and without having to install special client program [8]. It was very good for different regions, different levels of the fire department in China. Meanwhile, in this structure, the client did not interact with the database server directly and it improved the security of the database side. The system architecture was shown in Figure 1.

The first layer was the client's presentation layer, to achieve the interaction of the system and the client through browser, and to send various requests to the middle agent layer in accordance with the need. This layer contained the system's display logic, accessed to the client's information and behavior, the receiving of the task, processing the results of the feedback and so on through this layer to achieve the information Input, output. And the Students' End mainly distributed in fire departments of the country, Teachers End mainly distributed in The Chinese People's Armed Police Forces Academy of The Ministry of Public Security.

The second middle agent layer second layer of was the multi-Agent layer, and it was the most important part of the fire distance education system. It contained three classes Agents: client Agent (Student Agent, teacher and administrator Agent), Management Agent, and Task Agent (Learning Agent, Answer Agent, information query Agent, Information Retrieval Agent, etc). And the main function of Agent was as follows:

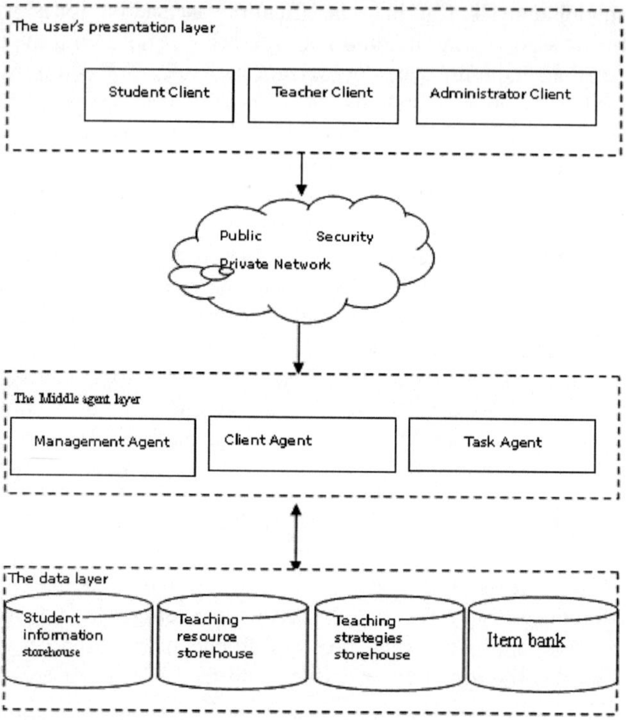

Fig. 1. Schematic diagram of system structure in the fire distance education system

3.1 Teacher Agent

It was the agent of the teacher, including lesson preparation Agent, teaching Agent and Q & A Agent. lesson preparation Agent was to help the teacher to complete daily lesson preparation, it might provide a platform for the teacher to be able to turn conveniently the course content to the Web page and put the courseware into the teaching website, and also could propose lesson planning suggestions for the teacher using the Agent characteristic intelligence. At the same time, it could also be possible to exchange information with students, understand the teaching form, the teaching media and other information that the students were more interested in, and adjust a way for preparing a lesson. Teacher Agent was mainly dealing with real-time teaching, besides could simulate face-to-face and real-time teaching; it might also exchange information with the student Agent mainly. Q & A Agent chose the teaching object target-oriented and dynamic through the understanding of the learner's study condition and hard problem.

3.2 Student Agent

It was the agent of the student, including learning Agent and collaborative Agent. Studies Agent was used to record learning process of the learners that logged on to the teaching server, carried on the inference dynamically according to the record of studies and organized the page of the hypertext structured teaching from the teaching database dynamically; when collaborative Agent exchanged the information, mainly enabled other Agents to have the destination to choose the exchange object, and managed the exchange object that the learner chose intelligently [9].

3.3 Management Agent

The user logged on to the system server, the system produced a Client Agent automatically, the management Agent accepted a unified for the management Agent including Client management, task management, information management, collaborative process management and so on through carrying on the registration. After carrying on the registration, if the Client Agent had any demand could request to Management Agent; the function of the behavioral analysis in Management Agent was used for determining the type of the task, then gave a particular task to the Duty Agent, such as knowledge learning, answering questions, information search, information search, etc . After completing the task, the Duty Agent returned to Management Agent and reported the implementation to the Management Agent at the same time; after making sure the thing, the Management Agent returned the determination to the Client Agent, simultaneously abolishing the Duty Agent. The Client exited and the Management Agent was to abolish the Client Agent.

The third layer was the data layer, and it was used to store various types of data resources. Including the teaching information storehouse, the teaching resources storehouse, the teaching strategy storehouse, and Item bank and so on. Teaching information storehouse was responsible for recording the personal information and learning of the students, such as study time, number of times, content, current level and learning capability and so on; The teaching resources storehouse was made of the concrete domain curriculum knowledge, which was taught. Its leading role was to organize the teaching material, deposit and manage all domains' knowledge that was taught; the basic constitution unit of the Item bank was the test question; the teaching strategy storehouse was mainly to record each kind of rule for software organization teaching, it provided the choice of learning environment, the amount of this learning task, the learning difficulty and whether to enter the review, the rules of the testing aspects, which provided basis for Teacher Agent to make decision.

4 Conclusions

The present information technology had provided the solid support platform for the distance teaching, but the artificial intelligence and the multi-Agent technology's introduction would further promote the humanization of modern distance teaching. In view of the learning demand and the education need for active duty fire officers and soldiers of The Ministry of Public Security, depending on third-level network construction of the public security firefighting troops and using advanced multi-Agent technology for achieving distance learning of fire, might carry on the free study for the students at any time, any place as well as the different study degrees, thus mobilized the students' self-interest in learning and improved teaching effectiveness, which played a positive promotion role in enhancing the information construction of the entire Fire Force and the quality of personnel. This article had a first attempt to the platform construction of the high efficiency and intelligent learning's research, proposed that one kind of model based on the multi-Agent fire distance learning, provided the reference for the next design and implementation of achieving the fire distance learning system that based on the multi-Agent technology.

References

1. The Main Points of 2008, Basic Education Resource Centre of Ministry of Education, http://www.ncet.edu.cn/html/news/work/20081024308.html
2. Zhang, H.: The Discussion of the Modern Distance-education of Armed Police Forces Academy. Journal of Chinese People's Armed Police Force Academy 25(1), 69–71 (2009)
3. Wang, R.-C., Xu, X.-L., Huang, H.-P.: pp. 287–301. Beijing University of Posts and Telecommunications Press (2006)
4. Wooldridge, M.: An Introduction to Multi-Agent Systems. Electronic Industry Press (2003) (Translation by SHI Chun-yi)
5. Zhou, Y.-Z., Chen, Y.-P.: Implementation of Agent in Distance Education. Journal of Xia'n University of Arts & Science (Nat. Sci. Ed.) 12(1), 24–26 (2009)
6. Yang, Y.-Q., Fu, Y.-Q., Liu, Y.-Y.: Individualized Distance Education System Based on Multi-Agent. Computer Science 34(9), 290–291 (2007)
7. Yu, L.-S., Peng, D.-W.: Research on An Intelligent Learning Platform Based on Agent. Computer Applications and Software 125(2), 99–102 (2008)
8. Ni, H.-M.: Collaboration and Intelligent Answering System Based on Multi-Agent. Journal of Yangtze University (Nat. Sci. Edit.) 6(1), 208–209 (2009)
9. Guo, P.: Research on the Multi- agent Technology in Distance Education. Journal of Jiangxi Science & Technology Normal University (6), 113–115 (2006)

NoC Router Power Macro-modeling at High Level

Fang Zhou[1,*], Ning Wu[1], Ying Zhang[1], and Xin Ge[2]

[1] Department of Electronic Engineering, Nanjing University of Aeronautics and Astronautics, Nanjing, China
[2] China Telecom, Nanjing, China
zfnuaa@nuaa.edu.cn

Abstract. This paper proposes a methodology for power extraction of the Network-on-Chip router. The router power model is devoted to overcoming the shortcoming of existing architecture-level power simulators with higher accuracy and providing a fast and precise power profile to enable power optimization such as power-aware compiler, core mapping, and scheduling techniques. Each component of the router is modeled by different methods according to different characteristics. Multiple linear regression is used to model the relationship between events occurring in the NoC and energy consumption. Using the EDA platform of Synopsys and SMIC 180nm standard cell library, we compare our power model to the gate-level power analysis by PrimeTime PX. Experimental results show that the average estimation error of the proposed power model is 5.0% against the gate-level simulation with 600 times speed up.

Keywords: NoC, router, power, macro-modeling.

1 Introduction

To solve the shortcoming of the traditional bus in communication efficiency, interconnect scaling and power consumption, etc., a new communication structure NoC (Network on Chip) has been proposed[1,2]. It aims to provide high performance communication for microprocessors, DSP, memory and other IP cores based on parallel computing and computer network design.

The high-performance communications in NoC increase at the cost of complexity of the structure, and an interconnection network dissipates a significant fraction of the total system power budget. Therefore, the interconnection power consumption has become an important constraint of designing the NoC[3]. It is desirable to get detailed trade offs in power and performance early in the design flow, preferably at the system level. Although the commercial EDA tools are accurate in gate level for power analysis, but there is limited ability in the system level to obtain more accurate power consumption data, and the power models are extremely time consuming and intractable. So establishing system level power model without going into the hardware details of components is needed. The router in NoC is the most important communication component, whose power model is the focus of the relevant research.

* Corresponding author.

D. Zeng (Ed.): Advances in Computer Science and Engineering, AISC 141, pp. 199–206.
springer link.com

In this paper, we propose a macro-model for power extraction of the Network-on-Chip router. Our router is decomposed into several components, and a model is built for each component according to their different characteristics in order to improve the power model's accuracy. Multiple linear regression is used to obtain the power models for the input buffer, the routing algorithm, the crossbar switch and the arbiter due to their linear characteristics. Using the EDA platform of Synopsys and SMIC 180nm standard cell library, we compare our power model against the gate-level power analysis by PrimeTime PX to evaluate its accuracy.

The rest of the paper is organized as follows: Section 2 surveys the related work. Section 3 introduces the architecture of our router. The model is described in Section 4 and Section 5. Section 6 presents the experimental results. Finally, Section 7 concludes the paper.

2 Related Works

Existing system level power models of the router can be divided into theoretical models and statistical models.

The basic idea of theoretical models is to estimate the effective capacitance of each circuit and the amount of bit-flips in the data stream. In [4], Wang et al. created a network simulator named Orion to estimate dynamic power of router components. This simulator was augmented in [5] to support leakage power. Ye et al.[6] proposed an energy estimation flow to derive bit energy models that are used to evaluate different switch fabrics in network routers. However, the models are tightly coupled with circuit implementations. As such, these models cannot be migrated to different technology libraries without a large amount of re-modeling. Moreover, low level of abstraction (i.e. gate and device level) and the extremely slow simulation make it definitely unsuitable to face with system level SW/HW exploration task.

Statistical models (i.e. power macro models) are usually based on multiple linear regression analysis to generate power estimation. Its basic idea is to establish the functional relationship between the statistical properties of the input data and energy consumed by routers through the implemented circuit. Power macro models usually are expressed by the look-up table or empirical polynomial. The authors in [7] automate the extraction of a power model for the STBus, a high performance industrial communication architecture supporting shared buses as well as crossbars, based on regression technique and presented an effective technology to minimize the Design of Experiments (DoE) in [8]. However, a packet switched router contains additional components that are not present in the STBus. Wolkotte et al.[9] created energy models for routers based on packet switching and circuit switching through calculating the average energy per bit traversing on single router by power analysis tool. Penolazzi et al.[10] presented an empirical formulation of the Nostrum NoC, which reflects the different input vectors on the power consumption. Meloni et al.[11] presented power model for xPipes switch[12] with average error of 5%. Chan et al.[13] built a cycle accurate power model by analyzing key signals which influenced power of each functional component. In [14] and [15], there are methods to obtain a good fit between the model parameters and energy consumption based on BP neural network.

3 Architecture of Noc Router

The router consists of five input/output ports, the routing algorithm, the arbiter and the crossbar switch. The input/output ports can be divided into input link controller, the virtual channel buffer and the transmission controller. The input link controller receives the request of communication from the last router, under the control of which the data is written to the virtual channel buffer. The transmission controller transmits the request to the arbiter and controls flits to be read from the virtual channel buffer. The routing algorithm judges the packet forwarding direction based on the routing algorithm. X-Y routing algorithm is used in the paper. The arbiter is responsible for steering flits towards appropriate output port. We use polling to complete the request of processing router signals in this paper.

4 Methodology

We create a power model for NoC router based on the theory of macro model, rather than gate-level circuit. Power macro model uses flits passing through a router to estimate power consumption, which consists of input data statistical property that has a strong correlation to power consumption. Firstly, power consumptions of different input vectors are get through the gate-level power analysis tool, and then power consumptions are used to obtain a good fit between input data statistical property and average power consumption.

The methodology used to create a power macro model for our work is illustrated in Fig. 1. Firstly, we use Synopsys Design Compiler to synthesize the router RTL description to the gate level net-list with technology library. In this step, physical information such as RC parasitic value files (SPEF), standard delay format (SDF) and design constraints file (SDC), are also generated for gate level power analysis and gate level simulation. Secondly, traffic patterns are generated to exercise the router under different conditions, which is named test bench. Configuration of different traffic patterns in test bench can improve the accuracy of power model. For the router, we can configure these traffic patterns by varying the traffic load and the amount of bit-flips in the data stream. Adjusting for the traffic load is in order to gain power consumptions under different congestion conditions. Meanwhile, the influence of input data stream on power can be accessed by changing the amount of bit-flips. Test bench and gate level net-list file are used for gate level simulation with the Synopsys VCS simulator. During the gate level Simulation, a Value Change Dump (VCD) file which has logic switching information for dynamic power measurement is generated. Thirdly, using the VCD and RC parasitic values, gate level power analysis is done with the Synopsys PrimeTime PX tool to create cycle accurate power waveform file stored as .out file. Finally, the energy consumption and signal values are used as observations in the regression analysis to obtain the power macro model.

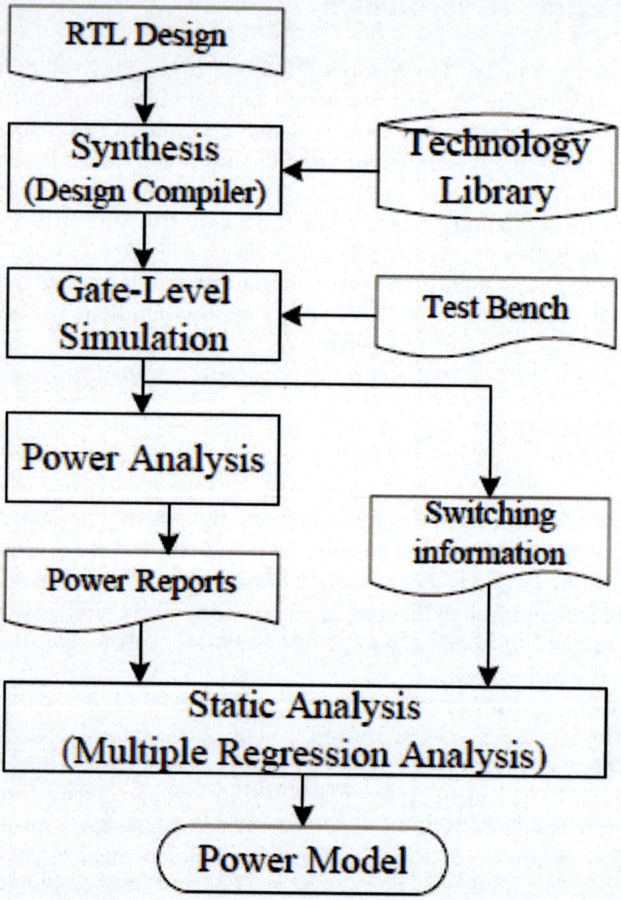

Fig. 1. Power model creation methodology

5 Power Modeling of Noc Router

The cycle energy consumption of our router can be expressed as:

$$P_{total} = \sum_{j=1}^{n} P_{ip,j} + P_{route} + P_{crossbar} + P_{arb} + P_{clk} + P_{leak} \quad (1)$$

where $P_{ip,j}$, P_{route}, $P_{crossbar}$ and P_{arb} are the power values derived from the component macro-models for the input/output ports, the routing algorithm, the crossbar switch and the arbiter respectively. Pclk is the clock power and Pleak is leakage power. n is the number of input/output ports, and n is equal to 5 in this paper.

5.1 Input/Output Ports

Dynamic power of input/output ports is mainly attributed to the reading and writing of the buffer and clock power. Hence, the power model can be formulated as:

$$P_{ip} = P_{write} + P_{read} \quad (2)$$

where P_{write} and P_{read} respectively are the power dissipated reading from and writing to the buffer; During writing to the buffer, both the frequency of write operation and the Hamming distance of data into the buffer have influence on the power consumption. Hence, the writing power macro model can be further decomposed as:

$$P_{write} = \psi_{write0} + r_w \psi_{w\_ctrl} + \alpha_F \psi_{store} \quad (3)$$

where r_w is the rate of write operation, for example, r_w=0.5 indicates writing to the buffer occurs once every two cycles. α_F is the quotient of hamming distance divided by bit-width. Ψ_{write0} is constant.

Similarly, the reading power macro model can be expressed as follows:

$$P_{read} = \psi_{read0} + r_r \psi_{r\_ctrl} + \alpha_F \psi_{retreive} \quad (4)$$

been running stably for some time, r_r can be considered to be equivalent to r_w. The following r will represent r_r and r_w.

The P_{clk} and P_{leak} are usually constant at a fixed clock frequency, so they can be merged into constant Ψ_{ip0} with Ψ_{write0} and Ψ_{read0}. Then the power of input/output ports can be calculated as:

$$P_{ip} = \psi_{ip0} + \psi_{ip1} r + \psi_{ip2} \alpha_F \quad (5)$$

The variables r and α_F can be obtained by simulation, regression coefficients Ψ_{ip0}, Ψ_{ip1} and Ψ_{ip2} are calculated by multiple linear regression.

5.2 Routing Algorithm

Route calculation takes place in routing algorithm receives the address information from input/output ports, which result in the power consumption. The power is related to the number of flit in packet header. Based on these observations, routing algorithm power can be given as:

$$P_{route} = \psi_{route0} + n_h \psi_{compute} \quad (6)$$

where n_h is the number of flit in packet header per cycle, $\Psi_{compute}$ is the regression coefficients of variable n_h, Ψ_{route0} is constant.

5.3 Crossbar Switch

Router used in the paper is fully connected crossbar network, and crossbar switch power is related to the hamming distance of the input data. Thus, given the bit-flips of input data $_c\alpha$, the following expression for the power of crossbar switch is resulted in:

$$P_{crossbar} = \psi_{crossbar0} + \alpha_c \psi_{dist} \tag{7}$$

5.4 Arbiter

The power of arbiter is caused by the arbitration operation, which is requested by the transmission of data packets. So its calculation is associated with the number of arbitration operation, and the power of arbiter can be calculated as:

$$P_{arb} = \psi_{arb0} + n_c \psi_{arb\_compute} \tag{8}$$

where n_C is the number of arbitration operation per cycle, $\Psi_{arb\_compute}$ is the regression coefficients of variable n_C, Ψ_{arb0} is constant.

6 Experimental Results

All router designs were synthesized using Synopsys Design Compiler and SMIC 0.18μm standard cell library were used for synthesis. Synopsys PrimeTime PX was used to obtain the power estimates and power waveform at an operating frequency of 200 MHz.

To estimate power more accurately, multiple random input datas were imported into router. We obtained the respective average power consumption of components on gate-level power simulations with a separate set of randomly generated traffic patterns (100 traces in total). 50 traces were used to establish the power models through multiple linear regression, and then we verified the accuracy of power models by all of the 100 traces.

Five separate experiments were performed to ensure the validity of our models under different traffic loads. We randomly characterized the injection rate and bit-flip of input data to generate 100 sets of traffic patterns.

Fig. 2 shows the expected power consumption and the actual one with our power model in five experiments as data flip rate is 50% and the flits injection rate is 10%. And Fig. 3 shows the error distribution of power model by statistical data on experiment.

From the figure, it is clear that the maximum power estimation error is 15.7% and the average error is 5.0%, which is in acceptable range. Fig. 3 also indicates that 70% errors are less than 6%, but errors are occasionally more than 15%. Experiments were performed in Inter Xeon 2.6GHz CPU and Red Hat AS4 operating system, and our power model is at about 600×faster simulation speed over gate-level power analysis by PrimeTime PX.

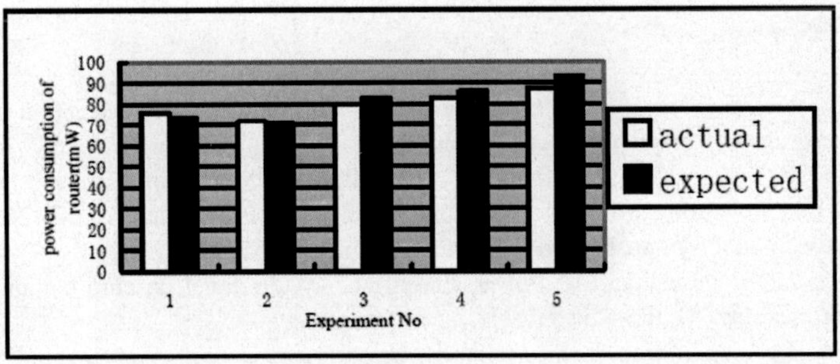

Fig. 2. Comparison between the expected power and the actual power with the power model

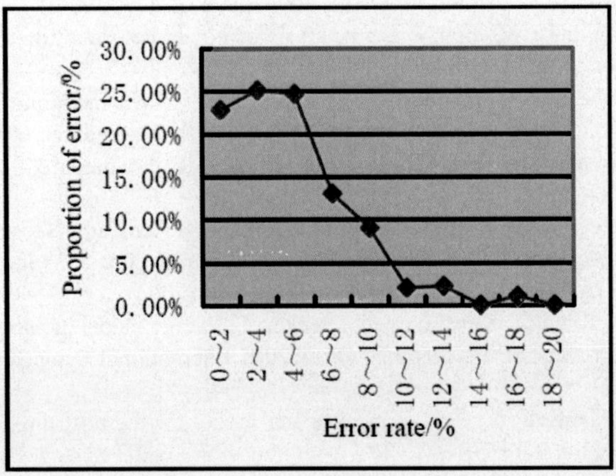

Fig. 3. The error distribution of power model

7 Conclusions

In this paper, we propose a heterogeneous macro-model for power model extraction of the NoC router. The model is built for each component according to their different characteristics: Linear regression is used to obtain the power models for the input buffer, the routing algorithm and the crossbar switch due to their linear characteristics. The power model for the arbiter is established based on BP neural network because of its nonlinear characteristics. Experiment results show that the model can accurately and efficiently evaluate the average power consumption of NoC router, and provide reliable and fast power simulation to designers in system level.

Acknowledgment. This work is supported by the National Natural Science Foundation of China (No. 61076019) and NUAA Research Funding (No. NS2010115).

References

1. Dally, W.J., Towles, B.: Router packets, not wires: on-chip interconnection networks. In: Proceedings of Design Automation Conference, Las Vegas, Nevada, pp. 684–689 (2001)
2. Kumar, S., Janstech, A., Soininen, J.-P., Forsell, M., Millberg, M., Oberg, J., et al.: A network on chip architecture and design methodology. In: Proceedings of the IEEE Computer Society Annual Symposium on VLSI, Pittsburgh, pp. 105–112 (2002)
3. Lahiri, K., Raghunathan, A.: Power analysis of system-level on-chip communication architectures. In: CODES+ISSS (2004)
4. Wang, H., Zhu, X., Peh, L.-S., Malik, S.: Orion: a power-performance simulator for interconnection networks. MICRO, 294–305 (2002)
5. Kahng, A.B., Li, B., Peh, L.-S., Samadi, K.: ORION2:A Fast and Accurate NoC Power and Area Model for Early-Stage Design Space Exploration. In: Design, Automation & Test in Europe Conference & Exhibition, DATE 2009, pp. 423–428 (2009)
6. Ye, T.T., Benini, L., Micheli, G.D.: Analysis of power consumption on switch fabrics in network routers. In: DAC, pp. 524–529 (2002)
7. Bona, A., Zaccaria, V., Zafalon, R.: System level power modeling and simulation of high-end industrial network-on-chip. In: DATE 2004: Proceedings of the Conference on Design Automation and Test in Europe, pp. 303–318. IEEE Computer Society, Washington, DC, USA (2004)
8. Bona, A., Zaccaria, V., Zafalon, R.: Low Effort, High Accuracy Network-on-Chip Power Macro Modeling. In: Macii, E., Paliouras, V., Koufopavlou, O. (eds.) PATMOS 2004. LNCS, vol. 3254, pp. 541–552. Springer, Heidelberg (2004)
9. Wolkottex, P., Smit, G., Kavaldjiev, N., Becker, J.: Energy model of networks-on-chip and a bus, system-on-chip. In: Proceedings of the 2005 International Symposium, November 17, pp. 82–85 (2005)
10. Penolazzi, S., Jantsch, A.: A high level power model for the nostrum noc. In: DSD 2006: Proceedings of the 9th EUROMICRO Conference on Digital System Design, pp. 673–676 (2006)
11. Meloni, P., Loi, I., Angiolini, F., Carta, S., Barbaro, M., Raffo, L., et al.: Area and power modeling for networks-on-chip with layout awareness. In: VLSI Design (2007)
12. Bertozzi, D., Benini, L.: Xpipes: a network-on-chip architecture for gigascale systems-on-chip. IEEE Circuits and Systems Magazine 4(2), 18–31 (2004)
13. Chan, J., Parameswaran, S.: Nocee: energy macro-model extraction methodology for network on chip routers. In: IEEE/ACM International Conference on Computer-Aided Design, ICCAD 2005, November 6-10, pp. 254–259 (2005)
14. Hsieh, W.T., Shiue, C.C., Liu, C.N.: An Efficient Power Modeling Approach of Sequential Circuits Using Recurrent Neural Networks. IEE Proceedings-Computers and Digital Techniques 153(2), 78–86 (2006)
15. Qiang, W., Cao, Y., Yan, Y., Gao, X.: Power Macromodel of CMOS Circuits by BP Neural Network. Journal of Wuhan University (Natural Science Edition) 52(3), 353–356 (2006)

Research on Customer Relationship Management Based on Web Mining

Wei Jianping

School of Economics & Management, Weifang University, Weifang, China
wfxywjp@163.com

Abstract. Identify, obtain and maintain customer is the main content of the customer relationship management. This paper adopts web mining technology, focused on the electronic commerce application research, and acquired of potential and loyal customer information from the data of users visiting the company's web site. With comprehensive understanding customer visit the company's web site of time, preferences and other relevant information, effectively to understand the demand of customer. To adjusted company website structure, market strategy, which makes e-commerce activities more focused on the customer.

Keywords: Web Mining, E-commerce Site, Customer Relationship Management.

1 Introduction

The traditional customer relationship management primarily through the establishment of call centers, sales records to establish a database to obtain information about the information of customs, then analysis and mining market trends and other useful knowledge, to develop strategic policy for the enterprise, sales programs to provide decision support. But with the rapid development of e-commerce, business sales site allows businesses to establish potential users all over the internet, then the business and customer information shows decentralized. The traditional customer relationship management can not meet the new demands of e-commerce businesses, through the Internet, by analysis the interest and value discovered of the majority of potential customs to have targeted marketing. This article will introduce e-commerce web mining to customer relationship management, and propose that the adaptation electronic commerce enterprise's customer relations management system, and the key part - the realization of site visitors' classification techniques are discussed in detail.

2 Customer Relationship Management E-Commerce

Customer Relationship Management (CRM) is a continuously strengthen communication with customers, the understanding of customer needs and continuously improve the products and services to satisfy customers and increase demand for continuous process [1]. The use of information technology and Internet technology to achieve operation of the integrated marketing to customers, based on customers as the core,

D. Zeng (Ed.): Advances in Computer Science and Engineering, AISC 141, pp. 207–214.
springer link.com

enterprise marketing implementation. Customer relationship management focus on the communication with customers, and the business is customer-centric, rather than the traditional product-or market center. To facilitate communication with customers, CRM can provide customers with a variety of communication channels. CRM includes three phases: customer acquisition, customer development and customer retention.

E-commerce environment, the most significant change is the business and customer communication changes. In addition to traditional call centers, media advertising, questionnaires, etc., customers understand the business and products mainly from the corporate website. These customers may be potential valuable customers, as e-commerce customer relationship management (E-CRM), should be able to identify and classify these customers, thus achieving customer development and retention. In general, e-commerce customer relationship management should be under the meet the following requirements:

- Improve the system integration. From the depth of integration that is divided into data integration, application integration, business process integration. The ultimate goal is to channel a variety of information gathering, analysis of the conclusions of other organic integration, optimization of customer relationship management effectiveness.
- Lay the foundation for personalized marketing. Internet technology makes the traditional relationship between business and customer-to-many relationship into a one on one, personalized marketing possible. E-CRM to customers to pattern recognition and classification, as lay the foundation for personalized marketing.

3 Web Mining

3.1 The Definition of Web Mining

Web mining is a comprehensive technical, involving web, data mining, computational linguistics, information science and other fields [2]. Web mining takes interesting, potentially useful patterns and hidden information from the web documents and web activities. Web mining can play a role in many aspects, such as mining the structure of the search engine to determine the authoritative pages, web document classification, web log mining, intelligent query, establishing data warehouse, etc. Web mining can be defined as follows.

Define 1: Web mining refers to found the hidden mode p from a large number of Web document structure and collection C [3]. If C as input, p as output, so Web mining process is from the input to the output of a mapping.

$$\xi : C \to p$$

Web mining comes from the data mining, therefore its definition is similar with data mining, that is carrying on the analysis to the massive data in the foundation, makes the induction inference, forecast the customers' behavior, helps enterprise to make policy-makers to adjust the market strategy, reduces the risk, makes correct policy-making the process[4]. But for web mining face great challenges, such as web information of excessive, web page complexity than any traditional text files, etc.

3.2 Web Mining Process

The typical web mining process is shown as figure 1.

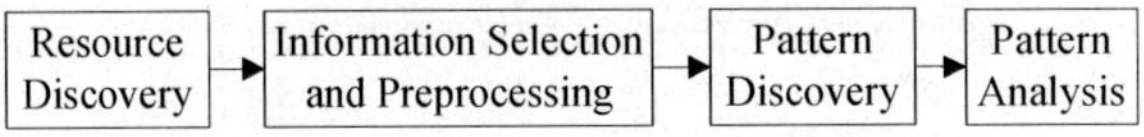

Fig. 1. Web data mining workflow

The main task of resource discovery from target web document obtained data (the source data). Information selection and preprocessing is to make according to excavate target from the source data removing the useless information and information purifying. Pattern discovery is to use various mining method, automatic mode discovered. Pattern analysis is to verify and explain a step up found model, to lay the foundation for the application of data mining results.

4 Web Mining Tool Architecture Based on E-Commerce Site

4.1 Web Mining Tool Module Design Based on E-Commerce Site

Web mining tool based on ecommerce site is auxiliary various types of e-commerce sites to understand customer needs, which operates in website user database or data warehouse [5]. Mainly includes the following function module:

- Data extraction and purification module: For ambiguity analysis according to Web server logs database or in data warehouse extracting relevant data, eliminating inconsistencies, provides the high quality for the next mining data source;
- Web mining module: This is a mining driver component. According to requirements to choose appropriate mining method from the Web mining algorithms library, and uses the method to perform the mining task;
- Web mining algorithms library: This is a comprehensive algorithms library of data mining analysis method. It organizes various mining algorithm by plug-in. Various methods can conveniently insert, realize scalability and selective, at the same time it also can keep into the new mining method, thus improving the mining efficiency;
- Evaluating the interface: With an intuitive way to represent data mining results, provide a and analysts interactive friendly interface;
- Mining results output module: Mining results with the most intuitionistic statements form output. Using exhumed useful information, to make corresponding work.

4.2 Web Mining Tool Diagram Based on E-Commerce

The diagram of web mining tool based on ecommerce shows as figure 2:

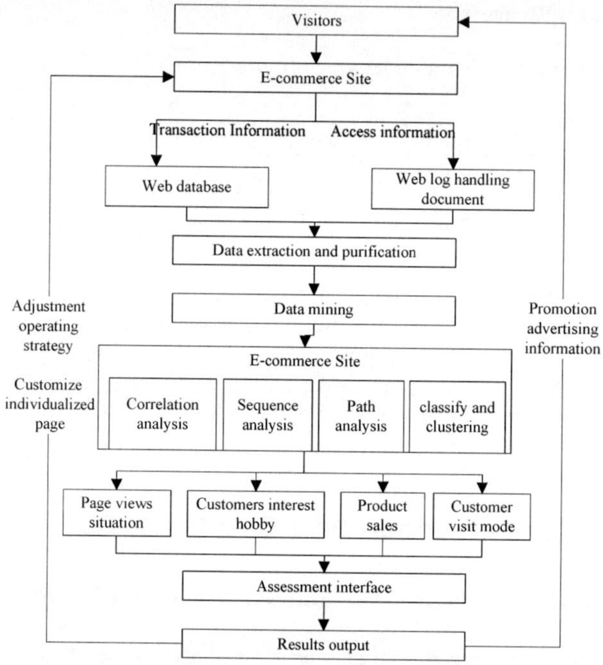

Fig. 2. Web mining tool based on ecommerce

5 Web Mining Tool Applications Based on E-Commerce Site

The web mining tool used in e-commerce sites, you can find a lot of useful information, thus can acquire new customers, retain the old customers and to customer upgrade purpose.

5.1 Acquire New Customers

There are lots of way to acquire new customers, this paper discuss it from finding potential customer group and found important pages.

1) Finding potential customer group

For an ecommerce site, if it can find the potential customer group from numerous visitors, then it can implementation of certain strategy for this type of customer, and causes those visitors to become listed customer group [6].

A Web site can be regarded as a directed graph, so we give following definition:

Definition 2: Web site refers to a web site that is a pair has the following forms of digraphs are:

$$G = (N, N_p, E, E_p) \tag{1}$$

Among them, N as node set, $N = \{Node \in N, \{(USERID, hits\}n\}, n \geq 1$, Record customer USERID and the number of visit Node, N_p as Node attribute sets, E as directed edge set, $E=\{(e \in E,\{number \ \ of \ \ path\}p)\},m \geq 1$, record directed edge and its path number, E_p as directed edge attribute set[7].

From Web site digraph G can get all URL of node set N, from the corresponding attribute set of N_p can gain access to every node USERID and the corresponding number of visits. In this with h_i j said customer j in a period of time visit the URL times. According to transaction database, if the customer not only made browsing and merchants with clinch a deal, and browse times more than specified threshold value, so, this is similar to potential customers customer group. Specific algorithm is as follows:

a) According to the formula to calculate the users visit each pages of the e-commerce sites, constitute the set $S=\{s_1,s_2,\ldots,s_n\}$, among which $j=1,2,\ldots,n$;

b) In set S take the values of $s_i > s_t$, constitute the set $C=\{c_1,c_2,\ldots,c_k\}$, among which s_t is according to the actual situation of the specified threshold;

c) Take out each USERID corresponding to c from C, and remove registered customer, the rest are potential customer group.

2) Found important pages

Through web mining tools, some important pages can be found in all of the electronic e-commerce sites pages, which user access to more often pages, so that important information and promotional information can be placed on these pages, thus to attract customers and converse potential customer groups into a registered customer group purpose[8].

Found important pages algorithm is as follows:

a) According to the formula to calculate the users visit each pages of the e-commerce sites, constitute the set $y=\{y_1,y_2,\ldots,y_n\}$, among which $j=1,2,\ldots,n$;

b) According to the formula to calculate weight value of y_j page, constitute a visit weight set $WEIGHT=\{weight1, weight2,\ldots, weightn\}$, among which $i=1,2,\ldots,n$;

c) For $i=1$ to n sort weight set, in front page namely for the main page.

5.2 Customer Maintain

1) Found the customer's interests, custom individualized visit space

a) Measure browses pages interestingness

Web site design generally follows a classification structure, namely a page of child pages are organized according to its page classes to arrange. Users accessing web sites, reflects the user's interests [9]. Usually users browse time of a web page and the ratio of the number of characters can be effectively reveal the user's interest. The access time of users stay in the page with no interest is shorter, and in the interest of the page to stay longer. We can use users browse route information and time information mining users' interest and their favorite commodities. The interest measure I of user UID on page can be definition by the following formula:

$$I(UID, Page)=\frac{The \ \ time \ \ spent \ \ by \ \ user \ \ UID}{The \ \ number \ \ of \ \ character \, sin \, page}\times Path \ \ factor \qquad (2)$$

b) Browse interested migration patterns

User access to web sites, which exists some orderly relationship that can reflect user access interest, which is user access interest and their visit sequence a strong correlation. Through interest association rules algorithm can excavate this interest patterns of migration. Specific algorithm is as follows:

Assume user access page sets $P=\{p_1, p_2, \ldots, p_n\}$, with weight said the possibility from a node to another, then triad akel $(p_i, weight, p_j)$ says the possibility from p_i to p_j, $0 \le weifht \le 1$.

- Calculated the value of weight of $(p_i, weight, p_j)$ according to the following formula:

$$weight = count(p_i - p_j) \Big/ \sum_{k=1}^{n} count(p_i - p_j) \tag{3}$$

- If $weight > S_t$, And for S_m users concern;
- Then $p_i \rightarrow p_j$ is interested migration pattern.

Weight said the possibilities from one entry (interest) to other entry (interest). Use it to combine the user currently visit page trace and users access to the current page can predict users might be visited link. Therefore, through the interested migration pattern can learn the user's interests and hobbies, according to its interests for their custom individualized visit space, timely provide users with his concern the product or service information, so as to achieve the purpose of retaining the customer.

2)Optimization link structure

For web site links structure optimization can be considered from two respects: on the one hand, according to the web log, found that the correlation of user access to page, so as to increase the close connection between web links, and user-friendly, on the other hand, according to the web log digging, found the expectations of the users position. If in the desired position access frequency higher than actual position access frequency, can consider to establish navigation links between the expected position and actual position, so as to realize the optimization of web sites.

*a)*To determine the correlation between the web pages

Set $P=\{p_1, p_2, \ldots, p_n\}$ is user access page set, S is specified threshold

Scanning web log file after sorting

While not. (eof).

{

if exist $P_k \rightarrow P_{k+1} \rightarrow \cdots \rightarrow P_m$ visit sequence;

Number+=1;

Skip,

}

If number > S, then P_k and P_m exists correlativity;

According to mining results, we can establish direct links between P_k and P_m, to provide convenient and quick service for users.

*b)*Found the location of the user's expectations
Web site links structure is designed according to certain strategy, which may exists gap between the links structures of user expectations. For example, as fig.3 shows, target page X under page A_1, but many users browse path is $F \rightarrow A_3 \rightarrow B_4$, so the user expectations page X position in A_3 or B_4, if we can understand beforehand to the expectations of the users place, and establish navigation from A_3 (or B_4) to page X, which is convenient for the user, so as to achieve the purpose of customer retention.

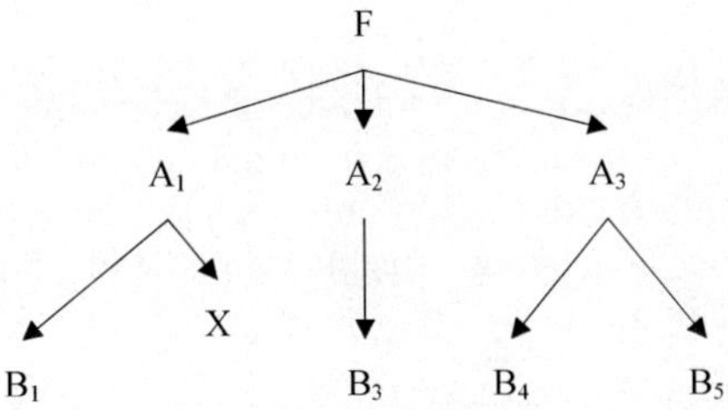

Fig. 3. Users access to WEB site map

Let E_1 as expects position, the algorithm is as follows:

- First building index of Web log file with user ID as primarily key, and time as secondary key;
- Scans web log index files, gather some pages sequence for each user ID;

i = 1;
While i < = n
{
Sort pages by support;
If support (P) > = S1;
Then P for was first expectations and occurred frequency than the system designer specified value page;
}

According to the mining results of this algorithm, we can adjust web site link structure; add navigation links between the desired position and actual position, so as to optimize web site links structure.

5.3 Upgrade Custom

1)Customer segmentation
Using this web mining tool clustering function can differentiate different user groups, according to different criteria, such as: customers' consumption psychology, consumption habits, purchase frequency, product demand and so on, which can realize the pertinent service and develop target products to customers, to enhance customer satisfaction. Through such fine classification helps enterprise provide the different characteristics of differentiation of service to customers.

2)Customer Tracking Service

Through web mining tool, it can timely understand customer requirements and satisfaction. Regarding customer change, it would make timely follow-up analysis for new consumption location, product name and performance, quantity, etc. Then timely analysis reasons why customer changes as soon as possible, and update the existing service quality, improve product performance, to prevent customer groups lost, thus help enterprises to obtain more competitive advantage.

6 Conclusions

Through web mining tool based on ecommerce site can discoverer lots of useful information of the site management, such as potential customers information, important page information, custom's interest hobby information, thus provide some references for enterprise in customer acquisition, customers maintain to promote enterprise the healthy and orderly development.

References

1. Goldman, L.: The Role of customer intelligence in successful CRM. DM Review, 12–14 (May 2004)
2. Rakesh, N., Gupta, M., Narasimhan, C.: Customer profitability in a supply chain. Journal of Marketing 65, 1–16 (2001)
3. Apte, C., Weiss, S.: Data Mining with Decision Trees and Decision Rules. Future Generation Computer Systems 13, 197–210 (1997)
4. Chen, M.S., Park, J.S., Yu, P.S.: Data mining for path traversal patterns in a web enviroment. In: Proc. of the 16th Intl. Conf. on Distributed Computing Systems, Hong Kong, pp. 385–392 (1996)
5. Gamma, E., Helm, R., Johnson, R., Vlissides, J.: Design Patterns: Elements of Reusable Object-Oriented Software, pp. 82–93. Addison-Wesley, Boston (2000)
6. Spiliopoulou, M.: The Laborious Way From Data Mining to Web Mining. International Journal of Computer System Science & Engineer Special Issue on Semantics of the Web 3, 105–113 (1999)
7. Niemi, T., Niinimaki, M., Nummenmaa, J., Thanisch, P.: Applying grid technologies to XML based OLAP cube construction. In: DMDW, pp. 204–207 (2003)
8. Winer, R.S.: A framework for customer relationship management. California Management Review 43, 89–105 (2001)
9. Lemon, K.N., White, T.B., Winter, R.S.: Dynamic customer relationship management-incorporation future consideratons into the service retention decision. Journal of Marketing 66, 1–14 (2002)

Improved Delay-Dependent Stability Criteria for Systems with Interval Delay

Dongsheng Xu[1] and Junkang Tian[2]

[1] School of Sciences, Southwest Petroleum University, Chengdu, Sichuan, 610500, China
[2] School of Mathematical Sciences, University of Electronic Science and Technology of China, Chengdu, Sichuan, 610500, China
{Dongshengxu09,btianjunkang}@163.com

Abstract. This paper is concerned with delay-dependent stability for systems with interval time varying delay. By defining a new Lyapunov functional which contains a triple-integral term, an improved criterion of asymptotic stability is derived in term of linear matrix inequalities. The criterion proves to be less conservative with fewer matrix variables than some previous ones.

Keywords: Delay-dependent Stability, Interval delay, time-varying systems, Linear matrix inequality(LMI).

1 Introduction

Time delays as a source of instability and poor performance often appear in many dynamic systems,such as, engineering system, biological system, chemical system and electrical networks. Since system stability is an essential requirement in many applications, much effort has been made to investigate stability criteria for various time delay systems during the last two decades. Recently, a less conservative stability criteria with fewer matrix variables was derived by Shao(2009). But in the Lyapunov functional of Shao(2009), the matrix Z_2 didn't change in whole delay interval, this is an important factor in leading the conservatism. By constructing a new augmented Lyapunov functional which contains a triple-integral term, an improved delay-dependent stability criterion is derived in Sun et al.(2009)by using the free weighting matrix methods. But the results may involve more computational complexity.

In this paper, the delay-dependent stability for systems with interval time-varying delay is investigated.By constructing a new Lyapunov functional which contains a triple-integral term with the idea of decomposing the delay interval into two equidistance subintervals ,on which different Lyapunov functionals are defined, then a novel criterion of asymptotic stability is derived in term of linear matrix inequalities. The criterion proves to be less conservative with fewer matrix variables than some previous ones.

2 Problem Formulation and Preliminaries

Consider the following system with time-varying delay;

D. Zeng (Ed.): Advances in Computer Science and Engineering, AISC 141, pp. 215–222.
springer link.com

$$\dot{x}(t) = Ax(t) + Bx(t - \tau(t)) \tag{1}$$

with the initial condition

$$x(t) = \Phi(t), t \in [-h_2, 0] \tag{2}$$

Where $x(t) = (x_1(t), x_2(t), \cdots, x_n(t))^T$ is the state vector, $A, B \in \Re^{n\times n}$, $\tau(t)$ are time-varying continuous functions that satisfy $h_1 \le \tau(t) \le h_2$, $0 \le \dot{\tau}(t) \le u$, in which h_1, h_2, u are constants. The initial condition, $\Phi(t)$ is a continuous vector valued initial function of $t \in [-h_2, 0]$.

Lemma 1. (Gu 2000):For any constant matrix $\Phi \in \Re^{n\times n}$, $\Phi = \Phi^T > 0$, scalar $\gamma > 0$, vector function $\dot{\omega} : [0, \gamma] \to \Re^n$ such that the integrations concerned are well defined, then

$$-\gamma \int_{-\gamma}^{0} \dot{\omega}^T (t+s) \Phi \dot{\omega}(t+s) ds$$

$$\le -(\int_{-\gamma}^{0} \dot{\omega}(t+s) ds)^T \Phi (\int_{-\gamma}^{0} \dot{\omega}(t+s) ds) \tag{3}$$

3 Main Results

For the asymptotical stability of the system described by (1) and (2), we have the following result.

Theorem 1. For given scalars $h_2 \ge h_1 > 0$ and u, the system described by (1) and (2) is asymptotically stable if there exists symmetric positive matrices $P = \left[P_{ij} \right]_{2\times 2}$, Q_i $(i = 1, 2, \cdots, 9)$ such that the following LMIs hold:

$$\begin{aligned}\Phi_1 = \Phi &- [0 \;\; I \;\; 0 \;\; -I \;\; 0 \;\; 0]^T Q_6 [0 \;\; I \;\; 0 \;\; -I \;\; 0 \;\; 0] \\ &- [I \;\; 0 \;\; 0 \;\; -I \;\; 0 \;\; 0]^T (h_{12} Q_8) [I \;\; 0 \;\; 0 \;\; -I \;\; 0 \;\; 0] < 0\end{aligned} \tag{4}$$

$$\begin{aligned}\Phi_2 = \Phi &- [0 \;\; I \;\; -I \;\; 0 \;\; 0 \;\; 0]^T Q_6 [0 \;\; I \;\; -I \;\; 0 \;\; 0 \;\; 0] \\ &- [I \;\; 0 \;\; 0 \;\; -I \;\; 0 \;\; 0]^T (h_{12} Q_8) [I \;\; 0 \;\; 0 \;\; -I \;\; 0 \;\; 0] < 0\end{aligned} \tag{5}$$

$$\begin{aligned}\Sigma_1 = \Sigma &- [0 \;\; -I \;\; 0 \;\; 0 \;\; I \;\; 0]^T Q_7 [0 \;\; -I \;\; 0 \;\; 0 \;\; I \;\; 0] \\ &- [I \;\; 0 \;\; 0 \;\; -I \;\; 0 \;\; 0]^T (h_{12} Q_8) [I \;\; 0 \;\; 0 \;\; -I \;\; 0 \;\; 0] < 0\end{aligned} \tag{6}$$

$$\begin{aligned}\Sigma_2 = \Sigma &- [0 \;\; I \;\; -I \;\; 0 \;\; 0 \;\; 0]^T Q_7 [0 \;\; I \;\; -I \;\; 0 \;\; 0 \;\; 0] \\ &- [I \;\; 0 \;\; 0 \;\; -I \;\; 0 \;\; 0]^T (h_{12} Q_8) [I \;\; 0 \;\; 0 \;\; -I \;\; 0 \;\; 0] < 0\end{aligned} \tag{7}$$

$$\Phi = \begin{bmatrix} \Phi_{11} & P_{11}B & 0 & P_{12}+Q_5 & -P_{12} & \Phi_{16} \\ * & \Phi_{22} & Q_6 & Q_6 & 0 & B^T P_{12} \\ * & * & \Phi_{33} & 0 & Q_7 & 0 \\ * & * & * & \Phi_{44} & 0 & P_{22} \\ * & * & * & * & -Q_3-Q_7 & -P_{22} \\ * & * & * & * & * & -\frac{2}{h_2^2-h_1^2}Q_9 \end{bmatrix} \tag{8}$$
$$-[A \;\; B \;\; 0 \;\; 0 \;\; 0 \;\; 0]^T U [A \;\; B \;\; 0 \;\; 0 \;\; 0 \;\; 0]$$

$$\Sigma = \begin{bmatrix} \Phi_{11} & P_{11}B & 0 & P_{12}+Q_5 & -P_{12} & \Phi_{16} \\ * & \Phi_{22} & Q_7 & 0 & Q_7 & B^T P_{12} \\ * & * & \Phi_{33} & Q_6 & 0 & 0 \\ * & * & * & \Phi_{44} & 0 & P_{22} \\ * & * & * & * & -Q_3-Q_7 & -P_{22} \\ * & * & * & * & * & -\frac{2}{h_2^2-h_1^2}Q_9 \end{bmatrix} \tag{9}$$
$$-[A \;\; B \;\; 0 \;\; 0 \;\; 0 \;\; 0]^T U [A \;\; B \;\; 0 \;\; 0 \;\; 0 \;\; 0]$$

Where

$$\Phi_{11} = P_{11}A + A^T P_{11} + Q_1 + Q_2 + Q_3 + Q_4 - Q_5 - \frac{2h_{12}}{h_2+h_1}Q_9$$

$$\Phi_{16} = A^T P_{12} + \frac{2}{h_2+h_1}Q_9$$

$$\Phi_{22} = -(1-u)Q_1 - 2Q_6 \qquad \Phi_{33} = -Q_4 - Q_6 - Q_7$$

$$\Phi_{44} = -Q_2 - Q_5 - Q_6$$

$$\Sigma_{22} = -(1-u)Q_1 - 2Q_7 \qquad h_{12} = h_2 - h_1$$

$$U = h_1^2 Q_5 + \frac{h_{12}^2}{4}(Q_6+Q_7) + \frac{1}{2}(h_2^2-h_1^2)(h_1 Q_8 + Q_9)$$

Proof: Define $\delta = \frac{h_1+h_2}{2}$ and choose a Lyapunov functional candidate as follow

$$V(x_t)=\begin{bmatrix} x(t) \\ \int_{t-h_2}^{t-h_1} x(s)ds \end{bmatrix}^T \begin{bmatrix} P_{11} & P_{12} \\ P_{12}^T & P_{22} \end{bmatrix} \begin{bmatrix} x(t) \\ \int_{t-h_2}^{t-h_1} x(s)ds \end{bmatrix}$$

$$+\int_{t-\tau(t)}^{t} x^T(s)Q_1x(s)ds+\sum_{i=1}^{2}\int_{t-h_i}^{t} x^T(s)Q_{i+1}x(s)ds$$

$$+\int_{t-\delta}^{t} x^T(s)Q_4x(s)ds+h_1\int_{-h_1}^{0}\int_{t+\theta}^{t} \dot{x}^T(s)Q_5\dot{x}(s)dsd\theta$$

$$+(\delta-h_1)\int_{-\delta}^{-h_1}\int_{t+\theta}^{t} \dot{x}^T(s)Q_6\dot{x}(s)dsd\theta$$

$$+(h_2-\delta)\int_{-h_2}^{-\delta}\int_{t+\theta}^{t} \dot{x}^T(s)Q_7\dot{x}(s)dsd\theta$$

$$+h_1\int_{-h_2}^{-h_1}\int_{\theta}^{0}\int_{t+\lambda}^{t} \dot{x}^T(s)Q_8\dot{x}(s)dsd\lambda d\theta$$

$$+\int_{-h_2}^{-h_1}\int_{\theta}^{0}\int_{t+\lambda}^{t} \dot{x}^T(s)Q_9\dot{x}(s)dsd\lambda d\theta$$

The time derivative of $V(x_t)$ along the trajectory of system (1) is given by

$$\dot{V}(t)\le 2\begin{bmatrix} x(t) \\ \int_{t-h_2}^{t-h_1} x(s)ds \end{bmatrix}^T \begin{bmatrix} P_{11} & P_{12} \\ P_{12}^T & P_{22} \end{bmatrix} \begin{bmatrix} \dot{x}(t) \\ x(t-h_1)-x(t-h_2) \end{bmatrix}$$

$$+x^T(t)Q_1x(t)-(1-u)x^T(t-\tau(t))Q_1x(t-\tau(t))$$

$$+x^T(t)Q_2x(t)-x^T(t-h_1)Q_2x(t-h_1)+x^T(t)Q_3x(t)$$

$$-x^T(t-h_2)Q_3x(t-h_2)+x^T(t)Q_4x(t)-x^T(t-\delta)Q_4x(t-\delta)$$

$$+h_1^2\dot{x}^T(t)Q_5\dot{x}(t)-h_1\int_{t-h_1}^{t} \dot{x}^T(s)Q_5\dot{x}(s)ds$$

$$+(\delta-h_1)^2\dot{x}^T(t)Q_6\dot{x}(t)-(\delta-h_1)\int_{t-\delta}^{t-h_1} \dot{x}^T(s)Q_6\dot{x}(s)ds$$

$$+(h_2-\delta)^2\dot{x}^T(t)Q_7\dot{x}(t)-(h_2-\delta)\int_{t-h_2}^{t-\delta} \dot{x}^T(s)Q_7\dot{x}(s)ds$$

$$+\frac{1}{2}h_1(h_2^2-h_1^2)\dot{x}^T(t)Q_8\dot{x}(t)-h_1\int_{-h_2}^{-h_1}\int_{t+\theta}^{t} \dot{x}^T(s)Q_8\dot{x}(s)dsd\theta$$

$$+\frac{1}{2}(h_2^2-h_1^2)\dot{x}^T(t)Q_9\dot{x}(t)-\int_{-h_2}^{-h_1}\int_{t+\theta}^{t} \dot{x}^T(s)Q_9\dot{x}(s)dsd\theta \tag{10}$$

Using Lemma 1 to obtain

$$-h_1\int_{t-h_1}^{t}\dot{x}^T(s)Q_5\dot{x}(s)ds \qquad \le -[x(t)-x(t-h_1)]^T Q_5[x(t)-x(t-h_1)] \tag{11}$$

$$-h_1\int_{-h_2}^{-h_1}\int_{t+\theta}^{t}\dot{x}^T(s)Q_8\dot{x}(s)dsd\theta$$
$$\le -h_1(h_2-h_1)\int_{t-h_1}^{t}\dot{x}^T(t)Q_8\dot{x}(s)ds$$
$$=-(h_2-h_1)[x(t)-x(t-h_1)]^T Q_8[x(t)-x(t-h_1)] \tag{12}$$

$$-\int_{-h_2}^{-h_1}\int_{t+\theta}^{t}\dot{x}^T(s)Q_9\dot{x}(s)dsd\theta \le \frac{2}{(h_2^2-h_1^2)}$$

$$\left(h_{12}x(t)-\int_{t-h_2}^{t-h_1}x(s)ds\right)^T Q_9\left(h_{12}x(t)-\int_{t-h_2}^{t-h_1}x(s)ds\right) \tag{13}$$

i) when $h_1 \le \tau(t) \le \delta$,we have

$$-(\delta-h_1)\int_{t-\delta}^{t-h_1}\dot{x}(s)Q_6\dot{x}(s)ds=$$
$$-\int_{t-\delta}^{t-\tau(t)}\dot{x}(s)(\delta-\tau(t))\dot{x}(s)Q_6\dot{x}(s)ds$$
$$-\int_{t-\delta}^{t-\tau(t)}(\tau(t)-h_1)\dot{x}(s)Q_6\dot{x}(s)ds$$
$$-\int_{t-\tau(t)}^{t-h_1}(\delta-\tau(t))\dot{x}(s)Q_6\dot{x}(s)ds-\int_{t-\tau(t)}^{t-h_1}(\tau(t)-h_1)\dot{x}(s)Q_6\dot{x}(s)ds$$

Define $\quad \alpha=\dfrac{\tau(t)-h_1}{h_{12}},$

Then we can obtain

$$-\int_{t-\delta}^{t-\tau(t)}(\tau(t)-h_1)\dot{x}(s)Q_6\dot{x}(s)ds$$
$$\le -(1-2\alpha)\int_{t-\tau(t)}^{t-h_1}(\tau(t)-h_1)\dot{x}(s)Q_6\dot{x}(s)ds$$

Therefore using Lemma 1 to obtain

$$\begin{aligned}
&-(\delta-h_1)\int_{t-\delta}^{t-h_1}\dot{x}(s)Q_6\dot{x}(s)ds\\
&\le -[x(t-\tau(t))-x(t-\delta)]^T Q_6[x(t-\tau(t))-x(t-\delta)]\\
&\quad -2\alpha[x(t-\tau(t))-x(t-\delta)]^T Q_6[x(t-\tau(t))-x(t-\delta)]\\
&\quad -(1-2\alpha)[x(t-\tau(t))-x(t-h_1)]^T Q_6[x(t-\tau(t))-x(t-h_1)]\\
&\quad -[x(t-\tau(t))-x(t-h_1)]^T Q_6[x(t-\tau(t))-x(t-h_1)],
\end{aligned} \tag{14}$$

$$-(h_2-\delta)\int_{t-h_2}^{t-\delta}\dot{x}(s)Q_7\dot{x}(s)ds \\ \le -[x(t-h_2)-x(t-\delta)]^T Q_7[x(t-h_2)-x(t-\delta)] \tag{15}$$

According to (10)-(15) yields

$$\dot{V}(t)\le \xi^T(t)\Phi\xi(t)-2\alpha[x(t-\tau(t))-x(t-\delta)]^T \\ Q_6[x(t-\tau(t))-x(t-\delta)]-(1-2\alpha)[x(t-\tau(t)) \\ -x(t-h_1)]^T Q_6[x(t-\tau(t))-x(t-h_1)] \\ -h_{12}[x(t)-x(t-\tau(t))]^T Q_8[x(t)-x(t-\tau(t))] \\ =\xi^T(t)[(1-2\alpha)\Phi_1+2\alpha\Phi_2]\xi(t) \tag{16}$$

Where Φ_1,Φ_2,Φ are defined in (4),(5),(8),respectively, and

$\xi^T(t)=\begin{bmatrix}x^T(t) & x^T(t-\tau(t)) & x^T(t-\delta) & x^T(t-h_1) & x^T(t-h_2) & \int_{t-h_2}^{t-h_1}x^T(s)ds\end{bmatrix}$, Since $[(1-2\alpha)\Phi_1+2\alpha\Phi_2]$ is a convex combination of matrices Φ_1,Φ_2 on α, $[(1-2\alpha)\Phi_1+2\alpha\Phi_2]<0$ for $0\le\alpha\le\frac{1}{2}$ can be handled by two less conservative LMIs(4),(5) respectively.

ii) when $\delta\le\tau(t)\le h_2$, we have

$$-(h_2-\delta)\int_{t-h_2}^{t-\delta}\dot{x}(s)Q_7\dot{x}(s)ds= \\ -\int_{t-h_2}^{t-\tau(t)}(h_2-\tau(t))\dot{x}(s)Q_7\dot{x}(s)ds \\ -\int_{t-h_2}^{t-\tau(t)}(\tau(t)-\delta)\dot{x}(s)Q_7\dot{x}(s)ds \\ -\int_{t-\tau(t)}^{t-\delta}(h_2-\tau(t))\dot{x}(s)Q_7\dot{x}(s)ds \\ -\int_{t-\tau(t)}^{t-\delta}(\tau(t)-\delta)\dot{x}(s)Q_7\dot{x}(s)ds \tag{17}$$

Let $\beta=\dfrac{h_2-\tau(t)}{h_{12}}$, then one can obtain

$$-\int_{t-h_2}^{t-\tau(t)}(\tau(t)-\delta)\dot{x}(s)Q_7\dot{x}(s)ds \\ \le -(1-2\beta)\int_{t-h_2}^{t-\tau(t)}(h_2-\tau(t))\dot{x}(s)Q_7\dot{x}(s)ds \tag{18}$$

$$-\int_{t-\tau(t)}^{t-\delta}(h_2-\tau(t))\dot{x}(s)Q_7\dot{x}(s)ds$$
$$\leq -2\beta\int_{t-\tau(t)}^{t-\delta}(\tau(t)-\delta)\dot{x}(s)Q_7\dot{x}(s)ds \tag{19}$$

$$-(\delta-h_1)\int_{t-\delta}^{t-h_1}\dot{x}(s)Q_6\dot{x}(s)ds$$
$$\leq -[x(t-\delta)-x(t-h_1)]^T Q_6[x(t-\delta)-x(t-h_1)] \tag{20}$$

According to (10)-(13), (17)-(20) yields

$$\dot{V}(t)\leq \xi^T(t)\Sigma\xi(t)$$
$$-(1-2\beta)[x(t-h_2)-x(t-\tau(t))]^T Q_7[x(t-h_2)-x(t-\tau(t))]$$
$$-2\beta[x(t-\tau(t))-x(t-\delta)]^T Q_7[x(t-\tau(t))-x(t-\delta)]$$
$$-\frac{h_1h_{12}\beta}{\tau(t)}[x(t)-x(t-\tau(t))]^T Q_8[x(t)-x(t-\tau(t))]$$
$$-h_{12}(1-\beta)[x(t)-x(t-h_1)]^T Q_8[x(t)-x(t-h_1)]$$
$$=\xi^T(t)\left[(1-2\beta)\Sigma_1+2\beta\Sigma_2\right]\xi(t) \tag{21}$$

Where Σ_1, Σ_2, Σ are defined in (6),(7),(9),respectively,

and $\xi^T(t)=\left[x^T(t)\quad x^T(t-\tau(t))\quad x^T(t-\delta)\quad x^T(t-h_1)\quad x^T(t-h_2)\quad \int_{t-h_2}^{t-h_1}x^T(s)ds\right]$,

Since $\left[(1-2\beta)\Sigma_1+2\beta\Sigma_2\right]$ is a convex combination of matrices Σ_1, Σ_2 on β, $\left[(1-2\beta)\Sigma_1+2\beta\Sigma_2\right]<0$ for $0\leq\beta\leq\frac{1}{2}$ can be handled by two less conservative LMIs(6),(7) respectively. Therefore, if the LMIs(4)-(7)are satisfied, then the system(1)is asymptotically stable. This completes the proof.

4 Conclusions

In this paper ,the delay-dependent stability for systems with interval time-varying delay is investigated.By defining a new Lyapunov functional which contains a triple-integral term with the idea of decomposing the delay interval into two equidistance subintervals ,on which different Lyapunov functionals are defined,then a novel criterion of asymptotic stability is derived in term of linear matrix inequalities. The criterion proves to be less conservative than some previous ones.

References

1. Gu, K.: An integral inequality in the stability problem of time-delay systems. In: Proceedings of 39th IEEE Conference on Decision and Control, pp. 2805–2810 (2000)
2. He, Y., Wang, Q., Lin, C., Wu, M.: Delay-range-dependent stability for systems with time varying delay. Automatica 43, 371–376 (2007)
3. Xu, S., Lam, J.: On equivalence and efficiency of certain stability criteria for time-delay systems. IEEE Transactions on Automatic Control 52, 95–101 (2007)
4. Sun, J., Liu, G., Chen, J., Rees, D.: Improved stability criteria for neural networks with time-varying delay. Phys. Lett. A 373, 342–348 (2009)
5. Shao, H.: Improved delay-dependent stability criteria for systems with a delay varying in arange. Automatica 44, 3215–3218 (2009)

Numerical Simulation Investigation on Flow Field of Axial Blood Pump

Lu Zhang[1], Yuemei Jia[1], Wei Zhang[1], Yuxuan Wang[2], and Qiuxia Zhao[1]

[1]Department of Mechanics, Taiyuan University of Technology, Taiyuan, P. R. China
[2]Department of Thoracic and Cardiovascular Surgery, The Second Hospital of Shanxi Medical University, Taiyuan, P. R. China
{zhanglu_tyut,wvzhang}@163.com,
{bjiayuemei,jiayuemei}@tyut.edu.cn, zqx55_2006@yahoo.com.cn

Abstract. To investigate the optimization of a certain type of axial blood pump, this article established the model of the axial blood pump with software PRO/E and simulates flow field in one axial blood pump with software FLUENT and GAMBIT. The simulation results are showed by the velocity vector graphs and pressure distribution contours, etc. The conclusion is drawn by analysis of the simulation, this axial blood pump meets human body, but there is some deficiencies need to improve.

Keywords: Axial blood pump, mechanical circulation support, flow field, numerical simulation, computation fluid dynamics.

1 Introduction

Among all kinds of heart diseases, heart failure is the leading cause of death. In recent years, the treatment of terminal heart failure has increasingly become a great challenge to cardiovascular clinical physicians. The limitations of routine medical therapy and surgical interventions, and the shortage of donor hearts have led to the rapid development of mechanical circulation support devices [1-3]. MCS, mechanical circulation support, is an artificial mechanism whose main component is blood pump, and it can substitute for cardiac pumping function that partially or completely to ensure blood supply in body tissues and organs [4]. According to the working principle of the blood pump can be divided into: roller pump, pulsating pump, rotary pump and total artificial heart. Rotary pump includes axial flow pump and Centrifugal Pump. Since the former is small in size, long-term into the human body, artificial auxiliary pump has become a research hotspot now. The conventional axial flow blood pump is based on the experimental design, the need for repeated experiments and amendments, and with the experience to determine the final design. The main drawback is the design cycle is long, costly, dependence on experience, efficiency is not too high. With computer simulation technology and the development of such disciplines as fluid mechanics recently, numerical simulation technology began to be used in the design of blood pump [5-8]. FLUENT, CFD software, is used for simulating flow field in one axial blood pump. The general structure of the blood pump shown in Figure 1.

D. Zeng (Ed.): Advances in Computer Science and Engineering, AISC 141, pp. 223–229.
springer link.com

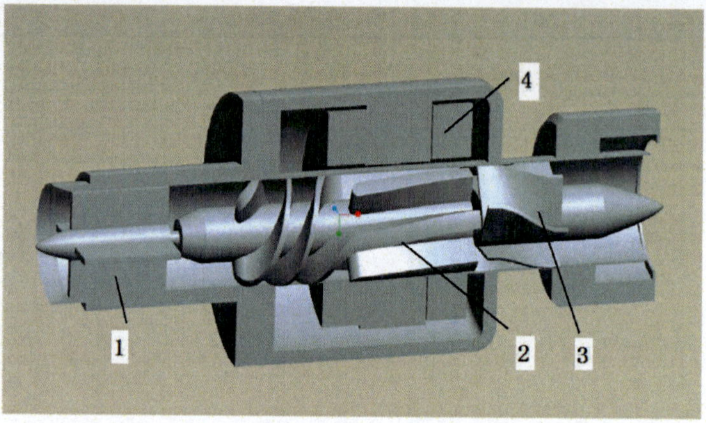

1.Front Guide vane 2.Rotor vane 3. Back Guide vane 4. Coil

Fig. 1. The general structure of the blood pum

2 Control Equation

2.1 Calculation Model

Since the standard $\kappa-\varepsilon$ model and the RNG $\kappa-\varepsilon$ model are based on isotropic turbulent viscosity to calculate the turbulent stress, these two models are difficult to take into account rotating flow and surface curvature of the flow direction. In order to reflect rotating flow of blood in flow field of the blood pump much better and ensure the accuracy and stability of the results, the paper chooses reynolds stress model (RSM) based on anisotropy.

2.2 Governing Equations

Incompressible continuity equations in constant temperature are

$$\frac{\partial u}{\partial x}+\frac{\partial v}{\partial y}+\frac{\partial w}{\partial z}=0 \tag{1}$$

$$\frac{\partial}{\partial t}(\rho u_i)+\frac{\partial}{\partial x_j}(\rho u_i u_j)=-\frac{\partial p}{\partial x_i}$$

$$+\frac{\partial}{\partial x_j}\left[\mu\left(\frac{\partial u_i}{\partial x_j}+\frac{\partial u_j}{\partial x_i}-\frac{2}{3}\delta_{ij}\frac{\partial u_l}{\partial u_l}\right)-\rho\overline{u_i'u_j'}\right] \tag{2}$$

Where $-\rho\overline{u_i'u_j'}$ is unknown reynolds stress component, $i, j=1,2,3$.

Turbulent equation is reynolds stress transport equation

$$\frac{\partial}{\partial t}\left(\rho\overline{u_i'u_j'}\right)+\frac{\partial}{\partial x_k}\left(\rho u_k\overline{u_i'u_j'}\right)=D_{i,j}+p_{i,j}+$$

$$G_{i,j}+\Phi_{i,j}-\varepsilon_{i,j}+F_{i,j}+S_{other} \tag{3}$$

Where $D_{i,j}$ is diffusion item, $p_{i,j}$ is generation item of shear stress, $G_{i,j}$ is generation item of buoyancy, $\Phi_{i,j}$ is stress and strain item, $\varepsilon_{i,j}$ is dissipative term, $F_{i,j}$ is generation item of system rotation, and S_{other} is user-defined source item.

$\kappa-\varepsilon$ equations in RSM are

$$\frac{\partial(\rho\kappa)}{\partial t}+\frac{\partial(\rho\kappa u_i)}{\partial x_i}=\frac{\partial}{\partial x_j}\left[\left(\mu+\frac{\mu_t}{\sigma_\kappa}\right)\frac{\partial\kappa}{\partial x_j}\right]+$$

$$G_k-\rho\varepsilon \tag{4}$$

$$\frac{\partial(\rho\varepsilon)}{\partial t}+\frac{\partial(\rho\varepsilon u_i)}{\partial x_i}=\frac{\partial}{\partial x_j}\left[\left(\mu+\frac{\mu_t}{\sigma_\varepsilon}\right)\frac{\partial\varepsilon}{\partial x_j}\right]+$$

$$\frac{C_{1\varepsilon}}{\kappa}G_k-C_{2\varepsilon}\rho\frac{\varepsilon^2}{\kappa} \tag{5}$$

Where κ is turbulent kinetic energy, ε is turbulent dissipation rate, $C_\mu, C_{1\varepsilon}, C_{2\varepsilon}$ are empirical constants, σ_κ and σ_ε is Prandtl number correspond to κ and ε, G_κ is generation item of κ that is cased by mean velocity gradient.

3 Numerical Simulation

3.1 Grid

Blood pump rotor model shown in Figure 2. Rotor diameter 6mm, inner diameter 2.6mm, the axial dimension of 27mm, the number of rotor vanes is 6, the speed is 12000rpm. The model of vascular can be represented simply with cylinder. Subtract rotor model from vascular model, a blood pump model of the internal flow field is obtained. Computational grid of model is divided by Gambit that is pre-processor of Fluent. Automatic mesh partition is used when division, and mesh type is unstructured grid. Grid number is 217328. Mesh model is shown in Figure 3.

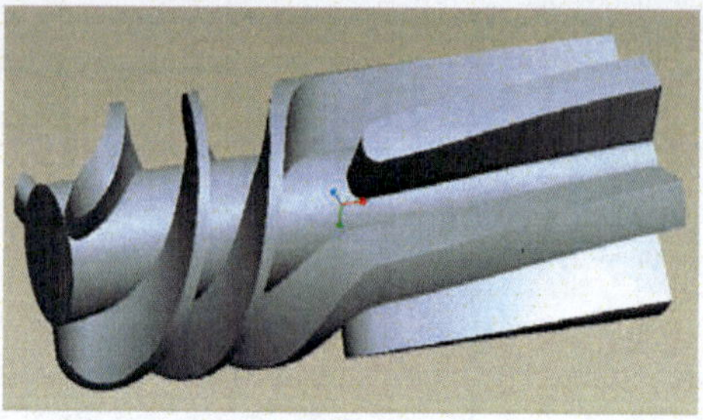

Fig. 2. Rotor model

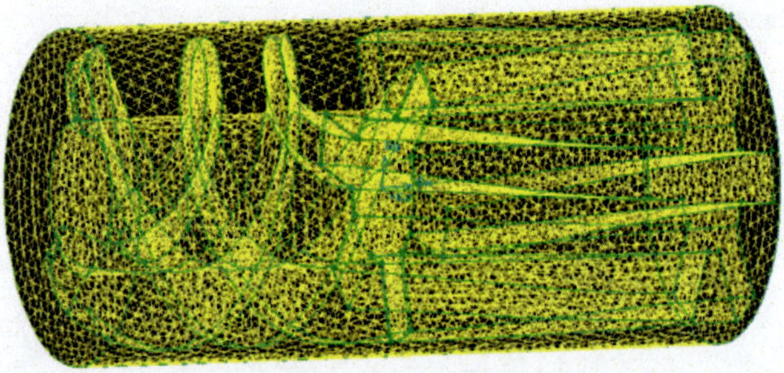

Fig. 3. Grid model

3.2 Boundary Conditions

- Inlet boundary conditions: medium is blood that is assumed incompressible Newtonian fluid. Density $\rho = 1.055 \times 10^3 kg/m^3$, viscosity $\nu = 3\times10^{-3} pa\cdot s$, blood flow rate is $Q = 5L/\min$, differential pressure between entry and exit is 90mmHg. Because blood is incompressible fluid, velocity-inlet is chosen as inlet boundary condition. Velocity $V = 0.6m/s$. Inlet turbulent is defined by turbulent intensity I and hydraulic diameter D_H, where $I = \frac{u'}{u_{ovg}} = 0.16(\mathrm{Re})^{-1/8}$, D_H equivalent to inlet equivalent diameter, Re is Reynolds number.
- Outlet boundary condition is outflow.
- Wall boundary condition: wall adopts rotational coordinate reference system, rotational speed $n = 12000 rpm$.

3.3 Simulation Results

(1) Figure 4 is outer wall surface shear stress contour of blood pump. Shear stress is essential reason of hemolysis. It can be seen from Figure 4 that high shear stress area locates in the rotor vane area where at the maximum shear stress of about 136Pa, and the area away from the vane at the lower shear stress values. According to analysis, the shear stress in the vane meet human needs.

Fig. 4. Outer wall surface shear stress contour of blood pump

(2) Pressure contour of blood pump flow field is shown in Figure 5, axial total pressure is shown in Figure 6. It can be seen from Figure 6 that outlet pressure is improved significantly from inlet, which is due to the promoting role of pump. In addition, in Figure 5, the pressure near the rotor vane surface is much smaller. This is because blood velocity value near the rotor vane surface is very high during the rotor rotation process, its pressure is smaller than the pressure at the other place inside the flow field.

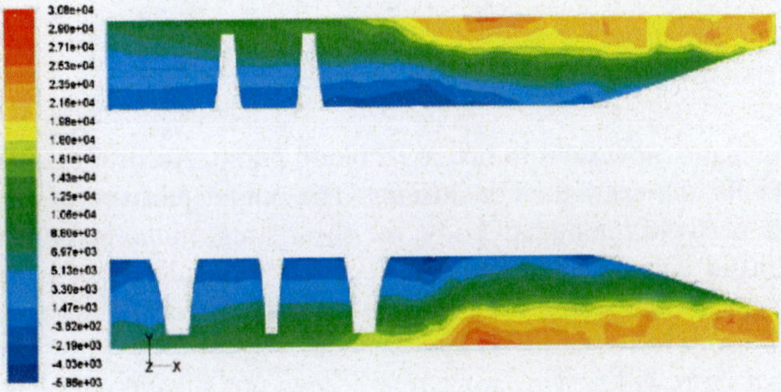

Fig. 5. Pressure contour of blood pump flow field

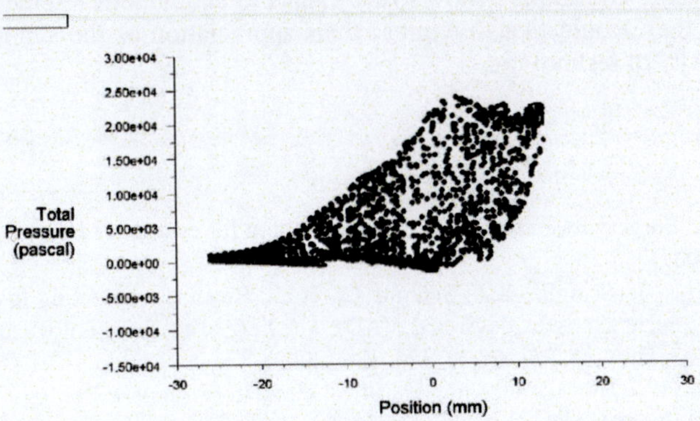

Fig. 6. Axial total pressure

(3) Fig. 7 shows velocity vector of blood pump flow field. Blood velocity in the flow field increases gradually along the channel direction and reaches its maximum at the outlet. Furthermore, in the back of the blood pump flow of blood appears obvious phenomenon of vortex and backflow. It follows that such a blood pump, the structure of the rotor vane is not reasonable, and it needs to be improved optimization.

Fig. 7. Velocity vector of blood pump flow field

4 Conclusion

The article simulates flow field in one axial blood pump. According to the analysis of simulation results, we draw the conclusions: The output pressure of the blood pump meet the basic needs of the human body; the shear stress in the pump flow is little, so the small hematic destructive power can reduce the incidence of hemolysis; blood pump outlet appears vortex and backflow, so the structure of the rotor vane is not reasonable, and it needs to be improved optimization.

These conclusions will provide some reference to optimization of blood pump in the future.

Acknowledgment. The authors show many thanks to anonymous referees' comments on this paper. they would like to express their appreciation of the support given by taiyuan university of technology.

References

1. Westaby, S.: Surgery for heart failure: now something for everyone? Heart Fail. Clin. 3(2), 139–157 (2007)
2. Wohlschlaeger, J., Schmitz, K.J., Schmid, C., et al.: Reverse remodeling following insertion of left ventricular assist devices (LVAD): a review of the morphological and molecular changes. Cardiovasc. Res. 68(3), 376–386 (2005)
3. Zhang, J.-M., Wan, F.: Advance in mechanical circulatory support. Beijing Medical 26(1), 60–61 (2004) (in Chinese)

4. Zhang, Y., Xue, S., Gui, X.-M., et al.: Digital simulation to the development of axial blood pump for artificial heart. Chinese Journal of Biomedical Engineering 26(1), 35–41 (2007) (in Chinese)
5. Wu, G.-H., Lin, C.-Y., Li, B.-Y., et al.: Developing and in vitro experiment about XZ-II A axial-flow blood pump. Beijing Biomedical Engineering 23(3), 215–217 (2004) (in Chinese)
6. Lin, C.-Y., Hou, X.-T., Wu, G.-H., et al.: The flow field analysis model XZ-II A axial flow blood pump. Beijing Biomedical Engineering 24(6), 405–409 (2005) (in Chinese)
7. Yun, Z., Tan, J.-P., Gong, Z.-L.: Research of CFD simulative optimization on impeller structure of axial blood pump. Journal of Machine Design 23(10), 6–9 (2006) (in Chinese)
8. Zhu, X.-R., Zhang, M.-Y., Liu, H.-N., et al.: Numerical investigation on flows and biocompatibility of an axial blood pump. Journal of Xi'an Jiaotong University 40(11), 1290–1294 (2006) (in Chinese)
9. Wang, F.-J.: Computational fluid dynamics analysis: the principle and application of CFD. Tsinghua University Press, Beijing (September 2004)
10. Anderson Jr., J.D.: Computational fluid dynamics: The basics with applications. McGraw-Hill, New York (1995)

Design and Implementation for the Node of Wireless Sensor Network Based on Zigbee Protocol*

Dai Chen[1], Shihui Jin[1], Yong Zhou[1], and Runtong Geng[2]

[1] School of Computer Science, China University of Mining and Technology
Xuzhou, China
[2] School of Electronic and Optical Engineering, Nanjing University of Science and Technology,
Nanjing, China
{chend,yzhou}@cumt.edu.cn, {xzjinshihui,grtzhuxian}@163.com

Abstract. This article compared several normal ZigBee solutions of the market based on full study of protocol in wireless sensor network. And choose the single chip solutions and selection of hardware platform with CC2430 as the core chip, using ZStack-1.4.3-1.2.1 as ZigBee protocol stack. Designed and produced a general-purpose wireless sensor network nodes can expand sensor and control mechanism module, and use the designed nodes build up a self-organizing wireless sensor network.

Keywords: ZigBee, CC2430, Wireless Sensor Network.

1 Introduction

Wireless sensor network (WSN) have been an active research topic for around a decade [1], which is component by lots of sensor nodes which have characteristics of miniature, low power, with communication and computation function from scattered in a particular area. ZigBee technology has low power consumption, low cost, low transmission rate, short range [2], short delay, high reliability, high expansibility, high security, free license band characteristics. ZigBee protocol stack can build up a star topology, cluster tree topology and mesh network topology. The ZigBee technology widely embedded in home, building and industrial automation, energy control, family control, medical monitoring, logistics and asset tracking, toys games and so on. The important point of the sensor design is to design the interface between sensor node and sensors. The general sensor interface design could expand simplifies and greatly shorten the product development cycle when make the future development design.

2 ZigBee Technology Overview

ZigBee protocol stack is based on the standard Open Systems Interconnection (OSI) seven-layer model. The ZigBee stack architecture is made up of a set of blocks called layers [3]. The physical (PHY) layer and the medium access control (MAC) sub-layer two lower layers were defined by IEEE802.15.4 standard. ZigBee alliance provides the

* Supported by "The Fundamental Research Funds for the Central Universities".

D. Zeng (Ed.): Advances in Computer Science and Engineering, AISC 141, pp. 231–239.
springer link.com

design of network (NWK) layer and the framework for the application layer. The application layer framework including application support sub-layer (APS), the ZigBee device objects (ZDO) and manufacturer-defined application object. The ZigBee stack architecture is shown in Figure 1[4].

The ZigBee wireless network nodes mainly include three logical device types: coordinator, router, end device. And the coordinator mainly responsible for build up the wireless sensor network. The router always used to extend the network. The main functions of the end device is gather the data of surrounding environment and control the device object when connected some sensors or control mechanism. Through these three kinds of wireless sensor network nodes can build up a self-organization and self-healing wireless sensor network. Each ZigBee network has one coordinator but can have multiple routers and end devices [5]. The ZigBee technology support three kinds of network: star topology, cluster tree topology, mesh network topology.

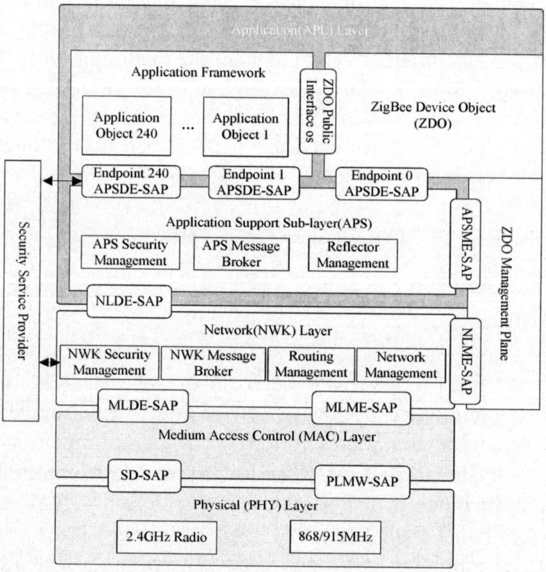

Fig. 1. ZigBee protocol stack architecture

3 System Design Scheme

3.1 Comparison and Selection of ZigBee Solution

At present the realization of build up a ZigBee network have three solutions for people to choose. These three kinds of solutions have their respective advantages. It's very important to choose a suitable solution to a product development. And then these three kinds ZigBee solutions were briefly introduced.

1) MCU + RF

The first kind of solution is the double chip solutions of MCU and RF transceiver separated. The ZigBee protocol stacks and the application program run on the MCU and the RF transceiver core mainly responsible for sending and receiving data. We need to consider MCU and RF transceiver in the design. But different types of MCU and RF transceiver could optional collocation, provides higher flexibility.

2) Single chip
The second kind of solution is the single chip solution used a RF transceiver and MCU integrated on a single chip. This kind of single chip solutions takes up the space is little because in a single chip integrated RF, memory, and micro controller and some other peripherals equipment and interface. Using the single chip to develop product more easily because simple circuit.

3) MCU + coprocessor
The third method for the ZigBee wireless sensor network to be set up is use ZigBee coprocessor and MCU double chip solution. The ZigBee coprocessor is generally responsible for ZigBee protocol stack operation and data transceiver. However, the specific applications run on the external MCU. Always won't need to spend too much time to care for the part of ZigBee development during developing new products. This solution can shorten the product development time and select the MCU optionally always have high flexibility.

The single chip solution with small size, low cost, low power consumption and development always need not consider the design of communication between the RF transceiver and MCU. The single chip solution is what we call the system on chip (Soc). Soc development in future development of the electronic technology is very important. Soc may cause the system to be more concise and need less peripherals circuit, PCB smaller, assembly and debugging easier and have higher reliability, higher anti-interference etc. From the low power consumption, low cost, low volume perspective, the single chip solution is most suitable for hardware platform of the wireless sensor network nodes.

3.2 The Design of System

ZigBee wireless sensor network is basic component as Figure 2 shows. The system contains a PC, a coordinator, several routers and more than one end devices. The end devices sensor scattered within a certain boundary through connections sensors and also control objects with some actuators connected. The three types of sensor nodes build up a mesh network topology. ZigBee coordinator choose the key network parameters and responsible for building network, by using the routers to extend the network. Coordinator and routers also act as a collector node to collecting around node of the data collected in the network. The sensor nodes in the network using USART interface communicate with computer by serial port connected. The end devices will monitor environment and acquisition environmental variables, the data collected sent to collector nodes through ZigBee wireless network, and then the computer shows the results after process. And also end devices can receive control commands from the computer to do configuration and modify the parameter of wireless sensor nodes or control the device objects which connected.

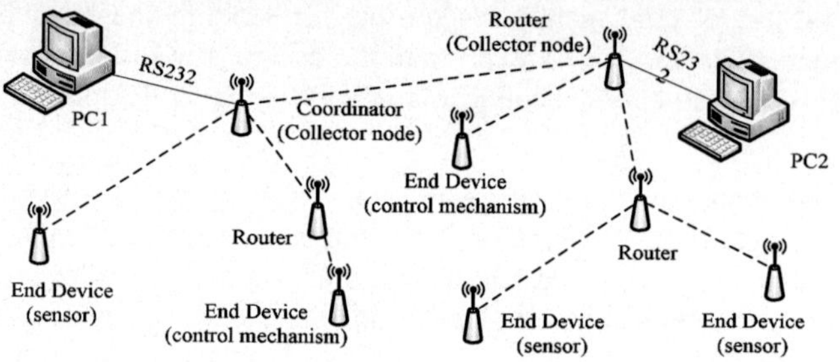

Fig. 2. System construction constitutes

3.3 The Design of the Communication Protocol

In order to make sure the universality and expansibility for end devices connect different sensors and control agencies, designed a kind of communication format. Once adopt the design standards as long as the sensor module or control module can be connected to the end devices and communication. The data packaged into a data frame to transfer when receiving sensor measurements and send control command to end devices to control the agencies. Figure 3 shows that is the data frame format of data communication between different devices in the system. Figure 4 shows, each sensor module has an 8-bits address, each sensor module on the device has a device ID. The address on each sensor module or control module in the same end devices and the ID on each sensor or controlled object in the same module are unique. In such a way, one end devices connected to sensors or control devices can be found in the wireless sensor network through the short network address, and then can be positioned to which particular sensor or a control agencies based on the device module address and device ID. Thus we can send command to a given agencies and know data PC received specific from which equipment.

Length of Data Frame (8b)	Destination Node Address(16b)	Source Node Address(16b)	Data

(a) The transmission format of the data frame between PC and Collector node

Length of Data (8b)	Device Address(8b)	Device ID (8b)	Command (8b)	Command Parameters

(b) The transmission format of the data frame between end device and device modules

Start Sigh0 (0x7E)	Start Sign1 (0xC3)	Length of Data Frame	Data	Stop Sign0 (0xE7)	Stop Sign1 (0x3C)

(c) The transmission format of the data frame added start sign and stop sign

Fig. 3. The transmission format of the data frame between devices

Data transfer between the wired data transmission part are transmitting used in the asynchronous serial interface UART. If the data transmission is interrupted unexpectedly, the data frame received is incomplete at the receiver without processing. Therefore, each data frame transmission give two tags as a start single and an end single. The head of data frame add two consecutive bytes 0x7E, 0xC3, with 0xE7, 0x3C two consecutive bytes added in the end of a data frame as the end single, so that we can read a data block to determine whether the complete data frame.

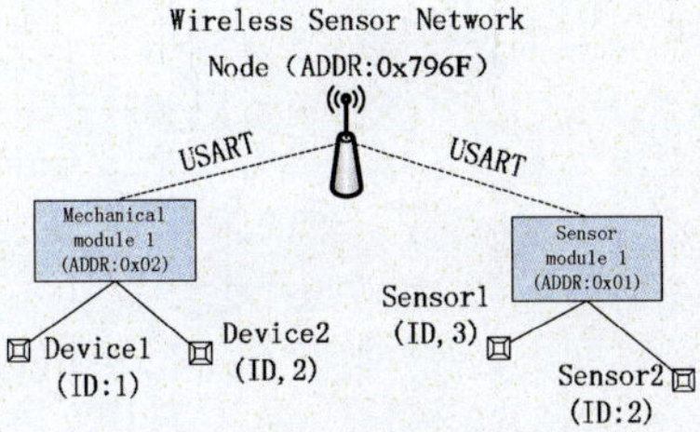

Fig. 4. The connection of end device and device modules

4 Hardware Platform Design

The hardware platform of system design using the TI's CC2430 as the core chip. The CC2430/CC2431 is Chipcon (already acquired by company of Texas Instruments in 2006) company's products used to implement for the ZigBee application system. The CC2430 is highly suited for systems where ultra low power consumption is required [6], which is a true System-on-Chip (SoC) solution specifically tailored for IEEE802.1d5.4 and ZigBee applications. It combines the excellent performance of 2.4GHz IEEE802.15.4 compliant DSSS (Direct Sequence Spread Spectrum) RF transceiver (industry leading CC2420 radio core) with an industry-standard enhanced 8051MCU, interior has optional 32/64/128KB of in-system programmable flash memory that correspond to three versions: CC2430F32/64/128, 12-bit ADC with up to eight inputs and configurable resolution, AES security coprocessor, one watchdog timer, one IEEE802.15.4 MAC timer, one general 16-bit timer and two 8-bit timers, the power on reset and brown out detection circuit, two powerful USART/SPI ports and 21 general I/O pins. Figure 5 shows is the basic CC2430 module circuit diagram.

The CC2430 as a core chip for running ZigBee protocol, additional nodes still designed a debug, RS232 and SPI, ADC interface etc, the modules also set up the buttons, LED lamp, power supply module etc. node module design for wireless sensor networks provides the hardware platform for running the ZigBee protocol.

5 Software Design

The system software is designed based on the free ZigBee 2006 protocol stack from TI, developed based on the SimpleAPP routine in ZStack-1.4.3-1.2.1 version. This routine provides a generic template of sensor data collection, through which we can build our own applications. We just need to modify the HAL (Hardware Abstraction Layer) according to our own hardware platform designed and then to add applications in APP (Application Layer).

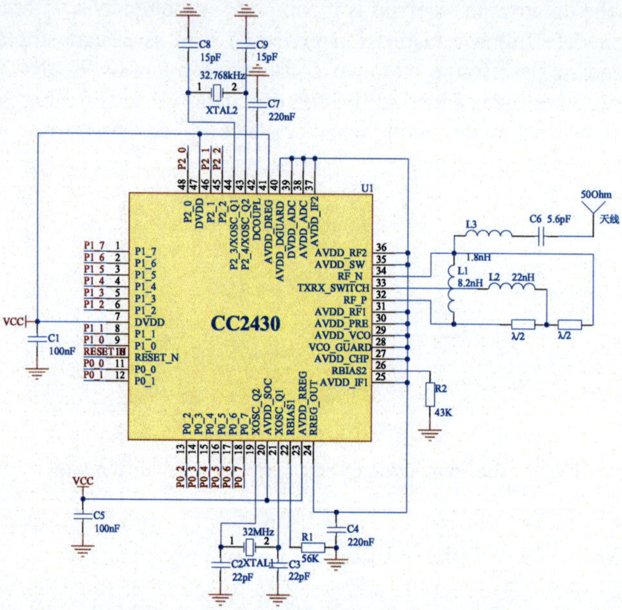

Fig. 5. CC2430 Application Circuit

The hardware abstraction layer is in HAL folder of the protocol stack. In TI's ZigBee 2006 protocol stack, the HAL defines all the hardware in accordance with TI's CC2430 ZigBee Development Kit, so must customized according to our own hardware circuit if we want the ZStack able to running on our hardware platforms, such as we must modify SPI, UART, ADC, Keyboard LEDs and other initialization.

The complete dada frame send and receive is continuously between collection node and PC and also the same between the end devices and control agencies modules or sensors modules. The data received from serial port is placed in a buffer. In order to get a command or a complete data packet from the receive buffer, we can judge the integrity of the data according to data length and the sign added in data beginning and ending. Figure 6 shows that the process of a full data frame received from the receive buffer.

A node may need to connect multiple devices module on the end device, here used a simple mechanism to avoid data transmission conflict. In addition, the serial port is can't receive data when the nodes or modules into sleep mode, so the devices or modules need to waked up while sending data.

Here used a simple way to deal with the two problems above. Between the module and node joined a wire, the I/O port connected is set to interrupt and input mode, falling edge trigger mode. The electrical level on the wire is always high, when you need to wake up the device and send data to it, another device make the level on the wire is low, so created a falling edge, the device in the sleep mode will waked up by interrupt. Device could receive data when awake. The electrical level on the wire is low all the way until the data transfer is complete, then the level on the wire turn to high. Thc dcviccs go to slccp again if no task to run. In this way, through dctccting thc level on the wire to determine if there are devices used for data transmission. Also can avoid conflict of communication effectively and can send data to sleeping device.

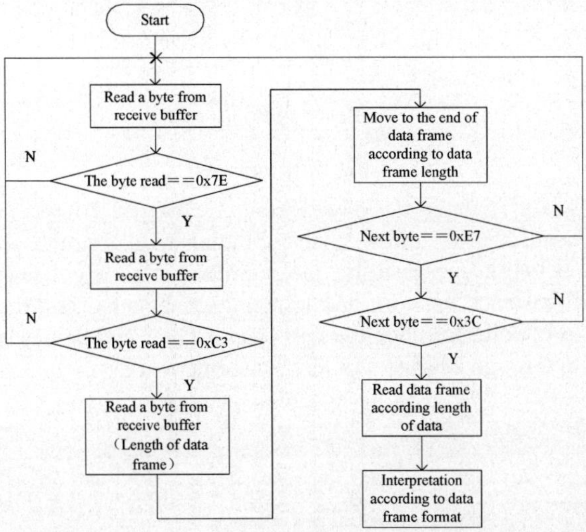

Fig. 6. The flow chart of read a complete data frame from the receive buffer

6 System Testing

Figure 7(a) shows that the design results of this topic, testing environment shown in Figure 7(b). The coordinator connected with PC as a collector node, the data collected from the sensors module given through the computer serial port, transfer the control information which sends to end device node to PC. Using serial debugging assistant to analyze data, and then verify that the correctness of system.

(a) ZigBee wireless sensor network nodes

Fig. 7. Designs and testing environment

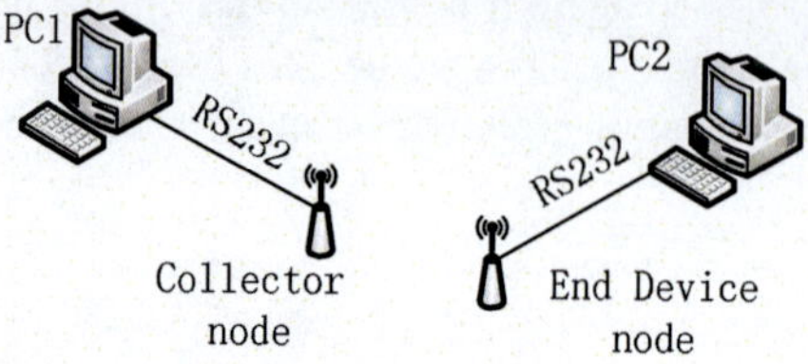

(b) Testing Environment

Fig. 7. (*continued*)

Tune on the coordinator, allow devices binding to it through press the key after established the network successfully, the coordinator as a collector node. The end device will find existence network and joined automatically, and then the end device send bind request to collector node the first found. Thus, end device and collector node can communicate through ZigBee wireless sensor network.

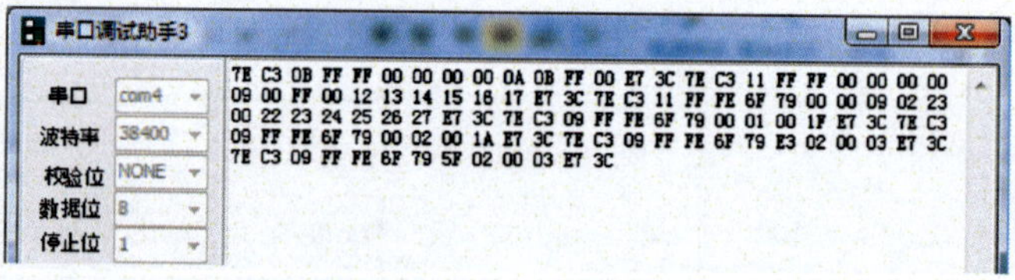

(a) The communications between PC and Collector node

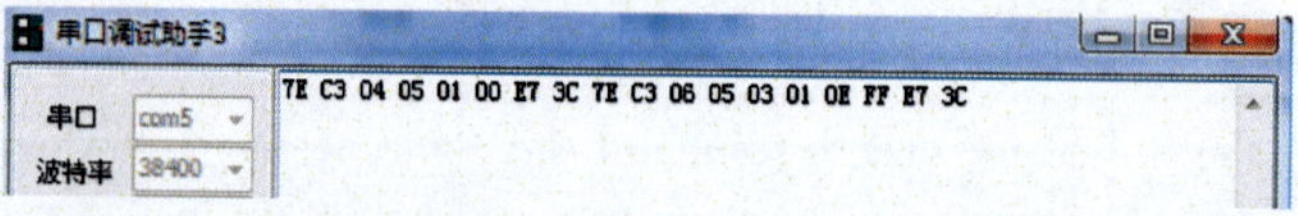

(b) The communications between end device and device module

Fig. 8. Test Results

Figure 8 shows that the serial debugging assistant received data frames between the nodes, on which one analysis of the data frame is as follows:

7E C3 09 FF FE 6F 79 00 01 00 1F E7 3C

Start sign		Length of data frame (9byte)	Destination Address		Source Address	
7E	C3	09	FF	FE	6F	79

Device Address	Device ID	Command	Command Parameter	Stop sigh	
00	01	00	1F	E7	3C

The contents of the above is a data frame from the end device node 0x796F, device address is 0x00 (the node itself), the devices ID is 0x01 (the on-chip temperature sensor), the command 0x00 (sensor values), the command parameter 0x1F (temperature 31°C).

7 Conclusion

This article is mainly use of ZigBee protocol to build up a mesh network topology wireless sensor network, designed and realized an universal and extensible sensor node which is used to conform a network and collect environmental parameters and control the objects. Finally the designs were tested and analyzed.

References

1. Andrew, W.: Commercial applications of wireless sensor networks using ZigBee. IEEE Communications Magazine 45(4), 70–77 (2007)
2. Li, C., Wang, Y., Guo, X.: The Application Research of Wireless Sensor Network Based on ZigBee. In: Second International Conference on Multimedia Information Technology (MMIT), April 24-25, vol. 2, pp. 89–92 (2010)
3. Li, P., Li, J., Nie, L., Wang, B.: Research and Application of ZigBee Protocol Stack. In: International Conference on Measuring Technology and Mechatronics Automation (ICMTMA), March 13-14, vol. 2, pp. 1031–1034 (2010)
4. ZigBee Alliance, ZigBee Specification Document 053474r17 (2008)
5. Wang, C., Sohraby, K., Jana, R., Ji, L., Daneshmand, M.: Voice Communications over ZigBee Networks. IEEE Communications Magazine 46(1), 121–127 (2008)
6. Texas Instruments, CC2430 Data Sheet (rev.2.1) SWRS036F (2007), http://www.ti.com/lpw

The Improved Location Algorithm of Apit Based on Midline Segmentation for Wireless Sensor Network

Fanrong Meng, Xiaocui Li, and Yong Zhou

School of Computer Science, China University of Mining and Technology
Xuzhou, China
{mengfr,yzhou}@cumt.edu.cn, xiaocuiworld@163.com

Abstract. Localization technology is one of the important supporting technologies in the Wireless Sensor Network. This paper promoted the original APIT algorithm by dividing the triangle into 3 small regions by lining from the centroid of the triangle to the mid-points of three sides of the triangle and determine the location of unknown node by testing the strength of the signal received by the three anchors of the triangle. Experiments prove that the algorithm improved the location accuracy by reducing the location area of original APIT algorithm. Through the centroid algorithm thoughts work out the unknown nodes' estimates coordinates that can improve the location efficiency.

Keywords: The Wireless Sensor Network, Localization technology, The improved APIT, Midline segmentation.

1 Introduction

Wireless sensor network (WSN) which consists of lots of sensor nodes constitutes the information collecting platform based on task through self-organization. Each sensor node is an independent data gatherers and routers, it can realize real-time monitoring, perceive and gather various environmental or monitoring information about objects, then process them[1]. It has many advantages that can remote monitor and real-time monitor, and it also can work in threatening or special environment. The WSN has been listed at the top of the 10 emerging technologies that has a far-reaching effect to human future life. The development and extensive application of Wireless Sensor Network will greatly influence on people's social life and industrial revolution and bring a huge boost.

Localization is base of many applications as one of the five support technologies of the WSN[2]. The data collected by node without location information has almost no application value. Since many applications require accurate localization. So in the past few years the Localization technology of wireless sensor network has been researching widely.

Wireless sensor network's localization system and performance of the algorithm directly affects its availability, their evaluation basically has the following standards [3].

D. Zeng (Ed.): Advances in Computer Science and Engineering, AISC 141, pp. 241–248.
springer link.com

(1) Location accuracy. It is that generally the distance between the estimate coordinate and actual coordinate (location error) divided by communication radius.

(2) Energy consumption. Since the sensor nodes' power is limited, wireless sensor network usually work in bad environment unreached by human, exhausting electrical energy means nodes failure. Computation, communication and awareness are all shortening nodes' lifetime, so many core issue of technical researching of wireless sensor networks focus on energy efficiency.

(3) The cost. The cost of location system contains temporal cost, special cost and capital cost. Time cost includes a system setup time, configuration time, time needed in localization. Special cost includes the necessary infrastructure, the numbers and the hardware size of the network's nodes needed in the location system or an algorithm, etc; Capital cost includes the total cost of the infrastructure and the nodes' equipments.

(4) Fault-tolerance and self-adaptability. The application background of Wireless Sensor network determines that the hardware and software of localization algorithm should also have very good fault-tolerance and self-adaptability to overcome the realistic environment's influence of communication, node failure, location errors, etc.

2 The Theory of APIT Algorithm

The principle of perfect point-in-triangulation[4] (short for PIT) shown in figure 1.

Proposition 1: if M locates in the triangle ABC, then, when M moves toward any direction, the new position will at least deviate from (or close to) triangle's three points A, B, or C.

Proposition 2: if M locates out of the triangle ABC, then, when M moves toward any direction, the new position will also deviate from (or close to) triangle's three points A, B, and C.

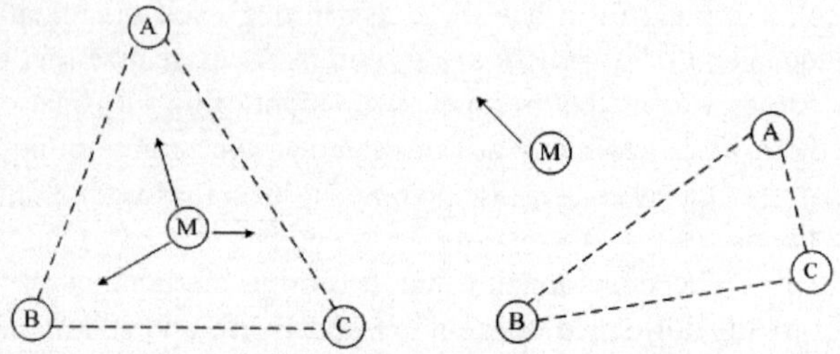

Fig. 1. The principle of PIT

The localization algorithm PIT is very suitable for that is relative lack of anchors. Compared to other similar localization algorithms, the location accuracy is on the optimal performance.

However, in the actual wireless sensor network, the unknown nodes are usually static. In order to realize the PIT algorithm in the static environment, Tian He, etc, put forward the APIT[4] (Approximation of the Perfect PIT Test) algorithm based on the

PIT. Use the relatively high density of nodes to simulate the node mobile, use wireless signals transmission to judge whether the node's deviating from or close to the anchor node. Usually the farther one node deviates from another, the weaker signal can be received. The unknown node can decide which one is more closer though exchanging received signal's strength of the neighbor nodes, thus mimic the nodes' movement in the PIT algorithm

Theory APIT: If there is no node M also deviate from (or close to) the triangle's three points A, B, and C, then M locates inside of the triangle. Otherwise, M locates outside of the triangle.

Use APIT algorithm to locate for the unknown nodes. Firstly find all the centroids of the regions that contain the unknown nodes and create a polygon using those centroids [5]· Then find the centroid of this polygon which seems to be the estimated coordinate of the unknown node. Details of location process are as follows:

Firstly, each node collects its neighbor anchors' information. Secondly select every three anchors to make a triangle. If there are n anchors, there will be in total C_n^3 triangles. Use the APIT theory to test whether the unknown node locates inside of these triangles. Thirdly, if the unknown node locates inside of one triangle, then find out in which small region the unknown node is and make a mark of it. Fourthly, find all the centroids of those small regions and create a polygon using those centroids. Finally, find the centroid of this polygon which seems to be the estimated coordinate of the unknown node.

The APIT algorithm can use the relatively high density to simulate nodes mobile and judge whether the unknown node is away from or close to the anchor by wireless signals. Usually in a given direction, the farther a node is away from another, the weaker it received another signal. Neighbor nodes through exchanging receiving signal's strength, judge the distance to a certain anchor, thus can mimic the nodes mobile.

3 The Improvement of APIT Algorithm

From the perspective of location accuracy, that using the APIT algorithm to locate can be divided into two stages. The first stage is the unknown nodes collect the information from the neighbor nodes. Then establish their own set of supporting data sets through the APIT algorithm. The second stage is to scan the data set by calling the appropriate scanning algorithm, and ultimately get the estimated coordinates of the nodes.

In the course of using APIT algorithm for localization, as shown in Figure 2: Let the three vertices of the triangle coordinates to be: $A(x_1, y_1)$, $B(x_2, y_2)$, $C(x_3, y_3)$. Divide the triangle into 3 small regions A', B' and C' by lining from the centroid of the triangle to the mid-points of three sides of the triangle. We can improve the APIT algorithm through comparing the strength of signal collect by M sent by the anchor A, B and C and reduce the location accuracy. Taking Figure 2 for example, by detecting the signal strength, we can know the signal strength received by M from A is stronger than form the point B and C. You can see that unknown node M should be in the region A'. Supposing M in the region A' is located in the geometric center O', according to the geometric relationship, the formula of obtaining the coordinates of O_A' is as 3.1:

$$O_A'(\frac{2x_1+\frac{x_2+x_3}{2}+\frac{x_1+x_2+x_3}{3}}{4},\frac{2y_1+\frac{y_2+y_3}{2}+\frac{y_1+y_2+y_3}{3}}{4}) \tag{3.1}$$

The same principle, when the M in the region B ' or C ', the centroid of region B ' or the C' can be obtained by the following formulas:

$$O_B'(\frac{2x_2+\frac{x_1+x_3}{2}+\frac{x_1+x_2+x_3}{3}}{4},\frac{2y_2+\frac{y_1+y_3}{2}+\frac{y_1+y_2+y_3}{3}}{4}) \tag{3.2}$$

$$O_C'(\frac{2x_3+\frac{x_1+x_2}{2}+\frac{x_1+x_2+x_3}{3}}{4},\frac{2y_3+\frac{y_1+y_2}{2}+\frac{y_1+y_2+y_3}{3}}{4}) \tag{3.3}$$

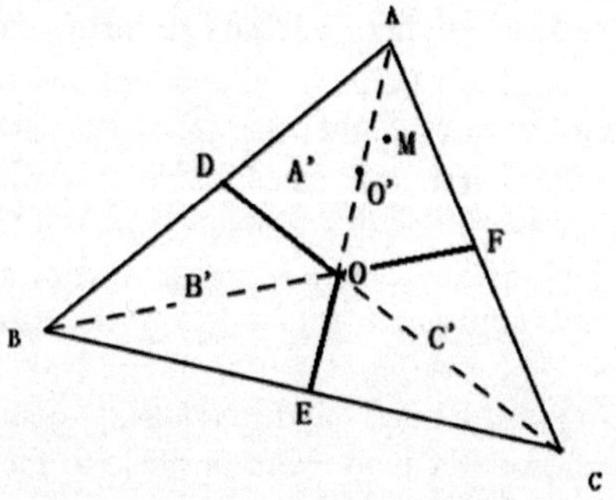

Fig. 2. The triangle divided by midline

Then exhaust by the same method for all the triangles constructed by anchors. Finally, average the centroids of all the small regions obtained, which is the centroid of a polygon constructed by all the centroids, as well as the estimated coordinates of M.

The following details are for improved APIT algorithm.

Step1: The unknown nodes collect information of neighbor nodes within a radius of unknown nodes, such as location, ID, the received signal strength, then the neighbor nodes exchange the information received within a radius of neighbor nodes.

Step2: Test whether the unknown node is within the triangle constructed by three different anchors by APIT theory. If there is no node M also deviate from (or close to) the triangle's three points A, B, and C, then M locates inside of the triangle. Otherwise, M locates outside of the triangle.

Step3: If you point is inside the triangle, using the center line, the triangle can be divided into three smaller regions. Through the comparison of the signal strength received by vertices of three anchors of a triangle, the node can determine in which small region the target node is.

Step4: Find all the centroids of those available small regions. Finally, average the centroids of all the small regions obtained, which is the centroid of a polygon constructed by all the centroids, as well as the estimated coordinates of M.

The improved APIT algorithm flow chart is as shown in Figure 3:

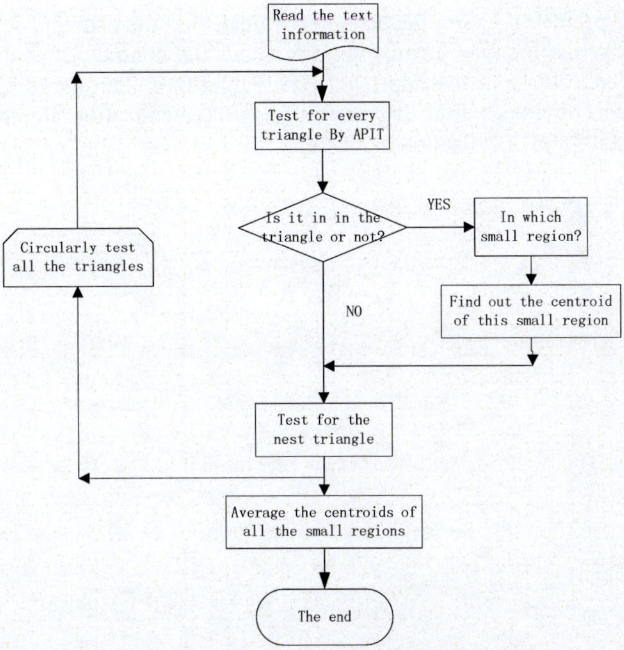

Fig. 3. The improved APIT algorithm flow chart

4 The Experiment

There are 34 anchors and 86 unknown nodes in this experiment, totally 120 nodes. The unknown nodes' communication radius is 100 and the anchors' communication radius is 200, the proportion of anchors is 28%. Figure 4 shows the contrast of actual coordinate and estimate coordinate in the original APIT algorithm, we can see that

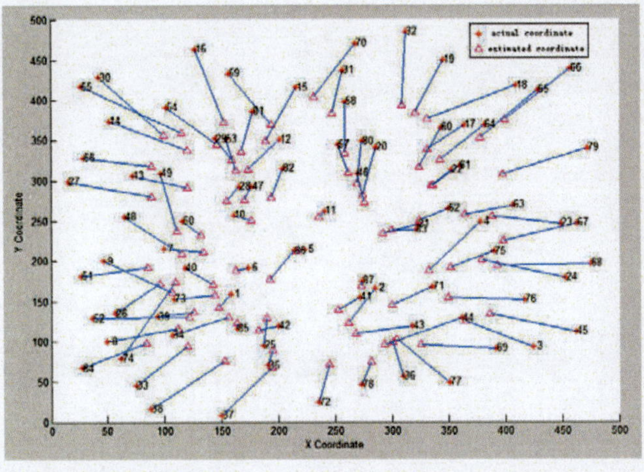

Fig. 4. The contrast of actual coordinate and estimate coordinate in the original APIT algorithm

there is some distance between the actual coordinate and the estimate coordinate, this distance we call the location error. Figure 5 shows the contrast of actual coordinate and estimate coordinate in the improved APIT algorithm. Through the figure 4 and figure 5 we can obviously find that improved APIT localization algorithm is better than the original APIT localization algorithm.

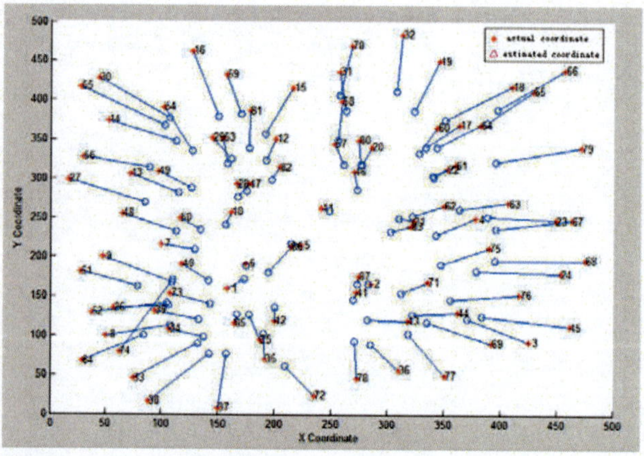

Fig. 5. The contrast of actual coordinate and estimate coordinate in the improved APIT algorithm

In order to observe the comparison clearly, we put the results of the original APIT algorithm and the improved APIT algorithm together as shown in figure 6, it can be seen that in the improved APIT algorithm the distance of actual coordinate and estimate coordinate is shorter than that in the original APIT algorithm. That explains the improved APIT algorithm performs better than the original APIT algorithm.

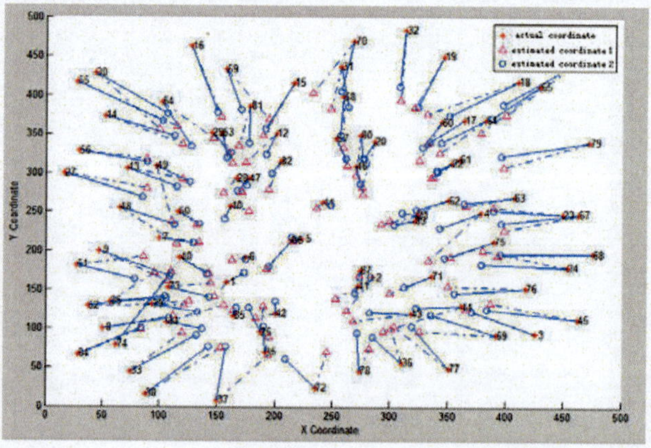

Fig. 6. The contrast of actual coordinate and estimate coordinate in the original and improved APIT algorithm

In order to better observe the distinction of the original algorithm and the improved algorithm. This paper will compare the location accuracy of these two algorithms as shown in figure 7:

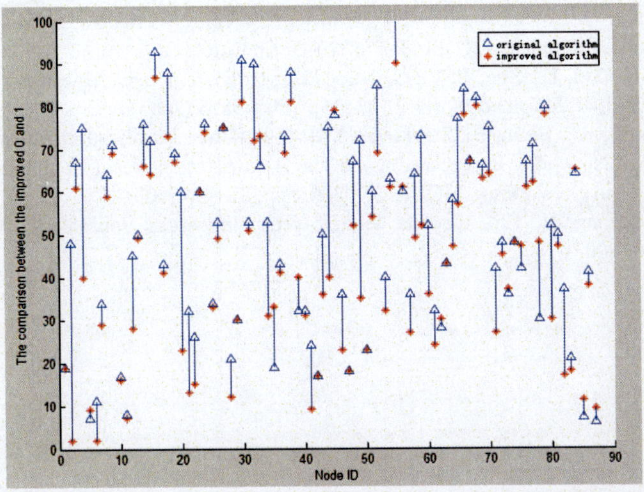

Fig. 7. The location accuracy of original and improved APIT algorithm

Through the above comparison, the average location accuracy of original APIT algorithm is 44.2% and 34.8% in the improved APIT algorithm, we can see that the improved APIT algorithm increased the average location accuracy by 9.4%.

5 Conclusion

This paper is focusing on the improved APIT algorithm based on the wireless sensor networks and location theory. By analyzing the advantages and the reason of the error occurred. This paper introduced the Improved APIT algorithm. Divide the triangle into 3 small regions by lining from the centroid of the triangle to the mid-points of three sides of the triangle and determine the location of unknown node by testing the strength of the signal received by the three anchors of the triangle. That is to determine which small region the unknown nodes locate at. Then with the same way, circularly execute all of the triangles formed by the anchors those are in the unknown node's communication range. Finally, find all the centroid of the small regions and form a polygon by them to find the polygon's centroid which is the estimate coordinate of the unknown node. Experiments prove that the algorithm improved the location accuracy by reducing the location area of original APIT algorithm. In the same time, it avoided the process of grid scan in traditional APIT algorithm, which can improve efficiency. Through the centroid algorithm work out the unknown nodes' estimates coordinates that can improve the location efficiency.

References

1. Sun, L., Li, J., Chen, Y., et al.: Wireless Sensor Networks. Tsinghua University Press, Beijing (2005) (in Chinese)
2. Cui, L., Ju, H., Miao, Y., et al.: Overview of wireless sensor networks. Journal of Computer Search and Development 42(1), 163–174 (2005) (in Chinese)
3. Wang, F.-B., Shi, L., Ren, F.-Y.: Self-Localization Systems and Algorithms for Wireless Sensor Networks. Journal of Software 16(05), 1000–9825 (2005)
4. He, T., Huang, C., Blum, B., Stankovie, J.A.: Range-Free Localization Schemes for Large Scale Sensor Networks. In: Proceedings of the Annual International Conference on Mobile Computing and Networking, MOBICOM 2003, pp. 81–95 (2003)
5. Feng, Y., Huang, H., Lin, C.: The wireless sensor networks. Journal of Software 14(7), 1282–1290 (2003)

Research on Computer Simulation of Mining Subsidence Mechanism with FLAC

Ji Hong[1] and Yu Xueyi[2]

[1] Network Center, Xi'an University of Science and Technology, Xi'an, China
[2] School of Energy Engineering, Xi'an University of Science and Technology Xi'an, China
modena@sohu.com, yuxueyi@xust.edu.cn

Abstract. This paper presents an analysis method of mining subsidence mechanisms of gob. The surface above gob appears displacement and deformation with the process of coal mining. The building foundation is affected seriously and the building is displaced, deformed and even destroyed. A plenty of data on the displacement and deformation is collected in Xinmi coalfield. By means of the data, the characteristics of the displacement and deformation and the effects on mining buildings are simulated with FLAC. After analysis, this paper gives the preventive measures of mining subsidence of gob and the protection measures of mining buildings.

Keywords: Numerical simulation, mining buildings, mining subsidence, FLAC, protechtion measures.

1 Introduction

With the process of coal mining, the surface above gob has displacement and deformation; moreover, it can seriously affect the building foundation above gob, and then, mining buildings appears displacement, deformation, and even destruction. Subsidence and deformation in large area surface cause serious damage to mining buildings and transport facilities; meanwhile, it is a serious threat to the ecological environment of gob and people's property. Therefore, it is very necessary to study on mining subsidence mechanisms and give reasonable control measures of mining environment protection based on computer numerical simulation.

2 An Engineering Project

The Xinmi coalfield is trapezoidal with slightly narrower in the northwest and wider in the southeast which includes five coal groups and the current project is the second coal group. The physical mechanical properties of the roof and floor are shown as follows: in the fine exploration search process, it takes 10 rock samples from the roof to the floor of the second coal group to analyze the physical mechanical properties. There is a

D. Zeng (Ed.): Advances in Computer Science and Engineering, AISC 141, pp. 249–254.
springer link.com

layer of $0.06 \sim 2.30m$ carbonaceous rock in the false roof; there is generally dark gray siltstone and sandy mudstone in the immediate roof with an average thickness of $11.7m$. The rock specific gravity is from $2.63KN/m^3$ to $2.81KN/m^3$; The bulk specific gravity is from $23.3KN/m^3$ to $26.6KN/m^3$; The vertical compressive strength is from $130.8KPa$ to $168.1KPa$; The shear strength is from $129KPa$ to $405KPa$ in normal stress and shear stress of dip angle 45°; The cohesion is from $4.6MPa$ to $20MPa$; The internal friction angle is from 27° to $33^\circ 30'$; The softening coefficient is from 0.42 to 0.48; The direct floor is made up with black sand, siltstone, sandy mudstone and mudstone with an average thickness of $11.4m$. The rock specific gravity is from $2.68KN/m^3$ to $2.74KN/m^3$; The bulk specific gravity is from $20.45KN/m^3$ to $26.20KN/m^3$; The vertical compressive strength is $110.9KPa$; The vertical tensile strength is from $0.57MPa$ to $7.20MPa$; The softening coefficient is 0.83; The old floor is composed of limestone with average total thickness of $9.36m$, and it is the main aquifer in this area. Because of the high hydraulic pressure of this aquifer, it can cause water gushing in the floor which is composed of the mudstone [1].

3 Design of Numerical Simulation

3.1 The Theory of Numerical Simulation

The Lagrangian method is generally used in fluid mechanics. The Lagrangian method is transformed to being used in solid mechanics. The study area is divided into grids, the grid nodes are equivalent to the fluid particles, and then it uses the Lagrangian time-step method to study the movement of mesh nodes. This method is called the Lagrangian method.

This method is most suitable for solving nonlinear large deformation problem. It uses dynamic relaxation method to solve the problem, without the formation of the stiffness matrix and without solving large simultaneous equations; moreover, it has small footprint and is easy to use computer to solve a larger project problem [2]. It is one of the reliable and feasible computer simulation methods.

3.2 The Analysis of Mining Deformation

In the context of gob, according to the borehole data near the working surface, the design model is designed as follows: the coal seam depth is $220m$, the mining height is $4m$, the loose thickness is from $30m$ to $70m$ and the terrane thickness above the coal is from $150m$ to $190m$. According to the physical mechanic test results of the rock in the research area, it can determine the rock's basic situation and the physical mechanic parameters of coal. It is shown in Table 1.

Table 1. The physical mechanic parameters of coal

Sequence	Lithology	Bulk Density (kg/m^3)	Modulus (MPa)	Poisson Ratio	Tensile Strength (MPa)	Cohesion (MPa)	Friction Angle(°)
1	Loess	1786	10	0.30	0.0018	0.019	14
2	Sandstone	2357	1740	0.23	3.13	7.27	42.0
3	Sand Mud Interbed	2395	1590	0.14	2.62	4.42	38.9
4	Sandy Mudstone	2431	7405	0.15	1.60	3.60	35.4
5	Coal	1307	6900	0.27	1.33	2.42	42.6
6	Siltstone	2482	18208	0.17	7.35	11.05	37.4

By means of FLAC numerical simulation, it is realized to simulate the process of mining subsidence in the working surface and to study on the characteristics of movement and deformation of the terrane and the surface above gob. In the process of numerical simulation, the design model grids before and after the coal mining is shown in *Fig. 1*, *Fig. 2*, the number from 1 to 6 are respectively as follows: loess, sandstone, sand mud interbed, sandy mudstone, coal and siltstone; the space means the gob area.

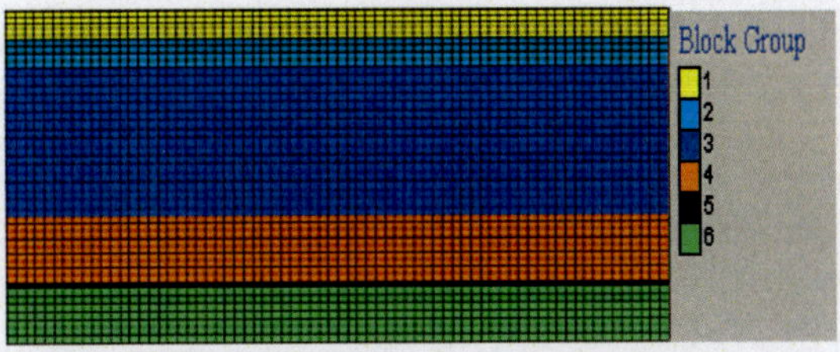

Fig. 1. The design model grids before the coal mining

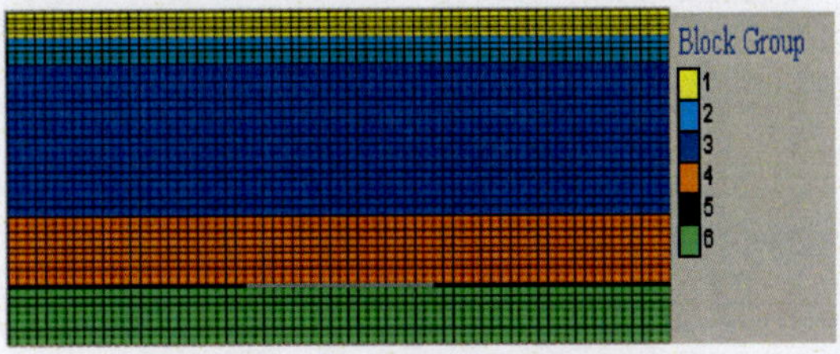

Fig. 2. The design model grids after the coal mining

The nephogram of subsidence and horizontal displacement after the coal mining are shown in *Fig. 3, Fig. 4*.

Fig. 3. The nephogram of subsidence after the coal mining

Fig. 4. The nephogram of horizontal displacement after the coal mining

The computer numerical simulation results show the characteristics of movement and deformation: The surface subsidence is from $2600mm$ to $3400mm$, and the surface horizontal movement is from $520mm$ to $1020mm$. It validates the accuracy of the results by means of probability integration method. It treats the normal distribution function as the influence function in probability integration method and uses integration method to indicate the movement and deformation. Based on its mathematical formula, it gives the characteristics of surface movement and deformation. The subsidence at any point is shown as follows:

$$w(x,y) = w_{\max} \cdot \iint_D \frac{1}{r^2} \cdot \exp\left\{-\pi[\eta - x]^2 + (\xi - y)^2 / r^2\right\} \cdot d\eta \cdot d\xi \tag{1}$$

The horizontal movement at any point is shown as follows:

$$ux(x,y) = u_{\max} \cdot \iint_D \frac{2\pi(\eta - x)}{r^3} \cdot \exp\left\{-\pi\left[(\eta - x)^2 + (\xi - y)^2\right] / r^2\right\} d\eta \cdot d\xi \tag{2}$$

4 The Effects on Mining Buildings

It destroys the balance of the terrane in the process of mining coal; the rock movement spreads to the surface, and then the surface begins to move. Meanwhile, it causes the building to damage. The main styles are given as follows: the ground subsidence damage, the ground horizontal deformation damage, the ground tilt damage and the ground bend damage.

The serious collapse area is in the western of the gob. The building in this area is severely damaged. The gob is $5.93km^2$ and subsidence area is $9.93km^2$ and the ratio of subsidence area and the mining area is 1.67. The subsidence value is $7055mm$ and the subsidence is greatly impacted on the surrounding villages. In this paper, based on the damage level of the brick structure buildings from "Coal Mining specifications under buildings, water and railway", it gives the classification of mining subsidence. It is divided into four levels of mining subsidence area: no subsidence area, the slightly collapsed area, the second seriously collapsed area, the seriously collapsed area.

The characteristics of different levels area are shown as follows: In the seriously collapsed area, the buildings are damaged seriously and the width of wall crack is from 12cm to 30cm. The windows and doors can't switch because of the serious deformation. And most buildings have been abandoned. The roads are collapsed seriously. The most land also have been abandoned for growing no crops; In the second seriously collapsed area, it has formed a small subsidence basin and the buildings are destroyed greatly, but they can live in by means of reinforcement. A small number of land subsidences are serious and the villagers have given up farming. The roads are collapsed greatly, but some of them can also use by means of reinforcement. In the slightly collapsed area, the buildings and roads mostly are destroyed slightly. In the no subsidence area, the buildings, roads and farmland have no effect by mining subsidence.

5 The Control Measures

5.1 The Preventive Measures of Mining Subsidence

It generally uses the blasting method, filling method and closed method to prevent mining subsidence. The blasting method is to drill deep holes or caverns in coal pillar and rock, and then it blasts coal pillar and rock, meanwhile it fills the whole mining area. The filling method is to use exotic materials to fill the whole mining area; moreover, it can support around surrounding rock and reduce the extent of damage; the closed method is to build enclosed strong walls in all roadways, then it closes the mining area and left it to collapse itself [3].

5.2 Layout of Buildings

Surface subsidence is caused by coal mining. The characteristics of displacement and deformation are different at any points. In the super-full extraction area, the final form of the static surface movement basin can be divided into three regions: the middle zone, the inner marginal zone and the outer marginal zone. It must avoid arranging the buildings in the inner marginal zone. The buildings axis and the working face direction should be parallel or perpendicular, otherwise it likely causes distortion and shear failure. In the middle zone, the buildings axis and the working face direction should be perpendicular. In the marginal zone, the buildings axis and the working face direction should be parallel [4].

5.3 The Protection Measures of Mining Buildings

By choosing the right structural protection measures, it can reduce the additional stress of mining buildings and enhance its ability to withstand ground deformation. Moreover it also prevents uneven settlement of building foundation. There are many effective measures to protect the buildings structure such as the buffer trench excavation, set deformation joint, set the steel rod, reinforced concrete ring beams and basic contact beams. [5].

6 Conclusion

In the process of coal mining under the buildings, it causes the surface above gob has movement and deformation; moreover, it also can seriously affect the building foundation. It gives the extent and scope of coal mining by means of computer simulation. According to the specific geological and mining conditions in mining area, it gives the preventive measures of mining subsidence and protection measures of mine buildings. It has important significance to sustainable development of the coal industry.

References

1. Tian, K., Zhang, Z.: Henan Province Coal Mine Investigation and Comprehensive Research Report. Henan Nonferrous Geotechnical Engineering Company, Henan (2006)
2. Li, Y.: Height prediction method and its application of fissure zone. Xi'an University of Technology, Xi'an (2007)
3. Guo, G.: Deformation mechanism and control of old building above gob. China University of Mining Press, Jiangsu (2001)
4. Yan, R.: Mining subsidence and surface construction. Metallurgical Industry Press, Beijing (1995)
5. Zou, Y., Hu, Y., Guo, Z.: Mining damage and protection. China University of Mining Press, Xuzhou (1996)

Existence of Solution for Boundary Value Problem of Impulsive Differential Equations in Banach Spaces

Dehong Ji[1] and Weigao Ge[2]

[1] College of Science, Tianjin University of Technology, Tianjin, 300384, China
[2] School of Science, Beijing Institute of Technology, Beijing, 100081, China
jdh200298@163.com

Abstract. In this paper, we discuss existence of solution for boundary value problem of impulsive differential equations in Banach spaces. The arguments are based upon the fixed point theorem of strict set contraction operator.

Keywords: Boundary value problem, Impulsive differential equations, Existence of solution, Banach spaces, Measureof noncompctness.

1 Introduction

The theory of ordinary differential equations in abstract spaces has become a new important branch, so boundary value problem in Banach Space has been studied by some researchers, we refer the readers to [7-12] and the references therein.

In [6], Feng and Xie considered the following boundary value problem (BVP for short)

$$\begin{cases} -u''(t) = f(t, u(t)), \quad t \neq t_k, \quad t \in (0,1), \\ -\Delta u'|_{t=t_k} = I_k(u(t_k)), \quad k = 1, 2, \cdots, n, \\ u(0) = \sum\limits_{i=1}^{m-2} a_i u(\xi_i), \quad u(1) = \sum\limits_{i=1}^{m-2} b_i u(\xi_i), \end{cases}$$

but paper [6] did not give the existence of solution for the above BVP in Banach spaces.

Very recently, Zhao and Chen in [4] studied the following m-point boundary value problem:

$$x''(t) + a(t) f(t, x(t)) = \theta, \quad t \in (0,1),$$

$$x'(0) = \sum_{i=1}^{m-2} b_i x'(\xi_i), \quad x(1) = \sum_{i=1}^{m-2} a_i x(\xi_i),$$

in Banach spaces E, where μ is zero element of E, $\xi_i \in (0,1)$ with $0 < \xi_1 < \xi_2 < \cdots < \xi_{m-2} < 1, a_i, \ b_i \in [0, +\infty)$ (i =1, 2, $\cdots$, m-2): By

D. Zeng (Ed.): Advances in Computer Science and Engineering, AISC 141, pp. 255–264.
springer link.com

using the fixed point theorem of strictset-contractions, they obtained some sufficient conditions for the existence of at least one or two positive solutions to the above problem.

Let E be a real Banach space with norm $\|x\|$ and $K \subset E$ be a cone of E: Motivated by the works mentioned above, the purpose of this paper is to investigate the existence of solution for the following boundary value problem of impulsive differential equations in Banach spaces

$$\begin{cases} u''(t) + f(t, u(t)) = \theta, & t \neq t_k, \quad t \in J, \\ \Delta u'|_{t=t_k} = -I_k(u(t_k)), & k = 1, 2, \cdots, n, \\ u(0) = \sum_{i=1}^{m-2} a_i u(\xi_i), & u(1) = \sum_{i=1}^{m-2} b_i u(\xi_i), \end{cases} \tag{1}$$

where $J = [0, 1]$; θ is the zero element of E; $t_k (k = 1, 2, \cdots, n$ where n is a fixed positive integer) are fixed points with $0 < t_1 < t_2 < \cdots < t_k < \cdots < t_n < 1$, $0 < \xi_1 < \xi_2 < \cdots < \xi_{m-2} < 1$, $\xi_i \neq t_k$, $i = 1, 2, \cdots, m-2$, $k = 1, 2, \cdots, n$,

$$\Delta u'|_{t=t_k} = u'(t_k^+) - u'(t_k^-), \tag{2}$$

where $u'(t_k^+), u'(t_k^-)$ denote the right limit (left limit) of $u'(t)$ at $t = t_k$, a_i, $b_i \in (0, +\infty)$, $i = 1, 2, \cdots, m-2$.

The cone K in E induces a partial order in E by $x \leq y$ if and only if $y - x \in K$. K is said to be normal if there exists a positive constant N such that $\theta \leq x \leq y$ implies $\|x\| \leq N\|y\|$, where θ denotes the zero element of E, and the smallest N is called the normal constant of K (it is clear, $N \leq 1$). K is called solid if its interior $\dot{K}$ is nonempty. If $x \leq y$ and $x \neq y$, we write $x < y$. If K is solid and $y - x \in \dot{K}$, we write $y \gg x$. For details on cone theory, see [1].

The operator $T : D \to E (D \subset E)$ is said to be a k-set contraction if $T : D \to E$ is continuous and bounded and there is a constant $k \geq 0$ such that $\alpha(T(S)) \leq k\alpha(S)$ for any bounded $S \subset D$; a k-set contraction with $k < 1$ is called a strict set contraction.

In the following, $\alpha(\cdot)$, $\alpha_{PC'}(\cdot)$ denote the Kuratowskis measure of noncompactness in E and PC' $[J,E]$, respectively.

The main tool of this paper is the following fixed point Theorems.

Theorems.

Theorem 1. [5] Let D be a bounded, closed and convex subset of the real Banach space E. If operator $A : D \to D$ is a strict set contraction, then A has a fixed point in D.

2 Prcliminaries

To establish the existence of solution in E of problem (1), let us list the following assumptions, which will stand throughout this paper:

(H_1) $f \in C(J \times E, E)$ and for any $l > 0$; f is uniformly continuous on $J \times B_l$, where $B_l = \{x \in E : \|x\| \leq l\}$;

(H_2) $I_k \in C(E, E)$;

(H_3) $\Delta < 0,\ \sum_{i=1}^{m-2} b_i\xi_i < 1,\ \sum_{i=1}^{m-2} a_i(1-\xi_i) < 1$ where

$$\Delta = \begin{vmatrix} -\sum_{i=1}^{m-2} a_i\xi_i & 1-\sum_{i=1}^{m-2} a_i(1-\xi_i) \\ 1-\sum_{i=1}^{m-2} b_i\xi_i & -\sum_{i=1}^{m-2} b_i(1-\xi_i) \end{vmatrix}$$

(H_4) There exist nonnegative constants η_l, d_l such that

$$\alpha(f(t,S)) \leq \eta_l \alpha(S), \quad t \in J,\ S \in B_l,$$

$$\alpha(I_k(S)) \leq d_l \alpha(S), \quad S \in B_l,$$

and

$$\Gamma_l = 2\eta_l + nd_l + \frac{1}{2}\left(\frac{-\sum_{i=1}^{m-2} b_i}{\Delta}\right)\eta_l + \frac{n}{4}\left(\frac{-\sum_{i=1}^{m-2} b_i}{\Delta}\right)d_l$$
$$+\frac{1}{2}\left(\frac{-\sum_{i=1}^{m-2} a_i}{\Delta}\right)\eta_l + \frac{n}{4}\left(\frac{-\sum_{i=1}^{m-2} a_i}{\Delta}\right)d_l < 1.$$

Let $PC[J,E] = \{x : x\}$ is a map from J into E such that $x(t)$ is continuous at $t \neq t_k$; left continuous at $t = t_k$ and its right limit at $t = t_k$ $x(t_k^+)$ exists for $k = 1, 2, \cdots, n\}$ be a Banach space with the norm $\|x\|_{PC} = \sup_{t\in J} \|x(t)\|$. $PC'[J,E] = \{x : x \in PC[J,E] : x'(t)$ is continuous at $t \neq t_k$, left continuous at $t = t_k$ and its right limit at $t = x'(t_k^+)$ exists for $k = 1, 2, \cdots, n\}$ be a Banach space with the norm $\|x\|_{PC'} = \max\{\|x\|_{PC}, \|x'\|_{PC}\}$

Let $J' = J\backslash\{t_1, t_2, \cdots t_n\}$ .In this paper, $u(t) \in PC'[J,E] \cap C^2[J',E]$ is called positive solution of problem (1) if $u(t)$ is a solution of (1) and $u(t) > \theta,\ 0 < t < 1$. In the proof of the main result, the following lemmas are fundamental.

Lemma 2: [6] Assume that (H_1) — (H_4) hold. Then $u \in PC'[J,E] \cap C^2[J',E]$ is a solution of problem *(1)* if and only if *u* is a solution of the following impulsive integral equation:

$$\begin{aligned} u(t) = \int_0^1 G(t,s)f(s,u(s))ds + \sum_{k=1}^{n} G(t,t_k)I_k(u(t_k)) \\ + t[A(f(\cdot,u(\cdot))) + B(I_k(u(\cdot)))] \\ + (1-t)[C(f(\cdot,u(\cdot))) + D(I_k(u(\cdot)))] \end{aligned} \tag{3}$$

where

$$G(t,s) = \begin{cases} s(1-t), & 0 \le s \le t \le 1, \\ t(1-s), & 0 \le t \le s \le 1, \end{cases} \tag{4}$$

$$A(\cdot) = \frac{1}{\triangle}\begin{vmatrix} \sum_{i=1}^{m-2} a_i \int_0^1 G(\xi_i,t)f(t,u(t))dt & 1-\sum_{i=1}^{m-2} a_i(1-\xi_i) \\ \sum_{i=1}^{m-2} b_i \int_0^1 G(\xi_i,t)f(t,u(t))dt & -\sum_{i=1}^{m-2} b_i(1-\xi_i) \end{vmatrix} \tag{5}$$

$$B(\cdot) = \frac{1}{\triangle}\begin{vmatrix} \sum_{i=1}^{m-2} a_i \left[\sum_{k=1}^{n} G(\xi_i,t_k)I_k(u(t_k))\right] & 1-\sum_{i=1}^{m-2} a_i(1-\xi_i) \\ \sum_{i=1}^{m-2} b_i \left[\sum_{k=1}^{n} G(\xi_i,t_k)I_k(u(t_k))\right] & -\sum_{i=1}^{m-2} b_i(1-\xi_i) \end{vmatrix} \tag{6}$$

$$C(\cdot) = \frac{1}{\triangle}\begin{vmatrix} -\sum_{i=1}^{m-2} a_i\xi_i & \sum_{i=1}^{m-2} a_i \int_0^1 G(\xi_i,t)f(t,u(t))dt \\ 1-\sum_{i=1}^{m-2} b_i\xi_i & \sum_{i=1}^{m-2} b_i \int_0^1 G(\xi_i,t)f(t,u(t))dt \end{vmatrix}, \tag{7}$$

$$D(\cdot) = \frac{1}{\triangle}\begin{vmatrix} -\sum_{i=1}^{m-2} a_i\xi_i & \sum_{i=1}^{m-2} a_i \left[\sum_{k=1}^{n} G(\xi_i,t_k)I_k(u(t_k))\right] \\ 1-\sum_{i=1}^{m-2} b_i\xi_i & \sum_{i=1}^{m-2} b_i \left[\sum_{k=1}^{n} G(\xi_i,t_k)I_k(u(t_k))\right] \end{vmatrix}. \tag{8}$$

Lemma 2: For $\forall t, s \in [0,1]$, we have

$$0 \le G(t,s) \le G(s,s) \le \frac{1}{4}. \tag{9}$$

Lemma 3: Suppose that (H_1) — (H_4) hold. We define an operator *T* by

$$\begin{aligned}(Tu)(t) = \int_0^1 G(t,s)f(s,u(s))ds + \sum_{k=1}^{n} G(t,t_k)I_k(u(t_k)) \\ + t[A(f(\cdot,u(\cdot))) + B(I_k(u(\cdot)))] \\ + (1-t)[C(f(\cdot,u(\cdot))) + D(I_k(u(\cdot)))]\end{aligned} \tag{10}$$

in which $G(t,s), A(f(\cdot,u(\cdot))), B(I_k(u(\cdot))), C(f(\cdot,u(\cdot))),\ D(I_k(u(\cdot)))$ are given by *(4)*, *(5)*, *(6)*, *(7)*, *(8)* separately. Then $u \in PC'[J,E] \cap C^2[J',E]$ is a solution of problem *(1)* if and only if *u* is a fixed point of *T*.

Lemma 4:[5] If $S \subset PC'[J,E]$ is bounded and the elements of S' are equicontinuous on each $J_k(k = 1,2,\cdots,n)$ then

$$\alpha_{PC'}(S) = \max\left\{\sup_{t\in J}\alpha(S(t)), \sup_{t\in J}\alpha(S'(t))\right\}.$$

Lemma 5: Assume that (H_1) — (H_4) hold. Then operator *T* is a strict set contraction from $B_l^{(1)}$ into $PC'[J,E]$, where

$$B_l^{(1)} = \{u \in PC'[J,E] :\ \|u\|_{PC'} \le l\}$$

Proof: By (H_1) — (H_4); it is clear that $T : B_l^{(1)} \to PC'[J,E]$ is continuous and bounded. Let $S \subset B_l^{(1)}$ be arbitrary given. So $T(S) \subset PC'[J,E]$ is bounded. Using (10), we get

$$\begin{aligned}\alpha((T(S))(t)) &\le \tfrac{1}{4}\alpha(f(t,S(t))) + \tfrac{1}{4}\sum_{k=1}\alpha(I_k(S(t_k))) \\ &+ \alpha(A(f(t,S(t)))) + \alpha(B(I_k(S(t_k)))) \\ &+ \alpha(C(f(t,S(t)))) + \alpha(D(I_k(S(t_k)))) \\ &\le \tfrac{1}{4}\eta_l\alpha(S(J)) + \tfrac{1}{4}\sum_{k=1}^{n}\alpha(I_k(S(t_k))) \\ &+ \tfrac{1}{4}\left(\frac{-\sum_{i=1}^{m-2} b_i}{\Delta}\right)\eta_l\alpha(S(J)) \\ &+ \tfrac{1}{4}\left(\frac{-\sum_{i=1}^{m-2} b_i}{\Delta}\right)\sum_{k=1}^{n}\alpha(I_k(S(t_k))) \\ &+ \tfrac{1}{4}\left(\frac{-\sum_{i=1}^{m-2} a_i}{\Delta}\right)\eta_l\alpha(S(J)) \\ &+ \tfrac{1}{4}\left(\frac{-\sum_{i=1}^{m-2} a_i}{\Delta}\right)\sum_{k=1}^{n}\alpha(I_k(S(t_k)))\end{aligned}$$

$$
\begin{aligned}
&\leq \tfrac{1}{4}\eta_l\alpha(S(J)) + \tfrac{1}{4}d_l \sum_{k=1}^{n} \alpha(S(t_k)) \\
&+\tfrac{1}{4}\left(\frac{-\sum_{i=1}^{m-2} b_i}{\Delta}\right)\eta_l\alpha(S(J)) \\
&+\tfrac{1}{4}d_l\left(\frac{-\sum_{i=1}^{m-2} b_i}{\Delta}\right)\sum_{k=1}^{n} \alpha(S(t_k)) \\
&+\tfrac{1}{4}\left(\frac{-\sum_{i=1}^{m-2} a_i}{\Delta}\right)\eta_l\alpha(S(J)) \\
&+\tfrac{1}{4}d_l\left(\frac{-\sum_{i=1}^{m-2} a_i}{\Delta}\right)\sum_{k=1}^{n} \alpha(S(t_k))
\end{aligned} \tag{11}
$$

the definition of $\alpha_{PC'}(S)$ in lemma 4 implies

$$\alpha(S(J)) \leq 2\alpha_{PC'}(S), \quad \alpha(S'(J)) \leq 2\alpha_{PC'}(S), \tag{12}$$

where $S(J) = \{x(t) : x \in S, t \in J\}$ and $S'(J) = \{x'(t) : x \in S, t \in J\}$

On the other hand, similar to the proof of Lemma 2 in [9], we can obtain

$$\alpha(S(t_k)) \leq \alpha_{PC'}(S), \quad \alpha(S'(t_k)) \leq \alpha_{PC'}(S)\; k = 1, 2, \cdots, n. \tag{13}$$

It follows from (11), (12), (13) that

$$\alpha(T(S)(t)) \leq \Gamma_r\alpha_{PC'}(S), \ \forall t \in J. \tag{14}$$

Similarly, in view of (10), (12), (13), we have

$$\alpha(T(S)'(t)) \leq \Gamma_r\alpha_{PC'}(S), \ \forall t \in J. \tag{15}$$

Noticing $\Gamma_r < 1$; we claim that T is a strict set contraction and the lemma is proved.

3 Main Result

In this section, we impose growth conditions on *f* which allow us to apply Theorem 1 to establish the existence of positive solutions of BVP (1). At the beginning, we define

$$\overline{\lim_{\|u\|\to\infty}}\left(\sup_{t\in J}\frac{\|f(t,u)\|}{\|u\|}\right)=\Lambda,$$

$$\overline{\lim_{\|u\|\to\infty}}\frac{\|I_k(u)\|}{\|u\|}=\Lambda_k\ (k=1,2,\cdots,n).$$

Theorem1: Assume that (H_1) — (H_4) hold. If

$$\begin{aligned}\delta=&\left[1+\tfrac{1}{4}\left(\frac{-\sum_{i=1}^{m-2}b_i}{\Delta}\right)+\tfrac{1}{4}\left(\frac{-\sum_{i=1}^{m-2}a_i}{\Delta}\right)\right]\Lambda\\&+\left[1+\tfrac{1}{4}\left(\frac{-\sum_{i=1}^{m-2}b_i}{\Delta}\right)+\tfrac{1}{4}\left(\frac{-\sum_{i=1}^{m-2}a_i}{\Delta}\right)\right]\sum_{k=1}^{n}\Lambda_k\\&<1,\end{aligned}\tag{16}$$

Then BVP *(1)* has at least one solution $u\in PC'[J,E]\cap C^2[J',E]$

Proof: By Lemma 5, operator *T* defined by (10) is a strict set contraction from $B_l^{(1)}$ into $PC'[J,E]$, and by Lemma 3, we need only to show that *T* has one fixed point $u(t)\in PC'[J,E]\cap C^2[J',E]$.

On account of (16), we can choose $\Lambda'>\Lambda,\ \Lambda_k'>\Lambda_k\ (k=1,2,\cdots,n)$ such that

$$\begin{aligned}\delta'=&\left[1+\tfrac{1}{4}\left(\frac{-\sum_{i=1}^{m-2}b_i}{\Delta}\right)+\tfrac{1}{4}\left(\frac{-\sum_{i=1}^{m-2}a_i}{\Delta}\right)\right]\Lambda'\\&+\left[1+\tfrac{1}{4}\left(\frac{-\sum_{i=1}^{m-2}b_i}{\Delta}\right)+\tfrac{1}{4}\left(\frac{-\sum_{i=1}^{m-2}a_i}{\Delta}\right)\right]\sum_{k=1}^{n}\Lambda_k'\\&<1,\end{aligned}\tag{17}$$

By the definition of Λ, there exists $r>0$ such that

$$\|f(t,u)\|<\Lambda'\|u\|,\quad\forall t\in J,\ u\in E,\ \|u\|>r,\tag{18}$$

so

$$\|f(t,u)\|<\Lambda'\|u\|+M,\quad\forall t\in J,\ u\in E,\tag{19}$$

where

$$M=\sup\{\|f(t,u)\|:\ t\in J,\ \|u\|\le r\}<\infty.$$

Similarly, we have

$$\|I_k(u)\|<\Lambda_k'\|u\|+M_k,\quad\forall u\in E\ (k=1,2,\cdots,n),\tag{20}$$

where M_k is positive constant. Now, (10), (18)-(20) imply

$$
\begin{aligned}
\|(Tu)(t)\| &\leq \tfrac{1}{4}\int_0^1 (\Lambda'\|u(s)\| + M)ds \\
&\quad + \tfrac{1}{4}\sum_{k=1}^{n}(\Lambda_k'\|u(t_k)\| + M_k) \\
&\quad + \tfrac{1}{4}\left(\frac{-\sum_{i=1}^{m-2} b_i}{\Delta}\right)\int_0^1 (\Lambda'\|u(s)\| + M)ds \\
&\quad + \tfrac{1}{4}\left(\frac{-\sum_{i=1}^{m-2} b_i}{\Delta}\right)\sum_{k=1}^{n}(\Lambda_k'\|u(t_k)\| + M_k) \\
&\quad + \tfrac{1}{4}\left(\frac{-\sum_{i=1}^{m-2} a_i}{\Delta}\right)\int_0^1 (\Lambda'\|u(s)\| + M)ds \\
&\quad + \tfrac{1}{4}\left(\frac{-\sum_{i=1}^{m-2} a_i}{\Delta}\right)\sum_{k=1}^{n}(\Lambda_k'\|u(t_k)\| + M_k) \\
&\leq \left[\tfrac{1}{4} + \tfrac{1}{4}\left(\frac{-\sum_{i=1}^{m-2} b_i}{\Delta}\right)\right. \\
&\quad \left. + \tfrac{1}{4}\left(\frac{-\sum_{i=1}^{m-2} a_i}{\Delta}\right)\right]\Lambda'\|u\|_{PC'} \\
&\quad + \left[\tfrac{1}{4} + \tfrac{1}{4}\left(\frac{-\sum_{i=1}^{m-2} b_i}{\Delta}\right)\right. \\
&\quad \left. + \tfrac{1}{4}\left(\frac{-\sum_{i=1}^{m-2} a_i}{\Delta}\right)\right]\sum_{k=1}^{n}\Lambda_k'\|u\|_{PC'} \\
&\quad + \left[\tfrac{1}{4} + \tfrac{1}{4}\left(\frac{-\sum_{i=1}^{m-2} b_i}{\Delta}\right)\right. \\
&\quad \left. + \tfrac{1}{4}\left(\frac{-\sum_{i=1}^{m-2} a_i}{\Delta}\right)\right]M
\end{aligned}
$$

$$+\left[\frac{1}{4}+\frac{1}{4}\left(\frac{-\sum_{i=1}^{m-2} b_i}{\Delta}\right)\right.$$
$$\left.+\frac{1}{4}\left(\frac{-\sum_{i=1}^{m-2} a_i}{\Delta}\right)\right]\sum_{k=1}^{n} M_k$$
$$\leq \delta'\|u\|_{PC'}+M', \tag{21}$$

where δ' is defined by (17) and M' is defined by

$$M' = \left[1+\frac{1}{4}\left(\frac{-\sum_{i=1}^{m-2} b_i}{\Delta}\right)+\frac{1}{4}\left(\frac{-\sum_{i=1}^{m-2} a_i}{\Delta}\right)\right]M$$
$$+\left[1+\frac{1}{4}\left(\frac{-\sum_{i=1}^{m-2} b_i}{\Delta}\right)+\frac{1}{4}\left(\frac{-\sum_{i=1}^{m-2} a_i}{\Delta}\right)\right]\sum_{k=1}^{n} M_k$$

Similarly, from (10), (17)-(20), we can get

$$\|(Tu)'(t)\| \leq \delta'\|u\|_{PC'}+M', \quad \forall t\in J. \tag{22}$$

It follows from (21) and (22) that

$$\|Tu\|_{PC'} \leq \delta'\|u\|_{PC'}+M', \qquad \forall u\in PC'[J,E].$$

Hence, we can choose a sufficiently large $r>0$ such that $T(B_r^{(1)})\subset B_r^{(1)}$.
On the other hand, by Lemma 5, T is a strict set contraction from $B_r^{(1)}$ into $B_r^{(1)}$. Consequently, Theorem 1 implies that T has a fixed point in $B_r^{(1)}$, and the proof is complete.

Acknowledgment. This work is sponsored by the National Natural Science Foundation of China (No. (11071014)) and the Tianjin City High School Science and Technology Fund Planning Project (No. (20091008)) and Tianyuan Fund of Mathematics in China (No. (11026176)) and Natural Science Foundation of Shandong Province of China (No. (ZR2010AM035)).

References

1. Lakshmikanthan, V., Leela, S.: Nonlinear Differential Equations in Abstract Spaces. Pergamon, Oxford (1981)
2. Gupta, C.P.: A note on a second order three-point boundary value problem. J. Math. Anal. Appl. 186, 277–281 (1994)

3. Ma, R.Y., Castaneda, N.: Existence of solutions of nonlinear m-point boundary-value problems. J. Math. Anal. Appl. 256, 556–567 (2001)
4. Zhao, Y.L., Chen, H.B.: Existence of multiple positive solutions form point boundary value problems in Banach spaces. J. Comput. Appl. Math. 215, 79–90 (2008)
5. Zhang, X., Feng, M., Ge, W.: Existence of solutions of boundary value problems with integral boundary conditions for second-order impulsive integro-differential equations in Banach spaces. J. Comput. Appl. Math. 233, 1915–1926 (2010)
6. Feng, M., Xie, D.: Multiple positive solutions of multi-point boundary value problem for second-order impulsive differential equations. J. Comput. Appl. Math. 223, 438–448 (2009)
7. Liu, B.: Positive solutions of a nonlinear four-point boundary value problems in Banach spaces. J. Math. Anal. Appl. 305, 253–276 (2005)
8. Guo, D.J.: Existence of solutions of boundary value problems for second order impulsive differential equations in Banach spaces. J. Math. Anal. Appl. 181, 407–421 (1994)
9. Guo, D.J., Liu, X.Z.: Multiple positive solutions of boundary-value problems for impulsive differential equations. Nonlinear Anal. 25, 327–337 (1995)
10. Guo, D.J.: Periodic boundary value problems for second order impulsive integro-differential equations in Banach spaces. Nonlinear Anal. 28, 983–997 (1997)
11. Guo, D.J.: Second order impulsive integro-differential equations on unbounded domains in Banach spaces. Nonlinear Anal. 35, 413–423 (1999)
12. Guo, D.J.: Multiple positive solutions for first order nonlinear impulsive integro-differential equations in a Banach space. Appl. Math. Comput. 143, 233–249 (2003)

Intelligent Optimal Control in Aluminum Electrolytic

Chengliang Wang[1], Juanjuan Chen[2], and Lianwei Wang[1]

[1] College of Computer Science, University of ChongQing, Chongqing, China
[2] College of Computer& Information Science, Chongqing Normal University, Chongqing, China
wangcl@cqu.edu.cn, cameroncjj@gmail.com, wlwxhl01@163.com

Abstract. Intelligent control is more and more important in aluminum electrolytic processes. This paper studies the operation of existing aluminum electrolytic control systems and expounds the problems confronting aluminum electrolytic control. It then applies Ontology theory to guide data analysis and modeling of intelligent control of aluminum electrolytes in the whole process. Through the application of data-driving and semi-supervised methods to analyze aluminum electrolytic processes, an aluminum electrolytic intelligent control expert knowledge was extracted. On the basis of combination of an expert system, neural network and intelligent time series data prediction algorithm, this expert knowledge base succeeds in real-time, adaptive controlling of aluminum electrolytic processes. It makes the concentration of alumina change in accordance with the current cell and working conditions, thereby allowing the conditions to maintain a proper level, or at least to develop in that direction. Finally, the feasibility and effectiveness of the intelligent control system was verified by long time practical operation.

Keywords: Intelligent system, electrolytic aluminum, ontology, expert system, intelligent control.

1 Introduction

The traditional aluminum electrolytic (AE) controlling system can barely meet the requirements of efficient production and energy saving, hence advanced integrated intelligent control techniques are very necessary. However, AE is a nonlinear industrial system, which is multivariable-coupling, has a time-varying delay, and is accompanied by complex physical and chemical reactions. The electrolytic cell (EC) operation is determined by many parameters and variables, which are strong coupling, uncertainty, and hard to continually predict. Therefore, the complex production process is difficult for mathematical modeling[1]. That is why there is a trend that many artificial intelligence techniques are being applied in AE production control[2], but the complete and accurate expert knowledge required by most intelligent controlling techniques are hard to be obtained. Although neural network is also introduced in AE[3], the parameters are

D. Zeng (Ed.): Advances in Computer Science and Engineering, AISC 141, pp. 265–274.
springer link.com

inadequate in physical meaning and so they barely understood by domain experts, which makes the machine learning hard to integrate with the expert knowledge[4]. Such problems are very common in complex industrial environment, not only typically in aluminum, but also in steelmaking and cement manufacturing etc. Therefore many scholars focus on seeking intelligent solving mechanism and related schemes.

This study combines Ontology theory [5], data-driving analysis[6] and various intelligent control together. Based on data-driving analysis and with the help of domain experts, AE production control Ontology Base is constructed, above which various artificial intelligent algorithms are integrated. Finally, the mechanism of hybrid intelligent AE control system is designed and realized.

2 Research Background

2.1 Essence of AE Controlling

The essence for the AE optimal controlling is:

1) Improve Data learning ability: based on a large amount of complex data analysis, obtain the accurate assessment of the stability and the working state of the AE cell and alumina concentration region;

2) Improve Controlling Intelligence: according to different working states of AE cell, appropriate controlling policies are applied to maintain the cell working at a healthy and stable state or prompt to the tendency.

2.2 Problem Facing Intelligent Algorithm in AE

According to the objectives in AE controlling mentioned above, several Artificial Intelligence technologies are introduced in this area, such as expert systems, neural networks and the wavelet theory. To some extent, these methods partially optimize AE controlling. However, these controlling algorithms have several limitations:

1) Failure of applying intelligent controlling technology throughout the whole process

2) Failure to effectively mining the hidden information

3) Failure to well adapt to both cell working condition and cell working state

2.3 The Solutions

To solve the problems mentioned in part B, we have to confront the reality that AE optimal controlling lacks complete conceptual model to define the working state; lacks corresponding mathematical model to identify the concept; and lacks a formalized parameter model to facilitate the effective computation of the mathematical model.

Therefore, this paper employs Ontology Theory to guide the data analysis and modeling in the AE optimal controlling process, namely with data-driving and semi-supervised pattern to analyze the AE process. Through adding the time scale changing operator and feature analysis operator in the traditional algorithm, while under the guidance of domain experts, this paper obtains the Ontology Base for AE intelligent control.

The whole control system is divided into production operation layer, data processing layer, data warehouse and computation control layer. Firstly, in the production operation layer slot-machines write real-time data into the real-time database through CAN bus. Secondly, the data processing layer uses ETL to regularly move the data from the real-time database to the data warehouse, while cleaning and integrating the data in the data warehouse for computational inference. Here ETL runs on the data migration server, and data cleaning and integration processes on the data processing server. Finally, the computational control layer conducts computational inference in accordance with different optimal control models. The structure of the 'intelligent information analysis and control systems for electrolytic process' is shown in Fig.1.

3 Expert Knowledge Base Construction

3.1 AE Ontology Base Construction Process

The basic idea of this section is: supervised by experts' prior knowledge in AE, and based on the massive data generated in the electrolysis process, we apply the semi-supervised flexible time windows mining algorithm to construct the Ontology Base, as is shown in Fig.2.

Because aluminum electrolytic voltage curve contains plenty of information, the semi-supervised flexible time windows mining algorithm might find some hidden concepts and attributes that reflect electrolysis production. Then experts evaluate whether these concepts and attributes have physical meaning, and furthermore help to obtain Ontology Base that reflects on electrolysis production and also data model [7].

3.2 AE Ontology

Based on the construction process described in part A, we classified AE Ontology in two categories, which are parameter Ontology and concept Ontology [8]. Part of the parameter Ontology is listed in Table 1 and the concept Ontology is shown in Table 2. The concept Ontology is composed by two key items: *cell working condition* (reflects the long-term capacity fluctuation of electrolyte) and *cell working state* (reflects the short-term performance of electrolysis ability).

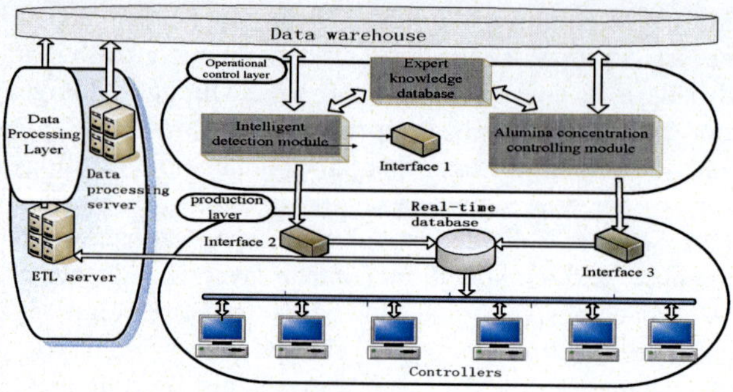

Fig. 1. General System Design Diagram

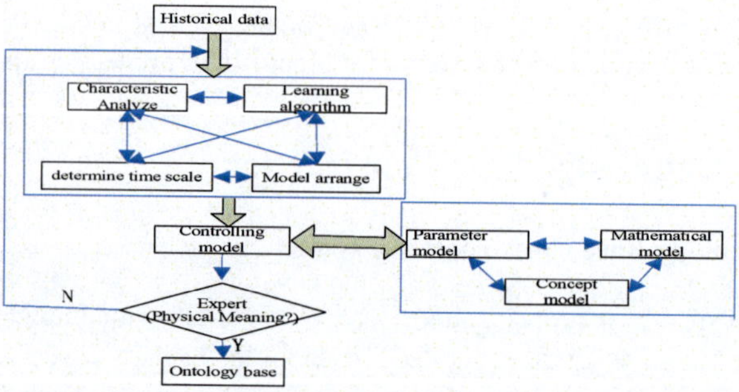

Fig. 2. The Process to construct the Ontology Base

Table 1. Parameter Ontology

Parameter Ontology	Parameter meaning	symbol	Equation
wave fluctuation	the ascent and descent wave range of real-time curve in certain time period	ρ	$\rho = \frac{\psi * \alpha}{\alpha + \beta} - \frac{\Omega * \beta}{\alpha + \beta}$
amplitude of wave	the amplitude of real-time curve in certain period of time	ϕ	$\phi = \frac{(\varepsilon + \eta)}{2}$
wave sudden change	the number and values of wave sudden changes of real-time curve	μ	$\mu = \delta * N_1$
fluctuations value	average value of voltage fluctuations(if $V_{i+1} - V_i \leq 30mv$ and $V_{i+1} - V_i = 0mv$)	χ	$\chi = \frac{[\sum_{i=1}^{N}(V_{i+1} - V_i)] * N_1}{N} (V_{i+1} - V_i > 30mv)$
condition ratio	ratio of working condition above normal(ω-the number of normal working condition, E- the total number of working condition diagnosed)	λ	$\lambda = \frac{\sum \omega}{E}$
Alumina concentration trends	the changing trends of alumina concentration in cell	$K(i)$	the detail of alumina concentration trends analysis is in section C

Table 2. Concept Ontology

Concept Ontology and Data Model	Characteristics of Real-time Curve	Collected Data	Data Model
Concept: *Cell working condition*	Sinusoidal or cosine voltage curve; fluctuating and irregular voltage curve; Regular in short term and proper in long term	Sampling Voltage Setting Voltage Anode-Rising (AR) AR in Aluminum flowing-out .... Anode-Dropping(AD) AD in Aluminum flowing-out	Very normal normal almost normal subordinate normal slight deterioration moderate deterioration serious deterioration
Concept: *Cell working state*		Normal Feeding Effect Feeding .... Normal Feeding in Busbar Rising Normal Feeding in Aluminum	healthy almost healthy sub-healthy subordinate healthy slight ill moderately ill severely ill
Concept: Alumina concentration		power input	low concentration area highly material-lacking area slightly excessive area moderately area low-right area high-right area

3.3 Knowledge Representation

1) Working condition and cells condition Knowledge

Based on the Ontology mentioned in part B, we use a prototype analysis method [10], combined with the working condition and cell working state diagnosis experience from electrolytic aluminum experts, we extract the prototype of working condition and cell working state diagnosis and formulate more than 200 working condition, more than 100 cell working state expert rules which are stored in the knowledge base of expert system. With the operation of the system, knowledge base constantly updates and upgrades to ensure continuity and consistency of knowledge and accuracy of the results of rules inference. The rules in knowledge base are production rules, some of which are shown in Table 3 and Table 4.

2) Alumina concentration Knowledge

a) Conceptual Layer of the Alumina concentration

During data mining, we need for the variable concept hierarchy and define it as a mapping sequence, which is the low-level concept mapping to the more general concept of a higher level. Concept hierarchy can be attributed to the data. Through such generalization of data, we can get simple and easily represent results for practical and knowledge level.

Through the repeated experiments of the process and analysis of aluminum electrolytic expert, conceptual models of all variables are given as follows:

Concentration: *Low, Highly Material-Lacking, Moderately Material-Lacking, Slightly Material-Lacking, Perfect Normal, Moderately Normal, Almost Normal, Slightly Material-Excessive, Moderately Material-Excessive, Highly Material-Excessive, Low-Right and High-Right.*

NB: *Normal, Less and Excessive*
Material-Lacking: *Mildly, Moderately, Highly, Strongly and Super-Strongly*
Material- Excessive: *Mildly, Moderately and Strongly*
Alumina resistance trends: *Steep Drop, Super-Strong Drop, Strong Drop, Moderate Drop, Mild Drop, Mild Rise, Moderate Rise, Strong Rise, Super-Strong Rise, Steep Rise*

Concept hierarchy can be manually provided by system users, domain experts and knowledge engineers, or generated automatically according to the statistical analysis of data distribution. However, when no standard definition is defined, or when there are knowledge discovery problems of data hierarchy, the entire data mining process requires repeated iteration among the concept hierarchy, data mining, mining results and experts' judgment[9].

b) Extraction of alumina concentration knowledge

Using the rules extracting process in figure 3, we got the alumina concentration knowledge, some of which are shown in Table 5 and Table 6.

Table 3. Diagnosis rules of cell working condition

Rules	Antecedent	Conclusion
Rules11	$\rho > 30$ & $4 < \phi \leq 5$ & $\mu > 1.0$	Slight deterioration
Rules13	$\rho > 30$ & $4 < \phi \leq 5$ & $\mu < 0.1$	Almost normal
Rules21	$\rho > 30$ & $\phi \leq 4$ & $\mu > 1.0$	Subordinate normal
Rules23	$\rho > 30$ & $\phi \leq 4$ & $\mu < 0.1$	Normal

Table 4. Diagnosis rules of cell working state

Rule	Antecedent	Conclusion
Rules11	$k \geq 200$ & $\lambda \geq 0.7$ & $\chi \geq 6$	Sub-health
Rules13	$k \geq 200$ & $\lambda < 0.3$ & $\chi \geq 6$	Subordinate health
Rules21	$k < 200$ & $0.3 \leq \lambda \leq 0.7$ & $3 < \chi \leq 6$	Slight ill
Rules23	$k < 200$ & $\lambda < 0.3$ & $\chi < 3$	Slight ill

Table 5. The parameters or threshold to determine the variable

No.	NB/K(i)	No.	NB/K(i)
1	$NB = 1 \rightarrow$ Normal	5	$K(i) < -3.5 \rightarrow$ Steep Drop
2	$NB > 1 \rightarrow$ Material -Lacking	6	$-3.5 < K(i) < -2 \rightarrow$ Super-Strong Drop
3	$0.9 < NB < 0.95 \rightarrow$ Moderately Excessive	7	$-2 < K(i) < -1.2 \rightarrow$ Strong Drop
4	$0.95 < NB < 1$	…..	………

Table 6. Calculation of Alumina Concentration Association Rules

Association rules to determine the concentration of alumina					
Concentration of last phase	$K(i)$	NB	$cd\_K(i)$		Current concentration
Low	0 ~ 1	<0.9	Positive		Mildly Lacking
Perfect Normal	≥ 2.5	≥ 1.0	Positive		Moderately Lacking
……	…….	…..	……		…….
Moderately Excessive	0.6 ~ 1.6	0.95 ~ 1.15	Positive	=>	Mildly excessive

4 System Constructions

4.1 Intelligent Detection Module

The aluminum electrolytic voltage curve contains a wealth of control information, such as significant mutations and fluctuations of the voltage curve that happen along with reaction and ascending and descending anode. Intelligent detection module analyzes the voltage curve and other indicators of real-time collection to mine the judging indicators of working and cell and uses forward inference mechanism to match the data set in database with the prerequisite of rules in knowledge base which is showed in Table 3, Table 4 and Table 5 to diagnose the current working condition, cell working state and alumina concentration as inference result only when the rules therein are all met.

4.2 Design of Alumina Concentration Control Model

At present, slot-machine control puts the alumina concentration control first, so accurate judgment of concentration is the basis of the whole concentration control process. The change of 'alumina concentration' in electrolytic cell has its own inherent characteristics; as many researches reveal, there is a qualitative relationship between the alumina concentration and cell resistance; but this relationship is a similar parabolic relationship that cannot be described in mathematical formula; Moreover, the actual production process has many uncertain factors. All these aforementioned result in the fact that the electrolyte cell resistance trends cannot tell the exact value of 'alumina concentration'.

Based on 'alumina concentration trends', 'alumina concentration changing process' and 'feeding interval coefficient', this paper combines neural network, association rules base and expert system to achieve accurate prediction and control of electrolytic cell resistance trends of aluminum electrolyte cells. Therein forms a new model for real-time prediction and controlling of alumina concentration of electrolyte cells, which might optimize aluminum electrolytic production control process. The basic control thoughts are mainly composed by the following steps:

1) Store data of aluminum electrolytic real-time data into data warehouse and use data mining techniques to form the 'aluminum concentration' judgments associated rule base that is mainly based on ' alumina concentration trends' and 'NB coefficient change value' in a certain period of time;

2) According to the target concentration range, the current concentration and control strategies from 'Module 1', infer control parameters;

3) Both control parameters and the corresponding neural network output serve as the input of association rules to predict the 'target concentration of alumina'; adjust the control parameters in accordance with the control strategy to best match the predicted concentration and target concentration;

4) Pass the determined control parameters to the real-time database to facilitate controlling and correct the control parameters according to the actual production output.

5 Experimental Results

Since January 2010, the system has begun in 6 sets of electrolysis electrolyte cells in the No. 2 branch factory of Guizhou CHALCO of China for trial operation. The main economic and technical indicators of system optimization before and after are shown in Table 7 and 8.

The main economic and technical indicators include effect coefficient, cell temperature, current efficiency and energy consumption. The effect mentioned here refers to a kind of chemical reaction that undermines electrolyte cell.

Effect Coefficient = effect number / cell *day;

Current efficiency (%) = (Actual production of raw aluminum) * 100% / (theoretical yield);

DC power consumption of electrolyte cell in the system has been reduced at some 100 ~ 150kwh/t-Al, which achieves the energy saving objection and further increases economic benefits for the enterprise. If an aluminum electrolytic company has an annual output of 650,000 tons of aluminum and power consumption reduced by 100~150kwh/t-Al, the annual cumulative electricity saving is 100kwh / tx650000t • 6 million kwh.

Fig.3 shows the operating process of No. 3325 electrolyte cell before implement Intelligent Optimal Control System. Fig.4 shows the operating process of No. 3325 electrolyte cell after implement Intelligent Optimal Control System. Obviously, after implementing the Intelligent Optimal Control System, the real-time voltage curve turns out to be a sine wave, the sine waveform leads to enhanced controllability of the electrolytic cell and a good cell working state.

Table 7. Main economic indicators before optimization

Cell No.	AE coefficient	Temperature (℃)	Current Efficiency	Energy Consumption(kwh/t)
3324	0.23	946.4	90.1%	14550 ～ 14600
3325	0.06	945.3	91.45%	
3326	0.29	943.5	91.8%	

Table 8. Main economic indicators after optimization

Cell No.	AE coefficient	Temperature (℃)	Current Efficiency	Energy Consumption(kwh/t)
3324	0.17	938.1	94.6%	14400 ～ 14500
3325	0.08	935.2	95.5%	
3326	0.08	936.5	95.9%	

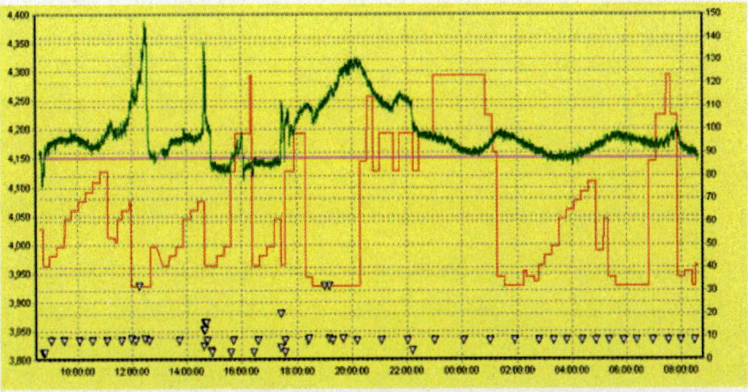

Fig. 3. The real-time curve of electrolytic cell before

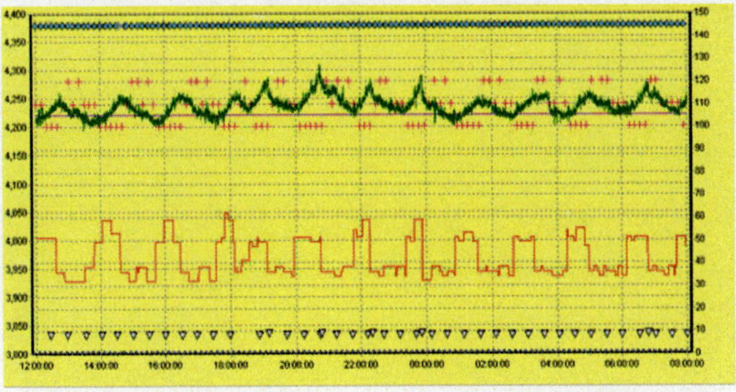

Fig. 4. The real-time curve of electrolytic cell after

Acknowledgment. The work described in this paper is supported by the national natural science foundation of china (61004112).

References

1. Homsi, P., Peyneau, J., Reverdy, M.: Overview of Process Control in Reduction Cells and Celllines. Light Metals, TMS, 223–229 (2000)
2. Meghlaoui, A., Bui, R.T., Thibault, J., Tikasz, L., Santerre, R.: Intelligent Control of the Feeding of Electrolytic Cells Using Neural Networks. Metall. and Materials Trans. 28B, 215–221 (1997)
3. Meghlaoui, A., Bui, R.T., Thibault, J., Tikasz, L., Santerre, R.: Predictive Control of Aluminum Electrolytic Cells Using Neural Networks. Metall. and Materials Trans. 29B, 1007–1019 (1998)
4. Tikasz, L., Bui, R.T., Cellocnik, V.: Aluminum Electrolytic Cells: a Computer Simulator for Training and Supervision. Engineering with Computers 10, 12–21 (1994)
5. Chen, P., Verma, R.: A query-based medical information summarization system using Ontology knowledge. In: 19th IEEE International Symposium on Computer-Based Medical Systems, pp. 37–42 (2006)
6. Cannone, R., Castiello, C., Mencar, C., Fanelli, A.M.: Data-driven Design of Fuzzy Classification Rules with Semantic Cointension. In: 2010 IEEE International Conference on Fuzzy Systems, p. 8 (2010)
7. A knowledge representation and inference mechanism for real-time adaptive process control. In: Proc. IEEE Workshop on Spatialand Temporal Interaction: Representation and Reasoning, pp. 190–209 (1994)
8. Sowa, J.F.: Conceptual Structures: Information Processing in Mind and Machine. Addison-Wesley, Reading (1984)
9. Das, G., Lin, K., Mannila, H.: Rule Discovery from Time Series. In: Proceedings of the 4th International Conference on Knowledge Discoveryand Data Mining, New York, U.S.A., pp. 16–22 (1998)
10. Chen, B., Xu, Y., Yang, J., Jiang, H.: A New Parallel Method of Smith-Waterman Algorithm on a Heterogeneous Platform. In: Hsu, C.-H., Yang, L.T., Park, J.H., Yeo, S.-S. (eds.) ICA3PP 2010. LNCS, vol. 6081, pp. 79–90. Springer, Heidelberg (2010)

Mathematical Model Study of Porous Media with Multi-branched Wells

Xiaoqing Wu[1], Xiao Guo[2], Yi Zhang[1], and Lijuan Wang[1]

[1] School of Sciences, Southwest Petroleum University, Chengdu, Sichuan Province, China
[2] School of Petroleum Engineering, Southwest Petroleum University, Chengdu, Sichuan Province, China
wuxiaoqing_swpu@163.com, guoxiao@swpu.edu.cn, Zhangyi331@126.com, 393295885@qq.com

Abstract. In the study of transient flow in porous medium, the precise point source solution for the transient flow. Described by initial value and boundary value problem of partial differential equations in the spherically symmetric indefinite infinite domain R^3, should be first considered. we can employ the precise point source solution to obtain the derivation of precise Mathematical model solution in vertical well, horizontal well and the line source, surface source, body source. This article's results was applied to the mathematical model of multibranch well.

Keywords: Partial differential equation, Point source, Multibranch well.

1 Introduction

In the study of transient flow in porous medium, the precise point source solution for the transient flow described by initial value and boundary value problem of partial differential equations in the spherically symmetric indefinite infinite domain R^3 should be first considered [1,2]. The initial value problem and boundary value problem of partial differential equations in the spherically symmetric indefinite infinite domain R^3 should also be taken into consideration at first.

$$\begin{cases} \frac{1}{r_D{}^2}\frac{\partial}{\partial r_D}(r_D{}^2\frac{\partial p_D}{\partial r_D}) = \frac{\partial p_D}{\partial t_D}, 0 < r_D < \infty, t > 0 \\ p_D\,|_{t_D=0} = 0 \\ \lim\limits_{r_D\to\infty} p_D = 0 \\ \lim\limits_{r_D\to 0^+} \frac{4\pi kL}{\mu}(r_D{}^2\frac{\partial p_D}{\partial r_D}) = -q(t) \end{cases} \tag{1}$$

where $p_D = p_i - p_f$, p_i is the initial pressure;

D. Zeng (Ed.): Advances in Computer Science and Engineering, AISC 141, pp. 275–282.
springer link.com

$$t_D = \eta t / L^2$$

$$r_D = \sqrt{(x_D - x_{\omega D})^2 + (y_D - y_{\omega D})^2 + (z_D - z_{\omega D})^2}$$

This paper introduces the approach of proper orthogonal transformation to turn the initial value and boundary value problem of partial differential equations (1) into initial value problem of ordinary differential equation. Then we can apply the theory of ordinary differential equations to have an easier access to the problem (1) of the exact solutions, which is called fundamental solution. The precise solutions of well test analysis model can be also obtained using a number of fundamental solutions. We can employ the precise point source solution to obtain the derivation of precise Mathematical model solution in vertical well, horizontal well and the line source, surface source, body source. The results was applied to the mathematical model of multi-branched wells.

2 Main Results

2.1 Fundamental Solution

Now we consider the eigenvalue problem

$$\begin{cases} \dfrac{1}{r_D^2}\dfrac{\partial}{dr_D}\left(r_D^2\dfrac{dE(r_D)}{dr_D}\right) = -\beta^2 E(r_D), 0 < r_D < \infty & (2) \\ \lim\limits_{r_D \to 0^+} r_D^2 \dfrac{dE(r_D)}{dr_D} = 0,\ \lim\limits_{r_D \to +\infty} E(r_D) = 0 & (3) \end{cases}$$

We can obtain the eigenvalue β^2 (β >0), the corresponding eigenfunction is,

$$E_\beta(r_D) = \frac{\sin \beta r_D}{r_D}$$

Eigenfunction system $\Lambda = \{E_\beta(r_D), \beta > 0\}$ constitute complete orthogonal system which has weight function r_D^2 in interval [0, ∞]. The orthogonality is,

$$\int_0^\infty E_\beta(r_D) E_{\beta'}(r_D) r_D^2 \mathrm{d} r_D = \frac{\pi}{2}\delta(\beta - \beta'), \beta > 0, \beta' > 0 \tag{4}$$

where $\delta(\xi)$ is a dirac delta function.

Using the complete orthogonality of eigenfunction system to introduce the corresponding orthogonal transformation.

Definition 1. we call

$$F[u(r_D)] = \frac{2}{\pi}\int_0^\infty u(r_D) E_\beta(r_D) r_D{}^2 \mathrm{d} r_D \tag{5}$$

as orthogonal transformation of $u(r_D)$,or $\overline{u}(\beta)$. Here we assume that the improper integral in the right of (5) is convergence.

We use expression (5) to obtain the inverse transformation formula [3,4,5]

$$u(r_D) = \int_0^{\infty} \overline{u}(\beta) E_\beta(r_D) \mathrm{d}\beta \tag{6}$$

Theorem 1. (differential properties)

Assume

$\lim\limits_{r_D \to +\infty} r_D u(r_D) = 0$, $\lim\limits_{r_D \to +\infty} r_D \dfrac{\mathrm{d}u(r_D)}{\mathrm{d}r_D} = 0$, and $F[u(r_D)]$ exists, so

$$F[\frac{1}{r_D^2}\frac{\mathrm{d}}{\mathrm{d}r_D}(r_D^2\frac{\mathrm{d}u(r_D)}{\mathrm{d}r_D})] = -\frac{2\beta}{\pi}(r_D^2\frac{\mathrm{d}u(r_D)}{\mathrm{d}r_D})\Big|_{r_D=0^+} - \beta^2 F\left[u(r_D)\right]. \tag{7}$$

We transform the problem(1) in orthogonal way on r_D , and record it as

$$F[p_D] = \overline{p}_D \tag{8}$$

Using differential properties, we have

$$F[\frac{1}{r_D^2}\frac{\partial}{\partial r_D}(r_D^2\frac{\partial p_D}{\partial r_D})] = \frac{\mu\beta q(t_D)}{2\pi^2 kL} - \beta^2 \overline{p}_D \tag{9}$$

$$\begin{cases} \dfrac{\mu\beta q(t_D)}{2\pi^2 kL} - \beta^2 \overline{p}_D = \dfrac{d\overline{p}_D}{dt_D}, t_D > 0 \\ \overline{p}_D \,|_{t_D=0} = 0 \end{cases} \tag{10}$$

So the precise solution of problem(10) is,

$$\overline{p}_D = \frac{\mu\beta}{2\pi^2 kL}\int_0^{t_D} q(\tau) \mathrm{e}^{-\beta^2(t_D-\tau)} \mathrm{d}\tau \tag{11}$$

Thus we obtain initial value problem of ordinary differential equations(1)

$$p_D(r_D,t_D) = \frac{\mu}{2\pi^2 kL}\int_0^{t_D} q(\tau)\int_0^{\infty}\frac{\beta\sin\beta r_D}{r_D}\mathrm{e}^{-\beta^2(t_D-\tau)}\mathrm{d}\beta\mathrm{d}\tau \tag{12}$$

The fundamental solution we obtained above is precise point source solution in the whole interval R^3 .But practically, the reservoir with a certain thickness, so we consider precise point source solution $h_D = (h/L)\sqrt{k/k_z}$ in region Ω :

$$-\infty < x_D < +\infty, -\infty < y_D < +\infty, 0 < z_D < h_D .$$

In $z_D = 0$ and $z_D = h_D$, if it is a no flow boundary, the boundary condition is

$$\frac{\partial p_D}{\partial z_D}\Big|_{z_D=0} = \frac{\partial p_D}{\partial z_D}\Big|_{z_D=h_D} = 0 \tag{13}$$

Now we consider the problem A

$$\begin{cases} \dfrac{\partial^2 p_D}{\partial x^2}+\dfrac{\partial^2 p_D}{\partial y^2}+\dfrac{\partial^2 p_D}{\partial z^2}=\dfrac{\partial p_D}{\partial t_D},(x_D,y_D,z_D)\in\Omega-\{(0,0,0)\},t>0 \\ p_D\,|_{t_D=0}=0 \\ \dfrac{\partial p_D}{\partial z_D}\,|_{z_D=0}=\dfrac{\partial p_D}{\partial z_D}\,|_{z_D=h_D}=0 \\ \lim\limits_{r_D\to 0^+}\dfrac{4\pi kL}{\mu}(r_D{}^2\dfrac{\partial p_D}{\partial r_D})=-q(t) \end{cases} \tag{14}$$

which

$$\Omega:-\infty<x_D、y_D<+\infty,\ 0<z_D<h_D.$$

Inspired by the image method, we extend the above area Ω on even cycle of z_D to the whole space R^3, and then apply the iterative point source solution to obtain the precise solution of problem A.

$$p_D(x_D,y_D,z_D,t_D)=\frac{\mu}{2\pi^2kL}\int_0^{t_D}q(\tau)\int_0^{\infty}\sum_{n=-\infty}^{\infty}[\frac{\sin\beta\sqrt{\rho_D^2+z_{D1n}^2}}{\sqrt{\rho_D^2+z_{D1n}^2}}+\frac{\sin\beta\sqrt{\rho_D^2+z_{D2n}^2}}{\sqrt{\rho_D^2+z_{D2n}^2}}]e^{-\beta^2(t_D-\tau)}\beta d\beta d\tau \tag{15}$$

which

$$\rho_D^2=(x_D-x_{\omega D})^2+(y_D-y_{\omega D})^2 \tag{16}$$

$$z_{D1n}=z_D-z_{\omega D}-2nh_D \tag{17}$$

$$z_{D2n}=z_D+z_{\omega D}-2nh_D \tag{18}$$

Similarly, if $z_D=0$ and $z_D=h_D$ are constant pressure boundary. The boundary condition should be

$$p_D\,|_{z_D=0}=p_D\,|_{z_D=h_D}=0 \tag{19}$$

We consider the problem B

$$\begin{cases} \dfrac{\partial^2 p_D}{\partial x^2}+\dfrac{\partial^2 p_D}{\partial y^2}+\dfrac{\partial^2 p_D}{\partial z^2}=\dfrac{\partial p_D}{\partial t_D},(x_D,y_D,z_D)\in\Omega-\{(0,0,0)\},t>0, \\ p_D\,|_{t_D=0}=0, \\ p_D\,|_{z_D=0}=p_D\,|_{z_D=h_D}=0, \\ \lim\limits_{r_D\to 0^+}\dfrac{4\pi kL}{\mu}(r_D{}^2\dfrac{\partial p_D}{\partial r_D})=-q(t), \end{cases} \tag{20}$$

The precise solution for problem B is,

$$p_D(x_D, y_D, z_D, t_D) = \frac{\mu}{2\pi^2 kL}\int_0^{t_D} q(\tau)\int_0^{\infty}\sum_{n=-\infty}^{\infty}\Big[\frac{\sin(\beta\sqrt{\rho_D^2 + z_{D1n}^2})}{\sqrt{\rho_D^2 + z_{D1n}^2}} - \frac{\sin(\beta\sqrt{\rho_D^2 + z_{D2n}^2})}{\sqrt{\rho_D^2 + z_{D2n}^2}}\Big] e^{-\beta^2(t_D-\tau)}\beta \mathrm{d}\beta \mathrm{d}\tau \quad (21)$$

2.2 The Application of Fundamental Solution to Line Source Mathematical Model

1. The Application of Fundamental Solution to Line Source Vertical Wells Mathematical Model

Using above results, we could easily obtain precise solution of line source vertical wells as well as horizontal wells. For instance, the line source vertical wells of the distance between $(x_{WD}, y_{WD}, z_{WD} - \frac{h_{WD}}{2})$ and $(x_{WD}, y_{WD}, z_{WD} + \frac{h_{WD}}{2})$, whose length is h_{WD}, is recorded as written as Λ. The parameter expression of vertical well is,

$$\Lambda : \begin{cases} x_D(s) = x_{WD}, \\ y_D(s) = y_{WD}, \qquad s \in [-1,1] \\ z_D(s) = z_{WD} + h_{WD}s, \end{cases} \quad (22)$$

Everywhere in the Λ line source satisfies the conditions of the flow rate

$$q(t, z_D) = -\lim_{\varepsilon_D \to 0^+}\lim_{\rho_D \to 0^+}\frac{2\pi kL}{\mu\varepsilon_D}\int_{z_D - \frac{\varepsilon_D}{2}}^{z_D + \frac{\varepsilon_D}{2}} \rho_D \frac{\partial p_D}{\partial \rho_D}\mathrm{d}z_D, \quad (23)$$

which

$$\rho_D = \sqrt{(x_D - x_{WD})^2 + (y_D - y_{WD})^2},$$

$$z_D \in (z_{WD} - \frac{h_{WD}}{2}, z_{WD} + \frac{h_{WD}}{2}).$$

Mathematical model of line source vertical wells

$$\begin{cases} \frac{\partial^2 p_D}{\partial x^2} + \frac{\partial^2 p_D}{\partial y^2} + \frac{\partial^2 p_D}{\partial z^2} = \frac{\partial p_D}{\partial t_D}, (x_D, y_D, z_D) \in \Omega - \Lambda, t > 0, (24) \\ p_D|_{t_D=0} = 0, (25) \\ \frac{\partial p_D}{\partial z_D}|_{z_D=0} = \frac{\partial p_D}{\partial z_D}|_{z_D=h_D} = 0, (26) \\ \text{Everywhere in the } \Lambda \text{ line source satisfies} \\ \text{the conditions of the flow rate (23).} \end{cases}$$

Definite problem (23)~(26) is a differential equation inverse problem which we want to know both p_D and $q(t_D, z_D)$. Then the solution of mathematical model of line source vertical well is

$$p_D(x_D, y_D, z_D, t_D) = \frac{\mu}{2\pi^2 kL}\int_0^{t_D}\int_{-1}^{1} q(\tau, z_D(s))\int_0^{\infty}\sum_{n=-\infty}^{\infty}\left[\frac{\sin(\beta\sqrt{\rho_D^2 + z_{D1n}^2})}{\sqrt{\rho_D^2 + z_{D1n}^2}} + \frac{\sin(\beta\sqrt{\rho_D^2 + z_{D2n}^2})}{\sqrt{\rho_D^2 + z_{D2n}^2}}\right] e^{-\beta^2(t_D-\tau)}\beta \mathrm{d}\beta \mathrm{d}s \mathrm{d}\tau \quad (27)$$

when $q(t_D, z_D)$ is determined. The solution of inverse problem (23)~(26) is determined by (23) and (27). We can get the solution of p_D and $q(t_D, z_D)$ by iterative solving (23) and (27).

3 The Application of Fundamental Solution to Line Source Horizontal Wells Mathematical Model

The line source vertical wells of the distance between

$$(x_{WD} - \frac{L_{WD}}{2}, y_{WD}, z_{WD}) \text{ and } (x_{WD} + \frac{L_{WD}}{2}, y_{WD}, z_{WD}),$$

whose length is L_{WD}, is recorded as written as Λ. The parameter expression of horizontal well is,

$$\Lambda : \begin{cases} x_D(s) = x_{WD} + sL_{WD}, \\ y_D(s) = y_{WD}, & s \in [-1,1] \\ z_D(s) = z_{WD}, \end{cases} \quad (28)$$

Everywhere in the line source satisfies the conditions of the flow rate

$$q(t, x_D) = -\lim_{\varepsilon_D \to 0^+}\lim_{\rho_D \to 0^+}\frac{2\pi kL}{\mu\varepsilon_D}\int_{x_D-\frac{\varepsilon_D}{2}}^{x_D+\frac{\varepsilon_D}{2}} \rho_D \frac{\partial p_D}{\partial \rho_D}\mathrm{d}x_D \quad (29)$$

which

$$\rho_D = \sqrt{(y_D - y_{WD})^2 + (z_D - z_{WD})^2} \quad x_D \in (x_{WD} - \frac{L_{WD}}{2}, x_{WD} + \frac{L_{WD}}{2}).$$

Mathematical model of line source horizontal well

$$\begin{cases} \frac{\partial^2 p_D}{\partial x^2} + \frac{\partial^2 p_D}{\partial y^2} + \frac{\partial^2 p_D}{\partial z^2} = \frac{\partial p_D}{\partial t_D}, (x_D, y_D, z_D) \in \Omega - \Lambda, t > 0, (30) \\ p_D |_{t_D=0} = 0, (31) \\ \frac{\partial p_D}{\partial z_D}|_{z_D=0} = \frac{\partial p_D}{\partial z_D}|_{z_D=h_D} = 0, (32) \\ \text{Everywhere in the } \Lambda \text{ line source satisfies} \\ \text{the conditions of the flow rate (29).} \end{cases}$$

Then the solution of mathematical model of line source horizontal well is

$$p_D(x_D, y_D, z_D, t_D) = \frac{\mu}{2\pi^2 kL}\int_0^{t_D}\int_{-1}^{1} q(\tau, x_D(s))\int_0^{\infty}\sum_{n=-\infty}^{\infty}\Big[\frac{\sin(\beta\sqrt{\rho_D^2 + z_{D1n}^2})}{\sqrt{\rho_D^2 + z_{D1n}^2}} + \frac{\sin(\beta\sqrt{\rho_D^2 + z_{D2n}^2})}{\sqrt{\rho_D^2 + z_{D2n}^2}}\Big]e^{-\beta^2(t_D-\tau)}\beta \mathrm{d}\beta \mathrm{d}s \mathrm{d}\tau \tag{33}$$

when $q(t_D, x_D, y_D, z_D)$ is determined.

The solution of inverse problem (29)~(32) is determined by (29) and (33).

4 The Application of Fundamental Solution to Line Source Multi-branched Wells Mathematical Model

The mathematical model of multi-branched well

$$\begin{cases} \frac{\partial^2 p_D}{\partial x^2} + \frac{\partial^2 p_D}{\partial y^2} + \frac{\partial^2 p_D}{\partial z^2} = \frac{\partial p_D}{\partial t_D}, (x_D, y_D, z_D) \in \Omega - \Lambda, t > 0, (34) \\ p_D|_{t_D=0} = 0, (35) \\ \frac{\partial p_D}{\partial z_D}|_{z_D=0} = \frac{\partial p_D}{\partial z_D}|_{z_D=h_D} = 0, (36) \\ q(t_D, x_{WD}, y_{WD}, z_{WD}) = -\lim_{\varepsilon_D \to 0^+} \frac{kL}{2\mu\varepsilon_D} \oiint_{\Gamma_{\varepsilon D}} \frac{\partial P_D}{\partial \rho_D} \mathrm{d}s, (x_{WD}, y_{WD}, z_{WD}) \in \Lambda, (37) \end{cases}$$

1) Geometrical description of multi-branched well

$$(x_{Dj1}, y_{Dj1}, z_{Dj1}) \in \Omega, (x_{Dj2}, y_{Dj2}, z_{Dj2}) \in \Omega,$$

$$\Omega = \{(x_D, y_D, z_D) \mid (x_D, y_D, z_D) \in \mathrm{R}^3, 0 < z_D < h_D\}$$

$\Lambda = \sum_{j=0}^{M} \Lambda_j$, Λ_j is a line segment, whose starting point is $(x_{Dj1}, y_{Dj1}, z_{Dj1})$ and ending point is $(x_{Dj2}, y_{Dj2}, z_{Dj2})$. $(x_{Dj1}, y_{Dj1}, z_{Dj1}) \in \Lambda_0$, $j \in \{1, 2, \cdots, M\}$.

2) The parameter expression of multi-branched well Λ : branch-j

$$\Lambda_j : \begin{cases} x_D = x_{Dj}(s) = x_{Dj1} + s(x_{Dj2} - x_{Dj1}),, \\ y_D = y_{Dj}(s) = y_{Dj1} + s(y_{Dj2} - y_{Dj1}), s \in [0,1] \\ z_D = z_{Dj}(s) = z_{Dj1} + s(z_{Dj2} - z_{Dj1}), \end{cases}$$

which $j \in \{0, 1, \cdots, M\}$.

3) Everywhere in the wellhead of multi-branched well satisfies the conditions of the flow rate

$$q(t_D, x_{WD}, y_{WD}, z_{WD}) = -\lim_{\varepsilon_D \to 0^+} \frac{kL}{2\mu\varepsilon_D} \oiint_{\Gamma_{\varepsilon D}} \frac{\partial P_D}{\partial \rho_D} \mathrm{d}s, (x_{WD}, y_{WD}, z_{WD}) \in \Lambda. \tag{37}$$

which

$$\rho_D = \sqrt{(x_D - x_{WDj})^2 + (y_D - y_{WDj})^2 + (z_D - z_{WDj})^2},$$

$$\Gamma_{\varepsilon_D} : \sqrt{(x_D - x_{WDj})^2 + (y_D - y_{WDj})^2 + (z_D - z_{WDj})^2} = \varepsilon_D, \qquad (x_{WDj}, y_{WDj}, z_{WDj}) \in \Lambda_j, j \in \{0, 1, \cdots, M\}.$$

The solution of mathematical model of multi-branched well is

$$p_D(x_D, y_D, z_D, t_D) = \frac{\mu}{2\pi^2 kL} \sum_{j=0}^{M} \int_0^{t_D} \int_{-1}^{1} q(\tau, x_{Dj}(s), y_{Dj}(s), z_{Dj}(s)) \cdot$$
$$\int_0^{\infty} \sum_{n=-\infty}^{\infty} \left[\frac{\sin(\beta\sqrt{\rho_D^2 + z_{D1n}^2})}{\sqrt{\rho_D^2 + z_{D1n}^2}} + \frac{\sin(\beta\sqrt{\rho_D^2 + z_{D2n}^2})}{\sqrt{\rho_D^2 + z_{D2n}^2}}\right] e^{-\beta^2(t_D - \tau)} \beta d\beta ds d\tau \tag{38}$$

when $q(t_D, x_D, y_D, z_D)$ is determined.

The solution of inverse problem (34)~ (37) is determined by (38) and (37) . which

$$x_{Dj}(s) = x_{Dj1} + s(x_{Dj2} - x_{Dj1}),$$
$$y_{Dj}(s) = y_{Dj1} + s(y_{Dj2} - y_{Dj1}),$$
$$z_{Dj}(s) = z_{Dj1} + s(z_{Dj2} - z_{Dj1}),$$
$$\rho_{Dj}^2(s) = (x_D - x_{WDj}(s))^2 + (y_D - y_{WDj}(s))^2,$$
$$z_{Dj1n}(s) = z_D - z_{WDj}(s) - 2nh_D, z_{Dj2n}(s) = z_D + z_{WDj}(s) - 2nh_D.$$

Note: The processing method and the results of multi-branched wells model in other boundary conditions is completely the same.

References

1. Ozkan, E., Raghavan, R.: New Solutions for Well—Test Analysis Problems: Part 1-Analytical Considerations. In: SPEFE, pp. 359–368 (September 1991)
2. Ozkan, E.: Performance of Horizontal Wells. PhD Dissertation (1) (1988)
3. Wu, X., et al.: A New Temperature Field and the Method for Designing Casing in Thermal Horizontal Well. In: SPE, 37540, pp. 251–256
4. Wu, X., et al.: A new volumetric source mathematical model for pressure behavior around horizontal wells in dual media reservoirs. Journal of Petroleum Science and Engineering 16, 301–314 (1996)
5. Wu, X.: Partial Differential Equations Theory and Practice. Science Press, Beijing (2009) (in Chinese)

The Study on Mathematical Model for Double Porosity Medium with Well of Curves

Rao Xiaotong[1], Xiaoqing Wu[2], Xiangjun Liu[3], and Fang Yang[2]

[1] School of Chemistry and Chemical Engineering, Southwest Petroleum University Chengdu, Sichuan Province, China
[2] School of Sciences, Southwest Petroleum University, Chengdu, Sichuan Province, China
[3] School of Petroleum Engineering, Southwest Petroleum University, Chengdu, Sichuan Province, China
xtrao2275980@sina.com, wuxiaoqing_swpu@163.com, liuxj@swpu.edu.cn, fangyangswpu2008@hotmail.com

Abstract. In the oil production and thermal recovery in the application of oil and gas field, well of curves are regarded as the crooked canal(namely not only the canal has horizontal section but also vertical section), it instead the horizontal well before. Thus establishmented the new elbow well model, the solution of the model not only suitable for theoretical study but also easy to calculate in practical production. This article considered the problem of indeterminate percolation of spherical symmetry infinite domain described by the initial boundary value problem of the system of partial differential equation and obtained the point-source accurate solution when research on the problem of indeterminate percolation for double porosity medium. Abtained the accurate solution about mathematical model of line source from appling the point-source accurate solution, and applied the results to mathematical model of well of curves, obtained the integral expression of it's accurate solution.

Keywords: Vertical well, Horizontal well, Well of curves, System of partial differential equations, Precise solution, Point-source accurate solution.

1 Introduction

This article considered the problem of indeterminate percolation of spherical symmetry infinite domain described by the initial boundary value problem of the system of partial differential equation and obtained the point-source accurate solution when research on the problem of indeterminate percolation for double porosity medium. Apply point-source accurate solution abtained the accurate solution of mathematical model about vertical well, horizontal well and line source, and applied the results to mathematical model of well of curves, obtained the integral expression of it's accurate solution.

In the study of the problem of indeterminate percolation for double porosity medium, first considered the point-source solution for the problem of indeterminate percolation of spherical symmetry infinite domain R^3 described by the initial boundary value

D. Zeng (Ed.): Advances in Computer Science and Engineering, AISC 141, pp. 283–292.
springer link.com

problem of the system of partial differential equation, namely considered the description problem of the initial boundary value problem of the system of partial differential equation for the problem of indeterminate percolation of spherical symmetry infinite domain R^3.

$$\begin{cases} \frac{1}{r_D^2}\frac{\partial}{\partial r_D}(r_D^2\frac{\partial \Delta p_f}{\partial r_D}) = \omega\frac{\partial \Delta p_f}{\partial t_D} + (1-\omega)\frac{\partial \Delta p_{ma}}{\partial t_D} & (1) \\ \lambda(\Delta p_{ma} - \Delta p_f) = -(1-\omega)\frac{\partial \Delta p_{ma}}{\partial t_D} & (2) \\ \Delta p_f \big|_{t_D=0} = \Delta p_{ma} \big|_{t_D=0} = 0 & (3) \\ \lim_{r_D \to +\infty} \Delta p_f = 0 & (4) \\ \lim_{\varepsilon \to 0^+} \frac{4\pi k_f L}{\mu}(r_D^2\frac{\partial \Delta p_f}{\partial r_D})_{r_D=\varepsilon} = -Q(t_D) & (5) \end{cases}$$

where

$$\Delta p_f = p_i - p_f, \Delta p_{ma} = p_i - p_{ma} \tag{6}$$

are the initial pressure;

$$t_D = \eta t / L^2 \tag{7}$$

$$\eta = k_f / \{[(V\phi C_t)_f + (V\phi C_t)_{ma}]\mu\} \tag{8}$$

$$\omega = (V\phi C_t)_f / [(V\phi C_t)_f + (V\phi C_t)_{ma}] \tag{9}$$

$$r_D = \sqrt{(x_D - x_{\omega D})^2 + (y_D - y_{\omega D})^2 + (z_D - z_{\omega D})^2} \tag{10}$$

$$e_D = (e / L)\sqrt{K_f / k_{ef}} \tag{11}$$

This paper introduces the approach of proper orthogonal transformation to turn the initial value and boundary value problem of partial differential equations (1)~(5) into initial value problem of ordinary differential equations. Then we can apply the theory of matrix differential equations to have an easier access to the exact solutions of the problem (1)~(5), which is called fundamental solution. The precise solutions of well test analysis model can be also obtained by using a number of fundamental solutions. Apply the point-source accurate solution we can employ the precise point source solution to obtain the derivation of precise Mathematical model solution in vertical well, horizontal well and the line source, surface source, body source. The results can be applied to the mathematical model of well of curves[2,3].

2 Main Results

2.1 Fundamental Solution

In order to transform the problem (1)～(5) into initial value problem of ordinary differential equation,we noticed the characteristic of equations and it's boundary condition, now we can consider the following characteristic value problem

$$\begin{cases} \frac{1}{r_D^2}\frac{\partial}{dr_D}\left(r_D^2\frac{dE(r_D)}{dr_D}\right)=-\beta^2E(r_D), 0<r_D<\infty & (12) \\ \lim\limits_{r_D\to 0^+} r_D^2\frac{dE(r_D)}{dr_D}=0,\ \lim\limits_{r_D\to+\infty}E(r_D)=0 & (13) \end{cases}$$

obtain the characteristic value β^2 ($\beta>0$), the corresponding characteristic function is,

$$E_\beta(r_D)=\frac{\sin\beta r_D}{r_D} \tag{14}$$

characteristic function system $\Lambda=\{E_\beta(r_D), \beta>0\}$ constitute the complete orthogonal system with the weight function r_D^2 on the interval [0, ∞]. The orthogonality is,

$$\int_0^{+\infty}E_\beta(r_D)E_{\beta'}(r_D)r_D^2\mathrm{d}r_D=\frac{\pi}{2}\delta(\beta-\beta') \tag{15}$$

where $\delta(\xi)$ is a Dirac delta function, $\beta>0, \beta'>0$.

Using the complete orthogonality of characteristic function system to introduce the corresponding orthogonal transformation.

Definition 1. we call

$$F[u(r_D)]=\frac{2}{\pi}\int_0^{+\infty}u(r_D)E_\beta(r_D)r_D^2\mathrm{d}r_D \tag{16}$$

as orthogonal transformation of $u(r_D)$, and also can be marked as $\overline{u}(\beta)$. Here we assume that the generalized integral in the right of (16) is convergence. Use the expression (15) to obtain the inverse transformation formula

$$u(r_D)=\int_0^{+\infty}\overline{u}(\beta)E_\beta(r_D)\mathrm{d}\beta \tag{17}$$

Do the orthogonal transformation for the problem (1)~(5) about r_D, and mark it as

$$F\left[\Delta p_f\right]=\overline{\Delta p_f}(\beta,t_D), F\left[\Delta p_{ma}\right]=\overline{\Delta p_{ma}}(\beta,t_D) \tag{18}$$

Using differential properties, namely

$$F\left[\frac{1}{r_D^2}\frac{\partial}{\partial r_D^2}\left(r_D^2\frac{\partial\Delta p_f}{\partial r_D}\right)\right]=\frac{\mu\beta\tilde{q}(t_D)}{2\pi^2 k_f L}-\beta^2\overline{\Delta p_f}(\beta,t_D) \tag{19}$$

Then obtained the initial value problem of ordinary differential equation

$$\begin{cases}\omega\dfrac{d\overline{\Delta p_f}}{dt_D}+(1-\omega)\dfrac{d\overline{\Delta p_{ma}}}{dt_D}=-\beta^2\overline{\Delta p_f}+f(\beta,t_D) & (20)\\ (1-\omega)\dfrac{d\overline{\Delta p_{ma}}}{dt_D}=\lambda\left(\overline{\Delta p_f}-\overline{\Delta p_{ma}}\right) & (21)\end{cases}$$

where

$$\overline{\Delta p_f}\Big|_{t_D=0}=\overline{\Delta p_{ma}}\Big|_{t_D=0}=0 \tag{22}$$

$$f(\beta,t_D)=\frac{\mu\beta Q(t_D)}{2\pi^2 k_f L} \tag{23}$$

from the formula (20),(21) subtracting the equations can be simplified as follows

$$\begin{cases}\omega\dfrac{d\overline{\Delta p_f}}{dt_D}=-\left(\beta^2+\lambda\right)\overline{\Delta p_f}+\lambda\overline{\Delta p_{ma}}+f(\beta,t_D)\\ (1-\omega)\dfrac{d\overline{\Delta p_{ma}}}{dt_D}=\lambda\overline{\Delta p_f}-\lambda\overline{\Delta p_{ma}}\end{cases} \tag{24}$$

Note

$$u=\left[\overline{\Delta p_f},\overline{\Delta p_{ma}}\right]^T \tag{25}$$

$$\varphi(\beta,t_D)=\left[\frac{1}{\omega}f(\beta,t_D),0\right]^T \tag{26}$$

$$A=\begin{bmatrix}-\dfrac{\beta^2+\lambda}{\omega},\dfrac{\lambda}{\omega}\\ \dfrac{\lambda}{1-\omega},-\dfrac{\lambda}{1-\omega}\end{bmatrix} \tag{27}$$

Write the ordinary differential equation as matrix differential equtaion

$$\begin{cases}\dfrac{dU}{dt_D}=AU+\phi\\ U\Big|_{t_D=0}=0\end{cases} \tag{28}$$

this is a nonhomogeneous matrix differential equation, A is 2×2 constant matrix.

First solving the corresponding initial value problem of the homogeneous matrix differential equation

$$\begin{cases} \dfrac{dX(t_D)}{dt_D} = AX(t_D) \\ X(0) = I \end{cases} \tag{29}$$

where I is 2×2 unit matrix, can use the method of characteristic value, characteristic vector to solve.

Form the characteristic equation

$$\det(A - \gamma I) = 0 \tag{30}$$

Obtained the two characteristic value of the matrix

$$\gamma_1 = \frac{-(\beta^2 + \lambda - \omega\beta^2) + \sqrt{G}}{2\omega(1-\omega)} \tag{31}$$

$$\gamma_2 = \frac{-(\beta^2 + \lambda - \omega\beta^2) - \sqrt{G}}{2\omega(1-\omega)} \tag{32}$$

where

$$G = (\beta^2 + \lambda - \omega\beta^2)^2 - 4\omega(1-\omega)\lambda\beta^2 \tag{33}$$

The characteristic vector α_i of the corresponding characteristic value $\gamma_i (i = 1, 2)$ can solved by linear algebraic equations

$$(A - \gamma_i I)\alpha_i = 0 \tag{34}$$

obtain

$$\alpha_i = \left[\frac{\lambda}{\omega}, \frac{\beta^2 + \lambda}{\omega} + \gamma_i\right]^T, i = 1, 2 \tag{35}$$

then obtain the matrix

$$B = [\alpha_1 \alpha_2] = \begin{bmatrix} \dfrac{\lambda}{\omega}, & \dfrac{\lambda}{\omega} \\ \dfrac{\beta^2 + \lambda}{\omega} + \gamma_1, & \dfrac{\beta^2 + \lambda}{\omega} + \gamma_2 \end{bmatrix} \tag{36}$$

and it's inverse matrix is

$$B^{-1} = -\frac{\omega^2(1-\omega)}{\lambda\sqrt{G}} \begin{bmatrix} \dfrac{\beta^2 + \lambda}{\omega} + \lambda_2, & -\dfrac{\lambda}{\omega} \\ -(\dfrac{\beta^2 + \lambda}{\omega} + \lambda_1), & \dfrac{\lambda}{\omega} \end{bmatrix} \tag{37}$$

So the solution of (29) is

$$X(t_D)=B\begin{bmatrix} e^{\gamma_1 t_D} & 0 \\ 0 & e^{\gamma_2 t_D} \end{bmatrix}B^{-1} \tag{38}$$

then use the theory of the linear nonhomogeneous matrix differential equation can obtain

$$U(\beta,t_D)=\int_0^{t_D} X(t_D-\tau)\phi(\beta,\tau)d\tau \tag{39}$$

By the simple matrix operation can obtain

$$\overline{\Delta p_f}=\frac{\mu\beta}{2\pi^2 k_f L}\int_0^{t_D}\left[A_1(\beta)e^{-\sigma_1(\beta)(t_D-\tau)}+A_2(\beta)e^{-\sigma_2(\beta)(t_D-\tau)}\right]Q(\tau)d\tau \tag{40}$$

$$\overline{\Delta p_{ma}}=\frac{\mu\beta}{2\pi^2 k_f L}\int_0^{t_D}\left[B(\beta)\left(e^{-\sigma_2(\beta)(t_D-\tau)}-e^{-\sigma_1(\beta)(t_D-\tau)}\right)\right]Q(\tau)d\tau \tag{41}$$

Where $a=\omega b,\ \ b=\dfrac{1-\omega}{\lambda}$ (42)

$$\sigma_i(\beta)=\frac{1}{2a}\left[1+b\beta^2+(-1)^{i+1}\sqrt{\left(1+b\beta^2\right)^2-4a\beta^2}\right] \tag{43}$$

$$A_i(\beta)=\frac{1}{2a}\left[b+(-1)^{i+1}\frac{b\left(1+b\beta^2\right)-2a}{\sqrt{\left(1+b\beta^2\right)^2-4a\beta^2}}\right] \tag{44}$$

$$B(\beta)=\frac{1}{\sqrt{\left(1+b\beta^2\right)^2-4a\beta^2}} \tag{45}$$

Use the inverse transformation formula (17) can obtain the accurate solution of the problem (1)~(5), namely the fundamental solution

$$\Delta p_f=\frac{\mu}{2\pi^2 k_f L r_D}\int_0^{t_D}Q(t_D-\tau)d\tau\int_0^{+\infty}\beta\left[A_1(\beta)e^{-\sigma_1(\beta)\tau}+A_2(\beta)e^{-\sigma_2(\beta)\tau}\right]\sin\beta r_D d\beta \tag{46}$$

$$\Delta p_{ma}=\frac{\mu}{2\pi^2 k_f L r_D}\int_0^{t_D}Q(t_D-\tau)d\tau\int_0^{+\infty}\beta\left[e^{-\sigma_2(\beta)\tau}-e^{-\sigma_1(\beta)\tau}\right]B(\beta)\sin\beta r_D d\beta \tag{47}$$

The fundamental solution we obtained above is precise point source solution in the whole interval R^3. But practically, the reservoir with a certain thickness, so we consider precise point source solution $h_D=(h/L)\sqrt{k/k_z}$ in region

$\Omega\ :\ -\infty < x_D < +\infty, -\infty < y_D < +\infty, 0 < z_D < h_D$.

In $z_D=0$ and $z_D=h_D$, if it is a no flow boundary, the boundary condition is

$$\frac{\partial p_D}{\partial z_D}\Big|_{z_D=0}=\frac{\partial p_D}{\partial z_D}\Big|_{z_D=h_D}=0 \tag{48}$$

We call the definite problem composed by (1)~(5), (48) as the problem A. According to the enlightenment of mirror method, Even periodic extend the above region Ω about z_D to whole space R^3, then use the iterative of point-source solution can obtain the accurate solution of problem A

$$\Delta p_f = \frac{\mu}{2\pi^2 k_f L}\int_0^{t_D} Q(t_D-\tau)d\tau\int_0^{+\infty}\beta[A_1(\beta)e^{-\sigma_1(\beta)\tau}+A_2(\beta)e^{-\sigma_2(\beta)\tau}]\times$$

$$\sum_{n=-\infty}^{+\infty}\left[\frac{\sin\left(\beta\sqrt{\rho_D^2+z_{D1n}^2}\right)}{\sqrt{\rho_D^2+z_{D1n}^2}}+\frac{\sin\left(\beta\sqrt{\rho_D^2+z_{D2n}^2}\right)}{\sqrt{\rho_D^2+z_{D2n}^2}}\right]d\beta \quad (49)$$

$$\Delta p_{ma} = \frac{\mu}{2\pi^2 k_f L}\int_0^{t_D} Q(t_D-\tau)d\tau\int_0^{+\infty}\beta[e^{-\sigma_1(\beta)\tau}-e^{-\sigma_2(\beta)\tau}]\times$$

$$B(\beta)\sum_{n=-\infty}^{+\infty}\left[\frac{\sin\left(\beta\sqrt{\rho_D^2+z_{D1n}^2}\right)}{\sqrt{\rho_D^2+z_{D1n}^2}}+\frac{\sin\left(\beta\sqrt{\rho_D^2+z_{D2n}^2}\right)}{\sqrt{\rho_D^2+z_{D2n}^2}}\right]d\beta \quad (50)$$

where

$$\rho_D^2=(x_D-x_{\omega D})^2+(y_D-y_{\omega D})^2 \quad (51)$$

$$z_{D1n}=z_D-z_{\omega D}-2nh_D \quad (52)$$

$$z_{D2n}=z_D+z_{\omega D}-2nh_D \quad (53)$$

As the same, if it is constant pressure boundary at $z_D=0$ and $z_D=h_D$, namely the boundary condition is

$$\Delta p_f\big|_{z_D=0}=\Delta p_f\big|_{z_D=h_D}=0 \quad (54)$$

The definite problem composed by (1)~(5), (54), it's accurate solution is

$$\Delta p_f = \frac{\mu}{2\pi^2 k_f L}\int_0^{t_D} Q(t_D-\tau)d\tau\int_0^{+\infty}\beta[A_1(\beta)e^{-\sigma_1(\beta)\tau}+A_2(\beta)e^{-\sigma_2(\beta)\tau}]\times$$

$$\sum_{n=-\infty}^{+\infty}\left[\frac{\sin\left(\beta\sqrt{\rho_D^2+z_{D1n}^2}\right)}{\sqrt{\rho_D^2+z_{D1n}^2}}-\frac{\sin\left(\beta\sqrt{\rho_D^2+z_{D1n}^2}\right)}{\sqrt{\rho_D^2+z_{D1n}^2}}\right]d\beta \quad (55)$$

$$\Delta p_{ma} = \frac{\mu}{2\pi^2 k_f L}\int_0^{t_D} Q(t_D-\tau)d\tau\int_0^{+\infty}\beta[e^{-\sigma_1(\beta)\tau}-e^{-\sigma_2(\beta)\tau}]\times$$

$$B(\beta)\sum_{n=-\infty}^{+\infty}\left[\frac{\sin\left(\beta\sqrt{\rho_D^2+z_{D1n}^2}\right)}{\sqrt{\rho_D^2+z_{D1n}^2}}-\frac{\sin\left(\beta\sqrt{\rho_D^2+z_{D2n}^2}\right)}{\sqrt{\rho_D^2+z_{D2n}^2}}\right]d\beta \quad (56)$$

2.2 The Application of Fundamental Solution to Well of Curves Mathematical Model

The mathematical model of well of curves

$$\begin{cases} \dfrac{\partial^2 \Delta p_f}{\partial x_D^{\ 2}}+\dfrac{\partial^2 \Delta p_f}{\partial y_D^{\ 2}}+\dfrac{\partial^2 \Delta p_f}{\partial z_D^{\ 2}}=\omega\dfrac{\partial \Delta p_f}{\partial t_D}+(1-\omega)\dfrac{\partial \Delta p_{ma}}{\partial t_D},(x_D,y_D,z_D)\in\Omega-\Lambda, t_D>0, (57) \\ \lambda(\Delta p_{ma}-\Delta p_f)=-(1-\omega)\dfrac{\partial \Delta p_{ma}}{\partial t_D}, (58) \\ \Delta p_f\big|_{t_D=0}=\Delta p_{ma}\big|_{t_D=0}=0,\ (59) \\ \dfrac{\partial \Delta p_f}{\partial z_D}\Big|_{z_D=0}=\dfrac{\partial \Delta p_f}{\partial z_D}\Big|_{z_D=h_D}=0, (60) \\ Q(t_D,x_{WD},y_{WD},z_{WD})=-\lim\limits_{\varepsilon_D\to 0^+}\dfrac{k_f L}{\mu|L_{\varepsilon_D}|}\iint_{\Gamma_{\varepsilon D}}\dfrac{\partial \Delta p_f}{\partial \rho_D}ds,(x_{WD},y_{WD},z_{WD})\in\Lambda\ (61) \end{cases}$$

Definite problem (57)~(61) is a differential equation inverse problem which we want to know { $\Delta p_f, \Delta p_m, q(t_D, z_D)$ }.

1) The geometric description of Well of curves $\Lambda=\sum_{j=0}^{3}\Lambda_j$,

$$\Omega=\{(x_D,y_D,z_D)\,|\,(x_D,y_D,z_D)\in \mathrm{R}^3, 0<z_D<h_D\}$$

Λ_j represents the line segment which take $(x_{Dj1}, y_{Dj1}, z_{Dj1})$ as starting point and $(x_{Dj2}, y_{Dj2}, z_{Dj2})$ as end point, j=1,3. Λ_2 represents the curve segment which take $(x_{D12}, y_{D12}, z_{D12})$ as starting point and $(x_{D32}, y_{D32}, z_{D32})$ as end point. $(x_{Dj1}, y_{Dj1}, z_{Dj1})\in\Omega, (x_{Dj2}, y_{Dj2}, z_{Dj2})\in\Omega$.

2) The parameters expression of the well of curves multilateral well:the j curve segment

$$\Lambda_1:\begin{cases} x_{WD1}(s)=x_{D11}, \\ y_{WD1}(s)=y_{D11}, \qquad s\in[0,1]. \\ z_{WD1}(s)=z_{D11}+s(z_{D12}-z_{D11}), \end{cases} \tag{62}$$

$$\Lambda_2:\begin{cases} x_{WD2}(s)=\varphi(s), \\ y_{WD2}(s)=\psi(s), s\in[0,1], \\ z_{WD2}(s)=\omega(s), \end{cases} \tag{63}$$

$$(x_{WD2}(0), y_{WD2}(0), z_{WD2}(0))=(x_{D12}, y_{D12}, z_{D12}),$$

$$(x_{WD2}(1), y_{WD2}(1), z_{WD2}(1))=(x_{D32}, y_{D32}, z_{D32}).$$

$$\Lambda_3:\begin{cases} x_{WD3}(s)=x_{Dj1}+s(x_{D32}-x_{D31}),\\ y_{WD3}(s)=y_{Dj1}+s(y_{D32}-y_{D31}), s\in[0,1].\\ z_{WD3}(s)=z_{Dj1},\end{cases} \tag{64}$$

3) The well of curves Λ is satisfy the flow condition at the well head everywhere

$$Q(t_D,x_{WD},y_{WD},z_{WD})=-\lim_{\varepsilon_D\to0^+}\frac{k_fL}{\mu|L_{\varepsilon_D}|}\iint_{\Gamma_{\varepsilon_D}}\frac{\partial\Delta p_f}{\partial\rho_D}ds,(x_{WD},y_{WD},z_{WD})\in\Lambda \tag{61}$$

$$\rho_D=\sqrt{(x_D-x_{WD})^2+(y_D-y_{WD})^2+(z_D-z_{WD})^2},$$

$$\Gamma_{\varepsilon_D}:\sqrt{(x_D-x_{WD})^2+(y_D-y_{WD})^2+(z_D-z_{WD})^2}=\varepsilon_D$$

$$K_{\varepsilon_D}:\sqrt{(x_D-x_{WD})^2+(y_D-y_{WD})^2+(z_D-z_{WD})^2}\le\varepsilon_D,$$

$$L_{\varepsilon_D}\equiv K_{\varepsilon_D}\cap\Lambda,$$

curve long of L_{ε_D} is definitioned by $|L_{\varepsilon_D}|$.

The solution of (57)~(60) is

$$\begin{aligned}&\Delta p_f(x_D,y_D,z_D,t_D)\\&=\frac{\mu}{2\pi^2k_fL}\sum_{j=0}^{3}\int_0^{t_D}\int_0^1 Q(t_D-\tau,x_{WDj}(s),y_{WDj}(s),z_{WDj}(s))\cdot\\&\int_0^{+\infty}\beta[A_1(\beta)e^{-\sigma_1(\beta)\tau}+A_2(\beta)e^{-\sigma_2(\beta)\tau}]\sum_{n=-\infty}^{+\infty}[\frac{\sin(\beta\sqrt{\rho_{Dj}^2(s)+z_{Dj1n}{}^2(s)})}{\sqrt{\rho_{Dj}^2(s)+z_{Dj1n}{}^2(s)}}+\\&\frac{\sin(\beta\sqrt{\rho_{Dj}^2(s)+z_{Dj2n}{}^2(s)})}{\sqrt{\rho_{Dj}^2(s)+z_{Dj2n}{}^2(s)}}]d\beta ds\end{aligned} \tag{65}$$

$$\begin{aligned}&\Delta p_{ma}(x_D,y_D,z_D,t_D)=\frac{\mu}{2\pi^2k_fL}\sum_{j=0}^{3}\int_0^{t_D}\int_0^1 Q(t_D-\tau,x_{WDj}(s),y_{WDj}(s),z_{WDj}(s))\cdot\\&\int_0^{+\infty}\beta[e^{-\sigma_1(\beta)\tau}+e^{-\sigma_2(\beta)\tau}]B(\beta)\sum_{n=-\infty}^{+\infty}[\frac{\sin(\beta\sqrt{\rho_{Dj}^2(s)+z_{Dj1n}{}^2(s)})}{\sqrt{\rho_{Dj}^2(s)+z_{Dj1n}{}^2(s)}}+\\&\frac{\sin(\beta\sqrt{\rho_{Dj}^2(s)+z_{Dj1n}{}^2(s)})}{\sqrt{\rho_{Dj}^2(s)+z_{Dj1n}{}^2(s)}}]d\beta ds d\tau \quad (66)\end{aligned} \tag{66}$$

when $Q(t_D, x_{WD}, y_{WD}, z_{WD})$ is determined. The solution of inverse problem (57)~ (61) is determined by (65) ,(66) and (61) . Where

$$\rho_{Dj}^2(s) = (x_D - x_{WDj}(s))^2 + (y_D - y_{WDj}(s))^2$$

$$z_{Dj1n}(s) = z_D - z_{WDj}(s) - 2nh_D, z_{Dj2n}(s) = z_D + z_{WDj}(s) - 2nh_D \quad .$$

References

1. Ozkan, E., Raghavan, R.: New Solutions for Well—Test Analysis Problems: Part 1-Analytical Considerations. In: SPEFE, pp. 359–368 (September 1991)
2. Wu, X.: Partial Differential Equations Theory and Practice. Science Press, Beijing (2009) (in Chinese)
3. Wu, X., et al.: A new volumetric source mathematical model for pressure behavior around horizontal wells in dual media reservoirs. Journal of Petroleum Science and Engineering 16, 301–314 (1996)

Research on Individual Tourism Service System Based on Web Mining

Liu Xiuhua

School of Economic & Management, Weifang University, Weifang, P.R. China, 261061
liuxiuhua001@163.com

Abstract. This essay puts forward the model of the individual tourism service system based on Web mining, and has a detailed introduction to the compose and function of this system, and has clusters analysis and association rules analysis to the individual tourism service, and at last, puts forward the design thought and achieving algorithm of the individual tourism service system.

Keywords: Web Mining, Individual, Tourism Service System, Central Point Clustering Algorithm.

1 Introduction

Thanks to the development of science of research, Individual ideas and technology has come into each and every field. By virtue of Individual ideas and service technique, customers can be tailored different services. This kind of new service is capable of studying customers' habit and behavior by gathering and analyzing necessary information so that corresponding recommendation can be made. In the field of Internet, the service can offer service of much higher quality, while in the tourism service WorldTravel agencies can benefit a lot from Individual service because of its capacity to make customers more satisfied. Current travel agencies, however, largely ignore customer's own like and disposition while recommending places of interest. To make things worse, customers' choice of traveling spots is to some extent influenced by their random decision. So they find their destination fall short of their expectancy after their journey. And it does harm to better satisfying the users.

2 Structure of the Personalized Tourism Service System

Individual tourism service system consists of user end and service end, and the latter is made up of identification of users, user model, matching data and recommendation. The four components can testify user identity legality, find out and their interest, make Individual recommendation and give final recommendation [1]. The user mainly works on sending customers' application and received information from the service end before rendering feedbacks to Individual recommendation. Each module in the system is introduced in detail as follow.

D. Zeng (Ed.): Advances in Computer Science and Engineering, AISC 141, pp. 293–298.
springer link.com

1) *Users requests.* It includes requirements and feedback information submitted by customers.
2) *User authentication.* It's a primary function lies in verification of user identity. Only customers who have passed the verification can have legal access to this system.
3) *User modeling.* Every customer may has vary in personal interest and need, so a user model is the core part of this system, it can meet users different demands. Building such a model can be subdivided into 2 steps:

- Register and recognize user targets on the basis of observation of their choice.
- Analyze their personal interest and bring up the basis on to make recommendation.

4) *Individual recommendation.* Take into consideration different user' models and interest-related inclinations, and search for matching information in the data base by certain rules. Then make a decision and give different customer different suggestions.
5) *Send recommendation information.* Pack up recommendation according to statistic structure and transmitting standard and send them to users.
6) *User feedback.* This section is responsible for gathering user's response's to the suggestions and for sending the information to service's en so that it can readjust customer's models and recommended tactics.

3 Individual Recommendation

3.1 Individual Recommendation Process

The process of Individual recommendation is as follow:

1) *Gather users information;*

2) *Analyze users interests,* according to the statistics and find out several popular feature spot then divide the users into several groups on the basis of their interests;

3) *Make another analysis of the grouped users to work out the association and rules of each group.* Summarize the common points of its members.

4) *Adjust user's weight,* in line with the changes of practices and highlight the increasing or declining trend of user's various interests, then reflected the recent circumstances of changes. As shown in figure 1.

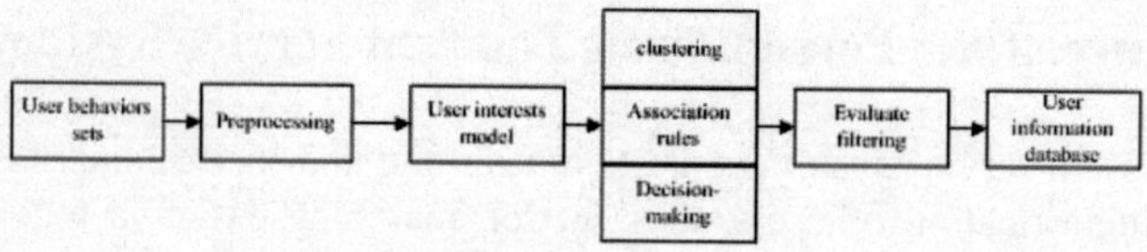

Fig. 1. Individual recommendation process

3.2 Clustering Algorithm Based on Central Point

1) K-Central Point Clustering Algorithm

Definition 1: Given D made up of n-tuple set, which aims to generate cluster of K. p is one tuple of set D, representing a given object of the cluster, O_j is a representative object of the cluster C_j [2]. The sum E_1 of absolute errors of all objects in collection D is defined as:

$$E_1 = \sum_{j=1}^{k} \sum_{p \in C_i} |p - 0_j|$$

the average dissimilarity between tuple of collection D and its reference is defined as:

$$E_2 = \sum_{j=1}^{k} \sum_{p \in C_i} |p - 0_j|$$

The division of k-central point is implemented based on the principle of minimizing the sum by difference between all objects and their reference point [3].

The basic strategy of k-central point clustering algorithm is: firstly, randomly select a representative object for each cluster, and according to their distance with the representatives, the remaining objects is assigned to the nearest cluster. Then repeatedly replace the representative object with non- representative object to improve the quality.

The quality of clustering results is inspected in a process the function with a price estimates the average dissimilarity between the measures object of this function and its reference [4]. To determine whether a non-representative objects O_{random} is a good alternative to the representative object O_j, the following 4 conditions needs to consider for each non-center object.

- p currently attaches to the center object O_j. If O_j is replaced O_{random} as a central point, and p is in the shortest distance from $O_i(i \neq j)$, then p is assigned by the center to $O_i(i \neq j)$.
- p currently attaches to the center object O_j. If O_j is replaced O_{random} as a central point, and p is in the shortest distance from O_{random}, then p is reassigned by the center to O_{random}.
- p currently attaches to the center object $O_i(i \neq j)$, If O_j is replaced O_{random} as a central point, and p is still in the shortest distance from O_i, then the membership of object does not change .
- p currently attaches to the center object $O_i(i \neq j)$, If O_j is replaced O_{random} as a central point, and p is in the shortest distance from O_{random}, then p is reassigned by the center to O_{random}.

When re-allocation is carried out, the poor of the absolute error function E_1 has impact on the cost function. Therefore, if the current representative object is not replaced non-representative object, the cost function calculates the poor of the absolute error [5]. The total cost of the exchange is the sum of the cost of all non-representative objects. If the total cost is negative, Absolute actual errors E will be decreased, and O_j can be replaced or exchanged by O_{random}. If the total cost is positive, the current representative object is acceptable, which has no change in this iteration.

2) *PAM Algorithm*

PAM Algorithm is one of the most famous central point algorithms[6].
Input: k-the number of results cluster; D- Tuple set;
Output: set with k- cluster.
Algorithm:
Begin:

a) Choose k objects arbitrarily from D as the original representing objects or seeds;

b) Repeat

c) Appoint every object which is left over to the nearest cluster which is stood for the object.

d) Choose an object O_{random} *which is not the representation randomly.*

e) Calculate the total cost S *by exchanging* O_{random} *to* O_j

f) In if $S<0$ *then, exchange* O_j *by* O_{random}*, and form a new gathering including k objects.*

g) until there to be nothing changed.

In 1995, Ester and Kriegel put forward the CLARANS Algorithm using effective space access about the differentiate method, then they improved the function of CLARANS Algorithm by using R*Tree and the technology of gathering focus.

3.3 Individual Recommendation Based on the Clustering Results

We recommend attractions to individual users based on the attraction choice staying behaviors of other users who are of the same type. For new users, we will not give Individual recommendations as the necessary interests degree cannot be calculated by their behavior.

Recommendation process: individual user U

a) Find category K that U belongs to;

b) Find in K category other users' records of the attraction choice, and calculate the total number of selections of each attraction;

c) List the number on descending order;

d) Recommend the first attraction that U has not chosen.

3.4 The Relative Rule Discovery of the Applied Individual Recommendation

Using the date excavating technology, we can carry on the pre-treatment to the log document, and then obtain the consumer conversation documents (Consumer conversation is about the feature spot and order the consumer choose in the appointed time.). All the consumers will make up the conversation document, so we can dig out the relative regulations which are as follows:

a) a% of the consumers who have chosen feature spot Placea1, and download Placec1 too.

b) b% of the consumers who have chosen feature spot Placea2, Placea3, Placea4 & Placea5, and choose Placec2 too.

c) c% of the consumers who start to choose feature spot from Placea6, and choose Placec3 next.

From the regulations above, Placeai $i = 1,2,3,4,5,6$ and Placecj $j = 1,2,3$ are the former term and the latter term of the relative regulation, a%, b% and c% are the believe degree of the regulation. According to the state form of the relative regulation, we can show them as follows:

a) Placea1→Placec1, the believe degree is a%; and the value of this regulation in the system is 75%;

b) Placea2 Placea3 Placea4 Placea5→Placec2, the believe degree is b%, the value of this regulation in the system is 50%, and the sliding window depth is 3;

c) Placea6→Placec3, the believe degree is c%, and the value of this regulation in the system is 60%.

The processes of the individual recommendation based on association rules are as follows:

a) The discovery of the relative regulation will be held at the latter of the quarter every year, and the time limitation of the log journal are the nearest two years.

b) Reject the consumer records which have less than twice choices of the feature spot in one year.

c) According to the system journals of two years, calculate the three regulations which surpass the value.

d) According to the discovery regulation, recommend the feature spot of the latter term regulation to the consumers who only satisfy the former term regulation.

3.5 Complementary Recommendation Policy of Association Rules and Cluster Analysis

Complementary recommendation method of association rules and cluster analysis are as follow: when mining, respectively by means of association rules and clustering analysis to get rules and clustering results; As for recommendation, if user's browsing patterns can match association rules well, results on the basis of association rules will be recommended first and those based on cluster analysis second; Otherwise, find out the similar cluster in line with the calculated matching point of user's browsing patterns and results of cluster analysis. Make a recommendation according to the results.

Compared with cluster analysis, association recommendation can provide more accurate results, but it can only make a recommendation for users whose behavior conforms to the former terms of association recommendation; though recommendation based on cluster analysis can provide recommended feature spot for all the users, it is not necessarily account.

4 Conclusion

As a new research field with great application prospect, individual information service makes good use of known behavior of uses and improves the ability to provide information service by mining user's interests. Thanks to the efforts of the researchers, it has made solid progress. At the beginning, the new model of Individual tourism service system was put forward, and then analysis the association between user behavior and their interests, thus advanced discovering tourists interests. At last, combined with tourists interest discovery suggests that the Individual recommendation ideas. Individual tourism service system is mainly composed of user interests mining and individual recommendation engine. Tourists interest mining mainly realizes user modeling, discovery user interests, provide decision-making basis for individual recommendation. Individual recommendation engine main comprehensive clustering analysis and association rules two methods achieve the recommendation process of feature spots.

References

1. Schafer, J.B., Konstan, J.A., Ried, J.: E-commerce recommendation applications. Journal of Data Mining and Knowledge Discovery 2, 115–152 (2001)
2. Freitag, D., McCallum, A.: Information extraction with HMM structures learned by stochastic optimization. In: Proceedings of the Eighteenth Conference on Artificial Intelligence, pp. 76–82 (2000)
3. Du, J., Guo, W., Wang, R.: Forecasting Model of Holiday Tourism Requirement on the Basis of RBF Neural Network. Intelligent Systems Design and Applications 1, 677–682 (2006)
4. Lin, Y.-S., Huang, J.-Y.: Internet blogs as a tourism marketing medium: A case study. Journal of Business Research 59, 1201–1205 (2006)
5. Conrady, R.: Travel technology in the era of Web 2.0. Trends and Issues in Global Tourism, 165–184 (2007)
6. Doris, S.: Blogs in tourism: Changing approaches to information exchange. Journal of Vacation Marketing 14, 99–110 (2008)

A Static-Error-Less Phase Detector for Delay Locked Loops

Gang Luo and Xianjun Zeng

School of Computer Science, National University of Defense Technology
Changsha, Hunan, China
ttywlhl@gmail.com, xjzeng1@163.com

Abstract. This paper presents a new phase detector circuit for a 30-phase 500MHz DLL. The new phase detector circuit solves the problem of dead zone in course of charge pump's switch transition and has advantages of a perfect linear PD with almost no static phase error which caused by current mismatch in the charge pump. The post-simulation results show that the proposed PD can detect the delay difference no less than 1ps, and RMS jitter of the DLL with the proposed phase detector decreased to 0.5629ps.

Keywords: Phase detector, static phase error, delay-locked loop, phase defence, RMS jitter.

1 Introduction

Phase detector (PD) is the circuit used to compare the difference of phase between two signals, output a pulse signal whose width correspond to the difference of phase, driving the charge pump then the DLL can track the phase. Because of not-ideal factor, such as not matching of input and dead zone of input signals, the output exist static errors inevitably [1]. In this paper, based on an analysis of the effects of static phase errors on delay-locked loop (DLL) and incorporates the feature of the Hogge PD with the feature of the Alexander PD, a high-resolution DLL phase detector is designed and presented for a 30-phase 500MHz DLL.

2 The Effects of Static Phase Error on DLL

Different from PLL (Phase-Locked Loop), DLL's phase results of the final output has especially direct effect to the result output ultimately, the existence of a static phase error in PLL loop will not cause the loss of PLL lock, the impact on the output frequency, simply increase the control voltage jitter. But in DLLs the static phase error will cause the output phase offset [2].

A period of time set to T, the static phase error is $\Delta\phi$, then ideally interval between every two adjacent phase is

D. Zeng (Ed.): Advances in Computer Science and Engineering, AISC 141, pp. 299–306.
springer link.com

$$PI_1 = \frac{T}{30} \tag{1}$$

As periodic time error of 1%, the time interval PI2 between two adjacent phase is

$$PI_2 = \frac{T}{30}(1 - \frac{\Delta\phi}{2\pi}) \tag{2}$$

Because no other non-ideal factors, The two cases, the first phase of the output phase difference is 0, the second phase is $PI_1 - PI_2$, the third phase is $2(PI_1 - PI_2)$, and so on. At this point you can calculate the RMS jitter of the output:

$$Jitter_{RMS} = \sqrt{\frac{\sum_{i=1}^{30}\left[(i-1)\frac{\Delta\phi}{2\pi}\frac{T}{30}\right]^2}{30}} \tag{3}$$

Let $\Delta\phi/2\pi = 0.01$ and T=2000ps, solve equation (3):

$$Jitter_{RMS} = \frac{2}{3}\sqrt{\frac{\sum_{i=1}^{29} i^2}{30}}ps \approx 11.258\,ps \tag{4}$$

RMS jitter at this time is close to the requirements for the bit error rate limit in a PCIE2.0 system, so a tiny loop in the DLL static phase error will cause a lot of jitter, so phase detecting Accuracy has an important influence on the loop output jitter and loop lock time.

In addition to the accuracy of the phase detector, charge pump current mismatch will also have a static phase error. Assuming charge the charge pump with current of I_{ch}, the discharge current is $I_{dis} = -(I_{ch} + \Delta I)$, the pulse width of output signal UP and signal DN is W_{UP} and W_{DN}, respectively. When phase locked, the control voltage is constant, for the loop filter the charging charge is equal to discharging charge. The following formula holds:

$$I_{ch}W_{UP} + I_{dis}W_{DN} = 0 \tag{5}$$

At this point the static phase error is:

$$W_{UP} - W_{DN} = \frac{\Delta I}{I_{ch} + \Delta I}W_{UP} = \frac{\Delta I}{I_{ch}}W_{DN} \tag{6}$$

It is now clear that the mismatch of the charge pump current is proportional to the static phase error. However, due to the charge pump current mismatch resulting from the switch transistor match degree, the output voltage, temperature, process and other aspects of the factors, and to completely eliminate the current mismatch is highly unlikely, therefore requested to consider the design of the phase detector can reduce the Charge pump non-ideal factors.

3 Traditional Phase Detectors

Common phase detector according to the phase detector output and the phase relationship between the phase detectors can be divided into linear phase detector[3] and the binary phase detector[4]. Most of the linear phase detector is based on Hogge structure, which outputs a phase difference proportional to the width of the HU or SD signal; the binary phase detector is based on Alexander the structure, which according to input data and clock signals between the lead or lag relationship, then output width equivalent LD or CAMP signal.

Linear phase detector performs a comparison to the phase relationships between enter data and feedback clock, and then generate an output signal whose width is equal to the phase error. For the conventional linear phase detectors, each cycle of the output signal is proportional to the phase difference; speed up the locking speed, and when the phase difference is reduced to certain range it can still guarantee very high phase accuracy. When DLL is locked, jitter of VCDL control voltage is very small. But there is also a drawback that both the charge signal for each cycle, there are discharge signals, and the pulse width of the output signal must be greater than a fixed value, to ensure that the output can avoid dead zone of the charge pump. By the derivation of formula (6) we can know in this case and the charge pump current mismatch will introduce a static phase error.

Different from linear phase detector, binary phase detector's output) can only characterize the lead / lag relationship between input data and feedback clock. The output of the binary phase detector can only reflect the leading or lagging phase relationship between clock and data can not reflect the phase difference in size, so the lock slowly, when the DLL Locked there will have a greater jitter . But the binary phase detector in each cycle will not have both charge and discharge, and this is equivalent to when the loop Locked the charge pump current mismatch do not have any impact, which resulting in a static phase error.

In order to improve the accuracy of phase detection current main methods are: In the case of the basic principles remain unchanged, in order to improve the accuracy of Hogge phase detector, [5] proposed a modified Hogge phase detector which can automatically adjust (clock edges) to the optimal sampling points; In order to achieve linear phase to digital conversion [6] proposed a modified Hogge phase detector with a small size and a low power consumption and other benefits. In addition, to improve the accuracy of Alexander phase detector, the researchers also conducted its structure a number of improvements, such as: In order to reduce the phase jitter to improve the accuracy, the literature [7] presents a bang-bang half-rate Alexander phase detector structure; the literature [8] proposed a linear half-rate Alexander phase detector. However, no matter Is the structure of the modified Hogge phase detector or optimized Alexander phase detector were unable to go beyond the basic structure of which it is based.

4 Design of the Hybrid Phase Detector

4.1 Circuits Description

The Hogge phase detector and the Alexander phase detector can not go beyond their inherent limitations on the accuracy increased restrictions. This paper proposed and designed a new hybrid phase detector with high accuracy, the phase detector combines the advantages of the linear phase detector and binary phase detector. Achieve a lower output phase RMS jitter, making it a higher sensitivity and accuracy. The phase detector circuit is shown in Figure 1.

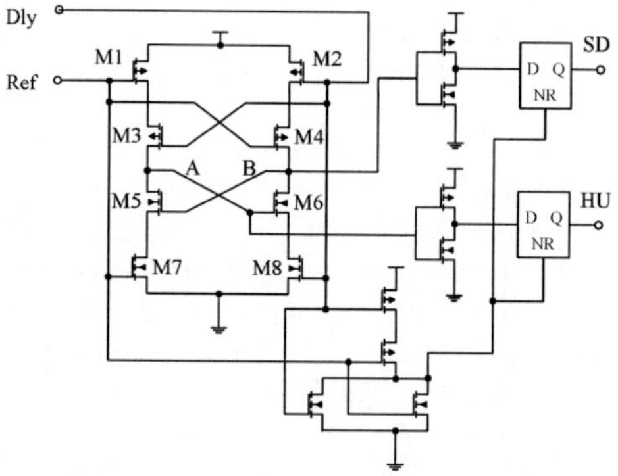

Fig. 1. The proposed phase detector

The circuit consists of two cross-coupled C2MOS inverter, an OR gate, two delay buffers and two negative edge-sensitive latch. The basic idea is that the first arrival of the input signal prevents the later input signal to change the output by controlling C2MOS inverter's clock control terminal. So that each cycle does not have the charge and discharge process at the same time, similar to the working mechanism of binary phase detector, avoided the charge pump current mismatch effects. At the same time phase difference control pulse width, greater phase difference lead to a wider pulse width to output , It draws on the advantages of linear phase detector, loop lock speed up and the control voltage jitter is reduced.

4.2 Operation of the Hybrid PD

The circuit in Figure 1 works as follows:

Before the arrival of the rising edge of Ref and Dly, transistors M1, M2, M3 and M4 are on, M7 and M8 are off, node A and B set to high by the power, M5 and M6 turns on. The nodes A and B connected to the latch through the buffer, so the latch input is high, as the latches is low level reset, so the output remains low, That is HU and SD are all Low.

If the signal of Ref arrives early than the signal of Dly, M1, M4 turns off, M7 on, node B is holding its previous value, M5 turns on, then node A is discharged, its level change from high to low, M6 followed turns off. At this point the first arrival signal of Ref given node A a falling edge, while the branch is blocked by M4 and M6 where node B is located, then node B into the high impedance state, regardless of how the input Dly, level of node B does not change.

AS the input signal of Ref change to high level , output of OR gate also change from low to high, At this point latch reset is released, its output will respond to input changes. It is worth noting that the delay between the rising edge of Ref and the falling edge of latch's node D must be longer than the delay between the rising edge of Ref and the rising edge of latch's node NR, to ensure that, reset signal has been withdrawn when data is inputting node D of the latch. That is the buffer delay should be large enough.

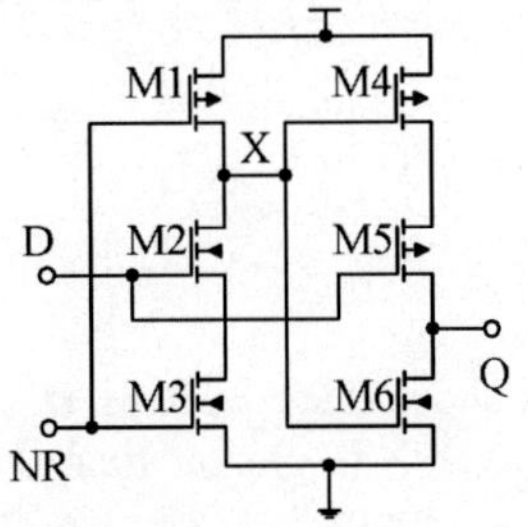

Fig. 2. Latch used in proposed phase detector

Figure 2 shows the negative-edge sensitive latch circuit diagram. When NR is low, M4 turns on and M6 turns off, the output is always low regardless of the value of node D. When NR is high, M1 off, M3 turns on, the input D is high, through the M2, M3 node X is discharged to low, then M4 turns on, M6 turns off, because M5 is off, output Q does not change. When the falling edge reach node D, M2 off, X nodes remains low, M5 turns on, through the M4, M5 output Q is charged to high, the state will hold until the NR active again.

Phase difference and the input signal pulse width joint decision output pulse width of phase detector. Time interval is usually very small when the phase difference between the input signals is small, if the phase detector output pulse width is too small then not enough time here to open the switch to charge and discharge of the charge pump. To solve this problem, we can increase the output pulse width of a fixed value such as half a clock cycle time.

In Figure 2, under the control of OR gate the reset signal will be delayed half-cycle. Shown in Figure 3, the falling edge of the output signal determined by the reset signal, while the reset signal comes from the input signal falling edge, so the range of the pulse width of output plus is T / 2 to T, where T is the cycle of input signal.

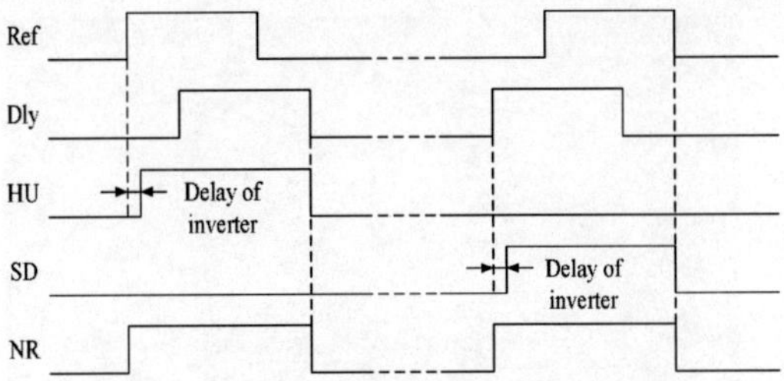

Fig. 3. Timing diagram of proposed PD

The cross-coupled C2MOS inverter used in the new phase detector to ensure accuracy of phase detect. But when the two phase of input signal is strict alignment to each other or phase difference less than resolution limit of the phase detector, node A and node B can not block each other's output using clock control node of C2MOS inverter, the rising edge of the input will produce a falling edge, thus output will be two positive pulse of equal width, which makes the phase detector still affected by the current mismatch when the phase-aligned. However, due to the high phase detect accuracy, the effects from static phase error caused by current mismatch is greatly reduced.

5 Simulation Result

Hspice pre-simulation results show that at 500MHz the phase detector can correctly identify the phase difference which is greater than 0.36 ° (2ps), Figure 4 shows the output waveform when phase difference between two input is 0.36 ° (2ps).

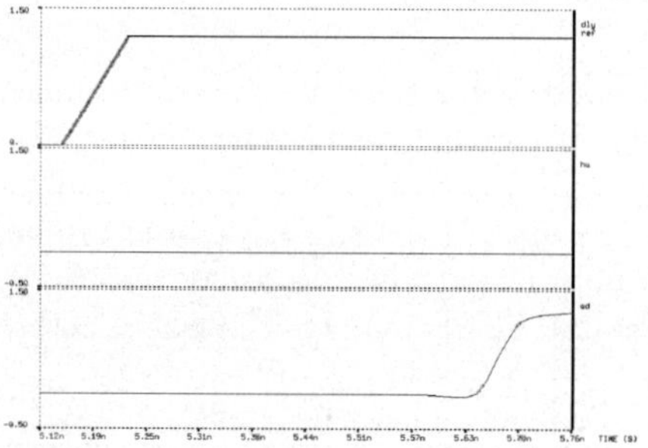

Fig. 4. Pre-simulation results of the PD

The layout of the PD was implemented in a 0.13um CMOS process. Power supply is 1.2V and the operation frequency of DLL is 500MHz. In order to ensure accuracy of the PD, the layout is strict geometric symmetry. In addition, wire routing is in the same metal layer as far as possible, for avoiding different delays of the signals that resulting in the generation of static phase error. Layout of the phase detector is shown in Figure 5.

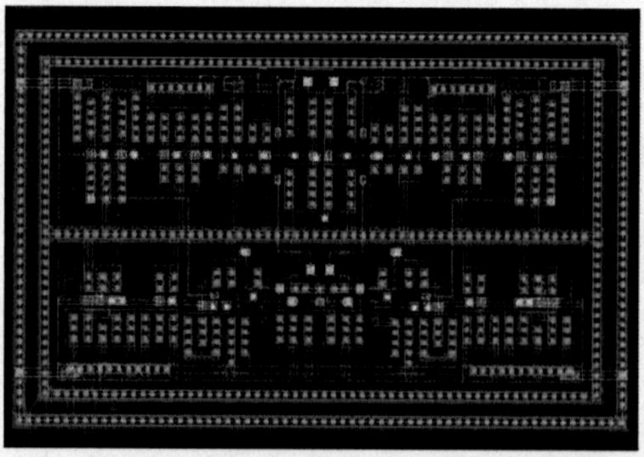

Fig. 5. Layout of the PD

Figure 6 shows the post-simulation result. It illustrates outputs of the phase detector In the typical case when the DLL is locked. Known by measuring, the phase difference less than 1ps can be identified correctly. It ensure that the static phase error almost does not Occur when the DLL locked.

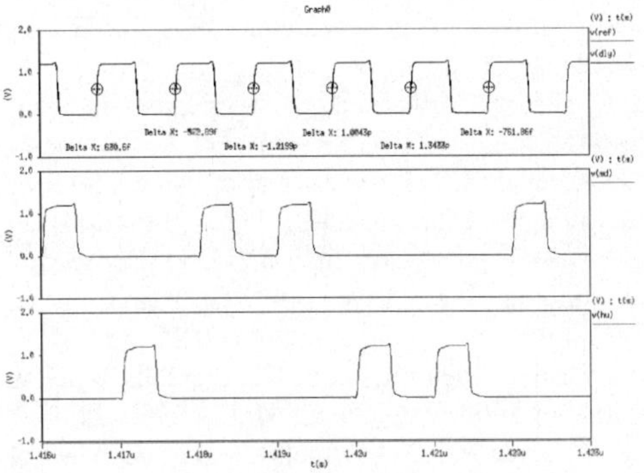

Fig. 6. Post-simulation result

Post-simulation shows the highest phase accuracy can reach 0.18 °,therefore the 30-phase 500MHz DLL with the proposed phase detector will not produce more than lps/2ns = 0.05% of the static phase error, According to the formula (4):

$$Jitter_{RMS} = \frac{2}{30}\sqrt{\frac{\sum_{i=1}^{29} i^2}{30}}ps \approx 0.5629p$$

RMS jitter of the DLL is only 0.5629ps.

6 Conclusion

A new hybrid phase detector is introduced. It combines the advantage of the linear phase detector and the binary phase detector, achieve a lower output phase RMS jitter and has high sensitivity and accuracy. By prevent the PD to charge and discharge the charge pump in a same clock cycle, it avoided the charge pump current mismatch effects, eliminates static phase errors caused by current mismatch in the charge pump. Loop locks speed up and the control voltage jitter is reduced also. The highest phase accuracy of the new type phase detector can reach 0.18 °, the static phase error of DLL with the proposed phase detector is lower than 0.05%, RMS jitter of the 500MHz DLL is only 0.5629ps. By using this hybrid phase detector for DLL, it is possible to meet the demand of high data rate serial links.

Acknowledgment. This work was supported by NSFC (No.60676016) in China.

References

1. Nurmi, J.: Processor design:system-on-chip computing for ASICs and FPGAs. Springer, Berlin (2007)
2. Farjad-Rad, R., Dally, W., et al.: A low-power multi-plying DLL for low-jitter multigigahertz clock generation in highly integrated digital chips. IEEE Joumal of Solid-State Cirauts 37(12), 1804–1812 (2002)
3. Razavi, B.: Design of analog CMOS integrated circuits(Gravure). Tsinghua Uruversity Press, Beijing (2005)
4. Dehng, G.-K., Hsu, J.-M., et al.: Clock-deskew buffer using a SAR-controUed delay-locked loop. IEEE Journal of Solid-State Circuits 35(8), 1128–1136 (2000)
5. Li, X., Gao, Q., et al.: The design of monolithic CMOS clock recovery. Journal of Electrorucs & Information Technology 29(6), 1496–1499 (2007) (in Chinese)
6. Perrott, M.H., Huang, Y., et al.: A 2.5-Gb/s multirate 0.25-μm CMOS clock and data recovery circuit utilizing a hybrid analog/digital loop filter and all-digital referenceless frequency acquisition. IEEE Journal of Solid-State Circuit 41(12), 2930–2944 (2006)
7. Perrott, M.H., Huang, Y., et al.: A 2.5-Gb/s multirate 0.25-μm CMOS clock and data recovery circuit utilizing a hybrid analog/digital loop filter and all-digital referenceless frequency acquisition. IEEE Journal of Solid-State Circuit 41(12), 2930–2944 (2006)
8. Kromer, C., Sialm, G., Menolfi, C., Schmatz, M., Ellinger, F., Jackel, H.: A 25-Gb/s CDR in 90-nm CMOS for High-Density Interconnects. IEEE Journal of Solid-State Circuit 41(12), 2921–2929 (2006)

Semipermeable Membrane Mass Transfer in Pressure-Retarded Osmosis Process

Yonghua You, Suyi Huang, Yi Yang, Chao Liu, Zhilin Wu, and Xiangfei Yu

School of Energy & Power Engineering, Huazhong University of Science & Technology
Wuhan, China
catia315@sina.com

Abstract. Pressure-retarded osmosis (PRO) is a feasible process in developing salinity power, which emerges in estuaries in large. Semipermeable membrane mass transfer plays an important role in PRO power generation. However, existing mass transfer models were hard to be validated due to the lack of suitable membranes and membrane modules. This paper presented a comprehensive PRO mass transfer model of semipermeable membrane, which included impacts of both internal and external concentration polarizations, as well as brine dilution and pressure loss. The model was solved by numerical method; and commercial asymmetric flat sheet membrane CA-3000 was taken as the instance. Concentration and hydraulic pressure, as well as brine flux were studied in the paper. Massive concentration gradient was observed in porous layer and effective concentration difference was much lower than expected; moreover, each brine concentration was found to have an optimal pressure for power density. An additional PRO process simulation was done with FLUENT software under the same condition. Two simulation curves approached closely. The paper could be referred for membrane optimization in PRO process.

Keywords: Mass transfer, semipermeable membrane, pressure-retarded osmosis, FLUENT simulation.

1 Introduction

Salinity power comes from chemical potential difference between solutions. When sea water is mixed with fresh water, salinity energy is released. Global salinity power in estuaries would be on the order of 2000 TWh per year [1,4]. They can be tapped by pressure-retarded osmosis (PRO) process invented by Israel scientist S. Leob[1]. A selective permeable membrane is utilized to create two chambers with different pressures. Pressurized brine flows in high pressure side and fresh water in low pressure side. Driven by chemical potential difference, water permeates through membrane into brine, and power is generated with diluted brine flowing through hydroturbine.

Scientists did PRO experiments with reverse osmosis (RO) membranes at first and found test data much smaller than projected. Leob and Mehta assumed hydraulic pressure and osmosis pressure had different water permeability and presented a two-coefficient transport equation [2]. Lee et al. regarded concentration polarization inside

D. Zeng (Ed.): Advances in Computer Science and Engineering, AISC 141, pp. 307–315.
springer link.com

porous layer (ICP) as the main reason of low water flux in PRO, and presented a mass transfer model including salt leakage and ICP[3]. Achilli et al. polished the model by adding the external concentration polarization (ECP) in their studies [4], which was proved with a custom-made laboratory-scale membrane module. Nowadays concentration polarizations are generally accepted as the reason of low water flux in PRO process [1].

Existing PRO membrane mass transfer models were based on some parameters determined by both RO and forward osmosis(FO) tests. They were deficient in experimental proof for lack of suitable membranes and membrane modules[4].

Seppala studied hollow fibre membranes in PRO power generation system [5]. He presented a witty optimization model, which included bulk brine dilution and pressure loss. However, the model ignored porous laryer ICP, and no test was given for verification.

This paper studied mass transfer of asymmetric flat sheet membrane in PRO process and presented a theoretical model, which included both ICP and ECP, as well as salt leakage, brine dilution and pressure loss. The model was solved with numerical method, and commercial RO membrane Toray CA-3000 was taken for the simulation instance. For the confirmation an additional FLUENT simulation was done under the same PRO condition. This paper studied sensitive factors of water flux and power density, such as brine concentration, hydraulic pressure and brine flux. Concentration distribution and power density were calculated; and results were showed in several figures.

2 Theoretical Model

A selective permeable membrane used in osmosis application allows water to permeate, but rejects salt. In real osmosis process, the majority of salt is rejected by semipermeable membrane, but a small amount still leaks through due to the concentration difference[3,4].

Semipermeable membrane is the key part in PRO power generation. Owning to high water permeability, an asymmetric membrane is widely applied in various osmosis applications. It usually consists of two layers, a thin dense skin and a thick porous support matrix. As Fig.1 shows, the dense skin faces pressurized brine in PRO process[4]. When water permeates through the membrane, salt is diluted on the brine side of the membranes, and accumulates inside the porous layer, which are called as ECP and ICP, respectively.

Both PRO and RO make use of water flux through semipermeable membrane, however, a big difference exists, i.e, their water flux directions are opposite. In PRO process, water permeates from fresh water to brine, and salt does in the opposite direction, which facilitates salt to accumulate inside porous layer and form severe ICP. ICP decreases the effective concentration difference across dense skin substantially, which results in water flux in PRO process much lower than projected.

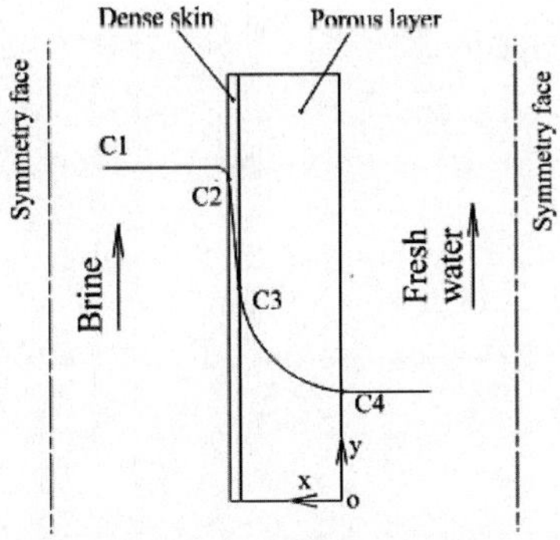

Fig. 1. Concentration distribution in membrane process

2.1 Species Transport through Membrane

Fig.1 demonstrates a typical asymmetric semipermeable membrane, and cross section concentration curve in PRO process. It can be seen the whole mass transfer resistances are made up of three parts, namely ECP, ICP and dense skin resistance, among which ICP acts as the most important one in PRO process.

1) Dense skin mass transfer

a) Water flux through membrane

As is known to all, osmosis is a process driven by chemical potential difference. Water flux through membrane is related to chemical potential difference determined by both concentration and hydraulic pressure.

Since both brine and fresh water are dilutive enough to be treated as ideal solutions, water flux J_w ($m^3.m^{-2}.s^{-1}$) can be calculated by (1).

$$J_w = A[RT(C_2 - C_3) - \Delta p] \tag{1}$$

where A is water permeability coefficient, $m^3.m^{-2}.Pa^{-1}.s^{-1}$;

R is constant, 8314 $kJ.kmol^{-1}.K^{-1}$;

T is Kelvin temperature, K;

C_2,C_3 are ion concentrations at two surfaces of dense skin, as showed in Fig.1, $mole.L^{-1}$;

Δp is pressure difference across membrane, Pa.

b) Salt leakage

In real PRO process, a selective permeable membrane rejects the majority of salt, but still a small amount permeates through into fresh water due to salt concentration gradient across membrane. Equation (2) can be used for salt leakage J_s ($kmol.m^{-2}.s^{-1}$).

$$J_s = -B(C_2 - C_3) \tag{2}$$

where B is salt permeabiltiy coefficient, $m.s^{-1}$.

Salt leakage decreases water flux, for it accumulates inside porous layer and strengthens the ICP, which reduces effective concentration difference.

2) Porous layer mass transfer

In PRO process, salt convects and diffuses in opposite direction, which facilitates salt to accumulate inside porous layer and creates relatively big concentration gradient normal to membrane surface.

The salt concentration gradient along bulk flow direction is trivial and can be neglected. According to salt species conservation in stable condition, salt leakage into porous layer equals to the convective plus the diffusive normal to membrane surface, as showed in (3).

$$J_s(y) = -D_s\varepsilon\frac{\partial C_s(x,y)}{\partial x} + J_w(y)C_s(x,y) \tag{3}$$

Where D_s is the salt mass diffusivity, $m^2.s^{-1}$;
ε is void fraction of porous layer;
C_s is salt concentration inside porous layer, mole.L^{-1};
x, y are coordinate axises, as showed in Fig.1, m.

ICP can't be minimized by stiring; it exerts much greater negative influences on water flux in PRO than in RO process due to the opposite flow direction of water and salt. ICP acts as the main reason of low water flux of PRO.

3) Skin surface mass transfer

In PRO process, salt is diluted at the dense skin surface on brine side, which is called as ECP. ECP decreases effective concentration difference and water flux drops.

ECP zone can be modeled with film theory, and salt concentration can be calculated by (4).

$$-J_w(y)C_3(y) = J_w(y)C_s(x,y) - D_s\frac{\partial C_s(x,y)}{\partial x} \tag{4}$$

Where boundary thickness $\delta = D_s/k$, m;
k is convective mass transfer coefficient, $m.s^{-1}$.

2.2 Brine Dilution

As brine flows forward, bulk brine is diluted by water flux through membrane. When water flux is appreciable to brine flux, dilution can't be ignored in the calculation.

Since salt leakage is very small, equation (5) and (6) can be used for bulk concentration after dilution.

$$\frac{dJ_d(y)}{dy} = J_w(y)U \tag{5}$$

$$\frac{dJ_d(y)C_1(y)}{dy} = 0 \tag{6}$$

where U is the wet circumstance, m; J_d is brine flux, $m^3.s^{-1}$.

2.3 Brine Pressure Loss

Brine pressure loss impacts both water flux and net power density. When brine flows between two flat sheet membranes, the pressure loss can be calculated by (7).

$$-\frac{dp}{dy} = 2f_f \frac{\rho}{d_h}\left(\frac{J_d}{S}\right)^2 \tag{7}$$

where p is hydraulic pressure, Pa;

ρ is brine density, kg.m^{-3};

d_h is the hydraulic diameter, m;

S is the channel cross section area, m^2;

f_f is the Fanning frictional coefficient.

2.4 Power Generation

The objective of PRO process is power generation, which depends on water flux through membrane. The flow loss inside the membrane should be offset. Assuming membrane width is one unit, the net power generation W(w) can be calculated by (8).

$$\dot{W} = \int_0^{exit} J_w(y)dy \times p_{exit} - \int_0^{exit} \frac{dp(y)}{dy} dy \times \dot{J}_{d,in} \tag{8}$$

where p_{exit} is pressure at the exit, Pa. ;$J_{d,in}$ is brine flux at the inlet, m^3.s^{-1}.

3 Numerical Computation

3.1 Computation Procedure Diagram

By solving the theoretical model of former chapter with numerical methods, concentration, water flux and hydraulic pressure in PRO process were all obtained, with which net power generation was calculated.

Fig.2 showed iteration flow chart of theoretical model presented in former chapter. A flat sheet asymmetric membrane was divided into discrete cells along bulk flow direction, each using an iteration cycle for cross section concentration. Concentration difference across dense skin C_2-C_3 was taken as iteration parameter.

Both brine dilution and pressure loss of cells were calculated before next length step.

3.2 Asymmetric Membrane CA-3000

Toray CA-3000 is a typical cellulose acetate asymmetric membrane studied by Loeb et al [6]. It consists of a thin skin and a porous layer of cellulose acetate, as well as a thick support fabric. The whole thickness is 0.26mm, and water permeability coefficient A, salt permeability coefficient B and salt resistivity K equal to 3.80×10^{-12} m^3.m^{-2}.s^{-1}.Pa^{-1}, 2.13×10^{-7} m.s^{-1} and 8.99×10^{6} s.m^{-1} respectively, with which void fraction of porous layer is calculated about 0.019.

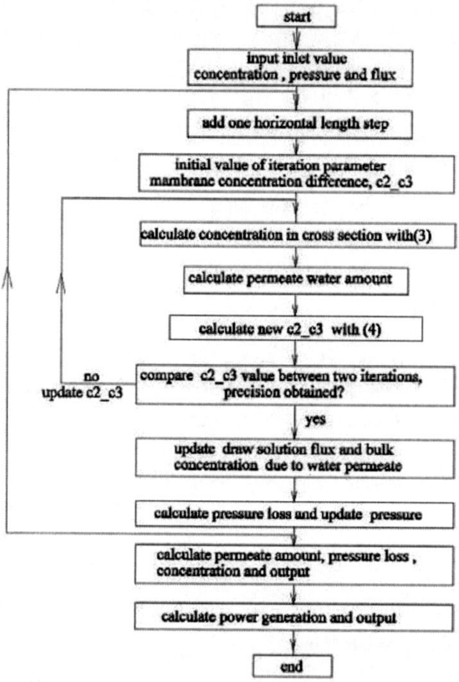

Fig. 2. Iteration flow chart of membrane osmosis process

The paper took CA-3000 flat sheet membrane as an instance for the PRO simulation.

3.3 CFD Simulation

Commercial software FLUENT is very popular for fluid simulation. In addition to being robust and precise, its solver has excellent flexibility owing to comprehensive and flexible User-Define-Functions(UDFs).

In this study FLUENT software was taken for PRO simulation under the same condition, and the results were used for the verification of former model.

The study used laminar species transport model, and created the mixture brine, which consisted of sodium chloride and water, assuming bulk flow fully developed. Three fluid zones were created and porous layer was treated as porous boundary, where Ergun Equation was adopted for pressure loss[7]. Both water flux and salt leakage through membrane were tackled by setting UDF sources.

3.4 Results and Analysis

Fig. 3 only showed the cross section concentration distribution inside the porous layer calculated in various hydraulic pressure by paper model (red curves), since concentration gradient in brine was really small. The internal concentration gradient was really big, taking up the majority of bulk concentration difference. FLUENT simulation under the same condition showed similar phenomena, which was also presented in the Fig.3 (blue curves). It was clear two sets of curves approached closely. The really big concentration gradient could be caused by severe ICP, and very thick porous layer of CA-3000 increased the severity.

Fig.4 showed how water flux changed along bulk flow direction. Five curves were presented, each representing one hydraulic pressure. All curves were consistent with theory. At the same concentration, low hydraulic pressure generated big chemical potential difference which allowed much water to permeate through; water flux dropped along bulk flow direction due to brine dilution, so curves were tilted down.

Fig.5 and Fig.6 showed power generation under various concentrations and pressures. It was observed the bigger concentration brine had, the bigger power density it generated, and each concentration had an optimal pressure for power density. The two figures could both be reasoned by (8). Under the same pressure, water chemical potential increased with salt concentration, so big concentration brought big water flux and power density. When hydraulic pressure increased, water flux would decreased. The product of pressure and water flux at the exit, and the power generation could rose and dropped according to specific circumstances, which appeared as a peak on the curve.

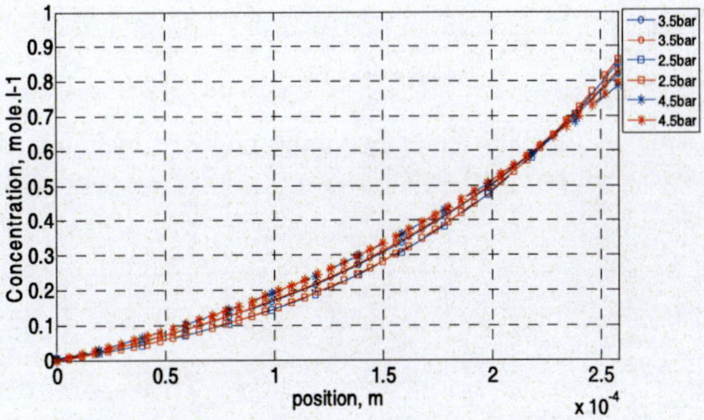

Fig. 3. Comparision of concentration in porous layer between paper model and FLUENT brine 1 mole l-1, deionized feed water.red: paper model,bule:FLUENT softmare.

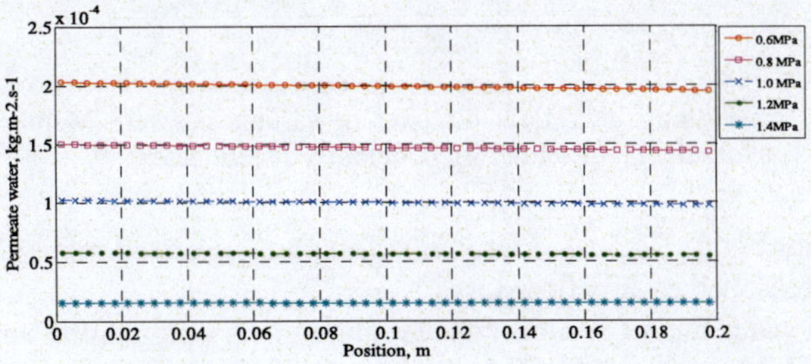

Fig. 4. Permeate water along bulk flow direcion in various pressure Membrane CA-3000, hydraulic diameter 0.16mm, brine 1.756 mole. 1-1, velocity 0.01 m.s-1, deionized feed water.

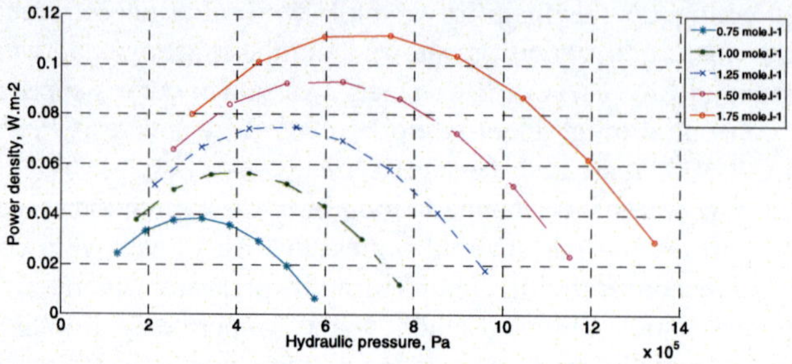

Fig. 5. Power generation of ca-3000 under hydraulic pressure Hydraulic diameter 0.16mm, brine velocity 0.01m.s, deionized feed water

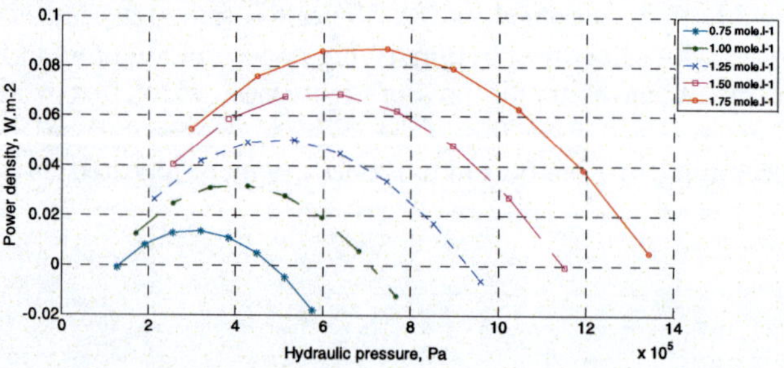

Fig. 6. Power generation of ca-3000 under hydraulic pressure hydraulic diameter 0.16mm, brine velocity 0.02m.s-1, deionized feed water

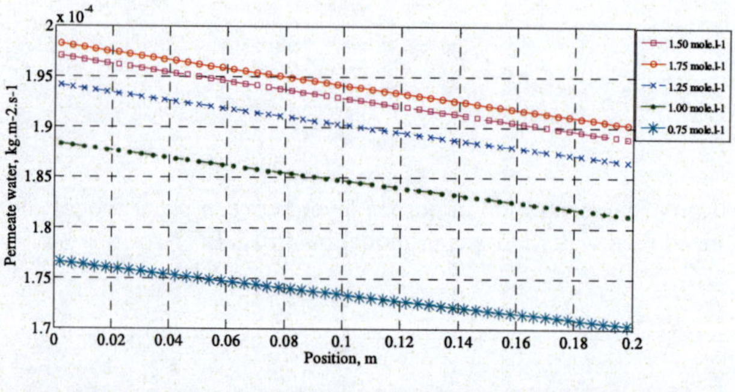

Fig. 7. Permeate water along bulk flow direcion in various pressure Membrane CA-3000, velocity 0.01 m.s-1, hydraulic diameter 0.16mm, deionized feed water

Compared with Fig.6, Fig.5 used less brine flux, but generated more power density, which could be reasoned by less flow loss.

Fig.7 showed water flux of various concentration under optimal pressure, where five curves were tilted down along bulk flow; these could all be caused by dilution of bulk brine.

4 Conclusion

The paper studied mass transfer of asymmetric flat sheet membrane in pressure-retarded osmosis process, and presented a theoretical model which included the impacts of both ICP and ECP, as well as pressure loss and brine dilution. Commercial RO membrane CA-3000 was taken for simulation. Most sensitive factors of water flux and power density were studied, such as brine concentration, hydraulic pressure and brine flux. Great concentration polarization was observed inside porous layer, which took up the majority of bulk concentration difference. The bigger concentration brine had, the bigger power density it generated, and each concentration had an optimal hydraulic pressure for power density. An additional FLUENT simulation was done for the verification of the model, and the two simulation results met well.

Acknowledgment. This work was supported in part by the national natural scientific foundation of China with No.51076057.

References

1. Achilli, A., Childress, A.E.: Pressure retarded osmosis: From the vision of Sidney Loeb to the first prototype installation — Review. Desalination 261, 205–211 (2010)
2. Loeb, S., Mehta, G.D.: A two-coefficient water transport equation for pressure-retarded osmosis. Journal of Membrane Science 4, 361–362 (1979)
3. Lee, K.L., Baker, R.W., Lonsdale, H.K.: Membranes for power generation by pressure-retarded osmosis. Journal of Membrane Science 8, 141–171 (1981)
4. Achilli, A., Cathb, T.Y., Childressa, A.E.: Power generation with pressure retarded osmosis: An experimental and theoretical investigation. Journal of Membrane Science 343, 42–52 (2009)
5. Seppala, A., Lampinen, M.J.: Thermodynamic optimizing of pressure-retarded osmosis power generation systems. Journal of Membrane Science 161, 115–138 (1999)
6. Loeb, S., Titelman, L., Korngold, E., Freiman, J.: Effect of porous support fabric on osmosis through a Loeb-Sourirajan type asymmetric membrane. Journal of Membrane Science 129, 243–249 (1997)
7. `http://www.fluent.com/`

A Fast Learning Algorithm of the Perception

Li Jinling

School of Management, Shanxi Normal University
Linfen, China
jll-li@163.com

Abstract. In this paper a new learning algorithm of perception with variable paces is given which adjusts its weights and threshold values according to the following formulation $W(K+1) = W(K) - 2W(K)X_K^T X_K$. A theorem shows that the algorithm must be convergence after finite steps.

Keywords: variable paces, perception, convergence, classification.

1 Introduction

In 1960, Rosenblatt first put forward a fast learning Algorithm of the perception, and its weight vector modified formula is $W(K+1) = W(K) + \alpha \frac{\varepsilon_K}{2} X_K$. It can be proved that the method above may get weight vector which can satisfy demand of classification, when α is small enough and the number of samples is big enough. Only if α is bigger, adjustment for the range is bigger, but the phenomenon of "over-adjustment" may appear. Therefore, someone proposed that α is bigger at the initial stage, and then, it gets gradually smaller. This idea is very good, but too general. How much is α actually "big" and how small does α mean "small" ? Furthermore, it may be big for a particular problem is, and for the other problems may seem small, thus, for different issues, specific changeable given α is the key to improving the learning speed.

2 Learning Algorithm of Perception with Variable Paces

Two ranges, R_1、 $R_2 \in R^n$ are separable, if hyper plane $WX^T + \theta = 0$, $WX^T + \theta > 0$. If $X \in R_1$, $WX^T + \theta \le 0$, if $X \in R_2$. $W \in R^n$ is called as weighted vector and $\theta \in R$ is threshold value. For an arbitrary group of training samples $X_1, X_2, \ldots, X_N \in R_1 \cup R_2$, training aim is to find weighted vector W and threshold value θ. When $X \in R_1$ or $X \in R_2$, they respectively have

D. Zeng (Ed.): Advances in Computer Science and Engineering, AISC 141, pp. 317–324.
springer link.com

$WX^T + \theta > 0$ or $WX^T + \theta \le 0$, as long as R_1 and R_2 are linear and separable. If $W = (w_1, w_2, \cdots, w_n)$, the problem above is transferred as

$$\sum_{i=1}^{n} w_i x_i - \theta > 0 \text{ if } X = (x_1, x_2, \cdots, x_n) \in R_1$$

or

$$\sum_{i=1}^{n} w_i x_i - \theta \le 0 \text{ if } X \in R_2$$

let $W = (w_1, w_2, \cdots, w_n, -\theta)$

$$Y = \begin{cases} (x_1, x_2, \cdots, x_n, 1) & \text{if} (x_1, x_2, \cdots, x_n) \in R_1 \\ (-x_1, -x_2, \cdots, -x_n, -1) & \text{if} (x_1, x_2, \cdots, x_n) \in R_2 \end{cases}$$ Thus when $(x_1, x_2, \cdots, x_n) \in R_1$,

$$\sum_{i=1}^{n} w_i x_i - \theta > 0 \text{ is equivalent to } WY > 0$$

When $(x_1, x_2, \cdots, x_n) \in R_2$,

$$\sum_{i=1}^{n} w_i x_i - \theta \le 0 \text{ is equivalent to } WY \le 0$$

Therefore, the problem of classification is to find the vector W, which makes $WY > 0$. Let $X_i (i = 1,2, \cdots, N) \in R^n$ be a group of training samples, and Y_i is the training sample obtained by X_i through the equation (1) 式.

The aim is to find W, which $WY_i^T > 0$, without loss of generality. Supposing $Y_0, Y_1, \cdots Y_N$ are all unit vectors.

Supposing $W(1) = Y_0$, adjusting $W(1)$ makes $Y_i (i = 0,1, \cdots, N)$ satisfy the inequality $W(1)Y_i > 0$ as far as possible. Thus, choosing the constant α and unit vectors $X_1 \in \{Y_0, Y_1, \cdots Y_N\}$, then as long as $\min_{1 \le i \le N} W(1)Y_i^T < 0$, thus

$$W(1)X_1^T = \min_{1 \le i \le N} W(1)Y_i^T \text{ and } (W(1) + \alpha X_1)X_1^T = 0$$

In the second equation, $\alpha = -W(1)X_1^T$

To ensure $W(2)X_1^T > 0$, supposing

$W(2) = W(1) - 2W(1)X_1^T X_1$, thus

$$
\begin{aligned}
W(2)X_1^T &= (W(1) - 2W(1)X_1^T X_1)X_1^T \\
&= W(1)X_1^T - 2W(1)X_1^T X_1 X_1^T \\
&= W(1)X_1^T - 2W(1)X_1^T \\
&= -W(1)X_1^T > 0
\end{aligned}
$$

Repeat the algorithm above till satisfy $W(k)$

$$\min_{1 \le i \le N} W(k)Y_i^T \ge 0$$

We get $W_1^* = W(k)$ (in the next part, we will prove the existence of $W(k)$)。The algorithm can be summarized as the following procedure:

(1) Input the group of vectors

$$S_1 \stackrel{\Delta}{=} \{Y_0, Y_1, \cdots Y_N\}, \text{ let } W(1) = Y_0, \ k = 1.$$

(2) Get X_k, let $W(k)X_k^T = \min_{1 \le i \le N} W(k)Y_i^T$, and $X_k \in S_1$.

(3) If $W(k)X_k^T \ge 0$, let it turn (4) , or

$$W(k+1) = W(k) - 2W(k)X_k^T X_k,$$

let $k = k+1$ turn to (2) .

(4) Output $W_1^* = W(k)$.

If $W_1^* Y_i^T > 0$ sets up for $i = 0,1,\cdots,N$, we choose $W = W_1^*$, then we get vectors and threshold value.

In the following, we consider the condition that there is at least a vector $X \in S$ which makes $W_1^* X^T = 0$. This incident happens only by small probability and we can adopt the method in the following to deal with it.

The first method is to change the initial condition $W(1)$. For example, let $W(1) = Y_1$. However, if the number of samples is too big, which means that huge computations will be abandoned. Therefore, another method can be used. If the group of vectors, $S_2 \subset S_1$, let any $Y \in S_2$, $W_1^* Y^T = 0$, to S_2, we repeat the algorithm to get W_2^*. From the algorithm, we can know W_2^* and W_1^* are orthogonal.

Repeating the process above, we can get a group of orthogonal vectors W_1^*, W_2^*, ..., W_m^* $(m \le n+1)$ and $S_1 \supset S_2 \supset \cdots S_m \supset S_{m+1} = \phi$. It is worth mentioning that such m must exist. In fact, if $S_{n+1} \ne \phi$, to $Y \in S_{n+1}$, $W_n^* Y = 0$. We notice that W_1^*, W_2^*, ..., W_n^* form a group of orthogonal vectors and each vector is the dimension of $n+1$. We know at most there is a vector in S_{n+1} (or S_{n+1} is one-dimensional unit vector, the only situation quantity S_{n+1} is formed of $Y=(X,1)$ and $-Y=(-X,-1)$. From (2.1) , we know $X_1 \in R_1$, $X_2 \in R_2$, and this is separable contradiction) with regard to $X \in S_{n+1}$, let $W_{n+1}^* = X$.

A lot of examples show that all we need to do is to compute W_1^*, but exceptional situation, may happen. For example, Y_1 is orthogonal to $Y_2,\ldots, Y_n$. We must compute W_2^*, and to make sure the integrity of algorithm, we provide the theorem as follows:

Theorem 1. Supposing the group of vectors W_1^*, W_2^*, …, W_m^* and $S_1 \supset S_2 \supset \cdots S_m \supset S_{m+1} = \Phi (2 \le m \le n+1)$ have been found, for any $Y \in S = \{Y_0, \cdots, Y_n\}$, as long as $W_m^* Y^T = 0 (i = 1,2,\cdots, m-1)$, thus $W_m^* Y^T > 0$, and let

$$m_0 = 1$$

$$m_i = \min \begin{cases} \min W_i^* Y \\ Y \in S, W_i^* Y^T > 0 \end{cases}$$

$$u_j = \prod_{i=1}^{m} m_i$$

$$W = \sum_{i=1}^{m} u_{i-1} W_i^* / 2^{i+2}$$

Therefore, $WY_i > 0 \quad (i = 0,1,\cdots, N)$

Prove: for any $Y \in S$, from the algorithm above, we know, in W_1^*, W_2^*, …, W_n^*, there is at least a vector which makes $W^* Y > 0$, and assuming that W_j^* is the first vector satisfying $W^* Y > 0$, we have

$$\begin{aligned} WY^T &= \sum_{i=j}^{m} u_{i-1} W_i^* Y^T / 2^{i+2} \\ &\ge u_{i-1} W_i^* Y^T / 2^{i+2} - \sum_{i=j+1}^{m} u_{i-1} / 2^{i+2} \\ &= u_{j-1} m_j / 2^{j+2} - \sum_{i=j+1}^{m} u_j / 2^{i+2} \\ &\ge u_j / 2^{i+2} - u_j / 2^{i+2} \\ &= 0 \end{aligned}$$

3 Convergent Theorem and Its Proof

Theorem 2. Supposing two models R_1, R_2 are linear separable, thus A learning Algorithm of the perception must be convergence after finite steps.

Prove:

Let $X_0, X_1, \cdots, X_N \in R_1 \cup R_2$ be a group of training samples. From the second part, we know R_1, R_2 are linear separable, and when, and only when $W^* \in R$ and $\delta > 0$ make $W^* Y_i^T > \delta$, there into

$$\overline{Y}_i = \begin{cases} (X,1) & \text{if } X_i \in R_1 \\ (-X,-1) & \text{if } X_i \in R_2 \end{cases}$$

Without loss of generality, supposing all the samples $Y_i (i = 0,1,\cdots,N)$ are unit vectors, and where there is the linear separable condition, there must be unit vectors W^* and positive number $\delta > 0$, which make any $i = 0,1,\cdots,n,\quad W^* Y_i^T > \delta$

Let $W(1) = \overline{Y}_0, W(k+1) = W(k) - 2W(k) X_k^T X_k$, thus

$$\begin{aligned} \|W(k+1)\|^2 &= \|W(k)\|^2 - 4(W(k)X_k^T)^2 + 4(W(k)X_k^T)^2 \\ &= \|W(k)\|^2 \\ &= \cdots \\ &= \|W(1)\|^2 \\ &= \|Y(1)\|^2 = 1 \end{aligned}$$

be a unit vector sequence.

$$\begin{aligned} W(k+1) &= W(k) - 2W(k) X_k^T X_k \\ &= W(k-1) - 2W(k) X_{k-1}^T X_k - 2W(k) X_k^T X_k \\ &= \cdots \\ &= W(1) - 2\sum_{i=1}^{k} W(i) X_i^T X_i \end{aligned}$$

Because

$$W^* W^T(k+1) = W^* W^T(1) - 2\sum_{i=1}^{k} W(i) X_i^T W^* X_i^T$$

$$W^* W^T(1) = W^* Y_0^T > \delta$$

$$W^* X_i^T > \delta, \quad -W(k) X_i^T > 0$$

so

$$W^* W^T(k+1) > 0$$

and

$$W^*W^T(k+1) = W^*W^T(k) - 2W(k)X_k^T W^* X_k^T$$

We know $\{W^*W^T(k)\}$ is a single minus positive sequence, and upper bound is 1, thus $\sum_{i=1}^{\infty} W(i)X_i^T W^* X_i^T$ is convergent series, thereby, $\sum_{i=1}^{\infty} \left|W(i)X_i^T\right| W^* X_i^T$ is also convergent series. Using $W^* X^T > \delta$, we have $\sum_{i=1}^{\infty} \left|W(i)X_i^T\right|$ convergence, so for any $\varepsilon > 0$, we have $N_1 > 0$, thus when $m > k \geq N_1$, $\sum_{i=k+1}^{\infty} \left|W(i)X_i^T\right| < \varepsilon/2$, we have

$$\|W(m+1) - W(k+1)\| = 2\left\|\sum_{i=k+1}^{m} W(i)X_i^T X_i\right\|$$

$$\leq 2\sum_{i=k+1}^{m} \left\|W(i)X_i^T\right\| \cdot \|X_i\|$$

$$= 2\sum_{i=k+1}^{m} \left|W(i)X_i^T\right| < \varepsilon$$

Therefore, $\{W(k)\}$ is a Cauchy sequence and must be convergent. Then, $\{W(k)\}$ is a convergent vector sequence. Supposing

$\lim_{k\to\infty} W(k) = W$, thus, $WY_i \geq 0 \quad i = 1,2,\cdots,n$, if $WY_i^T > 0$, then, $WY_i = (W(k) - W)Y_i^T + WY_i^T$, but because $\lim_{k\to\infty} W(k) = W$, we can find positive number K_i, which makes

When $K > K_i$, $\left|(W(k) - W)Y_i^T\right| < WY_i^T$, and then $k > k_i$, $W(k)Y_i^T > 0$.

Let $m = \max_{i\in N_0}\{K_i\}$, in which, $N_0 = \{i : WY_i^T > 0\}$, we know, for any $i \in N_0$, when $k > m$, $W(k)Y_i^T > 0$. But for $i \in \{0,1,\cdots,n\}$ —— N_0, $W(k)Y_i^T = 0$

So

$$WW^T(k+1) = WW^T(1) - 2\sum_{i=1}^{k} W(i)X_i^T WX_i^T$$
$$= WW^T(1) - 2\sum_{i=1}^{m} W(i)X_i^T WX_i^T$$

This equation shows that the angle of W and $W(k+1)$ is a fixed value when $k > M$. But $\lim_{k\to\infty} W(k) = W$, then $W(k+1) = W$, thus we prove $W(k)$ proved has convergence after finite steps W .

4 Example

In the two ranges $[0,1]^n$ and $[0,1]^{n+1} \times [-1,0]$, 100 samples are taken respectively, and with two kinds of methods, ten experiments are made, and the results are in the following table:

Table 1 : $n = 3$

Table 2 : $n = 4$

From the tables, we can see, sometimes, the paces of the learning algorithm with variable paces used may be much more, but in most cases, the pace number the method used is far fewer than that of Rosenblatt's iterations, thus, it can be said that the learning algorithm with variable is obviously superior to Rosenblatt's algorithm. learning algorithm.

5 Conclusion

A lot of examples show that the learning algorithm of perception with variable paces has faster convergent speed than the traditional learning method, and especially when the dimensional number is high, the convergent speed of the former is obviously faster than the latter. The original learning algorithm needs more learning times when the distance between two kinds of patterns is shorter, but the learning algorithm of perception with variable paces is not sensitive to this situation. The advantage of the new algorithm is reflected in the condition of samples' irregular distribution.

References

1. Gallant, S.I.: Perception-based learning algorithms. IEEE Trans. Neural Networks (2), 179–191 (1990)
2. Clark, D.M., et al.: A convergence theorem for Grossberg Learning. IEEE Trans. NN. 1(3), 87–97 (1990)

3. Shynk, J.J.: Performance surfaces of a single-layer perception. IEEE Trans. NN. 1(3), 267–274 (1990)
4. Widrow, B., Lehr, M.A.: 30 years of adaptive neural networks: Perception, Madaline and BP. Proc. IEEE 78(9), 1415–1442 (1990)
5. Jian, X., Zhang, H., Gong, X., YanDa, L.: A Nonlinear Perceptron Algorithm Based on Kernel Functions. Chinese Journal of Computers (July 2002)
6. Zhou, Y., Zhao, B.: A Faster Linearization Learning Algorithm For Nonlinear Weigh Function Perceptron. Computer Applications and Software (January 2004)
7. Cong, C., Lu, B.: Partition of Sample Space with Perceptrons. Computer Simulation (February 2008)

Analysis and Nonlinear Regression on Cyanobacteria Growth of Guishui Lake in Beijing Suburb

Weidong Li[1], Yan Meng[1], Lequan Min[1,*], Aijun Gong[2], and Hai Yan[2]

[1] School of Mathematics and Physics University of Science and Technology Beijing, China, 100083
[2] School of Chemical and Biological Engineering University of Science and Technology Beijing, China, 100083
minlequan@sina.com

Abstract. This paper analyzes the relationships among cyanobacteria number and water quality factors in the Guishui lake in Beijing remote suburb. The results show that the cyanobacteria growths are positively related with the concentrations of TP × P, TP/TN and TP/NO_2. Based on the tested data from six water quality monitoring sites, a nonlinear regression formulas on the growths of blue algae cells was established. The calculated results on the number of the cyanobacteria via the nonlinear regression formulas are well in agreeing with the tested data.

Keywords: cyanobacteria, total nitrogen, total phosphoro, phosphoro, nonlinear regression.

1 Introduction

One of the most serious environmental problems worldwide is the cyanobacteria blooms in natural waters. ([1][4]). Cyanobacterial blooms can produce toxic microcystins, which may cause acute and chronic hepatitis even hepatocellular carcinoma in human beings ([5],[6]). High densities cyanobacteria can also arise serious problems in domestic, industrial, and agricultural use of water.

Studying relationships among cyanobacteria blooms and water quality factors has important and effective roles for preventing cyanobacterial blooms.

Adequate amounts of water, air, sunlight, and nutrients like P and N, can make cyanobacteria grow very rapidly and form a bloom. Many researchers have been studied to find ways to control cyanobacteria blooms cyanobacterial blooms..

Havens et al. reported that Dual N and P input reductions are usually required for effective long-term control and management of harmful blooms [7]. Shukla et al. observed that high concentration of TN is the main factor arising cyanobacterial blooms [8].

Smith and Bierman [9] state that a mass TN:TP ratio of 22:1 provides a more distinct boundary between lakes dominated by N-fixing cyanobacteria and lakes with low occurrence of these algaes.

---

* Corresponding author.

D. Zeng (Ed.): Advances in Computer Science and Engineering, AISC 141, pp. 325–333.
springer link.com

Smith proposed a mechanism [10] to link cyanobacteria's dominance to a low TN:TP ratio. It states that all species of cyanobacteria are better able to compete for nitrogen than other phytoplankton when N is scarce.

Shukla et al. [8] hold that high concentration of N is the main reason of cyanobacteria blooms. However some researchers observed that low ratios of N:P was one of the main reasons for the formation of cyanobacteria ([7], [9], [11].

In a recent conference paper [12], we used nonlinear regression approach to evaluate the growth of the cyanobacteria number in Guishui lake in Beijing suburb. Although the cyanobacteria data calculated from the nonlinear regression formula were well agreement with in the ones tested from the water samples, the parameters of the nonlinear regression formula seem not to be reasonable for practical interpretations. In this paper we will propose a new simpler nonlinear regression formula to describe and interpret the growthes of the cyanobacteria of the water samples collected from 6 sties near the central line in Gishui Lake during June 29th 2007 to Oct.7th 2007.

2 Study Site and Data Source

2.1 Background

Guishui Lake is a narrow shallow nature lake and about 10 kilometers long located in the Yanqing Country — northwest of Beijing. Its tend toward is from west to east.

Now the water of Guishui Lake is mainly provided by rainfall and a sewage treatment plant. In recent years, the water pollution arising from cyanobacteria blooms in Guishui Lake has become one serious problem in this lake. The Yanqing Country is going to use biochemical approaches and artificial wetlands to solve water pollution problem of Guishui Lake.

2.2 Study Sites and Data Source

Water chemistry and biological data was collected at six water quality monitoring sites during June 29th 2007 to Oct. 7th 2007. The surface water samples were collected from the six sites during the monitoring period.

All sample sites in 2007 are on the central line of the lake. The distance between each two neighboring sites is about two kilometers. The #1 site is near to the original source part of the lake. The #6 site is near to the exit part of the lake. The #2 site is near to the sewage treatment plant.

Table 1. The average water temperatures of guishui lake during june -october, 2007

Month	Jun	July	August	September	October
Temperature	23.5	25.0	25.0	22.0	20.5

The average water temperatures (^{o}C) of Guishui Lake during June -Oct. 2007 are listed in Table 1.

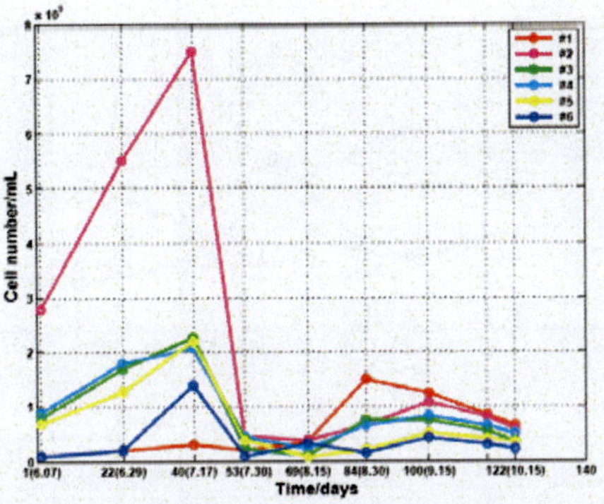

Fig. 1. The concentration evolution of algae cells at the six sample sites in 2007

The test items of water quality factors include: the concentrations of cyanobacteria, TN, TP, COD, NO^-, pH, Chl-a, ammonia-nitrogen (AN) and cyanobacterial toxins Microcystins (MC-RR, MC-LR).

Based on the China National Standard, all above items are tested by the authors' colleagues in the Chemistry Department at our University.

3 Results and Discussion

3.1 The Concentrations of Cyanobacteria

The curves of the concentration evolution of algae cells in the year 2007 sat the six sample sites are shown in Fig.1. We can obtain the following conclusions.

(a) On July 17th 2007, the cyanobacteria blooms occurred at #2~#6 sample sites except #1 sample site.

(b) On August 30th 2007, the cyanobacteria blooms occurred only at the #1 sample site.

(c) Temperature is an important factor for the growth of the cyanobacteria but not the key factor arising the cyanobacteria blooms.

3.2 The Concentrations of TN and TP

The concentration evolution of TN and TP, and TN:Tp at the six sample sites are shown in Fig.2 ~ Fig.3 , respectively. We can obtain the following conclusions.

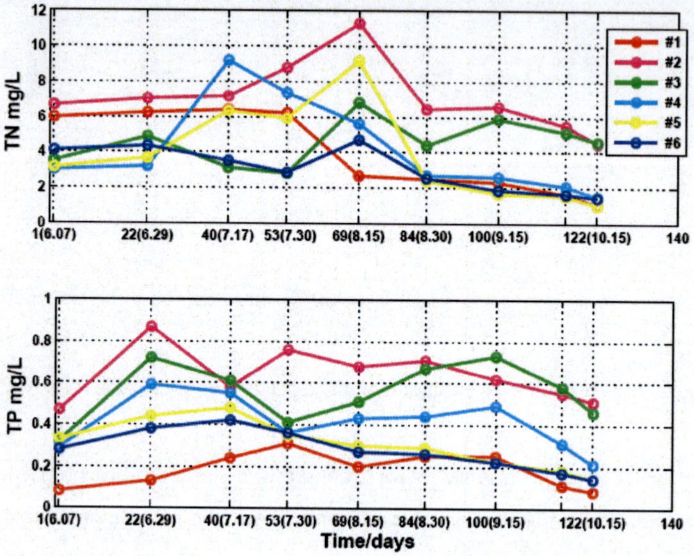

Fig. 2. The curve of the concentration evolution of TN and TP in 2007 at six sample sites

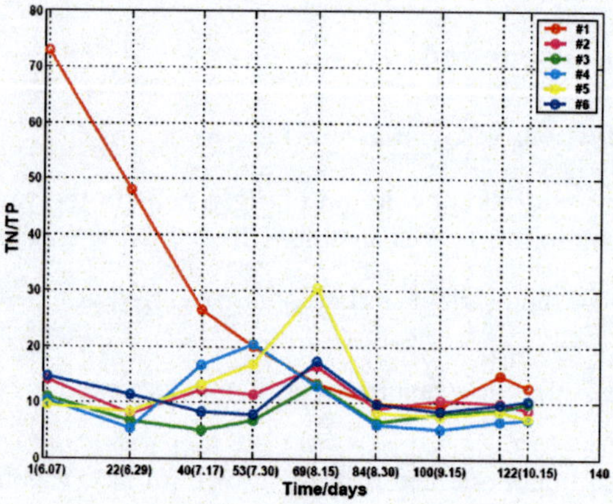

Fig. 3. The curve of the value of total TN:TP in 2007 at six sample sites

1) High ratios of TN:TP (> 20) can suppress the cyanobacteria blooms.
2) Low concentrations of TP (< 20) can constrain the cyanobacteria blooms.

3.3 The Concentrations of Some Other Tested Items

The evolution of AN, NO_2 and TP × P values are shown in Fig.4 ~ Fig.6, respectively. We can obtain the following suggestions which are different from those given in our previous paper [12].

1) The cyanobacteria growths are positively related with the concentration of TP × P if the value TP × P is the reason effecting the next time's test result on the number of cyanobacteria.
2) Low concentration of AN may arise the high concentra tion of cyanobacteria number in sample site #1 on August 30th.

Table 2. Calculated Parameters β's Of The Six Sample Sites

No.	# 1	# 2	# 3
β_1	1.2424×10^3	6.3705×10^5	1.5180×10^5
β_2	1.8903×10^7	2.5984×10^4	3.0864×10^5
β_3	3.7379×10^1	5.8940×10^3	-1.1532×10^3
β_4	3.0144×10^4	4.4902×10^2	9.0535×10^3
No.	# 4	# 5	#6
β_1	2.8012×10^4	-3.0619×10^3	2.5409×10^3
β_2	2.5543×10^6	4.9960×10^5	3.3258×10^7
β_3	-6.2840×10^2	-2.0928×10^3	-3.4687×10^1
β_4	0.1768×10^4	4.4079×10^3	3.1947×10^3

4 Nonlinear Regression

Based on the discussions addressed in Section 3, we can establish the following a new simpler nonlinear regression formula to describe the concentration evolution of the cyanobacteria at the sxie sample sites:

$$C = \beta_1\frac{TP^2}{TN^2} + \beta_2(TP \times P)^3 + \beta_3\frac{1}{AN} + \beta_4\frac{TP}{NO_2}$$,where C is the concentration of algae cells (cells/mL); TN and TP are the concentrations of total nitrogen and total phosphorus (mg/L), NO2 and AN are the concentrations of Nitrate ion and ammoniacal nitrogen (mg/L), respectively.

The calculated parameters β s of the 6 sample sites are shown in Table 2. The numerical simulations of the nonlinear regression formulas for the 6 sample sites are shown in Fig. 7. Observe that the numerical simulation curves of the nonlinear regression formulas are well in agreement with the practical tested concentrations of the cyanobacteria at the 6 sample sites.

5 Conclusions

(1) Keeping low concentrations of TP is the most important factor for controlling the growth of cyanobacteria. That TP < 0.3 mg/L, can restrict the concentration of blue algae cells below 1×10^5 /mL.

(2) The concentrations of TN seems not to be important to arise the growth of blue algae cells. In particular, the high ratio of TN/TP may control the growth of blue algae cells (see the red lines in Fig. 1 and Fig. 3).
(3) Compositive approaches should be established for controlling the growths of cyanobacteria and improving the water quality in Guishui Lake.

In summary, we need further experimental data from Guishui Lake to confirm our assumptions. There is a long way to solve the pollution problems in Guishui Lake.

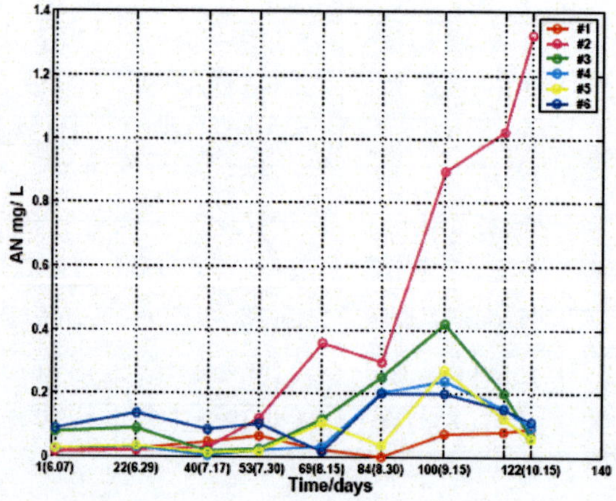

Fig. 4. The concentrations of AN at six sample sites

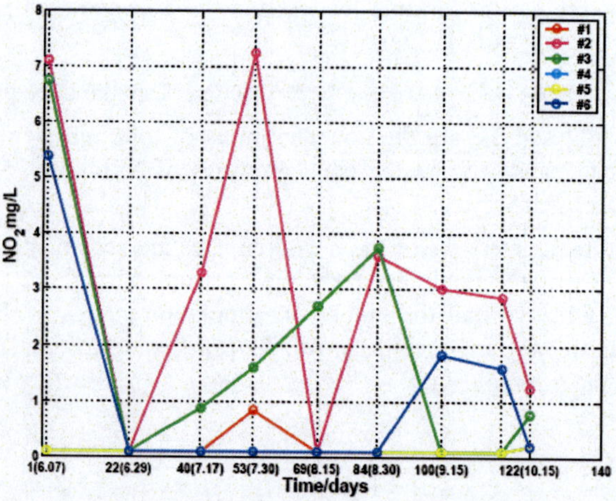

Fig. 5. The concentrations of NO2 at six sample sites

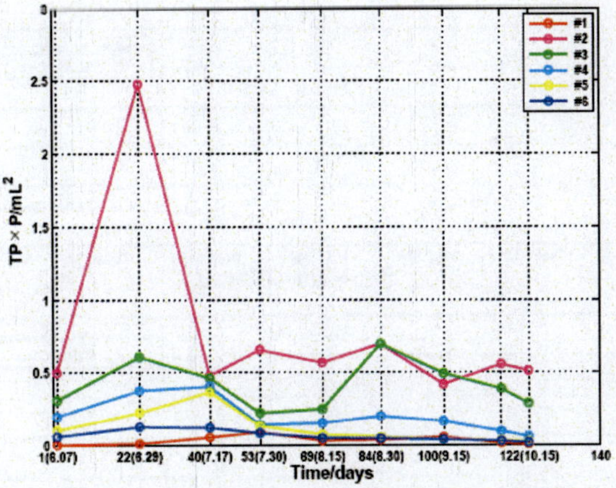

Fig. 6. The TP × P values at six sample sites

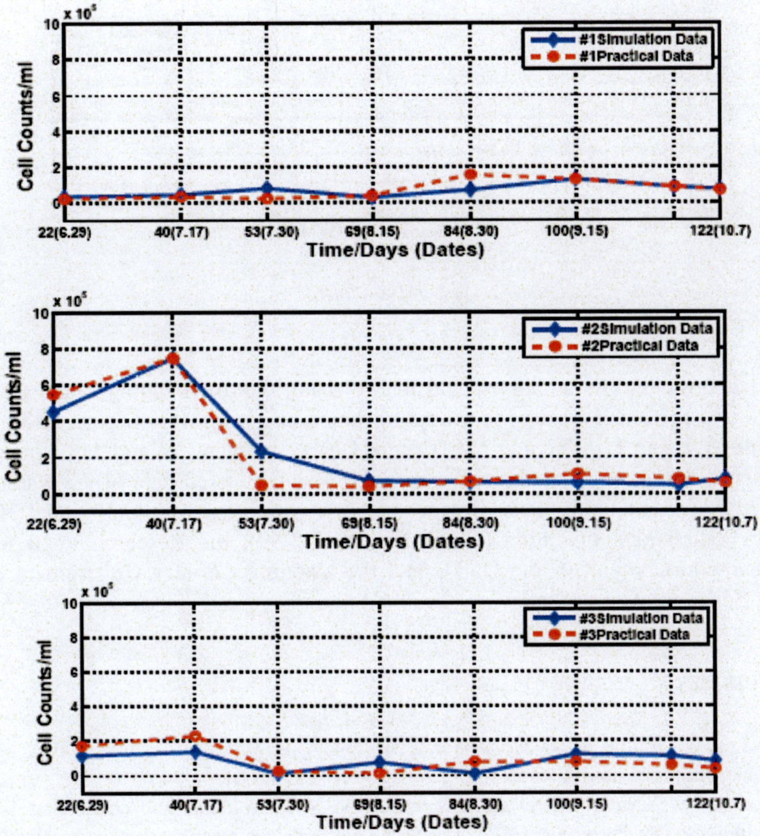

Fig. 7. The curves of the concentration evolution of the blue algae cells (blue) and the nonlinear regression (red) at six sample sites

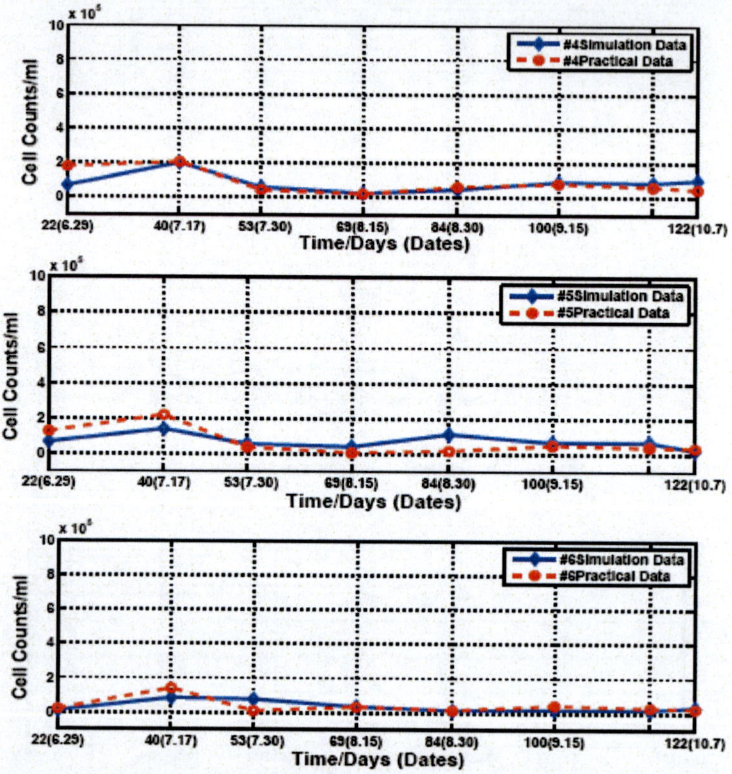

Fig. 7. (*continued*)

Acknowledgment. The authors would like to thank the the Yanqing Coun¬try Government and related Bureaus for providing helps and supports on our research, the colleagues Wenjun Li, Yanping Ye, Yanqiu Chao, Jianqiang Li and Jun Ning in the Chemistry Dept. at our University for providing us the tested data of the water samples in the Guishui Lake. This project is jointly supported by the National Natural Science Foundations of China (Grant No. 60674059), the Research Fund for the Co-construction Project of the USTB and the Yanqing Country Government of the Beijing Education Committee.

References

1. Paerl, H.W., Fulton, R.S., Moisander, P.H., Dyble, J.: Harmful freshwater algal blooms, with an emphasis on cyanobacteria. Science 1, 76–113 (2001)
2. Burch, M.D.: Cyanobacteria and water quality: New problems and challenges. In: Proceedings of Drinking Water Quality. Technical and Management Issues for the 1990's, Adelaide, October 12, pp. 18–24 (1993)

3. Haider, S., Naithani, V., Viswanathan, P.N., Kakkar, P.: Cyanobacterial toxins: a growing environmental concern. Chemosphere 52, 1–21 (2003)
4. Yao, Z., Fei, M., Li, K., et al.: Recognition of blue-green algae in lakes using distributive genetic algorithm-based neural networks. Neurocomputing 70(4-6), 641–647 (2007)
5. Carmichael, W.W.: Health effects of toxin-producing cyanobacteria: the cyanoHABs. Human Ecology Risk Assess. 7, 1393–1407 (2001)
6. Lora, E.F., Carlos, R., John, B.: Blue green algal toxins, surface drinking water, and liver cancer in Florida. Harmful Algae 6, 157–168 (2002)
7. Havens, K.E., James, R.T., East, T.L., Smith, V.H.: N:P ratios, light limitation, and cyanobacterial dominance in a subtropical lake impacted by non-point source nutrient pollution. Environment Pollution 122, 379–390 (2003)
8. Shukla, J.B., Misra, A.K., Chandra, P.: Modeling and analysis of the algal bloom in a lake caused by discharge of nutrients. Applied Mathematics and Computation 196, 782–790 (2008)
9. Smith, V.H., Bierman, B.L.: Historical trends in the Lake Okeechobee ecosystem IV, Nitrogen: phosphorus ratios, cyanobacterial dominance, and nitrogen fixation potential. Archiv. fur Hydrobiologie, Monographische Beitrage 107, 71–88 (2002)
10. Smith, V.H.: Low nitrogen to phosphorus ratios favor dominance by blue-green algae in lake phytoplankton. Science 221, 669–671 (1983)
11. Xie, L., Xie, P., Li, S.: The low TN: TP ratio, a cause or a result of Microcystis blooms? Water Res. 37, 2073–2080 (2003)
12. Min, L., Jiang, X., Fu, Y., Yi, Y., Meng, Y., Yan, H.: Nonlinear regression based analysis on cyanobacteria blooms and water quality in a Beijing shallow lake. In: Proceeding of the 2009 Int. Conf. on Computational Intelligence and Security, Beijing, China, December 11-14, pp. 498–504 (2009)

The Analysis on Mechanical and Thermal Property of Poly(Acrylic Acid)-Graft-Polypropylene Membrane Prepared by UV-Irradiation

Yanjun Liu, Chunying Lv, and Jia Yang

Environmental and Chemistry Engineering College of Tianjin Polytechnic University
Tianjin, China
yjliu72@126.com

Abstract. A series of pH-sensitive poly (acrylic acid)-graft-polypropylene microfiltration membranes were prepared by UV-irradiation. It was observed that the hydrophility and water permeability of grafted membranes was improved. After modification, the elongation of grafted membranes decreased significantly with the rise of grafting degree, and the glass transition temperature (Tg) of grafted membranes decreases as well. As a result, the grafted membranes have a great loss on its flexibility and become brittle. The grafting reaction weakens the thermal stability of grafted membranes. Both of the initial and the maximum decomposition temperature of 20% grafted membrane are 10°C lower than that of PP matrix. But grafted membranes still show good thermal resistance till 400°C. The result shows that the surface modification not only has great impact on water permeability of membranes but also affect their mechanical and thermal properties.

Keywords: UV irradiation, water permeability, mechanical property, thermal stability, glass transition temperature.

1 Introduction

UV-grafting reaction has been applied to modify membrane surface for serveral decades to improve membrane's seperation property. Some photosensitive polymer (eg polysulfone PES)[1,2] can directly produce free radicals under UV irradiation and induce grafting reaction, while more polymer needs photoinitiator (such as aromatic ketone) [3,4] to initate photo-grafting reaction. The membrane modification by UV-irradiation shows two prominent features: (1) UV light has low penetration capicity to material, so graftion can be strictly limited to the membrane surface or sub-surface and has little impact on the bulk properties of material; (2) UV radiation light sources are cheap and simple operated. The continuous UV modification was expected for great possibility[5].

Many studies on membrane modification by UV irradiation were reported to improve the hydrophilic and anti-fouling properties on membrane surface. The paper mainly discussed the influence of grafting degree on properties of poly(acrylic acid) (PAA) -grafted membranes. It shows that the grafting reaction not only leads to the

D. Zeng (Ed.): Advances in Computer Science and Engineering, AISC 141, pp. 335–342.
springer link.com

change in membrane permeability performance, but also may cause the loss on mechanical properties or thermal properties.

2 Experimental

2.1 Materials

Polypropylene (PP) hollow fiber microfiltration membranes supplied by The Institute of Seawater Desalination and Multipurpose Utilization China was used as the grafting substrate. The membrane had an average pore diameter of 0.4μm and a porosity of 58%, and was 50 μm thick with an internal diameter of 400μm. All other reagents and chemicals were of analytical grade and used as received. Water used in this work was treated with twice reverse osmosis.

2.2 Preparation of Poly (Acrylic Acid)-Graft-Propylene Membrane

A 1 kW high pressure mercury lamp (λ=254nm) was used as the UV supplier which was hung 0.4m high above the reactive quartz chamber. The PP membranes was dipped in 5 wt% benzophenone (BP) acetone solution for two hours and then dried in room temperature. The membranes coated by BP were put into quartz reactive chamber and irradiated by UV in the acrylic acid (AA) ethanol solution under nitrogen atmosphere and room temperature. After 30 minutes reaction time, the sample was taken out and washed with hot water repeatedly and then dipped in hot acetone for 24 hours to remove AA homopolymer. The membranes with different grafting degree were prepared by altering the concentration of AA. The grafting degree Dg was calculated according to the following equation:

$$Dg(\%) = \frac{W_1 - W_0}{W_0} \times 100\% \qquad (1)$$

where W_0 stands for the weight of original PP membrane and W_1 is the weight of grafted membrane after removal of homopolymer and unreacted AA.

2.3 Microfilitration

A mini crossflow microfiltration cell with a U-configuration module was employed to characterize water permeability and the BSA microfiltration performance of membranes. The external operation pressure was supplied by a magnetic drive pump. All measurements were carried out at a constant transmembrane pressure of 100 kPa and at 25±2°C with a cross-flow rate of 100 l/h. The measurement was carried out with an active membrane area of 33.3 cm^2 .The pH value of water was adjusted by HCl or NaOH. Except the special statement, prior to the measurement, the membranes were wetted with 70 wt% ethanol solution and washed with pure water repeatedly, and then equilibrated in the test solution under operation pressure until a stable flux was achieved. The permeation water flux J was calculated according to the following equation by measuring the mass of permeates (m) during a certain filtration time (t):

$$J = \frac{m}{S \bullet t} \tag{2}$$

where S is the area of the filtrating membrane.

2.4 Mechanical Property

The hollow fibers with a lenght of 20cm were fixed on the M350-5kN Fiber strength tester and their strength and elongth were measured with a rate of 200mm/min and a Gripper spacing of 100mm.

2.5 Dynamic Mechanical Thermal Analyses (DMA)

The membrane samples were tested by NETZSCH DMA242C Dynamic Mechanical Analyzer with the frequence of 1.0Hz, 5.0Hz, 10.0Hz respectively.

2.6 Thermogrametry(TG)

The clipped membrane samples with a weight of 10~20mg were tested by STA 409 Thermogrametry with a certain heating rate from room temperature to 600 °C at N2 and O_2 atmosphere respectively.

3 Results and Discussion

3.1 FTIR Characterization

Fig.1. demonstrates the IR spectra of virgin PP and grafted membranes. An absorption peak around 1720 cm^{-1} could be observed in the spectra of grafted samples compared with the original PP sample. The peak is assigned to the stretch of the C=O group. It proves that polyacrylic acid (PAA) chains had been grafted onto the PP substrate for grafted membranes. As expected, the strength of the characteristic peak (at 1720 cm^{-1}) of grafted samples increases with the increase of *Dg*.

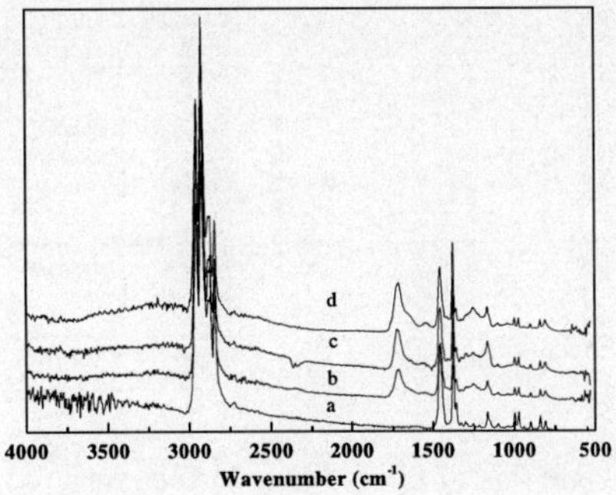

Fig. 1. FTIR Spectra of PP membrane (a) and the 5% (b), 11% (c) and 20% (d) grafted membranes

3.2 Pure Water Adsorption and Permeability

As shown in Fig.2, the flux of PP membrane without being wetted by ethanol is zero, and its water adsortption amount is 0.9%. With the rise of *Dg*, the water adsorption amout of grafted membrane increases gradually, and gets to the maximum at *Dg*=17%.

As *Dg*<10%, the small density of PAA chains on the membrane surface results in a zero flux though it also presents an slight increase of water adsorption amout than original membrane. Then the pure water flux ascends greatly with the increase of *Dg* and reaches the maximum of 8.6 $kg \cdot m^{-2} \cdot h^{-1}$ at *Dg*=17%. An inevitable result of membrane surface modification is that membrane pores can be blocked partially or completely. It causes a decline in pore size and porosity and an increasing hydraulic resistance to filtration. Therefore, the pure water flux increases till *Dg*= 17%. After that, the blocking effect leads to the decrease of membrane capacity to water. So the water adsorption amout decreases, and water flux declines as well.

3.3 The Influence of UV on Original PP Membrane

There is a large number of methyl tertiary carbon on polypropylene (PP). The hydrocarbon bond energy of methyl tertiary carbon atom is small (only 85Kcal/g). As a result, it is easy to form unstable free radical with a single electron when the high-energy ultraviolet light acts on it,. Free radical can promote the degradation of the chain severely.

UV grafting reaction is a kind of free radical grafting reaction. PP-based membranes first get –C-H bond rupture to form free radicals, which make graft chains grow. Thus UV irradiation is likely to have a certain impact on the mechanical properties of PP membranes. PP membrane samples dipped in 15% AA ethanol was irradiated by UV in N_2 and in the air respectively. And then they were vacuum dried after hot ethanol washed. The mechnical property of these samples were shown in Figure 3.

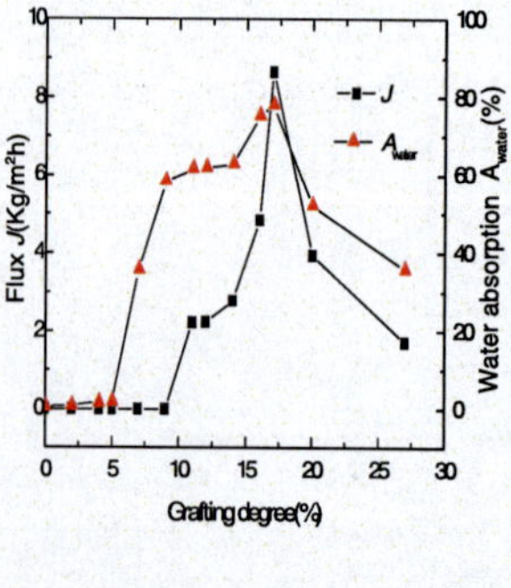

Fig. 2. The water absorption and flux of PP-g-AA membranes

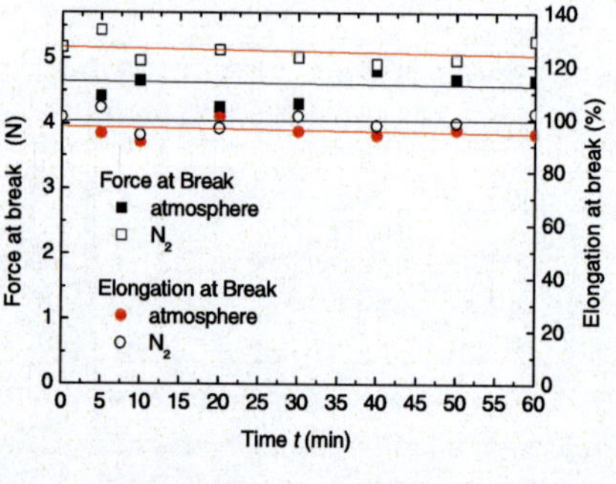

Fig. 3. The effect of UV irradiation on original PP membrane

As seen from Fig.3, UV irradiation in the N_2 had no significant effect on breaking strength and elongation of membrane. But the PP membrane UV-irradiated in the air shows a elongation at break between 95~105mm and a reduced breaking strength between 4.2~4.8N, while the original PP membrane had a breaking strength of 5.2N. It shows that the presence of oxygen in the air plays great role in promoting the degradation of PP membrane. Due to the grafting reaction had been carried out in N_2, UV had little impact on the mechanical properties of the PP matrix.

3.4 The Influence of Dg on Mechanical Property of Grafted Membranes

Fig.4 shows the tensile strength and elongation of grafted membranes. It can be found that the membrane breaking strength is about 5N which is similar as original PP membrane. Compared with the original PP membrane, the elongation of grafted membranes decreased significantly, which indicates that the grafted membranes have a great loss on its flexibility and lead to a crisp appearance. Grafting reaction generally occurs in the amorphous region of polymers[6], and does not significantly affect the crystallinity of the polymer itself. So the breaking strength of grafted membranes shows no major changes.

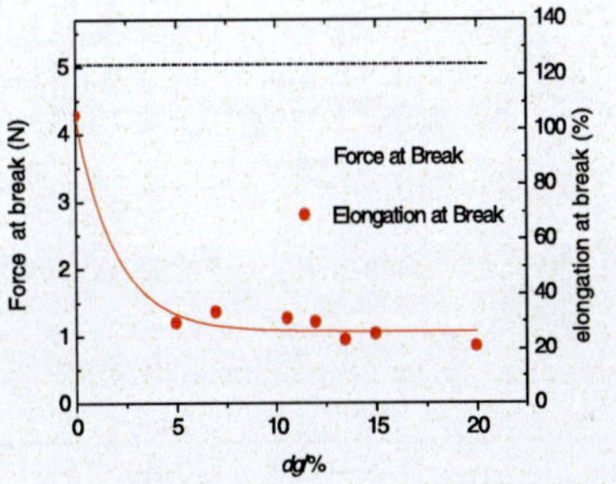

Fig. 4. The mechanical properties of grafted membranes

After grafted, there are strong hydrogen bonds among poly(acrylic acid) grafted chains to constraint the deformation of the PP main chain. Thus grafted membranes showed a significant decline in elongation, so that the grafted membranes show a reduction in their flexiblity, and they were prone to be damaged.

3.5 Dynamic Mechanical Thermal Analyses of Grafed Membranes

It can be seen from the Fig.5 that the grafted membranes has two loss peaks in the whole heating process. The loss peak at 30°C was caused by the glass transition of the PP matrix which was corresponding to the movement of PP macromolecular chain. The loss peak around 30°C decreases gradually with the increasing of grafting degree, while the glass transition temperature(Tg) of grafted membranes gradually shifts to high temperature. The loss peak at 150°C was caused by the crystalline phase change of the PP matrix corresponding to the movement of PP crystal chain. The peak area of PP matrix phase transition about 150°C increases and moves to the low temperature area gradually with the increase of Dg.

Table 1 listed the glass transition temperature(*Tg*) of original and grafted membranes. The *Tg* of poly(acylic acid) is at 106°C, while that of PP membrane is at 27.3°C. With increase of *Dg*, the *Tg* of grafted membrane increases. As *Dg* is low, most of the macromolecular chains of membrane PP are still in the free state and shows a decreasing loss peak around 30°C. With the increase of *Dg*, more and more PP chains are hydrogen bonded strongly by grafted chains. So the loss peak about 30°C of PP matrix reduces, and *Tg* increases obviously. Meanwhile, the phase transition of PP matrix coincides and the glass transition of poly(acrylic acid) grafted chain gradually, therefore, the phase transition move to the low temperature.

With the rise of *Tg*, the grafted membranes show reduced macromolecules motion capacity at room temperature and lower resistance to low temperature. It is the foundmental reason which cause the elongation of grafted membranes reduce.

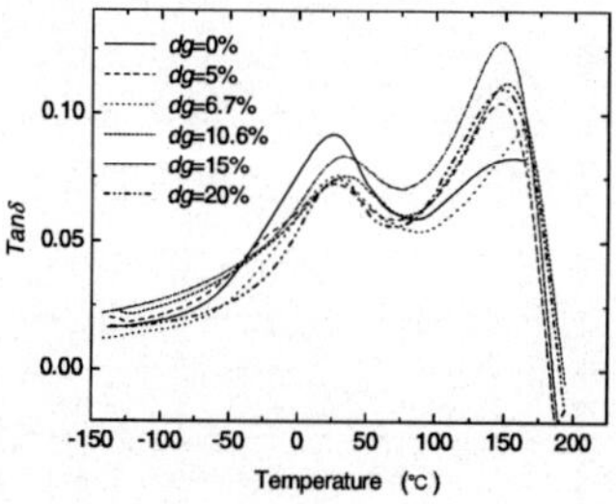

Fig. 5. The DMA spectrum of PP-g-AA membranes

Table 1. The glass transition temperature *Tg* of PP-g-AA membranes

dg(%)	0	5	6.7	10.6	15	20
Tg (°C)	27.3	27.8	28.0	29.8	33.3	35.9

3.6 Thermogrametry of Original and Grafted Membranes

Fig 6. shows the TG and the corresponding derivative curve (DTG) spetrums of membranes in the N_2 with the heating rate 10°C/min respectively. It can be seen that grafted membranes show good heat resistance and can remain stable till 400°C in N_2. Both original and grafted membranes show one-step decomposition from 430°Cto 600°C. It indicates that the grafted chains have good compatibility with PP matrix.

The main parameters from TG spectrum were listed in Table 2. In the N_2, the initial decomposition temperature(*Ti*) of the original PP matrix is around 435°C. *Ti* of grafted membranes with low *Dg* is similar as that of original membrane, while *Ti* of 20% grafted membrane is 10°C lower than that of original membrane. Meanwhile, the maximum decomposition temperature also dropped 10°C. After UV grafting, parts of –C-H bonds of PP matrix were boken to form tertiary carbon atoms and weaken the thermal stability of grafted membranes.

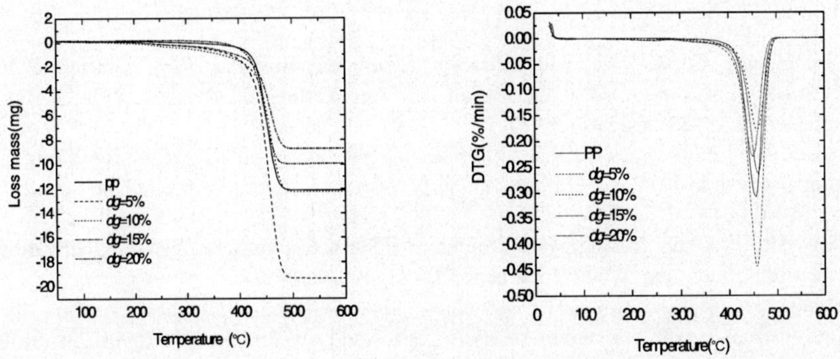

Fig. 6. The themogrametry and differential thermogrametry spectrum (DTG) spectrum of grafted membranes in N_2

Table 2. The data in the DTG spectrum of grafted membranes

dg(%)	*Ti*(°C)	*Tp*(°C)	*V*(%/min)	The total decomposition loss(%)
0	433.4	453.4	30.86	93.96
5	434.5	456.9	44.36	94.02
10	438.3	459.1	26.4	85.74
15	435.9	458.1	17.95	87.0
20	423.6	448.8	23.16	85.1

Ti, *t*he decomposition temperature; *Tp*, the fastest decomposition temperature; *V*,the fastest decomposition rate.

The final decomposition rates of original and low grafted membranes are 93 ~ 94%, but as the *Dg* is more than 10%, the final decomposition rate of grafted membrane shows a significant decline. It proves that the presence of polyacrylic acid grafting chains prevent the complete decomposition of PP matrix.

4 Conclusions

Membrane mechanical properties and thermal stability is determined by its intrinsic molecular structure. Although the surface modification occurs only in the thin membrane surface, but the interaction among the graft chains brings great changes both on water permeability and mechanical properties of grafted membrane. With increase of grafting degree, a significant reduction on membrane elongation could be found. Grafted membranes has good heat resistance. The introduction of polyacrylic acid grafting chain to PP matrix leads to the reduction both on glass transition temperature and on membrane's flexibility at low temperature, and causes the increasing brittleness at room temperature and a operational lifespan loss eventually.

References

1. Ulbricht, M., Riedel, M.: Ultrafiltration Membrane Surfaces With Grafted Polymer 'tentacles': Preparation, Characterization and Application for Covalent Protein Binding. Biomaterials 19, 1229–1237 (1998)
2. Kaeselev, B., Pieracci, J., Belfort, G.: Photoinduced Grafting of Ultrafiltration Membranes: Comparison of Poly (Ether Sulfone) and Poly (Sulfone). Journal of Membrane Science 194, 245–261 (2001)
3. Song, H., Wu, G., Zhong, H.: Progress is Surface Photografting Polymerization of Seperation Membrane. Water Treatment 30, 187–190 (2004)
4. Ulbricht, M., Matuschewski, H., Oechel, A.: Photo-Induced Graft Polymerization Surface Modifications for the Preparation of Hydrophilic and Low-Protein-Adsorbing Ultrafiltration Membranes. Journal of Membrane Science 115, 31–47 (1996)
5. Ma, H.M., Davis, R.B., Nbowman, C.: A Novel Sequential Photoinduced Living Graft Polymerization. Macromolecules 33, 331–335 (2000)
6. Feng, C., Zhou, C., Zeng, Q.: Progress on the Studies of Polypropylene Ion Exchange Fibers. Chemical Industry and Engineering Progress 22(6), 568–571 (2003)

Extracting River Networks of Lianyungang City from SRTM Data

Hongchun Peng and Haiying Li

School of Geodesy & Geomatics Engineering
Huaihai Institute of Technology
Lianyungang, China
{plateat,lihaiyingsd}@sina.com

Abstract. This article used SRTM data to extract the river networks in Lianyungang of China. Because of the defect of SRTM data, the data must be filled first. Through the filled data to extract DEM and then to obtain river networks. Different algorithm would get the river networks with different resolutions, so this article made a set of data processing methods to ensure the precision. Selecting different CSA thresholds would obtain differently detailed river networks. The analysis showed that the selection of CAS threshold is very important and will influence the statistic characteristics of whole basin river networks directly. The algorithm could simulate the real water system in the area with topographic relief. In contrast to the area with flat terrain, for example in Lianyungang, the algorithm will cause distinctive errors.

Keywords: SRTM, River Networks, DEM, Lianyungang.

1 Introduction

SRTM (Shuttle Radar Topography Mission) was implemented by NIMA of Defense of Ministry and NASA in USA, and provided a set of data about face of land and ocean in the world, and marked the new stage 2-Dimension into 3-Demension RS, which changed the methods getting topography information. SRTM produces 3-D high-resolution DEM maps, and solved the highly-difficult problems which was difficult to get the information of zone with complex and varied terrains and landforms. With the SRTM data, it achieved the work of all-weather terrain mapping[1,2,3].

SRTM data were precise to produce terrain maps, and the implement of 3-dimension simulation leads to the implements of stereoscopic map (3-Dimension map) which makes the different feature of the topography and terrain and water area more intuitionistic and visual[4,5]. On the base of these, digital river networks were extracted and the influence factors of precision extracting river networks were researched, which is possible to get a more precise algorithm.

D. Zeng (Ed.): Advances in Computer Science and Engineering, AISC 141, pp. 343–348.
springer link.com

2 Study Area and Data

2.1 Study Area

Lianyungang is situated between 118°24' -119°48' E and 34°11' - 35°07' N. She is a prefecture-level city northeastern Jiangsu province, People's Republic of China (Figure 1). She is the eastern end of the New Eurasia Continental Landbridge and the proposed Northern East West Freight Corridor. She administrates seven county-level divisions, including the three districts of Lianyun, Xinpu and Haizhou and four counties of Ganyu, Donghai, Guanyun and Guannan (Figure 2), and covers an area of 7 777 km². As one of the first 14 Chinese coastal cities opening to the outside world in 1984, she has developed about 20 years and has many changes in urban size or structure.

In this paper, the Lianyungang urban zone stand for the three districts, which is Haizhou, Xinpu and Lianyun from southwest to northeast. So, the Lianyungang include five parts: Lianyungang urban zone, Ganyu, Donghai, Guanyun and Guannan (Figure 2).

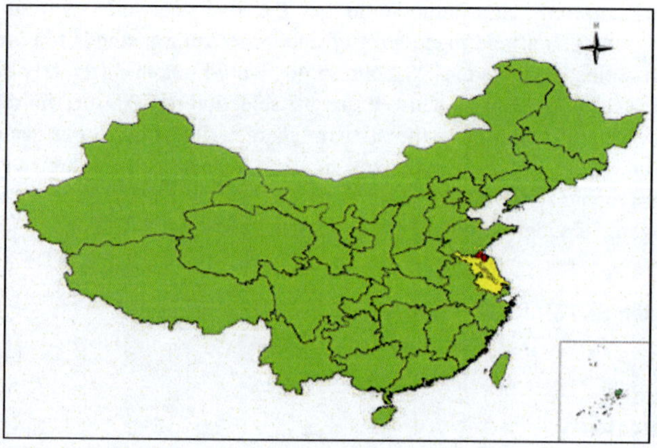

Fig. 1. The location of Lianyungang

2.2 Data and Methods

The data were downloaded from website http:// srtm.csi.cgiar.org/ SELECTION/ inputCoord.asp, and included SRTM data in Lianyungang area. The format of data was .tif and transformed to DEM which could be processed by MOCRODEM. In addition, 4 files downloaded from the website were combined into 1 file from which the area of Lianyungang as the center was extracted.

NASA provides users with the two resolutions of 1 and 3 arc known as SRTM-1 (30m) and SRTM-3 (90m). The users are allowed to obtain SRTM-3 data free. The elevation datum of SRTM-3 is the geoid of EGM-96, and the datum plane is WGS-84. In this article, SRTM-3 data were used.

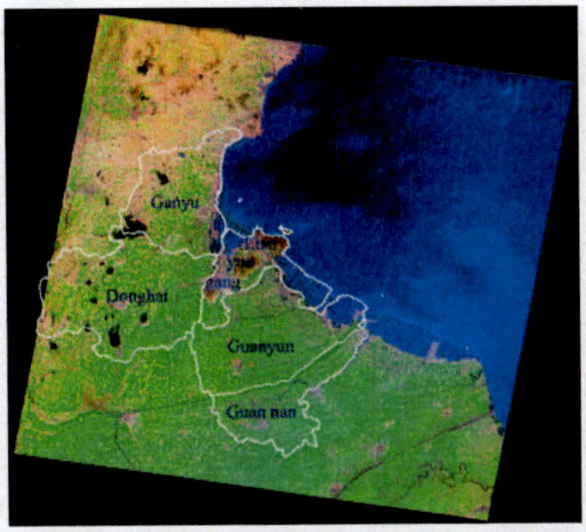

Fig. 2. The divisions of Lianyungang

The data in this article may be influenced by mirror reflection, radar shadow and echo lag in the course of obtaining data, which leads to the lack of data in some area and causes some null values. Although most of area were covered many times, it is impossible to avoid the currency of null values. To improve the Practicality of the data, the area with null values must be filled effectively before the application of the data. The methods included that the methods of filling existing contour, and of interpolating the grid of SRTM-3 directly, and of interpolating after extracting contour.

Up to date, researchers in the world tried to use varied methods to fill the null values of SRTM data. Because of different geographic range, technique used by users and auxiliary data, the course and result of filling were different. Which methods have the best or suitable effect and efficiency of filling? This question will be studied in this article.

In this article, the methods included that the methods of filling existing contour, and of interpolating the grid of SRTM-3 directly, and of interpolating after extracting contour.

3 Data Processing

SRTM data adopt the reference ellipsoid of WGS84 and elevation system of GM96. national fundamental geographic data in China adopted coordinates of Beijing-1954 and Yellow-sea elevation system of Krassowski ellipsoid. According the information of SRTM data and China's standards, we transferred the format of .tif into the grid of ArcGIS and then jointed them into one map. The vector boundary was transferred into the coordinates of WGS84. Using the vector boundary, the area of Lianyungang was cut from SRTM data.

Usually Raw DEM have Sinks or Peaks, so it is necessary to fill depression[5,6], which made Depressionless DEM (Figure 3) as input data. The input data would output the grid of flow direction. By the tool of flow direction, DEM data were used to the analysis of flow direction and got 8 flow directions.

Fig. 3. Depressionless DEM

The grid data from the data of flow direction were input, which would output confluence data. The confluence capacity map of grid would be generated by fixing the accumulation cell number of upstream (NIP) that the cell would flow into this cell[7, 8]. After the generation of the map, it would obtain the river networks by setting the threshold value of the number. The cell of value which was greater than the threshold became the point of gulch line, and connected the points and got the river networks[9].

Extraction of River networks all grid cells in the NIP which were greater than certain threshold would be extracted with the module of Hydrology in ArcGIS. The cells were assigned to 1 and the cells less than the threshold were assigned to 0, which would obtain a Binary Images. The thresholds were ranged from River Course Square Area (CSA) [10, 11, 12].

A threshold was in accordance with corresponding CSA (see Table 1).

Table 1. Correspondence of Threshold and CSA

Threshold	CSA	Threshold	CSA
20	0.162	90	0.729
30	0.243	100	0.810
40	0.324	110	0.891
50	0.405	120	0.972
60	0.486	130	1.053
70	0.567	140	1.134
80	0.648		

We selected different threshold to extract river networks with different density. Here we assigned threshold to the values of 90, 180, 1000 and 5000 (see Correspondence 4, 5, 6, and 7, respectively).

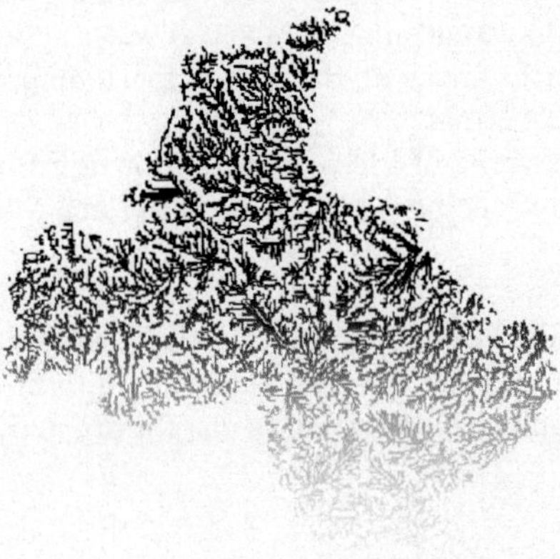

Fig. 4. River networks obtained by threshold of 90

Fig. 5. River networks obtained by threshold of 180

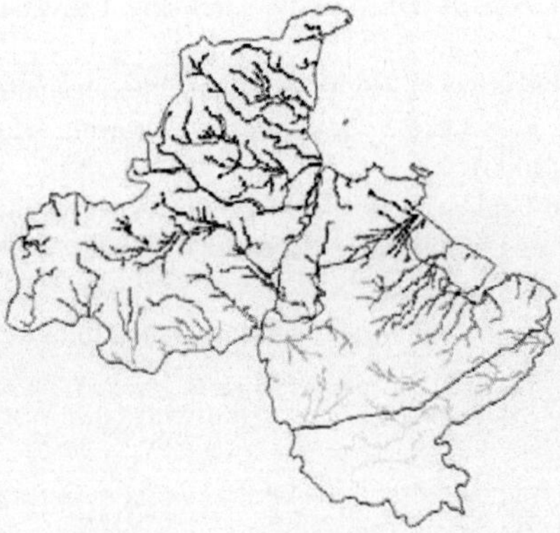

Fig. 6. River networks obtained by threshold of 1000

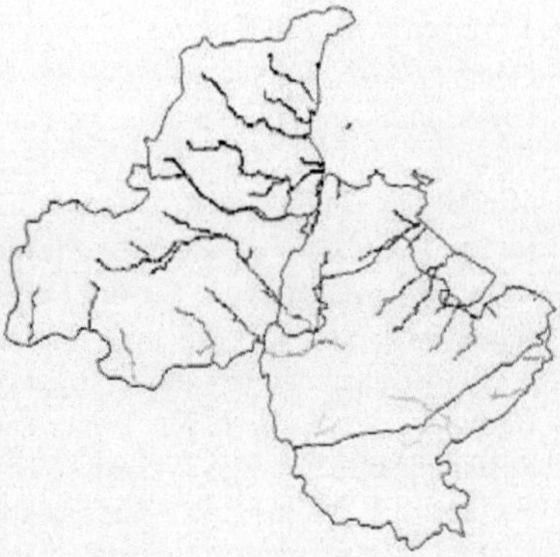

Fig. 7. River networks obtained by threshold of 5000

4 Results and Discussion

The results showed that more CSA could display the outline of river networks more, and the length of river networks was shorter, and the density of basin river networks was lower; less CSA could display the detailed features of basin river networks, and the length of river networks was longer, and the density of basin river networks was higher. The threshold of 5000 were gained the area of principal river in Lianyungang from the contrastive analysis.

The analysis showed that the selection of CAS threshold is very important and will influence the statistic characteristics of whole basin river networks directly.

The extraction of basin river networks based on DEM data has close relationship with the terrain feature in the study area. The algorithm could simulate the real water system in the area with topographic relief. In contrast to the area with flat terrain, for example in Lianyungang, the algorithm will cause distinctive errors.

The analysis showed that the basin river networks of Lianyungang extracting from data of SRTM-3 or the resolution of 90m could reflect the water system of the area when the threshold was assigned to 90.

Acknowledgment. This study is jointly supported by the fund of Huaihai Institute of Technology (grant Z2008011 and Z2008012) and the fund of Teaching Reform of Huaihai Institute of Technology (Bulding the multi-interdisciplinary plat for RS study). In addition, thanks Lü Ying's work.

References

1. Rabus, B., Eineder, M., Roth, A.: The shuttle radar topography mission a new class of digital elevation models acquired by space borne radar. ISPRS Journal of Photogrammetry & Remote Sensing 57, 241–262 (2003)
2. Szėkely, B.: On the Surface of the Eastern Alps-A DEMStudy. Germany University Tübingen Press (2001)
3. Hooper, D.M., Bursik, M.I., Webb, F.H.: Application of High-resolution, Interferometric DEMs to Geomorphic Studies of Fault Scarps Fish Lake Valley Nevada-California USA. Remote Sensing of Environment 84, 255–267 (2003)
4. Moore, I.D., Turner, A.K., Wilson, J.: GIS and land-surface-subsurface process modeling. In: Goodchild, M.F., Park, B.O., Steyaert, L.T. (eds.) Environmental with GIS, pp. 196–230. Oxford University Press, Oxford (1993); Analyst. Printed in USA, 104–107 (1998)
5. James Nelson, E.: WMS v6.1 Tutorials Environmental Modeling Research Laboratory, p. 236. Brigham Young University, Provo (2001)
6. Garbrecht, J., Martz, L.W.: Topaz Overview. USDA-ARS, Grazingland with GIS. Water Resources Bulletin 30(3), 45–46 (1994)
7. Garbrecht, J., Martz, L.W.: The assignment of drainage direction over flat surfaces in raster digital elevation models. Journal of Hydrology 193, 204–213 (1997)
8. Turcotte, R., Fortin, J.P., Rousseau, A.N.: Determination of the drainage structure of a watershed usinga digital elevation model and a digital river and lake network. Journal of Hydrology 240(3-4), 225–242 (2001)
9. Srinivasan, R., Arnold, J.G.: Integration of a basin scale water quality model 139(1/4), 263–293 (1992)
10. O'Callaghan, J.F., Mark, D.M.: The extraction of drainage network from digital elevation data. In: Computer Vision, Graphics and Image Processing, vol. 28, pp. 323–344 (1984)
11. Jensen, S.K., Domingue, J.O.: Extracting topographic structure from digital elevation data for geographic information system analysis. Photogrammetric Engineering and Remote Sensing 54(11), 1593–1600 (1988)
12. Tribe, A.: Automated recognition of valley lines and drainage networks from grid digital elevation models a review and a new method. Journal of Hydrology 139(1/4), 263–293 (1992)

Segmentation of LiDAR Point Cloud Based on Similarity Measures in Multi-dimension Euclidean Space

Qingming Zhan[1] and Liang Yu[2]

[1] School of Urban Design, Wuhan University Research Center for Digital City
Wuhan University, Wuhan, 430072, China
[2] State Key Laboratory of Information Engineering in Surveying, Mapping and Remote Sensing
Wuhan University, Wuhan, 430079, China
qmzhan@whu.edu.cn, yuliang@189.cn

Abstract. The segmentation of LiDAR point cloud is a key but difficult step for 3D reconstruction of architecture. Many researchers have tried to develop segmentation methods including edge-based, surface-based and cluster-based segmentation, etc. In this paper, we present a point data segmentation method based on similarity measures in multi-dimension Euclidean space. The main workflow of this method is made by calculating point normal vector, transforming color values, calculating Euclidean distance and angle in multi-dimension space, comparing the similarity among adjacent points, and segmenting the raw points set at last. The proposed method takes the both advantages of geo-metrical segmentation and color-metrical segmentation as compared with three different segmentation methods. It has been applied to LiDAR point data obtained by TLS (terrain laser scanner), the experiment results show that the segmentation method is promising.

Keywords: LiDAR, Point cloud, segmentation, Euclidean Space.

1 Introduction

Laser-scanning technology is increasingly widely applied in industrial and GIS fields, such as reverse engineering, surveying and mapping, 3D city reconstruction, etc. This technology is a non-contact measuring method. The raw data obtained by laser scanner are scatter points, usually called point cloud. The raw point data acquired by a Laser-scanning device generally require pre-processing operations such as noise filtering, smoothing, merging and data sorting in order to be useable in subsequent operations. Using the pre-processed point data, a 3D model can be generated. After data pre-processing, Segmentation is a key but difficult procedure for building 3D models.

1.1 Review of Existing Segmentation Methods

Segmentation of point cloud data is the process of labeling each point in a point cloud, so that the points belonging to the same surface or region are given the same label.

D. Zeng (Ed.): Advances in Computer Science and Engineering, AISC 141, pp. 349–357.
springer link.com

There are mainly three types of segmentation algorithms: edge-based segmentation, surface-based segmentation and hybrid segmentation.

The edge-based segmentation algorithms have a common stage: the points belong to surface edge always express different features. Those features include normal vector, gradients, principle curvatures, etc. Once finding out those edge points, tracking those points can obtain surface borders. Those points within borders are grouped together. Many researchers develop edge-based segmentation algorithms. Fan et al.[1] used local surface curvature properties to identify boundaries in the point data. In order to avoid the loss of localization, scale-space tracking, which convolves the entire data with Gaussian masks having different values of the spread parameter, was employed. Sappa and Devy[2] presented a technique consists of two stages. First, a binary edge map is generated; then. a contour detection strategy is applied for the extraction of the different boundaries. Meyer and Marin[3] presented a method to partition a polygonal network along the edges of a surface. In the method, the edge start point is precisely calculated by locally fitting a surface that has discontinuous tangent.

The region-growing methods proceed with segmentation by detecting continuous surfaces that have homogeneity or similar geometrical properties. BESL and Jain[4] used a piecewise-smooth surface model for point data that possesses surface coherence properties and developed an algorithm that simultaneously segments a large class of images into regions of arbitrary shape and approximates image data with bivariate functions. Pu and Vosselman[5] adopted the planar surface growing algorithm to segment building façade and automatically extract building features based on feature constraints such as size, direction and topology derived from convex hulls of segmented regions. Rabbani et al.[6] presented a method for segmentation of point clouds using smoothness constraint, which finds smoothly connected areas in point clouds. It uses only local surface normal and point connectivity which can be enforced by using either k-nearest or fixed distance neighbors.

The hybrid segmentation processes points with several kinds of point's properties. Those properties may be measure by geo-metrical or image-metrical, even texture-metrical. Lucieer and Stein[7] extracted spatial landform objects from a light detection and ranging (LiDAR) digital surface model (DSM), and the local binary pattern (LBP) operator, modeling texture, were integrated into a region growing segmentation algorithm to identify landform objects. Zhan et al.[8] presented a point cloud segmentation algorithm based on colorimetrical similarity and spatial proximity. The algorithm contains region growing, region merging and refinement processes. J Hernandez and B Marcotegui[9] used Top-Hat of hole filling algorithm to extract features from the connected components and a SVM machine learning method to segment and interpret the 3D point clouds from mobile LIDAR data. M Awrangjeb et al.[10] presented an automatic building detection technique fusing LIDAR data and multispectral imagery, in their experiment, the proposed technique can successfully detect urban residential buildings.

1.2 Motivation

Many point clouds segmentation methods are developed to segment industrial installation scanning data whose real world counterparts usually have relatively regular and simple geometric shapes. However, it is not easy for a computer to perform segmentation even for a simple object with quadric surfaces. The segmentation of an object with free-form surfaces, of course, is much more difficult and usually leads to ambiguous solutions.

In this paper, we present a point data segmentation method based on similarity measures in multi-dimension Euclidean space. The proposed method takes both advantages of geo-metrical segmentation and color-metrical segmentation. The main workflow of this method is made by calculating point normal vector, transforming color values, calculating Euclidean distance and angle in multi-dimension space, comparing the similarity among adjacent points, and segmenting the raw points set at last.

2 Proposed Segmentation Method

2.1 Overview of the Process

The point clouds obtained by laser scanners usually involve coordinates, color, and intensity etc. Isolated use of coordinates, color information may result in under- or over-segmentation problems. Since all these data are the components of the Euclidean space, we can utilize the spatial similarity to distinguish the point and segment the raw point set into several different point groups at last.

The criterion of similarity measures in the Euclidean space is based on two features, which are distance and angle.

The generalized Euclidean space distance calculation is given in Formula 1. The W_{1k} is the components of one element, while W_{2k} is the corresponding components of the other element.

$$DIST_{Euclidean} = \sqrt{\sum_{k=1}^{n}(W_{1k} - W_{2k})^2} \tag{1}$$

The angle cosine calculation is based on Formula 2. The $W_1 * W_2$ is the dot product, while $|W_1|$ and $|W_2|$ are respectively expressed the module osf the two vector.

$$\cos\theta = (W_1 * W_2)/(|W_1||W_2|)$$
$$\cos\theta = \sum_{k=1}^{n}(W_{1k} * W_{2k})/\sqrt{(\sum_{k=1}^{n} W_{1k}^2)(\sum_{k=1}^{n} W_{2k}^2)} \tag{2}$$

As the point clouds in three-dimensional space of discrete distribution, we use vectors instead of its coordinates. Meanwhile, due to normal vectors and point color existed in different dimension, we change the color range to the scope of any consistent normal vectors. If other feature, such as intensity, were used in the presented segmentation process, the feature must be adjusted consistently.

The main ideal of the presented segmentation method is that calculate the normal vector of each point, translate the *RGB* (red, green and blue) value of point color to the components of a vector. At last, we use the space similarity measurement to divide the raw point set. The details of these steps are given in algorithm 1 and further explained at Fig. 1.

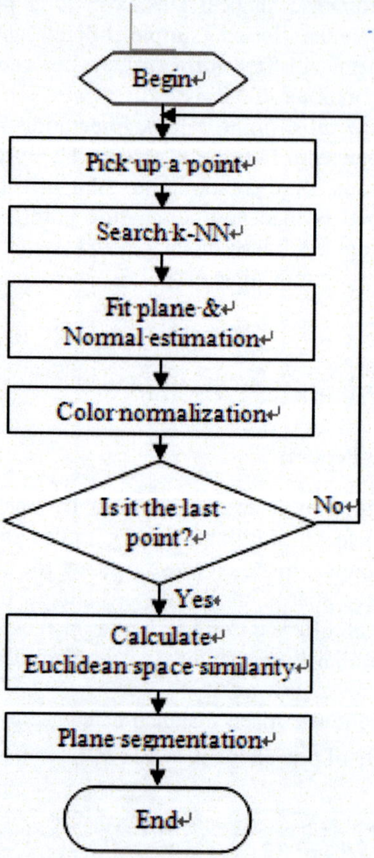

Fig. 1. Flowchart of the segmentation method

The 1st step, picking up a point from the point set and making it the current handle point;

The 2nd step, searching the k-nearest neighbors of the presented point.

The 3rd step, using the least square method to fit a plane of the k-NN search result, using the normal vector of the plane to represent the current point.

The 4th step, adjusting the color of the point and transforming the RGB value from range [0, 255] to range [0, 1].

The 5th step, handling all points with the same process to calculate the normal vector and color of the point set.

The 6th step, combining vector and color of the point to calculate the Euclidean space similarity.

The 7th step, using distance and angle tolerance to divide the raw point set into different group based on similarity measure.

2.2 K-NN Search

The k-nearest neighbors algorithm (k-NN) is a method for classifying objects based on closest training examples in the feature space. *K-NN* is a type of instance-based learning, or lazy learning where the function is only approximated locally and all computation is deferred until classification. The *k*-nearest neighbor algorithm is amongst the simplest of all machine learning algorithms: an object is classified by a majority vote of its neighbors, with the object being assigned to the class most common amongst its *k* nearest neighbors (*k* is a positive integer, typically small).

2.3 Normal Vector Estimation

The method of least squares is a standard approach to the approximate solution of overdetermined systems, i.e. sets of equations in which there are more equations than unknowns. The "least squares" means that the overall solution minimizes the sum of the squares of the errors made in solving every single equation [11].

The most important application is in data fitting. The best fit in the least-squares sense minimizes the sum of squared residuals, a residual being the difference between an observed value and the value provided by a model.

The least squares problems fall into two categories, linear least squares and nonlinear least squares, depending on whether or not the residuals are linear in all unknowns. The linear least-squares problem occurs in statistical regression analysis; it has a closed form solution. The non-linear problem has no closed solution and is usually solved by iterative refinement; iteration the system is approximated by a linear one, thus the core calculation is similar in both cases.

2.4 Color Transforming

The normal vector of the fit plane indicates the directions of the raw point. The components of vector are valued in range [-1, 1], while the color of corresponding point are valued in range [0, 255]. To avoid the difference between them that may influence the calculation result of Euclidean space similarity, we take a tactics to adjust their components to a similar range.

We adjust the *RGB* values of the point color by following formula.

$$R' = R/maxbyte$$
$$G' = G/maxbyte$$
$$B' = B/maxbyte$$

The $R\backslash G\backslash B$ are the raw components of the color and the $R'\backslash G'\backslash B'$ are the adjusted components. All *RGB* values will be adjusted to be in range [0, 1]. The transformed components of the color of the point will be used as input data in the remaining parts of the segmentation process.

2.5 Similarity Clustering in Euclidean Space

We use the Euclidean distance and angle to measure the difference between two elements: using formula 1 to calculate distance and using formula 2 to calculate angle.

In presented method, we assume a distance tolerance and an angle tolerance to judge two points whether they are belonging to the same group. We use the distance tolerance to roughly segment the raw point cloud, and then apply the angle tolerance to merge and refine the grouping result. If the space similarities of two points are within limitation, we consider that these two points are homogeneous and label them in the same region.

3 Application

The raw point set are acquired from a Chinese ancient architecture which contains 498,890 points. The Figure 2 is the perspective view of the raw data. The raw data contain points coordinate and color information, while they don't contain echo intensity; we can apply presented segment method to handle them.

Fig. 2. Raw point data with color

In presented method, we calculate each point's normal vector in the first step. We pick up a point of the point set and search it's nearest neighbors. The search result is used as the input for the plane-fitting. The normal vector of the plane is regarded as of the representative of the point. We define two parameters: neighborhood distance d=0.5m and number of neighbors for searching n=30. Then, we deal each point in the set with the same process. Furthermore, we adjust all points' color components to make their *RGB* values consistent with the vector space. In presented method, we simply assign transforming parameters as: maxbyte = 255. Through above transformation, the points' color values have been adjusted to the range [0,1]. At last, we use the Formula 1 and Formula 2 to calculate the space similarity, taking MaxDist = 1 as the threshold of $DIST_{Euclidean}$ and MaxAngle = 15 as the threshold of $\cos(\theta)$. The presented segmentation result in contrast with results produced by other geometric and color-metric segmentation methods are shown in Fig. 3, Fig. 4 and Fig. 5.

a: Angle threshold=15 b: Angle threshold=20

Fig. 3. Segmented results based on normal vector

We applied three point cloud segmentation methods to the raw point data. The first one is commonly used geometric segmentation method [6]. We define two parameters: neighborhood distance d=0.5m and number of neighbors for searching n=30. After calculating the normal vector of each point, we define another threshold: angle=15 or angle=20, the geometric segmented results are shown at Figure 3.

a:color threshold=50 b: color threshold=80

Fig. 4. Segmented results based on color

The second one is color-metric segmentation method[8]. We define two parameters: neighborhood distance d=0.5m and number of neighbors for searching n=30. Meanwhile, we define another color difference threshold: clrDiff=50 or clrDiff=80. The color-metric segmented results are shown at Figure 4.

a: distance threshold=0.8
angle threshold=10

b: distance threshold=1
angle threshold=15

Fig. 5. Segmented results based on similarity in color and geometric features

The Figure 5 shows the segmented results produced by the proposed method. From the results of the three segmentation methods, the geometric segmented method usually generates many small broken surfaces, while the color-metric segmentation method takes the similar color point into the same group but can't differentiate points that are located in different planes. The segmented results of the proposed method is laid between them, moreover, through adjusting the threshold of distance and angle, we can obtain much more reliable results. The comparison of different segmentation results is presented in Table 1. Based those figures, the proposed segmentation method is promising.

Table 1. Result comparison of three segmentation methods

Method	**Para-meters**	**Face number**	**Average point number**	**Max point number**	**Conclusions**
Normal vector	Ang:15	4552	110	22374	Over-segmentation
Point color	Clr:50	982	508	113853	Under-segmentation
Space similarity	Dist:0.8 Ang:10	2869	175	52317	Moderate-segmentation
Normal vector	Ang:20	3725	134	32176	Over-segmentation
Point color	Clr:80	402	1241	143853	Under-segmentation
Space similarity	Dist:1 Ang:15	2069	241	60852	Moderate-segmentation

4 Conclusions

A segmentation method based on the generalized Euclidean space similarity is proposed and implemented in this paper. The proposed segmentation method is performed by estimating the normal vector of point and transforming the components of color into vector space, then calculated the Euclidean distance and angle to segment the scatter point set based on the similarity measures.

The proposed segmentation method has been applied to LiDAR point data obtained by terrestrial laser scanner. As a result, the raw point cloud data are segmented with a series of surface. The segmented result clearly indicates the points which belong to roof, wall and basement are labeled in different groups. The proposed segmentation method has produced the best results.

Acknowledgment. This research is supported by the National Natural Science Foundation of China (No. 40871211) and the Hi-Tec Research and Development Project (863 project, No. 2006AA12Z151).

References

1. Fan, T., Medioni, G., Nevatia, R.: Segmented description of 3-D surfaces. IEEE Transactions on Robotics and Automation RA-3(6), 527–538 (1987)
2. Sappa, A.D., Devy, M.: Fast range image segmentation by an edge detection strategy. In: Proceedings of the Third International Conference on 3-D Digital Imaging and Modeling, pp. 292–299. IEEE Computer Soc., Los Alamitos (2001)
3. Meyer, A., Marin, P.: Segmentation of 3D triangulated data points using edges constructed with a C1 discontinuous surface fitting. Computer-Aided Design 36, 1327–1336 (2004)
4. Besl, P.J., Jain, R.C.: Segmentation through variable-order surface fitting. IEEE Transactions on Pattern Analysis and Machine Intelligence 10(2), 167–192 (1988)
5. Pu, S., Vosselman, G.: Automatic extraction of building features from terrestrial laser scanning. International Archives of Photogrammetry, Remote Sensing and Spatial Information Sciences 36(5), 25–27 (2006)
6. Rabbani, T., van den Heuvel, F., Vosselmann, G.: Segmentation of Point Clouds using Smoothness Constraint. International Archives of Photogrammetry, Remote Sensing and Spatial Information Sciences 36(5), 248–253 (2006)
7. Lucieer, A., Stein, A.: Texture-based landform segmentation of LiDAR imagery. International Journal of Applied Earth Observation and Geoinformation 6(2-3), 261–270 (2005)
8. Zhan, Q., Liang, Y., Xiao, Y.: Color-based segmentation of point clouds. International Archives of the Photogrammetry, Remote Sensing and Spatial Information Sciences, 248–252 (2009)
9. Hernandez, J., Marcotegui, B.: Point cloud segmentation towards urban ground modeling. In: 2009 Joint Urban Remote Sensing Event, May 20-May 22. IEEE Computer Society, Shanghai (2009)
10. Awrangjeb, M., Ravanbakhsh, M., Fraser, C.S.: Automatic detection of residential buildings using LIDAR data and multispectral imagery. ISPRS Journal of Photogrammetry and Remote Sensing (2010) (in Press)
11. Mitra, N.J., Nguyen, A.: Estimating surface normals in noisy point cloud data. In: Proceedings of the Nineteenth Annual Symposium on Computational Geometry, pp. 322–328. ACM, San Diego (2003)

Study on Soil Erosion in Zuli River Basin Using RS and GIS

Han Hui[1] and E. Youhao[2]

[1] Mathematics, Physics and Software Engineering School
Lanzhou Jiaotong University
Lanzhou 730070, China
[2] Chinese Academy of Meteorological Sciences
Beijing 100081, China
hanhui@mail.lzjtu.cn, eyh@cams.cma.gov.cn

Abstract. Combining Geographical Information System (GIS) and Remote Sensing (RS) with the Universal Soil Loss Equation (USLE), we try to extract the amount of soil erosion. Firstly we obtain the five factors that influence soil erosion severity. They are rainfall pattern, soil type, topography, crop system and management practices. Then the soil erosion distribution in the area is illustrated and we can analyze this disaster from spatial scale.

Keywords: Universal Soil Loss Equation (USLE), GIS, RS, soil erosion, Zuli river basin.

1 Introduction

China has the most serious soil erosion problem in the world. According to the national second Soil Erosion Survey in 2000, the area of soil erosion is 3,560,000 km2, which is more than one-third of the national territory area. Serious soil erosion not only deteriorates living condition and ecosystem, but also disserves the downstream area.

As the minimum branch of Yellow River, Zuli river which is located in the west of Loess Plateau has amazing sediment discharge and we select it as our study area. In Zuli river basin, land over cultivation and deforestation have changed the surface features, enlarged the soil erosion area and influences the balance of the ecosystem.

Combining GIS and RS with the USLE, we try to study on soil erosion in basin scale. By way of getting five factors' spatial distribution, such as rainfall pattern, soil type, topography, crop system and management practices, we extract the amount of soil erosion in Zuli river basin finally (See Fig. 1).

2 Use Model and Parameters' Extraction

In 1965, according to 30 years' observational data about more then 10,000 runoff plots in 30 states of eastern USA, USLE was brought forward by W. Wischmeier and D. Smith and revised by many researchers. Aiming at annual average soil

D. Zeng (Ed.): Advances in Computer Science and Engineering, AISC 141, pp. 359–364.
springer link.com

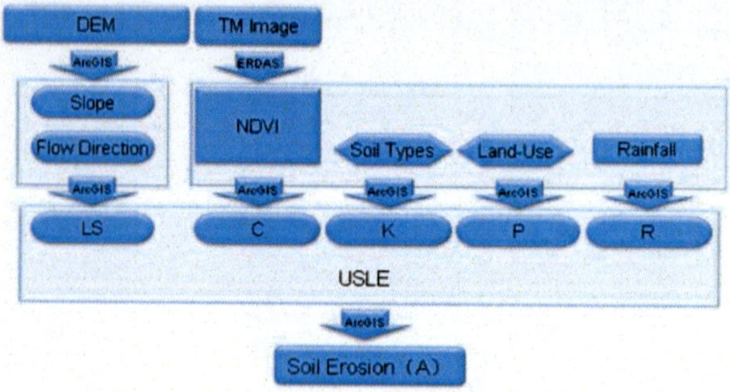

Fig. 1. Flow chart of the processes for calculating soil erosion using USLE

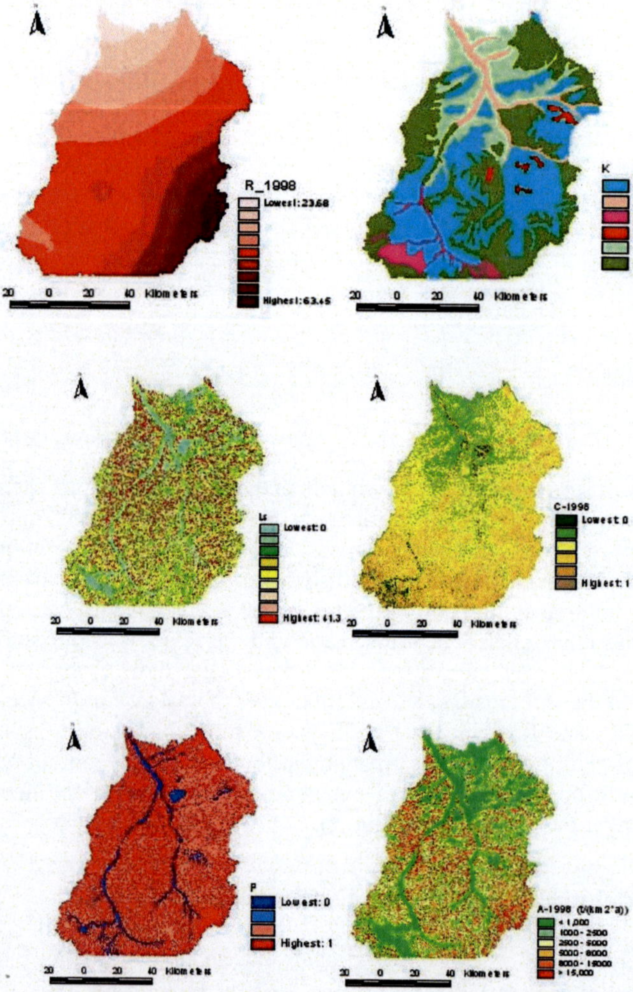

Fig. 2. Five factors and soil erosion distribution using USLE

erosion due to sheet erosion and rill erosion, this model considers five factors, such as rainfall pattern, soil type, topography, crop system and management practices. The equation is expressed as:

$$A=f*R*K*L*S*C*P \tag{1}$$

Where, A is soil erosion, R is rainfall erosion factor, K is soil erodibility, LS is slope length-gradient factor, C is crop/vegetation and management factor, P is support practice factor and f is a constant and equals to 224.2.

2.1 Rainfall Factor (R)

Rainfall erosion factor (R) expresses the latent influence to soil erosion by different climate and rainfall. In this article we use the below empirical formula to calculate R:

$$R=\sum_{i=1}^{12}[1.735\times10^{(1.5LgP^2{}_i/p-0.8188)}] \tag{2}$$

Where, P is annual rainfall and Pi is monthly rainfall. There are 17 rainfall stations in this area and using Kriging interpolation we get R in 1998.

2.2 Soil Erodibility (K)

Soil erodibility (K) can estimate the soil resistance for detaching, washing away and conveying. According to the experiences in loess areas, the formula of K is expressed as:

$$K=\left\{0.2+0.3\exp\left[0.0256SAN\left(1-\frac{SIL}{100}\right)\right]\right\}\left(\frac{SIL}{CLA+SIL}\right)^{0.3} \left[1.0-\frac{0.25C}{C+\exp(3.72-2.95C)}\right]\left[1.0-\frac{0.75SN_1}{SN_1+\exp(-5.51+22.9SN_1)}\right] \tag{3}$$

Where, SAN, SIL, CLA and C are contents of soil gravel, silt, clay grain and organic carbon, SN1=1-SAN/100. From the soil types distribution map, we calculate K factor in Zuli river basin.

Table 1. Main soil types and k factor in zuli river basin

Soil Type	Gravel (%)	Silt (%)	Clay Grain (%)	Organic Carbon (%)	K
Sierozem	0.415	0.402	0.183	0.034	0.447
Loessal Soil	0.4526	0.3459	0.2015	0.0052	0.436
Light Sierozem	0.7357	0.1803	0.0834	0.0054	0.446
Irrigated Desert Soil	0.6555	0.254	0.1594	0.0082	0.430

2.3 Topography (Slope and Slope Length) Factors (LS)

According to experiences of runoff plots in hill areas of Loess Plateau, LS can be expressed as:

$$LS = 1.07\left(\frac{\lambda}{20}\right)^{0.28}(a/10)^{1.45} \tag{4}$$

Where, λ is slope length and a is slope. Using Spatial Analysis Tools and Hydrology in ArcGIS, we get LS from DEM.

2.4 Crop/Vegetation and Management Factor (C)

Related to land-use and vegetation coverage, the C factor indicates the influence of vegetation to soil erosion. Using Regression Analysis the relationship between C and vegetation coverage can be set up like this:

$$C = \begin{cases} C = 1 & fc = 0 \\ C = 0.6508 - 0.3436 \lg f_c & 0 < f_c \leqq 783\% \\ C = 0 & fc > 78.3\% \end{cases} \tag{5}$$

Where, fc is vegetation coverage. TM image of 1998 is used to get fc according to the relationship between fc and NDVI.

2.5 Support Practice Factor (P)

The P factor represents the ratio of soil loss by a support practice to that of straight-row farming up and down the slope. Since the researches of P are relatively less, we use land-use map of 1997 to evaluate P for different land-use types according to relative references.

Table 2. Different land-use types and p factor in zuli river basin

Land-use Type	P	Land-use Type	P
Economic Forest	0.69	Commercial land	0
Woodland	1	Mining storage space	0
Shrub Land	1	Idle land	1
Other Forest	0.70	Non-irrigated farmland	0.7
Artificial Grassland	0.20	Water	0
Grassland	1	Saline	1

3 Discussion

According to the Chinese soil erosion magnitude classification standard for north-western Loess Plateau, there are six degrees for soil erosion.

Table 3. Standard soil erosion magnitude degrees in north-western loess plateau

Degree	Average Soil Erosion [t/(km^2·a)]
Slight	<1000
Light	1000-2500
Moderate	2500-5000
Strong	5000-8000
Extreme Strong	8000-15000
Violent	>15000

With the help of ArcGIS, soil erosion statistics are obtained.

Table 4. Soil erosion statistics in zuli river basin

Degree	Acreage (km^2)	Soil Erosion (t)
Slight	2824.51	35.36×10^4
Light	1073.07	187.54×10^4
Moderate	1658.09	616.91×10^4
Strong	1577.22	1011.78×10^4
Extreme Strong	1856.42	2014.65×10^4
Violent	790.13	1625.51×10^4

It shows that the main soil erosion degrees in this area are Extreme Strong and Violent, which are 66.3% of the total soil erosion, although their distribution acreage is relatively small, which is about 27%.

Including Strong, Extreme Strong and Violent degree, the serious soil erosion mainly happens to the southern basin (47.94% of the total amount) where the farm-land is widely distributed. In the northern basin, the main land-use is grassland and serious soil erosion is 36.98% of the total amount. The result is also influenced by R factor. In 1998, there is more precipitation in the southern basin.

Slight and Light soil erosion are distributed in the places where terrain is relatively flat and the summary area is 3897.58 km^2, which is more than one-third of the total acreage.

4 Conclusion

Using USLE model we can estimate the soil erosion in Zuli river basin. In this process RS and GIS can provide data and technical support and the result is based on spatial scale. After selecting suitable formulas in this area, we obtain the five factors that constitute USLE model and the soil erosion is calculated in ArcGIS environment.

From soil erosion distribution in 1998 we can see that the soil erosion situation is very severe in Zuli river basin. It is caused by the interaction of nature and human activity. How to reduce the influence of human activity in this area to keep ecological balance is very important for our government and researchers.

Acknowledgment. This research is supported by 'Qing Lan' Talent Engineering Funds by Lanzhou Jiaotong University.

References

1. Zhu, L., Feng, W., Zhu, W.: Progress of "3S" technique application in soil erosion study. Progress in Geography 27, 57–62 (2008)
2. Liu, G.: A review of the soil erosion model. Research of Soil and Water Conservation 10, 73–76 (2003)
3. Wang, D., Lu, Y.: Development of soil erosion models abroad. Science of Soil and Water Conservation 2, 35–40 (2004)
4. Shi, C.: Scaling effects on sediment yield in the upper Yangtze River. Geographical Research 27, 800–810 (2008)
5. Tong, X., Xu, Y., Li, H.: Effect of soil and water conservation measurements in the Xiangyanggou watershed. Bulletin of Soil and Water Conservation 26, 86–93 (2006)
6. Yang, G.: Study on soil erosion based on grid data in Karst mountainous region. Guizhou Normal University (2004)
7. Zhang, M.: Soil of gansu. Agriculture Press (1993)
8. Zhang, X.: Spatial analyze of soil erosion assisted by Remote Sensing data and GIS in Zuli river basin. Remote Sensing Technology and Applicaion 24, 772–776 (2009)

Predicted Effects of Nonuniform Beam Filling on Rainfall Retrieval for the Geostaionary Doppler Weather Radar*

Xuehua Li, Jianxin He, Zhao Si, and Xu Wang

Key Laboratory of Atmospheric Sounding,
Chengdu University of Information Technology,
Chengdu, China
{lixh,hjx}@cuit.edu.cn

Abstract. A novel instrument concept, Geostaionary Doppler Weather Radar, namely NEXRAD in Space (NIS), is developed for detailed monitoring of hurricanes, cyclones, and severe storms from a geostationary orbit., which has higher time-sampling and wider obversation area than those of low orbit weather radar such as TRMM and DPR radar. However, because of its too far distance away from the surface of the earth, even sharper beam width up to 0.019° ,it will lead to more than 12 kilometers horizontal resolution near surface, so that errors of nonuniform beam filling (NUBF) on rainfall retrieval would probably be fatal, especially for small scale convective rainfall. Due to no real data of NIS, an investigation using the high resolution data of the Second Generation Airborne Precipitation Radar (APR-2) to simulate low resolution NIS data is first started in this paper for the estimation of NUBF errors similar to the works before TRMM and DPR launched. Algorithms of rainfall retrieval and path-integrated attenuation(PIA) under uniform and nonunifrom beam filling are introduced to estimate these errors, and a statistic method is used to evaluate the error numerical distrubtion. Remakable results show that the errors due to NUBF are prospectively serious and the biases of PIA and reflectivity in the bottom of rainfall are larger and would lead to the rainfal retrieval to be overestimated or underestimated, their maximum biases could be larger than 15dB under small scale convective rainfall system.

Keywords: nonuniform beam filling, path integrated attenuation, APR-2, geostationary doppler weather radar.

1 Introduction

A spaceborne weather radar for precipitation intensity estimitation in quantity is usually used to measure the radar reflectivity(Z) and the path-integrated attenuation(PIA) of precipitation, and these two measurement parameters could be seriously effected by the horizontal resolution or footprint size near surface and the radar wavelength .when the footprint size is close to or larger than that of precipitation , the nonuniform beam filling will lead to large bias for the measurement of radar reflectivity and the PIA value, and finally result in the inaccuracy estimation of precipitation intensity.

* This work was partially supported by the National Science Foundation of China (41075010).

D. Zeng (Ed.): Advances in Computer Science and Engineering, AISC 141, pp. 365–372.
springer link.com

Therefore, the effects evaluation of nonuniform beam filling (NUBF) on rainfall retrieval is an important work before a spaceborne weather radar lauched. Nakamura [1] and S. L. DURDEN etc.[2] used ground-based and airborne data to simulate spaceborne radar data to investigate NUBF effects and found that biases in quantities measured by the PR could occur in coverage of intense rainfall, Following launch, some NUBF effects were indeed noted in TRMM PR data via direct comparison with airborne radar data acquired in TRMM under flights. For the first dual-frequency spaceborne radar (DPR) carried by Global Precipitation Measurement mission(GPR), S. L. DURDEN etc.[3] use the same technique as that of [2] to investigate potential NUBF effects on GPM DPR data using the data of the Airborne Second Generation Precipitation Radar (APR-2),and found that errors at ku-band are similar to those of found in[2] and errors in ka-band PIA value and radar reflectivity near surface are larger than at ku-band. The Geostationary Doppler Weather Radar, namely NEXRAD In Space (NIS) [4-5] is an advanced instrument concept and technology for future sapceborne atmospheric radar to hourly monitor life cycle of hurricanes and tropical storms from a geostationary orbit. And its operatation at ka-band(35GHZ) and in the distance of 36000 kilometers far away from the earth would bring to more serious precipitation attenuation and larger footprint size, thus probably result in more serious biases due to NUBF. In this paper, an investigation of potential NUBF effects aiming at NIS radar is first developed using the 2-dimension high horizontal resolution APR-2 data to simulate low resolution NIS data. This work is an essential evaluation for further development of NUBF correction or resolution enhancement.

2 Theory

The NIS Precipitation Radar (PR) footprint size of more than 12 km is larger than the typical size of raincells, for example, the median convective cell size for convective storms off the Virginia coast is only 1.9 km[2]. This implies that the rainfall within the PR footprint may not be uniformly spread when observing convective rainfall, and such nonuniform beam filling (NUBF) have been found biases in the estimated rainfall intensity [2-3]. Before further numeric evaluation through simulated NIS radar data , a reasonable contrastive method on rainfall rate retrieval under UBF and NUBF should be firstly established for the estimation of rain intensity from weather radar observations.

Typically, The rainfall intensity is retrieved from reflectivity measurement of weather radar obversation using the Z-R relation, which relate the radar reflectivity factor Z to the rainfall rate R by a power law $Z = aR^b$. Moreover, for the high frequency of ka-band in NIS radar, precipitation attenuation become a serious problem and K-R relation ($K = \alpha R^{\beta}$) must be added in. Thus, the rain intensity estimation from radar reflectivity measurement based on Z-R and K-R relation can be written [2-3]

$$Z\ (r) = aR(r)^b 10^{-0.2\int_0^r \alpha R(s)^{\beta} ds} \tag{1}$$

where r is range from rain top, α , β are the coefficients of K-R relation. if rain intensity across the antenna beam is uniform, the reflectivity measurement could be retrieved through average of rain intensity across beam in horizontal plane,

Thus equation (1) can be rewritten

$$Z_u(r) = a < R(r) >^b 10^{-0.2\int_0^r \alpha < R(s) >^\beta ds} \tag{2}$$

where u denotes uniform beam filling,< > denotes average in horizon. However, the radar apparent measurement is not the average of rain intensity R but radar reflectivity Z in fact. Thus, the radar apparent reflectivity measurement is

$$Z_a(r) = < aR(r)^b 10^{-0.2\int_0^r \alpha R(s)^\beta ds} > \tag{3}$$

where a denotes radar apparent measurement. From formula (2) and (3), we would found the two measurements above are equivalent if there is no precipitation attenuation and Z-R relation is linear. But unfortunately, the Z-R relation usually is not linear ($b \neq 1$) and precipitation attenuation in Ka-band is serious for NIS radar. Thus difference would inevitably occur between radar apparent measurement and uniform beam filling theoretical calculation and these difference will bring large biases to rainfall intensity retriveal.

Two typical position of rain top and rain bottom(near surface) along nadir-looking spaceboren radar range direction are chosed for further analysis of the measurement biases due to NUBF, which separately denotes the position of no precipitation attenuation and the most serious precipitation attenuation, through K-R relation these attenuation will lead to lots of effects to the reflectivity calculation. Near the top of rain column, where attenuation can be neglected, the apparent and uniform reflectivity can easily be showed as:

$$Z_{a\_top}(r) = < aR(r)^b > \tag{4}$$

$$Z_{u\_top}(r) = a < R(r) >^b \tag{5}$$

In the bottom of rain, due to the serious precipitation attenuation, an auxiliary method of path integrated attenuation (PIA) calculation through Surface Reference Technology (SRT) must be introduced to decrease the intrinsic instability of coefficient α, β. In SRT method, PIA value is calculated through the real radar reflectivity measurement in the clear rain area and in the rain area over ocean and removed the effects of various rainfall rate. the relation of apparent PIA and rainfall intensity is defined

$$PIA_a = < 10^{-0.2\int_0^L \alpha R(s)^\beta ds} > \tag{6}$$

where L is rain layer depth (from rain top to rain bottom). in order to simplify the computation complexity and considering that β is very close to unity in practice, a definition of path-averaged rain rate (PARR), R_p is introduced

$$R_p = L^{-1} \int_0^L R(s)\, ds \tag{7}$$

Thus formula (6) can be further simplified as

$$PIA_a = < 10^{-0.2L\alpha R_p^\beta} > \tag{8}$$

with the same simplification, the uniform PIA, the apparent and uniform radar reflectivity measurement in bottom of rain could be easily get:

$$PIA_u = 10^{-0.2L\alpha <R_p>^\beta} \tag{9}$$

$$Z_{a\_bottom} = a < R^b 10^{-0.2L\alpha R_p \beta} > \tag{10}$$

$$Z^m_{u\_bottom} = a < R >^b 10^{-0.2L\alpha <R_p>\beta} \tag{11}$$

After these contrastive methods have been established now, further analysis due to NUBF using simulated NIS data in the following text is mainly concentrated on those biases between PIA_a, Z_a and PIA_u, Z_u in rain top and rain bottom.

3 Data Simulated Technology

The NIS precipitation radar [4-5] is designed to operate in a geostationary orbit at analtitude of 36,000 km and work in 35GHz of radar frequency. A deployable, 35-m, spherical antenna reflector will be used to perform spiral scans up to 4° from boresight to cover a 5300-km circular disk on the Earth surface. This coverage is equivalent to 48° (±24°) in both longitude and latitude and the corresponding horizontal resolution ranges from 12 km at nadir to 14 km at 4° scan. The Airborne Second Generation Precipitation Radar [6] (APR-2) is a dual frequency precipitation radar, which operated at 13.4GHz and 35.6GHz with downward-looking, cross-track-scanning geometry similar to TRMM PR and GPM DPR, and was designed to pre-verify some key technology in ground before GPM launch. Table I shows some main radar parameters of NIS and APR-2. lots of reasons ,such as the same radar waveband, high resolution data, downward-looking scan and pulse compression system ,are worth us to use APR-2 data to simulate NIS radar data.

The simulated NIS data are generated from APR-2 data footprint by footpring on both vertical (range) and horizontal resolution units. Firstly, the APR-2 reflectivity data are sampled to a Cartesian grid along flight track, then a gaussian weight is added in range to simulate the response of transmitted pulse shape and receiver filter and in angular to simulate the function of antenna beam pattern which are usually used in resolution enhancement for gound doppler weather radar[2-3], After average of ten data units in range and thirty data units in horizon, the apparent reflectivity Z_a and path-integrated attenuation PIA_a data with 300 meters range resolution and 12 kilometers horizontal resolution for NIS radar are produced. Fig.1 shows an example of apparent reflectivity simulated results from APR-2 data. Different from those methods above, when to simulate PIA_u and Z_u data under uniform rainfall condition, gaussian weight in horizontal plane must be done on rainfall intensity R so that all reflectivity data must be first converted into R using the Z-R relation and finally return to reflectivity data Z after gaussian weight, in addition, in bottom of rainfall, the K-R relation need to be especillay used to acquire the path-averaged rain rate R_p and PIA_u ,then convert to Z_u .The coefficients in K-R and Z-R relation for this work are:

$$Z = 369.12R^{1.123} \tag{12}$$

$$k = 0.2336R^{1.021} \tag{13}$$

Table 1. Radar parameters comparison between NIS and APR-2

Parameter	*NIS*	*APR-2*
Frequency	*35GHz*	*35.6GHz*
Radar position	*36000km*	*< 10km*
3dB BW	*0.019°*	*4.8°*
Max scan angle	*±4°*	*±25°*
Scan model	*Spiral scan*	*Cross-scan*
Min. Zeq	*10dBZ*	*10dBZ*
Antenna sidelobe	*-30dB*	*-30dB*
Horiz.Res.(nadir)	*12km*	*400m*
Vertical Res.	*300m*	*30m*

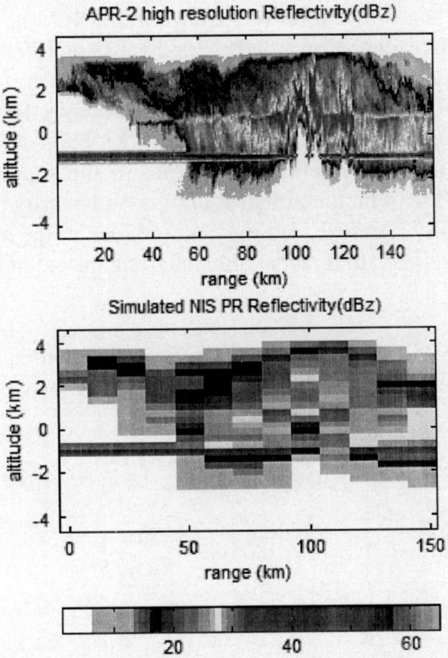

Fig. 1. APR-2 data collected in the Wakasa Bay AMSR-E validation campaign over the sea of Japan on board a NASA P-3 aircraft ,data filename used is "APR2.030119.040454.2. HDF". Vertical axis is altitude, and horizontal axis is a range position. Upper image is original APR-2 high resolution data, lower image is result of NIS PR simulation.

4 Results

The Airborne Second Generation Precipitation Radar (APR-2) data[5] used in this paper were collected in the Wakasa Bay AMSR-E validation campaign over the sea of Japan on board a NASA P-3 aircraft in January and February 2003. The raw APR-2 data were first saved in a unique APR-2 format, then calibrated to reflectivity Z, LDR, and velocity through a processor and finally created a geolocated Level 1B product using the aircraft navigation data and saved in a Hierarchical Data Format (HDF) format similar to the TRMM Precipitation Radar. The data used in this work were chosed from the HDF file under the condition of stratiform and convective rainfall over ocean. Removing the data with the PIA smaller than 3 dB and over land, 280 NIS radar footprint data have been simulated to investigated and analyze those errors of NBUF .

The NBUF errors were further analyzed among the PIA , the reflectivity Z in top of rainfall and in the bottom of rainfall versus their corresponding standard deviation of high resolution within NIS footprint which distribution could be further evaluated the relation between errors and precipitation system. Generally, small scale convective rainfall system will bring to larger standard deviation within NIS footprint in these simulation, while those for large scale straiform rainfall system are on contratry. Fig.2 to Fig.4 show scatter plots of the NUBF errors versus the standard deviation of the high-resolution measurements within the NIS footprint. In Fig.1 the errors of PIA are correlated with the standard deviation of PIA within NIS footprint, larger PIA value occurs when the standard deviation is large, and most of the PIA measurements in NIS were underestimated which directly result in large underestimation on rainfall intensity retrieval especially for. the small scale convective. The main factors maybe come from the exponential nature of attenuation of Ka band As shows in Fig3. with no precipitation attenuation the reflectivity errors in the top of rain are relatively smaller than those of PIA which mainly root in the no linearity of Z-R relation . the biases of rainfall intensity retrieval are most determined to the distribution of rainfall within NIS footprint. Noted that with the radar range extended to rainfall depth

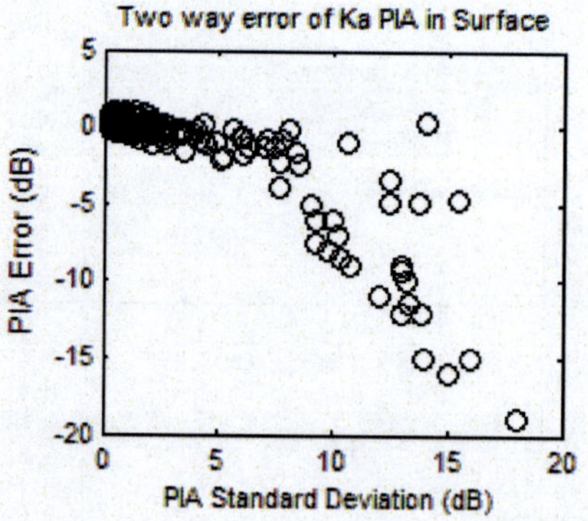

Fig. 2. Two-way PIA error versus standard deviation of high resolution two-way PIA within NIS footprint

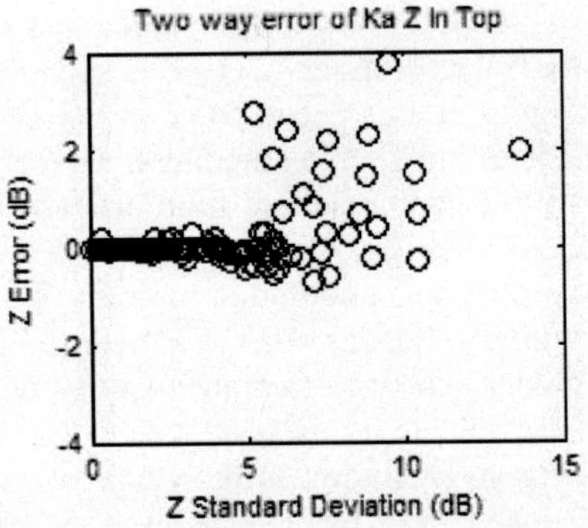

Fig. 3. Error in reflectivity at rain-top versus standard deviation of high resolution rain-top reflectivity within NIS footprint

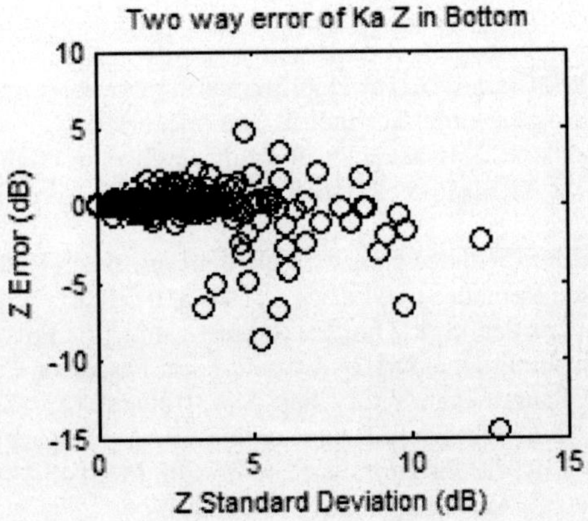

Fig. 4. Error in reflectivity at rain-bottom versus standard deviation of high resolution rain-bottom reflectivity within NIS footprint

larger biases and more instability will happen. The position in the bottom of rainfall (near surface) is a typical one where the path-integrated attenuation has reach maximum, from Fig.4 we can seen that the magnitude of the error increases with increasing standard deviation within NIS footprint, and the error sign can be positive or negative, large scatters may occur even in small standard deviation. All of those errors will result in larger bias and lots of instability on rainfall intensity retrieval and decrease the reliability of radar precipitation measurement.

5 Conclusion

Geostaionary Doppler Weather Radar is a novel instrument concept to provide detailed monitoring of hurricanes and the temporal information needed for creating advanced flood and hazard warning systems. However, low horizontal resolution more than 12 kilometers probably results in large biases on rainfall retrieval ,especially for small scale convective rainfall system. In this paper, some technology of PIA and radar reflectivity measurement calculation and data simulation are introduced to evaluate the effects of NUBF. lots of APR-2 data including stratiform and convective rainfall system over ocean were selected carefully to simulate NIS data and to calculate their PIA value and radar reflectivity under the condition of uniform or nonuniform beam filling in the top and bottom of rainfall . Results show that with the serious precipitation attenuation and nonuniform distribution of rainfall intensity within NIS footprint, the biases of PIA value and radar reflectivity measurement Z will be much larger than the rainfall measurement accuracy. Therefore, some NUBF correction measure as used in TRMM or resolution enhancement methods as used in spaceborne scatterometers must be adopted to decrease the effects of NUBF and improve the measurement accuracy on rainfall retrieval for NIS .

References

1. Nakamura, K.: Biases of rain retrieval algorithms for spaceborne radar caused by nonuniformity of rain. J. Atmos. Oceanic. Technol. 8, 363–373 (1991)
2. Durden, S.L., Haddad, Z.S., Kitiyakara, A., Li, F.K.: Effects of nonuniform beam filling on rainfall retrieval for the TRMM precipitation radar. J. Atmos. Ocean. Technol. 15(3), 635–646 (1998)
3. Durden, S.L., Tanelli, S.: Predicted Effects of Nonuniform Beam Filling on GPM Radar Data. IEEE Trans. Geosci. Remote Sens. Let. 5(2), 308–310 (2008)
4. Tanelli, S., Fang, H., Durden, S. L., Im, E., Rahmat-samii, Y.: Prospects for geostationary doppler weather radar. In: Proc. of RADAR Conference, Pasadena, CA, May 4-8 (2009)
5. Im, E., Smith, E.A., Chandrasekar, V.C., Chen, S.S., Holland, G.J., Kakar, R.K., Tanelli, S., Marks, F.D., Tripoli, G.J.: Workshop Report on NEXRAD-In-Space - A Geostationary Satellite Doppler Weather Radar for Hurricane Studies. In: Proc. of AMS 33rd Radar Meteorology Conference, Cairns, Australia, August 6-10 (2007)
6. Sadowy, G.A., Berkunand, A.C., Durden, S.L.: Development of an advanced airborne precipitation radar. Microw. J. 46(1), 84–98 (2003)

AOT Retrieval Based on CR Algorithm Integrated with BRDF Model in Urban

Lin Sun and Feifei Xu

Geomatics College,
Shandong University of Science and Technology,
Qingdao, China
Sunlin6@126.com, xufei048@163.com

Abstract. In this paper, the urban BRDF model was applied to AOT inversion with the improved contrast reduction (CR) algorithm. The measured data from AERONET stations has been collected to validate the algorithm and evaluate the accuracy. Results show the AOT inversion can reach a high precision.

Keywords: AOT, urban areas, CR algorithm.

1 Introduction

Aerosols play an important role in Earth's atmosphere, radiation budget, and global climate changes, which can be broadly classified into direct radiative effects and indirect effects[8,24]. Atmospheric aerosol particles can serve as condensation nuclei for the formation of both, atmospheric ice particles and cloud droplets [6,12]. Meanwhile, human health is directly endangered after the inhalable particulate, such as PM10 and PM2.5, are being inhaled and accumulated in the respiratory system[3]. Therefore, accurate monitoring of aerosol information has important scientific and practical significance in the research of climate changes and environmental pollution.

The radiance received by a satellite L can be described by:

$$L(\tau_a,\mu_s,\mu_v,\phi)=L_0(\tau_a,\mu_s,\mu_v,\phi)+F_d(\tau_a,\mu_s)T(\tau_a,\mu_v)\rho/[1-s(\tau_a)\rho] \quad (1)$$

where L_0 is the atmospheric path radiance, τ_a is the aerosol optical thickness, μ_s is the cosine of solar zenith angle, μ_v is the cosine of the view zenith angle, ϕ is the relative azimuth angel between solar and the instrument. F_d is the downward flux, T is the upward transmission, ρ is the surface reflectance, $s(\tau_a)$ is the spherical albedo. So the satellite measured signal is a composite of reflection of sunlight by the variable surface cover and backscattering by the semitransparent aerosol layer, if the reflectance of the surface is low, the radiance received by a satellite increases quickly with the loading of the aerosol increase. Therefore the errors in deriving the aerosol properties are expected to the smaller in this condition. Thus it is appropriate to use the dark pixels in the image to estimate the aerosol properties. Many authors have extensively investigated aerosol properties over ocean using satellite measurements as

D. Zeng (Ed.): Advances in Computer Science and Engineering, AISC 141, pp. 373–377.
springer link.com

well as over dark vegetated areas, and most of them have a high precision. The present approach for remote sensing of aerosol over the land from EOS-MODIS is based on detection of aerosol over dark surface covers. Yet in the urban area, for the bright surface and the complex structure, retrieving aerosol properties from satellite remote sensing is still a challenge. Because over these regions, the surface contribution to the radiance received by a satellite is larger than that over ocean and vegetated areas, yet in such area, it is hard to estimate the surface reflectance for the complex structure. Because of these difficulties, the resulting coverage of aerosol retrievals over urban areas is lacking in the current Moderate Resolution Imaging Spectroradiometer (MODIS) operational algorithm. However, remote sensing of aerosol over the urban area is important because Urban is the regions where a large part of the human activity takes place, and urban aerosol is considered as one of the main anthropogenic sources of aerosol.

To derive the aerosol properties over bright surface in high precision, Tanré et al provide a method called CR which determination aerosol properties based on atmospheric transmission. From (1) the difference of apparent radiance $\Delta L^*_{i,j}$ between two adjacent pixels (i, j) and $(i, j+1)$, where i and j are the geographical coordinates expressed in line and column numbers, is related to the actual ground reflectance difference $\Delta\rho_{i,j}$ by

$$\Delta L^*_{i,j}(\tau_a, \mu_s, \mu_v) = \Delta\rho_{i,j} \frac{T(\tau_a, \mu_v) F_d(\tau_a, \mu_s)}{1 - <\rho> s(\tau_a) \cdot 2} \tag{2}$$

Where $<\rho>$ is the mean reflectance of the two pixels.

If the method is applied to a group of images that includes a relatively clear image, the actual $\Delta\rho_{i,j}$ can be estimated from the clear image, and the optical thickness for each of the images can be derived from (2).

The CR method is applied to a group of images that includes a relatively clear image, in every image, the target is assumed to be invariant. The aerosol optical thickness can be calculated through the estimation of the variation in the transmission. The variation in the transmission is determined from the variation of the difference between the radiance from pixels located a specified distance apart. This method was developed and applied to TM(Tanré et al) and the AVHRR(Holben et al. 1992) images taken over arid regions, but untried in the urban area. In this paper, we will improve the method to adapt to the urban area.

Present Algorithms for retrieving the aerosol properties assume that the surface is Lambertian surface. The error resulting from this approximation is small for dark surfaces used to derive the aerosol optical thickness. But in the urban area, for the high reflectance and large heterogeneity, Lambertian assumption will lead to a large error. It is necessary, therefore, to develop BRDF model to estimate the urban areas reflectance accurately.

In this paper, the CR algorithm was improved and used to inverse aerosol optical thickness in urban areas from MODIS data. To reduce the affection of the character of bidirectional reflectance in aerosol optical thickness retrieving with CR algorithm, the BRDF model for urban areas is developed and applied in aerosol optical thickness inversing. To verify this approach, we compare the aerosol optical thickness derived from the MODIS data with ground-based sun photometer measurement and the API (air pollution index) measured by the environmental station.

2 BRDF Model Construction

Since 1980s', researches on the bidirectional reflectance characteristics have been gradually flourished, there has been a resurgence of interest in other kind of surface, such as water surface, soil and man made surfaces (Li, et al, 1995 ; Hu, B.et al, 1997). But in urban area, buildings, rivers, airports, train stations etc distribute irregularly, the structure is very complicated. It is quite a challenge to design an urban BRDF model. Meister et al (2001) assume the urban was composed by a street structure, and construct an urban BRDF model on a scale of more than 0.5km. The urban was defined of a street with buildings along the both sides, the roofs of the buildings are assumed to be flat, the street structure is straight and indefinitely long. To reduce the number of free parameters, here, assumption the sum of the road wide, a, and the building, b equal 1, a, b and the height of the building c are all relative value. To further simplify the approach, assume that all components (roofs, streets and walls) have the same reflectance. To avoid considering the trees complex 3-D shape, Meister restrict the determination of the BRDF of vegetated urban surfaces to grass (lawns). In this paper, based on the Meister's urban BRDF model, we form the Beijing city BRDF model on a scale of 0.25km, and apply it to the aerosol optical thickness deriving form the data of MODIS band1.

Based on the above assumption, the areas include the top horizontal areas, vertical areas and the bottom horizontal areas that was illuminated and viewed can be calculated by the geometrical optics, and then the reflectance of the pixel is determined.

In the urban BRDF model, the BRDF of the surfaces itself also was considering. The surface in the urban area was divide into two types, the vegetation and man made surface. Trees are difficult to integrate into the urban BRDF model, because of their complex 3-D shape and their unknown BRDF, thus restrict the determination of the BRF of vegetated urban surfaces to grass (lawns). A very convenient analytical empirical function was given to characterize the BRDF of the grass (lawns). The BRDF of the man-made surfaces was characterized by the database measured by the G.Mester's group and described in. Details can be seen in [G.Mester 2001].

3 AOT Retrieval Algorithm

To accurately express the contrast character of the target, structure function concept was proposed and the method was improved. In different area from different satellite, different structure function was defined. Holben et al defined the following structure function to derive the aerosol optical thickness over arid and semiarid areas from AVHRR data which has a spatial resolution of 1.1km. Gin-Rong Liu improved the structure function as formula 3 to increase the stabilization of the value of structure function. To adapt the CR method to retrieve the aerosol optical thickness over the urban area where is large heterogeneity, we provide a new structure function. Simulation indicates that it is more stabilization that the above two.

$$M^2(d)=\frac{1}{(d_{\max}-d_{\min})^2(n-d_{\min})(m-d_{\min})}\sum_{i=1}^{n-d}\sum_{j=1}^{m-d}\sum_{d_j=d_{\min}}^{d_j=d_{\max}}\sum_{d_i=d_{\min}}^{d_i=d_{\max}}\left[\left(\rho_{i,j}-\rho_{i+d_i,j+d_j}\right)^2\right] \tag{3}$$

To simplify the deriving process, a look up table was constructed to retrieve the aerosol optical thickness with MODIS channel 1. The look up table is a file in which multidimensional arrays of theoretical $\Delta L^{*}_{i,j}$ for all $\Delta\rho_{i,j}$, geometrical condition, and aerosol optical thickness are stored, here, and we used the radiative transfer model MODTRAN4 to compute the look up table. In the MODTRAN4, we chose a mid-latitude winter atmosphere and the aerosol model describe before for the region study.

4 Validation

Six MODIS images around Beijing city was selected, whose time is very close (2006/10/6, 2006/10/6, 2006/10/13, 2006/10/14, 2006/10/16, 2006/10/18). The selections have different geometrical condition. The images were matched with 0.3 pixel accuracy to have the same geographical area for different days. From the simultaneous ground measurements, we determined the image of 2006/10/13 as the clearest image, and use it as the reference day.

The retrieved aerosol optical thickness from MODIS data are compared with the ground measurement (Fig 1). The comparisons show that the aerosol optical thickness is retrieved within 0.1 when the aerosol optical thickness is low, and the relative error is less than 25% when it is high.

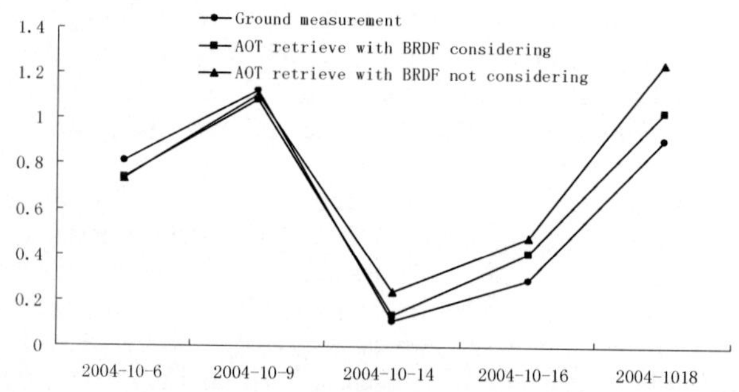

Fig. 1. Aerosol optical thickness at 550 nm obtained from the satellite data and from the ground

5 Conclusion and Discussion

An urban BRDF model was constructed, and applied to the aerosol optical thickness deriving in this paper. Compared with the Lambertian assumption, it can improve the aerosol optical thickness deriving. But for the structure of the urban surface is very complex, to simplify the process, we assume that the urban structure is a street structure and all components (roofs, streets and walls) have the same reflectance, this is a unrealistic assumption, and can lead to a significant error, and so affect the precision of the aerosol optical thickness retrieving.

Aerosol model which describes the range of the effective radius and single-scattering albedo, is also a main factor to cause errors in aerosol optical thickness retrieving. In this paper, we used the aerosol model measured by Jun Hua Zhang in Beijing. In fact, the aerosol model is very complex in the urban area, for the source is complicated.

In the aerosol optical thickness retrieving, using CR algorithm, a clear image should be selected first as the reference image. The precision of the aerosol optical thickness estimation and the atmospheric correction is directly affected the results deriving form other images.

Acknowledgement. Supported by National Natural Science Foundation of China (40701112); National Scientific and Technology support projects (2008BAC34B03).

References

1. Meister, G., Rothkirch, A., Spitzer, H.: Large-scale Bidirectional Reflectance Model for Urban Areas. IEEE Transactions on Geoscience and Remote Sensing 39(9), 1927–1942 (2001)
2. Levy, R.C., Remer, L.A., et al.: Remote Second-generation operational algorithm: retrieval of aerosol properties over land from inversion of Moderate Resolution Imaging Spectroradiometer spectral reflectance. J. Geophys. Res. 112, D13211 (2007)
3. Sunlin: Remote Sensing of Aerosols over Urban Areas. Ph.D. dissertation, Institute of Remote Sensing Application, Chinese Academy of Sciences (2006)
4. Tanré, D., Deschamps, P.Y., Devaux, C., Herman, M.: Estimation of Saharan aerosol optical thickness from blurring effects in thematic mapper data. J. Geophys. Res. 93, 15955–15964 (1988)
5. Hsu, C.N., Tsay, S.-C., King, M.D., Herma, J.R.: Aerosol Properties Over Bright-Reflecting. IEEE Trans. Geosci. Remote Sens. 42(3), 557–569 (2004)
6. Levy, R.C., Remer, L.A., Mattoo, S., Eric, F., Kaufman, Y.J.: Remote Second-generation operational algorithm: Retrieval of aerosol properties over land from inversion of Moderate Resolution Imaging Spectroradiometer spectral reflectance. J. Geophys. Res. 112, D13211 (2007)
7. Levy, R.C., Remer, L.A., Dubovik, O.: Global aerosol optical properties and application to MODIS aerosol retrieval over land. J. Geophys. Res. 112, D13210 (2007)

The BRDF Model Construction and Application in Urban Areas

Feifei Xu and Lin Sun*

Geomatics College,
Shandong University of Science and Technology,
Qingdao, China
xufei048@163.com, Sunlin6@126.com

Abstract. Albedo can be simulated through constructing the albedo calculation model in the three-dimensional urban structure. The paper utilizes the street structure BRDF model to simulate the BRDF in the urban areas. Based on the statistical methods, we determine the best close to the actual city building height, width and density and propose a new method to calculate the vegetation coverage which is an important parameter in simulating the BRDF in the study area. Based above, we simulate the reflectance in the actual area in the NIR band using multi-angular data. The simulated results compared with the MODIS surface reflectance product data in the same urban areas and the same band. The results show that the relative error is less than 6% with the angular change. It will be helpful to further study the BRDF in the urban area.

Keywords: BRDF, urban configuration, Geometrical optical models, vegetation coverage.

1 Introduction

In recent years, there is a primary topic of the urban area study that is improving the climates and the energy-balance in the cities. The bidirectional model can correct the surface bidirectional effects in time series of satellite observations, where both sun and viewing angles are varying [1]. The structures of the urban areas is very complex, there are many high and low buildings, criss-cross of roads. It is difficult to determine the surface reflectance precisely, especially the high buildings, make the characteristic of directional reflectance very clearly. All the above, makes the parameters of the urban area (such as AOT, LST .etc.) retrieval using remote sensing technology become difficult. To solve the problem, many efforts have been done to calculate the BRDF of the urban areas. The studies of bidirectional reflectance model of the surface features internationally have increased actively and the underlying surface types involving have also been growing continually [2]. The calculations of the urban albedo now almost based on the geo-optical model. There are Gerhard Meister,

* Corresponding author.

D. Zeng (Ed.): Advances in Computer Science and Engineering, AISC 141, pp. 379–386.
springer link.com

et al [3] establishing a large scale urban BRF model which is based on modeling on three different spatial scales: scattering on streetlike structures on a large scale, on the intermediate scale, and on the microscale and Parichart Chimklai, et al [4] establishing a albedo Calculator to calculate the BRDF of the urban through two types of urban configurations that buildings having the same height and buildings having two different heights.

So far, it is generally agreed that the urban components plays a key role in the urban albedo. However, Masaru Aida [5, 6] had made a model experiment to examine the effect of surface irregularity of an urban structure on the anomalous absorption of incident solar radiation seeing the urban albedo as a function of the urban structure and used a two-dimensional urban block-canyon array model to observe the effect of urban surface structure on albedo and performed numerical simulations getting calculated results of albedo. Kondo [7] based on Monte Carlo photon tracking method had determined the influence of the urban canopy configuration on albedo and net radiation flux using a three-dimensional radiation model. Because of the large heterogeneity, the homogeneity needed to properly define a BRDF [8] .Recent studies tend to combine muilt-scale structures in the urban configuration. Gerhard Meister [9] on a small scale with sample areas about 20cm×20cm,as well as the measurement and modeling of the BRDF of urban areas for spaceborne sensors with a pixel size of about 1km×1km. However, for simulation and application in the special urban area, there are some undetermined factors and the simulating BRDF of the urban area has low accuracy. Our goal is to improve these relative parameters to make the model apply to the actual area.

2 BRDF Model Description and Parameters Determination in Urban Areas

2.1 Model Description

Building BRDF model is based on Meister, et al [9] from the city streets structure to the surface of the grass and buildings built the BRDF model of the urban area on the different scales. It is mainly applying the results of BRDF model on the microscale in recent years to construct the BRDF model of the urban structure. And the model contains only three adjustable parameters describing the surface and can potentially be included in an algorithm of processing and correction of a time series of remote sensing data. The model considers that the observed surface bidirectional reflectance is the sum of two main processes operating at a local scale: a diffuse reflection component taking into account the geometrical structure of opaque reflectors on the surface, and shadowing effects; a volume scattering contribution by a collection of dispersed facets which simulates the volume scattering properties of canopies and bare soils. The core of constructed the model is to simply the main structure of urban area to the structure which has a street in the middle of two arrays building (e.g. Fig.1).

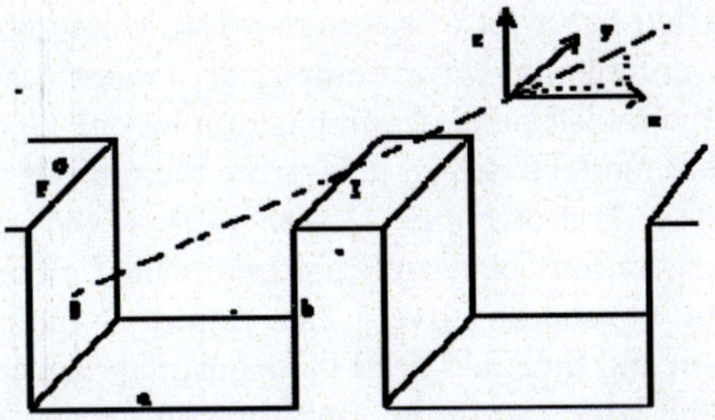

Fig. 1. The sketch of street structure In the paper, we did some improvements to the model. That made the model more practical.

2.2 Urban Building Height Determination

According to the characteristics of the urban configurations, based on the geometric optics principles, we calculate the BRDF in the urban areas. There are three adjusted parameters, including the width of the buildings and roads and the height of the buildings. In the practical urban areas there is different relative height of building. According to the characteristics of building distribution in Beijing, based on statistical methods, we determined the best close to the building height, width and density in Beijing. We assume the building height and street width as 0.2, 0.4, 0.5, 0.6, 0.8, five levels approximate.

Fig. 2. Adjust parameter to model street structure

Fig.2 show the parameters (street width a and building height b) various combinations model a city. From the figure we get that it is sufficient to simulate a city with the average value a=0.5, b=0.5. We conclude that in the solar Zenith angle 38°and the sensor Zenith angle 53°the simulated results compare with the MODIS surface reflectance data in NIR band, the relative error is less than 2% in the Beijing city.

2.3 The Method of Vegetation Coverage Determination

The vegetation cover is usually defined as the ratio of the vertical projection area of greenery and the area of the total surface. For the estimation of vegetation coverage area there has been carried out a lot of research work. Using remote sensing data to extract vegetation cover mainly have: empirical model law, vegetation index conversion model and the mixed pixel decomposition model. The empirical model by establishing the empirical model between measuring the ratio of vegetation cover area and the vegetation index. For example, Gratez [10] in the semi-acid soil region measured the ratio of vegetation cover area and established a linear regression model to estimate the ratio of vegetation cover. The empirical model law has a strong dependence on the region and time and need the measured data to special region, so it is difficult to promote the wide use in the long time and large scale[10][11][12]; the basic method of vegetation index conversion is through the analysis of the vegetation types and distribution characters in each pixel established the conversion of vegetation index and the ratio of vegetation cover area to extract directly the information about the ratio of vegetation cover area[13][14]. Use the normalized difference vegetation index (NDVI),

$$NDVI = \frac{\rho_{ir} - \rho_r}{\rho_{ir} + \rho_r} \tag{1}$$

ρ_{ir}, ρ_r Separately stand for the reflectance of NIR and red bands, Quamby [13] using AVHRR data established the linear mixed conversion model of vegetation index and the ratio of vegetation cover area. Gutman [14] from the vegetation distribution character in the pixel proposed the uniform sub-pixel model and the mixed sub-pixel model and utilized the NOAA AVHRR data to apply of the equal density model to estimate the ratio of global vegetation cover area. Although the vegetation index method didn't require measuring the large areas of ground samples, because the NDVI saturation is low [15], the largest NDVI value did not necessarily represent pure vegetation pixel. Mixed pixel decomposition is based on the reflecting of vegetation and background in different spectral bands to extract the vegetation cover. For example, Tang Shihao [16] proposed to make use of three-band gradient difference vegetation index (TGDVI) to estimate the ratio of vegetation cover area,

$$A = TGDVI \Big/ \left[\frac{\rho_{Vir} - \rho_{Vr}}{\lambda_{ir} - \lambda_r} - \frac{\rho_{Vr} - \rho_{Vg}}{\lambda_i - \lambda_g}\right] \tag{2}$$

Where,

$$\begin{cases} TGDVI = \dfrac{\rho_{ir} - \rho_r}{\lambda_{ir} - \lambda_r} - \dfrac{\rho_r - \rho_g}{\lambda_i - \lambda_g} \\ TGDVI = 0, \text{ if } TGDVI < 0 \end{cases} \tag{3}$$

Where ρ_{ir}, ρ_r and ρ_g separately stand for the reflectance of NIR, red, green band; ρ_v and ρ_s separately is the corresponding band's vegetation and soil reflectance; A is the ratio of the vegetation cover area, λ_{ir} λ_r and λ_g is the wavelength of corresponding band.

The current mixed pixel decomposition method is mainly based on two surface types of vegetation and soil to calculate the vegetation coverage. For urban area, the surface types mainly are building surface of cement, asphalt roads and other artificial surface, whose spectral characteristics have great difference with the soil. In red and

NIR bands, assume that the reflectance of the pixel is the weight of the area of vegetation and asphalt road in the pixel. Each surface reflectance in red and NIR bands is:

$$\rho_{ir} = A * \rho_{vir} + (1 - A) * \rho_{cir} \tag{4}$$

$$\rho_{r} = A * \rho_{vr} + (1 - A) * \rho_{cr} \tag{5}$$

Where, ρ_{ir} and ρ_r is separately the pixel reflectance of NIR and red bands; ρ_v and ρ_c is the reflectance of vegetation and asphalt roads or cement buildings of corresponding bands. So:

$$\rho_{ir} - \rho_r = A * (\rho_{vir} - \rho_{vr}) + (1 - A) * (\rho_{cir} - \rho_{cr}) \tag{6}$$

As in the red and NIR bands the surface reflectance of asphalt roads (or concrete buildings) are same, $\rho_{cir} - \rho_{cr}$ can be approximated zero, the ratio of vegetation coverage can be approximated:

$$A = \frac{\rho_i - \rho_r}{\rho_{vir} - \rho_{vr}} \tag{7}$$

In the calculation, we extract the pure vegetation reflectance in the red and NIR bands.

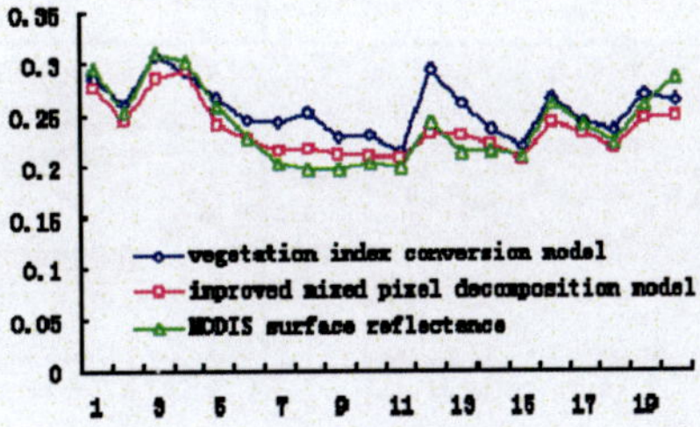

Fig. 3. Reflectance curve of vegetation coverage

Using the method of vegetation cover, we simulated the reflectance of NIR band, whose center wavelength is 0.86 nm. From fig.3 we see the results by improved mixed pixel decomposition model are closer to the MODIS surface reflectance. So using the determined vegetation coverage to calculate the BRDF apply to simulate the BRDF in Beijing city.

3 Application of BRDF Model

We choose six different scenes MOD09 surface reflectance data to compare with the simulating results. The daily angular situation showed in Table 1.

Table 1. The angular parameters situation in daily MOD09

days	Solar Zenith	Sensor Zenith	Relative Azimuth
244	38.7575	30.4060	43.2565
245	38.7575	53.4235	43.2565
246	34.4083	10.1957	75.1259
249	33.6867	51.4718	95.1794
324	59.8528	30.1910	94.9668
325	60.7098	53.5681	73.5576

We extract a representative row in each image, so the middle part stand for buildings in urban area and the both sides stand for the vegetation. Using the row data we analyze the reflectance vary in the simulating results. The simulating results comparing the MOD09 data showed in Fig.4, the ordinate axis stand for the reflectance and the horizontal axis stand for the pixel points of the row in the simulating image and MOD09 image in NIR band.

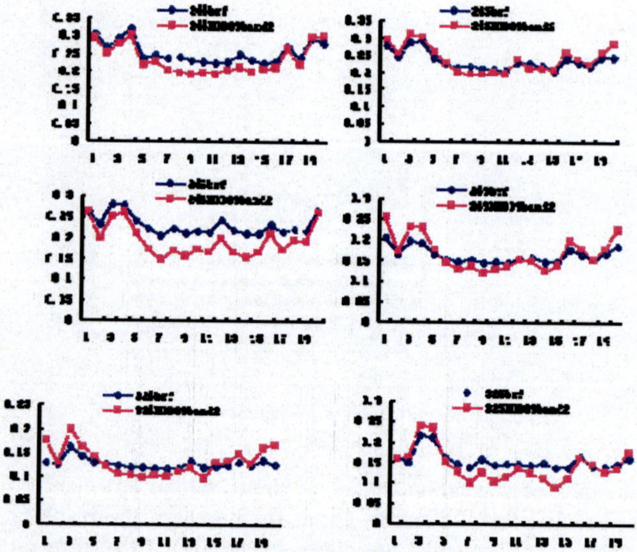

Fig. 4. Comparison of simulating BRF and MOD09 data from Beijing

4 Analysis

Fig.4 show in muti-angular conditions the simulating BRF results compare with the MOD09 data in the same area of Beijing. In the figure, we see the simulating reflectance is difference between the MOD09 surface reflectance. This is because in the simulating results we take into account about the directional reflectance, the

MOD09 surface reflectance data only simply consider. From 246, 244, 249, 245, we can see the correspondence of the reflectance of vegetation and buildings that in the solar zenith angle similar conditions, with the increase in the sensor zenith angle, the buildings error is less and less but the vegetation error is larger and larger. This can be caused by the hotspot.

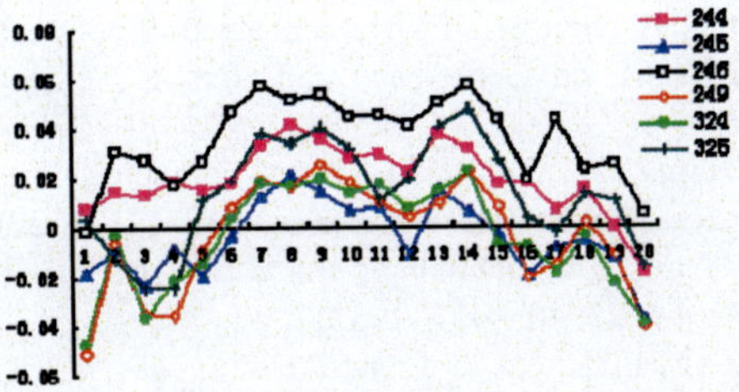

Fig. 5. The relative error with the angular change

Among the effects which occur as solar and observer geometry change is the hotspot effect. The canopy hotspots occurs when the solar and viewer are aligned so that there is no visible shadow, and thus the scene appears bright [17]. The magnitude and behavior of this phenomenon as viewing and illumination positions change provides important information about canopy structure. In particular, the angular change in brightness is a function of the relative size and spacing of objects making up the canopy. For the solar illumining, there is a block between the buildings. Hotpots effect will cast the shadow. With the angle increasing, the shadow between buildings will reduce, the reflectances strengthen, and the relative error will reduce (as showed in Fig.5). From the Fig.5 we see t using the multi-angel data to simulate the BRDF in the urban area; the relative error is less than 6% as the angles change.

5 Conclusion

The albedo determination in the urban area has been an issue in study. Through improving the BRDF street model, we simulate the BRDF in Beijing under multi-angular conditions in the NIR band. The simulated results compare with the MODIS surface reflectance product data. The comparing results show that using the multi-angel data to simulate the BRDF in the urban area is feasible and the relative error is less than 6% as the angles change. The further work we should focus on improving to establish a new model that can be more suitable for specific cities and improving the accuracy. It will be beneficial to serve the other quantitative inversion.

Acknowledgement. Supported by National Natural Science Foundation of China (40701112); National Scientific and Technology support projects (2008BAC34B03); Innovation Fund For Technology in Shandong University of Science and Technology (YCA100204).

References

1. Roujean, J.-L.: A bidirectional reflectance model of the earth's surface for the correction of remote sensing data. Journal of Geophysical Research 97(D18), 20,455–20,468 (1992)
2. Li, X., Strahler, A.H.: Geometric-optical bidirectional reflectance modeling of a conifer forest canopy. IEEE Trans. on Geoscience and Remote Sensing GE-24(6), 906–919 (1985)
3. Meister, G., Rothkirch, A., Spitzer, H.: Large-Scale Bidirectional Reflectance Model for Urban Areas. IEEE Trans. on Geoscience and Remote Sensing GE-39(9), 1927–1942 (2001)
4. Chimklai, P., Hagishima, A., Tanimoto, J.: A computer system to support Albedo calculation in urban area. Building and Environment 39, 1213–1221 (2004)
5. Aida, M.: Urban Albedo as a function of the urban structure - A model experiment. Boundary-Layer Meteorol. 23, 405–413 (1982)
6. Aida, M., Gotoh, M.: Urban Albedo as a function of the urban structure - A two dimension numerical simulation. Boundary-Layer Meteoral. 23, 416–424 (1982)
7. Kondo, A., Ueno, M., Kago, A., Yamaguchi, K.: The influence of urban canopy configuration on urban Albedo. Boundary-Layer Meteorology 100, 225–242 (2001)
8. Nicodemus, F.E., Richmond, J.C., Hsia, J.J., Ginsberg, I.W., Limperis, T.: Geometric considerations and nomenclature for reflectance, Monograph 161. National Bureau of Standards, US (1977)
9. Meister, G., Rothkirch, A., Spitzer, H., Bienlein, J.: BRDF field studies for Remote Sensing of Urban Areas. Remote Sensing Reviews 19(1), 37–57 (2000)
10. Graetz, R.D., Pech, R.R., Davis, A.W.: The assessment and monitoring of sparsely vegetated rang lands using calibrated Land sat data. International Journal of Remote Sensing 9(7), 1201–1222 (1988)
11. Dymond, J.R., Stephens, P.R., Newsome, P.F., Wilde, R.H.: Percent vegetation cover of a degrading rangeland from SPOT. International Journal of Remote Sensing 13(11), 1999–2007 (1992)
12. Wittich, K.P., Hansing, O.: Area-averaged vegetative cover fraction estimated from satellite data. International Journal of Biometerology 38(3), 209–215 (1995)
13. Quamby, N.A., Townshend, J.R.G., Settle, J.J., White, K.H., Milnes, M., Hindle, T.L., Silleos, N.: Linear mixture modeling applied to AHVRR data for crop area estimation. International Journal of Remote Sensing 13(3), 415–425 (1992)
14. Gutman, G., Ignatov, A.: The derivation of the green vegetation fraction from NOAA/AVHRR data for use in nurderical weather prediction models. International Journal of Remote Sensing 19(8), 1533–1543 (1998)
15. Xu, X.: Remote Sensing Physics. Beijing University Press (2005)
16. Tang, S., Zhu, Q., Wang, J., Zhou, Y., Zhao, F.: Principle and application of three-band gradient difference vegetation index. Chinese Science 48(2), 241–249 (2005)
17. Jupp, D.L.B., Strahler, A.H.: A hotspot model for leaf canopie. Remote Sensing 38, 193–210 (1991)

Spatial Distribution of Residential Prices Basing on RS and GIS in Harbin

Zhang Xinle and Liu Huanjun*

Resource and Environment Institute of Northeast Agricultural University Harbin, China
{huanjunliu,zhangxinle}@gmail.com

Abstract. To provide information and theoretical basis for government to formulate reasonable land use and urban planning, guide residential price, regulate real estate investment structure, consumers to choose resident, and to provide empirical instance for country macroscopical adjustment and control real estate operation, Harbin city was selected as study area, with ArcGIS and SPSS software as platform, basing on spatial analysis theory and geostatistics, RS and GIS were integrated to study residential location and price spatial distribution. The spatial autocorrelation of sampling points was analysed. Then, spatial interpolation was made, the spatial distribution of residential price was analysed. Finally, the results were classified to study spatial variation further. The results show that residential price in Harbin has significant spatial autocorrelation, the average decreasing order of districts is Nangang, Xiangfang, Daoli, Daowai, Songbei, Pingfang, Hulan; residential prices of Songbei rise significantly; the railway through the city has negative impact; spatial variation of residential price in Daoli district is remarkable and relatively equable in Xiangfang district; the original spatial pattern of residential price has been broken, and the city is developing from the single-core city to multi-core city. Spatial pattern of residential price will be further changed with the industrial layout adjustment.

Keywords: Harbin; residential prices; spatial characteristics.

1 Introduction

The rise of real estate is an important factor to stimulate rapid economic development in the 20th century. With the advance of China's urbanization process and the deepening of housing reform, residential construction is becoming a new growth point of national economy [1]. In recent years, China's commodity residential price continued to rise in the situation comprehensivly, residential price hit record high, has aroused widespread concern in industry and society, it is to become the focus of government, enterprises and individuals[2]. City commercial residential price, as regional conditions and other factors, the spatial distribution and time series has a strong difference. Revealing the spatial differentiation of residential price level can help the government manage grasp the distribution trend of residential price, macroscopical adjustment and control [3,4].

* Corresponding author.

D. Zeng (Ed.): Advances in Computer Science and Engineering, AISC 141, pp. 387–394.
springer link.com

In the past, studies of first-line cities' residential price are more, studies of second-tier cities' are relatively smaller. However, the second and third line cities has a huge development potential, residential price is rising faster in recent years, real estate developers and other real estate speculators began to enter these cities. Therefore, studing on urban housing price of second and third line has important scientific and practical significance to predict Chinese housing price trend and guide macroscopical adjustment and control for government to housing industry[5]. Harbin as a second-tier cities, its residential price is lower than average domestic price, but for the residents' purchasing power of Harbin is already a high burden on consumption. The super proportion of loans phenomenon is serious. And, residential price increased rapidly in the year of 2007 to 2009 in Harbin, which was well above expectations. In this paper, using the sales ordinary commercial housing data of Harbin City, in 2009, taking ArcGIS and SPSS software as platform, basing on spatial analysis theory and geostatistics [6,7], integrating RS and GIS technology to study residential location and price spatial distribution. Look forward to provide information and theoretical basis for government to formulate reasonable land use and urban planning, guide residential price, regulate real estate investment structure, consumers to choose resident; provide empirical instance for country macroscopical adjustment and control and real estate operation.

2 Study Area

Harbin is the capital of Heilongjiang province, located in 126.45° -126.81°E, latitude 45.56° -45.93°N, the center of Northeast Asia, and known as the pearl of the Eurasia continental bridge, the first Eurasia bridge and an important hinge for air corridor. Climate is temperate continental monsoon climate. Its winter is long, summer is short. It is called "city of ice". HarDa, SuiBin, Binzhou, Binbei, Labin are the mainly railways in this area connecting other cities. Songhua river flows through the city. Harbin is the largest provincial cities in China, the second largest population. The city's land area is 53,100 km^2, of which the urban area is 7086 km^2, built area is 302.41 km^2, including 8 section and 10 counties (cities). In this paper, Nangang, Xiangfang, Daoli, Daowai, Songbei, Pingfang, Hulan district of Harbin city was chosen as study areas (Fig.1).

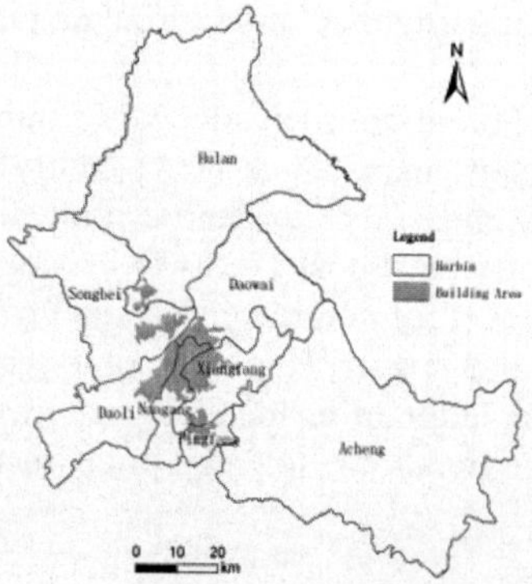

Fig. 1. Location of study area

3 Methods

3.1 Data Source

1) Sou Fang net data: According to Harbin Sou Fang net (http://hrb.soufun.com/) information, taking residential districts as units, operated average of the commercial residential price on December 2009, getting the average price of residential announced by the net. Using GPS for new real estate location, the price of 198 residential/real estate were collected. The sampling points distribution is shown in Fig.2.

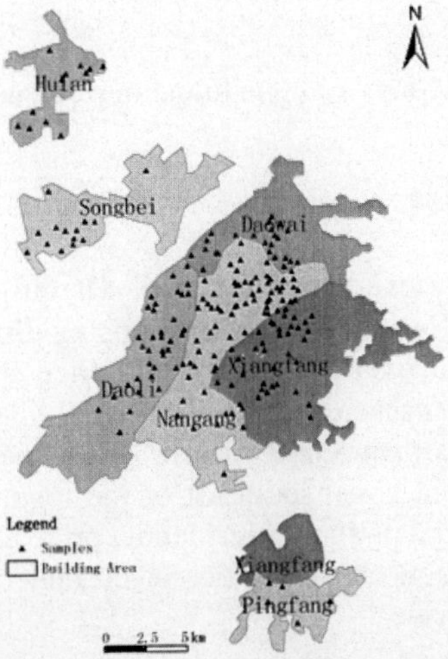

Fig. 2. Harbin city commercial housing prices spatial distribution of sampling points

2) Real estate agent transaction data: The real transaction price data was aqirde from the real estate services company, such as the 21st Century, Yitong conpany in Harbin.

3) Survey data: Based on field observation, interviewing residential communities, and neighborhood management and property management departments conversation and exchange methods, detailed survey was made for the 198 residential quarters.

4) Statistics: Demographic, socio-economic, real estate investment data were taken from statistical yearbook of Harbin and Harbin city statistics information network (http://www.stats-hlheb.gov.cn/).

3.2 Data Processing

1) Determine the residential price: According to the survey on the real estate services companies, there exists more stability relationship between network announced price and transaction price. Therefore, we took the price published by Sou fang net data as analysis target to study the spatial distribution of commercial housing price in Harbin.

2) Statistical analysis of sample data: Though statistics analysis with sample data, we found that the maximal sample price is 25,000 yuan/m^2, the minimum is 2,391 yuan/m^2, the mean is 6213.97 yuan/m^2, the standard deviation is 2011.23. The sampling points data approximate normal distribution.

3) Spatial Data: Using city electronical map, remote sensing (Landsat TM, Quikbird), city traffic map, of Harbin etc., ArcGIS software environment, setting Albers projection, digitized commercial housing, commercial centers, hospitals, schools, government departments, city's main roads, subway (under construction), water (Songhua river), parks, and built up spatial digital database.

3.3 Moran Index

The detailed formulaes of Moran's can be found in references[8-10].

4 Result and Analyse

Through analysis of historical data, by 2006, Harbin depends on the level of residential price from the "Nangang", "Daoli" shopping district in the distance; central avenue, friendship road near river area, residential price is 7000-10000 yuan/m^2, if in area near Dazhi street, Zhongshan road is 6000-8000 yuan/m^2, residential prices of other regions is mainly less than 5,000 yuan/m^2. Now the situation is that the Haxi, Qunli, Songbei, development zone, southeast of the regional formate numerous city sub center, the original spatial pattern of residential price has been broken. For further analysis spatial differentiation characteristics of current residential price in Harbin, the following studies was made.:

4.1 Spatial Autocorrelation Analysis of Residential Price

Spatial autocorrelation indexes of common residential price in Harbin were calculated with spatial autocorrelation analysis tool of ArcGIS. We can see that, Moran's index is 0.1, Z test value is 13.36, shows that residential price has significant spatial autocorrelation, that is, the price of real estate closer the similarity more significant.

Table 1. Spatial autocorrelation analysis of residential prices

Sampling Points	Moran's	Expectations	Z	Sig.
198	0.1	2.58	13.36	0.01

4.2 Characteristics of Residential Price Spatial Variation

1) Interpolation of residential price and analysis

Basing on geostatistical analysis of residential price sampling points, in the ArcGIS software environment, used GIS spatial interpolation to make interpolation of residential price. The residential prices of sampling points fitted the normal distribution from the normal Q-Q plot. And the spatial pattern of residential prices was significant correlation. So, Ordinary kriging[11,12] was used to acquire the spatial distribution of residential prices of the research area from sampling points. The best empirical fitted model of kriging is spherical model. Finally, the spatial distribution result of residential price in built-up area of Harbin City by December 2009 was acquired (Fig.3). Using zonal statistical method with ArcGIS 9.2 analyzed residential price in different districts of interpolation results (Table 2). The analysis results showed as the following distribution rules:

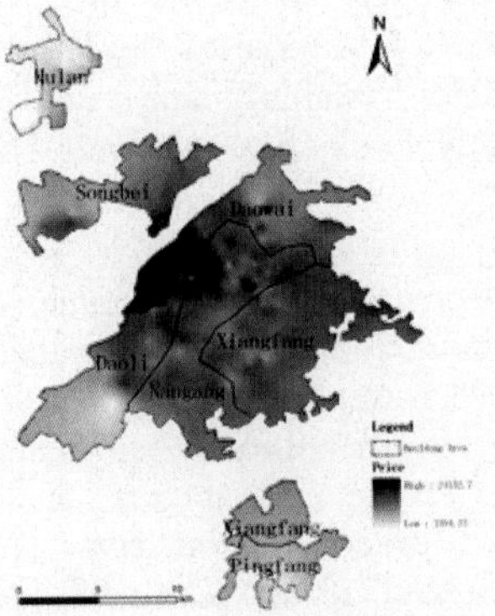

Fig. 3. The price basing on Kriging interpolation distribution of commercial housing

Table 2. Residential price of each district of interpolation results

Districts	Areas(km^2)	Min.	Max.	Range	Average	S.D.
Daoli(DL)	52.51	2812	24153	21341	5938	2448
Nangang(NG)	60.89	4110	11810	7700	6317	999
Xiangfang(XF)	63.16	4635	7314	2679	6099	358
Daowai(DW)	35.22	4228	11151	6923	5865	956
Songbei(SB)	38.64	3936	7975	4039	5248	730
Pingfang(PF)	17.30	2991	4693	1702	4026	392
Hulan(HL)	17.56	2404	5086	2682	3391	559

a) Overall distribution characteristic. Residential price in Harbin City in order to hang on central avenue and Qiulin Company located in Nangang district, museums and other major commercial areas is of high value for the center to the periphery gradient downward trend. The maximum value is found in the vicinity of the friendship road of Daoli District, mainly due to high-end communities such as Hagongguan caused by higher prices; sub-high regional distributs in Central Street business of Daoli District, east Dazhi Street and Fendou road vicinity of Nangang District, the price is more than 10,000 yuan/m^2.

b) Appeared several sub centers. Such as the exhibition center at the junction of Xiangfang and Nangang District, Haxi new region. With "north hopped, southern expanded, center blossomed, counties strengthened" development strategies, Harbin will form more sub-centers, and the role of Qunli new region, Haxi and other sub-centers will continue to be strengthened. We can see from Fig.3 that Harbin City, the area of high residential price distribute in Daoli district and Nangang district; residential price of Xiangfang district and Daowai district is medium; residential price of Hulan, Pingfang and Songbei district is lower.

c) Residential price of each district. Tab.2 shows that residential price in Harbin has significant spatial autocorrelation, the average decreasing order of districts is NG, XF, DL, DW, SB, PF, HL. Although the average of DL is lower than NG and XF, but its maximum value and standard deviation are the largest, denoting that spatial distribution of residential prices there has significant spatial heterogeneity. The residential price difference of XF is relatively smaller, relatively balanced regional development, denoting that it still need to develop regional business centers and actualize cluster effect also.

d) The residential price of the northern Songhua river has been increased obviously. Although the facilities such as schools, hospitals, commercial centers are not perfect, because of the recent construction of infrastructure facilities in Harbin, in particular the improvement of traffic conditions there residential price is taking on rise trend. The "north hopped" development strategies means that a modern new city the forecast will be built up in the northern Songhua river. Now and in near further several traffics being and will be constructed according the plans, such as bridges, tunnels across the river channel. These are bound to further promote SB's building, pushing residential price.

e) The railway through the city has negative impact. It is clear that residential price is lower than surrounding areas, which should arouse the city manager and planning department's attention.

2) Classification analysis of the spatial distribution of residential price

For quantitative analysis spatial distribution pattern of different residential price in different administrative regions, the results of space interpolation was classified. According to statistical analysis results of residential price in each sampling point and the per capita income in Harbin, took 3000,4500,5500,6500,8000,10000 as grading line, and divided residential price into seven levels(Fig.4). Then area of each residential price levels distribution within different district was analyzed.(Table 3)

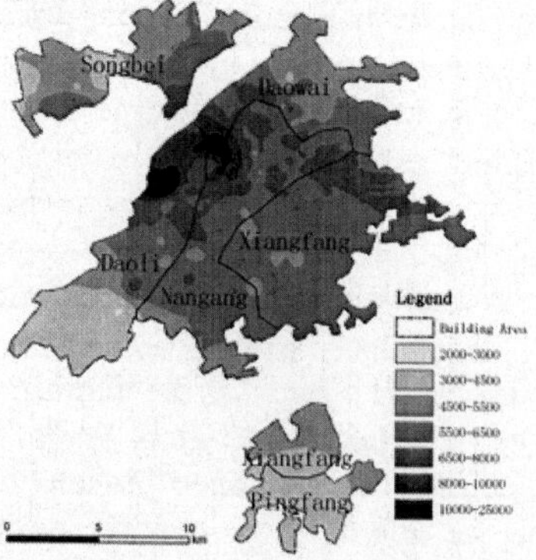

Fig. 4. Grade commercial housing price

Table 3 is the areas rate of residential price of different grades for each district of Harbin City. It can be seen that the high value areas of residential price are mainly in DL and NG. The seven levels of residential price are all distributed in DL district, showing strong spatial heterogeneity. The residential price is relatively uniform in XF, mainly in 5500-6500 yuan/m^2, the total area of 80.99%. Because residential price in the border area of DW with the NG and DL is higher, there are 47.45% of regional residential price in 6500-8000 yuan/m^2.

Table 3. The areas rate of residential price of different grades for each district for each district (%)

Price (1000 Yuan) / Districts	10-25	8-10	6.5-8	5.5-6.5	4-5.5	3-4	2-3
Daoli(DL)	0.24	17.93	30.73	21.46	13.73	10.72	5.20
Nangang(NG)		12.24	60.08	21.67	4.11	1.90	
Xiangfang(XF)			4.07	80.99	14.94		
Daowai(DW)			47.45	37.07	10.61	4.05	0.83
Songbei(SB)			0.08	67.75	27.81	4.35	
Hulan(HL)					22.10	61.93	15.97
Pingfang(PF)					0.03	43.71	56.27

With "north hopped, southern expanded, center blossomed, counties strengthened" development strategies, construction of west railway station, supporting facilities, songhua river bridge crossing the river and railway network, the same urbanization of Hulan and Acheng with the main city of Harbin, migration of provincial government and municipal government, the pattern of the economy, the spatial pattern of commercial residential price of Harbin will be significant changed. Harbin will be developed from the single-core city to multi-core city, namely DL and NG as cores, Huizhan center, Haxi new reagion, SB as sub cores.

In addition, due to residents have become increasingly demanding environment future, it will promote the industrial layout adjustment process. Some of companies of serious pollution sources, including chemical, pharmaceutical and other companies, will be moved out the around city. Spatial pattern of residential price will be further changed.

5 Conclusions

House prices in Harbin has significant spatial autocorrelation. Using spatial autocorrelation analysis tool of ArcGIS9.2 to calculate spatial autocorrelation indexes of common residential price in Harbin and acqired that residential price has significant spatial autocorrelation.

In ArcGIS9.2 software environment, used GIS spatial analysis mathod to make interpolation of residential price.The spatial distribution result of residential price in built-up area of Harbin City by December 2009 was acquired. The average decreasing order of districts is NG, XF, DL, DW, SB, PF, HL. The residential price of the northern Songhua River has been increased obviously. The railway through the city has negative impact.

Classifing the space interpolation results of residential price, to quantitatively analyse spatial distribution pattern of different residential price Spatial heterogeneity of residential price in DL district is strong; in XF is relatively uniform,and relatively balanced regional development, denoting that it still need to develop regional business centers and actualize cluster effect. Spatial pattern of residential price which will be developed from the single-core city to multi-core city has been broken up in Harbin, and with the industrial layout adjustment, its will be further changed.

Acknowledgment. This work was funded by "Science and Technology Research Project of Education Department of Heilongjiang Province (11551038)", Young foundation of Resource and Environment Institute of Northeast Agricultural University, and Project of Black soil Innotation Team of Northeast agricultural university, "National Natural Science Foundation of China (40801167)".

References

1. Liang, Z.: Study on Price Spatial Variation of Residential Price Based on Features Price. Beijing Traffic University, Beijing (2008)
2. Yuzhe, W.: Temporal and Spatial Evolution of Urban Housing Price Exploration Based on GIS and Its Application. Zhejiang University, Hangzhou (2005)
3. Zhiqing, Z.: Commercial Residential Prices Distribution of Guangzhou. Tropical Geography 3, 22–26 (2006)
4. Wei, Z., Jianwei, Z., Xinrui, L.: The Core of Urban Land Management – Management City. Yunnan Geographic and Environment Research 18, 56–59 (2006)
5. Chuanhua, L., Jinghu, P., Jun, Z.: Analysis on Spatial Pattern of Common Residential Price in Lanzhou City Basing on GIS. Yunnan Geographic and Environment Research (2007)
6. Yingjun, S., Jinfeng, W., Yanchen, B.: Study on Advances in methods of Geostatistics. Advances in Earth Science 2, 268–274 (2004)
7. Bin, M., Jingqiu, Z., Jinfeng, W., et al.: Spatial Analysis in The Study of The Real Estate Market. Geography Reserch 26, 956–964
8. Lu, Y., Shanyu, Z., Guifeng, L.: Spatial Analysis of Population Distribution in Shanghai 16, 83–87 (2006)
9. Songlin, Z., Kun, Z.: Comparative Study of Global Spatial Autocorrelation Coefficient Moran's Index and the G Index. Acta Scientiarum Naturalium Universitatis Sunyatseni 46, 93–97 (2007)
10. Sawada, M.: Global Spatial Autocorrelation Indices Moran's I, Geary's and the General Cross Product Statistic (2006), `http://www.Ipc.Uottawa.ca/publications/Moransi/Moran.html`
11. Zhixiong, M., Xia, L.: Spatial Structure of Residential Price in Dongguan City Based on ESDA and Kiging Methods 28, 862–866 (2008)
12. Yonghui, Y., Ziqiang, P., Yingjun, S., et al.: Practical Guide to Geostatistical Analysis of ARCGIS, pp. 8–47. Chinese Book Centre of ESRI, Beijing (2008)

Based on NetLogo Simulation for Credit Risk Management

Minrui Zheng

Faculty of Management and Administration,
Macau University of Science and Technology, Macau, China
Avenida Wai Long, Taipa, Macau
calla420@gmail.com

Abstract. With economic development, the credit card business has entered a stage of rapid development. Using quantitative methods of modeling to analyses and predicts the consumers use the credit card behaviors. It is one of hot project in the experimental economics, and it is important for avoiding credit risk. The essay based on a simple NetLogo simulation model to simulate individual consumers and economic activity, to observe the risk of credit card operation system. The experimental results is, in the different income levels, the cardholders' consuming and withdrawal behaviors would have no limits, then the issuing bank would be set line of credit is not higher twice than their income. In this situation, the cardholders could pay back the money on time. Otherwise, it will have risks for issuing bank. This method is common, so it can be used in many study areas.

Keywords: Credit, risk management, Netlogo, simulation.

1 Introduction

With the quick extension of the credit card business, it has brought huge business opportunities to China's banking industry. At the end of second quarter of 2011[1], the whole country issued 2.674 billion bank cards, including 257 million credit cards, an increase of 24.3%. With improve the environment for accepting bank cards, the cards becoming the most widely non-cash payment instrument used in our life, and the number will keep growing. However, the risk of credit card management is an unprecedented challenge, how to improve the bank's ability to control the risk and make the industry healthy development are our important missions[1].

Experimental economics is a subject based on a certain amount of practical rules, using stimulation method to create a laboratory environment like the real's, through changing the parameters to analysis the experimental results, and it uses in verify the past economic theory or hypotheses. Maybe the subject will find new theory, or giving some decision-makings theoretical analysis. In 2002, the Nobel Prize of Economics was awarded to Vernon Smith, an experimental economist, the experimental economics get more and more attention[2][3].

---

[1] http://www.pbc.gov.cn/publish/goutongjiaoliu/524/2011/20110808155824275687112/20110808155824275687112_.html, 2011-9-29.

D. Zeng (Ed.): Advances in Computer Science and Engineering, AISC 141, pp. 395–401.
Springerlink.com

For the risk of credit card management, some scholars use data mining to analyze one or more time sections, and find the potential risks[4] Also the quantitative model uses in credit cards' approval, in this model, we could segment the card holders[5][6]. We could use experimental economics to build the credit card's data model and predicate risks, to find out strategy of avoiding risks[7]. Zeng Wei[8] has been built a model based on Netlogo, in China's backgrounds to set up some parameters, and discussed whether the cardholder behaviors should be controlled. Pietro Terna with his students uses simulation models for economics to explore the practical application of Netlogo[9]. These reports show that if we use some quantitative methods to analyze and simulate the complex credit card operation system, the loss will be reduced. But in the process of simulate, how to choose best parameters, how to explain the result, that is still a problem.

In view of these facts, this paper will use quantitative model to analyze the commercial bank risk of credit card management, also use Netlogo and data mining. According to Netlogo platform, we using simulation models for economics to observe bank's potential risk in different cardholder's consumptive behaviors, and doing comparative experiment.

2 Designing for Experiment Model

Our simulation experiment is based on Netlogo modeling platform[10][11][12].

Netlogo is developed by the Northwestern University Center for Connected Learning and Computer-Based Modeling, CCL. The purpose is providing a useful and powerful computer-aided tool in scientific research and education. In 1999, Uri Wilensky with the help of the United States National Science Foundation began developing Netlogo. The version 1.0 was released in 2002, the version 2.0 was released in 2003, and the version 3.0 was released in 2005, now the version is 4.1.3. Netlogo is free software based on Java, and can be run in many different platforms. Netlogo is a multi-agent based simulation environment, especially suitable for time-varying of complex cases for simulation. These features of Netlogo make the possibility of connection between microscopical behavior and macroscopical mode becoming real. These macroscopical modes are composed by many individual interactions. Netlogo is a programming environment, and we could create our own models. As a powerful tool, Netlogo application covers natural and sciences fields, such as biology, ecology, geography, chemistry, architecture, computer science and economics and social psychology and so on [13][14].

The simulation experiment observes the risk of credit card operation system in different situations, and adjusts the system immediately. The experiment is proposed to provide a reference for credit risk management of commercial banks.

2.1 Designing for Experiment

2.1.1 The Agent of Experiment

The agent of experiment contains cardholder (consumer) and issuing bank. The numbers of cardholders set a constant number, and issuing bank's behavior as the whole financial system behavior.

According to the computer hardware setting the number of cardholders, but must be more than 100. This trial is 1000.

The cardholder's behaviors include: income, pay by card, withdrawal, repayment. Now we give some assumption: 1) income, cardholder has a steady income. 2) Pay by card, only use in consumption. 3) Withdrawal, in a credit limit. 4) Repayment, cardholder repays monthly, and the amount of repayment should be greater than the minimum payments. Otherwise, the credit card would be freeze. 5) Last-term debt should be pay in next month, also must pay interest. 6) Cardholder's other behaviors (e.g. saving, investment etc.) would exclude in this trail. Cardholder could use previous saving repay the debt.

We make market issuing banks as a whole, which is mean, the whole bank system face the risk from cardholder together. This setting considers that issuing credit card always bank own behavior in China's commercial bank. Cardholders could borrow other bank to repay a given bank's debt, so that their daily trading activities would be satisfied. If we take market issuing banks as a whole, this problem can adjust the total credit limit to be solved.

2.1.2 Experiment Flow

In this simulation experiment, each trail is based on real life credit card consumptive behaviors. Every month cardholder could get income, use credit card consumption, withdrawal and repayment. In use credit card consumption, we ignore the economic behavior between bank and seller. Withdraw is not a part of outflow of cash flow. In reality, the amount of minimum repayment should contain the money used in this month and last-term debt and its interest. When cardholder repays all the debt, these cash will return back to bank.

According to cardholder's consumption and withdrawal, bank use these information to calculate interest, accounts payable, the minimum repayment. Cardholder is based on the bill and his condition to repay the bill, and bank could confirm the amount of bad debt. Figure 1 shows the experiment flow.

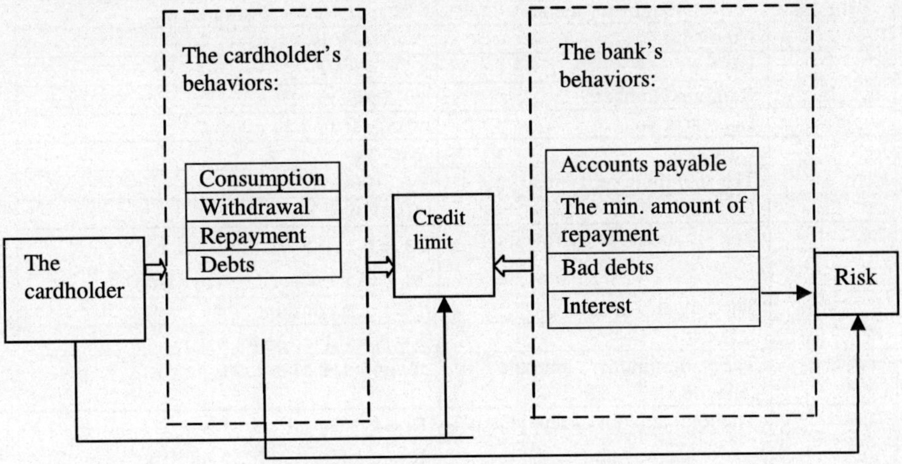

Fig. 1. The credit card's working flow chart

2.2 Assumption Condition

(1) The cardholder available money in each month must within the credit limit.
(2) The cardholder could draw the amount money must within the limit in each month.

(3) The daily interest is 0.5%, and it is compounded monthly. To ignored the credit card annual fee.
(4) The cardholder has a stable income.
(5) The cardholders keep their credit rating.

2.3 Building the Model

According to "Bank Card Business Management"[2], and these experiment assumptions, setting these parameters. We use these parameters on the Netlogo platform to build the model. The table 1 shows the main parameters.

3 Simulation Experiment and Results

3.1 Building the Model

We use the Netlogo platform to build the model. In this experiment, the variables are 1)the number of experiment agent, 2)daily interest, 3)the repayment part of income, 4)the ratio of the amount of accumulative unconsumed money, 5)the ratio of withdrawal part, 6)the ratio of the minimum amount of repayment, 7)cardholders initial money, 8)cardholders credit limit. The figure 2 shows the model interface.

3.2 The Experiment

Shown in table 2, we do cross-over experiment in income (I_t) and credit limit (V_t), and observes the repayment risk (Y_t). The figure 3 shows the risk curve.

Table 1. Netlogo parameters and formulas

Symbols	Parameter names	Formulas
V_t	Credit Limit	$V_t = V$, V is constant
C_t	Pay by Card	$C_t = b_1 \times V_t$
b_1, b_2, b_3	Random Number	(0, 100%)
d_1, d_2	Interest days	Range from 0 to 60
r_1	Cash rate	$r_1 = (V_t - C_t) / V_t$
r_2	The minimum repayment rate	$r_2 = 10\%$
r_3	Day rate	$r_3 = 0.05\%$
M_t	The amount of withdrawal	$M_t = b_2 \times r_1 \times Vt$, $(b_2 \leq 1 - r_1)$ $M_t = b_2 \times (V_t - C_t)$, $(b_2 > 1 - r_1)$
I_t	The holder's income	$I_t = I$, I is constant
P_t	Interest	$I_t = r_3 \times (C_t \times d_1 + M_t \times d_2)$
minR	The minimum amount of repayment	$minR = I_t + M_t + Y_t \times r_2$
R_t	The total amount of repayment	$R_t = I_t$, $(minR < I_t \leq T_t)$; $R_t = T_t$, $(I_t > T_t)$ $R_t = minR + b_3 \times (T_t - minR)$, $(minR \geq I_t)$
Y_t	Bad debts	$D_t = T_t - R_t$
T_t	Current accounts payable	$T_t = Y_{t-1} + C_t + M_t + P_t$
W_t	Available currency in short-term	$W_t = W_{t-1} + I_t + M_t - R_t$

[2] Bank Card Business Management, People's Bank of China [1999] 17.

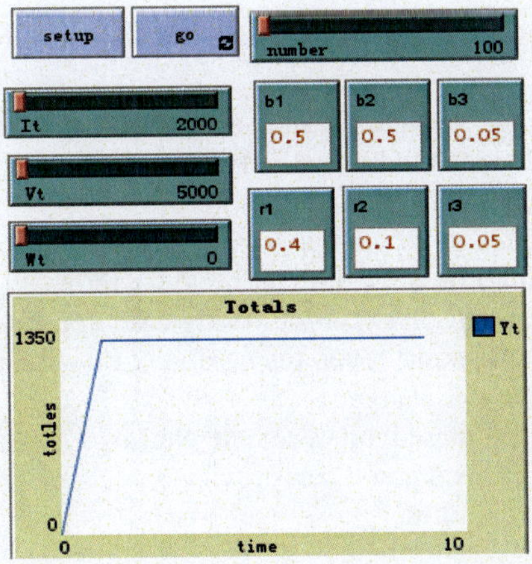

Fig. 2. Netlogo model interface

Table 2. Setting the parameters

Variables	Number of cardholders	b_3	r_1	r_2	r_3	W_t
Values	100	0.05	0.05	0.1	0.4	0

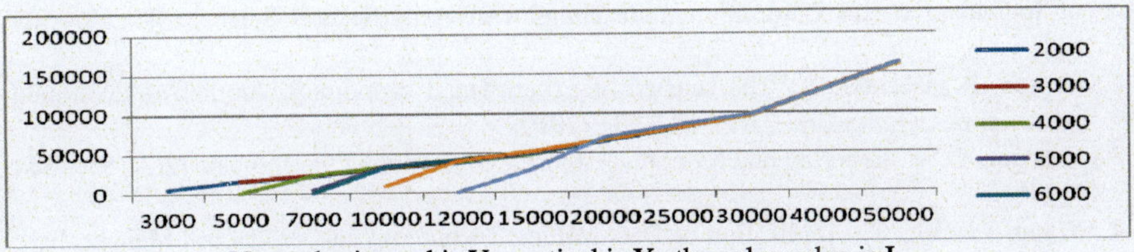

Notes: horizontal is V_t, vertical is Y_t, the color value is I_t.

Fig. 3. The relation map of I_t, V_t, Y_t

Through figure 3, when the credit limit gets a certain line, the potential risk curve in different income levels almost overlap. That is means the income has little negative influence for the bank potential risk curve. Then we concluded that between credit limit and cardholder's income has a special relationship. To observe this relationship, we use the model to find the relationship between I_t and V_t.

(1) It selected a set of I_t values, taking $V_t = 2I_t$, and doing separate trails.

(2) we set (I_t, V_t) model, use (2000,4100), (2000,4200), (2000,4300), ……, (2000,4800) to do trails.

Then we get the regression coefficient is 2.012067. The result shows that when the credit limit less than twice the cardholder's income, the bank has little risk.

4 Conclusion

This paper builds a simple simulation model of credit risk management, simulates individual behaviors influence credit card risk. This model will provide a reference to commercial banks to make their issuing card rules.

Through experiments, we observed that, if the repayment of credit is not higher than 2 times the income, cardholder's wealth can be the stability in the long-term, and be able to guarantee repayment, otherwise will bring risks for the issuer. We also found that when the credit line was some level, the cardholder's potential risks for banks and the cardholder's income do not matter.

Credit risk management based on research NetLogo is an effective method of experimental economics. This method can provide a decision-making reference and is suitable for further similar studies.

Acknowledgement. In this paper experiments used netlogo software, thanks for NetLogo provider.

References

1. Chi, G., Xu, W., Sun, X., et al.: Research on Evaluation System and Model for Credit Risk of Individual Credit Card. Journal of Tongji University(Natural Science) (4), 557–563 (2006)
2. Croson, R., Gächter, S.: The science of experimental economics. Journal of Economic Behavior & Organization 73(1), 122–131 (2010)
3. Harrison, G.W.: Experimental economics. Journal of Economic Psychology 12(3), 539–544 (1991)
4. Wang, Y., Cao, C.: Application of Data Mining Technology in Credit Card Management. Computer Engineering and Applications 10, 237–239 (2002)
5. Feng, C.Y., Sheng, D.Y., Qun, D.L., et al.: Comparison of Agent-based Simulation Platforms. Journal of System Simulation, 110–116 (2011)
6. Qiang, F.W., Dong, L.T., Xiamen, P.T., et al.: The prediction of the Default Risk of Credit Card Based on BP Neural Network. Computer Knowledge and Technology (10), 2348–2349 (2011)
7. Zheng, R.J., Feng, C.B., Li, Z.C., et al.: Estimation of the Behavioral for Credit Card Accounts by Markov Chain. Mathematics in Practice and Theory (9), 10–16 (2008)
8. Zeng, W.: Study on Credit Card on Net Logo. Tianjin University of Finance and Economics (2009)
9. Romulus-Catalin, D.: An agent-based computational study of wealth distribution in function of resource growth interval using Net Logo. Applied Mathematics and Computation 201(1-2), 371–377 (2008)
10. Wilensky, U.: NetLogo. Center for Connected Learning and Computer-Based Modeling. Northwestern University, Evanston (1999), `http://ccl.northwestern.edu/netlogo`

11. Sakellariou, I.: Net Logo BDI Extension, http://users.uom.gr/~iliass/projects/NetLogo/
12. http://ccl.northwestern.edu/netlogo/ (September 20, 2011)
13. Robertson Duncan, A.: Agent-based modeling toolkits NetLogo, RePast, and Swarm. Academy of Management Learning & Education 4(4), 525–527 (2005)
14. Elizabeth, S.: Software Review: Netlogo, A Multi-Agent Simulation Environment. Artificial Life 13(3), 303–311 (2007)

A Transmission Mechanism Analysis of Inflation's Affecting Stock Returns

Xiangdong Xu[1], Chaoyang Shi[1], and Xinbo Yu[2]

[1] School of Economics, Huazhong University of Science and Technology, No. 1037, Luoyu Road, 430074 Wuhan, China
[2] School of Business Administration, Hubei University of Economics, No. 8, Yangqiaohu Road, 430205 Wuhan, China
xuxiangdong28@163.com, shi-chaoyang@foxmail.com, 1076807054@qq.com

Abstract. This paper mainly analyzes transmission mechanism of inflation's affecting stock returns. And it shows that when inflation takes place, it will directly affect a country or region's macro-control policies, including money supply, interest rates, taxes and national debt. So, this paper, at the first time, describes transmission mechanism from the two aspects (monetary policy and fiscal policy), which directly impact on the stock market and stock prices. This resrearch also has an important practical significance to a number of macroeconomic policy-making.

Keywords: Inflation, stock returns, money supply, interest rates.

1 Introduction

The research of the relationship between Inflation and stock returns has been of great interest to financial economists. When inflation occurs, the general price level will be rising, which is sustained and significant , and purchasing power of money be decreasing. But at the same time, the stock price will be subtly changing, affected by the macroeconomic impact. Many empirical studies have shown that we can predict future stock returns through knowing the money supply.

This article mainly focuses on the analysis of transmission mechanism of inflation's affecting stock returns. Most of the literature agreed that: inflation can indirectly affect stock returns through which is affecting a country or region's macroeconomic policy, including monetary policy and fiscal policy. Loose monetary policy will be leading stock prices to rising (Lastrapes, 1998; Thorbeck, 1997; Patelis, 1997). Monetary policy not only affects stock prices independently, but also changes stock prices together with the state of business cycles (Jensen, Mercer, Johnson, 1996). Economists have also noted that the stock market, in turn, influence real economic activity and inflation (Kent, Lowe, 1997). Changes of stock returns make investors' financial wealth changing, which will affect their consumptions; and cost of capital also changes, which can affect the actual investment. Finally, stock prices will affect inflation (Campbell, 1996). Inflation occurs in varying degrees, which requires different macroeconomic policies. And then, the stock market will produce a different

D. Zeng (Ed.): Advances in Computer Science and Engineering, AISC 141, pp. 403–407.
Springerlink.com

response. When both inflation and stagnant economic condition accur(stagflation), the government will use various economic policies including monetary policy and financial policy, in order to maintain economic stability operation. At this time, the stock market's reaction is of diversity; changes of stock returns is more complex. Economists have agreed that inflation does not directly impact on the stock returns, but indirectly causes the stock price volatility by influencing national economic policy, including monetary and fiscal policies.

This paper is divided into four parts. The first part is introduction; the second shows the channels of the monetary policy about transmission mechanism of inflation's affecting stock returns;, including money supply, market interest rates; the third analyzes the channels of the fiscal policy, including tax, national debt, etc.; and the last part is conclusion.

2 Channels of Monetary Policy

Channels of monetary policy mainly include money supply and interest rates.

2.1 Money Supply

Inflation is mainly caused by excessive increase in money supply. The state's control of inflation is bound to control money supply. In general, money supply is positively correlated with stock prices, that is, increasing money supply in volume can raise stock prices (of course, there is a exception where the relationship between them is opposite).

There are three performances. First, increasing money supply can promote the prosperity of the stock market. Second, increasing money supply may make prices of commodities raise, and do shares of the company's sales revenue and profit increase, thus making the form of dividend money to rise at a certain level. All those will make stock prices rise accordingly. Third, inflation is often caused by a false boom. Meantime, the awareness of preservation tells people that investing stock market or the currency precious metals, the real estate or short-term national debt will be a good choice. So, the change of money supply is an important reason causing the stock prices of fluctuation.

When money supply increases, the excess part of the purchasing power will be put into the stock market, thereby raising the stock prices. If not, the stock market will come into a recession. In short, when inflation's stimulation is large to the stock market , the trend of the stock prices is consistent with the trend of inflation; but if its inhibition is large, the trend of the stock prices is negative correlation with the trend of inflation.

2.2 Interest rates

When inflation occurs, the real interest rates of bank deposits is lower or even negative. This time, the state will increase market interest rates.

In general, when interest rates fall, stock prices will rise; when interest rates rise, stock prices will fall.

We can use the basic model showing the relationship between interest rates and stock prices. r refers to the real interest rates, and CPI consumer price index (one index of inflation), and SR stock returns, and E external environment. We assume that CPI is an exogenous variable and dynamic.

The relationship between CPI and SR is as follows:

$$SR = f(E, r). \tag{1}$$

$$r = f(E, CPI). \tag{2}$$

Figures can be expressed as:

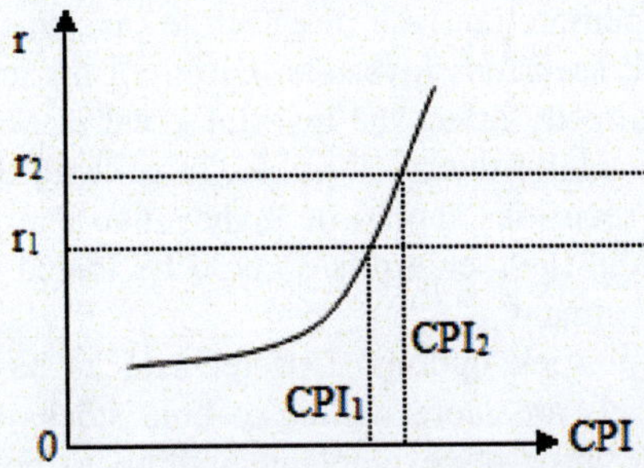

Fig. 1. E is constant.

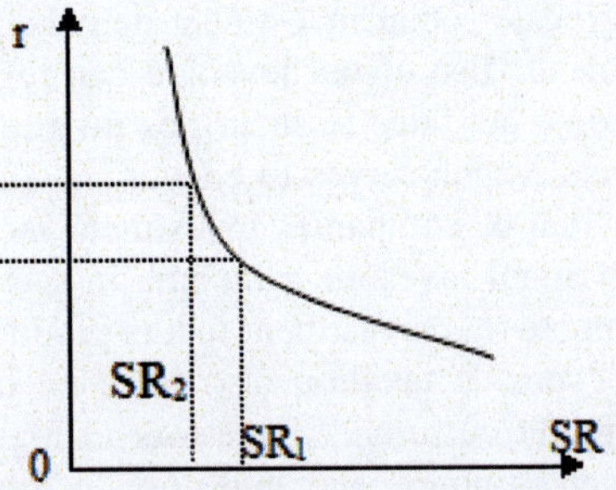

Fig. 2. E is constant.

From figure 1, we can see that when E is constant, r rises with the increase in CPI, indicating that the higher inflation, the higher the real interest rates (goverment regulation). Figure 2 shows that SR rises with the decline in r, showing that the higher the real interest rates, the lower stock returns. When inflation is CPI_1, stock returns should be SR_1; when inflation is CPI_2, stock returns should be SR_2.

The cause why Interest rates lead to the reverse movement in stock prices is mainly the following three aspects. (1) The rise of interest rates will not only increase the company's borrowing costs, but also make the company difficult to obtain the necessary funds. So the company will have to cut production scale, which is bound to reduce the company's future profits. And then, stock prices will fall. On the contrary, the stock price will rise. (2) When interest rates rise, the discount rate will rise by which investors assess the value of the stock. So the stock value will decline, and stock prices will make a corresponding decline. On the other hand, if interest rates decline, the stock price will rise. (3) When interest rates rise, a part of the funds will shift to the purchase of national debt and bank savings from the stock market. Because of reducing demands for the stock, the stock price should fall. Conversely, when interest rates decline, the profitability of savings is lower. This part of money is likely to return to the stock market, thus expanding the demand for stocks, so the stock prices will rise.

3 Channels of Fiscal Policy

Channels of fiscal policy mainly include tax and national debt.

3.1 Tax

Tax is not only means of forcibly obtaining financial income, but also a distribution way of a country's participating in the national income by virtue of its achieve political power, for the maintenance of its existence. Total amount of tax and national Government can adjust the size of securities investment and actual investmen through the changes of the structure and total amount of tax. That can curb the expansion of aggregate social investment demand, or compensate the lack of effective investment demand. Use of tax leverage can regulate the securities investors. Different tax and income tax rates to securities investors will directly affect the investor's real income levels, which serves to encourage, support or inhibit the stock prices. In general, the tax rate of companies investment income in securities should be higher than one of individual investment income in securities. And then, enterprises could be forced to promote real investment that is productive investment.

Types of taxation also have an impact on stock options. Different stocks have different customers. Investors in high tax levels are more willing to hold stocks of lower earnings, and investors in low tax levels or tax-exempt are willing to hold stocks with higher earnings. In general, if the more tax exists, the higher tax rates will thus have a negative impact on investors enthusiasm and stock investment. Because there are less surplus funds for development of the enterprises, and less dividends for investors that can buy shares of the funds. In contrast, low tax rates or tax relief can be appropriate to expand business investment and personal consumption, thereby stimulating the production and improving economic growth.

3.2 National Debt

National debt is a fiscal credit adjustment tool, which is different from bank credit. National debt can not be ignored of its impact on the stock market.

First, the debt itself is an important part of the total amount of financial assets on the stock market. As the national debt's credit is higher and its risk is lower, a large amount of national debt will lower the general level of earnings and risk.

Second, the change of T-bill rates may cause a serious impact on stock prices. When the rate of the national debt increases, investors will put money into national debt which is safe, and possesses high returns.

Therefore, national debt is competitive to other financial assets. When the capitals on the stock market are limited, too much debt is bound to affect the stock issuance and the stock prices. The above analysis shows that inflation would dampen the stock price, in particular, a serious inflation. Stock prices are negatively correlated with inflation.

4 Conclusion

This paper mainly analyzes transmission mechanism of inflation's affecting stock returns from the channels of monetary policy and fiscal policy. When Inflation occurs, it will inevitably make a country or region to chang macroeconomic policy. So, stock market will be affected, and stock prices will be caused to fluctuate. Different policy instruments have different results of market. In general, the increase in money supply will cause stock prices to rise; interest rates will cause stock prices to fall; tax cuts can stimulate stock prices to rise; a large number of issued national debt, to some extent, would repress stock prices. Therefore, the government should take into account all the implications in the choice of policy tools, in order to reduce the volatility of the stock market.

References

1. Campbell, J.Y.: Consumption and the stock market: interpreting the international evidence. Swedish Economic Policy Review 3, 251–299 (1996)
2. Kent, C., Lowe, P.: Asset-price bubble and monetary policy. Research Discussion Paper. Economic Research Department (December 1997)
3. Lastrapes, W.D.: International evidence on equity prices, interest rates and money. Journal of International Money and Finance 17, 377–406 (1998)
4. Jensen, G.R., Mercer, J.M., Johnson, R.R.: Business condition, monetary policy, and expected security returns. Journal of Financial Economics 40, 213–237 (1996)
5. Patelis, A.D.: Stock return predictability: the role of monetary policy. Journal of Finance 52, 1951–1972 (1997)
6. Thorbeck, W.: On stock market returns and monetary policy. Journal of Finance 52, 635–654 (1997)

Privacy Preserving Association Rules by Using Branch-and-Bound Algorithm

Chieh-Ming Wu[1] and Yin-Fu Huang[2]

[1] Department of Information, Taiwan Water Corporation, 404 No.2-1, Sec. 2, Shuangshih Rd., North District, Taichung City, Taiwan (R.O.C.)
[2] Department of computer science and information Engineering
National Yunlin University of Science and Technology
123 University Road, Section 3, Touliu, Yunlin, Taiwan 640, R.O.C.
g9210822@gmail.com

Abstract. Data mining techniques have been developed in many applications. However, they also cause a threat to privacy. In a very large database, there exists sensitive information that must be protected against unauthorized accesses. The confidentiality protection of the information has been a long-term goal pursued by the database security research community and the government statistical agencies. In this paper, we use the branch-and-bound strategies, which is one of the most efficient strategies to solve a large combinatorial problem. For take hidden every sensitiveness rule, hiding cost will be different from sensitiveness rule hidden order. We use the branch-and-bound algorithm to reach the best one hides order and minimum hiding cost. The experimental results showed the effectiveness of our approaches in terms of undesired side effects avoided in the rule hiding process. The results also revealed that in most cases, all the sensitive rules are hidden without generating spurious rules. First, the good scalability of our approach in terms of database sizes was achieved by using an efficient data structure FCET to store only maximal frequent itemsets instead of storing all frequent itemsets. Furthermore, in the framework, we combined the techniques for efficiently hiding sensitive rules with the transaction retrieval engine based on the FCET index tree. In particular, all research capitals before this, there is no order of considering every sensitive with hidden rule, because the interdependent complexity between the sensitive rules, cause the hidden order to influence the result seriously, so use branch-and-bound algorithm will make number of missing rules could be lowered as possible as we can.

Keywords: Data mining, privacy preserving, frequent itemsets, FCET, association rule, branch-and-bound.

1 Introduction

Privacy has become an important issue in Data Mining. Many methods have been brought out to solve this problem. In order to protect the privacy information, the objective of privacy preserving data mining is to hide certain sensitive information so

D. Zeng (Ed.): Advances in Computer Science and Engineering, AISC 141, pp. 409–416.
Springerlink.com

that they cannot be discovered through data mining techniques. The authors in [8] deal with the problem of association rule mining which preserve the confidentiality of each database. In order to avoided the privacy information broadcasted or been illegal used. The algorithm ISL to be divided into three categories: data hiding, knowledge hiding and mining results publishing. In [5] proposes an effective method for privacy preserving association rule. Trough the way of combining the basic strategies of PPARM algorithm and IMBA which are improved, making less-sensitive operations, a greater degree of hiding can be made while a less degree impact for the on sensitive rules can be made too. In [7], propose CRYPPAR, a novel, full-fledged framework for privacy preserving association rule mining based on a cryptographic approach and use secure scalar product protocols and public key cryptosystems in CRYPPAR to efficiently mine association rules over vertically partitioned data.

In [2][3], discuss when Data mining occurs on distributed data, privacy of parties becomes great concerns. They proposed a greedy algorithm, which unifies the input sets of parties without revealing any element's owner and has lower time cost than existing algorithms. They take the advantages of both in privacy-preserving Boolean association rules mining and in privacy-preserving quantitative association mining. In [4], uses a method which is effective to find frequent item sets on vertically distributed data. In [1], they propose a new approach that preserves privacy and maintains data utility in data mining. For k-anonymity model to preserve privacy while discovering and maintaining association rules through a novel algorithm, M3AR-member migration technique for maintaining association rules. In [6], also consider the problem of building privacy preserving algorithms for one category of data mining techniques from the association rule mining. In [9], they proposed algorithm clusters the sensitive association rules based on certain criteria and hides as many as possible rules at a time by modifying fewer transactions. Because of less modification in database it helps maintaining data quality.

2 Background and Related Work

The objective of data mining is to extract hidden unknown interesting rules or patterns from databases. However, the objective of Privacy Preserving Data Mining (PPDM) is to hide certain sensitive information so that they cannot be discovered through data mining techniques. The PPDM problem had to be demonstrated the solutions to this problem by reducing the support of the frequent itemsets via removing items from transactions; however, it was proved to be an NP-hard problem. In this paper, we discussed branch-and-bound algorithm to modify data in a database so that given sensitive rules cannot be inferred through association rule mining.

Branch-and-Bound algorithm is a general search method. Starting by considering the root problem, the lower-bounding and upper-bounding procedures are applied to the root problem. If the bounds match, then an optimal solution has been found and the procedure terminates. Otherwise, the feasible region is divided into two or more regions, these subproblems partition the feasible region. The algorithm is applied recursively to the subproblems.

The **PPDM** problem is an NP hard problem. Thus, the **PPDM** problem is hard to solve in worst case. But, as well be shown in this paper, the **PPDM** problem can be solved by using the branch-and-bound strategy. That is, if we are lucky, an exhaustive search through the solution space may be avoided. Now, we define the **PPDM** problem is to find tour, starting from any hidden sensitive rule, to hide every other sensitive rule and artificial rules.

Consider the cost matrix in Table 1. Where SR={sr1,sr2,sr3,…………srn}, where n is the number of sensitive rules. For example, from the Table 1, assume sr1=E=>B,C ; sr2=C,D=>B ; sr3=B=>E ; sr4=B=>D.

F(s_{ri},hide order)={(Zorder),(lemmaj),(selected order)}, the Zseq is an object function value, $lemma_i$ is the selected lemma to hide s_{ri}, the selected order is a boolean function, if 1 then the hide order is selected for hide s_{ri}. For example, F(s_{r3},1)={2,$lemma_1$,1}, that is to hide sr3 in first order then the cost is 2. In table 1, that is only the middle snapshot value from the execution of branch-and-bound algorithm. On Table 1, when the 3rd order, although hides sr3-> sr4-> sr1 may obtain the low cost, but in the 4th order will discover sr3-> sr4-> sr2-> sr1 is the best smooth solution.

Table 1. A cost matrix for the PPDM problem

hide order / sensitive rules	1	2	3	4
sr1	{4,lemma3,0}	{5,lemma3,0}	{7,lemma4,0}	{8,lemma1,1}
sr2	{8,lemma1,0}	{11,lemma1,0}	{8,lemma1,1}	NULL
sr3	{2,lemma3,1}	NULL	NULL	NULL
sr4	{2,lemma1,0}	{3,lemma3,1}	NULL	NULL

3 The Framework of Branch-and-Bound Algorithm

A. Framework of Branch-and-Bound algorithm, some important concepts used in proposed framework of Branch-and-Bound algorithm are as follows:

D: It is a transactional database.

D': It is a sanitized transaction database after sanitized algorithm processing.

Sensitive set: The strong association rules mining from original database D needed to be hidden by the data owner are called sensitive rule. The sensitive set stores all the sensitive strong association rules.

Artificial set: The false strong association rules mining from sanitized database D' that need not to be generated mining from the original database D. The artificial set stores all the false strong association rules.

Missing set: The missing strong association rules mining from sanitized database D' that need to be generated mining from original database D. The missing set stores all the missing strong association rules.

COST: cost(sri)=r*|artificial rules|sri+(1-r)*|missing rules|sri, where $0 \le r \le 1$. When we hide the sri strong association rule , the produce cost.

Objective Function: Zt=
$(\sum_{(sr_i,1)}^{t} \cos t(sr_i))$, *where* $1 \le i \le n$,
t is the level of the branch and sr_i is the selected hidden rule.

F(sri,hide order)={(Zorder),(lemmaj),(selected order)}, when to hide s_{ri} in selected hide order with $lemma_j$, then the cost is Zorder.

PPDM Constraints: All sensitive rules hide successfully, and the cost is minimal.

MONOTONE PROPERTY: The family Ψ of subsets of N={SR1, SR2……..SRn}, the cost(Ψ) is a monotone increasing function if $B \in \Psi$, $B' \subseteq B \rightarrow B' \in \Psi$, ∃ cost(B)>cost(B'). By the considering the power set P(N) as a symmetric probability space, one naturally defines the probability of Ψ by

$$\Pr(\Psi) = \frac{|\Psi|}{2^n}$$

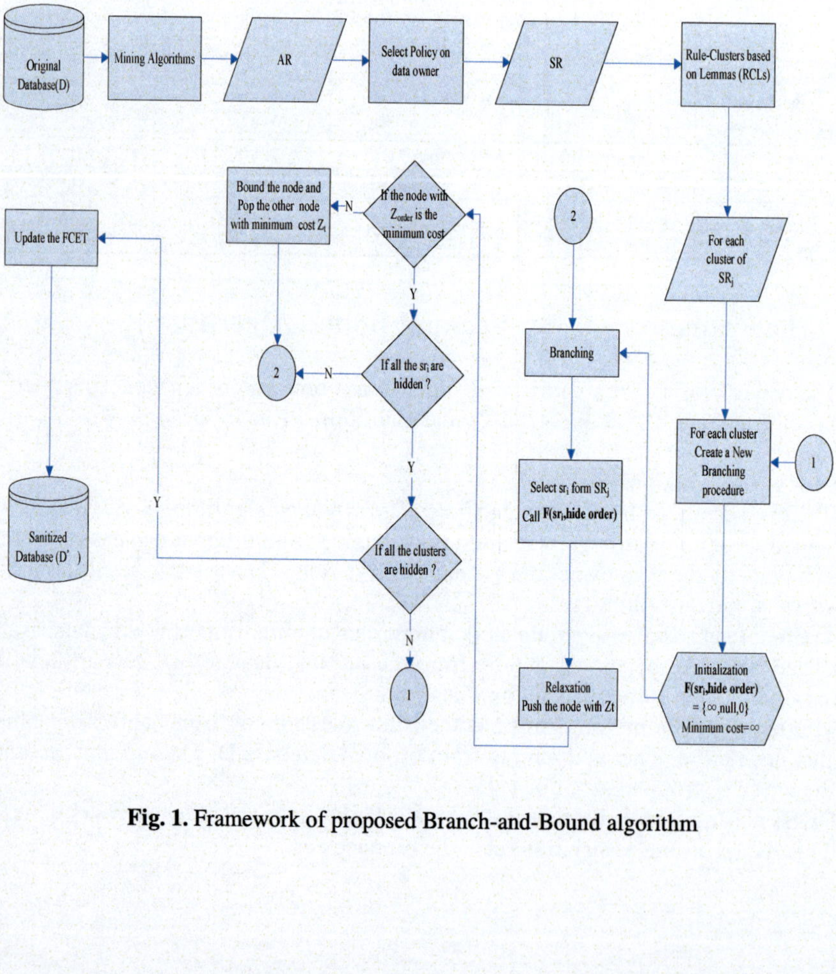

Fig. 1. Framework of proposed Branch-and-Bound algorithm

Thus, Pr(Ψ) is simply the probability that a randomly chosen subset of N lies in Ψ. If the hided sensitive strong association rules that are supported by disjoint large itemsets, then they are must satisfy the MONOTONE PROPERTY. Assume does not satisfy the MONOTONE PROPERTY, that is, when hiding a sensitive strong association rule will have a side effect on the rest of the other strong association rules . But if we try to hide overlapping rules, then hiding a rule may have side effects on the other rules to be hidden. This may increase the time complexity of the algorithms since hiding a rule may cause an already hidden rule to haunt back. Therefore, the algorithms should reconsider previously hidden rules and hide them back if they are no longer hidden. This will violate the sensitive rules are disjoint rules and they are not overlapping rules. Therefore, If the hided sensitive strong association rules that are supported by disjoint large itemsets, then they are must satisfy the MONOTONE PROPERTY.

The proposed framework of Branch-and-Bound algorithm is shown in Figure 1. Initially association rules (AR) are mined from the original database D by using association rule mining algorithms e.g. Apriori algorithm in [1]. Then sensitive rules (SR) are specified from mined rules. Selected rules are clustered based on some criterions of the rules. Rule-clusters are denoted as RCLs. Then for each Rule-cluster sensitive transactions are indexed in the FCET table. Sensitivity of each item (and each rule) in each Rule cluster is calculated based on proposed lemmas. The branch-and-bound algorithm process hides all the sensitive rules in sorted transactions for each cluster by using strategy mentioned in this section and updates the sensitivity of sensitive transactions in other cluster. Hiding process starts from highest sensitive transaction and continues until all the sensitive rules in all clusters are not hidden. Finally modified transactions are updated in original database and produced database is called sanitized database D' which ensures certain privacy for specified rules and maintains data quality.

4 The Cluster Criterions

According to above presented framework for hiding association rules in database, the proposed branch-and-bound algorithm is shown in Figure 1. By using given minimum support threshold (MST) and minimum confidence threshold (MCT), algorithm first generates the possible number of association rules from source database D. Now some of the generated association rules are selected as sensitive rule set by database owner. A strong association rule is a pair of disjoint item sets. If LHS and RHS denote the two disjoint item sets, the association rule is written as LHS → RHS. LHS and RHS are sets of items, the RHS set being likely to occur whenever the LHS set occurs. The support of the association rule LHS→ RHS with respect to a transaction set T is the ratio of support $|RHS \cup LHS|/|T|$. The Rule Cluster Algorithm based on lemmas in [3]. For a strong association rule RHS→LHS, the confidence is $|RHS \cup LHS|/|RHS|$. By using given minimum support threshold (MST) and minimum confidence threshold (MCT), algorithm first generates the possible number of association rules from source database D. Now some of the generated association rules are selected as sensitive rule set (set RH) by database owner. Then algorithm finds C clusters based on the some criterions to decide the sensitivity of each cluster. Finally algorithm updates all the modified transactions in original database. Proposed cluster criterions as follows:

CLUSTER CRITERION 1: To generate the clusters based on common RHS item of the selected rules.

For example: One can imagine that these sensitive rules E=>B; C,D=>B ; D=>B, which are selected for sensitive rules. They can be grouped according the cluster criterion 1. Because the confidence of rules |B,E|/|B|, |B,C,D|/|C,D| and |B,D|/|D| are less than MCT together, if we remove the item {B} from sensitive transaction.

CLUSTER CRITERION 2: To generate the clusters based on common LHS item of the selected rules.

For example: One can imagine that these sensitive rules B=>A,C ; B=>A ; B=>C, which are selected for sensitive rules. They can be grouped according to the cluster criterion 2. Because the confidence of rules |A,B,C|/|B|, |A,B|/|B| and |B,C|/|B| are less than MCT together, if we insert the new record with the item {B} into the transaction database.

CLUSTER CRITERION 3: To generate the clusters based on common item of {LHS∪RHS} of the selected rules under the condition of unsatisfied criterion 1 and criterion 2.

For example: One can imagine that these sensitive rules D=>A ; A,C=>B ; L=>J; K=>L, which are selected for sensitive rules. They can be grouped to two clusters { D=>A ; A,C=>B } and { L=>J; K=>L} according to the cluster criterion 3 and the cluster { D=>A ; A,C=>B } and { L=>J; K=>L} are independent, that is, when we use the branch-and-bound algorithm for sanitized processing , the hide order of these two cluster have no relationship.

5 Performance Evaluations

In the section, we evaluated the performances of six algorithms (i.e., greedy exhausted, greedy exhausted with branch-and-bound , greedy approximation, greedy approximation with branch-and-bound, algorithm 1, and algorithm 2) on a DELL GX 270 with Intel Pentium 4 3.2GHz and 1 GB main memory running Windows XP. All the experimental data were generated from the IBM synthetic data generator. The algorithm 1 and algorithm 2 was proposed by [9]. The performances of the developed algorithms have been measured according to two criteria: execution time and produced side effects. For execution time, we considered the time needed in each algorithm to hide a specified set of rules. For side effects, we considered the number of missing rules (MR) and the number of artificial rules (AR). In the experiments, given a sample database with MST 25 % and MCT 50%.

The figure 2 shows the performance on the overlapping degree. The algorithms with branch-and-bound strategies have the better performance than the others.

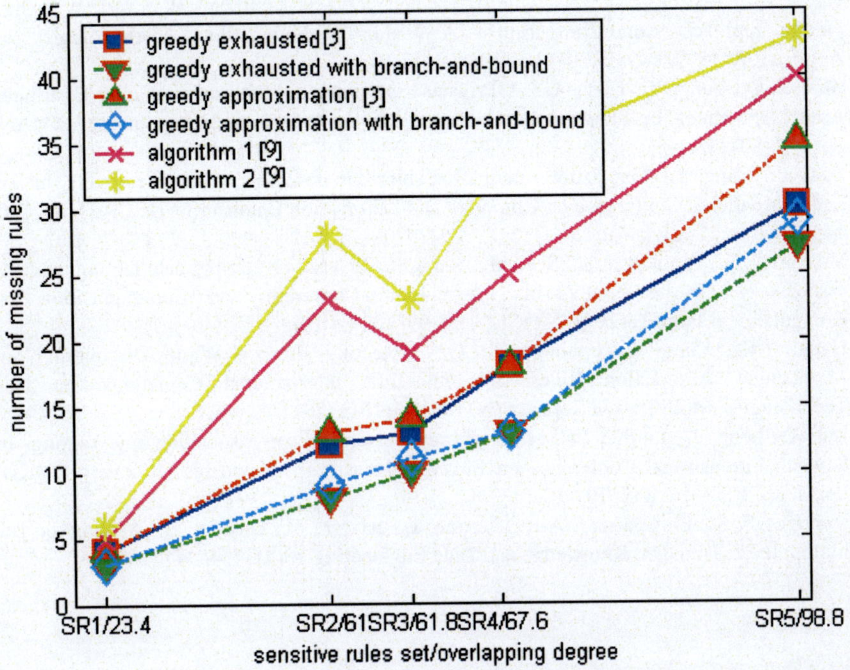

Fig. 2. Performance on the overlapping degree

6 Conclusions

In this paper, according to our experiments, it is very time-consuming to find out all associations. A kind of way to make changes according to specific conditions is to make the first order that is headed by sensitive rules. All research capitals before this, there is no order of considering every sensitive with hidden rule, because the interdependent complexity between the sensitive rules, cause the hidden order to influence the result seriously, In this paper, to use branch-and-bound algorithm will make number of missing rules could be lowered as possible as we can. From the experiment results, as expected, the greedy exhausted algorithm with branch-and-bound always performs better.

References

1. Agrawal, D., Aggarwal Charu, C.: On the design and quantification of privacy preserving data mining algorithms. In: Proc. of the 20th ACM SIGMOD-SIGACT-SIGART Symposium on Principles of Database Systems, pp. 247–255 (2001)
2. Huang, Y.-F., Wu, C.-M.: Mining generalized association rules using pruning techniques. In: Proc. of IEEE International Conference on Data Mining, pp. 227–234 (2002)

3. Wu, C.-M., Huang, Y.-F.: A cost-efficient and versatile sanitizing algorithm by using a greedy approach. Soft Computing - A Fusion of Foundations, Methodologies and Applications 15, 939–952 (2010)
4. Zhu, Y.-Q., Tang, Y., Chen, G.: A Privacy Preserving Algorithm for Mining Distributed Association Rules. In: International Conference on Computer and Management (CAMAN), pp. 1–4 (2011)
5. Sun, W., Wang, Y.: Association rule mining algorithm based on privacy preserving. In: The 2nd International Conference on Computer and Automation Engineering (ICCAE), pp. 140–143 (2010)
6. Vijayarani, S., Tamilarasi, A., SeethaLakshmi, R.: Privacy preserving data mining based on association rule- a survey. In: International Conference on Communication and Computational Intelligence (INCOCCI), pp. 99–103 (2010)
7. Dang, T.K., Küng, J., Phuong, H.V.Q.: Protecting Privacy While Discovering and Maintaining Association Rules. In: 4th IFIP International Conference on New Technologies, Mobility and Security (NTMS), pp. 1–5 (2011)
8. Ma, T., Leng, J., Li, K.: Full-scale privacy preserving for association rule mining. In: Seventh International Conference on Fuzzy Systems and Knowledge Discovery (FSKD), vol. 4, pp. 1668–1672 (2010)
9. Verykios, V.S., Elmagarmid, A.K., Bertino, E., Saygin, Y., Dasseni, E.: Association rule hiding. IEEE Trans. on Knowledge and Data Engineering 16(4), 434–447 (2004)

The Analysis of the SMEs Survival Strategy Based on the Ecological Niche Theory

Gu Lijuan[1] and Wang Min[2]

[1] Northwest Normal University, college of Geography and Environmental Science;
[2] Yang Zhou University, Yang Zhou Jiangsu 225009 China

Abstract. Small and medium enterprises play an important role in the economic and social development, however, short life span is an undisputed fact. This paper applies niche theory to survey the development strategies and competitive actions of SMEs. It points out that SMEs should not only select a suitable niche, but also need to regulate and optimize their niche properly according to the changes of the environment. It provides new ideas and methods for SMEs to survive and develop in the fierce market competition.

Keywords: Niche theory, SMEs, Survival Strategy.

1 Introduction

Small and medium enterprises (SMEs) in China's national economy plays an important role, the healthy development of SMEs, has an extremely important significance for China's sustainable economic and social development. As of 2008 June, our country small and medium-sized enterprise to already achieved about 36550000, occupy countrywide total of 99.8% enterprises, to create the final product and service value accounted for 60% of GDP, pay taxes accounted for 50% of the total national revenue, provided 77% of the post of town obtain employment, absorb 80% of the laid-off workers of state-owned enterprises, become alleviate employment pressure, eliminate poverty, maintain social stability of the important forces. Although China's small and medium enterprises have made great progress, but the overall situation is not optimistic. Many SMEs due to the small scale, weak competitiveness, financing difficulty, poor ability to resist risks and other reasons, in the fierce market competition, often at a disadvantage, and struggling to survive on the edge of. According to" scientific investment" interview, investigation and research, our country small and medium-sized enterprise average life generally between 3-4 years[1]. This international financial crisis impact on the domestic economy, small and medium-sized enterprises are the first affected and the greatest impact, the small and medium enterprises above Designated Size about 64000 homes damaged, cut-off semi-production enterprises accounted for 7.5% of the total number of. As the vulnerable groups, small and medium enterprises how to survive in the intense market competition, development, expand? With the idea of ecology gradually infiltrated into the field of management, with the help of niche theory revisited the small and

D. Zeng (Ed.): Advances in Computer Science and Engineering, AISC 141, pp. 417–424.
springer link.com

medium-sized enterprise development strategy and competitive behavior, for small and medium enterprises in the fierce market competition in the survival and development to provide new thinking. Biological community is a law of the jungle world, however, many of the apparently vulnerable species, and not because of fierce competition and extinction, but with a strong position in the species" coexistence", its reason is" niche" law plays a role. As organisms like, enterprise also deeply unified in the natural, economic, social, cultural and technological environment, enterprise niche right choice can also cause the enterprise to obtain the survival and development of space and time.

2 The Niche Analysis of Small and Medium-Sized Enterprise

1. The definition of niche and enterprise niche
Modern ecology niche is an important and abstract concept. For the current academic niche definition can be roughly divided into three categories: one is Green Neal's" Habitat niche", also known as spatial niche. He said the habitat niche divides spatial units, is the smallest species distribution unit, wherein the structure and condition to maintain the survival of species. One was Elton's" functional niche", also known as trophy niche. He mainly through food chain concept, ecological niche is the animal in the biological community in position, and the animals and their food and their relationship. Third was Hutcheson's hyper volume niche, he thinks the niche is every creature on the environment variable (temperature, humidity, nutrition ...) choice, because the environment variable is multi-dimensional (three-dimensional variable volume), so called super volume. All of these theories, this paper argues that ecological niche refers to a specific period of the specific ecosystems, biodiversity and environmental and other species is associated with a particular temporal position and function position. According to the impact of niche environmental factors (temperature, humidity, time) of a number, can be divided into one-dimensional niche, two dimensional niche, three-dimensional ecological niche and multidimensional niche.

The enterprise as a life characteristic of the tissues of the body, its survival and the competitive law is consistent with the niche theory. In order to survive, develop, the enterprise must be rooted in the specific natural, economic, and political environment, and with the surrounding environment and the interaction between arranged in a crisscross pattern. In interaction with the external environment interaction effect, catalytic role in the process of enterprises gradually formed unique products market, economic efficiency, technology, talent resource and asset size, possession of the survival and development of the necessary resources, adapt to changes in the external environment of ecological characteristics, these will constitute the enterprise niche. An enterprise's niche reflects the enterprise in a specific period, specific area, specific environment to occupy the space position, also mirror an enterprise to the external environment in a variety of resource selection and utilization capacity. We usually use the business niche width, overlap to describe an enterprise in the ecosystem occupies space size or use of resources many important quantitative index. Each enterprise should have its own niche, as enterprises in the ecosystem of a population, the small and medium-sized enterprise also has its own corresponding niche, also must occupy certain space and a considerable amount of resources.

2. The niche analysis of small and medium-sized enterprise

(1) The width of enterprise niche. Enterprise ecology niche width refers to an enterprise that utilizes a variety of market environment resources combined, is a measure of enterprise on the ecosystem environment resources such as technology, capital, and market and make use of the diversity index. Enterprise ecology niche has a width of. Wide niche means adapted to range expansion but adapted to the loss of efficiency, make the enterprise is universal or diversity trend; narrow niche means enterprise environmental fitness improvement but the range is limited, so that enterprises have the trend of specialization. On business niche width measurement, often uses Levins formula:

$$B_i = 1 / \sum_{j=1}^{R} P_{ij}^2$$

In the formula B_i represents enterprise I niche width, P_{ij} represents individual proportion of enterprise I uses resources J

For the majority of medium and small businesses, which can use a variety of resources limited? Technically speaking, small and medium enterprises technology into slow, development rate is low, have the technical staff is little, but most do not have the own development of the ability of new technology; from the level of management, the management of small and medium-sized enterprises relatively low quality of personnel, management concepts and management level is relatively backward; from the capital level speaking, small and medium-sized enterprise management Limited general funds, financing ability in general are quite poor, coupled with its financial credit with respect to be congenitally deficient, make small and medium-sized enterprises because of funding problems and difficult to seize the development opportunities, sometimes even normal operation due to shortage of funds by the collapse of doom; from the market perspective because the small and medium-sized enterprises, small scale, often only passive recipients of vicissitude of the market, product market and raw material supply influence is extremely limited; in addition to small and medium enterprises to obtain market information ability is weak, sometimes due to the lack of the necessary management basic information and lead to business decision-making errors, the impact of enterprise survival. The above aspects show that, small and medium-sized enterprise niche width is narrow, niche adaptation range. Small and medium-sized enterprises should adopt narrow niche model, so that they have a narrow niche area dominant position, in a large enterprise edge play their own unique expertise, from the fierce competition and challenges.

(2) The overlap of the enterprise niche. Enterprise niche overlap is refers to the enterprise niche similar degree or niche factors in the same proportion. Ecology niche theory that: when the two organisms use the same resource or common possession of other environmental factors, there will be the ecological dimension overlap. Similarly, if the space between two enterprises, time, market, technology and resources with other enterprises do not have the distinction, will form the enterprise niche overlap. In the condition of open market economy, the same target customers, common production resource consumption is likely to cause the niche overlap. Enterprise niche overlap used to measure as asymmetry of alpha commonly, the formula is:

$$O_{ij} = \sum_{a=1}^{n} P_{ia} P_{ja} / \sum_{a=1}^{n} P_{ia}^2$$

In the formula O_{ij} represents niche overlap of species i and species j; P_{ia} and P_{ja} represents use part of species I and j on species resource a (a = 1…n).

Ge Zhenzhong, Liang Jiahua (2004) argues that the competition between the enterprise, it is due to the niche overlap and cause, and in the same ecological environment to the long-term coexistence of enterprises can not be niche completely similar, they only have a limited similarity[2]. Xu Fang, Li Jianhua (2005) also hold similar views, they believe that if the two businesses target customers are similar, the two enterprise niche overlapping, if two enterprises have exactly the same niche, there will be one hundred percent overlap; if the two companies have a common part consumers, also each have different consumers, two enterprise niche overlap[3]. Niche overlap with the competition between the enterprise is proportional to the intensity of. The niche overlap causes the market share is relatively contractible, competition will be intense with each passing day, competition is the result of the realized niche differentiation. Niche differentiation is due to competition, enterprises from the part of their potential for survival and Development Zone exit, thereby eliminating the niche overlap. Two enterprise niche overlap either completely or partially overlapping, weak competitiveness will exit from overlapping portions, and strong competition in the final will remain in the niche. For the weak exit, may be passive, i.e., inferior position enterprise is strong company pushed out; also may be active, namely inferior enterprises in order to reduce the pressure of competition, active and strong enterprise niche separation, thus increasing the chance of survival.

Speaking of the small and medium-sized enterprise, niche overlap exists in two different situations, one is big enterprises and overlap, and even within the package from big business within the ecological niche [4]. As the disadvantaged group, active implementation of niche differentiation is to ensure its survival and effective method. Because if and large enterprises in the same market competition directly, because of the disparity between the strength of small and medium-sized enterprises adhere to each other, will soon be out of the suppression of large enterprises, merged or. Another is with the same or similar strength of the medium-sized and small enterprises partially or even entirely overlap. As a result of each other's strength of the gap is relatively small, in competition with one another on a long time, the enterprise should base on its own niche characteristics take the right competitive strategy, enhance competition ability ceaselessly, thereby the opponent out of niche overlap.

3 Chapter3 the Survival Strategy of Small and Medium-Sized Enterprise

Small and medium-sized enterprise wants to survive and develop in intense market competition, must produce oneself advantage adequately, avoid inferior position. Enterprises should choose according to their own conditions to give full play to the ability of the niche, and the formation of competitive advantage. At the same time as the enterprise environment resource to change with time, the enterprise also to niche timely, appropriate control and optimization. If the stick in the mud, hope to maintain the original ecological position unchanged, can only lead to strategic mistakes, missed opportunities, or even a threat to the survival of enterprises.

1. Concentration strategy
Small and medium-sized enterprises in the niche width and depth to fit enough, therefore unable to pass through a plurality to gain competitive advantage, therefore the small and medium-sized enterprises should abandon" small and complete " development train of thought, in the specialized division of labor and cooperation on the basis of enterprise activities, will focus on improving the core competitiveness of the value level, adhere to the " small and only"," small but excellent"," small and essence of life" and "small and strong" development train of thought. " Suncius strategies" is Chinese and foreign management theory to study the ancient Chinese classics, in its ninth part" the March of army" says: force is not in as much as possible, as long as no blind rash advance, and can concentrate power, take the enemy, internal unity, make good use of talents. On the strategy to use a centralized strategic principle, to play the small and medium-sized enterprise's competitive advantage has a special significance. Large enterprise human resources, although more, but also because of large institutions have redundant personnel is overmuch phenomenon, influences the efficiency, small and medium-sized enterprises organization if can maintain a lean, talents can be fully utilized, its efficiency is very high, is the same with large enterprises competitive basis. But if the small and medium-sized enterprises can use their own resources, it is impossible to establish their own competitive advantage. Small and medium-sized enterprises, especially small businesses there should be only one field of operation, otherwise it will disperse resources. Strategically despises the large enterprises, in the thought takes the large enterprises, establish a again small also will not admit defeat stubborn idea, thoroughly the pursuit of corporate personality, as long as the proper timing, centralized enterprise resource in essential points, concentrating, adventurous, there is success of the may. In the western developed countries, many small and medium enterprises rely on the weak overcame the strong, centralized strategic success, have proved this point.

2. Precise selection
For small and medium-sized enterprises, for the niche selection must be accurate, ensure the enterprise to obtain relatively ample space for development. Ecological selection refers to the material resources, technical advantages of information and the ability to choose the right products and market areas. Because of their own strength limitations, small and medium-sized enterprises and large enterprises are weak in the same niche on the fight, so if you want to survive and develop, we must take the initiative and enterprise niche differentiation. The realized niche differentiation is one way of finding and filling the enterprise ecosystem niche vacancy. These positions are also known as niche, Peter Drucker put these small bits are divided into: natural niche position, potential niche, niche and collaborative expertise Mini-Niche[5], these Mini-Niche large enterprises have been abandoned or neglected, is to provide SMEs survival, development space. Natural microhabitat are large enterprises do not want to get involved in some narrow the scope of the market, due to customer demand for products and services more unique, different require a higher degree, so many varieties, small batch, small profit on sales is the main mode of production, and this production

method is unable to obtain the scale economic effect. Potential niche, namely market neutral. Due to consumer demand preference transfer, technical innovation, regional economic differences and other reasons, there will always be some large enterprises has not been involved in the emerging market area or large enterprises have to actively explore the local region.Collaborative Mini-Niche is dependent on large enterprise niche position. Large enterprises in order to seek maximum profit or cost saving, avoiding the" big and complete" production of the drawbacks of the system, and the external enterprise collaboration, implementation of non-core business outsourcing, this collaboration for small and medium-sized enterprises to provide living space. Small and medium-sized enterprises can join the big enterprises of the ecological chain, and large enterprises to establish cooperative symbiotic relationship, so not only will not be big business threat, also with large enterprises co-evolution. Expertise Mini-Niche refers to with the invention patent of small and medium enterprises can use intellectual property to prevent large enterprise on its own proprietary knowledge to their own product market penetration, thus the legal system under the protection of the formation is beneficial to the growth of small and medium enterprises ' mini-niche. Of course is not only a technological development capability of small enterprises to enter, through the transfer of patent system, many small and medium-sized enterprises can get such a niche position themselves.

3. Invisible “monopoly”

Although small and medium enterprises to choose any one of these niche are able to avoid confrontation with large enterprises, increase the chance of survival, but in the niche position is also faced with competition. The enterprise wants to occupy the competitive advantage, must be in a related field to build the core competitiveness, in the niche position business areas to form a contact" monopoly", make yourself from the competition and challenge, in a large enterprise edge play their own unique expertise, for some special products and the level of technology to become a leader, the gradual accumulation of business resources, for the further development of SMEs to create a favorable opportunity. Natural selection Mini-Niche small enterprises to strengthen the analysis of demand and further meet the diversification of customers and scattered the ability of demand and according to the environment to take timely flexibility strategy. The selection of potential niche of small and medium-sized enterprises to strengthen a timely manner to meet market demand, to gain competitive advantage of quick response ability. On the future promising industries in early discovery, early preparation, early investment, earlier than competitors step introduced new products. Who first discovered the market gap and the first to enter, who can capture market empty shelves, to do exclusive business, opportunity benefit. Selecting cooperative Mini-Niche SMEs, mean that small and medium-sized enterprises face more and more important customers, not only for large enterprises to provide products and services, but also pay attention to and maintain good customer relations, because the loss of a large customers for enterprises by the impact is enormous. Selection of special knowledge niche position the small and medium-sized enterprises should cultivate the core technical capabilities, including R&D capability, product and process innovation, as the core technology capability determines the enterprises will be the technical resources to the technical advantages of conversion ability level. Only through ever fount technical innovation and technological innovation, constantly win is novel, advanced and practical technology inventions, or design a new structure, new specifications, the new style products, these can be as small and medium enterprises to open up new fractionize market, satisfy new social requirement, reduce the production cost of products, to expand the product difference method.

4. Optimization control
Small and medium-sized enterprises in the process of development, the external environment is not immutable and frozen, so small and medium-sized enterprises also have their own niche in control and optimization. Ecological control includes two aspects: the development of niche and the niche adjustment. A successful enterprise must be good at expanding the development of niche and adjust the niche, in order to transform and adapt to the environment. The small and medium enterprises to expand and adjust their own niche, must keep their common operation of the main line, namely the enterprise current products and market and future products and market exist between an inner contact, known as the" common operating line". Secondly, to master the niche expansion speed, avoid too fast and make the enterprise development loss of continuity, result in competition failure. Ecological optimization, which is optimized for utilization of ecological resource. Enterprise niche optimization is the enterprise in the development strategy under the guidance, system optimization and integration of enterprise external environment, internal structure and function of the organic process[6]. the appropriate effective enterprise niche optimization can improve the competitive advantage of the enterprise, the catalytic environment of enterprise inside and outside the communication and interaction between enterprise and environment, realize the cooperative evolution, so as to promote the sustainable development of enterprise niche, greatly enhance enterprises in ecological environment function status. The small and medium-sized enterprises should around development target, in accordance with the integrity, unity and the value of the optimal principle, effective organization and integration of enterprise internal structure of people, money, material and information elements, with the enterprise niche and ecological environment of continuous realization of material circulation, energy flow and information transmission function mechanism as the basis, relying on management, administrative and technical and other means, to the enterprise resource factor structure and ecological development space to optimize integration, implementation enterprise niche structure function complete, function maturity and size at a stable, and meet the demand of enterprise development evolution.

4 Conclusion

There are many reasons for the success of the enterprise," niche" should be one of them, because it requires that people and nature, people and society's harmonious development, the small and medium-sized enterprise success also lies in the right niche selection. As the United States famous management experts Peter Drucker said: "the small business success depends on it first place in a small field of ecology." In search of their own niche in the process, the enterprise must understand a truth: the only in its ecological niche is strong, the weak only in their ecological niche to free survival. If one does not have the ability to contend with big company of medium and small businesses ought not to do" tiger", and should be a flexible" monkey", in their own a small ecological field of play advantage. Small and medium-sized enterprise as long as they can find their niche, and can continuously according to the change of external environment to adjust and optimize its ecological niche, the successful reason lies in these.

References

1. Xin, B.: The Death of China's Private Enterprises. Investment in Science (11), 24–46 (2003)
2. Ge, Z., Liang, J.: Enterprise Niche and Competition of Modern Enterprises. Economic Management in East China (2), 113–116 (2004)
3. Xu, F., Li, J.: Principle and Model of Enterprise Niche Research. China Soft Science (5), 130–139 (2005)
4. Qian, Y., Ren, H.: The Analysis Based on the Niche Competition of enterprises. Finance and Trade Research (2), 123–127 (2006)
5. Drucker, P.: Innovation and Entrepreneurship. Shanghai Translation Publishing Company, Shanghai (1999)
6. Cao, Y., Liu, Z.: The Basic Connotation, Problems and Countermeasures of Enterprise Niche Optimization. Future and Development (6), 69–73 (2009)

Comparative Cost Model of Clothing Industry Transfer

Xiuchen Wang and Zhe Liu

Zhongyuan University of Technology, Zhongyuan Road 41,
Zhengzhou City, Henan Province, P.R. China
susongxinxiang@163.com

Abstract. A model of comparative cost for clothing industry transfer is proposed in this paper aiming at scanty economic cost basis about clothing industry transfer at present. The cost before transfer and after transfer is made a quantitative comparison by this model. In this paper, each cost factors influencing clothing industry transfer is given, and the calculation function of these factors is listed. Finally, the calculation method of cost model of clothing industry transfer is established. Empirical analysis results show that this model can evaluate the effect of clothing industry transfer better, can accurately measure the feasibility of enterprise transfer, and can provide the theory basis for enterprises transfer in China.

Keywords: Clothing, Industry transfer, Comparative cost, Model.

1 Introduction

China is a big production country with the biggest output and export in the clothing world. From 80's in the 21st century to now, those clothing enterprises once have drive the economic development in east coast of China. But there are some problems which are hard to overcome during these years. Various production costs are increasing, the sustaining power of development is lacking, the economic benefit is decreasing, and the competitive advantage is gradually losing.

Therefore, more and more scholars realize that these clothing enterprises in east coast must transfer to Midwest region which has rich resource, low cost of labor and wide land. In 2000, we proposed the idea of industry transfer in textile and clothing field, and published related papers. During these years, some scholars have done some researches: Yu cen [1] studied the enterprise strategy for the textile and clothing industry for Zhejiang province example; Chen yonghua [2] explored the method of internal industry transfer in the Yangtze River Delta and gave some advice. Lu peng [3] put forward existing problems and pointed out the route of transfer in manufacturing zone, and gave a view of power source of industry transfer from the globalization perspective; Li benben [4] analyzed the reason of slow increase of textile industry transfer in China, and proposed some advice and strategy; Zhong fang [5] gave some views of undertaking industry transfer.

Though scholars have put forward their own opinions of clothing industry transfer, they all limited to affirmation of this trend and only proposed some advises with no deep research. Therefore, this study will propose a new model to study the internal rule of clothing industry transfer from a completely new view.

D. Zeng (Ed.): Advances in Computer Science and Engineering, AISC 141, pp. 425–428.
springer link.com

2 Comparative Cost Model of Clothing Industry Transfer

The key of industry transfer is that the enterprise after industry transfer can obtain more benefits in the undertaking region. However, there are more factors related to enterprise management. Now the problem that the operation cost before transfer must be exactly analyzed and compared with the operation cost after transfer must sure. After a great deal of researches and deep investigations for many years, we construct a comparative cost model in China region. Two factors must be considered, they are direct cost and correlative cost. The need of industry transfer is evaluated by comparison of these two costs.

The direct cost is all estimation cost of single clothing production, they includes:

γ_1 : The resource cost of single clothing;

γ_2 : The production cost of single clothing;

γ_3 : The sale price of single clothing;

The correlative cost is the extra cost after transfer, they includes:

μ_1 : The transfer cost, it includes all cost of transfer, such as transportation cost, removing cost, installation cost.

μ_2 : The returning home cost, it includes total cost of returning home of staff;

μ_3 : The adaptation cost, it includes total cost of communication and relation.

The theoretical calculation formula of comparative cost is given as:

$$C = E \times 0.1 + \frac{P_O}{P_N} \times 0.4 + \frac{T_O}{T_N} \times 0.5 \tag{1}$$

Where, P_O denotes the direct cost of clothing enterprise in original region, P_N denotes the direct cost in new production region, T_O denotes the correlative cost in original region, T_N denotes the correlative cost in new production region. C denotes the comparative cost of transfer, E denotes the benefit between the land selling out and buying in. They are calculated as following:

$$P_O = \gamma' - (\gamma_1' + \gamma_2') \tag{2}$$

$$P_N = \gamma_3'' - (\gamma_1'' + \gamma_2'') \tag{3}$$

$$T_O = \mu_2' + \mu_3' \tag{4}$$

$$T_N = \mu_2'' + \mu_3'' \tag{5}$$

$$E = \frac{\mu_1' - \mu_2''}{\mu_1'} \tag{6}$$

Where, $\gamma', \gamma'_1, \gamma_2', \mu'_1, \mu'_2, \mu_3'$ denote various costs of original region, and $\gamma'', \gamma''_1, \gamma_2'', \mu''_1, \mu''_2, \mu_3''$ denote various costs of new production region. The relation of various cost factors is shown in Figure 1.

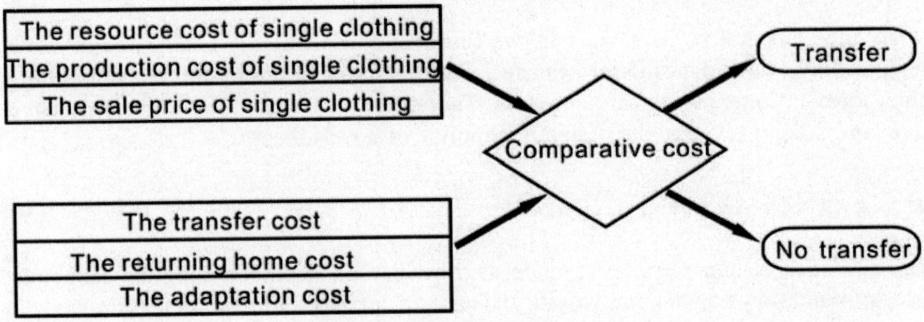

Fig. 1. Relation of transfer cost

3 Example

According to the cost model of transfer, we investigate many clothing enterprises in various regions and analyze and compare the transfer cost. Table 1 lists an example from these investigations.

Table 1. Cost comparison of investment production between Zhejiang province and Anhui province

Area	Resource of single clothing	Production cost of single clothing	Sale price of single clothing	Transfer cost	Returning home cost	Adaptation cost
Zhejiang province	γ_1=56	γ_2=16	γ_3=89	μ_1=5000	μ_2=3	μ_3=23
Anhui province	0.9 γ_1	0.8 γ_2	γ_3	0.6 μ_3	2 μ_2	2 μ_3

According to formula (2), formula (3), formula (4) and formula (5), we can calculate and conclude as:

$$C = 0.55 \tag{7}$$

The result denotes that the comprehensive profitability of this enterprise in Zhejiang province is 55 percent compared to that of the same enterprise in Anhui province. This is a successful case of enterprise transfer.

After a great deal of investigations, we give a standard to estimate the success of clothing industry transfer. The transfer effect is classified three classifications according to the transfer comparative cost, listed in Table 2.

Table 2. Transfer effect classification of clothing industry transfer

Transfer effect	better	good	bad	Worst (not suitable for transfer)
Classification standard	$C < 0.6$	$0.6 \le C < 0.8$	$0.8 \le C < 1$	$C \ge 1$

By the comparative cost analysis, we find that some clothing enterprises in some regions have advantages after transfer, but the transfer effect of some clothing enterprises in some regions are not good. Therefore, the transfer cost plays a decisive role, and doesn't relate to the preferential policy of government.

4 Analysis and Discussion

The aim of the comparative cost model is to reveal the relation among the economic factors of industry transfer and provide the quantitive basis for transfer measurement by enterprise and government. The accurate cost accounting of clothing industry transfer can provide reference for economic development strategy and provide powerful guidance for related policies by government. Otherwise, this model can provide the reference for transfer of enterprises from east to Midwest in China. So enterprises can search for the best time and the best method to carry out the transfer.

5 Conclusion

Each factors of clothing industry transfer is considered in the comparative cost model. The adapting cost is brought into the examination range first. This model can accurately measure the feasibility of enterprise transfer.

The comparative cost model can suitable for developing country with unbalance economic development and vast territory. This model has a theoretical reference value and academic value for clothing industry transfer in China.

References

1. Yu, C.: Going-out and Industry Transfer of Zhejiang Textile & Garment Industry. Silk 9, 5–8 (2008)
2. Chen, Y., Shan, X.: Empirical Analysis of Internal Industry Transfer in the Yangtze River Delta. Jiangsu Commercial Forum 8, 5–7 (2008)
3. Lu, P., Wang, Z.: Study of China's Transition of Manufacturing Zone in a Harmonious Economy: from the globalization perspective. Journal of Guangzhou City Polytechnic 2, 44–47 (2008)
4. Li, B., Li, J.: Reason and Measurement Analysis of Slow Increase of Textile Industry Transition in China. Shandong Textile Economy 2, 10–12 (2008)
5. Fang, Z.: Western Should Actively Carry on Industrial Move. China Business Update 9, 78–79 (2008)

Differential Analysis of National Economic Competitive Advantage between China and US Based on National Economic Security

Siyi Qin and Genhua Hu

School of Economics, Guangdong University of Business Studies, Guangzhou, China
qinsiyi2006@163.com, garyuser@hotmail.com

Abstract. This paper does a differential analysis of national economic competitive advantage between China and US from 1993 to 2009 based on national economic security using principal factor analysis of panel data. It is found that US has strong national economic competitive advantage and a more stable level of national economic competitive advantage, while it is weak and unstable in China. Besides, the level of national economic competitive advantage in China has been increasing in recent years, which helps to narrow the gap of national economic competitive advantage between China and US.

Keywords: National economic security, national competitive advantage, national economic competitive advantage.

1 Introduction

The concept of National Security was firstly proposed by Lippmann in 1943. And U.S. Government launched the first Act in the world, i.e. *The National Security Act of 1947*. According to the meaning of National Security, it can be understood from different viewpoints and fields, including politics, economy, culture and military, as well as diplomacy. In 1980, Japanese Government issued Report of National Comprehensive Security, which was the first official report referring to economic security and listed economic security in component elements of national security strategy. Economy had become a significant factor for national power strengthening, social steadiness and national security achievement as the ending of the Cold War. Therefore, many nations adjust their national security strategies for upgrading the position of national economic security in the strategies of national security. Since 1990s, economic crisis and financial crisis have been breaking out more and more frequently. In 2007, American subprime mortgage crisis broke out which remained people of rethinking about the issue of national economic security further.

In 2011, US debt default crisis has caused a great panic in the world. The crisis has brought a negative impact on national economic security for many nations and their national economic competitive advantages are facing a great challenge. With the development of China, the argument of "China threat" theory is running high. Under this international background, it is urgent to study the issue of national economic competitive advantage. China and US are the discussed focus, especially after China

D. Zeng (Ed.): Advances in Computer Science and Engineering, AISC 141, pp. 429–435.
springer link.com

became the second largest economy in the world in 2010. Therefore, the paper does a differential analysis of national economic competitive advantage between China and US based on national economic security.

2 Literature Review

Nowadays, national economic security has become a key aspect of national security. There exists different comprehension to it but without an authorized definition so far. Neu, New and Wolf (1994) [1] suggest that economic security can protect economic interests when a nation faces the events which could threat economic interest or hinder economic running, and economic security endows a nation with the capability of national economic environment which is in accordance with itself interests. However, Zhao, Xu and Xing (1994) [2] propose that national economic security is economic competitiveness, capability of resisting interference, threat and aggressiveness at home and abroad, and internal and external environment for economic development of a nation. However, the factors influencing national economic security have changed dynamically as the development of the world economy. With the development of the world economy and economic globalization, the factors influencing national economic security have changed dynamically. Richards (2009) [3] suggests that economic security refers to global capital flows and the capital and commodities markets. Besides, he proposes that providing true economic security requires an analysis of the economic system with an approach of econophysics. Pankov (2011) [4] analyzes the problems of national economic security and international economic security in the world economic crisis of 2008-2010. He reminds of people that particular attention should be paid to the current problems, e.g. food, finance, energy and environment.

In fact, national economic competitive advantage is a vital part of competitive advantage of nations. Besides, it is the outward manifestation of national economic security. Therefore, to study national economic competitive advantage is the key aspect of analyzing national economic security.

Porter (1990) [5] advances the theory of Competitive Advantage of Nations and proposes whether a nation can obtain economic competitive advantage in the international competition is the fundamental element to determinate prosperity or recession for a nation. Stigliz and Walsh (2002) [6] review the developmental process of economic growth achievement and competitive advantage of nations upgrade firstly, then address the concept of absolute advantage and regard technology innovation as a nation's competitive advantage. Stone and Ranchhod (2006) [7] build a Quantitative Model of Competitiveness Advantage of a Nation Determination based on Porter's Diamond Model and analyze the competitiveness advantage of nations in the UK, USA and BRIC nations using a quantitative model. Aleshchenko and Nedelea(2008) [8] analyze the competitive advantages of Russian economy under conditions of open markets. Zhang (2010) [9] suggests that structure upgrading and strategic transition can strengthen China's economic competitive advantage and economic competitiveness. Qin and Hu (2011) [10] propose the low-carbon factor intaking energy efficiency as one of the major elements in the diamond model of national economic competitive advantage based on national economic security and analyze the path how the national economic competitive advantage is upgraded by low-carbon factor and how to protect national economic security.

At present, few literatures are found on comparison research of national economic competitive advantage in different nations based on national economic security. Therefore, this paper tries to do a differential analysis of national economic competitive advantage between China and US based on national economic security using the method of factor analysis of panel data.

3 Variables, Data and Empirical Results

3.1 Variables and Data

In this section, we choose some factors influencing national economic competitive advantage based on national economic security and apply the method of principal factor analysis to calculate the level of national economic competitive advantage. Then, we do differential analysis of national economic competitive advantage between China and US based on national economic security. These factors include growth rate of GDP (X1), inflation rate (X2), unemployment rate (X3), growth rate of M1 (X4), ratio of dependence on foreign trade (X5), ratio of external debt (X6), rate of financial deficit (X7), growth rate of fixed assets (X8), balance of foreign exchange reserve (X9), growth rate of energy input per GDP (X10) and high-tech-product-GDP ratio (X11).

The data is from IMD World Competitiveness Yearbook (2002-2010), China Statistical Yearbook (1994-2010) (in Chinese), International Statistical Yearbook (1992-2009) (in Chinese), World Economy Yearbook (1992-2009) (in Chinese) and some websites, including UN data website and www.bea.doc.gov.

3.2 Empirical Results

3.2.1 Comparison Analysis Based on IMD World Competitiveness Report

Table 1 reports the scores and ranking of world competitiveness in US and China. It is found that US scores 100 and it ranks the first position from 2002 to 2009, while US world competitiveness is a little weakened and scores 99.091 in 2010, ranked the third position. The potential influencing factors might be as follows: underperformed economy, inefficient government and business, and slow renewed infrastructure, impacted by American subprime mortgage crisis in 2007. Although the ranking of China's world competitiveness is not high, its scores have been increasing in general, indicating that China's world competitiveness has been strengthening during the sample years. In recent years, Chinese Government has been sticking to proactive fiscal policy and prudent monetary policy, which keeps a good economic state in China. Further, American subprime mortgage crisis has few impacts on the economy of China because its capital markets have incompletely been opened. What's more, China continues to proactive policies to strengthen the capacity of risk resistance for the economy.

Table 1. Scores and Ranking of World Competitiveness in US and China

Year	US Score	US Ranking	China Score	China Ranking
2002	100	1	51.980	12
2003	100	1	50.813	12
2004	100	1	70.725	24
2005	100	1	63.219	31
2006	100	1	71.554	19
2007	100	1	79.484	15
2008	100	1	73.759	17
2009	100	1	76.595	20
2010	99.091	3	80.182	18

Source: IMD World Competitiveness Yearbook (2002-2010).

In 2010, China became the second largest economy in the world. However, its world competitiveness isn't high and ranks the eighteenth position, indicating that China has to make great efforts to upgrade its world competitiveness. Figure 1 presents the differential graph of world competitiveness between China and US. It shows that there exists a large gap of world competitiveness between China and US while the gap is narrowing in general.

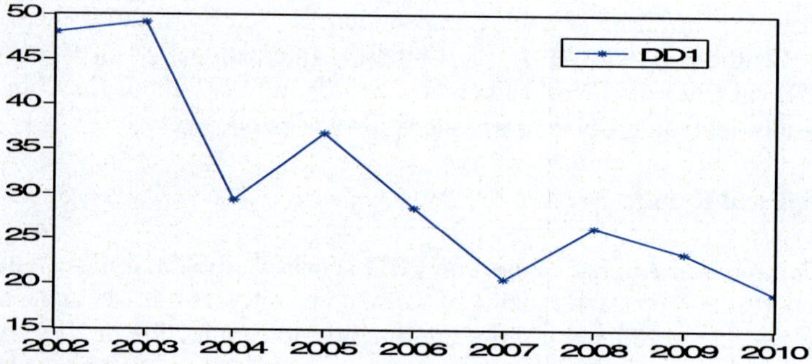

Fig. 1. Differential Graph of World Competitiveness between China and US

Note: DD1 represents the gap of world competitiveness between China and US.

3.2.2 Comparison Analysis Based on National Economic Security

Table 2 displays the scores of national economic competitive advantage in US and China. The method is Principal Factor Analysis of panel data. The detailed results are as follows. Firstly, US scores are generally larger than those of China from 1993 to 2006, indicating that national economic competitive advantage in US is stronger than that of China. Secondly, China's score is larger than that of US in 2009. After American subprime mortgage crisis breaks out, US economy is badly impacted and it keeps in the shadow of the crisis in the next few years. On the contrary, China is less impacted and its economy keeps fast growth which helps to narrow the gap of national economic competitive advantage between China and US.

Table 2. Scores of National Economic Competitive advantage in US and China

Year	US	China	Year	US	China
1993	2.865	3.926	2002	2.616	-0.508
1994	3.084	1.961	2003	2.656	0.270
1995	3.118	0.381	2004	2.671	0.448
1996	3.053	0.034	2005	2.793	0.450
1997	2.876	-0.672	2006	2.615	0.368
1998	2.765	-0.407	2007	2.647	1.301
1999	2.500	-1.376	2008	4.575	2.601
2000	2.498	-1.112	2009	2.583	3.132
2001	2.298	-0.569			

Figure 2 is the graphs of national economic competitive advantage in US and China. It shows that the changing trend is more stable for US national economic competitive advantage than that of China, while jump appears in 2008. The possible reason is that, to cope with the crisis in 2007, the U.S. Government carried out a series of economic incentive policies, including cutting interest rates, implementing auction operations of short-term loans and reusing discount window.

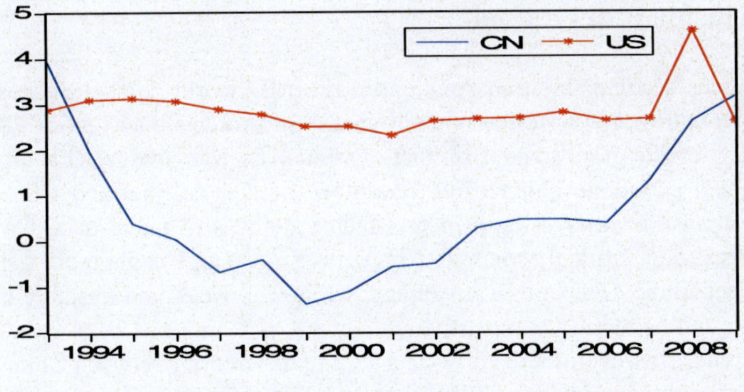

Fig. 2. Graphs of National Economic Competitive advantage in US and China

In Figure 2, it can be seen that the national economic competitive advantage of China has been decreasing from 1993 to 1999 because of a new round of inflation in 1993 and 1997-1998 Asian Financial Crisis. Since 2000, China's economic competitive advantage has been exhibiting an increasing trend. However, there still exists large differential between China and US on the national economic competitive advantage based on national economic security, see Figure 3.

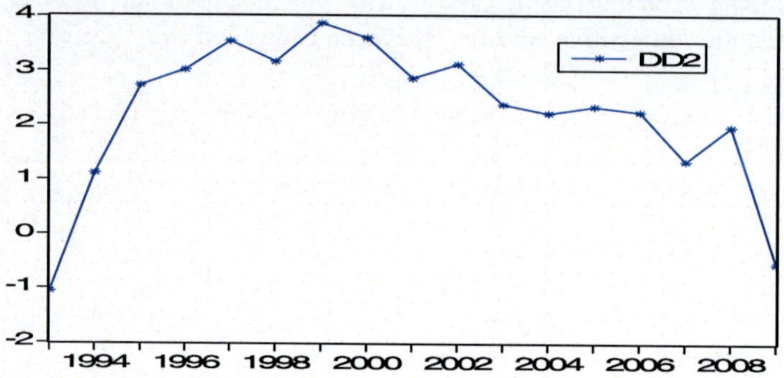

Fig. 3. Differential Graph of National Economic Competitive Advantage between US and China

Note: DD2 represents the gap of national economic competitive advantage based on national economic security between US and China.

Figure 3 presents the differential of national economic competitive advantage between China and US based on national economic security. The detailed results are: firstly, the differential has been enlarging from 1993 to 1999 and it reaches the peak point in 1999; secondly, the gap begins to be narrowed from 2000. Some possible reasons are listed in previous sections.

4 Conclusion and Outlook

This paper does a comparison analysis based on IMD world competitiveness report firstly. It is found that US world competitiveness is stronger than that of China and US changing trend is more stable than that of China. Then, it analyzes the differential gap of national economic competitive advantage between China and US based on national economic security using principal factor analysis of panel data. On the one hand, US has strong national economic competitive advantage and a more stable level of national economic competitive advantage, while it is weak and unstable in China; on the other, the economic competitive advantage of China has been exhibiting an increasing trend. However, there still exists a large differential between China and US on national economic competitive advantage based on national economic security.

In 2011, US debt default crisis has caused a great panic in the world and the crisis brings a negative impact on national economic security. At the same time, the national economic competitive advantage is also facing a great challenge. Therefore, we propose that more relevant researches should be done in the future. Further, more attention should be paid to the relevant problems on national economic competitive advantage based on national economic security.

Acknowledgments. Our work is supported by grants from the General Program of Humanity and Social Science, Ministry of Education, China (No. 10YJAGJW012) and the Eleventh Five-year Planning Program of Guangdong Philosophy and Social Science (No. 09E-15).

References

1. Neu, C.R., Wolf, C.J.: The Economic Dimensions of National Security. RAND (1994)
2. Zhao, Y., Xu, H.P., Xing, G.R.: Risks Faced by Chinese Economy-Theory of National Economic Security. Yunnan People's Publishing House (1994) (in Chinese)
3. Richards, J.G.: Economic Security and National Security: Interaction and Synthesis. Strategic Studies Quarterly 3(3), 8–49 (2009)
4. Pankov, V.: Economic Security: Essence and Manifestations. International Affairs 57(1), 192–202 (2011)
5. Michael, E.P.: Competitive Advantage of Nations. The Free Press (1990)
6. Stiglitz, J.E., Walsh, C.E.: Economics, 3rd edn. W. W. Norton & Company (2002)
7. Stone, H.B.J., Ranchhod, A.: Competitive Advantage of a Nation in the Global Arena: a Quantitative Advancement to Porter's Diamond Applied to the UK, USA and BRIC Nations. Strategic Change 15(16), 283–294 (2006)
8. Aleshchenko, V., Nedelea, A.: Competitive Advantages of the National Economy under Conditions of Open Markets. The Annals of the "Stefan cel Mare" University of Suceava, Fascicle of the Faculty of Economics and Public administration 8(1), 17–21 (2008)
9. Zhang, Y.G.: Strengthen China's Economic Competitive Advantage under the Background of Strategic Transition. Economic Vision (1), 10–11 (2010) (in Chinese)
10. Qin, S.Y., Hu, G.H.: Diamond Model of National Economic Competitive Advantage Based on National Economic Security. Communication in Computer and Information Science 208(1), 250–257 (2011)

The Technology Selection for Realizing the Campus E-Commerce System

Zhaochuan Qi, Qianghua Li, and Shuoling Dong

Institute of Finance and Economics Hebei Normal University of Science & Technology
Qinhuangdao, Hebei, China
Qzch423@163.com

Abstract. It should adopt to the users' habit and the actual requirement of the system itself when selecting the technology to realize the campus Electronic commerce system due to the special users, special characteristic and special environment of the system. The article will discuss the selections of the technology of the campus Electronic commerce system from the sides of the development tools, the Web testing server and the software of the terminal.

Keywords: Campus Electronic commerce system, B/S, Tools and Technology.

1 The Development Tools and Technology

This The campus E-commerce system is a network system based on B/S; two sets of the development tools should be prepared, one set is for the foreground of the website, the other set is for the background of the website.

1.1 The Selection of the Development Tools for the Foreground

The selection of the development tools for the foreground is mainly to select tools for designing the web, the software Dreamweaver or Frontpage could be chosen.

FrontPage has been many users' favors as it is simple and easy to use for web making, just like the description of someone: powerful, simple and easy. It is very convenient to edit and preview the web by operating the tab edit, preview and HTML in the work window. With the help of the editor of GIF animation and graphic, FrontPage support CGI and CSS. The most characteristic of FrontPage is its management function to the website. Not needing to create or change files and directories if website in the server needs to be updated during system maintenance. Not only the files in the local but also could the files in the remote server connected to internet be operated via FrontPage.

Dreamweaver has the software package of visible programming, HTML and code editing; it also well supports ActiveX, JavaScript, Java and Flash. To design dynamic HTML and to make dynamic HTML animation by mouse dragging in the whole course could be done in virtue of Dreamweaver. Adopted technology of Roundtrip HTML, Dreamweaver makes the conversion of the web free between the code editor of Dreamweaver and HTML; furthermore it is convenient to expand its functions,

D. Zeng (Ed.): Advances in Computer Science and Engineering, AISC 141, pp. 437–441.
springer link.com

which is based on the opened designing idea. Dreamweaver also supports the program development language, the basic language of ASP. NET and connecting and operating database.

The campus E-commerce system should be designed to an open system, and should support dynamic HTML designing and convenient function expanding, for the handsome tempo of E-commerce and correlation technology. So ASP.NET is selected to be the basic language of the system, and Dreamweaver is chosen to be the development tool for the foreground designing of the system.

1.2 The Selection of the Development Tools for the Background

For the campus E-commerce system, the J2EE platform was used mostly in the past, which is realized by Java+Jsp, for now the technology of .Net is applied and developed greatly. Commonly J2EE is thought more difficult, but the .NET platform of Microsoft more simple and convenient.

C# is an Object-oriented programming language which was published by Microsoft Corporation as one of latest production; it is applied on the Framework. C# is resembled to Java greatly, such as the single inheriting, the interface and so on. It is also different from Java; it has become the protagonist of .NET windows network frame from Microsoft Corporation, for the realization of direct integration with COM referred on Delphi.

To ensure the information security, confidentiality, integrality and non- repudiation is the basic requirement to the campus E-commerce system. With the feature of security, stabilization, simpleness and elegance, C# is a fit language for the background development. C# has powerful operating ability with innovative language, and supports convenient Object-oriented programming, which makes it become the first selected language for the system development of the campus E-commerce system based on ASP.NET.

Using C# language that a simple development technology decreases the pressure and obstacle to develop, the development of the campus E-commerce system could be developed fast and effectively with low cost for C# system runs on the Window platform but Web system runs on IIS server.

The users of the campus E-commerce system depend on Windows very much. C# makes Windows programs and COM integrate without any gap, which is the advantage of it. A class compiled by C# could make an existing COM component become a child class; the child class can also be used as a COM component. So after a running environment created, the components in the environment providing service for the network can child class any .NET language.

C# has many data types, such as int, uint, long, ulong, short, ushort, float, double, decimal and so on. There will be some money data in the campus E-commerce system; 28 bit decimal digits can be stored using C# data type, which makes the system increase more practicability. So C# is selected as the development language for the system.

ASP application is easy to developed and modified, which can make HTML page, ActiveX component and script command assembled, then the dynamic and alternant Web server and Webs with powerful functions can be created and run.

PHP like ASP is a HTML embedded language, it can execute dynamic web page faster. PHP and ASP are used broadly at home. Another tool for Web development has to be mentioned is JSP which is a relatively new technology. It has become the Web technology standard that was applied broadly. JSP is very popular overseas; especially it is adopted to develop E-commerce websites.

The technologies mentioned above, each has its strong point. Some famous website for example: Sina and Chinaren adopted to PHP. One of reasons for PHP used broadly is that it is supported well in database. Because of the shortcoming of PHP such as disunion interface of database visiting, complex installation, short of supporting from enterprises and formal commerce, the users have to be familiar with UNIX environment, and the platform should be FreeBSD/Linux/UNIX operating system with Apache server, and the project to be developed should be finished quickly with little cost. Then this technology is not fit to be used in a large E-commerce Website, but it is fit for small website.

Relatively ASP and JSP have no the shortcomings that PHP has. ASP can be supported not only by ActiveX which can be got via COM/DCOM of Microsoft Windows, but by DCOM and Transcation Server in configuration. JSP was spark plugged by SUN Corporation, it could get Java Class and EJB scale supporting, and also get framework support of Application Server from many manufacturer.

In three ones mentioned above, JSP which represent the development trend of technology can provide solutions for some large E-commerce platform, that's to say it more suits for the construction of high end E-commerce platform.

The campus E-commerce system requires not only security and steady, but the course of development should be simple and easy to operate, and the running platform should be based on Microsoft Windows operating system. The campus E-commerce system facing to the users in campus should be a medium or low end E-commerce platform, so C# and ASP.NET was selected as the development tool for the background.

ASP.NET which is a translating and editing and running program based on currency language can run preferably on the Web application platform. The campus E-commerce system which is based on B/S structure runs on the Web application platform, ASP.NET suits to the requirement for its system development, so the basic database of currency language and the data interface can be processed and the information mechanism can be integrated. ASP.NET with the characteristic of language independence which supports C# can protect the programs developed with COM+, furthermore the programs can be transplanted whole.

As we know, the identities of the users of the campus E-commerce system have to be validated; ASP.NET can finish this work by running some very common tasks, such as form submitting, furthermore the configuration of the website will be simpler. The users can define legible and convenient profile their own by ASP.NET page constructer.

The continuing running has to be ensured as soon as the campus E-commerce system starts to run, which is a basic requirement for E-commerce system. The server need not have to be restarted when ASP.NET applications are installed for these files are only copied to it.

The developer can define "plug-in" module or add any components defined by him when he develops the campus E-commerce system because ASP.NET has user-defined and expansibility, which makes the work of the development of Website program simpler, avoids the bad effect because of the system expanding.

The system security of E-commerce is the first. The authentication technology basing on Windows and the configuration on every application ensure the original program of the campus E-commerce system security absolutely.

1.3 Database Modeling Tools

Although there are many database modeling tools to select, such as Sybase PowerDesigner, Rational Rose Enterprise, Visio Professional, Dezign, ERD Tool List and so on, Rational Rose Enterprise is selected to finish the database modeling for repeating modification is necessary in the course of database modeling for the campus E-commerce system, and the modeling tools should be convenient to modify, the codes that should not be changed should be unchanged after the model is modified. Actually Rational Rose Enterprise usually is used with power designer together.

2 The Selection of Server for Web Testing

To ensure the concision and shortcut of the development, jboss+tomcat can be selected to be the testing server in the course of developing JSP and Servlet.

Tomcat which was developed by Apache and Sun as the core of the Jakarta project of Apache foundation is a free or saving Servlet container, criterion always could be seen as the Sun participated in and supported.

3 The Selection of the Terminal Ection of Server for Web Testing

3.1 Microsoft IE6.0 Is Selected as the Browser

There are many browsers such as 360 Explorer, saayaa Explorer, Maxthon Explorer and Sougou Explorer and so on. Microsoft Corporation also keeps to publish new version of IE. The browser selected should support IE6.0, for the users are more accustomed to use IE, for Windows is also the mainstream operating system.

3.2 Weblogic Is Selected as Application Server

Application server has becomes a very pivotal technology to be applied in the E-commerce. User's interface, business logic and back-end service could be separated by using application server. Application server is in the middle layer in the distributed three-tier environment, which is among the front interface of user, back-end data and enterprise applications.

The followings are the popular mainstream application servers: IBM Websphere, Sybase Enterprise Application Server, Oracle Internet Application Server, BEA WebLogic.

Among the application servers above, IBM WebSphere with the characteristic of integration database, application development and message service which supports multiple platforms, which accords with the standard of latest Java could be selected as the credible platform for the E-commerce application development.

The particular of Oracle Internet Application Server is mutually compatible with other productions of Oracle. Oracle developed plug-in unit module for Apache to process small Java service programs. The enterprise version of Oracle iAS configuring Oracle Portal provides the tools for users to deploy enterprise information.

BEA WebLogic application server with the high capability and credibility which Website requires more accommodates the requirement of the E-commerce, furthermore it follows up the latest standard. The tool of EJB Deployer could manage multiple EJB.jar files.

Each server mentioned above has its merits and advantages on the function and capability, but BEA WebLogic should be focused on for its high capability/price ratio. The application server of the campus E-commerce system requiring the characteristic of unification, convenience and being able to expand should support MS SQL Server database well, so BEA WebLogic should be the first selection as the application server of the campus E-commerce system.

3.3 The Selection of Database Server

Oracle, IBM DB2 and MSSQL Server are the three mainstream database servers in the current market of database management system, and the difference of their essential capability is little. Although Oracle and IBM DB2 have a good capability of crossing platform, the capability of mutual operating is weakness.

Because the campus E-commerce system is build on the Windows platform, MS SQL SERVER which can be integrated close with Windows system and has excellent capability is the most familiar database server on the Windows platform. The majority system administrators are also familiar with it. So the campus E-commerce system selects MS SQL SERVER as the default database server.

References

1. Peipei, H.: Design of Campus E-commerce System Based on Web. Computer & Telecommunication (1) (2010)
2. Pan, S.: The Design and Construction of Campus Ecommerce Information Platform. Meteorology Journal of Inner Mongolia (3) (2010)
3. Business Office, `http://Services.htmljhsph.edu/businessoffice/StudentAccountsandBusiness`
4. School Charge, `http://www.SchoolCharge.com/SchoolChargeinstantonlinepaymentsystemforeducation.html`

Analysis for Innovation Performance of the Enterprise in Industrial Cluster Based on the Network

Min Cui and Xiujian Wei

Number 74, West Yanta Road, Xi'an, Shaan'xi Province, China
Xi'an Jiaotong University, 710061
mmmm_111@126.com

Abstract. With the high development of economy in era of knowledge economy, enterprise's innovation should be achieved by the network. In a network, the innovation of an enterprise is affected by many factors. Through the exploitation, test, correlation analysis and regression analysis of tables, the article show that there is a positive correlation between network density and technological innovation, a negative correlation between network strength and technological, a positive correlation between network members(Suppliers, partners, financial institutions, intermediaries, government)and enterprise's innovation. It is also give some conclusion: Enterprises in cluster should explore and make good use of the value of the innovative network, for the network members may become potential resource to promote enterprise innovation. The basis of enterprise's innovation is to manage competition and cooperation during the enterprises in cluster, to create trust and cooperation atmosphere. Besides, the government can use policy to guide the formation of a network structure and also take advantage of the innovation to enhance the competition in industry cluster area.

Keywords: Network, Industry cluster, Innovation of the enterprise.

1 Introduction

With the more fierce competitive environment, the diversity and dynamism of people' lifestyle, the technological change of the enterprises become more faster, the product lifecycle gets more smaller, and the competition become more intense in today's modern market, it is difficult for a company to research and innovate by itself. Therefore, network is very important for the enterprises to achieve their innovative goal. Because of the computer and communications technology, the trend of economic globalization become more and more obvious, the production factors such as capital, technology, information and labor force flow freely on the global scale at a high speed. Beside of the economic globalization, the phenomenon of the world economy localization also appeared, that means many similar industries gathered together in a same area, which is called industrial cluster. In an industrial cluster, enterprise use contract, social relationship and information networks as a bond to cooperate with the university, company, research institution, government, capital market, intermediaries. After these resources integrated together, the enterprise

D. Zeng (Ed.): Advances in Computer Science and Engineering, AISC 141, pp. 443–450.
springer link.com

network formed gradually. In the network structure, knowledge and information can be rapidly transmit and spread, if the enterprises can learn, communicate and share the technological knowledge from the other organization. All these actions can accelerate the formation of the technological innovation system, and then promote innovation performance of a enterprise. So, analyse and establish a enterprise's innovation system based on the network is very necessary for a corporation to promote it's competitive and developmental ability.

2 Definition and Feature of the Network

2.1 Definition of the Network

The concept of the network originate from 1960s and 1970s, it has been described as a synthetic gut, metallic thread and some other similar things looked like a "net ". Since 1980s, the conception of network came into vogue. According to Hakansson' viewpoint, network should include three elements---actors, activity and resources. Actors not only contain individual, enterprise and enterprise clusters, but also embody government, intermediaries, educational and training organizations. The action in network include transferring knowledge within organization, external transaction and internal transition activity of the enterprise, as well as the flowing of some important factors such as information, knowledge and technology in a whole network. The resource consist of material resource, financial assets, human resources and so on.

Innovative network mainly analyse the relationship behavior between organizations during enterprise's innovative process and also the innovative efficiency, it primary focus on the flowing of innovative resources. The paper is from such point of view, using empirical analysis to discuss influence of the basic variable on enterprise's innovation in a network.

2.2 Feature of Network's Fundamental Structure

Hakansson and Snehota give the basic variable (actor, action, resource) which influent the network and the connection with the organization (enterprise, relationship, network).The nodes of network mainly include enterprise, university, research institute government and intermediaries and so on (Figure 1), the route of network connected by the nodes become the channel to exchange the information and resource between the actors, and when these nodes (actors) located in different position, they can control the different network information and resource.

Because the network density and intensity determine the impact on the enterprise's innovation performance in an industrial cluster, it is necessary to introduce the concept of the network density and intensity here.

Network density describe the proportion that the real connection account on possible connectionIn coleman' point, high density network can produce a large number of connection between these enterprises, so that the information and resource can flow more rapidly. Besides, high density network can easily develop a trust relationship, a normal principle and a joint action model.

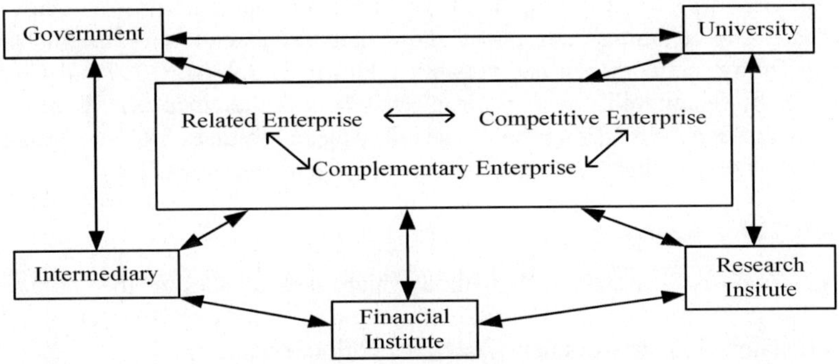

Fig. 1. The Basic Structure of Network

Network intensity describe the frequency of actor's connection and the commitment level the organization resources offer the connection. Granovetter classified network intensity into strong connection and weak connection, and there are two different point of views about the impact of network intensity on the innovation performance. Some scholars think that a enterprises which have a intensive network can form a shared attitude and a common principle, so that it can promote a deep communication for knowledge and information in a network, exchange the valuable information between each other. However, Granovetter think that most of the information in a intensive network is repetitive and similar, in which less innovation opportunity and information can generate, while a weak network can provide an abundant information, so that the weak network is more important compared with the intensive one.

3 Empirical Analysis of Enterprise's Innovation in a Industry Based on the Network

Here, the we hand out and take back 150 pieces of questionnaires from the enterprises which have the innovative ability to some degree, and got the empirical analysis scales. Then we use the SPSS software to give the reliability test on scales, the correlation analysis and regression analysis on innovation factors of the enterprise.

3.1 Reliability Test

Reliability test refers to the consistency of measurement data, it measured wether there exist consistent result on the testers when they accepted repeated test on the same paper, and check wether the measurement itself is stable. There are several measuring method, here the paper adopt Cronbach's α to test. On the measurement of variable, the paper adopt Likert' Five-point analysis, which means we can use strongly disagree, disagree, neutrality, agree and strongly agree to stand for one point, two points, three points, four points and five points. Then we use SPSS to analyze these questionnaires, and get test result about the enterprise's innovation reliability, the impact of network density and network intensity on enterprise's innovation performance. The reliability test result show that Cronbach's α of theenterprise's innovation ability, network density, network intensity is 0.9004, 0.7972 and 0.8034 respectively. In an reliability test, if Cronbach's α is higher than 0.7, it means the scales have high reliability. From the test result, we can see that all of the Cronbach's α is higher than 0.7, so this variable measurement pass the reliability test.

3.2 Correlation Analysis

Correlation analysis is a statistical method which used to analyze the intimacy of variable. The paper use Pearson Correlation method to analyze the correlation during enterprise's innovation, network density and network intensity.

3.2.1 Correlation between Enterprise's Innovation Performance and Network Density

As the result given above, correlation coefficient between enterprise's innovation performance and network density is 0.851, and significant level is less than 1%, which means there is a significant and positive correlation between these two factors.

Table 1. Correlation between enterprise innovation and network density

		Enterprise innovation	Network density
Enterprise innovation	Pearson Correlation	1	0.851**
	Sig.(2-tailed)	.	0.000
	Sum of Squares and cross-products	9.488	5.075
	covariance	0.727	0.175

Note: ** Denotes significant at the 1% level

3.2.2 Correlation between Enterprise's Innovation Performance and Network Intensity

According to the result, we can see that the correlation coefficient between enterprise's innovation performance and network intensity is -0.406, and significant level is less than 5%, which means there is a negative correlation between these two factors. This result correspond with the idea represented by Burt and Granovetter, who think that the a weak connective network can provide abundant information and the information exchange is more valuable.

Table 2. Correlation between enterprise innovation and network intensity

		Enterprise innovation	Network intensity
Enterprise innovation	Pearson Correlation	1	-0.406*
	Sig.(2-tailed)	.	0.026
	Sum of Squares and cross-products	9.488	-2.900
	covariance	0.372	-0.100

Note: ** Denotes significant at the 5% leve

3.2.3 Correlation between Enterprise Innovation Performance and Formal Network

The formal network in an industrial cluster refers to a formal cooperative relationship which formed through a contract made by actors. Such formal network include horizontal relationship, vertical relationship of the maket, and a external relation of the non-market. Horizontal relationship embody the relationship in cooperative enterprises and competitive enterprises, vertical relationship involve the relationship during the enterprise, customer and supplier, while external relation of a non-market mainly mention about the relationship in enterprise, intermediary, research institution, financial institution, government and so on.

Table 3. Correlation between enterprise innovation and formal network

		Enterprise innovation	cooperator	competitor	supplier	customer	financial institute	research institute	intermediary	GOV.
Enterprise innovation	Pearson Correlation	1	0.637**	0.081	0.754**	0.675**	0.523**	0.500**	0.491**	0.759**
	Sig.(2-tailed)	.	0.000	0.669	0.000	0.000	0.003	0.005	0.006	0.000
	Sum of Squares and cross-products	9.488	6.760	0.800	6.000	7.160	4.600	4.400	4.240	6.640
	covariance	0.327	0.233	0.028	0.207	0.247	0.159	0.152	0.146	0.229

Note: ** Denotes significant at the 1% level

As we can see from the table, in these relationships, the significant level between enterprise and other network factors are all less than 1%, which means there is a positive correlation between enterprise innovation performance and other factor in the formal network. Among them, correlation coefficient between enterprise innovation performance and some factors such as supplier, customer and government is quite higher, while there is no significant correlation between enterprise innovation performance and it's competitors.

3.3 Regression Analysis

In order to further analyze the relationship in enterprise innovation performance, network density, network intensity and formal network, it is necessary to use regression model to analyze the causality in these variables, and check the result conclude from correlation analysis.

3.3.1 Regression Analysis in Enterprise Innovation, Network Density and Network Intensity

From the table 4, we can get the regression equation:

$$\text{Enterprise innovation performance} = 0.805*\text{Network density} - 0.138*\text{Network intensity} \quad (1)$$

As we can see from result, the fit of model R^2 is 0.74, which means the degree of fitting is much higher, the value of F is 38.494, which means the model can better explain the enterprise innovation ability. Therefore, through regression analysis, we have further verified that the more dense network can short average route of information transmission, accelerate the flowing of the information, rapidly spread of new knowledge and achievement, which is benefit for the enterprise innovation. Besides, the significant level of network intensity is less than 0.05 and t value is -1.329, which means the t test is successful. The result show that there is a negative correlation between network intensity and enterprise innovation, the more intense the network is, the more disadvantage the innovation will have.

Table 4. Regression analysis in enterprise innovation, network density and network intensity

	Regression coefficient B	Standard error of regression coefficient	Standard regression coefficient	t	Sig.	R^2	Adjusted R^2	F
Network density	1.279	0.165	0.805	7.738	0.000	0.740	0.721	38.494
Network intensity	-0.183	0.138	-0.138	-1.329	0.195			

3.3.2 Regression Analysis between Enterprise Innovation and Formal Network

In this part, we use forward stepwise method to analyze which group are more benefit for the enterprise innovation in a network relationship. Forward stepwise method means the variable will enter in to the regression equation step by step. The variable which enter into the regression equation firstly has the most higher correlation with dependent variable, then the second one will enter into the eguation, in which the correlation is lower than than first one, and the rest can be done in the same manner. According to the result, the regression eguation is:

$$\text{Enterprise innovation performance} = 0.405*\text{Government} + 0.374*\text{supplier} + 0.264*\text{customer} \quad (2)$$

As the result given above, we can see that in the formal network, there are three variables entered into the regression model, which are government, supplier and customer. The fit of model R^2 is 0.711, which means the degree of fiting is quite well. The F value is 29.186, which means the whole equation has a much higher significant level. Besides, the significant level of these three variables are all less than

1%, and all of them are pass the t test. On the other hand, from the order the three variables follow by when they entered into the regression egiation, the government factor is the first one, the supplier is the second on and the customer is the last one. This result show that all of these variables have a positive correlation with enterprise innovation, and the government as a factor has the highest correlation. Therefore, government is the main mechanism which influent the enterprise innovation most, and it use innovation distribution to impact the speed, orientation and scale of the enterprise innovation. Supplier and customer are the next influential factors which only next to the government, which is due to the enterprise, supplier and customer are on a same supply chain.

Table 5. Regression analysis between enterprise innovation and formal network

The entered variable		Regression coefficient B	Standard error of regression coefficient	Standard regression coefficient	t	Sig.	R^2	Adjusted R^2	F
1	government	0.844	0.133	0.769	6.357	0.000	0.591	0.576	40.104
2	government	0.543	0.138	0.494	3.928	0.001	0.726	0.706	35.739
	supplier	0.547	0.150	0.459	3.648	0.001			
3	government	0.445	0.136	0.405	3.276	3.276	0.771	0.754	29.186
	supplier	0.446	0.147	0.374	3.041	3.041			
	customer	0.236	0.104	0.264	2.266	2.266			

4 Conclusion and Enlightenment

The empirical analysis show us that the innovative network structure can stimulate the enterprise innovation in an industrial cluster, promote the efficiency and the success of innovation, keep the motivation of enterprise's consistent innovation, and accelerate the transmission of the innovative achievement in an industrial cluster. Through the conclusion, we can get some enlightenment for the enterprise innovation in an industrial cluster:

First, the enterprise in an industrial cluster should explore and make full use of innovation network, expend and strengthen the connection and interaction with it's cooperator, competitor, supplier, customer, research institution, intermediary, government and so on. The enterprise should should bases on the mutual benefit when they construct the innovation network, so as to ensure the flowing of the information and the stability of the network structure. Furthermore, the enterprise should focus on the diversification of network members to make sure the abundance of information and resource in the network.

Second, on the cultivation and development of an industrial cluster aspect, government can use the policy and market to develop the typical industry and guide a completed network, so that the industrial cluster can promote the core competitiveness through it's innovation. Government also can provide some preferential policy and encourage the technological enterprise, innovative intermediary, education institution to join in the network structure. Through training, guiding and cooperating supplied by government, enterprise can improve it's innovative ability and make full use of the network structure to carry on it's innovation.

Third, regulating the competitive and cooperative relationship of those enterprises in an industrial cluster, creating a trusting and a cooperative atmosphere. Wether the market system and environment are normal or healthy can directly effect cohesion of the whole industrial cluster. In an industrial cluster, there are the competitive and collaborate relationship between each enterprises, in order to prevent the cutthroat competition and the trust losing, the government can enhance it's supervision and the enterprise can adopt the self-discipline to protect the property right of the innovation and the intellectual property right, so that competition can be transparency and the members in the industrial cluster can regulate each other on the mutual interest and benefit.

References

1. Gai, Q.: Innovative network—A new thinking for the development of regional economy. M.Peking University (2002)
2. Zhang, S.: The research of network impact on the enterprise competition in a geographical cluster—An empirical analysis on XinZhu technological industrial area. D. A master thesis in ChaoYang technological University (2002)
3. Simsek, Z., Lubatkin, M.H., Floyd, S.W.: Iterfir networks and entrepreneurial behavior: A structutral embeddeness perspective. Journal of Management (2003)
4. Etikowitz, L.: University and global knowledge economy. M.Jiang Xi education (1999)
5. Huang, W.: An analysis of innovation system and innovation performance in a industrial cluster based on the network. Journal of Ningbo University (Liberal Art and Humanities) (2004)

The Impact of Humanities on Science Education

Yanmin Wu

Xi' an University of Technology, Xi' an, Shaanxi Province, China

Abstract. The study is to talking about the correct understanding of the relationship between science and humanities is a combination of the necessary means to promote quality education. Especially in the humanities and the humanistic spirit of the role of science education. It is found that is very realistic for university to establish the educational concept integrating science education and humanistic education in the case that economy and culture are rapidly developing.

Keywords: Humanities. Science education. Impact.

1 Introduction

Scientific rationality as a tool, but also an integral part of human culture as a whole, therefore, all scientific activities embodies the meaning of human. Conversely, human culture can not be divorced from the reality of life and independent existence, it should be and must be compatible with scientific rationality, united in the practice of human activities. Symbiotic interaction of science and humanities, the same exchange. However, after all, science and humanities, both closely related to each other, but also significant difference in it. Related, that can communicate, can complement each other, distinguish differences that should be interactive and should complement each other, seek common development of harmony, in order to help qualified senior personnel training.

2 Science and Humanities Have Common, But also Different

2.1 Science and Humanities Are Different, But Can Complement Each Other

Science and humanities led to two important human dimension of understanding of the world. Human purpose is to achieve "good", it is to people-centred, rational human values, develop moral character and ideals of human personality, and strive to enhance the spiritual realm. The purpose of science is to seek "truth", emphasis on human intelligence training to improve people's ability to conquer and transform nature. In understanding the complex external world, the two dimensions are essential, because both constitute the basis of human wisdom, if only knew one, it can only be "half a person." Science and humanities dominated the two basic ways of human understanding the world. Scientific thinking mainly refers to the strict logic, it is the basis for the formation of correct understanding. However, scientific thinking are not the only way of thinking, the irrational factors in the humanities ,such as imagination,

D. Zeng (Ed.): Advances in Computer Science and Engineering, AISC 141, pp. 451–454.
springer link.com

intuition, inspiration, etc, plays an important role in the scientific understanding and development.

2.2 Science and Humanities to Promote and Penetrate Each Other

Humanities guide the correctness of scientific development. The goal of science is "truth", but science itself does not guarantee the correctness of its direction, the scientific need human guidance. For example, the early 1970s, the success of recombinant DNA technology, means that human genes of a species can be transplanted to another species or even to create a new species. From a scientific point of view, this is a very promising area of research. Its pioneer, Professor Berg was keenly aware of that if technology unchecked may result in unpredictable consequences of human. In his appeal, the 1975 International Conference to develop the norms and treaties, to ensure safe and reliable testing, genetic recombination experiment was re-launched. This case shows that must concerned about the impacts of science for man's own, natural and social, can make science real for human happiness and effectiveness.

Humanities Provide the basis for the development of science. The goals of humanities is "good", which requires proceed from reality, concern for human survival situation, the human quest for knowledge must comply with objective laws. At the same time, scientific and cultural, scientific knowledge of human culture results, survival and development of human knowledge, to provide support.

3 The Impact of Humanities on Science Education

3.1 Humanistic Education Is an Important Part of Education

With the rapid development of science and technology, education, there have been not overcome its own problems: human control of the means of conquering nature, but did not learn how to live in harmony with nature; people to create powerful weapons, these weapons failed to prevent harm human World War itself; people able to "clone" technology continues to explore, but can not avoid the resulting confusion of the ethical relationship. When we are faced with environmental degradation, the dysfunctional relationship between man and nature, decay of moral values ... Education is also a heavy reflection. There is no scientific education, human development and progress, humanities education, human civilization and prosperity, or even destruction. Therefore, education must take the harmony of science and human development, science and engineering university-trained scientists, not just an "expert" but a knowledgeable all-round development of people.

Humanistic education is to provide value orientation for science education. Science Education is the base to establish the world, which deeply affects people's scientific thinking and changes people's life. Today, for a country, not modern science, there is no advanced technology, is left behind, thus will be beaten. However, the development of science and technology also led to the so-called "development syndrome." These issues are professional and technical personnel with the expertise can not be resolved. Misuse and blind abuse of technology are often due to the lack of nature and a comprehensive understanding of human society. Human needs science, and needs to harness the spirit of health science. Natural science is motivated by the humanities passion, ideals and exploring. Meanwhile only the humanistic ideals as the ultimate concern and goal can we promote human, natural, social harmony and progress. In terms of the scientific and technological development, more attention must be given to value. Only when scientific and technological research and application benefit the humanity as a whole, the long-term survival and development, conducive to their full human development and the full realization of their value, the technology is meaningful for the real progress. If the scientific and technological development of human neglect or deviate from the value of comprehensive care and the comprehensive development of this fundamental purpose, which will damage the natural environment for human survival and the rich man's own spirit and personality development, thus leading to mental wither and alienation of human nature.

3.2 Humanistic Spirit Is an Important Kernel of Scientific Education

Humanistic spirit is benefit to students in mining technology in the value inherent in human. We focus on science and technology learning in science and engineering. Students learn to think of human values of science and technology and also think about how to make every day count are engaged in research and future work, which can be undertaken to promote technological development and social progress. This is a powerful weapon more than a springboard and tool for individuals to meet their material needs and to enjoy the privileged life. Scientists and even scientists, if we can not recognize the humanistic value of technology, science and technology can only be a simple materialized, science, science education will be materialized in external manifestations, then the emergence of science and technology is abused, misused with all the technological lead over the defects appear not hard to understand.

In science education, the devotion of the humanities provides science and engineering students with the spirit of a powerful spiritual force, so that they can grasp the right direction, continue to stimulate creative thinking sparks burst, so that they can reach the heights of science. Meanwhile, the human spirit in the emphasis on human values, human destiny, the meaning of existence and other issues of reflection and exploration, will encourage people to actively explore the human spirit of science and technology. For example, scientific research in practical, practice-oriented hard work, such as unity, cooperation and innovative spirit, the same life, work, scholarship, etc. are required, that they are universal human spirit. Meanwhile, the pursuit of science and technology of the sublime also reflects the highest state of human spirit. Because human spirit of science and technology is hidden in the learning, research and production process, they are not the dominant lead and, therefore, it is necessary to convert them into the minds of college students to become conscious behavior, humane education is precisely played a very important role in imparting knowledge and through human transmission of the human spirit, students think about various social problems, found their own study of science and technology, achieve the wealth scientific spirit and focus on the goals of science ultimate human care and other issues, thus make science education through the development of human knowledge, ability, wisdom, teaching people to resolve the relationship between man and nature.

4 Conclusion

Educational concept is the idea and spirit pursuit guiding educational behavior, which must be determined according to the special educational object, target and task. There is no doubt the impact of humanities on science. It is very realistic for university to establish the educational concept integrating science education and humanistic education in the case that economy and culture are rapidly developing.

References

1. Che, L.-N.: Integration of unified of scientific education and humanistic education. The Guide of Science & Education 3 (2010)
2. Zhou, C.-C., Zhu, J.: Integration and uniform of science education and humanistic education. Modern Education Science 7 (2008)
3. Values in Education and Education in Values. University of London Press, London (1996)

Game Behavior between Institutional and Individual Investors in Chinese Stock Market

Zheng Yan[1] and Rukai Gong[2]

[1] School of Economics, Huazhong University of Science and Technology, No. 1037, Luoyu Road, Wuhan, 430074, China
[2] School of Economics, Shanghai University of Finance & Economics, No. 777, Guoding Road, Shanghai, 200433, China
{advanced,gongrukai}@163.com

Abstract. Based on KMRW reputation model proposed by Kreps, Milgrom, Roberts and Wilson(1982), in the light of the trading behavior characteristics of institutional and individual investors in Chinese stock market, this paper built a model that individual investors play game with institutions according to the observed changes of stock price. The game results show that the behavioral strategies which can achieve utility maximization are not able to be carried out by all institutional and individual investors due to special economic environment in which Chinese stock market is. Therefore, strengthening supervision and enhancing the transparency of information should be the most important tasks.

Keywords: KMRW Reputation Model, Institutional Investors, Individual Investors, Dynamic Game of Incomplete Information.

1 Introduction

The main investors in Chinese securities market are divided into institutional investors and individual investors. Fewer institutional investors have more funds and more individual investors have fewer funds. Although the institutional investors have generous financial advantage and information superiority and may be showing the trend of stock price manipulation, in the game, they need to try to figure out the individual investors' preferences and beliefs before making investment decisions. The institutional investors wish to achieve their utility maximization by taking effective first-mover strategies which can trigger the individual investors' following strategies. The other hand, as a group with information disadvantage, the individual investors must choose their strategies according to institutional investors' first-mover strategies and speculating the type of each institution. In the process of making investment strategies, the individual investors' prior beliefs will be amended according to Bayesian rule, and then individual investors will take their own optimal strategies according to their modified posterior beliefs. In securities transactions, a major strategy of individual investors is modifying the prior belief about the type of institutions according to the change of the price of stocks, and then taking profitable strategy in order to achieve their utility maximization.

D. Zeng (Ed.): Advances in Computer Science and Engineering, AISC 141, pp. 455–462.
springer link.com

Game theory is used as an analytical tool to analyze the behavioral strategies of participants in Chinese securities market mainly from the late 1990s. Lin Guochun (1997) use dynamic game of incomplete information to analyze the game relationship between institutional investors and individual investors in Chinese stock market. Jiang Xiaogan et al (2001) established an asymmetric information dynamic game model about institutional investors and individual investors. He proposed to solve this problem by making institutional innovations which will make the manipulative institutions pay a heavy price for manipulating market. Wang Jiankun (2000) made the analysis of the game relationship between dealers and retail investors, dealers and dealers, dealers and government, dealers and stock exchange, respectively. Li Hua, and Du Li (2002) established a model based on the game relationship between listed companies, institutions and individual investors, and made a simple analysis of the operation and equilibrium condition of the stock market. Based on the structure of Chinese stock market investors, Xiao Xinrong and Tian Cunzhi (2002) used a game framework of Laffont and Maskin (1990), under the assumption of the negative exponential utility function, to research separate pricing strategy and confused pricing strategy of institutions and investors, and proved that Chinese stock market was inefficiency. Through making an in-depth analysis and the evolutionary game of the group behavior of stock market, Xie Shiyu (2002) pointed out that the key measures for achieving stable development of Chinese stock market were allowing credit trading, expansion and prohibiting the behavior of manipulating stock market. Wang Jiankun (2000) established a game model of investors, and pointed out that the behavior of investors would cause the stock market bubble under the condition of incomplete information.

The stock market is a game arena which contains government, stock exchange, listed companies, institutional and individual investors, in order to maximize their own utility, each party is bound to observe the behavioral strategies of the other parties and develop one's own action strategy accordingly. In addition, Chinese stock market is not yet mature, so the study of behavioral game of relevant actors in Chinese stock market is even more necessary for protecting the interests of individual investors and maintaining market stability.

Based on KMRW reputation model proposed by Kreps, Milgrom, Roberts and Wilson (1982), in the light of the trading behavior characteristics of institutional and individual investors in Chinese stock market, this paper focused on the transaction process, built a game model of individual and institutional investors, and made a detailed analysis of their game behaviors.

2 The Model

2.1 Basic Assumptions

(1) Supervision department has perfect functions in the management of market, namely keep stock market just, fair and relatively stable in order to make sure no severe concussion during development process of stock market.

(2) Both two subjects of participants in the stock market, institutional and individual investors have enough rationality to strive for maximizing their own utility under certain constraints.

(3) The price manipulated by institutions fluctuates within a reasonable range, that is to say, the change of a separate stock's price is in the fluctuation range of the whole stock market.

2.2 Game Design

The KMRW reputation model proposed by Kreps, Milgrom, Roberts and Wilson (1982) brought incomplete information into repeated games and proved that the incomplete information what one participant has about the payment function or strategy space of other participants, has great effect on the equilibrium result. As long as the game repeated enough times, not necessary to be infinite, cooperative behavior will appear in a limited times game. At the same time, it shows that a "bad person" maybe behave as a "good person" for quite a long time.

In general, the institutional and individual investors are in extremely unbalanced status in the information acquisition and interpretation, market analysis ability and many other aspects. By a large number of professional and extensive information collection channels, institutional investors have the professional ability to analyze the economic situation, macro policies and situation in production and management of listed companies comprehensively, have the inside information what individual investors can't obtain. In contrast, individual investors usually only know some public information which is almost "Common Knowledge" for all the investors. Admittedly, there is obvious "unequal trading status" between institutional and individual investors. In this case, the two players play the game obeying path as follows.

First, virtual "nature" chooses the signal, strong or weak, as the type of institutions. Then, institutions take action according to their types. After that, individual investors infer the types of institutions by watching the reaction of institutions' action on stock price and wait a chance to adopt their strategy. Therefore, the game between institutional and individual investors belongs to dynamic game of incomplete information category. The optimal strategy for both two players is to reach the refined Bayesian Nash equilibrium.

Since institution's action and its type depend on each other, "natue" chooses the type of institutional investors first, institutions are aware of their types ahead but individual investors don't. Assume the prior probability that individual investors make subjective judgment on the type of institutions is η_1 (strong institutions) and η_2 (weak institutions), so $\eta_1 + \eta_2 = 1$. If one buys stocks manipulated by strong institutions, he will get rich return, while it will gain little and even get loss, if buy stocks which weak institutions manipulate. After "nature" chose, institutions begin to act. Individual investors speculate the action of institutions through the observation of the specific conditions of the stock price rising and falling and then modify the prior beliefs on the type of institution and wait a chance to take the optimal action. If individual investors induce the type of institutions is strong based on posterior faith, the individual investors will take the buy strategy; Otherwise, the individual investors will take wait-and-see (not buy in) strategy. At the same time, institutions forecast that their actions are observed by the individual investors and they speculate the type of institutions correspondingly. Then they try to relay good news to themselves. In order to enhance their reputation, increase the value of their own assets and prepare for the future selling, strong institutions often hope that the individual investors follow when they lift the stock price. And weak institutions also make up themselves. They try their best to disguise as a strong institution to attract individual investors to follow. which is good for them to clear their stock rapidly and leave with profit in time and avoid the loss. Therefore, both palyers need to revise prior beliefs constantly according to the Bayesian law. And they will wait and see the best time to choose perfect action strategies.

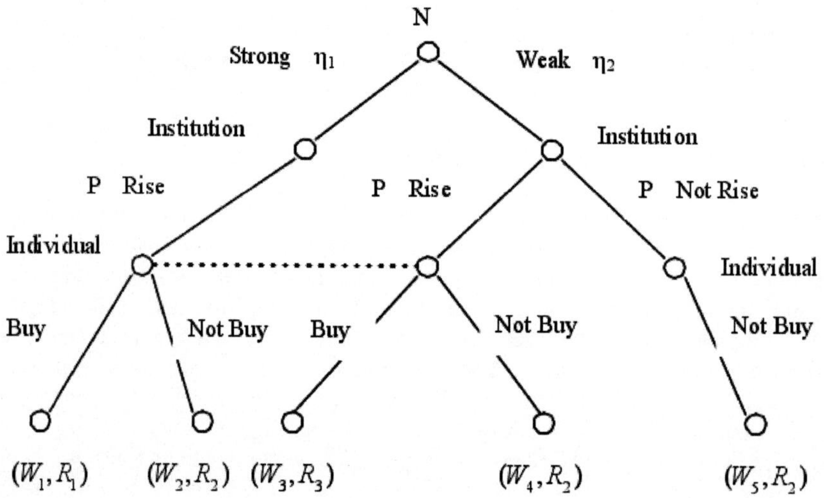

Game Map

There are two types of institutions, η_1 and η_2, but only one type of individual investors, and η_1 and η_2 are also prior probability that individual investors judge about the types of institutions. Let P stand for the stock price observed by individual investors. Assume, W_i is the revenue of institutions, $i = 1, 2, 3, 4, 5$, and $W_1 > W_2 > W_3 > W_5 > W_4$, R_j is the revenue of individual investors, $j = 1, 2, 3$, and $R_1 > R_2 > R_3$.

Specific game processes are as follows. If strong institutions induce individual investors to follow and buy successfully, the profits W_1 are biggest. Strong institutions lift the stock price, but aren't able to let individual investors to follow correspondingly, then the profit is W_2, which is higher than the profit W_3 that weak institutions attract the individual investors' follow successfully. This is due to the capital strength of strong institutions is much stronger than weak ones. And this will form the "sing and play solely" situation of strong institutions in market. Even if a few individual investors follow, strong institutions can still maintain the high stock prices. After a long time of sending stocks and sharing out bonus, they dilute share prices eventually and clear market successfully. Weak institutions push up the stock price and don't make the individual investors follow. At the same time for their weakness of own financial strength and the pressure of clearing, their profit W_4 is the least. At the moment it is possible for weak institutions to be back to a prototype and form a series of bad circulation.

At the same time, individual investors will also get benefits in the game. The individual investors get the biggest profit R_1 by following up stock controlled by strong institutions. And no matter the judgment about the type of institutions, individual investors don't follow up and get the profit $R_2 = 0$. At this point there is no transaction. If individual investors misjudge the type of institutions, weak institutions imitate strong institutions lift stock price and take the follow strategy, they get the least profit R_3 ($R_3 < 0$). This is because that with the weak institutions' successful attracting, individual investors will face lower stock prices and suffer a loss, even be locked up for long.

2.3 Strategies

Suppose the utility function of institutions in Single stage is:

$$W = -(1/2)\,v^2 + a\,v - b\,(v - v^e) \tag{1}$$

In the function, a is the highest growth that stock price can reach objectively for stage game, $a > 0$. b is the type of institution. When $b = 0$, namely a strong institution. At this time, the individual investors follow up, stock prices would continue rising. When $b = a$, namely weak institutions. There is no probability that the stock price continues rising. At the moment if individual investors follow up, stock price will drop rather than rise; individual investors have risk of getting loss and being bound. v represents the actual stock growth rate that institutions control. v^e represents the stock growth rate that individual investors expect.

By formula (1), If $b = 0$, it is strong institution, so $W = -(1/2)\,v^2 + a\,v$. By first order condition is 0; we can get the optimal strategy of institution, $v^* = a$, correspondingly $W = -(1/2)\,a^2$

When $v > a$, W decreases with v increasing; when $v \geq 2\,a$, $W \leq 0$, which means that when share prices rises too fast and it exceeds $2\,a$, institutions may get loss. The reason is when individual investors observe the rapid rise in stock price, they may think it as institutions are ready to "pump and dump", and instead more reasonable individual investors dare not follow up easily. At this time, there is a lack of persons who chase after and make stock price rises further, unless the strong institution have super capital strength to control the trend of stock completely. Otherwise the probability that this situation happens becomes little with the development of the stock market and the government's management. At this time, the optimal strategy for the organization is $0 \leq v \leq a$.

If $b = a$, the institution is weak, $W = -(1/2)\,v^2 + a\,v^e$

We can get $v^* = 0$, namely the expected stock growth probability that institutions control is 0. If individual investors is rational, then its $v^e = 0$, so at the moment the utility of institutions is $W = -(1/2)\,v^2 + a\,v^e = 0$.

At this time, the weak institutions know individual investors will see through its face, the optimal strategy is not controlling prices, in order to avoid the situation that there is nobody chasing after and getting loss.

But individual investors' group is often hindsight, and there is a certain delay in the process of amending beliefs, for information of participants on the stock market is asymmetry. At the same time, the institutions will make noise information consciously to make individual investors misjudge the situation. If individual investors expect mistakenly $v^* > 0$, then $W > 0$, hence the weak institutions will get their way, but the individual investors will suffer losses.

3 Dynamic Game Equilibrium

Suppose that in the initial period, namely $t = 0$, the prior probability when $b = 0$ is p^0, so the prior probability when $b = a$ is $(1 - p^0)$, that is, individual investors think that the prior probability of the type of institutions is strong is p^0, is weak is $(1 - p^0)$. Suppose that y_t is the probability of weak institutions choose the strategy of lifting stock price in the period of t.

This game is a T-stage repeated game. Suppose that x_t is the probability that institutions choose the strategy of continuing lifting stock price when individual investors think institutions are weak. In the equilibrium conditions, $x_t = y_t$. Based on the Bayes rule, if in the period of t, individual investors observe the stock price continues rising, $v_t = a$, now the posterior probability that individual investors deduce the institutions are strong in the period of $(t + 1)$ is:

$$p_{t+1}(b = 0 \mid v_t = a) = (p_t \times 1) / [p_t \times 1 + (1 - p_t)\, x_t] = p_t / [p_t (1 - x_t) + x_t] \geq p_t \qquad (2)$$

In the function:

p_t is the probability that institutions are strong in the period of t. The probability that strong institution make stock price rise is 1. If individual investors observe stock price rising, they think the probability that the institutions are strong increasing. When $x_t < 1$, the inequation in formula (2) is correct strictly; when $x_t = 1$, the equation in formula (2) is correct.

If Individual investors observe stock price not rising in the period of t, $v_t = 0$, individual investors speculate that the probability of the institutions are strong is:

$$p_{t+1}(b = 0 \mid v_t = 0) = (p_t \times 0) / [p_t \times 0 + (1 - p_t)\, x_t] = 0 \qquad (3)$$

That is, when individual investors speculate the institutions are weak, the optimal strategy of individual investors is not buying (wait and see). Therefore, weak institutions should choose lift stock price as the best strategy until the last period of T, in order to attract investors to follow up the strategy and control the stock price to make a profit. In the last period of the game, the stock price will reach the maximum, and then institutions don't have the necessity to continue lifting the stock price. Meanwhile, the optimal strategy of the institutions is $v_T = 0$, and the expectancy of individual investors is $v_T^e = a\, p_t$. The utility of the institutions is:

$$W_T = -(1/2)\, v_T^2 + a\, v_T - b\,(v_T - v_T^e) = a\, v_T^e = a^2\, p_T$$

According to the first order condition, we get $a^2 > 0$. It indicates that the utility of weak institutions is a increasing function of its reputation. It is the internal reason that weak institutions package themselves and build the external image like strong institutions reputation. The institutions prop the stock price up more powerfully, the more individual investors believe that the institutions are strong. The longer the weak institutions disguise, the more the profits are. If the weak institutions continue pushing shock price before the stage $(T - 1)$, it will make individual investors buy in, at the price of $p_{T-1} > 0$, by formula (2), and the probability that the individual investors expect the stock price will rise is: $v_{T-1}^e = p_{T-1}\, a + (1 - p_{T-1})\, x_{T-1}\, a$. Make δ $(0 < \delta < 1)$ as the discount factor of institution, which indicates its patience degree. Consider the basic strategy of weak institutions $y_t = 0, 1$. If the weak institutions choose to cast stock in stage $(T - 1)$, namely $v_{T-1} = 0$, $y_{T-1} = 0$, then $p_T = 0$. The total revenue of institutions is:

$$W_{T-1}(b = a) + \delta\, W_T(b = a) = -(1/2)\, v_{T-1}^2 + a\, v_{T-1}^e + \delta\,[-(1/2)\, v_T^2\; a\, v_T^e] = a\, v_{T-1}^e$$

If weak institution continue rising the stock price in the period $(T - 1)$, namely, $v_{T-1} = a$, $y_{T-1} = a$, its total revenue is:

$$\begin{aligned} &W_{T-1}(b = a, v_{T-1} = a) + \delta\, W_T(b = a, v_{T-1} = a) \\ &= -(1/2)\, v_{T-1}^2 + a\, v_{T-1}^e + \delta\,[-(1/2)\, v_T^2\; a\, v_T^e] \\ &= -(1/2)\, a^2 + a\, v_{T-1}^e + \delta a^2\, p_T \end{aligned}$$

In order to make $v_{T-1} = a$ better than $v_{T-1} = 0$, it requires,

$$-(1/2)\, a^2 + a\, v_{T-1}^e + \delta a^2\, p_T \geq a\, v_{T-1}^e$$

So we get,

$$p_T \geq (1/2\,\delta).$$

in the equilibrium state, $x_{T-1} = y_{T-1} = 1$, by formula (2), we have,

$$p_T = p_{T-1}, \text{ and } p_{T-1} \geq (1/2\,\delta)$$

The formula above indicates that, in the period $(T - 1)$, if individual investors consider the probability that institutions continue raising stock price is $p_{T-1} \geq (1/2\,\delta)$, so weak institutions would also pretend to raise prices aggressively. Hence, the equilibrium is that institutions will continue raising stock price in the period $(T - 1)$ to attract individual investors to buy in, and sell off in period T, get the maximum utility. The prerequisite is $p_{T-1} \geq (1/2\,\delta)$, which means that the institutions' patience of raising stock price is sufficient.

4 Conclusion

Based on KMRW reputation model proposed by Kreps, Milgrom, Roberts and Wilson (1982), this paper build the dynamic game of incomplete information to capture the game behavior among institutional and individual investors in Chinese stock market, Since short-mechanism is still not set up in Chinese stock market and the small number of shares is in short supply, institutions involved in almost all of the stock transactions. The game results show that, using a number of advantages of information and funds and lifting stock prices to lure individual investors to follow to buy on the upswing is institutions' rational choice of the optimal strategy. For the disadvantage of asymmetric information and less funds, individual investors' posterior belief can be amended according to the observed change of stock price, but they are easily misled by institutions. Even if the individual investors could accurately determine the type of institutions, they also may miss the optimal invest opportunity because of a long delay in this process. In such case, actually the optimal strategy in game is very difficult to implement. If the transparency of information can not be effectively increased and supervision can not be effectively strengthened, individual investors will undoubtedly be the biggest victims in the market.

References

1. Kreps, D.M., Milgrom, P., Roberts, J., Wilson, R.: Rational Cooperation in the Finitely Repeated Prisoners' Dilemma. Journal of Economic Theory 27, 245–252 (1982)
2. Laffont, J.J., Maskin, E.S.: The Efficient Market Hypothesis and Insider Trading on the Stock Market. Journal of Political Economy 98, 70–93 (1990)
3. Laffont, J.J., Maskin, E.S.: A Differential Approach to Dominant Strategy Mechanisms. Econometrica 48, 1507–1520 (1980)
4. Wang, J.: Game Analysis of Chinese Stock Market. Finance & Trade Economics 217, 54–57 (2000)
5. Xiao, X., Tian, C.: Institutional and Individual Investors in Stock Market: A Speculative Game Analysis. The Journal of World Economy 285, 62–68 (2002)
6. Jiang, X., Wu, Q., Han, Y., Jiang, S., Cheng, G.: Stock Market Game Model under Asymmetrical Information. Systems Engineering-Theory & Practice 154, 24–28 (2001)
7. Xie, S.: Group Behavior Game and Systematic Scientific Analysis for Speculation In Stock market. The Journal of Quantitative & Technical Economics 223, 69–72 (2002)
8. Li Hua, Du Li: Game Analysis of Chinese Stock Market. Shanghai Finance 263, 29–31 (2002)
9. Lin, G.: Speculative Game between Institutional and Individual Investors in Stock Market. Economic Science 100, 44–48 (1997)

Industry Stock Price Effect and Its Influencing Factors of Cash Dividend Distribution: Based on Chinese Real Estat Listed Companies

Rongcheng Yan[1] and Chi Xie[1,2]

[1] School of Business Management, Hunan University, 410082 Changsha, China
[2] Center of Finance and Investment Management, Hunan University 410082 Changsha, China
yanrongcheng111@163.com, xiechi@hnu.edu.cn

Abstract. This paper examines industry stock price effect of cash dividend announements in Chinese listed real estate companies and finds that the own-impact of announcers experience a significant decrease during the announcement period. This findidng, coupled with a significant abnormal returns in the same direction of nonannouncing firms, suggests the existence of industry stock effect to cash dividend. Further study of its influencing factors indicates that the abnormal return and firm size of announcers, the degree of homogeneity between announcing and nonannouncing firms, and the level of asymmetric information in nonannouncing firms are important determinants in explaining nonannouncers' abnormal returns.

Keywords: Cash dividend, Industry Stock Effect, Influencing Factors, Real Eatate.

1 Introduction

As a core of profit distribution in contemporary corporation, cash dividend policy has always been a hot topic at home and abroad. Previous studies make great achievements on cash dividend from micro and macro level. In micro level, firms' characteristic and quality, such as the negotiability and concentration of corporate ownership structure and business performance, are considered as main factors influencing cash dividend[1]; Based on catering theory, agency hypothesis and signal transfer hypothesis, the motivation and market reaction of cash dividend are studied. The results show that, inorder to cater for investors' demands, the level of cash dividend is different according to the different stage of listed companies. Meanwhile, affected by the randomness of cash dividend distribution and the speculative of Chinese securities market, agency hypothesis and signal transfer hypothesis are hardly work. The findings of most prior studies document that the distribution of cash dividend is utilized by large shareholder for cash and isn't welcome for investors[2]. In macro level, policy guidance of cash dividesd is the main factor considered. Related researches show the change of income tax, the specific stipulations of cash dividend put forward by Securities Regulatory Commission have produced significant policy effect[3].

D. Zeng (Ed.): Advances in Computer Science and Engineering, AISC 141, pp. 463–469.
springer link.com

Industry factor is an important external governance mechanism for listed companies, however, there are few researches of cash dividend from industry level. This paper focus research object on real estate which has colse relation with economic growth. As a capital-intensive and resource-integration industry, real estate is slow in cash folw, large in cash demand, and sensitive to cash dividend distribution. Studying the industry stock price effect and its influencing factors to cash dividend distribution in real estate industry, can enrich and expand related research of cash dividend with the purpose of helping investors and listed companies to make effective descision.

2 Theoretical Analysis

2.1 Industry Stock Price Effect of Cash Dividend

Cash dividend announcement reveals information about the industry in general and affects market performance of nonannouncing firms, therefor prosuces industry stock price effect. Industry effect includes contagion effect and competitive effect. Contagion effect to cash dividend means the stock price reaction for nonannouncing firms in the industry moves in the same direction as that of the announcing firm. Because of asymmetric information, external investors cann't make accurate judgement for firms' real value. The homogeneity of cash flow and management model among companies in same industry is significant in a extent, so the management state of a company probably implies that of other compamies. Cash dividend announcement is a process of information leakage in certain sense. During this process, both self-value and other firms' real value in the industry are partly exposed. Therefore, the stock price effect of announcers may have a contagion effect acorss other firms in the industry.

However, real estate is an industry with high competition and the market concentration is gradually strengthened. Cash dividend announcements break the competition equilibrium in the industry, which may benefit to its industrial rivals and cause a shift in the stock price to the rivals. That's the competitive effect. The prior literature documents that cash dividend usually becomes a carrier for controlling shareholder to transfer interests and damages coporate value. Thus, the distribution of cash dividend is good news for rivals at a certain extent in the industry, which may mean new chances to transfer investment opportunity and market profit. In other words, when facing with new investment opportunity, there is more sufficient cash flow for nonannouncing firms to relize value added.

What needs to piont out is that contagion effect and competition effect are not mutually exclusive. The observed industry stock price effect is the sum of that two effects with opposite orientation, therefor it may insignificant. The actual conditions still need further empirical test.

2.2 Influencing Factors of Industry Stock Price Effect

The Degree of Homogeneity. The greater degree of homogeneity between announcing cash dividend firms and nonannouncing firms always means the larger probability the two affected by common factors. Cash dividend announcements will produce a greater effect on nonannouncing firms with higher degree of homogeneity due to the considerable information transfer efficiency among them. Firth suggests that firms that are similar to each other have a high earning correlation[4]. So, the paper uses stock earning correlation between announcing firms and nonannouncing firms over one year period preceding the cash dividend announcement to measure the degree of homogeneity.

The Level of Asymmetric Information for Nonannouncing Firms. Chinese security market is filled with lots of speculative transactors. After announcing cash dividend, nonannouncing firms with higher level of asymmetirc information are more likely to be speculated. That means that kind of nonannouncing firms will suffer more significant abnormal returns from the announcement than will other firms in the induatry. Krishnaswami and Subramaniam suggest that the dispersion in stock returns captures the level of asymmetirc information well, therefore, the return variance is used as a proxy for the level of asymmetirc information[5].

The Stock Reaction of Announcing Firms. Cash dividend announcements that are companied by significant price reaction for the announcing firms are, other things equal, more likely to transfer industry related information and affect the stock price of nonannouncing firms. Therefor, industry effect will be more significant. It is considered that there is a positive association between cumulative abnormal returns of announcing and nonannouncing firms.

The Size of Announcing Firms. Larger firms are aways more dominant and representative and likely to reveal industry relevant information, which will probably affect nonannouncers' valuations. The proxy of size is the natural logarithm of the market value of the announcers one month prior to the cash dividend announcement.

The Dividend Payout Ratio. Dividend payout ratio is the ratio of cash dividend per share and earnings per share. High ratio of dividend payout means a high profit for investors, while also indicates a high possibility of ultra-ability dividend payout which may result in the loss of investment opportunity because of lacking funds and go against long term development. However, considering the significant negtive market reaction of cash dividend documented by prior literature, we predict there is a negtive relation between the dividend payout ratio and abnormal returns for nonannouncing firms.

3 Empirical Analysis

3.1 Sample Selection and Data Sources

The sample used in this study is obtained form Chinese listed real estate companies that finish shareholder structure reform and issue net cash dividend announcement in Shenzhen and Shanghai stock exchange from 2007 to 2011. Inorder to avoid the influence of outliers and new stocks, we eliminate ST or PT samples and samples listed less than one year when distributing cash dividend; Besides, those samples lacking transaction data are also excluded. The final samples consist of 123 cash dividend anncouncements.

The matching sample, that is the nonannouncing firm, satisfies the following criteria: (1) They have transaction data during the announcement period; (2) They do not announce any kind of dividend and do not polluted by other important events during the announcement period. Considering the announcement period for a given cash divided is the same for all nonannouncing firms in the industry, these industry counterparts are formed into an equally weighted portfolio to study the reaction of nonannouncing firms[6].

The announcements of cash dividend announcement, stock returns, market returns, market value and earnings per share of all samples are collected and arranged from RESSET data source(www.resset.cn).

3.2 Statistical Characteristics of CAR

We adopt event study to examine whether the cash dividend announcement produces industry effect. The day issues the announcement is defined as event day(t=0), and the stock returns 240 days preceding the cash dividend announcement are used to estimate the parameters of market model as follows:

$$R_{it} = \alpha_i + \beta R_{mt} + \varepsilon_{it} \quad (1)$$

Accordingly, the cumulative abnormal returns(CAR) during the event window T(−1, 1) are calculated. The cumulative abnoamal return T-test results for both announcing and nonannouncing firms are presented in Table 1.

Table 1. Cumulative Abnormal Return T-test Results

Announcing Firms		Nonannouncing Firms	
N	123	N	123
Mean	−0.0107	Mean	−0.0030
Sig	0.009	Sig	0.057

It is showed in Table 1 that, for announcing firms, cash dividend announcements cause a large decrease to average cumulative abnormal return of −1.07%, significant at the 1% level. The result is consistent with the researches prior to shareholder structure reform that cash dividend is unwelcome for investors.

For nonannouncing firms, the mean cumulative abnormal return is −0.3%, and significant at the 10% level. This suggests the existance of intra-industry reaction at the announcement of cash dividend. That means, nonannouncing firms is affected by both contagion effect and competitive effect of announcing firms, but the contagion effect is overrun competitive effect at the mean level, producing negtive and significant abnormal return to matching samples.

The reasons are showed as follows. From macro level, the increasing strength of macro control to real eatate industry and the frequent promulgation of property-purchasing limitation restrict real eatate's overdevelopment. Meanwhile, affected by continuous deflation of credit policy, capital flow of real eatate is short in supply, resulting the growth slows down. From industry level, it is a seeking and developing period for real eatate accompanied with uncertain return and risk during 2007-2011. Captial is in great demand but tight for real eatate at the growth stage, and the distribution of cash dividend influences the listed companies' net profit and cash flow directly. Therefore, cash dividend affects the value expection of both announcing and nonannouncing firms. It can be said that the stock price effect manifested in real eatate industry is a instinct response inflecting by related administrative stipulation and the development degree itself. From company level, in post shareholder structure reform era, controlling shareholders keep the same interests with minority shareholders due to the overall circulation of shares, and the motivate to transfer intrest by distributing cash dividend for controlling shareholders is not tenable in theroy. However, the fact showed in the study documents that the phenonmenon of unwelcome for cash dividend is continued. There is a delay to achieve expected target for shareholder structure reform.

3.3 The Regression Anaysis of Influencing Factors

Based on the former theoretical analysis, we can specify the model as follows:

$$CAR = \alpha + \beta_1 CAR_d + \beta_2 DPR + \beta_3 COR + \beta_4 VAR + \beta_5 SIZE \tag{2}$$

Where

CAR= one day, value-weighted cumulative abnormal return of nonannouncing firms;

CAR_d= one day cumulative abnormal return of announcing firms;

DPR= the ratio of cash dividend per share and earnings per share;

COR= earning correlation, representing the degree of homogeneity between announcing and nonannouncing firms;

VAR= return variance, representing the level of asymmetric information for nonannouncing firms;

SIZE= the logarithm of market value of the announcers one month prior to the cash dividend announcement.

The regression estimates are presented in Table 2.

Table 2. Multivariate Regression Results

Variables	Unstandardized Coefficient		Standardized Coefficient	t	Sig.
	B	*Std. Error*	*B*		
(Constant)	−0.228	0.114		−2.005	0.047
CARd	0.070	0.034	0.189	2.080	0.040
DPR	−0.007	0.005	−0.135	−1.501	0.136
COR	0.025	0.014	0.177	1.747	0.083
VAR	1.687	0.767	0.202	2.200	0.030
SIZE	0.022	0.011	0.198	1.915	0.058

From Table 2 we can find that the CAR_d variable is positive and significant at 5% level, reflecting that the cumulative abnormal return is associated with a counterpart positive reaction to matching samples. This finding reinforces the earlier conclusion drawn in relation to Table 1 that cash dividend has significant contagion effect.The coefficient for COR is also positive and significant as predicted, implying that cash dividend announcement transfers more information for nonannouncing firms with a higer homogeneity to the announcing firms than for those that are lower homogeneity. The VAR exhibits a significant posotive role at 5% level in the regression, supporting the hypothesis that nonannouncing firms with high level of asymmetric information experience a greater return. Besides, the coefficient on size of announcer firms is positive and significant, suggesting that the larger in size, the more likely their cash dividend announcements are to be industry relevant. DPR has no role to play in the regression, which may related with the lower level of dividend payout in real estate industry.

4 Conclusion

This paper examines whether cash dividend announcements affect stock prices of nonannouncing firms in real estate industry. The results indicate that, with regard to the own-impact of announcers, firms react negtive and significant arround the announcements; with regard to the cross-impact on non-announcers, they expreience a small but significant decrease during the announcement period. This finding suggests that the information carried by cash dividend announcements conveys ralated information of nonannouncing firms and results in industry effet.

The regression results indicate that announcing firms with higher CAR and dividend payout ratio obtain higher returns; nonannouncing firms with higher asymmetric information or similar to announcing firms can earn higher abnormal returns.

The study results can supply a more informed view from industry effect for investors to get related information and adjust portfolios. Meanwhile, considering that the firms' action can have wide ranging impact on transfering information within the real eatate industry, it is an effective strategy for listed company to construct adaptive dividend distribution mechanism according to industry and firm characteristics.

Acknowledgments. We are grateful to the editors for their review and recommendations on the article. Research reported here is partially supported by the National Social Science Foundation of China(07AJL005), Doctoral Fund of Ministry of Education of China (20070532091) and Humanities and Social Sciences Project of the Ministry of Education(09YJC630063), thanks for the sponsors.

References

1. Yan, D.Y.: The Propensity of Large Ownership and Cash Dividend Behavior in Chinese Stock Market: An Empirical Analysis. Nankai Economic Studies 6, 94–100 (2004)
2. Lv, C.J., Xu, J.J.: Dividend Signaling Effect: From the View of Dividend Change Announcement. Nankai Business Review 13, 90–96 (2010)
3. Li, Z.F., Zhang, S.F.: Can the Policy of Reducing Dividend Income Tax Increase Cash Dividends that Chinese Listed Companies Pay: A Study on the Effect of the Taxation No. 2005102. Finance & Trade Economics 5, 34–36 (2010)
4. Firth, M.: Dividend changes, Abnormal Returns, and Intra-industry Firm Valuations. J. Fin. Qua. Ana. 31, 189–211 (1996)
5. Krishnaswami, S., Subramaniam, V.: Information Asymmetry, Valuation, the Corporate Spin-off Decision. J. Fin. Eco. 53, 73–112 (1999)
6. Tawatnuntachai, O., Harrisburg, P.S.: Intra-industry Reaction to Stock Split Announcements. J. Fin. Reas. 25, 39–57 (2002)

Low Carbon Economy and New Energy Applications in China

Liming Gong[1], Qiang Wu[1], and Wanbin Bi[2]

[1] School of Economics, Huazhong University of Science and Technology, No. 1037, Luoyu Road, Wuhan, China. (430074)
[2] School of Management, Wuhan University of Science and Technology No.947, HePing Street, Wuhan, China. (430081)
jiuarthur@163.com

Abstract. Investigations have been carried out on the relationship of energy, environment problem and economic growth, however, little attention has been paid on background of low carbon economy. It is the purpose of this paper to interpret the connotation of low carbon economy, analyzing how to achieve a low carbon economy, and effects of new energy application on China. With the method of unity of history and logic, we find that energy is the driving force of economic growth, and decides the scale and speed of economic growth. Moreover, the growth rate of environmental pollution should be less than zero and growth rate of energy stock should be greater than zero. At last, we analyze the development of low carbon economy and new energy applications to the Chinese economy.

Keywords: Low carbon economy, new energy, sustainable development, economic growth.

1 Introduction

The emergency of energy and environment issues makes resources and environment to be included in the production function. When resource elements are concerned, you will find the energy consumption and economic growth is closely related; economic growth depends on their level of satisfaction of energy needs. Energy is the driving force of economic growth, and decides the scale and speed of economic growth. From the perspective of long-term economic growth, environmental quality and energy as endogenous growth factors are set in the analysis framework, when consumers maximize their intertemporal utilities, optimal dynamic process of system shows that, in the long run of economic growth, in order to achieve sustainable economic growth and ensure sustainable use of energy, the ratio of growth rate of energy and growth rate of human capital should be less than that of output elasticity of human capital and energy output elasticity.

In order to make growth rate of environmental pollution less than zero and growth rate of energy stock greater than zero, we should develop low carbon economy, using of new energy technologies to supplement and replace traditional fossil energy sources. The low carbon economy is a low power pollution-based economy, it creates a new

D. Zeng (Ed.): Advances in Computer Science and Engineering, AISC 141, pp. 471–478.
springer link.com

energy structure to reduce energy consumption and pollutant emissions by technological innovation and institutional innovation to mitigate climate change and promote sustainable human development.

At last, analyze development of low carbon economy and new energy applications to the Chinese economy.

2 The Relationship between Energy and Economic Growth

Energy is always being paid much attention by many governments. When adding resources into production of a national economy, you will find the energy consumption and economic growth are closely related. Generally speaking, in the point of view of level and extent of national economic development, the national economy growth rate of a country or a region embodies a positive relationship in its growth rate of energy consumption. In the research of causal relationship between energy consumption and GDP, many different characteristics of countries have been found. The time series of energy consumption and GDP are not stationary, while been differenced. Co-integration of these two factors has been found in many countries. In the case of China, they are stationary after two order differences in the period of 1990-2009 when logarithm processed, and a co-integration relationship between them can be found under the level of 5% significance, there is a long-term and stable relation between them. On the one hand, economic growth has a dependence on energy, on the other hand, sustainable energy development also need economic growth as premise. In the end, the achievement of economic growth depends on their level of satisfaction of energy needs, that's why long-term and stable relations of energy supply has been widespread concerned. Energy is the driving force of economic growth, and decides the scale and speed of economic growth.

3 China's Energy Structure in the Perspective of Low Carbon Economy

As the largest developing country in the world, China is a big producer and consumer of energy, its energy production is just after the United States and Russia, ranking third in the world, basic energy consumption occupies the world consumption of 1 / 10, after the United States, ranking second in the world. China is a country with coal as main energy, so the contradiction of developing economy and protecting environment is prominent. Energy issue has been hot and difficult in China's national economic and social development for a long time. As a large country with vast land area, China has abundant energy resources. The conventional energy resources of China shows that, total volume of coal resources is 1.1 trillion tons, proven recoverable reserves is 114.5 billion tons, accounts for 11.6% of the world. Total volume of oil resources is 78.7 billion tons, proven recoverable reserves is 2.5 billion tons, accounts for 1.7% of the world. Natural gas resource is estimated at more than 38 trillion cubic meters, proven reserves are 10 trillion cubic meters. China is one of the countries which have ample hydropower resources reserves. Its theoretical reserves is 67.6 billion kilowatts, water resources can be developed is 378 million kilowatts, the corresponding annual power generation is 1.92 trillion kWh. For large number of people, China's per capita energy share is very low, its per capita consumption of oil is only 11% of the world's index, natural gas is less than 5%, per capita of fossil resources (including coal, oil, natural gas) is only 58% of the world's index, ranking fifty three in the world.

From the perspective of equilibrium of energy supply and demand, the date of the trend of total energy consumption over the total energy production is still spreading since 1992 for the first time. Although in recent years the total energy consumption decreased, its range is lower than that of total energy production. This indicates that on one hand with the rapid economic development, energy demand continues to grow, not only the total amount in the growth, but also the speed is accelerated, on the other hand, it reflects the production of energy is lack of stamina, cannot well adapt to the energy demand of economic and social development, the gap of energy consumption and production mainly relies on imports to balance. Since oil import reached 15.67 million tons in 1993, the trend increased year by year, it reached 36.61 million tons in 1999, doubled in the next year to more than 72 million tons. This level had soared to 218.885 million tons by the end of 2009. This not only increased the country's foreign exchange expenditure, while increasing hidden danger of energy security. From perspective of environment, coal production is in a dominant position in China's energy structure. So2, co2 and dust which are produced from coal are major sources of air pollution and greenhouse gases. China's emissions of so2 and co2 have ranked first and second in the world. Environment problem caused by China's energy development and utilization in large scale, especially the greenhouse gas emissions like co2, have been widespread concerned by many countries. China will endure more and more international pressure on the problem of greenhouse gas emissions at the same time of the effect of "Kyoto Protocol" and the beginning of the negotiations of "Post-Kyoto Protocol".

4 The Path of Sustainable Development of Economy

The outbreak of the oil crisis shocked the world economy in the nineteen seventies, unemployment, recession and energy crisis had been widespread concerned by the world. According to the analysis of world energy authority, oil recoverable period remains only 41 years, which accounts for 40.5% of world's annual total energy consumption. In the condition of oil consumption dominants energy consumption, energy crisis will occur unless the structure of energy consumption changes. Natural gas recoverable period remains 61.9 years, which accounts for 24.1% of world's annual total energy consumption. Coal is not inexhaustible although it is more than oil; its recoverable period remains 230 years, which accounts for 25.2% of world's annual total energy consumption. When the energy stock drops to a certain extent, the development of residual reserves will increase in difficulty, and gradually loses its exploitation value. At the same time, the large scale of development and use of fossil energy is one of the main reasons which caused environmental pollution and ecological destruction.

When macroeconomic growth theory explores the long-term economic growth, it always ignores environment, the potential assumption is that economic growth on environmental quality is lossless. However, from the perspective of long-term economic growth, environmental quality and energy as endogenous growth factors are set in the analysis framework, when consumers maximize their intertemporal utilities, optimal dynamic process of system shows that, in the long run of economic growth, in order to achieve sustainable economic growth and ensure sustainable use of energy, the ratio of growth rate of energy and growth rate of human capital should be less than that of output elasticity of human capital and energy output elasticity. Energy security and global climate warming promote the worldwide researches on energy saving, emission reduction and renewable energy. The wave to meet the low-carbon economy and build a new structure of energy has been raising world widely.

British Prime Minister Tony Blair delivered a speech entitled "Our energy future --- to create a low carbon economy" in February 24, 2003---a white paper, first proposed the concept of low-carbon economy, planning to reduce CO2 emissions by 20% on the level of 1990 by the end of 2010, and reduce CO2 emissions by 60% when in 2050, building a low carbon economy society. The low carbon economy is a low-power pollution-based economy, it create a new energy structure to reduce energy consumption and pollutant emissions by technological innovation and institutional innovation to mitigate climate change and promote sustainable human development. Its foundation is to build low carbon energy system, low carbon technology system and low carbon industrial structure, set up a mode of production ,a consumption pattern, encourage international and domestic policy, legal system and market which are adapted to the low carbon development. Visible, the mode of production of low carbon economy is consistent with the model of long-term sustainable development which was concerned above in the growth theory. Otherwise, the economic system due to the constraints imposed by energy and environmental issues can not converge to the balanced growth path.

5 The Impact of Low Carbon and the Application of New Energy on China's Economy

In order to change the traditional mode of economic growth which with high energy consumption and high pollution, many countries vigorously promote energy conservation, develop low carbon economy which with the symbol of low power and low emissions, for the aim of sustainable development. In order to control the global greenhouse gas emissions, the " Kyoto Protocol" had been signed by 141 United Nations Member States , were went into effect in February 16, 2005. this movement will push development and planning of renewable energy to a new stage. EU claimed that renewable energy in a percentage of primary energy should be increased from 6% in 1997 to 12% in 2010, 20% in 2020, would achieve 50% by2050. The United States put forward that renewable energy production will be two times that of 2000 by the end of 2025 except for hydropower, the biomass power 45 million kilowatt, wind power 10 million kilowatts, 3 million kilowatts of solar photovoltaic power generation, Thermal power of 20 million kilowatts. It is estimated that by 2020 China's renewable energy generating capacity will reach 120 million kilowatts, accounts for 12% of the total generating capacity, annual power generation of about 40 billion kWh, accounts for 8% of the total power generation. Renewable energy generating capacity is equivalent to fossil energy of about 1.2 tons of standard coal. The influence on China's economy of developing low carbon economy and new energy technology is mainly reflected in the following aspects:

5.1 The Promotion of Energy Structure Optimization

The development of low carbon economy is the important means to optimize energy structure, the connotation of it reveals there are two ways to achieve low carbon economy, one is to adjust structure of energy, reducing the emissions intensity of CO2, the other is to improve energy efficiency. China's coal reserves are located in the top of all petrochemical resources. Resource endowments determine the core of China's energy consumption is coal, and it is the highest carbon containing energy in all petrochemical resources. In addition, China is still in industrialized metaphase, heavy industry accounts for a relatively heavy proportion, extensive mode of production has been given priority, energy utilization efficiency is not high, and therefore the situation of energy shortage and environmental pollution is particularly serious. Development of low carbon economy and increased use of new energy which represented by solar, wind and biomass can adjust energy structure, reduce emissions of CO2. The difference of processing and consumption efficiency of world energy is so big, that the potential energy efficiency can be improved. Since China's energy utilization efficiency is nearly half of that of world, we have much room for improvement.

5.2 Sustainable Development of Economy

The development of low carbon economy is to reduce greenhouse gas emissions, to maximize the reduction of coal and oil and other high carbon energy consumption, for the purpose of building an economy based on low energy consumption and low pollution. The principle of development of low carbon economy is reduction; it requires more consideration on energy substitution and saving in the beginning of production, improvement of comprehensive utilization of energy, prevention of waste generation, rather than focusing on the terminal management of production process. Currently, it is an effective way that we develop new energy, achieve substitutions not only between traditional energies, but also between conventional energy and new energy to develop low carbon economy, solve the bottleneck of energy supply and demand and relieve the pressure on the environment, further more, it is a guarantee to achieve circular economy. China will actively develop hydropower, nuclear power, encourage the development of wind energy, solar, biomass and other renewable energy sources, and actively exploit and utilize geothermal energy and ocean energy. In this way, the environmental pollution in economic development of our country can be effectively reduced, and the growth rate of energy stock will significantly increase, ensuring that the economy converge to the balanced growth path in the long term and achieve sustainable development.

5.3 The Impact of Energy Price Fluctuations on the Macroeconomic

With the increasing of rapid development economy after war, the global energy resources were inadequate, and had become the bottleneck of economic development. The sharp rise of energy price not only increased consumption expenditure but also led the rise of overall price and eroded the income of consumers, for energy had been the main cost of many commodities. The rapid fluctuations of energy price had directly conducted to the real economy, caused stagflation and downturn of the entire economy, intensified risk of rapid fluctuations of macro economy when capital bubble burst. The development of low carbon economy and increasing new energy application can effectively weaken the adverse effects by fluctuations of petrochemical energy price, is conducive to the long-term economic stability and development.

5.4 Emerging Industry to Be a New Economic Engine

Low carbon economy has long industrial chain; the effect of output is obvious. It requires the development of new energy, renewable energy, the development of low carbon technology, developing carbon capture technology, using of low energy contained products, and these movements have a significant effect on output. Each movements talked above can be derived from a complete industrial chain when they are market oriented, further more, they can drive the overall low carbon economy, stimulating demands, promoting economic growth and creating a large number of employment opportunities. From a historical perspective, countries which occupied the high ground of technique in different technical revolutions would lead an era, the low carbon economy is considered to be a new technological revolution which after two industrial revolutions, information revolution and revolution of biotechnology. So, if advanced technology of new energy were mastered, the national competitiveness would be enhanced. In order to establish the world low carbon technology and hegemony of low carbon economy, Europe and the United States are actively competing for control of low carbon economy, have proposed plan of emission reduction, established carbon trading market, launched the " low carbon" technology and product standards, fought for the discourse right concerned with low carbon. China must promote the strategy height of development of low carbon economy to a nation, how to seize the opportunity in low carbon technology revolution is not only a chance but also a challenge. We will lose this great opportunity to overtake if not examine the development of low carbon economy problem on a higher perspective.

5.5 The New Trade Barriers

In recent years, in order to adjust their economic structures, Europe and the United States have put forward new concept of carbon tariff in the context of the world financial crisis, is about to launch a new round of economic game theory. Carbon tariff which in the name of environmental protection is actually the reply of United States of America for the developing countries including China, especially having a great effect on the improvement of trade deficit with China, the creation of new manufacturing industry and increase of domestic employment opportunities. Since late start of development, the industrial structure and technical progress of developing countries are at a disadvantage compared with developed countries; carbon tariff will enable developing countries shouldering the responsibility of cutting emissions, but also undertaking the risk of economic contraction, it is unfair for developing countries because developed countries have experienced the stage which with a large number of industrial energy consumption required and damage to the environment when industrialization, having caused damage which is difficult to recover on the human environment, should bear the primary responsibility for climate change. The carbon dioxide emission of United States of America is 54.05 tons in 2009; the value of average per capita is 17.55 tons which is much higher than China for 4 tons. The Copenhagen conference closed on December 19, 2009, had just reached a non-legally binding file ---"Copenhagen agreement", did not determine the binding emission reduction targets. So, the situation of emission reduction will become more uncertain which is in first commitment period under the framework of "Kyoto Protocol". United States, the world's largest emitters of carbon dioxide has not yet joined the" Kyoto Protocol", therefore, China should have reservations about the carbon tax transaction.

5.6 The Rhythm of Low Carbon Economy Development and Application of New Energy

China as a developing country has not finished industrialization process is difficult to cut large number of carbon emissions in a short time because of the labor-intensive industries, huge energy consumption and rough machining. At the same time in terms of resource endowment, coal as the main energy source in China, is not possible to exit the main structure of energy in a short period of time. Now we are in a particular phrase of heavy industrialization and urbanization, so the low carbon industry is a gradual process, but not accomplished at one stroke. According to historical experience, the development of low carbon economy and promotion of new energy application of our country will experience two stages: the first stage is to introduce and imitate the frontier technology of low carbon, including the imitation and innovation of low carbon policy mechanisms of developed country. The feature of this stage is that the utilization efficiency of fossil energy rising ceaselessly, environmental impact of production and consumption will gradually reduce, while the production of high carbon and high carbon industry still hold the dominant position, the gap of low carbon technology between China and developed countries will gradually narrow. The second one is the phase of innovation, we will usher in a comprehensive technical innovation through continuous investment of R & D on low carbon technology, in the end, China should strive to become the birthplace of low carbon technology, leading the technology and industry development of low carbon, achieving the complete substitution of low carbon industry and production mode with that of high carbon, and finally get rid of the dependence on fossil energy.

6 Conclusions

It is reflected of a new relationship between energy and environment with economic growth in the context of low carbon economy. It claims that the growth rate of environmental pollution should be less than zero and growth rate of energy stock be greater than zero. So we should develop a low carbon economy, using of new energy technologies to supplement and replace traditional fossil energy sources. At the same time, we should pay more attentions to the rhythm of low carbon economy development and application of new energy, while avoiding the new trade barriers that produced in development.

References

1. Ahmad, N., Wyckoff, A.: Carbon Dioxide Emissions Embodied in International Trade of Goods. OECD STI Working Paper (15) (2003)
2. Copeland, B.R., Taylor, M.S.: North-South trade and the environment. The Quarterly Journal of Economics (8), 755–787 (1994)
3. Solow, R.M.: A contribution to the Theory of Economic Growth. Quarterly Journal of Economics 70, 65–94 (February)
4. Weber, C.L., Peters, G.P., Guan, D., Hogback, K.: The contribution of Chinese exports to climate change. Energy Policy (9), 3572–3577 (2008)
5. Ross, S.: Use of List Cycle Assessment in Environmental Management. Environmental Management 29(1), 132–142 (2002)
6. Stokey, N.: Are there Limits to Growth? International Economic Review

Interactive Mechanism of Agricultural Growth and Urbanization in China

Qiang Wu[1], Liming Gong[1], and Wanbin Bi[2]

[1] School of Economics, Huazhong University of Science and Technology, 1037 Luoyu Road, Wuhan, China
[2] School of Management, Wuhan University of Science and Technology, 947, Heping Street, Wuhan, China
Victorwool1125@gmail.com

Abstract. The dual structure of urban and rural areas has been manifested inch by inch at the advent of booming economy in China. Currently, our inescapable issue is to coordinate the development between urban and rural areas as well as eliminate their contradictions as far as possible. In view of this, this essay thoroughly illustrates the interactive mechanism of agricultural growth and urbanization from economic view, thus makes a conclusion that the two elements are mutually facilitated and restricted.

Keywords: urbanization, agricultural growth, mechanism.

1 Introduction

China, an agricultural country with 700 millions peasants, should unswervingly take agricultural as their top priority either to attain a sustainable economic growth or political stability. At present, Chinese economy is in transition, but the years of rapid growth has made its economy enter into demand dominant one. In the near future, the demand, especially the domestic demand, will become a major constraining factor. Meanwhile, the policy choice and system arrangement by Chinese government must adapt with the transformation of dual economic and social structure.

Urbanization was affected by many factors. Its generation and development is related to a series of economic and social variables. in turn, the degree of urbanization could also have an influence on those variables, among which conclude direct and indirect factors. Two typical direct factors are the change of demand and employment transformation. In a word, urbanization comes from the change of demand and results in labor migrating from agriculture to other industries.

Throughout the history of economic development worldwide, urbanization is a dispensable component in economic development. Then, how to deal with the development of agriculture in the process of urbanization? How to balance them when facing increasingly exhausted resources? Are they equally vital or not? These issues above need deliberation of the authorities. The paper elaborates the interactive mechanism of urbanization and agricultural development from economic perspective.

D. Zeng (Ed.): Advances in Computer Science and Engineering, AISC 141, pp. 479–485.
springer link.com

2 Mechanism of Agricultural Growth to Urbanization

The level of urbanization can be promoted by economic development through industrialization which is an intermediate variable. With the development of industrialization, population, industry and resource elements are gathering in the city. So we can say that economic growth can improve the level of urbanization. But the prerequisites for the development of industrialization are agricultural development. If agriculture cannot be grew up abundantly, there will not supply desired products, capital, market, resource elements for the industrialization development. More precisely, Agricultural development is a prerequisite to improve the level of urbanization.

2.1 Positive Effect on Agricultural Growth to Urbanization

2.1.1 Direct Mechanism

First, direct mechanism which agricultural growth to promote the process of urbanization is embodied in the urbanization of rural areas, namely agriculture development will improve the transformation between part of rural areas which economy were better than others and city gradually. Agricultural growth can be caused the accumulation of agricultural surplus. Pursuing to more profits, people will increase investment to various sections in agricultural production in order to get more production surplus. And then market scale of agricultural production will be expanded. The expansion of this can promote a certain degree of agglomeration economy effects, which making agriculture developed rural areas gradually to the urban transformation. Expansion of agricultural production scale can bring division of labor, and then direct communication and transportation cost are more important. So that industry chain which related agricultural sector has been extended, and spurred on upstream and downstream industries around the agriculture industry, also can make this area converse to urban.

Secondly, the growth of agriculture makes the rural population transfer to the city. Fei and Ranis (1961) dual-structure model and Jorgenson (1961) dual economic theory model are pointed out, the agricultural surplus is a necessary and sufficient condition which transferred labor from the agricultural sector to the industrial sector. If agricultural surplus can not meet the increasing demand for agricultural products in the expansion of industries, labor force migration will be block. Only agricultural production expansion and agricultural residual increases, that will be realized which making the rural population is transferred to town. Therefore, from this perspective, agricultural growth can promote the process of urbanization.

In addition, income on agricultural practitioners will be increased by agricultural growth. Pursuing more comfortable and convenient living environment in the city, a lot of people "escape from" rural areas to the city. With the rising in income and improvement on education situation, according to incomplete statistics, the majority of graduates from rural areas lived in the city after graduation. It also accelerates the process of urbanization in China.

2.1.2 Indirect Mechanism

The indirect mechanism which agricultural growth improves the process of urbanization is mainly manifested that agricultural growth can accelerate the industrialization, thereby indirectly promote the level of urbanization.

The promotion of industrialization can improve level of urbanization through two aspects which conclude industrial structure and employment structure. The first is variation in industrial structure. As the propulsion of industrialization, people income will rise. The share of the first industry will drop significantly, and the share of second industry and the third industry will be increased. The change of this kind of structure can alter domestic demand structure, production structure and trade structure rapidly. That will be better to the level of urbanization. The second is transformation in employment structure. Along with the change of industrial structure, large amount of labor force migrate from agriculture sector to non-agriculture sector. Then employment structure correspondingly changes. With the development of economy and the increase of per capita income, according to "Petty - Clark theorem", labor force will transfer from the first industry to the second industry; and with the further development of economy, labor force will gradually migrate to the third industry. Industrialization can make population gradually migrating to the city.

The growth of agriculture will contribute to the development of industrialization, such as resource element, product, market, capital. Agricultural products can provide food, raw materials for industries development in addition to meet their needs. The promotion of agricultural departmental efficiency can provide more labor force which required for industrial development. The market of industrial products will be more vast, if increasing income on agricultural practitioners. And its further expansion of production will be realized. The growth in agricultural sector can provide strong financial support for industries. Therefore, the growth of agriculture is a condition for development on industries, and the development of agriculture can promote the process of urbanization in indirect.

2.2 Negative Effect on Agricultural Growth to Urbanization

Of course, agricultural growth is not entirely improve the process of urbanization. From the above, there is a certain time lag which agricultural growth improves urbanized process. That is to say, from a long-term perspective, agricultural growth can promote the process of urbanization, but in the short term is not obviously. It is possibly, to some extent, hinder the process of urbanization.

Todaro (1969) pointed out that the income difference on urban and rural is a required condition which population want to migrate from rural to city in his population flow model. That is to say, without considering the migration cost, whether transferring to the city depends on the comparison on their expected income and opportunity cost. The population will migrate, when the expected income is more than the opportunity cost. In the near further, because per capita income improved in rural areas which caused by agricultural growth per capita, the opportunity cost on transferring to the city will rise. Comparing with their expected income, opportunity cost is more than their expected income. The number of population who transfer to the city will reduce, and thus hinder the process of the urbanization.

Secondly, with the development of agriculture, the living environment of rural areas was improved. So, that can reduce the number of population who pursuit more convenience and comfortable living environment in city to some extent. That will slow down the process of urbanization.

To sum up, agricultural growth may have a certain ripple effect on the process of urbanization in the short term. That is to say, it may slow the process of urbanization and also may promote the course. It depends on the compare on its positive and negative effect. But in the long run, agricultural growth may promote the process of urbanization.

3 Mechanism of Urbanization to Agricultural Growth

The economic process of urbanization is indivisible in some extent. But from concept of it, this continuous process can be divided into two sequential processes: labor forces migrate from agriculture to non-agriculture sectors and those population are concentrated in the city area. Along with those two successive processes, the population who were supported by those workers coming from rural areas will also be transferred to the city from the rural areas. Aiming to the urban agglomeration economic effect, those farmers migrate to the city. As a result, that accelerates labor force transferring from the agricultural sector to non-agricultural sector. One reason is that their wages rise steadily, namely relative to the return rate of capital, the wage rate is on the rise. The other is due to the weakness of agricultural products. The income elasticity on commercial and industrial products and service is greater than agricultural products in the city. That will cause the prices of agricultural products dropping, namely the two relative prices are falling. That can" push" rural labor flowing to the city.

In this case, agriculture practitioner will reduce production cost by taking capital to substitute of labor. And, along with the reduction of the rural population, large-scale and mechanized production can be realized. In the process of urbanization, the phenomenon of technology spillover permeate into the agricultural production appears gradually. That will improve the efficiency of agricultural production, increase the output of agriculture and improve agricultural development. But that has brought certain negative effect which high quality talents from the rural flow to the city in the process of urbanization.

3.1 Positive Effect on Urbanization to Agricultural Growth

3.1.1 Direct Mechanism

First of all, with a large number of agricultural labor force transfer from the countryside to the city in the process of urbanization, land is increasingly concentrated and agricultural productive scale is continued to expand. Mechanization and scale management can be realized. That will improve labor productivity, and then promote agricultural development.

Secondly, technology innovation and diffusion in the city are favorable for agricultural production, and constantly improve the level of agricultural modernization. That will directly affect the agricultural production, which will bring improvement on the agricultural technology. There are two forms manifested on agricultural technology: one is the mechanical technology, another is the biochemical technology. This technological progress will stimulate the development of agriculture in a certain extent. As in the 1960s, some developing countries carried out "green revolution". The output of crops increases substantially through a large area and new varieties of wheat and rice.

With the improvement of living standard, the consumption of agricultural products increased rapidly. That will improve the development of agriculture. The increasing population in the city, for their survival, will inevitably lead to the demand for agricultural products increasingly, which will lead to the expansion of agricultural production.

In addition, the development of urbanization can be" feed back" agriculture. It will increase the investment in agriculture and accelerate agricultural capital accumulation. According to Mellor's (1967) three stage theory of agricultural development, the third stages of agricultural development is capital intensive technology progress. In order to obtain continuously and stable development, we must increase investment. That will take capital and technology to substitute of labor. Then labor productivity improved continuously, and agriculture can obtain steady growth.

3.1.2 Indirect Mechanism

The indirect mechanism which urbanization will contribute to the agricultural growth is mainly reflected in the following three aspects. First, urbanization and industrialization are complementary. The demand for manufactured goods will increased with the process of urbanization. That will cause the expansion of production, and then the demand for raw material will increase. In the supply and demand action, that will bring about expansion to the agricultural production. So the agricultural growth can be realized. Secondly, urbanization is conducive to innovation and diffusion on technology, and that will reduce the cost of some industrial production which is use for agriculture, such as chemical fertilizer, agriculture machinery. With its price falling, the farmer can use a large quantity of those. Then labor productivity will be effectively improved, and it will stimulate the agricultural growth. Finally, the process of urbanization can improve regional development through" development pole". It can bring about splendid development in rural areas surrounding the city, so as to promote the growth of agriculture.

3.2 Negative Effect on Urbanization to Agricultural Growth

To some extent, the process of urbanization hindered the development of agriculture. The negative effect is reflected in the following respects:

(1) The population mobility has seriously affected the quality of the rural labor force in the process of urbanization. Due to some institutional constraints, in the early stages, population flow is mainly manifested as a "flow of migrant workers" which usually have some knowledge. In the countryside, they tend to be of high qualities. The proportion of males is more than females; undergoing secondary or above education is more than accepting only primary school education; unmarried is more than married; smart, flexible, has a certain expertise is more than general honest farmer. These population movements led to the quality of labor force in rural areas

improving slowly in a long-term. This will seriously affect the efficiency of agricultural production. And then this will bring tremendous obstacles on agricultural development and seriously affect the sustainable development of agriculture.

(2) Urbanization will make the rural factor of production loss amount. With the increasing of migrant workers and some national policy adjustment, those farmers who have proficiency in a particular area or find a stable job in the city emigrate from the rural areas to the city. After many years of accumulation, these farmers have accumulated a certain amount of capital. When they migrate to the city, large amounts of wealth will flow to the city, such as tangible monetary capital, basic household facilities and intangible ability to obtain wealth. Thus it is easy to cause agricultural resources using on non-agricultural field. That will seriously affect the development of society and economy in rural areas in a long run. At the same time, when those township enterprises gathering in the city, their wealth and resources also transfer to the city. And, some policies, in a certain degree, assist city in "plundering" resources and productive elements in rural areas, Such as combining city and county, township merger, city circle, city group construction, etc.. In order to increase cooperation in urban and rural, and then the city has assistant to the development of rural areas which it can be played the role of developmental pole, those policies were introduced. But often get the opposite of what one wants. In order to get more politic achievements, some" politicians" often prefer to urban development by the deprivation of agricultural resources in the distribution of labor, capital and other materials. This will result that more resources are speeding up rural areas to the cities.

4 Conclusion

From an economic point of view, the internal mechanism of agricultural growth and urbanization is agglomeration economy: agricultural growth caused the city gathered. The expansion of the scale of production stimulates the formation of city, expansion of city scale and increase the level of the city. The higher level of urbanization promote agricultural growth through its effects of agglomeration economies and spillover effects of developmental pole.

To some extent, the process of urbanization has negative impact on agriculture. But they are a mutually reinforcing relationship in general.

References

1. Black, D., Henderson, J.V.: A Theory of Urban Growth. Journal of Political Economy 107(2), 252–284 (1999)
2. Evenson, R.: Indian urbanization and economic growth since 1960. Journal of Economic Literature 31(3), 1488–1489 (1993)
3. Kevin, Zhang, H., Song, S.: Rural-urban migration and urbanization in China: Evidence from time-series and cross-section analyses. China Economic Review 14(3), 386–400 (2003)
4. Bertinelli, L., Strobl, E.: Urbanization, Urban Concentration and Economic Growth in Developing Countries. Core discussion paper, No. 03/14 (2003)

5. Bertinelli, L., Black, D.: Urbanization and growth. Journal of Urban Economics 56, 80–96 (2004)
6. Pencavel, J., Abramovitz, M.: Economic growth and urbanization in developing areas. Journal of Economic Literature 29(3), 1285–1285 (1991)
7. Rosenthal, S.S., Strange, W.C.: Geography, Industrial Organization, and Agglomeration. Review of Economics and Statistics 85(2), 377–393 (2003)

The Study on the Effects of Bank Employees' Perceptions of Organizational Politics to Job Performance

Shouming Zhang and Yanmei Yang

Zhejiang Sci-Tech University, Hangzhou City, Zhejiang Province
imjlfe@126.com

Abstract. For bank employees, it's more sensitive to organizational politics because of the special industry and work. The study investigates bank employees' perceptions of organizational politics and analyzes its effects on job performance by the questionnaire survey. The conclusions show that: two aspects of perceptions of organizational politics, self-serving behaviors and coworker relation, have significantly positive correlation with task performance. But pay and promotion policies has significantly negative correlation with both task performance and relationship performance.

Keywords: Perceptions of organizational politics, job performance, task performance, relationship performance.

In the knowledge economy era, enterprise development depends more on excellent human resources. When each enterprise strives for product market, the keen competition for talents tends to be intensified. Banking is the industry which is risk intensive, information intensive and knowledge intensive. The domestic banking is in increasingly fierce competition because of the influx of foreign banks and the competition between the same industry. And the bank employees are facing more working pressure and learning load because of the widely application of new technologies, the uncertainty of internal reforms and important job responsibilities. These situation factors will make members produce organizational political behaviors which will affect or may affect the distribution of organizational limited resources, then they have significant influence on rewards related to work such as job performance and promotion opportunities. In this case, organization members will judge and interpret these behaviors according to their own value and conception, and then the perceptions will significantly affect their work attitudes and results.

This paper aims to study the bank employees' perceptions of organizations politics and their dimensions' influences on job performance, so that the banks can fully mobilize their employees' work enthusiasm and initiative, improve job performance and then enhance organizational performance.

1 Research Hypothesis

(1) Perceptions of organizational politics. As a special employee group, Chinese bank employees possess relevant professional skills and relative high personal quality, they

D. Zeng (Ed.): Advances in Computer Science and Engineering, AISC 141, pp. 487–493.
springer link.com

enjoy a high compensation level but undertake high pressure at the same time, they have strong willing of self value achievement but blocked by the realistic situation, the above contradictions make bank employees sensitive to the organizational politics. So in the background of characteristics of national banking and bank employees, I support the classification by Ma Chao, Ling Wenquan and Fang Liluo in theory and propose research hypothesis 1: Bank employees' perceptions of organizational politics contain three aspects, self-serving behaviors, pay and promotion policies and coworker relation.

(2) Job performance. Motowidlo and Borman firstly put forward the concepts of relationship performance and task performance, and they considered job performance as synthesis of behaviors and results, then they proved the two dimensional structure. Chinese scholars also supported the model. Therefore, the research hypothesis 2 is: bank employees' job performance can be divided into two dimensions, task performance and relationship performance.

(3) Perceptions of organizational politics and Job performance. Be based on the perceptions of organizational politics model of Ferris, lots of domestic and foreign scholars began the empirical study of it. Bank employees are a kind of special research objects, and their sensitivities to organizational politics can make them be accustomed to treat people and things in organization according to their own standards. Once they regard others' practices as organizational political behaviors which have damaged their benefits and threatened their deserved opportunities, they may have negative reactions, such as reducing work enthusiasm and satisfaction, and then affecting individual job performance. Therefore, the research hypothesis 3 is: perceptions of organizational politics has significantly negative correlation with job performance.

2 Study Design

2.1 Study Objects

This study adopted the method of questionnaire survey, and the objects refer to the employees who were from Bank of China, Industrial and Commercial Bank of China, China Construction Bank, Agricultural Bank of China and so on. 400 questionnaires have been given out, a total of 374 questionnaires were effective, and the effective rate was 93.5%.

A breakdown showed that 52.4% of the sample were male while 47.6% female; the sample who were at the age of 26 to 30 years old were the most and the proportion was 50.8%; 69.5% of the sample's educational background were undergraduate degree or above; the general level employees accounted for 69.5% by occupation. The above demographic characteristics were consistent with bank employees' structure.

2.2 Measures

(1) Perceptions of organizational politics. This variable was measured by Perceptions of Organizational Politics Scale (POPS) that was first developed by Ma Chao, Ling Wenquan and Fang Liluo, and the scale included 16 items. According to factor analysis, KMO measure of sampling adequacy was 0.911 and significant level of Bartlett's Test of Sphericity was less than 0.001, which could explain this scale have

good structural validity. Reliability of the scale and each factor respectively was 0.875, 0.809, 0.826, 0.745, which were higher than 0.7, so the scale also had ideal reliability. This study only changed six-point Likert scale into five-point scale. It ranged from 1 (strongly disagree) to 5 (strongly agree).

(2) Job performance. The objects were bank employees, so this study adopted the enterprise knowledge workers' performance scale designed by Wang Che. This scale included 12 items. According to factor analysis, KMO measure of sampling adequacy was 0.814 and significant level of Bartlett's Test of Sphericity was 0.000, which could explain this scale have good structural validity. Reliability of each factor respectively was 0.896 and 0.901, so the scale had ideal reliability.

3 Data Analysis

This study utilized SPSS13.0 to analyze the survey data.

3.1 Dimensional Analysis of Bank Employees' Perceptions of Organizational Politics

Firstly, I conduct exploratory factor analysis about perceptions of organizational politics. The result shows that, the KMO value is 0.830, the value of Bartlett Test is 1602.776, and significant level is 0.000. The questionnaire has a pretty good validity.

Finally I gain three factors. I name them after perceptions of self-serving behaviors, pay and promotion policies, and coworker relation separately. Self-serving behaviors is the way of act that employees often take when the individuals or the small teams in the organization want the result they expect. Pay and promotion policies is whether the formulation and execution of pay and promotion policies is fair and reasonable, or political. Coworker relation is the relationship between employees. And in order to make the analysis results more clear, I define them respectively as X1, X2, X3. In addition, because the research object of this article is different and bank employees answer the questions according to their own understanding, the distributions of questions about perceptions of self-serving behaviors and coworker relation are different from original questionnaire. The result supports the hypothesis 1.

3.2 Dimensional Analysis of Bank Employees' Job Performance

I conduct factor analysis about questions of job performance and the result shows that the KMO value is 0.884, the value of Bartlett Test is 1900.212, the significant level is 0.000. The questionnaire has a pretty good validity and I finally gain two factors.

The factor analysis of this research fits a question which belongs to relationship performance into the task performance questions. This article supposes that bank employees answer this question according to current task performance level, so this question should measure task performance. Therefore, in the two factors we gained, the first factor, which is named after task performance, mainly describe the situation directly relevant to employees' current task; the second factor, which is named after relationship performance, mainly describes the behavior indirectly relevant to the task. I define them respectively as Y1, Y2. The results support the hypothesis 2.

3.3 Difference Analysis

We can use difference analysis to test whether different gender, ages, degrees, marital status, tenure and duties have significant influence on the dimensions of perceptions of organizational politics.

Firstly, I conduct independent-Sample Test about employees who have different gender and marital status. The result is illustrated in the form below.

Table 1. Independent-Sample Test

	X1		X2		X3		Y1		Y2	
	t	Sig.	t	Sig.	t	Sig.	t	Sig.	t	Sig.
Gender	3.482	0.001**	-2.983	0.003**	-2.214	0.027*	1.862	0.063	-0.839	0.402
Marital status	3.371	0.001**	-0.895	0.371	1.396	0.163	1.403	0.162	-1.840	0.067

** Correlation is significant at the 0.01 level (2-tailed).
* Correlation is significant at the 0.05 level (2-tailed).

The result shows that, bank employees who has different gender have significant difference in the dimensions of self-serving behaviors, pay and promotion policies, and coworker relation. Male employees have higher perceptions of self-serving behaviors and female employees have higher perceptions of pay and promotion policies and coworker relation. Employees who have married have higher perceptions of self-serving behaviors.

Then, about bank employees of different gender, degrees, tenure and duties, we need one-way analysis of variance to test their influences to the two variables. The result is as below.

Table 2. One-Way AVOVA

	X1		X2		X3		Y1		Y2	
	F	Sig.	F	Sig.	F	Sig.	F	Sig.	F	Sig.
Age	3.415	0.018*	10.052	0.000**	0.127	0.944	3.464	0.016*	2.205	0.087
Degree	0.286	0.835	3.144	0.025*	1.116	0.342	0.554	0.646	3.508	0.016*
Tenure	3.556	0.015*	4.205	0.006**	2.527	0.057	5.633	0.001**	3.074	0.028*
Duty	2.578	0.053	9.814	0.000**	0.241	0.867	2.719	0.044*	3.045	0.029*

The result shows that, employees who have different ages have significant differences in the dimensions of self-serving behaviors, pay and promotion policies, and task performance. The employee's perceptions of self-serving behaviors will be higher and perceptions of pay and promotion policies will be lower when he gets older. I find employees who have different degrees have significant differences in the dimensions of pay and promotion policies and task performance. Furthermore, employees who have bachelor's degree have higher perceptions than those who have master's degree or above in these two dimensions. Also I find employees who have different years of work have significant differences in all these three dimensions. Employees who have 2 to 5 years of work have the most significant differences than the other ones. They have the lowest perceptions of self-serving behaviors, the highest perceptions of pay and promotion policies, the lowest task performance, and the highest relationship performance. Employees who have different duties have significant differences in the dimensions of pay and promotion policies, task performance and relationship performance. Common employees have the highest perceptions of pay and promotion and the lowest task performance and relationship performance.

3.4 Correlation Analysis of Bank Employees' Perceptions of Organizational Politics and Job Performance

I conduct correlation analysis about perceptions of organizational politics and job performance and Pearson correlation coefficient is illustrated in table 3.

Table 3. Pearson Correlations

	X1	X2	X3	Y1	Y2
X1	1				
X2	0.000	1			
X3	0.000	0.000	1		
Y1	0.197**	-0.247**	0.183**	1	
Y2	-0.012	-0.213**	-0.083	0.000	1

The result shows that, perceptions of self-serving behaviors and coworker relation are significantly positively correlated with task performance, and pay and promotion policies is significantly negatively correlated with task performance and relationship performance. In order to research the relationship of perceptions of organizational politics and job performance further and prediction function of perceptions of organizational politics to job performance, this article conduct regression analysis.

Table 4. Regression analysis results

	Y1				Y2			
	B	Beta	.t	Sig.	B	Beta	.t	Sig.
X2	-0.247	-0.247	-4.903	0.000	-0.213	-0.213	-4.201	0.000
X1	0.197	0.197	3.997	0.000				
X3	0.183	0.183	3.779	0.000				
R^2	0.133				0.045			
AR^2	0.126				0.043			
F	18.912**				17.646**			

The result shows that all dimensions of perceptions of organizational politics have the function of prediction to task performance and total explained variance is 13.3%. Perceptions of pay and promotion policies have the strongest prediction function. The regression equation is:

$$\text{Task performance} = -0.247\times\text{perceptions of pay and promotion policies} + 0.197\times\text{perceptions of self-serving behaviors} + 0.183\times\text{perceptions of coworker relation} \quad (1)$$

If we use perceptions of pay and promotion policies to predict criterion variables from the angles of five questions, then the variables enter the equation will be "pay and promotion criterion is complete in our organization", " employees clearly know the policy of pay and promotion criterion in their organization", " employees who do excellent job will be promoted in our organization".

Only perceptions of pay and promotion policies have the function of prediction to relationship performance. The regression equation is as below.

$$\text{Relationship performance} = -0.213\times\text{perceptions of pay and promotion policies} \quad (2)$$

If we predict criterion variables again, the variables which enter the regression equation will be "managers encourage employees to speak frankly, even if they criticize the authority", "employees who do excellent job will be promoted in our organization".

4 Study Conclusions and Suggestions

4.1 Study Conclusions

According to the study results, we can get the following conclusions: in the background of China, (1) Bank employees' perceptions of organizational politics can be divided into three dimensions, self-serving behaviors, pay and promotion policies and coworker relation; (2) Bank employees' job performance contains two dimensions, task performance and relationship performance; (3) self-serving behaviors and coworker relation have significantly positive correlation with task performance. But pay and promotion policies has significantly negative correlation with both task performance and relationship performance; (4) different gender, ages, degrees, marital status, tenure and duties have significant influence on perceptions of organizational politics.

4.2 Management Suggestions

Some suggestions can be put forward on the basis of the above conclusions:

(1) Improve the performance appraisal standards and make the compensation policies open and transparent, so as to embody and follow the principle that work and efforts proportionable to gains and returns, and that can offer good physical benefits to employees; In addition, personalized compensation management for different groups of employees should be put in practice, for example: younger employees generally have higher degree, shorter tenure, lower positional level, and they always have more desire for compensation and benefits than other older employees, correspondingly their perceptions of pay and promotion policies are stronger, so the organization should adopt appropriately pay adjustment policy: accelerate salary adjustment frequency according to the employees' individual performance, but each adjustment can be relatively small, so as to show respect and recognition to them, but also can take into account the old employees' psychological feeling.

(2) Provide fair occupation development channel, decrease organizational political color such as "Guanxi", and establish fair competition mechanism, so as to make people who are fit for the positions do the work; Encourage employees to participate in the decision-making and management, and the cooperative working atmosphere will make employees understand and support the formulation and implementation of organizational policies more clearly; In addition, provide personalized learning and training opportunities to improve employees' skills and competitive power, so as to adapt to the working pressure went with the technological progress and the organizational changes. For example: establish the organization's talent pool and encourage competition of internal employees; carry out coordination meetings in department or between different departments, encourage employees to offer advice and suggestions; build organizational knowledge networks and training teams, develop reasonable training plans according to the different training demands and so on.

(3) Apply organizational politics reasonably. Most researches mainly focused on the negative effects of organizational politics, but every coin has two sides, organizational politics also has positive effects. The high perceptions of self-serving behaviors and coworker relation can form pressure to employees' psychology, under such organizational atmosphere, employees will transfer pressure to motivation and restrain themselves better, working hard in order to avoid severe punishment due to organizational politics, as a result, task performance largely increased, besides, the formulation and implement of organizational politics can not deviate from the management, so the formulation and perceptions of organizational politics are highly related to the management style of leaders, in order to promote the quality and management skills of the management and apply skills of organizational politics reasonably.

References

1. Ferris, G.R., Fedor, D.B., King, T.R.: A political conceptualization of managerial Behavior. Human Resource Management Review 4(1), 1–34 (1994)
2. Ma, C., Ling, W., Fang, L.: Construct Dimension of the Enterprise Staff's Perceptions of Organizational Politics. Acta Psychological Sinica 38(1), 107–115 (2006)
3. Ferris, G.R., Russ, G.S., Fandt, P.M.: Politics in Organizations. Impression Management in the Organization, 143–170 (1989)
4. Wang, C.: The Empirical Research on the Impact that Psychological Contract to Enterprise Knowledge Workers' Performance. Shandong University (2008)

Research on Resource Integration of New Venture

Chenglin Rong

School of Management. Hebei University of Technology Tianjin, China
lieren130@yahoo.com.cn

Abstract. The acquisition, usage and integration of resource are the key to decide the core competitiveness of new venture, which plays great important role for enterprise. This paper starts from the classification of resource type needed by the new venture, and analyzes its resource identification, acquisition and developing process, then constructs resource integration process model, at last discusses the factors of influencing enterprise's resource integration.

Keywords: New Venture, Resource Integration, Resource Acquisition, Information.

1 Research Background and Theory Overview

1.1 Basic Theory of New Venture and Its Resource

Entrepreneurship is the process of commercial opportunity with market value from initial concept to enterprise formation[1]. Howard H .Stevenson（1990）defined entrepreneurship as a series of behavior process of integrating resource developing opportunity. These opportunities may not exist in present application range of resource, but in the future it may create value. New venture is the entity with legal personality which takes use of commercial opportunity through resource integration to create, and it can provide product or service. It takes vitality and growth as the goal and can create value[2].

Resource is the foundation of new venture's establishment. How to use resource to form enterprise's core capacity has become the fundamental problem of enterprise development[3]. First, the meaning of the concept enterprise resource must be clear and definite. From the perspective of management, resource is that enterprise, as an economic entity, owns or governs the combination of each factor to realize enterprise's strategic objective in the process of providing product or service for society (Bamey, 2001) [4].

The resource needed by enterprise is not only the accumulation of amount, but the integration and usage of resource. How to use the present resource to form core competitiveness is important for new venture to survive in sharp market competition.

1.2 Overseas Research Situation

On the research of new venture, there are some scholars who pay attention to the research of the entrepreneurship resource. Brush has researched on enterprise resource allocation based on different stage of life cycle, and has compared the key resource and

D. Zeng (Ed.): Advances in Computer Science and Engineering, AISC 141, pp. 495–502.
springer link.com

allocation process of new and old venture, then has summarized the key resource for new venture's survival and development[5]. On this basis, Brush, Greene, Hart, Hailer has carried on deep analysis on the construction of enterprise resource foundation, dividing the main challenges into four parts: focus resource, attracting resource, integrating resource and transformation resource[6]. Lichtenstein and Brush researched on the resource identification of new venture and the process of resource acquisition based on resource basic theory, put forward the key resources relayed by the new venture in the stage of entrepreneurship, and confirmed the influence of resource conversion for new venture's performance.

1.3 Domestic Research Situation

For the domestic research of new venture, Lin Chong(2004) and others analyzed the formation process of new venture and introduced the simple model, complex model[7] and dynamic adjustment model. He compared and reviewed these three typical models and pointed the "complex" problems in the empirical research of new venture. Yang Jun and others (2003) maintained that entrepreneurship opportunity actuation behaviors based on entrepreneurs resources endowment evolution and pointed that entrepreneurs resources endowment evolution is the primary cause resulting in the heterogeneity of the micro-level entrepreneurial behavior[8]. Cai Li and others researched on the integration of new venture and proposed the general model of resource integration[9] and some factors of influencing new venture's resource integration.

2 Resource Types of New Venture

New venture is in the initial stage of formation, and needs the support of all kinds of resource to promote its development. On the classification of these resources, researchers have proposed different ideas. Dollinger divided new venture resource into six classification, including material resource, reputation resource[10], organization resource, financial resource, intelligence and human resource and technology resource. However, Barney thought new venture resource can be divided into human and technology resource, financial resource and other production business resource. Social network proposed by new economics is formed in the way of dividing new venture by resource acquisition path. Entrepreneurs and its team's all social relations form new venture's social network. Social network is the important path for the entrepreneur to acquire resource. Combining all the research results, this paper divides the resource needed by the new venture into the following six types, including human resource, reputation resource, material resource, financial resource, technology resource and organization resource.

2.1 Human Resource

Human resource is a wide concept, not only including human element, but intelligent element. In specific, human resource includes entrepreneur or its team and its employers' knowledge, ability, experience and personal social network. It covers each individual's judgment, insight, creativity, vision and intelligence, even including social skills and social relations. Human resource is the core of new venture. All enterprise's operations are inseparable from human and intellectual support. In addition, entrepreneurs' entrepreneurial ability is an important and strategic resource for new venture.

2.2 Reputation Resource

Reputation resource is the perception and acceptance of people for the new venture who are in the environment of new venture. Reputation resource is an invisible resource. It can exist above the level of product, appearing as the form of brand loyalty and it can also exist in the level of company as the form of the company's overall image.

2.3 Financial Resource

Financial resource is a basic resource for new venture's survival, which is the visible property of using in the process of creating and operating the new venture. In specific, it includes factory, machinery equipment, all kinds of office infrastructure, raw materials and energy. Material resource is valuable and is the basis for one enterprise's existence, which can make profit for new venture. It's especially important of the initial stage of new venture,

2.4 Financial Resource

Financial Resource is the currency assets of the new venture, including the amount of debt, the amount of capital which can be raised in the internal operation and cash flow. For entrepreneurs, the acquisition of financial resource is an important factor for the beginning of operation. Each stage in the enterprise's development can't separate from the support of financial resource.

2.5 Technology Resource

Technology resource is composed of crafts, system or physical transformation method. It includes lab, research, developing equipments, testing and quality control technology. It also includes the patent formed by knowledge generated by research. The unique formula, trademark, copyright and franchise are also the important parts of technology resource.

2.6 Organization Resource

Organization Resource includes structure, process and system of new venture. In general, organization resource means formally reporting system, information processing transfer and decision-making system. The design of new venture's organization structure and system should accord with its industry and characteristics. Therefore organization resource is unique. It needs to choose the suitable organizational model according to its specific environment. For example, the entrepreneurs committing IT industry can design flat organizational structure and reduce level as to flexible communication; the entrepreneurs committing manufacturing tend to divide organization structure by the nature of functions, ,making each functional department perform their duties, clear their division of labor and responsibility in place.

3 Resource Integration Process of New Venture

Most new ventures start with poor resource, little management and industry experience. How to more efficiently and enduringly ensure the entrepreneurial opportunities to realize the needed resource needs to establish a set of perfect enterprise resource integration mechanism as to achieve high efficient utilization of resource, which can realize the long-term development of new venture. Resource integrating process model is shown as in the following figure 1:

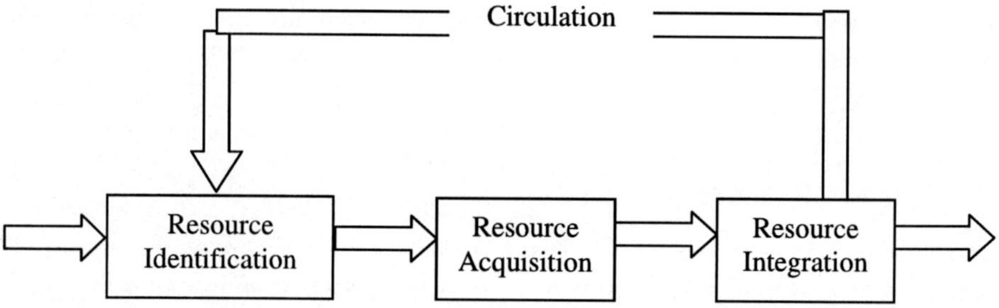

Fig. 1. New Venture Resource Integration Process Model

3.1 Resource Identification

The primary task of resource integration is to conduct resource identification. The manager of new venture should identify the specific resource classification needed according to the macro environment and the microscopic demand. As the above-mentioned six types of resource, material resource, human resource, organization resource, financial resource, technology resource and reputation resource, the manager of new venture should identify them, not only make clear the specific type of these resources, but evaluate the amount of the needed resource and the special requirements. Besides these basic resources, enterprise should focus more on identifying the resource that can help new venture realize its core competitiveness. From the strategic perspective of new venture, the enterprise needs valuable, scarce, hard to imitate, irreplaceable and sustainable resource to help it form core competitiveness, which can keep a foothold in the sharp competition of mature enterprises. As shown in the following figure 2.

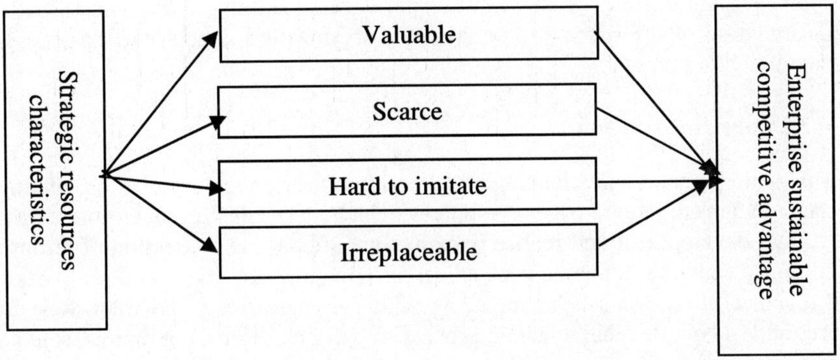

Fig. 2. Strategic resources characteristics and Enterprise sustainable competitive advantage

3.2 Resource Acquisition

New venture has various ways to acquire resource, including the primitive accumulation of entrepreneurs, resource purchase, invest resource of investors, enterprise alliance, finance lease, operating lease and other ways.

The acquisition way of human and intelligent resource includes the entrepreneurs' accumulation of knowledge, ability, experience, social relations, and the continuous study of new knowledge and successful experience. New venture can find suitable talents through ordinary recruitment or head-hunting company, besides, it can seek intelligent support through consultation, management planning company.

The acquisition way of reputation resource includes create excellent enterprise culture and value, shoulder social and environmental duty, provide product and service with high quality, keep healthy financial index, establish creative and positive image of enterprise. No matter which way the reputation resource needs to accumulate over a long period, thus for new venture, the acquisition of reputation resource is not easy.

The acquisition way of material resource includes the past accumulation of entrepreneurs, the acquisition by inheriting, bought deal, lease and merger. For entrepreneurs, the acquisition of material resource is easier.

Relatively speaking, in the environment of economic development globalization, the acquisition of financial resource is various. There are many financing means, including the primitive accumulation of entrepreneurs, partner funding, debt, bonds, private equity and other means.

The acquisition way of technology resource includes the technology which the entrepreneurs have already owned and the technology which will be formed through continuous development. The research and development of technology is usually time-consuming, so the new venture can acquire technology resource through trade, or acquire the right to use the technology by paying certain fee to other enterprise.

For most new ventures, organization resource can't be formed immediately like other tangible resource, but be formed through gradual evolution in certain period. Besides, organization resource can also come from the initial design and continuous adjustment of entrepreneurs or its team; in the meanwhile it includes the adaptation to environment and the study of successful experience. One thing to mention is that the simple imitation of the successful enterprise's organization structure can't make the new venture gain success.

3.3 Resource Integration

After the entrepreneurs experiencing resource identification and acquisition, the most important thing is to integrate these resources, which makes the resources utilize by the enterprise's development and realize its value, not simply the possession of resource. Integration means the component of adjusting company and the process of making them together. It's normal behavior along with the enterprise's operation. Resource integration is a complex and dynamic process. It chooses, learns, configures, activates and integrates for the resource of different origin, different level, different structure and different content, which make them more flexible[11], rational, systematic and valuable. It also reconstructs the original resource system and gives up the invaluable resource as to form new core resource system. Resource integration of new venture is usually conducted in the following 4 parts:

(1) Integration of Individual Resource and Organization Resource
Through the arrangement of the dispersed resource, it will be rational and organized, which makes it permeate into organization and transform to organization resource for better utilization. In the meanwhile, organization resource can be quickly integrated into the vehicle of individual resource, which can stimulate the potential of individual resource vehicle and increase the utilized value of individual resource.

(2) Integration of New Resource and Traditional Resource
The combination of new resource and traditional resource can not only realize the efficiency of new resource for traditional resource, but the rational utilization of traditional resource can stimulate new resource in turn, and promote the renewal of new resource.

(3) Integration of Horizontal Resource and Vertical Resource
Horizontal resource means the relevancy of one certain type of resource to other related resource. Vertical resource means the resource of certain type's width and depth. Through their integration, it does good to set up the three-dimensional structure of horizontal resource and vertical resource for new venture.

(4) Integration of Internal Resource and External Resource
On the one side, it needs to identify, choose, and learn valuable and other scarce resources which are suitable with the internal of enterprise such as the implicit technology and knowledge, and to take them into the enterprise's own resource system; on the other hand, it can realize the suitable connection and stimulate the internal and external resource as to fully develop its efficiency of internal and external resources.

For the resource integration of new venture is a dynamic process with continuous adjustment and is the process of transforming the present resource to systematic resource, entrepreneurs need to maintain the resource basis of the new venture through individual ability and the power of the whole team. The best condition the new venture can get is to realize the transformation of present resource through the above-mentioned set of resource integration process, which makes it transform into the core competitiveness as to realize the competitive advantage of new venture. Through the above-mentioned sets of integration, these complex resources transform into ordered and complementary as to be utilized by new venture and achieve its strategic goal

4 Influencing Factors in the Process of Resource Integration for New Venture

4.1 Information

The information asymmetry restricts the new venture from acquiring resource, on the one hand, because the new venture exists a relatively short time, in the process of groping, a series of new problems, such as dealing resource, rules and regulations, product, technology, market and environment, are highly uncertain. The lack of credit and rationality causes that the investors are unwilling to invest to the new venture for getting not enough information. On the other hand, if the entrepreneurs lack the information of the resource owner, it's hard to approach the resource owner and to acquire resource. Information has become an important influencing factor of identifying and acquiring resource. New venture should set up unobstructed access to information and fully utilize all kinds of ways to collect useful market information. Second, new venture should increase the ability to manage and utilize information and establish perfect information collecting, evaluating and utilizing mechanism. Filtering the useful information from the mass information will help the enterprise acquire all kinds of needed resources.

4.2 Enterprise Ability

Enterprise ability of new venture is another key factor of influencing resource integration. Enterprise ability means the ability to allocate resource, develop its productive and competitive function. The level of enterprise ability decides the efficiency and quality of resource integration. Enterprise should overall improve its organizational management ability, research and developing ability, productive management ability, marketing ability and financial ability as to achieve the high level of management for resource. In addition, in the dynamic operating progress, the improvement of the organization's whole study ability is of great importance for promoting resource integration.

5 Conclusion

Resource integration is the key step for the new venture to gain the core competitiveness. Entrepreneurs should fully develop his entrepreneurship and lead its members to utilize the internal and external resources to acquire all kinds of needed resources. In the process of integration, it should focus on the utilization efficiency of resource and the problem of cost-control and fully develop its creativity to stimulate the vigor of new venture. In the meantime, there is no existing model for new venture to follow, and the environment is changing day by day, so there are a lot of new problems in the dynamic process of integration for the new venture to solve.

References

1. Alvarez, S., Barney, J.B.: Resource- based theory and the entrepreneurial firm, Strategic entrepreneurship: Creating a new mindset, pp. 89–105. Blackwell Publishers, Oxford (2002)
2. Li, L.: Intergrating Structure of Venture Theory. Industrial Technology & Economy (2), 89–91 (2010)
3. Barney, J.: Firm resources and sustained competitive advantage. Journal of Management 17(1), 99–120 (1991)
4. Cai, L., Xiao, J.: Research on the Effects of Entrepreneurial Orientation on the Resource Utilization of New Venture based on Resource Development Process. Science of Science and Management of S. & T (1), 102–105 (2008)
5. Brush, C., Greene, P., Hart, M.M.: From initial idea to unique advantage: The entrepreneurial challenge ofconstructing a resource base. Academy of Management 15(1), 64–81 (2001)
6. Lichtenstein, B.M.B., Brush, C.G.: How do ' resource bundles' develop and change in new ventures6A dynamic model and longitudinal exploration. Entrepreneurship Theory and Practice 25(3), 37–58 (2001)
7. Lin, C., Zhang, W., Qiu, Q.: The Discussion and Development Trends of Venture Creation Proces. Nankai Business Review (7), 47–50 (2004)
8. Yang, J., Zhang, Y.: Analysis of Venture Behavrior Process Based on Entrepreneuship. Foreign Economics and Management (2), 2–6 (2004)
9. Cai, L., Liu, Q.: Resource Combination Process Model during New Venture Creation. Science of Science and Management of S. & T (2), 95–97 (2007)
10. Dollinger, M.J.: Entrepreneurship: Strategies and Resources, pp. 45–48. Irwin, Boston (1995)
11. Danny, M., Jamal, S.: The resource- based view of the firm in two enviroments: The Hollywood film studios from 1936 to 1965. Academy of Management Journal 39(3), 519–543 (1996)

Efficacy Coefficient and Fuzzy Triangular Based on the Performance Evaluation of Science and Technology of Scientific Research Institutes in the Energy and Chemical Industry*

Ling Li[1], Lixiang Zhang[2], and Jinsuo Zhang[1]

[1] School of Management, Xi'an University of Science and Technology, Xi'an, 710054
[2] Research Institute of Shaanxi Yanchang Petroleum (Group) Co., Ltd, Xi'an 710075
llldou@yahoo.com.cn, zwshnyt@163.com, mark56zhang@163.com

Abstract. Performance assessment of science and technology in scientific research institutes is an effective means in guiding theirs technology innovation, promoting technical progress and supervising management. This paper constructs evaluation index system on performance assessment of science and technology in scientific research institutes by the method of investigation and study based on the theoretical and practical analysis. According to quantitative and qualitative indicators' properties of the evaluation index system, the paper calculates the index value, through the efficacy coefficient method and fuzzy triangular comprehensive evaluation method. The results have certain direct sense for performance management of science and technology of scientific research institutions in energy chemical enterprise.

Keywords: Energy and chemical industry, scientific research institutes, performance evaluation, efficacy coefficient method, fuzzy triangular comprehensive evaluation.

1 Introduction

Performance assessment is closely related to the management and incentive measures. In recent years, it develops from a single evaluation system to systematic and comprehensive evaluation and gradually forms distinguished performance management theories in all fields. For the scientific research institute' s scientific performance assessment, mainly focuses on national non-profit scientific research institutes: Our national scientific institute is founded and subsided by nation, it includes the Academy of science, Research Center, Central government's Research institutions, National Laboratory. Since 1980, our national scientific institutes have introduced performance assessment. Dengjie researched the application of BSC in public beneficial scientific research institute's performance assessment[1]. Jin Zhinong constructed performance

* Supported by Shaanxi province Science & Technology Fund Planning Project (Program NO. 2010KRM71), Scientific Research Program Funded by Shaanxi Provincial Education Commission (Program NO. 2010JK169).

assessment index system according to the office management, research development and characteristics of different departments in local scientific research institute[2].What' more, analyzed Dolphi method and AHP method in local scientific research institutes performance assessment to determine the index weight. Zhang Qianwei aimed at the choice of index, handling of data, calculating of weight coefficient and the way to taking account of the advantages and disadvantages of the assessment object, thus he could introduce a combination assessment model based on linear an nonlinear assessment[3].

We can see clearly that as the different aims and main emphasizes of research, at present, it is hard to see scientific performance assessment of a scientific research institute of an enterprise. The aim of scientific assessment is multiply, but also ever-changing. In a specific assessment, the aim of assessment has different emphasizes according to different objects and the main bodies of assessment. So we should construct a series of index system which is intrinsically linked and complementary to adequately reflect the scientific research institute's characteristics of an energy and chemical industry .choose and design a specific method and steps.

2 Design of Evaluation Index System

This research mainly uses logical approach and empirical approach to construct the scientific performance assessment of energy and chemical enterprise's scientific institute:

Firstly, collecting and carding some papers about the existing performance management and the technical innovation of scientific research institute. Combining its characteristics, we can design the open type interview outline of this research .At the same time, come up with some alternative indexes to form a index set of an energy and chemical enterprise's scientific research institute.

Secondly, we made a further interview with the senior mangers of Petrol China, Shenhua energy and coal group in April 2011 .In the interview, we mainly pay close attention to the special need, the main method, questions and difficult existing. Thirdly, according to the analytical structure of the aim, principle and index in systematic analysis. Discussion group also combine the result of the theory analysis and practical research clearly to determine the goal which the scientific performance assessment of an enterprise should to get to .They break the aim down into different principles and design the messier indexes which correspond to them.

Fourthly, we choose five scholars who are related to this research and make them judge this primary index system. We mainly judge whether the system leaks some important indexes and whether they are complete. Whether the express of the index is clear and whether the discrimination of these indexes is obvious.

Finally, we should adjust and process the index system with professions' advice to form the final performance assessment index system of the energy and chemical enterprise `s scientific research institute. Apply this index system into practice and dynamically correct the index in time according to the feedback of the application.

3 Construction of the Assessment System

The technical innovation performance assessment index system of the energy and chemical enterprise's scientific research which made up of general goal, specific principle, and measure index is a three levels hierarchical structure.

3.1 Energy Chemical Enterprise Institution Innovation Target

The current world oil prices volatile, with petrol-chemical based energy chemical industry faces the new pattern of world energy and new technology revolution brings challenge. In science and technology, focusing on the introduction of absorption and innovation, based on the set of technologies to speed up oil refining, ethylene, polyethylene, polypropylene and other key technologies and equipment related to the localization process. While the energy chemical industry by strengthening the original innovation and integrated innovation, energy and chemical build a complete system of independent innovation, to achieve the sustainable development of energy and chemical. Energy and chemical business research institutes of technology ahead of reserves, technology research and technology application should be launched around this goal.

3.2 The Principle Level

Energy chemical enterprise scientific research institutions of science and technology innovation appraisal of specific standards include innovation investment, innovation and management innovation output three aspects.

Scientific research institutions, the most basic job is daily scientific research tasks, the main task is in the scientific research achievements, out on the performance, and can promote applied to production . By the papers, monographs, patents, scientific and technological achievements such as the main form of technological innovation output, innovation and research institutions is reflected in the performance of the visual. In order to eliminate the size different produces the science and technology innovation mechanism innovation output absolute difference phenomenon; we choose the per capita amount.

Current national, scientific research institutions to assess the focus has gradually shifted attention to the academic and research institutions in the economic and social aspects of a comprehensive medium-term and long term results, such as the training and mobility, technological innovation, overall contribution to society, international cooperation, etc. Science and technology inputs not only indicate the potential strength of innovation of science and technology is also the guarantee of the normal operation of the foundation of scientific research institutions. But energy chemical enterprise subordinate to the scientific research institutions, its investment in science and technology are often by enterprise directly transfer, for personnel recruitment and selection may to some degree subject to enterprise. Combined with this characteristic, we use technology training and development of the evaluation of scores as one of the criteria.

A sound management system of science and technology, technological innovation can be implemented on all the factors in the process control and coordination of activities to ensure smooth progress and completion of research tasks. Combination of field surveys, we believe that the degree of realization of this norm can be done through science and technology plan, the operation of science and technology institutions, research and information management, financial management of funds to specific measures in four areas.

3.3 The Index Level

Based on the analysis above we get the energy shown in Table 1 chemical science and technology research institutes performance appraisal index system (shown as table 1).

4 The Performance Assessment Method of an Energy and Chemical Enterprise's Scientific Research Institute

The index in the index system constructed above can be divided into quantitative and qualitative; we can respectively choose effective coefficient method and fuzzy triangle comprehensive evaluation method.

4.1 Use the Efficacy Coefficient Method to Calculate the Quantitative Index in the Index System

The efficacy coefficient method has been propounded on the basis of multiple objectives programming theory. It gives every index a satisfaction value and a not allowed value. Makes the satisfaction value as the upper limit and makes the not allowed value as the lower limit .It determines the score of every index through calculating its satisfaction level and then synthesizes its weighted mean to assess the research object's comprehensive condition[4].

(1) Determined the quantitative index's satisfaction value and not allowed value for the scientific performance assessment of the energy and chemical enterprise.

The satisfaction value means that every index is possible to get to the highest level that is to say the upper level in the participated assessment units. Not allowed value means that every index should not be as the lowest level that is to say lower level in the participated assessment units. There are some common methods such as high standard location method industry standard method, intranet business location method, Delphi method. When assess the performance, we can combine the annual plan of the enterprise science and the best, worst value to determine them.

Table 1. Performance assessment index system of energy and chemical enterprise's scientific research institute

Aim	Principle	Index	Index level	Caption
Solve the key technology in construction, develop new technical research, production and launch soft science research. Promote technical progress of an enterprise, Strengthen Innovation (S)	Output of technical innovation (U_1)	Paper and monograph (X_{11})	capita paper and monograph of scientific research(X_{111})	quantitative
			proportion of paper recited into four retrieval system(X_{112})	quantitative
		patent(X_{12})	capita patent number of scientific member(X_{121})	quantitative
			proportion of authority for applied patent (X_{122})	quantitative
		Awards for science and technology (X_{13})	number of receive awards for science and technology(X_{131})	quantitative
			proportion of national awards (X_{132})	quantitative
	Technical training and input of supporting scientific research (U_2)	The training of technical staffs (X_{21})	times of learning and training member participte(X_{211})	quantitative
			proportion of members participate in learning and training(X_{212})	quantitative
		External exchanges and cooperation (X_{22})	times of gonging and coming to visit of the scientific members(X_{221})	quantitative
			cooperation with other offices($X_{22}2$)	qualitative
		Input of technical research (X23)	gain supporting funds for scientific research(X231)	quantitative
			improving condition of research(X_{232})	qualitative
	The management of innovation (U_3)	Performance of technical plan(X_{31})	completion rate of important projects (X_{311})	quantitative
			completion rate of technical projects(X_{312})	quantitative
		Management of technical message (X_{32})	trail of industrious knowledge(X_{321})	qualitative
			trail of technical needs(X_{322})	qualitative
			exchanging and sharing of the inner message(X_{323})	qualitative
			the storing and secret of the technical data (X_{324})	qualitative
		Operation mechanism of scientific research office(X_{33})	construction of technical management system(X_{331})	qualitative
			executing of technical management system (X_{332})	qualitative
		Management of funds for scientific research (X_{34})	the obtaining of the scientific research funds(X_{341})	qualitative
			regularly management of the funds for scientific research(X_{342})	qualitative

(2)Calculating every single index's score
In the formula, s_i is index's specific evaluation value, x_i is the actual value, x_{max} is the most satisfaction value, and x_{min} is the not allowed value.

$$S_i = (\frac{x_i - x_{min}}{x_{max} - x_{min}} \times 40 + 60) / 100 \tag{1}$$

(3) The calculating of comprehensive score of the quantitative part.
The important degree of every index is different in the index system, in order to reflect every factor's important degree, we give every factor some weight .Thinking about the condition of the energy and chemical enterprise's scientific management, we choose point-to-point comparison method of the subjective evaluation method to make the assessment has some operability.

4.2 Use Fuzzy Triangle Comprehensive Method to Calculate the Qualities Index of the Index System

The general fuzzy evaluation is hard to quantities or can't precisely measure the index, so it also use "best, better, middle, bad" such fuzzy words to assess indexes, thus leads much useful message lost in the calculating process ,but fuzzy triangle comprehensive method can broaden the scope and the quantity of useful message. So that makes the result more actual and it also improves the possibility and the precision of the evaluation[5]. Fuzzy triangle method uses 3numbers to express the index's evaluation value .They can be recorded as $\tilde{A}(a,b,c)$
Its membership function is

$$\mu_{\tilde{A}}(x) = \begin{cases} 0 & x < a \\ \frac{x-a}{b-a} & a \le x < b \\ \frac{c-x}{c-b} & b \le x < c \\ 0 & x \ge c \end{cases} \tag{2}$$

Determine scientific performance assessment's qualitative evaluation value of the energy and chemical enterprise's scientific institute.

Express the qualitative index's evaluation value of the energy and chemical enterprise `s scientific institute as $C = \{C_1, C_2 \ldots, C_n\}$

Its corresponding weight is $\tilde{\omega} = (\tilde{\omega}_1, \tilde{\omega}_2, \ldots, \tilde{\omega}_n)^T$ the index c_i 's evaluation value is x_i. They can be given as variables by the professor according to the remarks set (shown as table 2) which is designed in advanced, and we can exchange the variables to corresponding fuzzy triangle numbers.

Table 2. The weight of the evaluation index and the fuzzy triangle numbers of the evaluation index's remarks

Rank	Weight set of the index	Remark set of the index	The corresponding fuzzy triangle numbers
A	lowest	worst	(0, 0, 0.1)
B	low	worse	(0, 0.1, 0.3)
C	Medium-low	bad	(0.1, 0.3, 0.5)
D	middle	middle	(0.3, 0.5, 0.7)
E	Medium-high	good	(0.5, 0.7, 0.9)
F	high	better	(0.7, 0.9, 1.0)
G	highest	best	(0.9, 1.0, 1.0)

Fuzzily calculate l professors\` evaluation values, we can get the evaluation value of the index x_i

$$\tilde{x}_i = (1/l) \otimes \left[\tilde{x}_i^{(1)} \oplus \tilde{x}_i^{(2)} \oplus \ldots\ldots \oplus \tilde{x}_i^{(l)} \right] \tag{3}$$

In the formula "$\otimes$"和"$\oplus$" respectively express fuzzy numbers \`s multiplication and addition.

The calculating of every index's weight

Synthesize the weights given by every professor; we can get the index's weight vector. $\tilde{\omega} = (\tilde{\omega}_1, \tilde{\omega}_2, \ldots, \tilde{\omega}_n)^T$,

$$\tilde{\omega}_i = \left(1/l\right) \otimes \left[\tilde{\omega}_i^{(1)} \oplus \tilde{\omega}_i^{(2)} \oplus \ldots\ldots \oplus \tilde{\omega}_i^{(l)} \right], \quad i = 1, 2, \cdots, n \tag{4}$$

Calculating of the expectation of the fuzzy importance and fuzzy donation

Index weight $\tilde{\omega}_i$ and index value $\tilde{x}_i$ calculated above are all fuzzy triangle numbers. It is hard to compare them directly, so we need to calculate the corresponding expectation. Take the evaluation value $\tilde{x}_i$ as example, according to its function, $\tilde{x}_i$ \`s left membership function and right membership function is

$$f_{x_i}^{\ L}(x) = (x - \alpha_i)/(\beta_i - \alpha_i) \text{ , } f_{x_i}^{\ R}(x) = (x - \gamma_i)/(\beta_i - \gamma_i) \tag{5}$$

Their inverse functions are

$$\varphi_{x_i}^{\ L}(y) = \alpha_i + (\beta_i - \alpha_i)\, y \text{ , } \varphi_{x_i}^{\ R}(y) = \gamma_i + (\beta_i - \gamma_i)\, y \tag{6}$$

Obviously, they are continuous and respectively increase and reduce progressively. So we can calculate $\tilde{x}_i$ \`s expectation by integrating its left and right expectation.

$$I(\tilde{x}_i) = \eta I_L(x_i) + (1 - \eta) I_R(x_i), \quad 0 \le \eta \le 1, i = 1, 2, \cdots, n \tag{7}$$

In the formula, η is optimistic-pessimistic coefficient.

4.3 Calculate the Comprehensive Evaluation Value

Calculate the value and value weight of the evaluation index with the two methods mentioned above and use the method of weighted mean to get the final comprehensive evaluation value can be used to further analysis the existing problems of the scientific performance management of the scientific institute.

5 Conclusions

This paper constructs scientific performance assessment index system through theory analysis at the levels of its total aim scientific actions, principle and specific balance index according to the scientific assessment situation of energy and chemical enterprise's scientific research. What's more, it chooses and designs effective coefficient and fuzzy triangle number to take the specific evaluation data. But as the condition is limited, the research can't design the management steps and operational system according to assessment result. This also is our main effort.

References

1. Jin, Z.: A Study of Performance Appraisal Index System and and Their Weights of local research institutions. Science and Technology Management Research (12), 103–106 (2009) (in chinese)
2. Deng, J.: Balanced Score Card's Applying in the Performance performance appraisal of the commonweal research institutes. Communication of Finance and Accounting (19), 58–59 (2009) (in chinese)
3. Zhang, W., Miao, Y.: The Application of Combined Evaluation Model on the Performance Evaluation of China's National Research Institutes. Science of Science and Management of S. & T. (4), 36–40 (2008) (in chinese)
4. Li, W.: Competitiveness theory and method of evaluation. China Market Press, Beijing (2009) (in chinese)
5. Zhang, S., Su, Y.: Study on measurement of enterprises' technology innovation capability based on innovation process under the network environment. Science of Science and Management of S.& T. (1), 101–105 (2007) (in chinese)

Study on the Transformation Strategy for Resource-Exhausted Cities in View of Sustainable Development

A Case Study in Baiyin City, Gansu Province

Haifei Wang

Geography and Environment Science College, Northwest Normal University,
Lanzhou, Gansu, China, 730070

Abstract. Industrial and sustainable development of resource-exhausted cities are China's current economic and social harmonious development problems to be solved nowadays. In the course of China's industrialization and urbanization, there are many cities which are currently facing industries transformation, pay close attentions to the transformation of these cities and study the means, mechanisms, measures and methods,It is significant and critical for building a harmonious society with Chinese characteristics, In this paper, take the resource-exhausted city ,Baiyin Gansu province for example, analyzed the status of urban development and transformation of the main problems, through positioning the industrial transformation of the city , proposed restructuring of the relevant strategic vision, point out that baiyin city should try to integrate itself into the Lanzhou-baiyin urban economic circle as early as possible and make joint development with lanzhou; Actively cultivate alternative industries, implement industrial diversification; Optimize the investment environment of the region and absorb industries from other developed regions; Develop circular economy, improve use efficiency of resource and energy ; Make use of two resources and two markets and expand the space of using resources. These ideas are beneficial for other resource-based cities' sustainable development with a certain reference value.

Keywords: Resource-exhausted city, Strategic orientation, Development strategy.

1 Introduction

The economic transformation of resource-exhausted cities is the important content of coordinated and sustainable development for a country. Exploring the answer of it is a worldwide problem, in our country it is also a subject with certain universality. The resource cities grew up with the development of resource, expand with the prosperity of resource industry and decline with the depression of resource industry. Part of the countries in the world have experienced or are facing the phenomenon of resource exhaustion which make the cities' economic development blocked. Such as the Lorraine of France, the Ruhr of Germany and the Montreal of Canada, etc. In the

D. Zeng (Ed.): Advances in Computer Science and Engineering, AISC 141, pp. 511–520.
springer link.com

development history of urban resources at home and abroad, there are two very different situations, because since resources exhaustion some regions couldn't develop any longer, they left there, finally the cities formed the so-called empty "ghost town. Others similar to Houston and Los Angeles such comprehensive city developed more and more flourishingly. In the consideration of China's special national conditions, the resource-exhausted cities of China could not go the way of abandoning city to solve the problem. We must base on the regional economic and social sustainable development, through establishing relevant development mechanism to build new industries , making the industries diversified, and strive to make regional economic and social development enter the benign orbit again.

Sustainable development has become the direction of a modern city development, the resources city as a type of the city, in order to realize the sustainable development, it is necessary to achieve transition of industry[1]. In China there are 426 all kinds of mining cities (towns), including 178 mining cities. These cities and towns provide more than 80% of the mineral resources and energy resources to the national economic development, they are the strong power of our national economy development and they also once made the glorious history. Such as China's Baiyin of Gansu, Kelamayi of Xinjiang, Daqing of Heilongjiang, Fuxin of Liaoning, Jiaozuo of Henan, ZaoZhuang of Shandong, Datong of Shanxi, Panzhihua of Sichuan and other cities. Since its high dependence of resources it formed the highly dependent economic structure. This structure easily makes resource cities get into the crisis of resource exhausted and city ruined [2]. But the reality is that China now has more than 50 cities are faced with this threat, the potential development of some resource cities in the future is worrying. As the first construction project of five-year plan and earliest non-ferrous metals industry base, BaiYin has gone through 50 years development process, the main nonferrous metal mineral resources have a sharp decline, the self-sufficiency degree of copper resources decreased from 100% to less than 20%, the sustainable development of society and economy are facing great challenges. In 2003, BaiYin became one of the 18 typical national resource-exhausted mining cities. In March of 2008, BaiYin was approved as one of 12 national pilot cities of resource transformation, the state, local government and related departments paid close attention to the sustainable development of it.its development is facing opportunities and filled with challenges. Discuss its transformation development is of great significance to the region economy coordinated and sustainable development of Gansu province.

2 The Inevitability and Necessity of Industry Transformation for Resources-Exhausted Cities

Industry is the core of the regional economic development; the area is the supporter of industry. Thus, the sustainable development of industries is the guarantee of regional lasting vitality. Industry transformation is the only way which must be passed to get out of difficulties and step on the way of sustainable development for resources city.

2.1 Development of Recycling Economy Is a Big Choice Which Resources City Is Facing

Recycling economy is the conserving resource and intensive economy. The sustainable development theory emphasizes in the process of economic development do not pursue the expansion of quantity too much; it shall make the depletion speed of the non-renewable resources under the speed of looking for renewable resources as a substitute [3]. The development of resources cities are built on the excessive dependence on the original and specific resources, and these particular resources are non-renewable and irreversible. If the resource cities blindly follow traditional development road, its development will be unsustainable. Its development mode must make the transition. Therefore, change the way of the economic growth and the exploitation of resources, achieving resource saving and improving efficiency of resource utilization. The sustainable utilization developments of specific resources are the objective requirement of promoting the sustainable utilization of resources, and the urgent reality choice of promoting sustainable development of resource industry and the resource cities.

2.2 Cultivating and Developing the Continued Industry Is the only Way to Get Out of Difficulties for Resource Cities

To solve the increasingly prominent economic and social development problem, the resource-exhausted cities can only rely on industry transformation, cultivate and develop alternative industry, realizing the diversity of industrial structure and abating the excessive dependence on the traditional industry. The area of Most of the domestic resources-exhausted cities is too larger and the population is concentrated, their rise and fall have great externalities in the social and economic aspects. Different from many small resource-exhausted towns abroad, we shouldn't take the way which close down mines and abandon city, people leave and the city ruin, we can only go the way of regional "industry reengineering", and use the internal and external resources of the region, capital, technology and talent to put industry transformation into effect, making the urban economic and social development to stage a comeback.

2.3 The Economic and Social Transformation of Resource Cities Will Promote the Change and Innovation of the Regional Economic System

The unreasonable use and development of resources are the fundamental factors that impact the sustainable development of resource cities, while the development of the resource-exhausted city is the important component of the national economic and social development. The resource cities who under the guidance of our country's sustainable development strategy, according to the characteristics of local resources and the situation of social economic development set industry development and transition strategy that fit the requirements of local development., surely resulting in promoting a series of change and innovation of resource cities` economic system, and in a higher starting point to promote the sustainable development of the region. Besides, the resources-exhausted cities deal with the relationship between the economic development and improvement of ecological environment in the long term, through the effective implementation of the sustainable development strategy of economic transformation.

3 The Current Situation of Baiyin's Industry Transformation and the Problems of It

3.1 The Current Situation of Baiyin's Transformation

BaiYin is a typical resourceful city of China. Its copper mine which as one of national 156 key projects in the period of first five-year began to be built from 1954. In 1960 the copper and sulfur system had been finished and put into production, becoming the largest non-ferrous metals enterprise and production base of copper and sulfur of our country. In 1980 s, we had the construction of "two factory and a mine" (i.e. the northwest smelter of lead and zinc, Baiyin aluminum manufacturer and Changba lead-zinc mine), thereby, forming a sound industrial development pattern which is integrative in mining, beneficiation, smelting, processing, scientific research and trade. It took the copper, lead and zinc as the main non-ferrous metal resources. At present, the production capacity of BaiYin`s industrial non-ferrous metal is 400000 tons and can produce more than 1000 tons of coal. The installed capacity of electric power plants built and under construction is 3.3 million kilowatts; the production capacity of TDI is 5 tons. For 50 years, BaiYin`s nonferrous metal industry has accumulated to provide more than 5 million tons nonferrous metal to our country and pay over 15 billion RMB tax to the government. It once had made a achievement of the copper production first in the country in the consecutive 18 years. However, after decades of mining, the proven reserves of the copper, lead, zinc and other main resources that Baiyin`s industrial economy is dependent on has been dry up. The copper mine which is built in 1954, from the 1980 s is forced to close down mine and take deep mining underground, greatly reducing the amount of resources; Changba lead-zinc mine built in the period of seventh five-year to the eighth five-year because of l the massive and excessive mining result in the collapse of open-pit, which influence the mine's sustainable development badly; the supply of non-ferrous metal resources is in seriously shortage; Part of resources of Jingyuan coal company in BaiYin also began to dry up. It already has had several mines closed. The development of mineral resources also faces a lot of problems. And it has not been able to get breakthrough and progress in geological prospecting work. Under the double restriction of Resource exhaustion, and system transformation, its system, structure and social contradictions become increasingly prominent. In the period of ninth five-year, the city's economic growth went to the bottom. Thus accelerating the economic transformation of cities is becoming the major historical subject for BaiYin.

3.2 Problems of BaiYin`s Transformation

BaiYin from putting transition into effect to it was identified as the national resource-exhausted city in the March 2008, the main line of the development strategy has two. One is the development of substituted industry; the other is the cultivation of continued industry. But the recent years economic situation of BaiYin shows that although the city maintains a high economic growth, there exist some outstanding problems in the industrial structure, the quality of economic growth, the restructuring of state-owned enterprise, operational efficiency and ecological environment. They main refer to: (1) the economy of BaiYin is still very depend on mineral resources and non-ferrous industry , and the contribution of the traditional industries to BaiYin`s economy as much as 70% ; the proportion of new and high technology industries is small; Agriculture foundation is weak ; the advantage of characteristic industrial is not obvious; The way of secondary industry development still extensive ; the restructuring pace of state-owned enterprises who as the support of the city's economy is small; The development of tertiary industry is slow; (2) industries in their region are not close, the industrial chain is short, the development of continued industry is slow, the cultivation of substituted industry is not enough; (3) the large state-owned enterprise who as the leading power of BaiYin`s economy don't have high operational quality, its growth of profit margins is too slow ,even some enterprises have not make up deficit and get surpluses; (4) the innovation of technology is weak, the way of economic growth is comparatively extensive, the strength and depth of traditional industry transformation are not enough, the development of recycling economy lack internal force, the contradiction of ecological construction is more prominent. Worse still, Baiyin didn't make use of its own advantages of characteristics and location, when they cooperate with outside world, the pace and intensity are small, such as the regional cooperation, industrial interaction, investment attraction and industry transfer .etc. The Baiyin high-tech industry and the Chinese academy of sciences was once the representative project of the city, but the energy and potential function of transformation did not play out. It is obvious that the transformation of BaiYin is still difficult, it is a long journey. The difficulties that the transition is facing, including objective factors, such as, the endogenous force is not strong, the resources prospect is not good, technology, capital and talents are short, but more factors are subjective ones, such as the change of concept is slow, system constraints, the strength of policy guidance is not strong, etc. After investigation, the state-owned enterprises of BaiYin widely have the problems and vulnerabilities of inefficient and poor management. The space is larger for management to enhance enterprise performance.

4 The Localization and the Transformation Strategy of the Resource-Exhausted City of Baiyin

4.1 The Localization and Ways of the Resources City's Transformation

Industry can not be automatically or completely depend on the market mechanism to realize the optimization of the structure, coordination of proportion, perfection of layout and organization spontaneously, so the government must take some proper management and regulation. The localization of resources-exhausted city's transformation is the key problem, that is, to determine which industry as its dominant industry. The leading industry in one area is who has promising market prospect in the region economic development, strong innovative ability and diffusing effects; they are stimulus to the other industry's growth; they play a role of direction to the regional economic development. the process of resources-exhausted city's transformation is to determine and develop the dominant industry, this is the important strategy problem of regional economic development.

4.1.1 The Power of City's Transformation: The Technical Innovation

Science and technology are the first productive forces. The traditional industries who take the resources development as the guiding industry have disadvantages in the regional economic development. If these traditional industries can't to be transformed and upgraded in time, the development of resource-exhausted city will be blocked by its insufficient driving force, resulting in economic decline, and even causing serious social problem. We must transform and upgrade traditional industry with the high and new technology and extend the industrial chain, which has important strategic significance. The economic development of resources-exhausted city must depend on science and technology innovation instead of resources exploitation leading economic growth to promote industrial development. This includes two aspects; the one is around resource development and processing to introduce a group of new high technology industry, take over industries from the developed regions and take use of the applicable technology; The other one is to transform traditional industries with the high and new technology, benefiting by science and technology. The economic development should focus on the five new high technology industries, they are fine chemical engineering, new material of non-ferrous metal, new energy technology, materials and technology of ecological restoration and eco-friendly materials. We should take the cooperation of producing, studying and researching as the important driving force to realize the economic transformation.

4.1.2 The Guarantee of the Transformation: System Innovation

The government leading industry is still the main way of resources city's transformation in our country. The government should give full play to the policy guidance, system innovation and optimization of regional development environment. System innovation is not only a part of transformation strategy, but also the system guarantee of strategic transformation. To put the economic transformation into effect, we should eliminate institutional obstacles first. System innovation is the big breakthrough in establishing management and operation system that adapt to the market economy, setting up a platform which is helpful to the supernormal development of economy and enhancing the power and energy of development. It is necessary to deepen the reform of enterprises in all round way, establish the modern enterprise system, develop the shareholding system economy and the non-state ownership economy and take substantial steps in achieving diversity of property rights. Concentrate on supporting reform and development of large and medium-sized state-owned enterprise and focus on cultivating a batch of competitive enterprises and large blocs. Make non-ferrous, coal, electric power and the chemical industry be the leading industries, driving development of regional economic.

4.1.3 The Support of City Transformation: Diversification in Industry

According to the comparative advantage and disadvantage factors of local economic transformation, BaiYin is suitable to cultivate following pillar industries: non-ferrous metal and new materials industry of rare earth, integration industry of fine chemical, recycling industry of mining and resources, energy and new energy industry, machinery and special equipment manufacturing, non-metallic minerals industry, deep processing of characteristic farm and animal products industry and cultural tourism industry of the Yellow River, etc. Economic transformation is the urgent task for

Baiyin to speed up the development. Developing and cultivating diverse pillar industries is the effective way to promote the economic transformation. The only way for the resource-exhausted city to realize sustainable development is following: to cultivate diverse industries, to speed up the development of the continued industry, to extend the industrial chain and to improve the additional value of the products. BaiYin should expand space of economic development , develop diverse industries, forming a new situation that the traditional industry, new industries and new and high technology industries compete to develop.

4.2 The Strategic Ideas of Resource-Exhausted City's Transformation in BaiYin City

4.2.1 Through Lanzhou-Baiyin Economic Circle, Baiyin Develop together with Lanzhou

Combining the transformation of resources-exhausted city and the construction of city circle is the innovation of transformation [4]. The construction of lanzhou-Baiyin urban economic circle is the objective requirement of economic geography's law. Urban economic circle is a development strategy to realize the regional economic integration and promote the whole competitiveness of regional economy through the coordinated development. Under the framework of regional economic integration, plan the regional resource allocation and distribution of productive forces well; actively adjust the industrial structure to avoid separating development and the waste of resources, magnifying the efficiency and quality of regional economic growth. According to the "center drive" theory, giving full play to LanBai urban economic circle's the leading role is helpful to expand the space of Lanzhou`s development, promote the economic integration of LanBai and surrounding towns and improve the whole competitiveness of city circle; it is helpful to make a combination of petrochemical industry, non-ferrous metallurgy, equipment manufacturing and the high and new technology industrial, optimize the industrial chain, foster new innovative industrial chain and promote the value chain. Its rise and development is not only of great significance to Lanzhou and BaiYin`s ecological construction, but also to the regional economy development of Gansu province. The LanBai economic circle will be the growth pole of Gansu province's development and will make a "core engine" to the Gansu`s development. Lanzhou and BaiYin through complementing each other's advantages, resource integration and mutual influence realize the synergistic effect of 1 + 1 > 2 and promote the rapid development of BaiYin`s social and economic transformation. BaiYin and lanzhou can completely develop together in space in future, realizing the spanning of heavy industry, equipment manufacturing, mechanical electronic, biological medicine, ecological agriculture and food processing, tourism and the modern service industry. Develop and expand the LanBai economic circle, making it become the economic center of Gansu province even the western region [5]. At present, it is the key period of Baiyin`s transformation. The construction of "LanBai economic integration" will inevitably produce new driving force in regional economic integration, which not only form a benign interaction between Lanzhou and Baiyin; it is helpful to accelerate the pace of Baiyin`s transformation, cultivate economic growth pole and promote the regional core competitiveness which will become the new "engine" and "booster" of transformation. The construction of "dual-core" (Lanzhou and Baiyin) economic circle makes Baiyin`s transformation into the saltant stage.

4.2.2 Based on the Characteristic Resources to Implement Diverse Industry and Advance New Industrialization and Urbanization

According to their existing resources characteristics, talents and technology advantage enhance strong points and avoid weaknesses.Vigorously develop alternative industries, and actively cultivate a new economic growth point. Continue to write the mining articles and do it better and better. For some fundamental industries, expand the industrial chain as long as possible. The transformation of traditional industries drives the cultivatation and development of alternative industries; the cultivatation and development of alternative industries promote the transformation and upgrade of the traditional industries. At present, the traditional industries in the BaiYin occupy a quite large proportion of GDP, they are still one the main strong point of the city's economic and social development. Therefore, we should reform the tradition non-ferrous metal, energy, chemical industries.etc. it is very important to to benefit from science and technology. At the time when approved as " the first group of the resources-exhausted transformation city in the country " strive for the relevant national policies;strengthen the echnical modification and investment of old enterprises,such as the Baiyin non-ferrous company, Yinguang company, Changtong Cable Plant, Jing coal company, making them full of vitality.

Advancing new industrialization and urbanization is the basic direction. The state-owned enterprises are still the main body of Baiyin`s industry,they universally exist vulnerabilities of inefficient management and poor management. The space is larger in the enhancement of enterprise performance. So, further to perfect enterprises` operation mechanism and management innovation, making the enterprises to establish the modern enterprise system which" property rights are clear, rights and liabilities are definite , separate enterprises from administration and take scientific management ". which make large and medium-sized state-owned enterprises get out of troubles, this is the basic guarantee to gain development power for regional economy. Industrialization and urbanization are the two big trends of leading the future economic development. According to the requirements of the new industrialization, strive for national related policies of resource-exhausted city. In process of city`s transformation adhere to drive industrialization through informationization and promote informationization through industrialization; promote traditional industries and develop alternative industries through the application of high technology and information.

4.2.3 Open Wider to the Outside World, Improve the Environment and Pay Attention to the Projects

BaiYin`s location advantage is not obvious, so we need to rely on the government and policies to create a good investment environment. As the western remote city, Baiyin should grasp the opportunity and actively create a condition to absorbe and take over industries of the developed area. Environment is the productive force, so we should continue to improve the investment environment, actively attract the multinational companies and famous enterprises at home and abroad to invest; combine the capital, technology and brand superiority of the developed area with our advantages of land, raw materials, energy and labor force to take technological renovation, joint cooperation and achieve win-win situation. According their own functional planning, focus on developing several industrial clusters.

4.2.4 Develop Recycling Economy and Promote the Efficient Use of Energy and Resources Conservation

Circular economy is the basic requirement of the Scientific Outlook on Development. we must take energy conservation and consumption reduction as the main target of transforming economic growth pattern; insist on saving energy ,reducing consumption and developing together;advance the low consumption and efficient use of resource and energy. Really put the sustainable development idea into the regional economic development and the industrial development. According to the principle of "reducing, reusing and recycling " realize the the maximum use of resources and the minimum production of pollutants. Develop and perfect technology and craft of resources recycling, set up the recycling and supply system of waste resources, realizing recycling of resources. Establish ecological process, realizing resources recycling and multilevel utilization and making them turn into the maximum products, and gradually realize the ecological development.

4.2.5 Should Be Good at Using Two Kinds of Resources and Two Markets and Expanding Resources Space

Two kinds of resources namely are the internal resources and external resources; two markets, namely are the domestic market and overseas market. The development of the city, it especially means the sustainable development of resources-exhausted city. We should not only base on our limited regional resources, also must go outside for the rich resources, providing power for the fundamental industry in the future. Currently, for the development and utilization of mineral resources, we should keep our eyes on other domestic places, even the international market, and learn to cooperate with others, building strategic alliances. At the same time, continue to prospect the deep existing mining area and the surrounding area and strive to expand resources. Actively use the external resources and develop a market of raw material supply. We must pay attention to the outside and modestly learn the experience from the similar enterprises in developed area. Some enterprises in BaiYin have already made some achievements on technology and product innovation, but this does not means that it will have a good market performance. Enterprise shall set up the idea that take the market as the direction guide, establish scientific and standard management system and draw up strong marketing strategy, making a batch of competitive products go to the domestic and international market and rapidly improving sales of products. The important contents for the resources city to draw up and implement economic transformation strategy of the sustainable development are to make a long-term development planning of rational utilization of resources. Resources industry and the long-term sustainable development of resources economy are the strategic goal of transformation.

5 Conclusion

The important contents for the resources city to draw up and implement economic transformation strategy of the sustainable development are to make a long-term development planning of rational utilization of resources. Resources industry and the long-term sustainable development of resources economy are the strategic goal of transformation[5].

The Implementation of industry transformation is the only way for the resource-exhausted city to promote economic development and maintain social harmony and stability; especially the orientation of transformation is the key point. The situations of resources-exhausted cities are varied, so there is no fixed mode. According to the principle, adjust measures to local conditions to make an orientation. The principles are" to take advantages, to base on characteristics, the science and technology innovation, the development of diverse industry and the increase of employment and the promotion of development". To find a combining point between their own advantages and the market demand is the joint direction. In the process of transformation we must deal with the relationship between the industrial upgrade and the employment growth, the relation between the traditional industries and the alternative industries and the relation between the industry transformation and the system reformation. "To gradually transform" is the core content of the transformation; it means dynamically adjust the orientation of industry transformation, realize the industrial upgrade, economic growth, sufficient employment and social harmony. At the moment when Gansu province implements the strategy of "the center drive, developing together and advancing overall" Baiyin depend on outside power and develop with Lanzhou together. We should Use all kinds of way to promote economic and social transformation of resources-exhausted city, such as through science and technology innovation to cultivate diverse industry, promoting the common development of rural and urban areas and compacting characteristic culture, making it on the virtuous and sustainable development road.

References

1. Sun, Y.J.: The transformation and development ways of resources city, pp. 2–5. China Economic Press, Beijing (2006)
2. Zhao, T.S.: Research of China's economic transformation of resources city, pp. 2–3. China Economic Press, Beijing (2007)
3. Liu, L.G.: The sustainable development strategy of resources city, pp. 37–38. Economic Management Press, Beijing (2006)
4. Wang, H.F.: Baiyin's sustainable development strategy study in the view of LanBai city circle. J. Economic Geography 4, 566–571 (2011)
5. Zhang, Z.Q., Gao, F., Feng, Y.M.: City's innovating development-Baiyin's innovative development strategy research. Gansu Science and Technology Press, Lanzhou (2009)

On the Coordinated Development of Regional Economy in Gansu Province

Haifei Wang[1] and Weixin Kong[2]

[1] School of Geography and Environmental Science, Northwest Normal University, Lanzhou, Gansu, 73007

[2] School of Economy and Management, Zaozhuang University, Zaozhuang, Shandong, 277160

Abstract. The unbalanced development of regional economy is common in the world and it worthies attaching great importance, regional economic unbalance and difference expansion has its inevitable condition, while the coordinated development is a way to weaken the impacts of these conditions. Through the analysis of the current situation and the influencing factors of the region economy in Gansu province, in this paper, the counter measurements of coordinated development of Gansu regional economy were put forward according to writer's investigation. Given the premise of regional economic differences, the implementation of non—balanced regional development strategy and the industrialization of undeveloped area is proposed to promote the reconstruction of regional ecology, and helps to the attract the investment.

Keywords: Regional economic difference, coordinated development, industrialization, ecological reconstruction.

1 Introduction

The difference of regional economy is a general economic phenomenon; all countries in the world are facing the problem of unbalanced development. As the world's largest developing country, China is facing a more prominent regional difference, which leads to the unbalance of its economic and social development. The problem of regional economic difference has become the focus of our economic and social life in China. Regional economic disparity refers to the non-equalization of the economic growth of each region in many aspects in the process of economic development. The phenomenon of the differentiation of regional economic development, whose cause of is very complex, can never be neglected. With regard to the economic and social development of Gansu province, this article mainly analyzes the reasons of the regional economic difference in Gansu province from the natural, economic and social aspects, puts forward some countermeasures, aiming to offer policy proposals for the coordinated development of the regional economic in Gansu province.

D. Zeng (Ed.): Advances in Computer Science and Engineering, AISC 141, pp. 521–527.
springer link.com

2 The Current Situation and the Influencing Factors of the Regional Economic Development in Gansu Province

2.1 The Current Situation of the Regional Economic Development in Gansu Province

With a poor natural foundation, Gansu province is underdeveloped, but in the national strategic function it is very important [1]. The regional economic development in Gansu Province presents the gradient pattern. The western area, with Jiayuguan as its representative, is rich in resources and has a higher level of economic development, so the industrial competitive power is strong. Zhangye City and Wuwei city are the typical agricultural cities, which have a better condition of the agricultural production, while their industrialization levels are relatively low. The central area—Lanzhou city, as the center of the economy and politics and also the capital city of Gansu Province, it occupies a favorable position in the economic and cultural development. Baiyin is a typical resource industrial city, its economic level ranks the front row, and the location advantage is prominent, but the industrial economic benefit is not good enough. The resources in that city are at the edge of exhaustion, which leads to the deficiency of its follow-up industry. Due to the limitations of the historical and natural conditions, the economic development of Dingxi city lags far behind the average provincial level. The eastern region, with its low industrialization level inconvenient traffic, poor living and production conditions, low science and technology education level, prominent poverty problem, is the bottleneck of the coordinated regional economic development in Gansu Province.

2.2 The Factors Influencing the Regional Economic Development

2.2.1 Environmental and Resource Factors

As the material basis of all the economic activities, the natural condition determines the levels of the whole economic activities. Gansu province is located in the northwest inland area in our country. Gansu province is located in the upper reaches of the Yellow River. The plateau regions such as Hexi Corridor, Longzhong Loess Plateau, Longnan Loess Plateau, Longnan Mountain and Longnan plateau area distribute from the west to the east. Most areas have a low amount of precipitation, the annual rainfall decreases from the southeast to the northwest. The complex geographical and natural condition of in Gansu province determines the differences of the provincial economic development. The metallogenic condition in Gansu province is superior, but its distribution is quite asymmetrical. The coal resources are mainly distributed in Lanzhou, silver in Pingliang; copper in Baiyi (now close to the edge of depletion); iron ore are mainly distributed in Jiayuguan; nickel ore are mainly distributed in Jinchang, gold resources are rich in Gannan district. The reserve of mineral resources in Gansu is not so rich, but it formed a relatively complete industry base relying on the mineral resource mining, smelting and processing. The mineral resources in Linxia, Dingxi city are rare. Due to the low level of its development, other natural resources are not effectively utilized. Longnan city has rich mineral resources, and is especially rich in high grade lead zinc mine and high quality antimony ore reserves, but due to the underdevelopment of mining industry and raw materials industry, and the processing chain is short, the resources there have not been given effective value-added processing, which leads to the relatively backward economic development. All in all, the resources factor is the important factors of the cause for its regional economic difference [2].

2.2.2 Social Factors

The development of the society is the premise of its economic development, and in turn the economic development accelerates and promotes the development of the society to a certain extent. The process of regional economic development is in fact the process of regional urbanization. The level of the urbanization, the number and scale of cities and the infrastructure construction level play a vital role in the regional economic development and the enhancement of its comprehensive strength.

2.2.3 Level of Urbanization

Urbanization is the spacial transfer and movement process of continuous development of the productive forces in human society and those of deepening geographical division of labor[3]. Located in the inland northwest of China, Gansu province is a relatively backward area in economy, which is mainly due to the horizontal inequality of the urbanization level of different areas. In Hexi City, the urbanization development in three cities out of five is higher than the average level of the province, and the city of Jiayuguan ranks first in Gansu province. And in the middle region of Gansu Province, except Lanzhou City, Baiyin is close to the average level of the province, Dingxi city is far lower than that of the two cities. In Longnan area, except Longnan city, which is close to the average level of the province, other areas are far below the average level of the province. The regional urbanization level in Gansu province shows asymmetry in its regional distribution. Thus, the level of urbanization is also an important factor affecting regional economic development.

2.2.4 Infrastructure Construction

The infrastructure construction level is also an important factor influencing the level of the regional economic development. The improvement of infrastructure needs the support of government. Thus, large-scale investment and construction is necessary for the gradual realization of the urbanization. The Jiayuguan ranks the first place in the infrastructure level in Gansu province, while Longnan City is the lowest. Except the tap water production capacity and the number of communal steam (electronic) car every million people are less than that of Lanzhou city, the other indicators of Jiayuguan City are higher than that of Lanzhou[4]. In addition, mainly due to the higher rate of its gas penetration, the level of city infrastructure in Qingyang is also higher. From the above analysis of the difference in infrastructure construction, it is clear that the infrastructure construction level differences and slow development is also key indicators of the regional economic difference in Gansu province.

3 Suggestions on Regional Economic Development

3.1 Strategy of Coordinated Development of Unbalanced Region

Unbalanced development theory believes that economic progress cannot be reached in all areas at the same time. In a free competitive market, once there is a gap among different regions in regional development and conditions, the areas with good condition and rapid development speed would attract such production factors as labor force, capital, information, technology and so on from the areas with low conditions, and at last leads to *polarization effect*. At the same time, through the promotion of new technology and processing of raw materials and products, the center regions play a diffusing effect, and lead the poor areas to the path of becoming prosperous. Therefore, unbalanced development theory advocates that the investment should be specially put in some areas or departments, so as to breed the "growth pole", and at last achieve the overall development through the diffusion effect. Thus, the difference of region economy could only be narrowed but is impossible to be completely eliminated. Therefore, the economic development of Gansu province should also allow the existence of area difference, the developed area motivates the underdeveloped areas, and at last reach the goal of common prosperity and overall progress.

As one of the three new economic belts in west development, Xilonghai-Lanxin economic belt bring a new opportunity for the economic development of Gansu province. Gansu province has 9 cities, 30 counties in the economic belt. Therefore we should put forward the "Eagle" development strategy with a "One Body Two Wings" development strategy. "One body" refers to the establishment of the economic entity with the capital Lanzhou as the center, with the main body consisted of Lanzhou, Baiyin and Yongjing along the Yellow River basin. With the advantage of the manufacture and location of Lanzhou—Baiyin economic region, the “One body” could be constructed as the 1st growth pole of the development of Gansu economy, and in turn construct the Lanbai urban economic circle. The two wings" refers to the construction of the growth pole with Tianshui, Qingyang and Pingliang as the center. On the basis of advantages in machinery and electronics, tourism and other industries, Tianshui could be built into a central city in the east wing which is rich in tourism, electromechanical industry and fruit processing. Qingyang and Pingliang would be constructed into a new energy resources city. The scope of “east wing" could be expanded to Longnan. As for "the West Wing", relying on the industry of the sources of energy, resource industry, defense industry, agriculture, tourism and other industries, we could build another growth pole with Jiuquan and Jiayuguan as the center of the growth pole. We could carry out the integration strategy of combining Jiuquan - Jinchang - Wuwei Jiayuguan. Through the" one body with two wings" strategy, the central city and central town, city agglomeration area and rural areas, remote and impoverished area would be connected together, which in turn makes the industry spread from the fast developing cities to the relatively undeveloped and impoverished areas, make lots of rural areas, especially impoverished area economic integrate with the process of city and countryside integration, which is conducive to the harmonious development of regional economy.At the same time, we should also note that, on the imbalanced condition, to realize the coordinated development of regional economy is a complex and arduous project. The regional economic development can truly be achieved through the government's guidance and support, the matching of administrative financial price system and the reform of political system. The coordination of government should be mainly from the following aspects: the economic support and preferential policy to backward area, government’s construction of regional infrastructure and the development of public utilities, the construction of perfect authoritative regional economic development macro control mechanism through guiding the elements to flow to the areas needing support.

3.2 Rural Industrialization and Poverty Shake-Off

On the unbalanced development of regional economy, government can't just sit back and wait for "becoming co-rich" and should actively take measures to minimize differences. Because of the large proportion of agricultural population and remarkable difference between urban and rural areas, the emphasis and difficulty of poverty shake-off in Gansu Province lies in rural areas. When helping underdeveloped areas in economic development, government should choose some core areas which can drive the development of the whole regional economy, make them key development zones for special capital investment, guide them in enterprise investment space by implementation of preferential policies and then breed them as growth point to drive the development of regional economy. Government should also pay great attention to town enterprises, promote the process of rural industrialization, focus on agricultural management and agricultural and sideline products processing industry and cultivate a rational structure of the township enterprises group. Meanwhile, government should further strengthen the management of township enterprises loan, expand the size of the loan under the condition of the national financial resources allowance, promote and guide the loans from other provincial financial organizations to township enterprises and speed up the construction in small towns and rural industrial areas to provide a necessary infrastructure and a favorable external environment for the development of township enterprises.

3.3 Industrial Bases, Technological Advantages, Industrial Zone and Circular Economic Zone

Relying on previous industrial bases, industrial advantages and personnel resources in science and education, attracting domestic and foreign capital, technology, constructing economic supporting industries with new systems, high-tech, extroversion, ecology, reasonable structures will provide Gansu industrial zone a new approach to industrial chain and industry integration development[5]. Gansu province has a solid traditional industrial base and rich technical personnel. Making good use of these favorable conditions is of great importance to create a new success in the development of Gansu industry. According to Gansu regional development strategy (one center, two wings, group development and whole advance) and the state approved the test zone of circular economy, industrial zones and industrial clusters should be planned rationally.

The following circular economy bases should be focused on for their regional resource features: non-ferrous petrochemical centered "Lanzhou - Baiying" circular economy base, clean energy and metallurgical materials centered "Jiuquan - Jiayuguan" circular economy base, nonferrous metal new materials centered Jinchang circular economy base, coal chemical industry and petrochemical industry centered "Pingliang - Qingyang" circular economy base, manufacturing and electronic information centered Tianshui equipment manufacturing industry base, agricultural and sideline products processing centered "Zhangye - Dingxi - Wuwei" circular economy base, and forestry and Prataculture centered Gannan - Linxia - Longnan circular economy base. In order to enhance the status of regional core cities and build a Lanzhou-Baiyin metropolitan area as the focus, with the Lanzhou economic and Technological Development Zone, Lanzhou High-tech Development Zone, the Chinese Academy of Sciences Baiyin high-tech industry and Lanzhou New Area as the carrier, government should emphasize the high-tech industry, undertake the transfer of industries, promote regional development with industries and projects and make Lanzhou-Baiyin economic zone become the engine of regional economic development in Gansu.

3.4 Implementation of Regional Ecological Reconstruction

Being in the upper reaches of the Yangtze River and the Yellow River and located in the convergence of three plateau areas of Qinghai-Tibet, Huangtu and Meng Xin, Gansu province lies in northwest arid area of China and in the intersection zone of the Tibetan Plateau and the eastern monsoon region and other three natural regions, so its natural condition is abominable. Most economic underdeveloped regions are in this area, and the development of local economy is restricted. Therefore, in the process of exploitation and development, the ecological environment should be protected; ecology should be reconstructed so as to strive to follow a path of economic development and ecological civilization.

Firstly, we should adhere to the principle of taking measures according to local conditions and do a good job in the natural ecological environment protection and construction. Natural ecological environment construction should take the sustainable development as the guidance, stick to the restoration of the vegetation construction, and gradually restore ecosystem benign loop system.

Secondly, ecological agriculture demonstration model should be established. Establishing and perfecting the construction of ecological agriculture demonstration zone drives similar areas of ecological agriculture construction and development. According to the requirement of "enhancing farming, increasing income, decreasing burdens, lying fallow," in the 21 Century Gansu Agriculture and Rural Economic Development Ideas and Countermeasures, we should carry out retreating cultivated land to forest and grassland to planting area for crops and fruits

Thirdly, the comprehensive management of urban and rural environment should be strengthened. Industrialization and urbanization should promote the ecological environment protection and construction. City environmental protection should make improving environmental quality as its core to accelerate city environmental protection infrastructure construction and enhance city function. Counties and key towns should do well in the construction of wastewater sewage discharge and waste sanitary landfill facilities. The countryside should focus on controlling agricultural pollution, promote organic and ecological agriculture implement green civilization life and guide the establishment of sustainable consumption model.

4 Conclusion

Regional economic difference is a widespread phenomenon in the process of economic development all over the world. In fact, the existence of regional economic development difference is the reflection of general pattern in regional economic development and the result of economic growth and development imbalance. Generally speaking, during its economic growth and industrialization, a country or a region has a process in which regional economic development difference changes from gradual expansion to gradual reduction, namely, from non-balanced development to relatively balanced development. Therefore, the existence of economic differences is a normal economic phenomenon, so there is no need to completely eliminate the regional economic differences. The industrialization development process in Gansu province is from the initial stage to metaphase stage [6], so the amplification of area difference is difficult to eliminate in a short period of time. But the implementation of coordinated and sustainable development strategy will undoubtedly accelerate the process of regional coordinated economic development in Gansu province.

References

1. Expert Advisory Group of Technical Assistance Project from Asian Development Bank: Gansu Provincial Development Strategic Research, pp. 1–2. Science Press, Beijing (2009)
2. Wei, X.H.: The Status Analysis of Gansu Regional Economic Competitiveness, pp. 101–102. Gansu Science & Technology, Lanzhou (2005)
3. Nie, H.L., Lu, D., Quan Li, Q.: A General Theory of Modern Regional Economics, pp. 8–9. China Social Science Press, Beijing (2008)
4. Wei, T.J.: Western Large Development and Regional Economic Development in Gansu. The Journal of Gansu Administration Institute, 63–64 (2004)
5. Wei, S.W., Zhu, Z.W.: An Analysis and Forecast of Gansu Economic Development from 2009 to 2010, pp. 202–203. Gansu People's Press, Lanzhou (2009)
6. Wang, H.F.: Study of Developing Low Carbon Economy in Gansu Province. Regional Research and Development, 64–68 (2011)

Empirical Study on the Trans-regional Financial Flows between Northwest and Eastern in China

Zhang Qiuliang, Bai Yongping, and Wang Peian

School of Geography and Environment, Northwest Normal University,
Lanzhou 730070, China

Abstract. The Northwest China is one of the most undeveloped regions, whose economic and financial situation have lagged far behind the developed Eastern China. The paper uses relative quantitative analysis to study the reality of the financial flows from northwest China to Eastern China. And based on Co-integration analysis, Granger causality test and model VEC, this article finds out that these is a stable causal relationship between the economic gap and financial gap of two regions. And the financial elements escape out of Northwest China to Eastern China, which has hampered the development of economic of Northwest China and aggravated this gap. At last, this article finds out the appropriate policy recommendations.

Keywords: financial flows, VEC model, Granger-Causality test, Northwest and Eastern of China.

1 Introduction

The financial flows is specifically refers to monetary form or value form of finance factors in circulation field or circulation process, including the circulation of currency, financial assets and financial instrument and derivatives. The famous economist Lucas-1995 Nobel winner in economics-wrote a classical paper in 1990-"Why do not the capital flow from the rich to the poor countries"[1] and drew opposite conclusion that the financial flows is often from developing to developed countries, this is the famous "Lucas paradox" or "Lucas puzzle". In fact, the phenomenon also exists within different areas of the same country. In other words the financial flows would from developed areas flow to developed areas in a country, which will make developed areas more developed and underdeveloped areas due to the lack of effective financial support and even further behind. As refer to the flow problems of regional financial flows have been deeply discussed by many foreign scholars and the formed a theoretical system and research methods, and Feldstan Martin and Horioka Charles wrote a classic paper is the most notable. Firstly they proposed the thesis about estimation and test the degree of capital spatial mobility and adopted "Inter-temporal saving-investment Model"-is called F-H model that estimated the degree of regional capital flows by measuring the correlation between investment and savings and known as the savings - investment correlation test (F-H method)- to examine the relationship between domestic savings and domestic investment of OECD countries, what's more they also calculated the correlation coefficient between this two variables. Since then, many

D. Zeng (Ed.): Advances in Computer Science and Engineering, AISC 141, pp. 529–537.
Springerlink.com

foreign scholars have carried out empirical studies about regional capital flows by F-H method [2-7] to provide a large number of literatures and theoretical methods for domestic research and many domestic scholars have proved this point. Gou Jinlong and Wang Hongwei for the financial flows' situation and economic development disparity of Midwest three key inter-regional carried out multi-angles comprehensive investigation, such as investment in fixed assets, bank fund, financial transfer payment and foreign investment. It is concluded that Midwest fund flow to Eastern through various channels, and the trans-regional financial flow have increased regional economic disparities. Hu Yongping, Zhang Zongyi and Zhu Jiejin examined long-term co-integration relationship of the various regional savings-investment and calculate the savings retention coefficient of the F-H test model β by adopting ARDL-ECM model, and concluded that generally the Eastern is net capital inflow region, the Western is net capital outflow region and central region is essentially flat, they also confirmed the inter-regional capital flows' basically trend of capital flow to Eastern. Ji Fajun and Bai Yongping et al. confirmed that china exit the Lucas paradox from the side and they used institutional economics, game theory and other related theoretical approaches to explore the reasons for this phenomenon and propose countermeasures. Thus, it is an indisputable fact that financial flows from China's Northwest flow to Eastern coastal areas. However, is it gradually widened or narrowed that the economic gap between Eastern coast and Northwest since the reform and opening up? And the study focuses on whether some relationship exists between the economic gaps and financial flows.

2 Research Region

As the Northwest inland is typical of less developed area, while the Eastern coast is the most developed core area and the two areas is the most representative in economic and financial developments, so we choose the five Eastern coastal provinces and five Northwest interior provinces as research regions. Selected for comparison of the two areas to better characterize our country inter-regional financial flows and reveal the deep-seated reason. The five northwest inland provinces consist of Shaanxi Province, Gansu Province, Qinghai Province, Ningxia Hui Nationality Autonomous Region and Xinjiang Uygur Autonomous Region. The economic financial development of our Northwest have been seriously restricted by some problems, such as the total less productive, the inefficient use of funds, the low-income population, especially the poor population, the allocation inefficient of raising funds, the escape of financial elements seriously, more difficult to accumulate capital and financial repression. The five eastern coastal provinces consist of Shanghai city, Jiangsu province, Zhejiang province, Shandong province and Fujian province. It is located in our country's economically developed eastern coastal region that is the core area of economic and financial development.

3 Direction of the Financial Flows

Since the reform and opening up, along with the development of society and economy the different areas of our country's total amount of economy and finance is growing, that is, it is very difficult to analyze the flowing of financial flows trans-region dependent on the actual total amount of economic finance. However, if consider our country's total finance as unit "1", the ratio of a region of the total finance and the national is called relative financial amount, and it is an effective method that use the changes of relative financial amount replacement the direction of financial flows. Similarly, the ratio of a region's total economy and national total economy is called relative economy quantity. We call this kind of method as relative quantity analysis, the Chinese economic and financial total as reference frame and the computation formula of a regional relative GDP quantity is: $RelGDP = \frac{lGDP}{tGDP}$, $lGDP$ point to the regional GDP and $tGDP$ point to the national total GDP. Similarly, a regional relative savings quantity computation formula is: $\mathrm{Re}lDep = \frac{lDep}{tDep}$ and loan relative quantity computation formula is: $\mathrm{Re}lLoa = \frac{lLoa}{tLoa}$. On the three decades of 1978-2010, the Eastern coastal areas' relative GDP increasing year by year. While the change of relative savings can be divided into two stages, the first stage is 1978-1990 and in this period it showed obviously down trend-in the early stages of the reform and opening up the market economy is at the exploratory stage, the planned economy still plays an important role, the state set up special economic zone and be directed by policy the funds of the East and Midwest flow to pearl river delta region, which may be the main reason of Western and Eastern deposit are periodically relative reduction. The second stage is 1990-2010 and in this period the Eastern coastal area relatively savings continue to rise while the Eastern is declining. It is the main reason that our country established the core status of market economy and increased economic and social reforms, during this period three major economic zones-the Bohai, the Yangtze River Delta and the Pearl River Delta are gradually becoming the engine of economic development and this has led to a obvious phenomenon of the "pull money" effect from West to East. Although the main trends of between the amount of relative loan and the relative deposit in the Eastern and Western generally maintain a consistent, the following analysis is focused on the relationship of between the two regions the amount of relative loan. During 1978-1982, the eastern relative deposit far higher than relative loan and this reflected the fact of the high savings. During 1982-1990, as the East to increase investment in the early days of reform and opening up the amount of relative loan and the relative deposit generally maintain a consistent. But in the period of 1990-2002, the relative deposit exceeded the relatively loan, which indicating that the market economy status to be consolidated and the Eastern be of sustained economic development in this stage. During 2002-2010, the amount of relative loan is obviously higher than the amount of relative deposit because the Eastern economy is relatively active, increasing investment and stimulating economic development. The Northwestern total relative deposit amount continued atrophy, indicating that this region long-term "bleeding" and seriously hindered the healthy development of the economy (fig.1). Since three decades of the 1978 reform and opening up, the economic gap between the Eastern and Western have been growing. Although the deposit and loan gap is narrow between the Eastern and Western in 1978-1990, after 1990 the

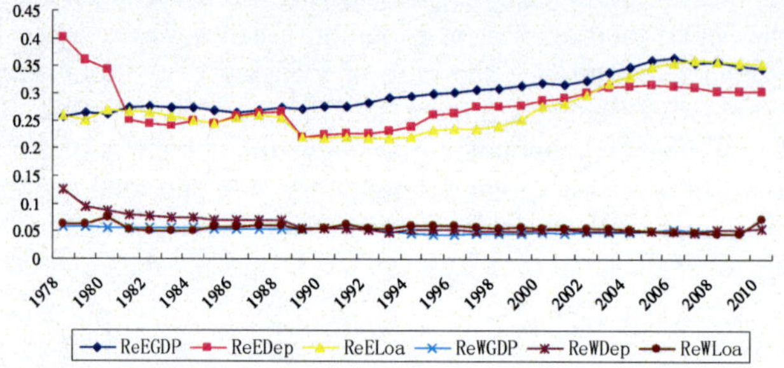

Fig. 1. The running condition of economical and financial indicators(1978-2010)

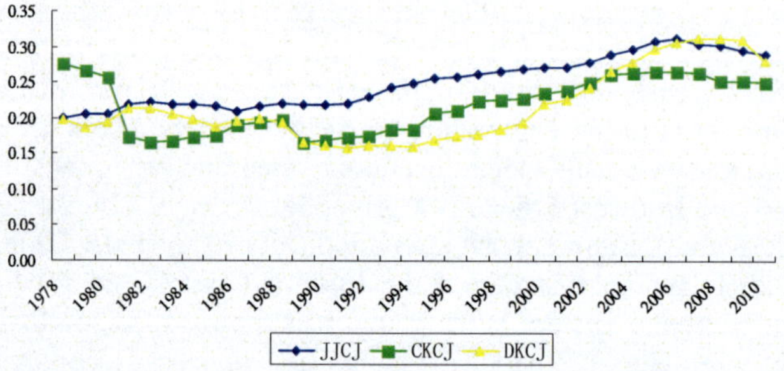

Fig. 2. Gap between the two regional economical and financial situation(1978-2010)

relative deposit and loan gap is gradually widened (fig.2). In general, there is a long-term stable relationship between the relative GDP gap and the relative deposit and loan gap.

Based on the above analysis, the analysis of financial flows' direction between northwest inland and Eastern coast was done using the relative amount analysis method. See northwest inland, Eastern coast and other areas as three inter-connection pools and see financial flows as stream, then water on free movement in the three pools. If there isn' t interference from outside, the pool water level are higher must have from others pools and the pool water level are lower must have to others pools. However, due to the country's financial system, the water levels of three pools constantly rising, therefore we can not analyze the absolute direction of stream flow of the three pools. Then see each of the three pools' total amounts of water as a unit "1", and at this point each pool's water account for the proportion of the total amount of water is relative water amount. If the relative water amount increased considers there is water inflow, conversely, considers have water outflow. We base on such an idea and combined with the above analysis, we can draw a conclusion that financial flows from Northwest areas flow to Eastern coastal areas since three decades of reform and opening up (fig.3).

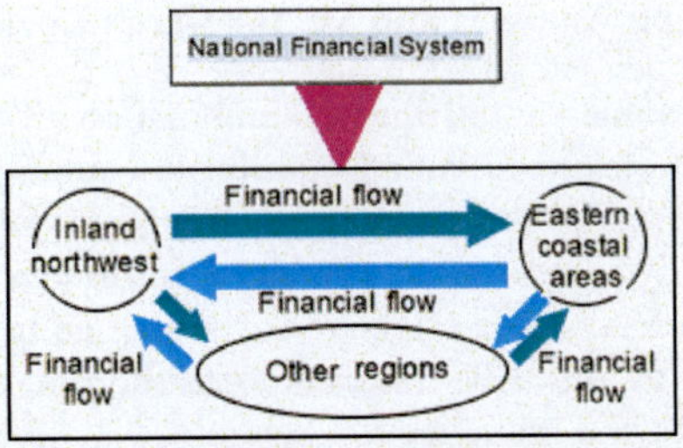

Fig. 3. Schematic diagram of regional financial flow

4 Empirical Analysis

The financial flows from the Northwest areas flow to the Eastern coastal areas, and economic gap between the Eastern and Western also gradually widened, then and what sort of relationship exists between the widening economic gap and financial flows' direction? Whether the widening economic gap led to financial flows from the Northwest flow to the Eastern or the financial flows lead to widening economic gap.

4.1 Unit Root Test

Since most economic variables is unstationary, in order to guarantee the regression results unbiased ness, effectiveness, and the optimization, firstly the economic and financial variables should be analyzed by the stability analysis of ADF (augmenter's base-fulmar) inspection. Before the test, data should be processed, defined as follows: LJJCJ = ln (JJCJ * 100), LCKCJ = ln (CKCJ * 100), LDKCJ = (DKCJ * 100). This treatment can eliminate the heteroscedastic, facilitate data processing at the same time and do not affect the development trend of each index and relationship. According to take tendency item and intercept term model, take intercept term model, don't take trend and intercept model of three models, inspection process should be taken gradually. Test results show that all the time series is a one order of each single whole sequences (table 1), therefore, we can use co-integration method to analyze relationship between them.

Table 1. The unit root test of time series

Variable	Test form •c,t,k•	ADF Statistics	Critical value•1%•	Critical value •5%•	AIC	SC	Test result
LJJCJ	•c,0,1•	-2.203357	-4.284580	-3.562882	-4.654948	-4.469918	unstable
DLJJCJ	•c,0,0•	-3.381959*	-2.641672	-1.952066	-4.605341	-4.559084	stable
LCKCJ	•c,0,8•	-2.603577	-4.394309	-3.612199	-3.745463	-3.205521	unstable
DLCKCJ	•c,0,7•	-4.640434*	-2.641672	-1.952066	-2.056398	-2.010141	stable
LDKCJ	•c,0,1•	0.538077	-2.641672	-1.952066	-3.092235	-2.999719	unstable
DLDKCJ	•c,0,0•	-2.931064*	-2.641672	-1.952066	-3.146817	-3.100559	stable

Note: (1) in the test forms (c, t, k), c with a constant, t said with trend, k said the lag order number; (2) the base of choice of period of lag k is AIC information criterion and SC information criteria. (3) * and * * express in the significant level of 1% and 5% to refuse non-stead conjecture respectively.

4.2 The Co-integration Analysis and Vector Error Revised Model

It may be smooth for some of the linear combination of economic variables; the long-term stability relationship is called equilibrium relationship in the non-steady economic variables. For many the co integration analysis of variable used Johansen inspection mostly. It is based on a complete information maximum likelihood estimation model of the VAR, it is effective for giving co integration vector estimate parameters. According to the AIC information criterion and the SC information criterion it is certain that the sequence unconstrained VAR model optimal lag order number 7, so Johansen inspection for the optimal lag order number is six. In order to make the conclusion more accurate effectively, the conclusion here to choose sequence co-integration equation has intercept model but no certainty trend, and at the same time the mark test and the maximum eigenvalue test results should be considered, the test results are shown (table 2).

Table 2. The Johansen test of three variables'Co-integretion relation

null –hypothesis, Co-integration vector	characteristic value	Trace statistic	Critical Value (5%)	Great characteristic value	Critical Value (5%)	Test Conclusion
System(LJJCJ,LCKCJ,LDKCJ)						
0*	0.690384	59.82931	29.79707	30.48299	21.13162	three Co-integration relationship exists
no more then one *	0.616061	29.34631	15.49471	24.88906	14.26460	
no more then two *	0.157543	4.457253	3.841466	4.457253	3.841466	

The test results show that obvious co-integration relationship is existed between the economic difference and financial difference in the eastern and the western, it explain long-term equilibrium relationship is existed between the economic difference and the financial different in the eastern and the western, the relationship would be showed by vector error correction model (VEC) model, and the matrix form be showed behind:

$$\Delta Y_t = \begin{bmatrix} -0.077 & 0.199 & -0.081 \\ 2.209 & 1.399 & 0.371 \\ 1.227 & 0.691 & 0.682 \end{bmatrix} \Delta Y_{t-1} + \begin{bmatrix} -0.523 & 0.145 & 0.089 \\ -2.700 & 0.884 & 1.091 \\ -1.197 & 0.689 & 0.422 \end{bmatrix} \Delta Y_{t-2}$$

$$+ \begin{bmatrix} 0.440 & -0.044 & -0.040 \\ -0.340 & 0.494 & 0.667 \\ 0.063 & 0.331 & 0.328 \end{bmatrix} \Delta Y_{t-3} + \begin{bmatrix} -0.588 & 0.072 & -0.133 \\ -1.188 & 0.693 & 0.476 \\ -0.893 & 0.432 & 0.401 \end{bmatrix} \Delta Y_{t-4}$$

$$+ \begin{bmatrix} 0.009 & 0.058 & 0.198 \\ 0.138 & 0.235 & 1.098 \\ 0.360 & 0.183 & 0.802 \end{bmatrix} \Delta Y_{t-5} + \begin{bmatrix} 0.077 \\ 2.209 \\ 1.223 \end{bmatrix} VECM_{t-1} + \begin{bmatrix} 0.008 \\ -0.021 \\ -0.029 \end{bmatrix}$$

$$VECM = [1 \;\; -1.412 \;\; 0.159] \Delta Y + 0.590$$

$$\Delta Y = [LJJCJ \;\; LCKCJ \;\; LDKCJ]$$

Based on the above analysis, the economic and financial difference between the eastern and the western exist co-integration relationship which may appear unbalance phenomenon in the short term and the error correction mechanism will link up this short-term equilibrium and long-term equilibrium of the relationship. The AIC and SC of VEC model were -12.40721 and -9.815541, and they can't meet the effectiveness of the model requirements.

4.3 Granger Causality Test

Granger causality test is used in examining whether the sequence X is the causes of sequence Y or not: Firstly estimate the current y value is its own lag phase value can explain, then the extent of validation by introducing the lag of the x whether can improve the value is explain degree. Y If it is, it says sequence x is y Granger reason (Granger Cause), at this time the x lag phase coefficient with a statistical significant. Similarly, also can examine whether granger x sequence y reason.

Table 3. The Granger causality test of all series

Null Hypothesis:	Obis	F-Statistic	Probability
LCKCJ does not Granger Cause LJJCJ	28	2.46213	0.10742
LJJCJ does not Granger Cause LCKCJ	28	2.31820	0.12101
LDKCJ does not Granger Cause LJJCJ	28	0.76744	0.47572
LJJCJ does not Granger Cause LDKCJ	18	7.84310	0.00253
LDKCJ does not Granger Cause LCKCJ	18	5.45044	0.01155
LCKCJ does not Granger Cause LDKCJ	18	1.08261	0.35535

Granger causality test results show that in the 85% of credibility, economic gap and deposit gap between the eastern and western exists the two-way causality; Economic gap is the granger reason for loan gap, but loan gap is not the granger reasons for the economic gap; Loan gap is the granger for the deposit, but deposit gap is not the granger reason for the loan gaps. To sum up, the gap of total loan between the eastern and western is the result of the economic gap of growth between the eastern and western; there exist causality between economic gap and deposit gap; The gap of total loan between the two areas led to The gap of total deposit between the two areas. It shows that the economic gap between two areas cause financial elements flowing from less development northwest region to the more development east, due to economy of northwest area is relatively backward and lack of the investment demand, makes the credit capital can't invest, instead, escape to the economic and financial active areas of east, this hinder the economic growth of northwest regions further, and formed such a vicious cycle. The east, by contrast, the economic growth attracts financial flows inflow largely, and abundant financial elements promote the economic growth further. In certain controllable scope, this is a kind of benign circulation. The two cycles makes the gap become larger between the eastern and western.

5 Conclusion and Suggestion

5.1 Main Conclusion

To sum up, our country also exist "Lucas paradox" phenomenon, so-called financial flows which flow from less developed areas in northwest to the developed area in coast of eastern. For the northwest regions, lack of money is a serious problem, the small deposit amount suppresses total loan of the market, and even more inhibit the economic growth of the northwest region. To the east, the problem is not the funds but how to use the abundant money to create greater economic and social benefits. The economic gap of development between the eastern and western is the main reason to cause direction flow of finance, meanwhile the financial flow wide the economic gap further. In addition, financial environment and investment opportunity is far behind the eastern coastal developed regions, due to financial elements of the benefit-tendency, make financial flows from northwest to the eastern region, the financial flows exacerbate the lack of funds in the northwest area and depress economic growth of the northwest region further.

5.2 Policy Suggestions

(1) Countries which formulate regional development policy for the backward area northwest continue to make appropriate tilt and formulate appropriate finance, tax and other aspects of the incentive mechanism to guide finance to flow to the northwest reasonably. (2) To improve financial environment of investment, the northwest region should promote the economic and financial systemic reform by creating a favorable environment of policy in northwest. (3) Combined with its actual situation the northwest region should utilize the more developed eastern regional model of economic development to explore the scientific path of development.

Acknowledgements. This study was supported by the National Natural Science Foundation (Grant No. 40771054), the Joint Specialized Research Fund for the Doctoral Program of Higher Education (Grant No.20106203110002), and the Fund for the Knowledge, Science and Technology Creative Research Groups of Northwest Normal University (Grant No. nwnu-kjcxqc-03-50).

References

1. Lucas, R.E.: Why doesn't Capital Flow from Rich to Poor Countries? American Economic Review 80(2), 92–96 (1990)
2. Bayoumi, T., Klein, M.W.: Aprovineial View of Capital Mobility. NBER Working Paper No.5115 (1995)
3. Bayoumi, T., Klein, M.W.: Aprovineial View of Economic Integration. IMF Staff Paper WP/97/41 (1997)
4. Helliwell, J.F., Mekitriek, R.: Comparing Capital mobility across Provincial and national boards. NBER Working PaPer No.6624 (1998)
5. Hiroshi, F., Ktamura, Y.: Feldstein-Horioka Paradox Revisited. Bank of Japan, Monetary and Economic Studies 13(1), 1–16 (1995)

6. Engel, C., Rogers, J.H.: How Wide is the Border? American Economic Review 86(5), 1112–1125 (1996)
7. Ottaviano, G., Tabuchi, T., Thisse, J.-F.: Agglomeration and Trade Revisited. Intemational Economic Review 43(2), 409–435 (2002)
8. Guo, J., Wang, H.: China's Regional Capital Flows and Regional Economic Gap Research. Management World 7, 45–58 (2003)
9. Hu, Y., Zhang, Z., Zhu, J.: An Analysis of Inter-regional Capital Flows: Saving and Investment. China Soft Science 5, 130–134 (2004)
10. Ji, F., Bai, Y., Niu, J.: The Game Analysis between Institutional Improvement and Financial Flaw-a Case Study of Northwest China. Economic Geography 29(9), 1502–1506 (2009)
11. Zhang, X.: Econometrics. Nankai University Press, Tianjin (2000)
12. Granger, C.W.J.: Modeling Economic Series. Oxford Clarendon Press, London (1990)

Research on Protection and Renovation of Historic Streets of YuQuan District in Hohhot City

Zhang Qiuliang, Bai Yongping, and Zhao Xiuqing

School of Geography and Environment, Northwest Normal University,
Lanzhou 730070, China

Abstract. Historical streets preserved the information of city culture in the city's development and reflected the style and features of city. Today urban construction fleetly increases, so it is very important to protect historical streets in favor of transmitting the information of city culture, exhibiting the style of city, enhancing the city's quality and improving people's life. This article based on the actuality researching of YuQuan District in cohort, expounded the street's history and culture resource, analyzed the problem of the historical streets, put forward the concept how to protect and renovate the historical area.

Keywords: Historical streets, protection and renewal plan, YuQuan District in Hohhot City.

1 Introduction

Yuquan District is former city of Hohhot and it was called the "Old City". It has lots of places of interest, for example, the Big Temple, the Small Temple, the Wuda Temple, the Chinese Buddhist Temple which merge Mongolian history and the religious cultural and other key units to be protected, it also has "Zhaojunmu" mausoleum and other cultural landscape which are enjoyed a good reputation as a testimony of the marriage of the Han nationality and Hu nationality. Therefore, today the urban renewal is gradually deepened. The purpose are expressing the historical information of the city relatively, showing the history of the urban development and cultural characteristics. It has important meanings to conduct preservation plan of the historical street in Yuquan District, protect the historical inheritance in Yuquan Streets, then it could form a more complete historical and cultural features of the environment.

2 Survey of Historical Street Profile in Yuquan District

Yuquan District is one of four districts in Hohhot, which is the capital of Inner Mongolia Autonomous Region. Yuquan district is in the southwestern side of Hohhot, and it has one rural town, one town and seven offices. The total area is 270 square kilometers, and the total population is 270,000.Many nationalities are live in this place, and the Mongolian is the main nationality. Yuquan District is the birthplace of Hohhot, and it already have 400 years of history. In the long history of development, people of all ethnic groups lived in harmony, created a splendid culture and unique customs of a

D. Zeng (Ed.): Advances in Computer Science and Engineering, AISC 141, pp. 539–545.
springer link.com

Yuquan District. According to research, the first urban city is the Ministry of Tumote Alatan Khan, which built in the nine years of Emperor Wanli of the Ming Dynasty (1581). That time it is called "Hohhot." This original green city is very big, and it gain the name of "Guihua city" from the Ming Dynasty with strong nationalistic discrimination. The city destroyed in war at the end of Ming Dynasty. A small city was built at the ruins in the early of Qing Dynasty, and it was just two kilometers and only two city gates in the north and south. This small city is the predecessor of "Old City".

"Guihua city" took the distribution of freestyle in general, and the temple was held prominent position .In general there were markets in front of the temple, which was the industrial and commercial center, so the commerce and handicraft are very developed. Though the wall and the gate of "Guihua city" has been demolished, the district still own lots of temples ,like the Wudasi temple, the big temple, the small temple and so on, so today the historical and cultural atmosphere of temples and believers still exist . The following passage is a brief introduction of key cultural relic in the historical street of Yuquan District.

2.1 The Big Temple

The Big Temple located in former street in Yuquan District of Hohhot. It is called "Yike temple" in Mongolian, the meaning is big shrine. The Chinese name was "Hongci Temple", later the name was changed by "Wuliang Temple". The Big Temple is the earliest temple in Inner Mongolia Autonomous Region, and it has the highest status and the greatest impact. It is the head of the temple in the south desert, and the samples of the temple in north desert and Tian mountain, so it has been called "the first temple in south desert". The temple possess hundred of precious "Gan Zhuer" classics of 600 years ago; and its sculpture and mural are very glorious. It was the temple which own by feudal lord to enshrine and worship, the feudal upper paid more attention to it.

The Big Temple's construction pattern can be divided into three vertical, it has the Decorated Archway, the Entrance, the Kings Temple, the Puti Hall, the Daxiong Hall, the Tibetan Classics House, the Buddhist Hall, the Jiujian Floor, the Subsidiary Hall and so on along the axis from the south to the north .The Big Temple is the largest and most complete wooden structure, and is still intact and full of magnificent after several hundreds of years. In addition to the Great Hall, the Big Temple's construction patterns are both Chinese-style temple layout. The Great Hall is common-Tibetan lamaseries shapes, which composed of three parts. There is a pair of stone lions in front of the Great Hall

2.2 The Xilitu Temple

The Xilitu Temple located in the north of the Rock Avenue in Yuquan District of Hohhot, about 100 meters from the Big Temple. It was built in 13 year of the Ming Dynasty (1585). At the beginning it was a little shrine, blocked north to south, covered 13,160 square meters, and the area of temple was 5,000 square meters. Now the temple has become the largest temple in Hohhot and has good reputation in history and the history of ancient architecture, which was enlarged by the Qing Emperor Yongzheng, Xianfeng, and Guangxu.

The Xilitu Temple is a series of buildings which combine Tibetan and Chinese pattern. Now the buildings inside the Xilitu Temple adopt the type of traditional Chinese Buddhist temples. It means that the axis is from the entrance to the mosque. Along the axis were distributed the Decorated Archway, the Entrance, the Temple, the Mosque, the Hall and other buildings. At the two side of the axis the Clock Tower, the Drum Tower, the Subsidiary Temple, the kiosks, the storehouse and the room are built. The Xilitu Temple is very magnificent and unique. There are 81 luxurious and great Hall of the Tibetan and Chinese architectural structure. The roof of the Hall paves green glazed tiles, the spine decorates gilt bronze monastery, Buddhist wheel, the dragon and the lucky deer, the courtyard of the Hall towers the monument which establish Mongolian, Tibetan and Chinese aged four words engraved on the Kangxi's personal expedition, at the east of the courtyard was a tower which is the most perfect type of Lama tower in Inner Mongolia. The Xilitu Temple are one of the most beautiful temples in Hohhot, the Buddist meeting and other religious activities held here each year.

2.3 Sheli Pagoda with Adamant Base

In Mongolian the Sheli Pagoda with Adamant Base was called "Ben tower", "Sipuriga" or "Wuda Temple". It is in Wudasi Back Street of Yuquan District of Hohhot. It was built in five year of Emperor Yongzheng of the Ching Dynasty(1727).At the beginning the pagoda was a building which was in "Cining Temple", now it doesn't exist. The tower is 16.5 meters high, has the adamant base, and is composed of three parts, the floor, the adamant base and the roof. There are five small towers on the roof, which are straight up into the sky, and the shape is unique and elegant. More than 1,560 Buddhist statues carved with the entire tower, so it is also called "1000 pagoda". The Sheli Pagoda with Adamant Base is not only a building, but also an enormous sculpture art.

The Sheli Pagoda with Adamant Base is made up of adamant base and five sheli pagodas, this form of temple appeared later in China and relatively rare. The temple's is composed of brick and stone, about 13 meters high, and it was a rectangular plane. The sculpt is graceful and precise, and it has a high status in architectural history.

3 The Main Problems of Historical Streets

3.1 The Historical and Cultural Outlook Was Destructed Constructively

With the development of Hohhot, the historical and cultural outlook of Yuquan District was destructed constructively. The temple in Guihua City has been destroyed; especially a large-scale transformation of the old city in recent years, which made the original pattern of the Old City and the unique spatial scale had gradually disappeared. In the important protected areas, the new buildings don't coordinated with the historical environmental features.

3.2 The Shrine Are Coming under Destruction and Threaten

At present the shrine of the historical heritage streets are still facing difficulties in protecting. Apart from the cultural protection units on the national and regional level, other shrine are not protected well. Some shrine didn't repair immediately. Meanwhile, some high value house and commercial did not identified as historical buildings, so at the transformation of the city they didn't acquire proper protection, the result would be demolished.

3.3 Lack of Legal Protection Planning

There is no long-term, operational and legal planning to guide and regulate the urban construction.

4 Planning Strategy

4.1 Guiding Ideology

(1) The shrine should be protected fully , and the historical and cultural blocks , the environment, and the continuity of cities and urban civilization should be protected, too. (2) Protection and the city construction should develop harmoniously. Urban space development, land distribution, and function in areas should be macro-controled and guided. The scale of the ancient city should be strictly control, and the city population and traffic congestion should be eased. (3) Strengthen protective measures and management, enhance characteristics study, based on the protection, create conditions to promote economic development and tourism development.

4.2 Protection Principles

(1) Authenticity: protect historical and cultural blocks to reflect the true value of the historical original item, protect all historical information. (2) Overall: protect the whole pattern and features of historical streets, protect the environment and its historical and cultural heritage, protect cultural relics and comprehensive material and non-material cultural heritage. (3)Readability: protect the various cultural heritage which reflect its specific historical and cultural backgrounds at the heritage of different historical periods. (4) Sustainability: it is a long-term undertaking to protect historical and cultural heritage, achieve the unified development of social, environmental, economic and cultural benefits.

5 Planning Idea

5.1 Protect Blocks: The Protection of Heritage Buildings-the Protection of Historical Environment

Change from the protection of monomer and scattered heritage to the protection of historical overall environment.

5.2 Update Blocks: History Retained-Value Returned

For most historical architectural and historical environment, distinct from the protection and maintenance of individual heritage, regard as the model of a large number of museums and memorials, it would need to consider the renewal of blocks, which is how to protect the value of their historic outlook and accomplish value. Update the blocks should be through the collation and revilement of historical information, and record and display city's traditional culture objectively and completely, then meet modern requirements through the use of multi-mode modern materials and techniques.

6 Planning Objectives and Development Orientation

6.1 Development Orientation

Bulid the integrated blocks, which possess the commercial services, recreation and tourism, and living of the unique cultural and historical characteristics.

6.2 Planning Objectives

Protect the outlook of blocks, improve the historical environment, continue the historic context and restore the value of neighborhoods. (1)Protect the whole pattern and historical outlook of blocks, change it to become an outstanding example of the ancient city of Hohhot. (2) Protect the true and rich historical and cultural heritage, retain the authenticity of the historical blocks, change it to become a witness of the history of urban development. (3) Manage environment of blocks, improve infrastructure, optimize the living environment, change it to become the neighborhoods which maintain the traditional cultural. (4) Use history and culture resources of blocks reasonably, promote the first-class quality of the tertiary industry, change it to become a unique cultural landscape of urban vitality Strip. (5) Deal with the relationship of core and the background. Designate the protected area of conservation and protection of historical and cultural blocks at all level, that is to designate the core of protected areas and coordinate district of the outlook in a targeted manner.

6.3 The Area of Core

Designated the outlying areas and within 100 meters of the conservation and protection. The area of key reserve of key department is designated by the protection of cultural relics, approved by higher-level departments in charge of cultural relics, based on the scope approved.

6.4 The Area of Coordinate District of the Outlook

The area of west of the Fluent Street, east of the Park Road, south of Lü Dongbin Temple Street and north of West 50 Street should coordinate with the core protection zone and the spatial pattern of the traditional style. Plan the core of protected areas to be "backstage" areas, the so-called "background" is not only the visual background, but include natural "background" and historical "background".

7 Planning Method

7.1 System Protection

(1) Tablet (protection of the entire film) : Protect and control the historical streets overall. Layout the main planning areas to key conservation area, residential land, construction sites and public green space. (2) Line (construct perspective along the line) : Protect and renovate landscape and prosperous environment of the Big South Street, the East Street, make up the history and culture scenes axis. (3) Points (key repairs) : Repair monuments at all levels strictly, control and protect historical buildings and other historical elements, improve historical building.

7.2 The Protection of Blocks Keeps Pace with the Ecological Construction

Green System Planning emphasis on the coordination of urban Green System. The green space which is at the south of the Wuta Temple Street will be extended to the east Wuta Temple street, green space along the green space is an important part of urban transportation. The green space around the Big Temple and Xilitu Temple will be formed to groups, and they will infiltrate to neighboring city. They will eventually become an integral part of urban green network. Plan to design visibility corridor through restricting the height of the surrounding building. Then the landscape system in and out the planning district will be agreement, and they will strengthen ecological effects. The planning follow point to an area, and combine points, lines and sections, achieve an organic combination of green landscape. Rely on the main focus of conservation areas, and form a landscape system with one coalition, four tapes, and many green. Pay attention to three points of diversity of plants, native, and the ecological capacity. Adopt ecological cultivation of natural communities, focuses on the using of plants in the perennial color, fruits and other changes.

7.3 Analysis the Architectural Space

(1) Control Height: Control the height of the building in the core protected district . The height of the building shall not exceed 6 meters, which is inside the 50 meters of the key protected district of the key protected unit. The height of the building shall not exceed 12 meters, which is inside the 100 meters of the key protected district of the key protected unit. (2) Control Outlook: Combine the geographical and historical cultural conditions, control the new, reform, and expansion projects of protected district strictly, and the form, height, weight, color, decorative materials of building and construction, scale, proportion must be coordinated with the traditional architectural style. Reform and change the common buildings, which impact on the general outlook of the construction history, reduce the negative impact of long-term then remove and update the buildings in the future.

8 Safeguards

8.1 Funding Policies

The renovation projects of historical streets include the repair and construction of houses, roads, and bridges and the refurbishment of riverside, improvement of facilities, control of termite, projects of transform into specialized pipeline. In all construction projects, household will hand out the funding with the government in the house repairing and sewage pipes placing, the proposed ratio is 55% for government and 45% for households.

8.2 Policy of Population Dispersal

In order to reduce population density, improve living conditions, according to the actual situation, the government has allocated some economic housing indicators that is taking care of the poverty households with per capita area of 0.8 square meters, the former housing will repurchase by the government. In addition, encourage residents of the resettlement currency. Through the implementation of this policy, 20% of the population Street area will be evacuated.

8.3 The Policy of Vacant Housing

In the process of protecting blocks, disposal the structure of the dangerous house, control termite and lay pipes, suspend the power and water. There are a lot of unexpected risks and consider the safety of the residents, so in principle, ask residents vacated during the renovation.

8.4 Policy of Demolishing Illegal Buildings

In order to maintain the coordination of historical streets outlook, the illegal construction, temporary construction must unconditionally remove in the specified period of time, and there is no compensation, if not it will be removed according to the law.

Acknowledgements. This study was supported by the National Natural Science Foundation (Grant No. 40771054), the Joint Specialized Research Fund for the Doctoral Program of Higher Education (Grant No.20106203110002), and the Fund for the Knowledge, Science and Technology Creative Research Groups of Northwest Normal University (Grant No. nwnu-kjcxqc-03-50).

References

1. Architecture and Urban Studies, Tsinghua University. Transformation of the Old Design, Planning and Research. Tsinghua University Press, Beijing (1993)
2. Yang, J., Wu, M.: Modern urban renewal. Southeast University Press, Nanjing (1999)
3. Wang, J., Xiao, Y., Wei, C.: Historic district renovation planning research and application. In: Proceedings of 2006 Annual Meeting Urban Planning in China. China Construction Industry Press, Beijing (2006)

The Property Tax Mechanism Analysis from the Perspective of Economics

Chaoyang Shi[1], Xiangdong Xu[1], and Dan Wang[2]

[1] School of economics of Huazhong University of Science and Technology, No.1037, Luoyu Road, Wuhan, 430074, China
[2] Local Taxation Bureau of Jiaozuo, 454100, China
shi-chaoyang@foxmail.com, xuxiangdong28@163.com, 120274829@qq.com

Abstract. From the year 2001,within the rapid growth of economy,as the pole of growth of our country,realty business has made a more and more important role in our country development. However ,following the hypergrowth of the real estate investment ,the problem within the realty business has became outstanding day by day.This paper analyzing from the economic theory of supply and demand of real estate tax, to mine real estate taxation policy function theory on real estate prices and adjustment.

Keywords: The real estate market, Property tax, Equilibrium.

1 Introduction

The impact of taxation on real estate mainly through acting on both of the supply and demand of the real estate market to impact the Equilibrium of the real estate market.Therefore, researches on the property tax reform is inseparable from the theory research of supply and demand market.As an important part of the market economy, the real estate market follow the same operation of supply and demand theory. Doing research on the real estate market supply and demand theory,is the theoretical basis of our study on the real estate market equilibrium of supply and demand behavior ,and also is the beginning of our research on the role of the tax on the real estate market.

2 General Economic Theory of the Real Estate Supply and Demand

Real estate has many different characteristics from the general goods, such as its non-mobility, the construction and the development cycle is long, the land resources which is an important part of its are scarce and so on.These characteristics make the real estate market not only has the common characteristics of the general demand and supply goods market, but also has its own unique characteristics.

The supply of real estate has the characteristics of diversity, price rigidity and long life cycle and so on. In reality,the supply of real estate requires two conditions: first, the seller has the ability to provide real estate; Second, the seller would like to provide real estate products.There are many important factors affecting the supply of real estate,such

D. Zeng (Ed.): Advances in Computer Science and Engineering, AISC 141, pp. 547–551.
springer link.com

as:price, financing, interest, tax burden, land development construction costs, macro-economic situation, government policy, the original number of real estate and the building management capacity of the real estate seller.

The demand for real estate is the number of real estate goods that in a period of time, under corresponding price level, the real estate buyers are willing and able to purchase. According to the buyer's different demand motives of buying housing, we can divid the needs of housing into:consumer demand, investment demand and speculative demand. The consumer demand means buyers purchase a housing is to meet the housing needs of normal life; the investment demand is the demand which is generated from taking the real estate as an asset of the holding pattern, which is a long-term investment to achieve the value of assets preservation and increasing;the speculative demand is the short-term real estate transaction which is generated from the purpose that to gain benefit from the changes in the real estate prices. The latter two demand is an important part of more suites and large suites demand.

The price of the real estate market is determined by the power of buyers and sellers, and in the meanwhile, affects the two sides in the market. Because of different income levels, family structure, population structure and buying motivation, which producing different requirements, so there exist various kinds of houses with different areas, classes and structures, thus leading to the diversity of kinds of the real estate markets in details. Due to the long construction cycle and limited urban land, the short-term price elasticity of the real estate supply is relatively small and the supply curve is steep, thus making it impossible to decreasing the price of real estate by the means of increasing the supply . The price of real estate is mainly determined by the demands. In the paper, we choose the way of restraining demands to study the problem of real estate. In details, with the help of property tax reform, reducing tax burden during the circle of development and circulation and increasing the cost of holding, we can limit the speculation or investment of suite, thus influencing China's demand structure and eventually easing contradictions between the supply and demand of real estate. The ideas and studying methods are suitable for the country with large population but limited land, rigid land supply and rapid development of population and income level, such as China.

3 The Action Principles of Taxation on the Real Estate Supply and Demand

The real estate tax is the free, mandatory imposing activities by the state on the real estate property or the wealth of the land gained by engaging in business activities with real estate ownership of social non-public subject.it is an important means of the State to regulate the real estate economic operation.Taking different tax policies on different taxation objects and linkages will make a profound impact and effect on the resources allocation of real estate market,social wealth distribution and market restructuring and many other aspects.

For different objects of taxation, tax on real estate market will have different effects.

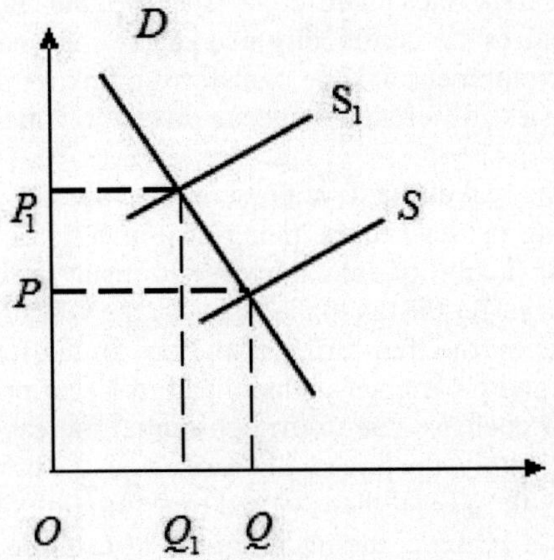

Fig. 1.

When imposing on the supply side:1.Imposing tax on the producer makes the supply curve moving;2.Because of imposing tax on the producer, it makes the cost of goods rise, reduces the profitability and the supply curve to move left .It makes the market equilibrium prices rise and reduce the Equilibrium yields.As shown in **Fig. 2.**,Tax increase makes the supply curve moving from S to S1, the equilibrium price from P up to P1, Equilibrium output from the Q decreased to Q1.

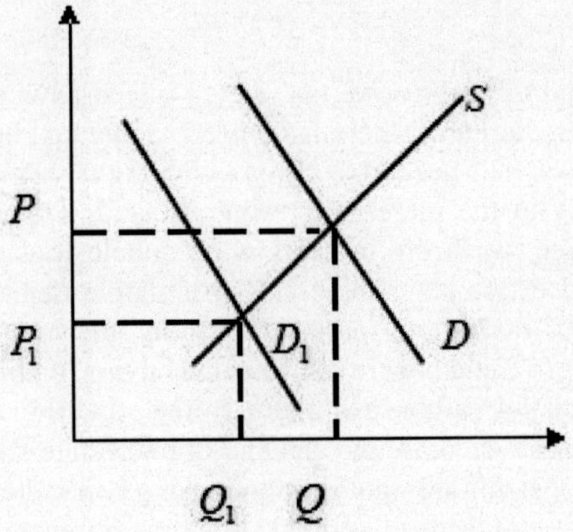

Fig. 2.

When imposing on the demand side:1.Imposing tax on the consumer makes the demand curve moving;2.Because of imposing tax on the comsumer,it makes the attractiveness of goods smaller,after adding tax factor in the price determination reduce the number of demand, the demand curve moves left.It makes the market equilibrium prices decline and reduce the equilibrium yields.shown in **Fig. 2.**.

Taxation plays different roles on different aspects of the real estate. According to the differences of the real estate commodity and marketing chain, we separate the real estate market into development linkage, transferring linkage and keeping linkage. The demand elasticity of these different linkage are different, which leads to different effect of taxation.

In the linkage of the real estate development, the supplier directly bear the taxation burden, so that levying tax lead to the rising of production cost. In short run, the demand curve is lack of flexibility, which is easy to result in the transfering of burden. The developer could transfer the tax to the buyer by the way of increasing the price, and the price will increase by the transfering of the tax. In the long run, the buyers could change their choices and preferences, which lead to larger price elasticity of demand. Housing demand will decrease due to the substantial increace in house prices. In the condition of constant supply, the price will decrease, so the price rise or drop extent and duration is decided by the adjust intensity and time duration of demand.

In the tax linkage of keeping, the main objects of taxation are lands and properties owned by the taxpayers. The direct impact of the increasing tax is the rising of holding costs, which is not prone to pass to others. In details, due to the different types of demand elasticity of consumers, including small flexibility of consumers and large flexibility of investors and speculators. This taxation mainly plays a role in consumption and then effectively reduces the excessive demand, thus moving demand curve to the left, reducing the price, and stimulating the third class market of real estate, which eventually lead to optimal allocation of real estate.

In the tax linkage of transfer, the tax is mainly indirect, which is apt to transfer the tax burden to others, causing higher transaction cost born by the consumers. The role of taxes is similar to that in the linkage of real estate development.

4 Conclusion

Property tax is a kind of progressive tax, which increases with the increasing area of house. Below the standard area, there is no need to pay tax, but if the area of house is bigger, the excessive area is needed to pay tax with higher tax rate and at the same time to pay tax according to the increase in value of land. Through supply and demand analysis of real estate, we bring out following conclusions. First, property tax will effectively curb real estate investment and speculative demand, which is helpful to decrease the price of real estate market. Secondly, the levying of property tax can reducing real estate price and benefit low-income buyers, thus leading to the increasing of the demand of small-area-houses.Finally, In the condition of constant land supply, levying property tax can decrease the demand of real estate and large suite houses and increase the supply of small-area-houses, thus easing contradiction between supply and demand, which in the end leads to optimal allocation of resources.

References

1. Alvarez, L., Kanniainen, V., Sdersten, J.: Tax Policy Uncertainty and Corporate Investment: a Theory of Tax-Induced Investment Spurts 1(7), 17–48 (1998)
2. Hamilton, B.W.: Zoning and property taxation in a system of local governments. Urbanstudy 12(2), 205–211 (1975)

3. Judd, K.L.: The impact of tax reform in modern dynamic economics. In: Hassett, K.A., Hubbard, R.G. (eds.) Ransitional Costs of Fundamental Tax Reform, pp. 5–53. AEI Press, Washington (2001)
4. Shi, C.: Property tax reform theory and empirical analysis. Huazhong University of Science and Technology (2009)
5. Zhong, F., Chu, X.: The empirical analysis between the real estate progressive tax and the supply and demand. Finance & Trade Economics (6), 83–87 (2008)
6. Wang, Y., Qin, C., Cao, Y.: Tax Regulate Mechanism of the real estate market. Modern Business Trade Industry (7), 193–194 (2008)

The Research on Impact of International Trade on China's Economic

Aiqin Xu

Zibo Vocational Institute, Zibo China 255314
xuaiqin66@163.com

Abstract. In this paper, according to logarithmic difference sequence of China's economic growth, import and export, and the establishment of the VAR model test and the relativity of China international trade and economic growth; From 1984 to 2008, this paper discusses the measures impact effect on economic growth, the contribution of the import and export of change caused by impulse response function and variance decomposition (IRF). The results show that in the first phase, variance decomposition GDP by their own fluctuation, and only in the second period from the influence of the effect of export is fully reflect tend to rise, thereafter. After about the sixth period, it gradually becomes stable. From the influence of the effect of export is reflected in the second half, the lakers started in the third quarter, in order to reduce to the greatest extent. After about the sixth period, it gradually becomes stable. The influence from the import of GDP is relatively weak in the first two times, and then rising. After about the sixth period, it often is stable.

Keywords: International trade, economic growth, China.

1 Introduction

Since the reform and open policy, our country international trade policy is greatly increased; At the same time, the interest rates of economic growth to a higher level. The relationship between international trade and economic growth, therefore, became the theory hotspots.

On the relationship between international trade and economic growth, most scholars of empirical studies support export-led growth (ELG) assumption that the main idea, the proposition to go along two lines, some scholars, use, multinational (or area) parts (or panel) data, using the correlation coefficient or ordinary least squares (OLS) test ELG assumptions, some scholars use a single country time series data validation ELG hypothesis.

Balassal (1978), use the relevant (RC), using the data of the rand 11 industrialized countries from 1960 to 1966, from 1966 to 1973, testing and analyses the actual average growth of the relationship between real gross domestic product (GDP) average growth of exports, and draw the conclusion, export played a positive role in promoting economic growth [1]. Feder (1982), using the average percentage of OLS method together with such variables as investment to GDP average growth of population, the proportion of GDP, and make an analysis of the relationship between

D. Zeng (Ed.): Advances in Computer Science and Engineering, AISC 141, pp. 553–560.
Springerlink.com

foreign capital occupies actual GDP average growth and export average rate of GDP in an overall production function, and support ELG hypothesis [2]. Dollars (1992) using OLS method, this paper analyzes the 92 countries panel data from 1976 to 1985 [3], McNab and Moore (1998) using OLS method, this paper analyzes the level 3 data 41 developing countries from 1963 to 1973, from 1973 to 1985, two conclude: exports to stimulate growth [4].

Jung and Marshall (1985) to analyze the relationship between and real gross domestic product growth (GNP), and the growth of exports and developing countries and the area from 1950 to 1981 use granger causality test model, but found VARD five 37 countries and regions support ELG assumptions and 11 this phenomenon's export-oriented economic growth. Israel has bilateral granger causality, while in the other 20 DuoGe country does not exist granger causality exports and economic growth [5]. Karunaratne (1994), based on time series data, validation ELG Australia. He found the granger test method is to use double varieties of support, and IRFs ELG test methods and FEVDs using 6 variables no causality (or weak granger causality) [6].

Chinese scholars research mainly focus on the problems related to the test time series data of our country. LiWen (1997), the paper makes an empirical analysis using economic growth model, and finds that the export sector productivity is higher than the total factor productivity the export department conclusion, thus drawings export growth driver of China's economic growth [7]. Jiaqin Chen (1999), it can be found that the export to stimulate growth [8]. Peng Fuwei (1999) can be found, foreign trade on economic growth needs little effect [9]. ZhaoJinLing Shaohua, song song and Hongming (2001) for granger causality test in 3 variables error correction model using annual data from 1978 to 1999 and the relationship between the quantitative analysis and real GDP exports actually, but found that China's export growth on the economic growth of the driving effect will last a short-term, a very long period of time, it is not obvious [10]. Xu Qifa, jiang zemin Cuixia analysis (2002) said that, while there are the intense correlation international trade and economic growth, not simple the import or export of lead to economic growth, and export total volume of the cause of the economic growth is very [11].

Go to the above-mentioned documents, I found that most scholars focused on exports and economic growth of the research on the relationship between the impact on the economy of imports is very small. Consider factors in the process of building model are single, even ignoring some important variables. On this basis, this paper aims to select import and export to illustrate influence the rapid development of China's economy.

2 Model and Analyzing Method

This paper goes as follows:

Firstly, the paper has the stationary test of China's economic growth, import and export by employing unit root test techniques.

Time series of many economic indicators do not have the feature of stable process. For the time series formed in non-stationary process, traditional mathematical statistics and econometrics methods seem powerless. Besides using sequential autocorrelation analytic chart, modern econometrics judges the stationary of time series by a more formal approach, that is, to have statistical tests. Unit root test is one of the statistical tests that are universally applied. This approach judges the stationary of a certain time series through judging whether it has unit root. Commonly-used hypothesis testing approaches include DF test, ADF test and PP test. This paper, by employing ADF test [12], gives a stationary test of time series.

According to Schwert (1989)[13], the ADF test with long lags is superior to the others. The three differencing AR models of ADF are expressed as the following forms:

$$\triangle y_t = \phi y_{t-1} + \sum_{i=1}^{p-1} \beta_i \triangle y_{t-i} + \varepsilon_t \tag{1}$$

$$\triangle y_t = \alpha + \phi y_{t-1} + \sum_{i=1}^{p-1} \beta_i \triangle y_{t-i} + \varepsilon_t \tag{2}$$

$$\triangle y_t = \alpha + \phi y_{t-1} + \gamma t + \sum_{i=1}^{p-1} \beta_i \triangle y_{t-i} + \varepsilon_t \tag{3}$$

Model (1) is a pure random walk with the lag terms. Model (2) possesses a drift. Model (3) includes a drift and a time trend. The null hypothesis for ADF test is: H_0: $\phi = 0$, with the alternative H_1: $-2 < \phi < 0$.

Secondly, this paper builds VAR model [14]. VAR model is a kind of modeling idea doing research on the dynamic relationship among multivariable. It builds the model with each endogenous variable in the examined economic system considered to be the lagged value's function of all the endogenous variables in the system, without any restrictions beforehand.

To regard each variable as endogenous variable can avoid the problem of regression of the lagged item for every endogenous variable of the system in structure modeling approach. Thereafter, this model can be used to make an analysis of the shock on variant system from relevant time series and random perturbation, and further makes an explanation of various economic shocks causing the formation of economic variant. This paper supposes VAR(p)model to be:

$$y_t = A_1 y_{t-1} + A_2 y_{t-2} + \cdots + A_p y_{t-p} + \varepsilon_t \qquad t = 1, 2, \cdots, T \tag{4}$$

Where, y_t is two-dimension endogenous variable vector T stands for the number of samples, p is the hysteretic exponent number, ε_t is the two-dimension perturbation vector, $A_1, A_2, \cdots, A_p$ are parameter matrix.

Thirdly, the paper further makes a convincing explanation of the built models on the basis of vector autoregression by virtue of impulse response function (IRF) and variance decomposition [14]. The impulse response function is the impulsive effect on an endogenous variable from a follow-up system, while variance decomposition is to decompose the system's forecasted mean square error into shares of each variable impact in the system.

Impulse response function (IRF) is adopted to examine the effect on the present or future value of endogenous variable by a standard deviation impact of random perturbation item. In the models built in this paper, we examine the effect taken by a standard deviation's random perturbation of the lagged value of two variables, that is, export and import, with economic growth being endogenous variable, and the route change of it.

Through the medium of the dynamic structure supplied by VAR model, the shock of a singer variable can not only affect its own alteration but also affect the alteration of other relevant variables. From Equation4, we can get vector moving average model (VMA):

$$y_{it} = c_{ij}^{(0)}\varepsilon_{jt} + c_{ij}^{(1)}\varepsilon_{jt-1} + c_{ij}^{(2)}\varepsilon_{jt-2} + \cdots + c_{ij}^{(p)}\varepsilon_{jt-p} + \cdots,\ t = 1,2,\cdots,T \quad (5)$$

Suppose during the base period (t=0), y_1 is given impulse with one unit, that is, $\varepsilon_{1t} = 1$, then the result goes, $c_{ij}^{(0)}, c_{ij}^{(1)}, c_{ij}^{(2)}, \cdots, c_{ij}^{(p)}, \cdots$ representing y_i's response function caused by its impulse. Variance decomposition analysis supplies another approach describing system dynamics.

Variance decomposition is to decompose the system's mean square error and analyze the contribution degree of each structure impact on the alteration of endogenous variable. As this model meets stationary, variance decomposition model employs approximate relative variance contribution (RVC):

$$RVC_{j\to i}(s) = \frac{\sum_{q=0}^{s-1}(c_{ij}^{(q)})^2\sigma_{jj}}{\mathrm{var}(y_{it})} = \frac{\sum_{q=0}^{s-1}(c_{ij}^{(q)})^2\sigma_{jj}}{\sum_{j=1}^{k}\left\{\sum_{q=0}^{s-1}(c_{ij}^{(q)})^2\sigma_{jj}\right\}},\ i,j = 1,2,\ldots,k \quad (6)$$

Where c_{ij}^{q} is impulse response function, σ_{jj} is the root-mean-square error of the jth variable, and y_{it} is the ith variable of autoregression vector. $RVC_{j\to i}(s)$ shows the degree of the jth variable's effect on the ith variable according to the jth variable's relative contribution to y_i on the basis of the impact's variance. The higher value of $RVC_{j\to i}(s)$ is, the greater the jth variable's effect on the ith variable will be.

3 Empirical Test

3.1 Data and Variables

Data used in this paper are annual data, with samples from 1985 to 2008. The data are all taken from China Statistical Yearbook of every year. As to the selection of the variables, import and export both use those with the unit of Renminbi converted according to the average rate of foreign exchange of the very year. In this paper, GDP stands for the level of economic growth, EX stands for export, and IM stands for import. To eliminate the effect of inflation, it is necessary to make amendments to GDP, EX and IM by using CPI. In order to eliminate the heteroskedasticity appearing in time series, the variables will be transformed to natural logarithm in the form of lnGDP, lnEX, and lnIM. All data will be manipulated on a computer by means of software Eviews 5.1[15].

3.2 Stationary Test

The three selected variables tend to be rising steadily as time varies, that is, time series is non-stationary. This kind of building models directly will cause the problem of pseudo linear regression. The first thing to do before building the model is to realize the stationary of time series; thus we transform each variable to logarithm and the first order difference. Having ADF unit root test of each variable, we will get such results as are shown in Table 1 [16]:

Table 1. Unit root test

Variable	ADF Test Value	Critical Value	Conclusion
lnGDP	-2.153	-3.269***	reject null
Δ lnGDP	0.697	-1.607***	reject null
Δ^2 lnGDP	-4.762	-2.692*	cannot reject
lnEX	-2.042	-3.249***	reject null
Δ lnEX	-0.882	--1.607***	reject null
Δ^2 lnEX	-6.822	-2.686*	cannot reject
lnIM	-2.462	-3.255***	reject null
Δ lnIM	-0.866	-1.607***	reject null
Δ^2 lnIM	-5.981	-2.680*	cannot reject

Notes: 1. *,** and *** in the above table respectively stands for the unit root test under the significance level of 1%, 5% and 10%; Δ stands for difference symbol; 2. p is defined by the minimum of the value of AIC and SC; 3. The critical value of test comes from Eviews5.1

It is shown from the test results that under the significance level of 1%, the absolute value of ADF statistic quantity are always higher than that of relative critical value. It is obvious that these three series both belong to second order integration I (2).Therefore, VAR model can be used to describe the impact relationship between variables.

3.3 VAR Model Building and Estimation

Firstly, it is necessary to define the lag interval p of VAR model. This paper comprehensively takes AIC (Akaike Information Criterion)and SC (Schwarz Criterion) test approach into consideration to measure and define the best lag interval.

Table 2. Definition of the lag order P in var model

Lag	1	2	3	4	5
AIC	-7.555	-7.548	-8.711	-10.932	-15.029*
SC	-6.963	-6.507	-7.219	-8.991	-12.643*

Notes: * stands for the selected lag interval at each standard.

Define the lag interval to be 5 according to the minimum principle of AIC and SC. Secondly, build VAR5 Model with lnGDP, lnEX and lnIM regarded as endogenous variable simultaneously.

$$\ln GDP_t = 0.902\ln GDP_{t-1} - 0.236\ln GDP_{t-2} - 0.046\ln GDP_{t-3} - 0.016\ln GDP_{t-4}$$
$$+0.264\ln GDP_{t-5} + 0.013\ln EX_{t-1} + 0.092\ln EX_{t-2} - 0.043\ln EX_{t-3} + 0.025\ln EX_{t-4}$$
$$-0.024\ln EX_{t-5} + 0.114\ln IM_{t-1} - 0.111\ln IM_{t-2} + 0.087\ln IM_{t-3} - 0.050\ln IM_{t-4}$$
$$+0.006\ln IM_{t-5} + 0.566$$
$$R^2 = 0.9998 \quad F = 1216.355 \quad \text{AIC=-5.411,} \quad \text{SC=-4.616}$$

Next, have an AR boots graph of VAR model. Test results show that there is at least one root outside the unit circle. VAR does not satisfy the stability condition. Thereafter, it cannot be analyzed as an impulse response function (IRF).

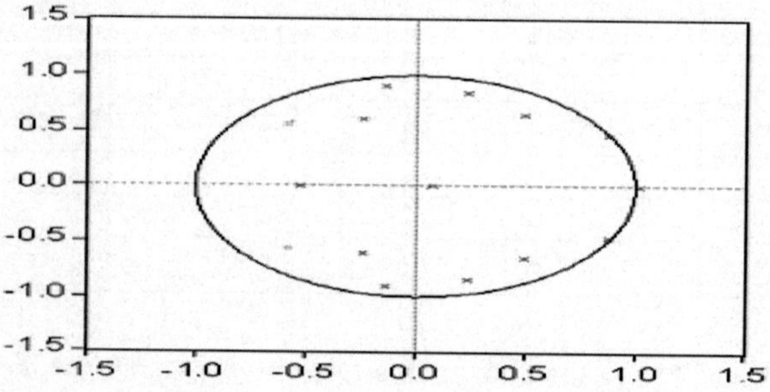

Fig. 1. Inverse roots of AR characteristic polynomial

3.4 Variance Decomposition

Contribution degree of each part in GDP fluctuation is predicted for 15 periods, and the variance decomposition is divided into such three parts as GDP, foreign trade export and foreign trade import. The percentage of each impact factor in the formation of predictive standard deviation is listed in Table ċ.

It is shown from the results of variance decomposition that for a long period the alteration of China's GDP is mostly affected by its own alteration, denoting that GDP has the problem of self-accumulation and self-reliance. Its contribution always sustains above 97%. The second greatest influential factor is foreign trade export and remains a relatively steady percentage. However, the influential power of import is rather low with the highest contribution equaling 0.299%, showing import's effect on China's economic growth is not yet remarkable.

Table 3. Variance decomposition of GDP variation

Period	S.E.	lnGDP	lnEX	lnIM
1	0.018	100.000	0.000	0.000
2	0.035	97.768	2.138	0.094
3	0.051	93.339	3.532	0.130
4	0.063	97.039	2.816	0.145
5	0.072	97.639	2.176	0.186
6	0.078	97.771	1.986	0.244
7	0.082	97.852	1.865	0.283
8	0.085	97.912	1.789	0.299
9	0.087	97.942	1.759	0.299
10	0.088	97.972	1.735	0.293
11	0.088	97.996	1.714	0.291
12	0.089	98.014	1.700	0.287
13	0.091	97.969	1.753	0.277
14	0.094	97.900	1.838	0.263
15	0.098	97.949	1.804	0.247

4 Model Explanation

From the results of VAR model approach and variance decomposition approach we know the analyses of the relationship between foreign and GDP by using these two approaches are comparatively similar: for a short period, export has a rather strong impact on China's GDP and the effect is a generally positive, while the effect coming from foreign trade import is relatively weak; for a long period, export' impact on China's GDP grows weaker and weaker and the effect tends to be steady, while the effect coming from import grows stronger and stronger and tends to be steady.

References

1. Balassa, B.: Exports and Economic Growth: Further Evidence. Journal of Development Economics 5, 181–189 (1978)
2. Feder, G.: On Exports and Economic Growth. Journal of Development Economics 12, 59–73 (1982)
3. Dollar, D.: Outward-oriented Developing Economies Really Do Grow More Rapidly: Evidence for 95 LDCs, 1976-1985. Economic Development and Cultural Change 40, 523–544 (1992)
4. McNab, R.M., Moore, R.E.: Trade policy export expansion, human capital and growth. Journal of International Trade and Economic Development (7), 237–256 (1998)

5. Jung, S.W., Marshall: Export, Growth and Causality in Developing Countries. Journal of Development Economics 18(1), 1–12 (1985)
6. Karunaratne, N.D.: Growth and Trade Liberalization In Australian: a VAR Analysis. International Review of Economics and Business (41), 625–643 (1994)
7. Li, W.: Quantitative Analysis of the Contribution of Export to Economic Growth in China. Economy & Audit Study (5), 12–15 (1997) (in Chinese)
8. Chen, J.: The contribution of international trade to China's economic growth and Basic Orientations of international trade Supporting Policies Adjustment. Finance & Trade Economics (6), 49–59 (1999) (in Chinese)
9. Peng, F.: How to view the effect of present international trade on China's economic growth. International Trade Journal (1), 15–19 (1999) (in Chinese)
10. Zhao, L., Song, S., Song, H.: The Experiential Analysis on China's Export-oriented economic growth. World Economy (8), 14–20 (2001) (in Chinese)
11. Xu, Q., Jiang, C.: An Analysis of Correlation Relationship between External Trade and Economic Growth. Forecasting (2), 14–18 (2002) (in Chinese)
12. Dickey, D.A., Fuller, W.A.: Distribution of the estimators for autoregressive time series with a unit root. Journal of American Statistical Association 74, 427–431 (1979)
13. Schwert, G.W.: Tests for Unit Roots: A Monte Carlo Investigation. Journal of Business and Economic Statistics 7, 147–159 (1989)
14. Li, Z.-N., Pan, W.-Q.: Econometrics. Higher Education Press (2005) (in Chinese)
15. Eviews 5.0 User's guide, Quantitative Micro Software (2004)
16. Yi, D.: Data analysis and Eviews application. China Statistics Press (2002) (in Chinese)

Influence of New Accounting Standard to Reserve for Property Insurance Companies

Yongming He and Xinguo Xi

Hubei Branch, Taiping General Insurance Co.,Ltd.
No.145, Xianggang Road, Jiang'an District, Wuhan, China
hbheym@tpi.cntaiping.com

Abstract. Because of the implement of new accounting standard, the premiums' measurement basis of insurance company is changed, then the earned premiums and the underwriting profit are also influenced. The different measurement of UPR between the traditional accounting standard and the new accounting standard is analyzed with one case. Study shows that the implement of new accounting standard results in the change of UPR's measurement principle, meanwhile it strengthens the development confidence of insurance company, shortens the profit period of company and improves the professional standards and valuation of the property insurance industry, which is the theoretical basis for the further development of property insurance industry.

Keywords: new accounting standard, UPR, underwriting profit, influence.

1 Preface

According to Interpretation No.2: Measurement principles of Reserve for property insurance companies in *Accounting Criteria for Enterprises*, the China Insurance Regulatory Commission requires that the measurement principles under the best estimate guideline should be observed for every insurance company, in which the operating results in the insurance industry is accounted objectively, the differences reports both domestic and abroad for the publicly listed company is eliminated and the professional skill level and value assessment capability in the insurance industry is improved[1]. But the influence is reflected in the unearned premium reserve under the new measurement principles. The calculated result of earned premiums is influenced directly by its adjustment, as a result, some index such as the comprehensive loss ratio, the comprehensive expense ratio and the combined ratio, are all influenced significantly[2]. By the case analysis, the influence of new accounting standard to the undue responsibility reserve and financial indicators of the overall is discussed.

2 Case Analysis

Unearned Premium Reserve(UPR) is the property insurance policy within one year during the profit and loss accounting period, which is the premium reserve for the cross

D. Zeng (Ed.): Advances in Computer Science and Engineering, AISC 141, pp. 561–567.
Springerlink.com

period. According to the accrual basis principle, to account the profit and loss account during the fiscal year, the premium across the profit and loss accounting period is divided as the income for the later period of profit and loss accounting by the form of premium reserve. For the preliminary underwriting business, there is a part of insurance liability to this period. The premiums from the preliminary is transferred to this policy period as the UPR, which is the premium income responsible for the unexposed risk and is the reserve for the unexposed risk[1-5].

2.1 Calculation for UPR on the Condition of Traditional Standard

For the traditional standard, there are different calculation methods such as 1/365 method, 1/24method,1/8method and 1/2 methods etc., in which the 1/365 method and the 1/24method are the main calculation method.

For example, there is a policy from the July 1,2010 to the June 30,2011 and the insurance premium is 1000.0 Yuan (RMB), which is retained totally without the reinsurance processing. For the 1/365 method the UPR is 550.0 Yuan (UPR=1000.0×181/365), but it only applies to the case by form calculation. If there are no policy and the data of endorsement, the 1/24 method is recommended, shown in table 1.

Table 1. UPR calculated by 1/24

month	insurance premium	Unexpired proportion	UPR
1	120	1/24	5.00
2	150	3/24	18.75
3	180	5/24	37.50
4	150	7/24	43.75
5	180	9/24	67.50
6	240	11/24	110.00
7	180	13/24	97.50
8	210	15/24	131.25
9	150	17/24	106.25
10	180	19/24	142.50
11	240	21/24	210.00
12	150	23/24	143.75
total	2130		113.75

There are two preconditions, first the policy period is one year, second, the density of signing is uniform distribution. The calculated results shows that, there is a very small error between the 1/24 and method the 1/365 method, which implies the former is the best method to prove the 1/365 method.

2.2 Calculation for UPR on the Condition of New Accounting Standard

For the new accounting standard, there are some new concepts such as discount and marginal and there is no fixed model for the UPR, therefore the related data could not obtained by the financial system, it requires that the unbiased estimation of future cash flows should be obtained by the endeavor of different departments such actuarial and finance, the reasonable amount of money is estimated by the future net cash flows, then the UPR is calculated by the cash flow method according to the current market condition, shown in fig.1.

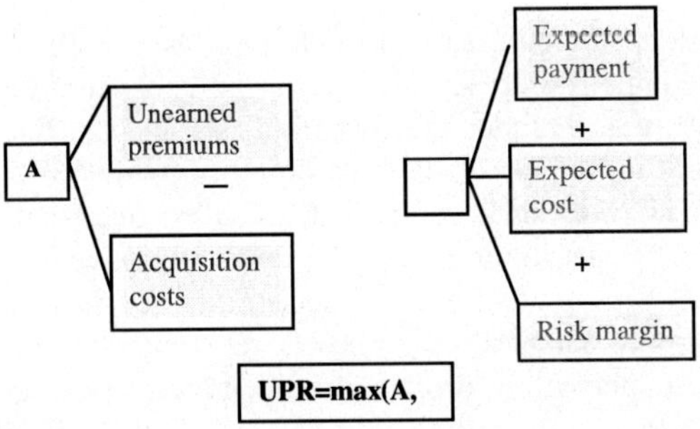

Fig. 1. UPR by cash flow method on the condition of new accounting standard

For example, there is a policy from the July 1,2010 to the June 30,2011 and the insurance premium is 1000.0 Yuan (RMB), which is retained totally without the reinsurance processing. Its expected loss ratio is 60.0% and the acquisition costs including variable costs is 23.5%, the fixed cost rat is 12.0%, then the UPR on December 31, 2010 is calculated as the following:

(1) Unearned premiums is 500.0 Yuan and related acquisition cost is 118.0 Yuan (=500×23.5%), then A=500.0-118.0=382.0 Yuan.

(2) the predicted compensation cost in the future is 300.0 Yuan (=500.0×60.0%)

(3) the cost related to the (2) is 60.0 Yuan(=500.0x12.0%)

(4)the cash flow in the future is 360.0 Yuan(=300.0+60.0)

(5) the risk margin: for B, it is 3.0% of cash flow in the future with the cash flow in the future, about 371.0 Yuan(=11.0+360.0). Now A is 382.0 Yuan and B is 371.0 Yuan, it is obvious that A is larger than B, therefore UPR=382.0 Yuan.

2.3 Underwriting Profit Influenced by Adjustment of UPR on the Condition of New Accounting Standard

According to the new accounting standard, the result of earned premiums is influenced directly by the adjustment of UPR, then the index such as comprehensive loss ratio, comprehensive expense ratio and combined ratio is also influenced.

Now the underwriting profit both of traditional and new accounting standard is discussed on the above example.

For this policy, the variable expenses includes sales charge about150.0 Yuan, taxes about 65.0 Yuan and sales performance about 20.0 Yuan, the fixed expenses is 120.0 Yuan, payment for 2010 and 2011 is 300.0 Yuan per year, total 600.0 Yuan. Therefore the equation of underwriting profit of this policy is as the following: income subtracts the total costs of variable costs and fixed costs and compensation cost (Underwriting profit = 1000.0-(150.0+65.0+20.0)-120.0-(300.0+300.0)=45.0 Yuan).

The underwriting profit on account in 2010 is the following: the underwriting profit equals the total costs of the net earned premiums and the comprehensive compensation and the comprehensive costs[3].

In this equation, the net earned premiums is 500.0 Yuan(1000.0-500.0), the comprehensive compensation is 300.0 Yuan and the comprehensive costs is 295.0 Yuan(235.0+60.0) that includes the variable costs and the fixed costs. Therefore, the underwriting profit in 2010 is -95.0 Yuan(500.0-300.0-295.0)

The underwriting profit on account in 2011 is the same equation on the above. In the equation, all data is the same except the comprehensive costs, which includes the variable costs(0.0 Yuan) and the fixed costs(60.0 Yuan), therefore the underwriting profit in 2011 is 140.0 Yuan(500.0-300.0-60.0).

Table 2 is the underwriting profit on the conditions of traditional, new accounting standard. It shows that on the traditional accounting standard, the underwriting losses in 2010 is 95.0 Yuan, the underwriting profit in 2011 is 140.0 Yuan, which indicates that there is no balance between income and expense on the condition of traditional standard. But on the condition of new accounting standard, UPR is 382.5 Yuan, underwriting is 22.5 Yuan for each year, which balances during the underwriting period.

From the above analysis, the underwriting profit on the condition of new accounting standard balances during the underwriting period and the report forms is reflected realistically the actual operation results compared to the condition of traditional standard.

There is another question: whether it still has the same influence for the underwriting profit on the condition that it is the underwriting losses when the earned premium is influenced by the accounting standard? The above example is still discussed only the compensation ratio is from 60.0% to 70.0%, in this condition the policy is the underwriting losses.

Table 2. Underwriting profit on different accounting standards

(a) the traditional accounting standard

index	2009	2010	2011
earned premiums	500.0	500.0	1000.0
+premium income	1000.0	-	1000.0
-UPR drawing	500.0		
+UPR reversal	-	500.0	
compensation expense	300.0	300.0	600.0
comprehensive costs	295.0	60.0	355.0
+sales charge	150.0	-	150.0
+taxes	65.0	-	65.0
+sales performance	20.0	-	20.0
+fixed costs	60.0	60.0	120.0
underwriting profit	-95.0	140.0	45.0

(b) the new accounting standard

index	2009	2010	2011
earned premiums	617.5	382.5	1000.0
+premium income	1000.0	-	1000.0
-UPR drawing	382.5		
+UPR reversal	-	382.5	
compensation expense	300.0	300.0	600.0
comprehensive costs	295.0	60.0	355.0
+sales charge	150.0	-	150.0
+taxes	65.0	-	65.0
+sales performance	20.0	-	20.0
+fixed costs	60.0	60.0	120.0
underwriting profit	22.5	22.5	45.0

On the condition of traditional accounting standard, UPR is 500.0 Yuan(1000.0×181/365), because it is the fixed model for UPR, it is fixed value during the whole policy period, as a result, the earned premium is the fixed. If there is a higher compensation ratio, the total policy is reflected as the underwriting losses. But because all the expenses happens in the first year, so the losses is also reflected in the first year, and the second year it is reflected as the profit, shown in table 3. On the condition of new accounting standard, the future cash flow and marginal is considered in the UPR, calculated as the following: A is 382.0 Yuan, B is 422.0 Yuan(500.0×70.0%+500.0×12.0%), UPR is the bigger one in both of A and B, so UPR is 422.0 Yuan, shown in table 3.

Table 3. Underwriting losses on different accounting standards

(a) the traditional accounting standard

index	2009	2010	2011
earned premiums	500.0	500.0	1000.0
+premium income	1000.0	-	1000.0
-UPR drawing	500.0		
+UPR reversal	-	500.0	
compensation expense	350.0	350.0	700.0
comprehensive costs	295.0	60.0	355.0
+sales charge	150.0	-	150.0
+taxes	65.0	-	65.0
+sales performance	20.0	-	20.0
+fixed costs	60.0	60.0	120.0
underwriting losses	-145.0	90.0	-55.0
underwriting profit	-95.0	140.0	45.0

(b) the new accounting standard

index	2009	2010	2011
earned premiums	578.0	422.0	1000.0
+premium income	1000.0	-	1000.0
-UPR drawing	422.0		
+UPR reversal	-	422.0	
compensation expense	350.0	350.0	700.0
comprehensive costs	295.0	60.0	355.0
+sales charge	150.0	-	150.0
+taxes	65.0	-	65.0
+sales performance	20.0	-	20.0
+fixed costs	60.0	60.0	120.0
underwriting losses	-67.0	12.0	-55.0
underwriting profit	22.5	22.5	45.0

Table 3 shows that for those policy that is the one of underwriting losses, the losses happens in the first year and shares those undefined risks. Therefore on the background of rapid increasing of casualty insurance premiums, the Interpretation No.2: Measurement principles of Reserve for property insurance companies in *Accounting Criteria for Enterprises* plays an important role to check objectively the operating results and the profit and losses in the year for property insurance companies, which helps the investor in the insurance industry to understand the state of profitability of those companies and is provided as the reference for the companies' operating decision and performance review.

3 Influence of Insurance Company by the Change of Premiums' Measurement[4]

The above case shows that there are some influences for property insurance company by the change of premiums' measurement, which is resulted from the implementation of new accounting standard.

Strengthen the development confidence of insurance company. Because the cost of policy is not deferred and belongs to the current profit and losses, on the condition of traditional accounting standard, there is no balance between income and expense in the annual policy period. The annual profit decreases with the rapid increase of business, even it is losses. So there is unreasonable situation: the more development the property insurance company, the harder the profit is obtained, which discourages the company's confidence and hinders its development. But the earned premiums increases with the decreasing of UPR calculated by the new accounting standard, which is higher than the one calculated by the traditional accounting standard and decreases the comprehensive expense ratio and increases the annual profit, as a result, the company's confidence is strengthened.

Shorten the profit period of company. On the condition of traditional accounting standard UPR is measured by statutory premiums calculating standard in the financial report, in which UPR is higher the annual actual debts of company, and the profit has the different tendencies during the policy period which increases slowly in the first part period then rapidly in the second part period, as a result, the time value of investors' money devalues. On the condition of new accounting standard the premiums calculating standard is transferred to fair value, which is reflected objectively the true financial strength and profitability of insurance company. If there is no change for the other conditions except that the accounting standard the earlier stage of losses of company is shortened, which indicates that it help the insurance company satisfy the listing requirement of profit within 3 years[5] and widen the financial channels.

Improve the professional standards and valuation of the property insurance industry. The fixed model is the basis for the principle of premiums measurement on the condition of traditional accounting standard, but the premiums measurement requires the estimation of future cash flow of insurance liability meanwhile the factor of risky marginal is considered on the condition of new accounting standard, which involves more risky models and risky parameters, which improve the professional standards and valuation of property insurance company and is the basis for the industry's further development.

It is the positive impact for property insurance company by the implement of new accounting standard, its negative impact is also considered for its involving more risky models and risky parameters with higher uncertainty. Particularly the intervention of UPR's estimation for the management of nonstandard operating insurance company, non-independence of professional assessment of the actuarial and auditors, increased risk for manipulation financial results, all these hinders the supervision of insurance.

4 Conclusion

Study shows that the implement of new accounting standard results in the change of UPR's measurement principle, meanwhile it strengthens the development confidence of insurance company, shortens the profit period of company and improves the professional standards and valuation of the property insurance industry, which is the theoretical basis for the further development of property insurance industry.

References

1. Interpretation No.2: Measurement principles of Reserve for property insurance companies in Accounting Criteria for Enterprises. The Ministry of Finance of the People's Republic of China (August 2008)
2. Zhao, Z.: Influence and Its Countermeasures of Insurance Company by Implement of New Accounting Standard. Financial Accounting (8), 35–39 (2007) (in Chinese)
3. Fu, B., Li, X.: Evaluation Index of Property Insurance Company's Operation Influenced by New Accounting Standard. Shanghai Insurance (10), 8–10 (1999) (in Chinese)
4. Influence Analysis of Property Insurance Industry in China by Implement of New Accounting Standard (in Chinese), http://money.163.com/10/1021/14/6JHAPR500025335M.html
5. Discussion of Property Insurance Industry Influenced by Implement of New Accounting Standard. International Finance News (2010) (in Chinese)

Study of Influencing Mechanism of Waste Clothing Recycling

Zhe Liu and Xiuchen Wang

Zhongyuan University of Technology, Zhongyuan Road 41,
Zhengzhou City, Henan Province, P.R. China
xyliuzhe@163.com

Abstract. A force vector model of influencing factors is proposed in this paper aiming at scanty further research of recycling mechanism for waste clothing at present. This paper put forward that main influencing factors are economy environment, clothing output, consumption ability, consumption concept, fashion update, special need. A calculation function of acceleration is given to describe the influencing rule of waste clothing recycling by these factors. Empirical analysis results show that this model can describe the mechanism of waste clothing recycling by influencing factors better, can provide a new method for rationally evaluating recycling and reusing of waste clothing, and provide a new index for estimating the trend of waste clothing recycling and reusing.

Keywords: Waste, Clothing, Recycling, Influencing mechanism.

1 Introduction

With the development of the national economy and the improvement of people's living standard, the clothing consumption is increasing; various change of clothing style is emerging. Therefore, waste clothing is increasing explosively during these years. If these waste clothing could not be distributed reasonably, it will produce three main disadvantages: 1) A great deal of limited resource in the earth will be wasted because waste clothing materials are not reused; 2) All kinds of ecological pollution will be produced because waste clothing is disposed unsuitably; 3) domestic demand Expanding that can drive the economic development will bring many negative effects caused by overstock of waste clothing.

Therefore, the problem of waste clothing recycling has increasingly become the focus of people's attention. The important influences of ecological pollution, rational utilization of resource and stimulation of consumption growth are gradually approved by people. It is not only an urgent problem need to resolve in china but also a subject need to explore in academic field to study recycling steps of waste clothing and analyze the influencing factors and establish an evaluating model.

However, there is little information about further research of waste clothing in the world at present. Zhang Zheng et al. [1] studied the recycling and redesigning of waste clothing in some countries, and concluded some measures and evaluating system. Tao Hui et al. [2] described the reusing and recycling methods of waste clothing. BIRTWISTLE [3] explored the future of waste clothing and proposed his own opinion.

D. Zeng (Ed.): Advances in Computer Science and Engineering, AISC 141, pp. 569–573.
Springerlink.com

Zu Yidan [4] investigated storage of waste clothing and put forward corresponding disposal suggestion. Zhao Guodong considered the waste clothing were the plastic bags that they could not fly, and pointed out the harmfulness of environment.

Above literatures have provided different suggestion for waste clothing recycling and reusing and have important reference value. But thorough analysis and research of recycling mechanism of waste clothing are little from theoretical aspects. More researches are focused on recycling mechanism of waste electrical equipments. There is difference between electrical equipment and clothing from property to price. Therefore, the research results of waste electrical equipment could not be used in clothing recycling.

This study uses all kinds of ways and research methods to analyze influencing factors of waste clothing recycling and try to describe the influencing mechanism of factors by constructing surfing correlation model. This model can provide a theoretical reference for the research of strategy mechanism and recycling model of waste clothing.

2 Establishment of Surfing Model of Waste Clothing Recycling

2.1 Factors Selection and Quantization

We consider that the main influencing factors of waste clothing recycling are economy environment, clothing output, consumption ability, consumption concept, fashion update and special need. The key problem how these factors influence the waste clothing recycling is to quantitate these factors. We propose a method to measure the quantization of factors. All influencing factors are expressed by unified unit.

Let the basic promoting force of waste clothing recycling is Φ, its value is estimated by the recycling condition, then quantization results of factors can calculated as following:

The economy environment is determined by GDP growth rate, it is expressed as:

$$F_G = g_G \times \Phi \tag{1}$$

The clothing output is determined by the growth rate of annual output of clothing, it is represented as:

$$F_F = g_F \times \Phi \tag{2}$$

The consumption ability is determined by the price expense index, it is denoted as:

$$F_C = g_C \times \Phi \tag{3}$$

The consumption concept is denoted by the consumption confidence index, it is expressed as:

$$F_B = g_B \times \Phi \tag{4}$$

The fashion update is denoted by the fashion index, it is represented as:

$$F_T = g_T \times \Phi \tag{5}$$

The special need is denoted by the domestic demand index, it is expressed as:

$$F_I = g_I \times \Phi \tag{6}$$

2.2 Force Vector Model of Influencing Factor

Combining above influencing factors, we consider that the development mechanism of waste clothing recycling is influenced by external force of various factors. These factors effect on the waste clothing recycling is just like external forces act on an object and produce acceleration. When the comprehensive force of factors is larger than a positive value, the velocity of waste clothing recycling will be accelerated. Therefore, we construct a force vector model. Its schematic diagram is shown in figure 1.

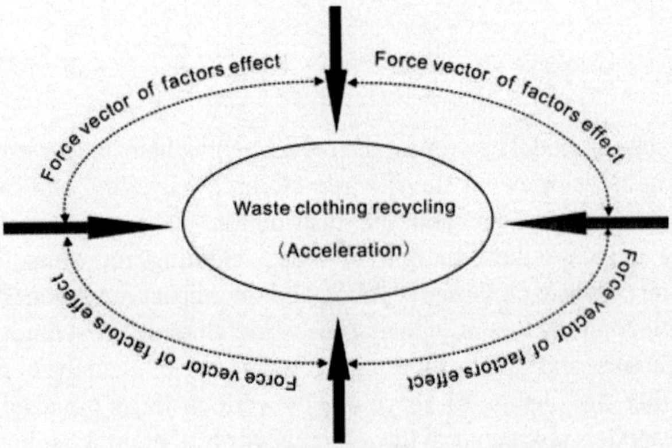

Fig. 1. Force vector model of influencing factors

In order to describe the development velocity of waste clothing recycling, an acceleration index is given to measure the velocity and trend of waste clothing recycling. The acceleration index can calculated as:

$$Va = F_G + F_F + F_C + F_B + F_T + F_I \tag{7}$$

3 Example

We take the Guangdong province which is a developed province in China for example. We collect firsthand data and information by a great of investigation and operation with enterprises and authoritative departments. Various factors are selected to calculate using Formula (1) – (6). The calculation results is listed in Table 1, where the value of Φ is 1.

Table 1. Factors evaluation of waste clothing in Guangdong province

Factors	F_G	F_F	F_C	F_B	F_T	F_I
Evaluation value	12	17	8.3	4.4	13	1.1

According to the results from table 1, we calculate the acceleration index by formula (7) and the result is obtained as:

$$V_a = 12\text{-}17\text{-}8.3\text{-}4.4\text{+}13\text{+}1.1\text{=-}3.6 \quad (8)$$

The value show that the recycling effect of waste clothing in Guangdong province is not good, the storage of waste clothing is increasing, and the distribution is decreasing.

4 Analysis

The surfing correlation model can reveal the recycling mechanism of waste clothing, can further explore the development rule of waste clothing recycling, and can describe the correlation between the storage and the distribution. This theoretical research can comprehensively evaluate the situation of waste clothing recycling, and grasp the development trend of waste clothing recycling and the impact on national economy. The model describe the main factors influencing the waste clothing recycling and the relation between these factors and waste clothing recycling, also it analyze the influencing mechanism, so that the change of recycling by various factors can be pre-evaluated rationally. These models also can provide reference for further establishing profit strategy model of waste clothing recycling, and can provide reference basis for making policy and measurement by government and related departments.

5 Conclusion

The force vector model of influencing factors constructed in this paper can describe the recycling influence and action by various factors. Government and enterprises can evaluate the status of waste clothing recycling by this model, and can rationally understand the present situation of waste clothing in china.

The acceleration function given in this paper can estimate the reference index for waste clothing recycling, can accurately judge the situation of waste clothing reusing, and can guide the enterprises to make some counter measurements for waste clothing recycling.

References

1. Zhang, Z., Wang, X.: The status of recycling and redesign of waste clothing. Tianjin Textile Science & Technology 191, 5–8 (2010)
2. Tao, H., Wang, X.: The research of recycling and reuse of waste textile and clothing as well as rethink. Mellind-China 12, 43–46 (2009)
3. Birtwistle, G.: Fashion clothing-where does it all end up? International Journal of Retail Distribution Management 3, 210–216 (2007)
4. Zu, Y., Li, X., Cui, S.: Investigation on Storage and Treatment of Family Waste Clothing. Ecological Economy 10, 180–182 (2010)

Research on the Factors Affecting Consumers' Willingness to the Use of Mobile Payment

Chenggang Li and Miaomiao Zhang

Ningbo Institute of Technology, Zhejiang University, South Qianhu Road 1, Ningbo, China
chglizju@163.com

Abstract. In this paper, we want to explore the factors which influence consumer's adoption of mobile payment. Firstly, according to the theoretical study, we put up the research model and hypothesis. Then, we got data through questionnaires(N=229) and it showed that the data had good validity and reliability. Lastly, the research model and hypothesis were verified by statistical analysis. The results showed that mobile payment's comparative advantage, compatibility, complexity, image, social influence and individual innovative of consumer affect the consumers' adoption of mobile payment.

Keywords: Mobile payment, willingness to pay, factors.

1 Introduction

With the coming of the 3G era, mobile e-commerce based on credit card payment has been greatly developed, mobile payment as part of the core of mobile e-commerce stand out conspicuously. China has a huge population of mobile users and of bank cards users, which provided a huge mass base for the growing of the mobile payments. As of 2010, Chinese mobile phone users reached 738 million and mobile Internet users increased to 233 million accounting for 60.8% of total Internet users. Estimated to 2013, China's mobile payment market will reach $ 20 billion and China is expected to become the world's second-largest payment market. HC's findings show that there are currently 97.5% of users are very interested in mobile payments, the will of nearly 80% of users using the mobile payment for public transportation, shopping and dining is much strong, 60 percent of mobile phone users query transaction details and features of interest, 20% of users willing to use recharge, refuel, to pay high tolls and other functions. Thus, it is extremely potential for China's mobile payment market, which will become a major e-gold payment system of China.

However, China's mobile payment system is still in the market cultivation stage. The survey shows that, less than 15% of mobile phone users to pay full trust, about 40% of mobile consumers who lack confidence in payment security, and 65% of mobile users refused to send their credit card information via mobile network, and large most users can only take small mobile payment. Therefore, to explore why mobile payment service is limping in such a huge market in China and the factors which limit consumers to take such a convenient way, has important theoretical and practical significance.

At present, most studies on mobile payment focused on the characteristics and the restriction factors of mobile payment industry and the study on the will for mobile

D. Zeng (Ed.): Advances in Computer Science and Engineering, AISC 141, pp. 575–580.
springer link.com

payments was only focused on the technology acceptance model and planned behavior model. In this paper, we regarded mobile payment as an innovative technology, based on innovation diffusion theory, studying factors affecting innovative behavior .on this basis, the policy recommendations are proposed, which is practically significant for the popularity of mobile payment in China.

2 Literature Review

Rogers pointed out that the main factors impacting on individuals or organizations to use innovative products is comparative advantage, compatibility, complexity, operability and observability [1]. Siau and Shen considered that whether the consumer use an innovative depends on the following five factors [2]: (1) the consumer's personal characteristics; (2) Other consumer influencing on him / her; (3) innovative products features; (4) innovative vendors marketing activities; (5) the level of competition in the market. Among them, the company's marketing activities and the level of competition in the market are the external influencing factors, which is not to be included in this study. How the innovative characteristics of the product, the consumer's personal characteristics influence on the innovation adoption process and affecting the diffusion model in a particular market is the focus of this study. In addition, Davis pointed out that the use of innovative products associated with the perception of risk affect innovation adoption, the greater the risk, the more spread slowly [3].

Heijden studied the factors affecting successful promotion of mobile payment system [4]. From the aspects of the reception from consumers and merchants, six factors that affect the acceptance of mobile payment were proposed, namely the cost of consumer awareness, consumer perception of ease of use, perceived risk of consumers, businesses perceived cost, perceived ease of business and businesses perceived risk. It was also pointed out that consumer acceptance of mobile payment and business acceptance of mobile payment have a positive correlation.

Zhao et al investigated consumer psychology of mobile payment through empirical study [5]. They believed that there were two factors affecting consumer psychology, namely internal and external factors. Internal factors include individual attitudes, subjective norm of mobile payment, perceived behavioral control; external factors include user age, educational level, the use of mobile phones, network skills, environmental policy factors and etc. Min et al pointed out that in the context of mobile commerce, consumer behavior intentions are positively affected by perceived trust [6]. Gimun et al considered that in the process of building trust, reliability and security are very important for mobile commerce; security is a very important factor for a customer to refusing mobile payment [7].

3 Models, Hypothesis and Data

3.1 Models

According to the theoretical study, mobile payment is deemed as an innovative consumer payment instruments in this paper. The personal use of the innovation process was not only affected by the personal impact of the level of innovation, but also by the innovative features of the product itself. Meanwhile, the social impact will be also introduced as an important factor in research model. Therefore, whether consumers use mobile payment or not was mainly affected by three factors: consumers' awareness of the innovative characteristics of mobile payment, social impact and the consumer's personal level of innovation. Among the three factors, consumer awareness and evaluation of mobile payment innovation will play a critical impact on the use of mobile payment, which will be focused on in this paper. Research model was shown in Figure 1.

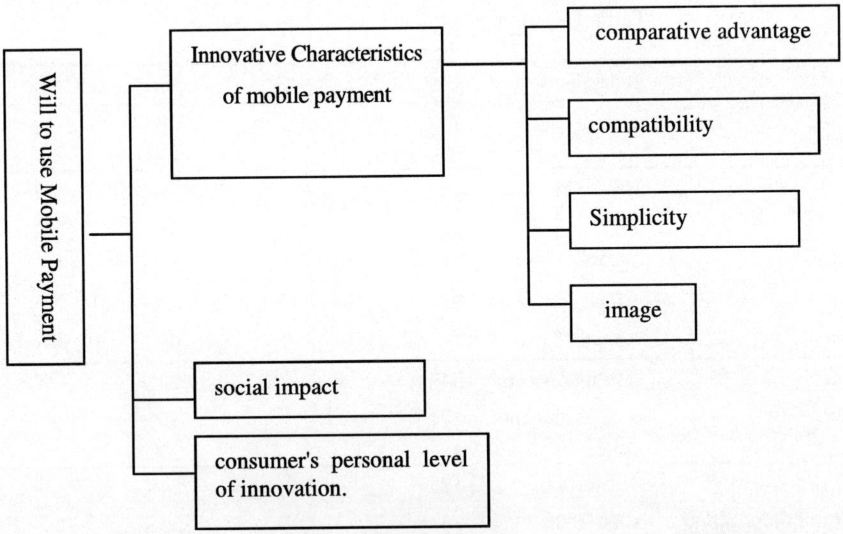

Fig. 1. Research Model

3.2 Hypothesis

Based on the research model and previous research results, combined with the relevant theory about the will to use mobile payment, the paper made the following hypothesis:

H1: The comparative advantage of mobile payment and consumers' willingness to pay has positive correlation.

H2: The compatibility of mobile payment and consumers' willingness to pay has positive correlation.

H3: The simplicity of mobile payment and consumers' willingness to pay has positive correlation.

H4: The image of mobile payment and consumers' willingness to pay has positive correlation.

H5: The social influence of mobile payment and consumers' willingness to pay has positive correlation.

H6: Consumers' personal level of innovation and consumers' willingness to pay has positive correlation.

3.3 Data

This study investigated through interviews and questionnaires. Online and paper questionnaires were issued at the same time. The object of this investigation is divided into two categories: college students and serving officers. 276 questionnaires were returned of the total of 300 questionnaires with 92% of recovery. The number of valid questionnaires is 229, of which 74 were online questionnaire, 155 paper questionnaires; the questionnaire response rate was 83%.

Table 1. Respondents' Fundamental Information

Index	Index value	Quantity	Proportion(%)
gender	male	132	58%
	female	97	42%
age	Below 18	2	1%
	18-24	202	88%
	25-29	14	6%
	30-34	9	4%
	Above 35	1	0%
profession	serving officers.	54	24%
	college students	171	75%
	others	4	2%
Monthly disposable Income(yuan)	below1000	123	54%
	1000-2000	76	33%
	2000-4000	27	12%
	Above 4000	3	1%
Monthly phone bill(yuan)	Below 50	118	52%
	50-100	90	39%
	100-200	19	8%
	Above 200	2	1%

4 Result and Discussion

4.1 Reliability and Validity Analysis

Cronbach's α coefficient is used in this study to measure the questionnaire's reliability. The Cronbach's α coefficients of the variables was shown in Table 2.

Table 2. Reliability Statistics

Variable	Cronbach's Alpha	Cronbach's Alpha Based on Standardized Items	N of Items
comparative advantage	.806	.807	7
compatibility	.803	.803	3
Simplicity	.766	.768	2
image	.754	.754	2
social impact	.836	.836	4
consumer's personal level of innovation.	.855	.856	4
Will to use Mobile Payment	.904	.905	5

Seen from the table 2, Cronbach's α coefficient of the variables is mostly more than 0.8, only simplicity (0.768) and the image (0.754) is below 0.8, which means that the questionnaire of this study has high reliability and can meet the requirements of empirical study.

The six factors of measuring willingness to use mobile payment were checked by KMO and Bartlett's chi-square test. The result showed that: KMO value is 0.914 (Table 3), significance probability of Bartlett's chi-square test is 0.000, indicating that the questionnaire has good validity.

Table 3. KMO and Bartlett's Test

Kaiser-Meyer-Olkin Measure of Sampling Adequacy.		.914
Bartlett's Test of Sphericity	Approx. Chi-Square	3539.415
	df	351
	Sig.	.000

4.2 Correlation Analysis

This study used the Pearson correlation coefficient to measure the degree of correlation between the willingness to use mobile payment and various factors, ** indicates a significantly positive correlation at the 0.01 level, * indicates a significantly positive correlation at the 0.05 level.

Table 4. Correlation Analysis

Factors,	Will to use Mobile Payment
comparative advantage	.317**
compatibility	.566**
Simplicity	.554**
image	.171**
social impact	.552**
consumer's personal level of innovation.	.621**

Seen from the results of the above analysis (Table 4), the correlation coefficient among the relative advantage of mobile payment (that is, its convenience and quality of service), compatibility, simplicity and image, as well as social impact and personal innovative level and mobile payment were 0.317,0.566,0.554,0.171,0.552,0.621, respectively. This indicated that six factors and willingness to use mobile payment has a positive correlation, which has a significantly positive correlation at the 0.01 level. Therefore, the proposed six hypotheses have been verified.

In conclusion, consumer's willingness to use mobile payment in China is affected by three main effects: Firstly, consumers awareness for four innovative features of mobile payment, including the relative advantages of mobile payment, compatibility, simplicity , the image; the second is social environment; thirdly, the innovative level of individual consumers .

References

1. Rogers, E.M.: The Diffusion of Innovations. The Free Press, New York (1983)
2. Siau, K., Shen, Z.: Building Customer Trust in Mobile Commerce. Communications of the ACM 46, 91–95 (2003)
3. Davis, F.D.: Perceived Usefulness, Perceived Ease of Use, and User Acceptance of Information Technology. MIS Quarterly 13, 319–340 (1989)
4. Heijden, H.: Factors affecting the successful introduction of mobile payment systems. In: 15th Bled Electronic Commerce Conference, pp. 430–434 (2002)
5. Zhao, Y.L., Zhang, Y.H.: Study on development of mobile payment in China: application of TPB. Computer Engineering and Design 30, 2311–2314 (2009)
6. Min, Q.F., Ji, S.B., Meng, D.C.: Trust in mobile commerce adoption factors. Management World 12, 184–185 (2009)
7. Gimun, K., Bong, S.S., Ho, G.L.: Understanding Dynamics Between Initial Trust and Usage Intentions of Mobile Banking. Info. Systems J. 269, 283–311 (2007)

Principles of Emergency Management

Weihai Yuan

Electronic Commerce School of Management, HeFei University of Technology, 230009, Hefei, China
Anhui Administration Institute, 230059, Hefei, China
yuanweihai@sina.com.

Abstract. Due to the imbalance of economic development, the increasing wealth gap between rich and poor, the ongoing damage to eco-environment, and the accumulated social conflict, emergencies occur constantly in China, a country which is entering the transformation period. The highest level of emergency management is to eliminate the risk in the cradle.

Keywords: Emergency, emergency management, Emergency Response Law.

1 Introduction

As Sun Tsu says in *The Art of War*, "In the practical art of war, the best thing of all is to take the enemy's country whole and intact; to shatter and destroy it is not so good. So, too, it is better to recapture an army entire than to destroy it, to capture a regiment, a detachment or a company entire than to destroy them. Hence to fight and conquer in all your battles is not supreme excellence; supreme excellence consists in breaking the enemy's resistance without fighting. Thus the highest form of generalship is to balk the enemy's plans; the next best is to prevent the junction of the enemy's forces; the next in order is to attack the enemy's army in the field; and the worst policy of all is to besiege walled cities." By the same token, as to emergency management, the "supreme excellence" is not to mobilize the whole society and to muster nation-wide strength, but to harbor the power of people, take precautions beforehand, and to eliminate the risk in the cradle. In this regard, we must follow the principles as below.

2 Focus on Prevention, and Combine Constant Precautions with Emergency Responses

To think of danger in times of safety, focus on precautions and be constantly prepared is the foundamental guideline to emergency responses. As the beginning phase of emergency management, the goal of precautions is " to spend a small sum of money on prevention rather than a large sum of money on treatment". This is the priority of emergency management and we should take it seriously. The effective precautions could possibly avoid crises, save large amount of responding and recovering work; or save a lot of manpower and material resources, to lay a solid foundation for rehabilitative measures. In other words, only to get prepared beforehand, combine constant precautions with emergency responses, enhance inspection and treatment on

D. Zeng (Ed.): Advances in Computer Science and Engineering, AISC 141, pp. 581–585.
Springerlink.com

potential safety hazard, strengthen monitoring and early-warning system, can we effectively eliminate the risk in the cradle. And only to raise the awareness of evading risks, take full precautions, can we assure effective responses to emergencies and to decreasing the damage and hazard in maximum. Thus, to some extent, effective prevention is more meaningful than treatment, and is the most powerful way to eliminate the risks. In contrast, no sucessful rescue could possibly totallly eliminate the influece of emergencies.

There is not necessarily no danger when there is preparedness, while there is surely big danger when there is no preparedness. As the Chinese saying goes, "to repair the house before it rains", and to combine precautions with emergency responses, these are basic rules for emergency handling. Thus we should enhance strategic planning, set up and amplify the emergency responding plan and early-preparing mechanism, raise the risk awareness of the whole society, strengthen crisis monitoring, reserve material supplies, create accident fund, implement insurance plan, carry out emergency reponding rehearsals, and increasingly improve the emergency management ability. Only by doing these, can we get fully prepared and stay calm when emergency occurs and carry out rescue action in an orderly way, thus effectively control the development of the situation. According to relevant overseas research results, whether or not setting up the emergency responding plan can determine the direction of emergency. If there is no responding plan, an emergency could cause consequences 2.5 times bigger than otherwise as to the lasting perild of the accident itself and the post-accident complications. In this regard, article 5 of *Emergency Response Law* declares, "emergency responding work follows the principle of focusing on precautions and combing prevention with treatment, to get a comprehensive evaluation of possible emergencies, in order to decrease the amount of major accidents and reduce the influence of major accidents in maximum. "

3 Implement Humanity-Oriented Measures, and Take into Account Both Social Safety and Economic Profits

To preserve social stability and ensure the safety of the people's life and property is the unshakable obligation of the government, and is the starting point and the objective of emergency management work. People's life cannot be duplicated. Thus we must follow humanity-oriented concept while dealing with emergency management. Rescue action must be implemented according to the rule of "life is above all" and put people's life before all the other materials. As another Chinese says goes, "Where there is life, there is hope". When dealing with life-relevant emergencies, especially when there is hostage concerned, we must put the safety of people's life above all and to avoid uncessary casualties. As Premier Wen Jiabao said at the press conference of the 7·23 significant traffic accident of Wenzhou, "When dealing with accident, Our basic principle is to rescue people. After the accident happens, President Hu Jintao immediately ordered relevant parties to put saving lives to the first place. And I also called the responsible ones of Ministry of Railways when I heard the news. I only said two words, 'save people'. " This accident reminds us that to put humanity-oriented concept into action, we must not only ensure the safety of the accident victims, but take every measure to protect the emergency responding staff; we must not only meet the basic needs of victims, but carry out actions in aspects of social safety, medical treatment and psychological relief so as to ensure social stability, reduce the negative influence of emergencies to a minimum, and create the image of a responsible government.

Scientific and reasonable emergency management should be to realize maximum public safety with the minimum social cost. Thus we must take both social safety and economic profits into account, avoid both response deficiency and response excess. In general, there are two points worth mentioning. On the one hand, when emergency severely harms the interests of the overall society, to avoid response deficiency, we must put efficiency first, authorize the administrative agencies to effectively pool all kinds of social resources and power, control or eliminate the urgent threat posed by crisis. On the other, emergency management must be carried out in legal scale, the measures or ways taken should be appropriate, necessary and proportional so as to avoid over-response or power abuse. To be more specific, there are three levels of meaning. Firstly, actions taken in emergency management should help effectively control and eliminate real threat, which means "appropriate actions". Secondly, on the basis of appropriateness, measures chosen in emergency management should help protect people's rights, which means "necessary measures". Thirdly, goals pursued by laws or public power and the means taken should not cause excessive burdens to prople, which means "damage to people's rights cannot be unproportionate". As article 11 of *Emergency Response Law* declares, "The measures taken by the concerned government and its departments to deal with emergencies should be in proportion to the nature, degree and scale of possible damage to the society. In face of many options, the concerned government and its departments should choose measures that could help protect interests of citizens, legal persons and other organizations to the greatest extent." In this regard, the so-called wording of "at all costs" is open to question.

4 Let the Government Take the Leading Role, and Coordinate Social Mobilization with International Cooperation

Government is not only the stakeholder of emergency management, but also the leading and core role of it. That is determined by the nature, obligation and power of the government. First of all, public safety is the basic value of social civilization. Preserving public safety is the basic requirement of government. When public safety is theatened by emergencies, as the reprensentative of public interests, the protective umbrella which the public could count on, the government should initiatively lead the public to actively deal with and defeat crisis, which is determined by the nature of government. And then, to provide public safety service and protect the safety of people's life and property, that is one of the basic obligations of service-oriented government. In the situation of crisis, whether or not the emergency management is sucessful not only poses a challenge to the public-managing capability of government, but also concerns the public identification of the administrative capability of the government, even concerns the life and death of the government itself. Lastly, the government not only has the mission of emergency management, but also has the capability to realize it. This reflects both in the advantages that government has in organizing, information and finance and in generality of government's effective managing of the public. The government not only masters huge amount of resources and powerful capability of organizing, mobilizing and managing, but also enjoys legacy of system arrangement and implementation, and the coersive power provided by laws. For this reason, to let the government take the leading role in emergency responses is both reasonable and obligational.

Apart from the leading role taken by the government, the participation of all the other forces is also necessary. Emergency has the features of urgency, difussiveness and unstability. It could cause negative influence on a large scale and severe social damage which cannot be reversed by government alone. Emergency cannot be perfectly responded without the active participation and support of the whole society. In this regard, we should mobilize the whole society to provide all sorts of manpower, material and finance so as to create a joint force to deal with emergencies. When necessary, international force is also a primary option. Since China is a government-guided country, social mobilization is a top-down government action, which means the mobilizing subject is the government while the non-governmental forces are mobilizing objects and therefore a helpful and necessary supplement to governmental emergency staff. And it is also a tendency to enhance global cooperation and use international power to deal with all kinds of crises. There are two reasons for this tendency. On the one hand, under the background of economic integration, the cause and result of an emergency often involves the whole world. Through international cooperation, crisis can be effectively solved, social order recovered and global civilization rebuilt. On the other hand, internatinal coordination and cooperation can raise emergency managemnet efficiency and reduce rescue cost, which is even more outstanding and essential when faced with significant natural disasters. Aften the 911 event happened in 2001, through cooperation with the United Nations and most of the countries in the world, the United States started the anti-terrorist war that has far-reaching influence.

5 Open Administration, and Unify Scientific Decision with Quick Response

As one of the basic rules for emergency management, open administration includes three aspects. Firstly, besides the confidential ones, emergency management-related matters, such as administrative policies, administrative legislation, administrative enforcement of law, administrative adjudication and administrative reconsideration, should be made public so as to avoid abuse of administrative power and erosion of national body. As article 10 of *Emergency Response Law* declares, "The relevant government and its departmentss should make public its decisions and orders about emergencies. " That is so-called "open administration". Secondly, in face of emergencies, administrative subject should follow the procedure of taking administrative counterpart's opinions, informing administrative counterpart of its rights and explaining reasons and basis of administrative actions. And then, the administrative subject should open its decisions and the manners, steps and orders taken in their implementation, and accept press interviews, reports and comments, be open to people's review and supervision. In this way, objectivity of emergency management can be assured. That is so-called "open administrative procedure".

Thirdly, when emergency happens, to ensure people's right to know and avoid public panic, the administrative agency should follow the legal procedure and open to the public the information acquired during the emergency management process in a convenient and easy way. As article 53 of *Emergency Response Law* declares, "The government responsible of unified leadership or organizing emergency responses should publicize information about the relevant emergency and its responding work in a unified, precise and timely way according to related regulations. " And that is so-called "open administrative information".

When emergency happens, how timely the leaders make decisions and orders can determine the developing trend of the event itself. And it is also essential that those involved in the emergency responding process can take rescue actions promptly. But due to the complexity of the cause of emergency and imbalance of information in the crisis situation, the risk of emergency response decision does always exist. *"In face of an unstable situation, the government's action could shows features of gambling."* In order to achieve scientific decision, efficient solvemnt, we must follow scientific concepts and gain support from scientific ways and technologies such as Comprehensive Information Platform and Satellite Remote Sensing Techinque, respond quickly and take actions promptly. Therefore, when faced with emergencies, we should take all the factors and conditions into full consideration so as to take actions accordingly. We should send rescue staff to save victims as quickly as possibly, investigate the cause of event, figure out the nature of event, and also actively get in touch with the press, elaborate the truth, declare the responding measures, try to obtain understanding from all the involved parties. Not only should we make prompt decisions and take emergency measures to control the situation in the shortest period of time, but also mobilize all forces of the society, especially the emergency response experts, so as to make scientific decisions and professional responses. Through this way, we can make full use of every second for rescuing and combine scientific decisions with quick responses. And that is the best way to reduce the damage caused by crises to a minimum.

References

1. China Government Gateway Website. Emergency Response Law of the People's Republic of China (Augest 30, 2007), http://www.gov.cn/ziliao/flfg/2007-08/30/content_732593.htm
2. Yuan, W.: The Openness of Government Information on Emergency Management. Chinese Public Administration (1), 66–69 (2011)
3. Yuan, W.: Emergency Response Information System and Management Framework of China. Journal of Anhui Administration Institure (4), 12–16 (2011)
4. Yang, B.: Preliminary Discussion on the Media Roles on Public Emergent Events. Theory Horizon (7), 56–58 (2009)

A Design of the Third Party Payment for Reducing the Risk of the Remittances

Yaping Li

Dept. of Computer Engineering, Anhui Economic Management Cadres Institute, Hefei 230059, China
ypliflyr@126.com
School of Management, Hefei University of Technology, Hefei 230009, China

Abstract. In this paper, we designed a new transaction process to solve the risk of online trading program that is brought by a large number of remittances deposited on the third-party (TTP) payment platform for a long time.The process is a new TTP program based on the SET protocol to reduce the risk and solve the problems of the certification verified repeatedly,the third-party lack of credibility by re-seting participants and changing the way to square the account.Furthermore, the new process enhances the incorporation of the TTP program and the Secure Transaction Technology (SET) protocol, and improves the overall efficiency of the online trading process.

Keywords: SET protocol, TTP program, Remittances, Information flow.

1 Introduction

Security Electricity Transaction(SET) protocol [1] is based on the merger of the Secure Electronic Payment Protocol(SEPP) of MastCard and the Secure Transaction Technology (STT) of VISA and Microsoft. It provides a better authentication scheme for the security of the online payment based on the bank card. However, there are also many drawbacks [2]. In 2001 Zhang dalu proposed to improve the efficiency of SET transactions by using new encryption algorithms and a new certificate of authentication process [3]. Gan yuanju did same research to improve the processes from the guarantee transaction atomicity point[4]. However, the improved trade process based on the SET protocol, could not solve the problem of credit risk that non-receipt of payment and receivables are not shipped.Thus, The Third Party(TTP) payment program began to use in the process of online payment. After that, some researchers such as Xu Xiaoping, Zhong Hongshan, Sun Ling-fang improved the design of the TTP payment program for the security, the transaction risks and the economic recovery, etc [5-7]. Moreover, some scientists designed a online payment process on the combination of the TTP program and the e-cash,because of the less efficient of certificate transfer and validation, higher-cost problem of construction for SET protocol [8]. In practical usage, Domestic third-party payment platform is generally a complex that combines TTP program with Secure Sockets Layer (SSL) protocol. In this paper, the author analyses the information flow of the SET protocol in detail, and the problems of the SET protocol

D. Zeng (Ed.): Advances in Computer Science and Engineering, AISC 141, pp. 587–593.
Springerlink.com

combination with the TTP program, In order to make the combination of the SET protocol and the TTP program more effectively, some new trading participants, a new Electronic Transaction Center (ETC) and the corresponding information flow of purchase, payment and requested have been designed. Moreover, this design can not only accelerate the pace of capital flows ,but also reduce risks in the transaction of the Business. The new process is more effective.

2 Analysis of the Transaction Flow Based on the SET Protocol

In this paper, the tags involved in the following description, for the further analysis of the process from the perspective of the information flow.

Table 1. Description of the tags

tags	Description	tags	Description
CH	CardHolder	CaP	Certificates of the payment gateway
MC	Merchants	$Sign_u()$	Digital signature of the Cardholder
PG	Payment Gateway	$Sign_m()$	Digital signature of the merchant
InitReq	Initial Request	$Sign_p()$	Digital signature of the payment gateway
InitRes	Initial Response	H()	Operate of the message digest
OI	Order Information	$ES_u()$	Digital envelope of the Cardholder
PI	Payment Information	$ES_m()$	Digital envelope of the merchant
OP	H(OI)+H(PI)	$ES_p()$	Digital envelope of the payment gateway
ATR	Authorized Request	EK1{}	Encrypte information using key k1
ATRes	Authorized response	EKi{}	Encrypte information using key ki
CT	Capture Token	SRes	Shopping response
CaC	Certificates of the cardholder	CapR	Capture request
CaM	Certificates of the merchant	CapP	Capture response

The article will describe the transaction process based on the SET protocol from There are eight main steps in the flow of transaction which is based on the SET protocol [9].the perspective of the Information transmission as follows:

(1) $\mathrm{CH} \to MC : Init\,\mathrm{Re}\,q$
(2) $MC \to CH : Init\,\mathrm{Re}\,s + CaM + CaP + Sign_s(H(Init\,\mathrm{Re}\,s))$
(3) $\mathrm{CH} \to MC : CaC + EK1\{OI + H(PI) + Sign_c(H(OP))\} + EK2\{PI + H(OI) + Sign_u(H(OP))\} + ES_m(k1) + ES_p(k2)$
(4) $MC \to PG : CaC + CaM + EK2\{PI + H(OI) + Sign_u(H(OP))\} + ES_p(k2) + EK3\{Sign_m(H(ATR))\} + ES_p(k3)$
(5) $PG \to MC : CaP + EK4\{AT\,\mathrm{Re}\,s + Sign_p(H(AT\,\mathrm{Re}\,s))\} + EK5\{CT + Sign_p(H(CT))\} + ES_p(K5)$
(6) $\mathrm{MC} \to CH : CaM + S\,\mathrm{Re}\,s + Sign_m(H(S\,\mathrm{Re}\,s))$
(7) $\mathrm{MC} \to PG : Sign_m(H(CapR)) + EK6\{CapR\} + ES_p(k6) + CaC + CaM + EK5\{CT + Sign_p(H(CT))\} + ES_p(K5)$
(8) $\mathrm{PG} \to MC : EK7\{CapP\} + ES_m(k7) + CaP + sign_p(H(CapP))$

From the Information flow above, we can see that the transaction process ensure the confidentiality of transactions, authentication, data integrity and non-repudiation through the techniques of symmetric encryption, digital signatures and digital envelopes and so on, but the operation of transfer and validation of the certificate is excessive. In this transaction process, the first problem is the existence of duplicate certificate sent,the second problem is the repeated verification certificate is a kind of redundancy while the two sides in the exchange of information mutual trust.

Therefore, the introduction of TTP program for online transactions, will provide a better solution to the credit risk of trading. With the introduction of TTP program has made some defects of online transaction process. In the transaction process of TTP program,each transaction funds will precipitate a certain period of time on the third-party platform [10], The current period, the third-party payment platform is lack of financial supervision, it will bring about greater financial risks and lower obviously the use efficiency of the business funds.

To sum up, the main defects of the TTP program based on the SET protocol are as follows: (1) the financial risks of Remittances and the efficiency decline of the use of the merchant funds; (2) the redundancy of transfer or verify of certificate, the transaction process more complexand and more inefficient ; (3) Third-party payment platform is lack of credibility, to some extent, hinder its wider application.

3 The Improvements and Analysis of the Online Transaction Process

In response to these three major defects of TTP program based on the SET protocol, this article provide a improved design prgram taken to improve the design of the following methods

(1) The funds request operations of the business in the TTP traditional program will chang from once to twice,and change the conditions of the funds request. On the one hand, to reduce the amount and time of the precipitation funds, and the risks associated with precipitation of funds, on the other hand, to reduce the original TTP cardholder the possibility to extend maliciously the payment;

(2) Certificate in one easily accessible, one certificate in the transaction process, as well as the two sides have been building up mutual trust these three cases, the re-design the information flow and reduce transmission of the certificate to verify the operation of the redundanc.

(3) The reason for the lack of credibility of the third party payment platform is Lower the threshold of market access and lack of financial supervision.So, in this paper, the third party platform we designed authorized by the Government,the new design will improve the market access conditions, thereby reducing the number of payment platforms, and the chaotic situation of the existing third-party payment platform, thus increasing the credibility of third-party payment platform.

We design a new third-party payment flow based on the SET protocol,attempt to improve the main defects of the existing TTP program. The improved model of the transaction process as shown in Figure 1.

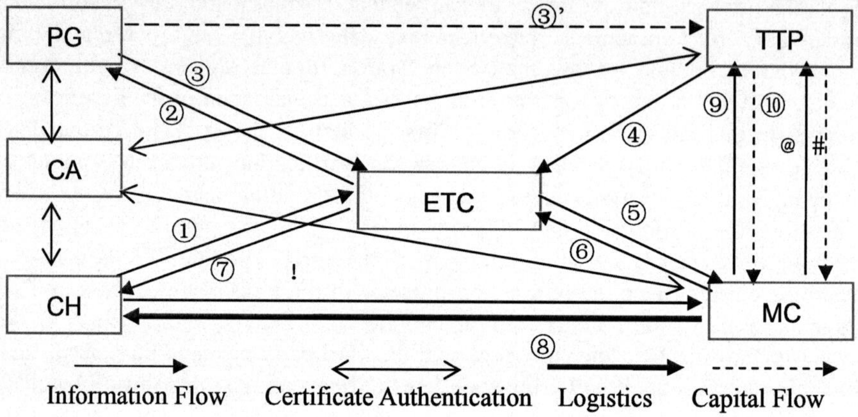

Fig. 1. The improved process online transactions

Figure 1 shows a new transaction process from the perspective of the Participants, the entities involved in the transaction process include: cardholders (CH), Merchants (MC), Payment Gateway (PG), e-trading center (ETC), third-party payment platform (TTP), Certificate Authority (CA). The improved process online transactions specifically described as follows:

① Cardholders or Customers (CH) view the electronic trading center (ETC) web site, and place an order.

②The electronic trading center send to the payment gateway center (PG) a authorization request (ATR)information.

③ Payment gateway to send a authorize response(ATRes) to the electronic trading center, and payment to the corresponding third-party payment platform account.

④The Third-party payment platform receivables, and sent to the electronic transaction Center collection information(Pinfo).

⑤ the Electronic Transactions Center to inform the businessmen (MC) shipments until get the authorization of the payment gateway and the collection information of the third-party payment platform(CSInfo).

⑥Business center to send response information of shipments request to electronic transactions information (RInfo).

⑦Electronic trading center to respond their customers to send a shopping information （SRes） .

⑧Merchant shipping and retain the shipping certificate (SCer).

⑨Merchant send a funds request(the money of the Logistics) to the third party payment platform with the shipping certificate （CapR1） .

⑩ The third party payment platform send a response information to the funds request of Merchant （CapP1).

!The cardholder send to the merchant a confirmation information after receipt the goods(Cinfo).

@ Business to balance the request of third-party payment platform once again by the confirmation information of the cardholder(CapR2).

#The third party payment platform send a response information to the residual funds request of Merchant (CapP2).

According to the transaction steps above-mentioned,the corresponding information flow is designed as follows:

(1) $\mathrm{CH} \rightarrow ETC: CaC + EK1\{OI + H(PI) + Sign_c(H(OP))\} + EK2\{PI + H(OI) + Sign_c(H(OP))\} + ES_e(k1) + ES_p(k2)$

(2) $ETC \rightarrow PG: CaM + CaE + EK2\{PI + H(OI) + Sign_m(H(OP))\} + ES_p(k2) + EK3\{Sign_e(H(ATR))\} + ES_p(k3)$

(3) $PG \rightarrow ETC: EK4\{AT\mathrm{Re}s + Sign_p(H(AT\mathrm{Re}s))\} + ES_e(k4)$

(4) $TTP \rightarrow ETC: CaT + EK5\{PInfo\} + Sign_t(H(PInfo)) + ES_e(k5) + EK6\{CT + Sign_t(H(CT))\} + ES_t(k6)$

(5) $ETC \rightarrow MC: CaE + EK7\{CSInfo\} + Sign_e(H(CSInfo)) + EK6\{CT + Sign_t(H(CT))\} + ES_t(k6) + ES_s(k7)$

(6) $\mathrm{MC} \rightarrow ETC: CaS + EK8\{RInfo\} + Sign_e(H(RInfo)) + ES_e(k8)$

(7) $ETC \rightarrow CH: EK9\{S\mathrm{Re}s\} + Sign_e(H(S\mathrm{Re}s)) + ES_u(k9)$

(8) $MC: SCer$

(9) $MC \rightarrow TTP: Sign_m(H(CapR1)) + EK10\{CapR1\} + ES_t(k10) + CaC + CaM + EK6\{CT + Sign_t(H(CT))\} + ES_t(k6)$

(10) $TTP \rightarrow MC: EK11\{CapP\} + ES_m(k11) + sign_t(H(CapP))$

(11) $\mathrm{CH} \rightarrow MC: CaC + EK12\{CInfo\} + Sign_c(H(CInfo)) + ES_m(k12)$

(12) $MC \rightarrow TTP: Sign_m(H(CapR2)) + EK13\{CapR2\} + ES_t(k13) + EK6\{CT + Sign_t(H(CT))\} + ES_t(K6)$

(13) $TTP \rightarrow MC: EK14\{CapP2\} + ES_m(k14) + sign_t(H(CapP2))$

4 The Efficiency Evaluation of the Improved Process

To ensure the safety of the premise, the flow of the process based on the SET protocol and the TTP program has been reinstituted. Now, the efficiency evaluation of the payment process is lagging behind. This article compared the efficiency of the improved process (B) and the process (A) based on the SET protocol by a combination approach of quantitative and qualitative from the risk of funds in transit,the redundancy of transmission information, the credibility of the third-party payment platform and the safety of the Transaction process.

(1)The risk of funds in transit. Online transactions, the emergence of third-party payment platform, effectively alleviate the problem of credit risk transactions, but also makes the transaction of funds sediment down to the third-party platform.it occupy the liquid capital and reduce the utilization factor of the funds, At the same time , there is a big financial risks because of the lack of an effective financial supervision.The process

B design twice request payout, the first section requested (step 8 to step 10) is a request for logistics money business by shipping certificate, The second paragraph (the first step 12,13) is a request for the remaining funds by the confirmation certificate of receipt of the cardholder. It allows the quantity and the time of precipitation funds getting corresponding reduction,thus reducing the financial risks of the program B.

(2) The redundant of information transmission. The principle of the program B to reducing the redundant of certificate transmission and of authentication is that the transfer is unnecessary a transaction the certificate has been received or have easy access through other channels (For example, the Step 3, 7, 10, 12, 13 in program B) .In specific applications of E-commerce systems, on the one hand, the program B to reduce the operation of the transmission and verify to formation certain economies of scale.

(3)The credibility of the Third-party payment platform. Currently, there are more than 100 third-party payment platforms in China, but the legislation of the third-party platform is lagging behind. The low threshold of the market access of thehird-party platform cause the the user's risks confidence. The third-party payment platform in program B has been limited to be the nature of a government, quasi-governmental nature or authorized by the Government. The restrictions can increase the credibility of third-party payment platform effectively and to promote a wider range applications of the Third-party payment platform.At the same time, this restrictions also provides a degree of protection for online transactions regulatory capital. Because the different nature of the third parties in program B, making the supervision of capital flow and information flow is reasonable, it also provide a possibility to the taxation of the internet transactions.

(4) Security. This articles give a comparative analysis of the program and B mainly from the perspective of information security. From the information flow of the above design, Program B is still using the same technologies in the transmission of information of the SET protocol, such as digital signatures, digital envelopes, digital abstracts. Thus program A and program B within the same security in the process of information flow. At the same time, the literature[11] pointed out that the risk of the payment system client, can be considered to be the largest electronic payment risk factors, And the risk is directly related to the number of sensitive information operation by the user. The number of user Send operation from the above information flow analysis,the program A is 4 times and the program B is 2 times. Program B reduces the input risk by reduce the number of sensitive information operation. So the authors believe that the safety of program B did not lower.

5 Conclusion

SET protocol and the TTP program provide solutions for the payment and transaction in different aspects, which alleviate the problem of information security and trust issues effectively. However, the efficiency of the SET protocol and the funds risks of the TTP program, inhibit the integration and application of the two programs. In order to achieve a more effective combination of them, in this paper, the authors design the new programs that focuses on in-transit funding of the TTP programs and other issues based on the SET protocol.First, change the number of the get funds from once to twice,in order to improve risks of the deposition of funds. Second, to reduce the redundancy of the transmission and verification of certificates under certain conditions, and improve the overall efficiency of combination of the SET protocol and the TTP program. Third, enhance the credibility of third-party payment platform by increasing the transaction participants and set its properties.

References

1. Visa and Master Card: SET Sepecification. Books (May 1997)
2. Alexandris, N., Burmester, M., Chrissikopoulos, V., et al.: Secure linking of customer; merchants and banks in electronic commerce. Future Generation Computer Systems (16), 393–401 (2000)
3. Zhang, D., Li, E.: The analysis and improvement measures of the SET payment process. Micro Computer Applications 17(8), 8–10 (2001)
4. Gan, Y.: The deficiencies and improvement program SET protocol system. Computer Engineering and Applications (20), 137–138 (2003)
5. Xu, X.: Research on the Secure Electronic Payment Model based on the safety of third-party. Computer Engineering and Design (6), 2137–2139 (2006)
6. Zhong, H.: Optimal Design of e-commerce payment processes. Business Times (32), 73–74 (2007)
7. Sun, L., Xu, J., Zhun, Y.: The design and analysis of the secure payment protocol based on the trusted third party. Computer Application 26(12), 2910–2912 (2006)
8. Magdalena Payeras-Capellà, M., Ferrer-Gomila, J.L., Huguet-Rotger, L.: Anonymous Payment in a Fair E-Commerce Protocol with Verifiable TTP. In: Katsikas, S.K., López, J., Pernul, G. (eds.) TrustBus 2005. LNCS, vol. 3592, pp. 60–69. Springer, Heidelberg (2005)
9. Su, C., Hu, Q., Zhao, F.: The analysis and improvement of the SET protocol. Computer Times (3), 20–21 (2004)
10. Li, E.: The funds in transit of the Third-party payment. Sociologist 3, 163–164 (2006)
11. Zhang, S.: Discussion of the client security of the electronic payment. Computer and Modernization 118, 120–122 (2005)

The Evolution of Enterprise Profit-Making Paradigms

Yong Wu

School of Economics and Management, Huazhong Agriculture University,
1# Shizishan Road, Wuhan, China
enice@163.com

Abstract. This research divides enterprise profit-making paradigms from the aspect of evolutionary economics with clear definition. Also, further analysis of the defined four main paradigms will be demonstrated under specific economic conditions of each period, with both positive and negative effects on the evolution of enterprise profit-making paradigms, in order to guide companies' share-oriented expansion and marketing.

Keywords: Profit-making paradigm, Share-oriented, Customer-oriented, Value.

1 Introduction

Paradigm, is the inherent disciplines and evolution trajectory of each specific subject. (Kuhn, 1962) As for enterprise profit-making paradigms, they are specific forms of marketing through the administration and management of companies. Since Adam Smith put forward "Wealth of Nations", the academic analysis about enterprise market paradigm has never stopped, from the perspective of the evolution of the economy, corporate profitability paradigm's development can be divided into four stages as production-oriented , share-oriented, customer-oriented and value-oriented.

2 Production-Oriented

For a considerable period of time, the whole society was of extreme material shortage, which is often referred to as "Seller's Market". During this period, the management of companies was rather simple. For the majority of enterprises, as long as they produced, there would be customers. In China, especially, people were most impressed by such profit-making paradigm, for they had experienced purchasing goods with limited "coupons" in time of centrally planned economy----- which is a typical profit-making paradigm oriented by production.

From the economic aspect, the production-oriented period was also the time when the capital of society rapidly accumulated. During this period, the increase of production, due to the limitation of technology, was within the applicable sphere of scale effect, which means the more they produced, the more they profited, and so was the wealth of the society. However, with the development of science and technology, the profit-making practice oriented by production lost its advantages gradually. When scale effect reached to its maximum, the on-going increase of production scale would lead to deficiency and waste of resources in the society. It is worth mentioning that even

D. Zeng (Ed.): Advances in Computer Science and Engineering, AISC 141, pp. 595–599.
springer link.com

today, there are still many domestic enterprises remaining at the production-oriented stage-----some of them may be able to make profit with the support from the government, but to perceive from the aspect of substitution effect, it would squander public resources severely.

3 Share-Oriented

The transition from the stage of production-oriented to the market share-oriented, to a certain extent, may be triggered by the transition from sellers' market to customers' market. The theory of market share originated from the DuPont formula suggested by Brown, the Chief Financial Officer of DuPont Company in the early 1900s. Then it was specially highlighted in 1970 by the Boston Consulting Group (BCG) of United States when formulated the combined evaluating methods for investment. Yet the popularity of pursuing market share as a basic strategy was due to the BCG model in the milestone report by the Boston Consulting Group (BCG) in the US in 1975. The BCG model divides investment projects into two categories: market growth rate and the related market share of a company. Market growth rate is synonymous with market appeal. The greater the demand of the targeted market, the higher the market growth rate and thus the more opportunities for profitability. Market share represents a company's competence in the market.

In the meantime, academics also conducted in-depth researches in terms of market share. Robert D • Buzzell (1974) revealed 37 key factors affecting profits, among which market share was the most significant. In 1981, he suggested that the companies who successfully gained larger market share usually outperformed their peers in the following three areas: activeness of new products, relative product quality and expenditure on marketing. Upon this point, the Institute of Strategy Planning of the United States compared market share and profitability in thousands of companies.

Although there were different voices towards the development of market share theory in the academic field, it is undeniable that market share theory has played a guiding role in terms of profitability in the 1970s and 1980s. As Marx pointed out in 'The Theory of Surplus Value', "For industrial production, the continuous expansion of the market is its living condition, because the continuous expansion of massive production is not restricted by the existing demand boundaries, but only by the amount of existing capital and the development level of labor productivity. It constantly floods the existing market with commercial products and continuously drives the market to expand and change." In the development course of profitability paradigms, market share theory is of great importance.

Specifically for China, since the implementation of market economy, some companies have been pursuing market share and insisted that "the bigger market share, the more profits" (BCG diagnostic report about British motorcycle industry in 1975). Compared to most of the production-oriented enterprises, these enterprises had absolute competitive advantages, and therefore earned huge profits. However, as the social and economic developed, many of the market share leader dominators found out that they had involved in a "black hole" with no profits.

As market share became increasingly important for business, this black hole went even deeper. The decline of color TV leading enterprises, the downfall of bidding kings on CCTV, the nightmare of dumping tycoons... People started to think about the shortage of the market share theory: market share did not equal profit. The reason why companies were so interested in increasing their market share was to gain competitive advantages in the market, in the hope that someday it would bring in huge profits, but in fact, the increase of the market share does not necessarily mean the growth of profit.

Enterprises compete solely on market share would result in a depleting war, which would even end up in damaging each other. Such situation has been happening in China in the manufacturing industry of television, refrigerator, air-conditioner and DVD-player. Market share is like a rearview mirror. Managers tend to evaluate the performance of an enterprise according to the market share of its products or service. However, market share is merely a 'rearview mirror' in regards to accessing. It evaluates the achievement of an enterprise in the past rather than the to-be performance in the future. Market share as an index could only reflect a static state. If we rely on such an index to guide the running of an enterprise, we will undoubtedly end up in a blind operation. Market share tends to be suppositional; it is at the most the appearance or referential statistic of business competition, which is suppositional to a certain standard. It cannot reveal the real competence of an enterprise; therefore we can't dogmatically judge whether the management of an enterprise is successful or not solely according to its market share.

4 Customer-Oriented

Though scholars'question to share-oriented theory started since the theory had been brought out., it was the new theory which had been brought out by Frederic from Bei Tu company and Suss from Harvard Business School that took place of share-oriented theory. In 1990, they studied the relationship between share-oriented and profitability and they discovered that there was no direct connection between them by research to the service firms.

In the study of other factors of profitability, they found out that in service enterprises higher customer loyalty lead to higher profitability. And those loyal customers would bring the enterprises more profits in a certain period. In 1996, Frederic also wrote an article pointing out that, companies which enjoyed extremely high customer loyalty and continuously loyalty were able to keep their profit at a very high level and had fast growth rate. Suss further put forward "Service profit chain" in 1997: the success and growth of enterprise depended on the link between value, profits and customer loyalty. The cost of gaining new customer was huge and increasing growing. It spent far less on new service or product promotion to exciting customers than on the promotion to new customers. The reason was very simple, to those familiar customers, little communication is needed in marketing and there is no need for new credit check, which can save a lot of time on both sides. In addition, due to the customer price sensitivity is low, profits will improve.

At the same time, rapid technological advances and venture capital inflexion reduced the cost of market entry barriers and operation. The margin in yesterday became no profit quickly. As the rules of the game changed, innovators had been concerning about how the profit changed? Where was the profit zone today and where it would be in tomorrow? Initially, they started to realize that the environment of economic was now moving from focusing on product and share-oriented to emphasizing customer and profitability.

The characteristic of new economic is not balance but mobility. Profitability is always changing. Profit has been accompanied with suppliers with highest share-oriented in yesterday while profit goes with suppliers with best business model. This model targets the right customers and has very high profit rate. When customer change, profit change as well. As a result, enterprise design must be updated to be in step of these changes.

Customer-oriented theory introduces a new order of economic perspective which reflects those extremely successful enterprises' profit model and further explain how customer replace share-oriented in bringing in margin. When opportunities arrive, these companies and their innovative leaders are always able to seize the chance. Due to being capable to create continuously excellent achievement to surmount the economic products and technology cycle, every company succeed to keep innovating enterprise design to maintain business margin.

The successful enterprise design must be established in the" customer-centric" principle. And this principle should be reflected by meeting customers' demand in the every section of enterprise product design, production, marketing and internal management and tap into the potential profit by seizing the share-oriented with high value.

The customer-oriented theory is not only the enterprise profit Paradigm development summit period, but also the recent focus of theoretical research. As countless scholars' participation, the achievement of theory has been greatly enriched with further relative extended theories such as relationship marketing, one-to-one marketing, business marketing. The success of numerous enterprises such as DELL, American express and the Ritz-Carlton Hotel illustrate the importance of customer-orientated from the point of view of demonstration. However with two ultimate questions having been put forward, it gives rise to our new thinking that: what is our target market and who is our competitor?

5 Value-Oriented

No matter in the class or in practice, we have a basic consensus: Target market is target customer which product have serviced. There is no doubt whether it is share-oriented driven or customer-oriented. When we place the target customer into in the target market, peer competitor becomes our business competitor involuntary. so the final outcome is obvious: the malignant competition .Although after a certain period, via repeating the game, it returned to the state of virtuous, but it's cost too much for scarcity social resource, so that it's can't afford such waste.

If we known customers in-depth, it's easy to found, the basic requirement of enterprise is create value for customers. And how can the basic requirements of the enterprise as a business target market? At that moment, the basic responsibility of enterprise, it have to do, and it can be avoid. What is our target market? In fact, in the study of Coase's talk about the essence of social shows that point .Coase believes that there is a certain degree of business's existence is to market alternative; it's to save transaction costs. So the profitability of business paradigm, it should be based on transaction cost control. For marketing, is the exchange of cost savings? Therefore, the conclusion is: The target market is the business enterprise's entire value chain, competitors are value chain costs. It requires grasp the value-driven of corporate design art, to focus on value chain management, in order to achieve a significant growth in shareholder value.

Value-oriented theory is built on the basis of the customer-oriented theory, it acknowledge the importance of customer value, and propose following point base on that point: the enterprise have to becomes value chain manager , to hole initiative position of the value chain; optimize and reorganize the channels to create value for customers as a strategic control, control transaction costs from the source; identify and develop core business, streamline the value chain; from seeking share-oriented and customer share into increasing shareholder value.

From the view of evolutionary, there will be a circuitous process that value-oriented instead of customer-oriented, and it's a falsification process which presents the value-oriented. Changing business ideas. We need to care about the customer's differences, rather than differences in sales; we should be thinking about the value of customer needs in order to determine the value of the product definition. Enhance services value. We should focus on products used in the process of transformation of services to enhance product value, committee to service of customer value contribution. And whether they can bring the performance dedicated to the growth of marketing services.

Focus on value-delivery. Positioned to deliver value during the most critical point is to ensure those products, sales methods and the establishment of consistency between the value orientations. Marketing is not a business of department, the company should be focus on value-delivery that from the product design, production to sales, distribution, pricing and service.

References

1. Kotler: Armstrong Principles of Marketing, 9th edn. Prentice-Hall, Inc. (2001)
2. Kotler, P.: Marketing Management, 10th edn. Prentice-Hall, Inc. (2000)
3. Payne, A., Chritopher, M., Peck, H.: Relationship Mketing For Competitive. CIM/ Butterworth-Heinemaan, Oxford (1998)
4. Perreault, M.: Basic Marketing, 12th edn. Richard D Irwin (1996)
5. Keegan, W.J.: Global Marketing Management, 5th edn. Prentice-Hall, Inc. (1995)

Hedging or Speculation?—From Intensive Derivative Users of CHINA Listed Companies*

Weiying Jia and Min Li

School of Business, Beijing Wuzi University, 101149, Beijing, China
Jiawy2007@yahoo.com.cn

Abstract. This paper examines the influence of derivative usage on their return volatility using data from intensive derivative users of CHINA listed companies from 2006 to 2010. We find that the derivative usage by CHINA listed companies does not realize the goal of avoiding performance risk, and what is more, it increases the return volatility. The results disclose that some CHINA listed companies violate the original intention of hedging, and speculation is hidden behind the derivative usage to some degree.

Keywords: Derivatives, Hedge, Speculation, Return Volatility.

1 Introduction

With the rapid development of derivatives such as foreign exchange and interest rate in 1970's, more and more companies have engaged in the derivatives market to conduct the risk management. It is showed by the 24th annual meeting held by the International Swaps and Derivatives Association that 94% of fortune 500 companies use derivatives to manage and manage both the business and financial risks. As the development of the global capital market and the full liberalization of China's financial market, more and more domestic enterprises are increasingly affected by international economic and financial environment to be involved in the application of the derivatives to hedge the risks.

However, since 90s of the 20th Century, derivatives market likes a rising wind and scudding clouds. Western developed countries there had been a spate of huge losses case due to engage in derivatives trading, and almost every economic crisis and financial crisis in the world had a close relationship and derivative transactions. So development of derivatives market and its risk is remarkable especially. It turned out that derivative is a "double-edged sword ", which means it can avoid the risks, but bring a certain degree of risk. Some of China's listed companies used derivatives hedging to avoid risk, but its profits and losses gyrated violently, and then sensational case followed. So both in academia and practice, it is controversial about limiting and encouraging enterprises to use derivatives, which government departments have no agreement.

* This paper is supported by Funding Project for Academic Human Resources Development in Institutions of Higher Learning Under the Jurisdiction of Beijing Municipality (PHR200907134, PHR201008224) and Funding Project for Business Innovation of Beijing Wuzi University(WYJD200904).

D. Zeng (Ed.): Advances in Computer Science and Engineering, AISC 141, pp. 601–608.
springer link.com

2 Literature Review

Western risk management theory (Smith and Stulz, 1985; Mayers and Smith, 1990; Myers, 1977 ; Froot, Scharfstein and Stein, 1993) argues that the application of derivatives aims at lowering the business performance volatility .This theory provides four typical explanations: (1) hedging maybe result from the tax incentive. Hedging can help companies reduce expected tax cost when the companies confront the convex function of tax. (2) hedging can also bring down the expected cost of financial crisis. (3) hedging can increase the borrowing capacity of the companies. (4) hedging is able to slow down the problem of information asymmetry between the over-investment and insufficient investment.

But the managers and the stock holders have the intention to raise volatility of the business performance. Black and Scholes (1973) point out that the stock holders' claim to the leveraged company's cash flows is similar to the reward for the call option. Jensen and Meckling (1976) notes that as the call option compensation rises when the business performance changes, so the stock holders who seek for the maximization of the stock holders' value have the intention to transfer wealth from the outside creditors by adding the risks of the company. Nonetheless, the stock holders are likely to have difficulties in making the managers who are risk averters to invest on the projects which would increase the business performance volatility.

Even if there is not conflict between stock holders and creditors, the managers still have the motive to employ the derivatives in increasing the volatility of the business performance. Guay (1999) indicated that the value of the employees' preemption rights to buy stocks rises as the volatility of the stock price increases, which leads the managers are motivated to take part in the activities which will bring the risks to the company. Similarly, when the company's earnings are just in or near the lower limit of the award scheme, the managers' compensation plan is just like the compensation of a call option. However, the compensation plan in the pursuit of the maximization of the stock holders' value would take the motive of the risk lovers into consideration as well as reduce the adverse effects.

Finally, the managers can utilize derivatives to speculate on interest rates, exchange rates or the fluctuations of the commodity prices. As Dolde (1995) pointed out that nearly 90% of the users of the derivatives would take a look at the fluctuations in the financial market when deciding the traits of their own portfolio of derivative instruments. Generally speaking, speculation is not related to the enterprises' basic risks. Hence, the application of the derivatives is supposed to increase the volatility of enterprises' business performance instead of reducing that.

The results of the foreign research are inconsistent with each other in the research of the experiences. The majority believe that non-financial companies can make use of the derivatives to reduce the company's stock price sensitivity to the financial risks (Smithson and Simkins, 2005) . Domestic academic research in this area is little. The aim of this article lies in examining hedging is for preservation or speculation by empirical study on the relationship between derivative usage and the return volatility.

3 Derivative Usage and Company Characteristics

3.1 Samples and Data Sources

China's Ministry of Finance published The Enterprise Accounting Standard on February 15, 2006, to make specifications on derivatives accounting treatment on China's enterprises. It is implied in the listed companies since January 1, 2007, and non-listed companies are encouraged to use the new standard. It breaks through traditional financial accounting framework, and redefines the accounting elements. Derivatives are accounted in the financial statement which were disclosed in the notes to the financial statements according to the old standards, and measured with the fair value, which sound value flexible loss and profit is confirmed in the current period, in order that financial statement reflects hidden risks of company's derivatives business and its effect for the enterprise financial position and operating results, meanwhile it makes this study possible.

The selects all of A share non – financial listed companies in Shanghai and Shenzhen Stock Market from 2006 to 2010 as initial samples which hold derivative financial asset or derivative financial liabilities. The information about engaging in derivatives transactions' type and amount in their balance sheet and notes is accessed respectively. And the data about the company use of derivatives and the use of reason is collected. Due to the derivatives market in China underdeveloped, and strictly limited offshore derivatives trades. Most listed companies use derivatives at a tentative stage, whose size of usage is far lower than the level of developed countries, resulting in difficulty of investigation of its effect. So the samples are processed as follow, if the end of fair value of derivative financial asset or derivative financial liabilities is greater than or equal to RMB100 million in annual reports, it will be redefined intensive derivative user. At last, 23 companies are selected as samples. Considering better explaining samples company characteristics compared with other companies, we select the paired samples according to the rules followed for every sample. First, paired companies are same industry as samples, which takes three - digit SIC code from Guidance on the Industry Classification of Listed Companies announced by The China Securities Regulatory Commission (CSRC). Second, paired sample's total assets and the non- paired sample's ones are most similar. Last, paired companies do not use derivatives.23 companies are paired sample which are selected in turn according to the rules. 230 effective observations are received as well.

The derivatives data involved in the article is received by manual collection and arrangement from Listed Companies' Annual Report one by one. The other data is from REESET（www.resset.cn）,Cninfo（www.cninfo.com）and CSRC （http://www.csrc.gov.cn）publicly disclosed Listed Companies' Annual Report. To ensure the reliability of data, the data is sampled and then double-checked. We use SPSS and Excel for the data processing, Excel is for basic data processing, and SPSS is for descriptive statistics, T-Type Test and Non - Parameter Z – Test.

3.2 Return Volatility

To fully reflect the sample company's operating performance, this article adopts two absolute indicators—total profit（Profit）and Net Cash Flows and two relative indicators—Earnings per Share（EPS）and Return on Equity（ROE）to measure samples company's operating performance. Return volatility is reflected by standard deviation of managerial achievement index above from 2006 to 2010.

3.3 Derivatives the Sample Companies Hold

Table 1 lists kinds and types of sample companies use derivatives of China's listed companies. From the kinds in the table, it can be seen 11 companies use one kind derivatives, about 47.8% of all, 9 companies use two kinds derivatives, about 39.1%, 3 companies use three kinds derivatives, about 13.1%. From the types in the table, it can be seen that most derivatives used are foreign exchange derivatives, about 65.2% of all, the second type is commodity derivatives, about 56.5%, the third is interest derivatives, about 43.5%. According to the survey on the World's top 500 enterprises use derivatives from ISDA in 2009, they use foreign exchange derivatives most, about 93.6%, the next is interest derivatives, about 88.3%, the last is commodity derivatives, about 50.9%. The kind of derivative usage of China's listed companies is single, which there has been gap compared with international level.

Table 1. Kinds and Types of Sample Companies Use Derivatives

Kinds	Quantity	Rate(%)	Types	Quantity	Rate(%)
1	11	47.8	Foreign Exchange	15	65.2
2	9	39.1	Interest	10	43.5
3	3	13.1	Commodity	13	56.5

Table 2 lists the size of the sample companies' use of the derivatives. In general, the size of China's listed companies' derivatives use is small and its change has been little from 2006 to 2010, except that there had been large amounts of derivative financial liabilities and sound value flexible loss and profit because of global financial crisis in 2008. Southwest Airlines, the second biggest airline company, is the only one which has remained profitable every year since 1973, and profit increase rate is the top (Zhang liang and Jan yin, 2008) . Table 3 lists that hedging benefits makes up the most of pre-tax earnings since 2004, even exceeded pre-tax earnings in 2005.It means that Southwest Airlines had been a enterprises running in the red, if it never transfer risk of fuel cost. So there is big gap in the size of derivative usage between China's companies and the ones in developed countries.

Table 2. Size of Derivative Usage of Sample Unit: 10,000 yuan

Projects	2006	2007	2008	2009	2010
DFA	13794	26387	20299	7008	13567
DFA /total asset(%)	0.15	0.60	0.41	0.33	0.32
DFD	6695	6149	87371	20149	12231
DFD/total asset(%)	0.12	0.18	1.19	0.50	0.51
PLFV	-1312	14117	-86354	41487	14581
PLFV/total profit(%)	-0.33	2.50	-41.65	12.38	2.19

Note: DFA (Derivative financial assets),DFD (Derivative financial debts), PLFV(profits and losses on the changes in fair value).

Table 3. Earnings Obtained from Hedging Unit: $1,000,000

Year	2001	2002	2003	2004	2005	2006	2007
EBT	631	417	483	554	779	790	1058
EFH	80	45	171	455	892	634	686
EBT/EFH	12.68%	10.79%	35.4%	82.13%	114.51%	80.25%	64.84%
Net profit	511	241	442	313	484	499	645

Note: EBT (Earnings before tax),EFH(Earnings from hedging).

3.4 Financial Characteristics of Sample

Table 4 lists financial characteristics indicators of intensive derivative users and paired samples. As shown in Figure 1, total asset of both intensive derivative users and paired samples increase every year, which reflect good growth. In addition, total asset of intensive derivative users is higher than paired samples in each year, which is consistent with the research findings of most scholars abroad on experience, or the bigger companies more likely to use derivative. As shown in Figure 2, analyzing assets liabilities ratio of two groups, assets liabilities ratio of intensive derivative users is higher, which is consistent with the research findings abroad. In order to reduce the cost of financial crisis, the company holding higher ratio use derivatives more. Judging from the operating income growth rate, as shown in Figure 3, these indicators of intensive derivative users is lower than paired samples from 2006 to 2008, however it is higher during 2009 and 2010. The average the operating income growth rate of intensive derivative users is 26.7328%, while the rate of paired samples is 22.7198%. This finding is same as the theory that derivatives can reduce the lack of potential investment. Generally speaking, the issue that a lack of investment is more possible in high growth company, which resulting in losing good opportunities, and reduce return volatility by derivatives usage. By the above analysis, it is concluded that the financial characteristics of intensive derivative users from China's listed companies are basic anastigmatic with the theory about derivatives abroad.

Table 4. Comparison of financial characteristics of sample

Firm	Project	2006	2007	2008	2009	2010	Total
User	Assets	23.7940	24.0964	24.2262	24.4685	24.6604	24.2491
	Leverage	60.4016	60.4087	64.4519	66.3669	65.4559	63.4170
	Growth	22.1578	25.0899	15.6209	14.4583	56.3372	26.7328
Non-User	Project	22.7401	23.1482	23.3384	23.5454	23.7267	23.2998
	Assets	60.5543	58.2126	58.0735	59.0640	58.4991	58.8807
	Leverage	26.2766	32.4637	18.0747	0.7838	36.0001	22.7198

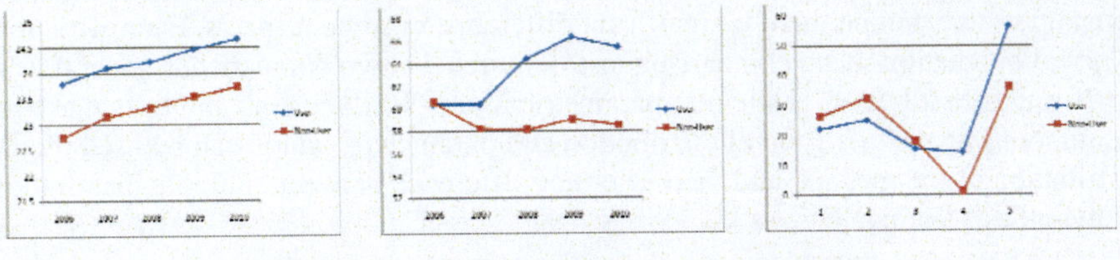

Fig. 1-3. Comparison of Assets, Leverage , Growth

4 Derivatives Use and Return Volatility

4.1 Operating Performance of Sample Company

Table 5 lists the indicators of operating performance of intensive derivative users and paired companies. Judging from the absolute indicators of operating performance, as shown in Figure 4 and 5, every year both Profit and NCF of derivative users are higher than those of sample companies, because the asset size of companies holding derivatives is generally bigger than the paired companies, which this finding is consistent with the research findings of most scholars abroad on experience. At the same time, the volatility is bigger on Profit and NCF of intensive derivative users. Judging from the relative indicators of operating performance, as shown in Figure 6 and 7, there is little difference between the two groups on EPS and ROE. It is obvious that volatility of intensive derivative users is bigger than the other.

Table 5. Comparison of performance of sample

Firm	Project	2006	2007	2008	2009	2010	Total
User	Profit	40.3381	56.4320	20.4949	33.5027	66.5040	43.4543
	NCF	53.1589	54.3978	54.6084	63.4736	77.4489	60.6175
	EPS	0.4548	0.7786	0.2440	0.4342	0.6984	0.5220
	ROE	11.6221	21.9341	0.4957	11.3252	17.4083	12.5571
Non-User	Profit	10.5043	15.4253	6.9032	5.8330	15.6130	10.8558
	NCF	13.9618	13.9469	17.3188	18.6726	18.2382	16.4276
	EPS	0.4657	0.7111	0.4113	0.4101	0.5119	0.5020
	ROE	12.9947	19.5196	9.4184	9.9209	12.7296	12.9166

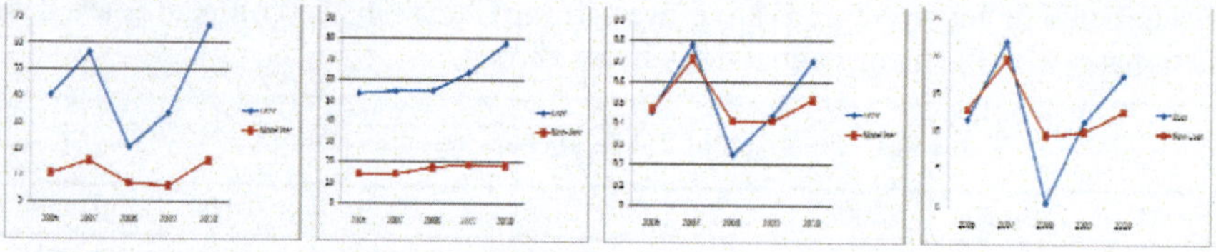

Fig. 4-7. Comparison of Profit, NCF, EPS and ROE

4.2 Return Volatility

This article adopts parametric and non-parametric tests, by pair testing of a series of characteristics of variables between the companies holding derivatives and paired companies, to examine there is significant difference in some respects. Parametric test (Normal population mean comparison test) requires the two comparisons general is in the Normal Distribution, while non-parametric Mann-Whitney rank and tests make no assumptions against its general distribution and parameters, which can be used on the distribution of the median, and there is or not difference between multiple distribution by hypothesis testing.

As showed in Figure 6, no matter parametric tests or non-parametric tests, there is significant difference in standard deviation of Profit σ_{Profit} on a confidence level of 1% between two groups companies, and the Profit of intensive derivative users is significantly higher than the paired samples. Judging from the standard deviation of NCF, the value of intensive derivative users is more than the paired samples. It is significant on a confidence level of 5% of T-Type Test and 1% of Z – Test, which implied the volatility of NCF of intensive derivative users is obviously more than the other. The standard deviation of EPS of intensive derivative users is higher than the other. It is significant on a confidence level of 10% of T-Type Test and 5% of Z – Test as well, which implied the volatility of EPS of intensive derivative users is obviously more than the other. At last, according to the standard deviation of ROE, it is higher of intensive derivative users than paired samples. It is significant on a confidence level of 10% of T-Type Test and 15% of Z – Test as well, which implied the volatility of ROE of intensive derivative users is obviously more than the other. It is concluded that, comparing with paired companies, rather than reduce return volatility, derivative usage increase the volatility.

From the results of the article, it is concluded that derivative usage significantly increase the volatility, not reduce the volatility as Western financial theory and practice has proved. We consider the possibilities as follow. First, some derivative users violate the aim of hedging, in other words, the intension of speculation is concealed behind of hedging. Second, Domestic undeveloped derivative market limits the companies, especially offshore derivatives trades are greatly restricted. The kind of derivative available is little, and hedging companies involved in is limited. The companies engaging in hedging are hedge some of the risks. Last, China's listed companies often deviate from the main line of the risk control, such as operation against the regulations, weak risk control and deficiencies of Internal Control Mechanism, resulting in loss for companies.

Table 6. Results of T-Test and Z-Test

Indicators	Groups	Mean	T-test	Sig.	Median	Z-test	Asymp. Sig.
σ_{Profit}	User	35.0790	3.684	0.001	15.6528	3.406	0.001
	Non-User	8.9790			3.5430		
σ_{NCF}	User	34.2063	2.721	0.012	17.7851	2.889	0.004
	Non-User	13.8487			6.7436		
σ_{EPS}	User	0.4466	1.974	0.061	0.3755	1.977	0.048
	Non-User	0.2924			0.2659		
σ_{ROE}	User	14.7865	1.757	0.093	9.6637	1.642	0.101
	Non-User	8.6704			7.0572		

5 Conclusion

This article examines whether firms systematically reduce or increase their return volatility with derivative usage using data from intensive derivative users of China's listed companies from 2006 to 2010. We find that, first, types of derivative usage of

China's listed companies are single, and the size of China's listed companies' derivatives use is small, which there has been gap compared with international level. Second, the relationship of derivative usage and the financial characteristics is basic anastigmatic with the theory derivatives abroad, or derivative usage is associated with company size, the cost of financial crisis and avoid underinvestment. Last, firms using derivatives display more return volatility.

The aim of derivative usage is to avoid the risks, is not speculation arbitrage. This article shows that return volatility of intensive derivative users of China's listed companies is significantly higher than the firms that do not use derivatives. This implies that some derivative users violate the aim of hedging, in other words, the intension of speculation is concealed behind of hedging.

References

1. Black, F., Scholes, M.: The Pricing of Options and Corporate Liabilities. Journal of Political Economy 81, 637–659 (1973)
2. Dolde, W.: Hedging Leverage and Primitive Risk. Journal of Financial Engineering 4, 187–216 (1995)
3. Guay, W.R.: The Impact of Derivatives on Firm Risk: An Empirical Examination of New Derivative Users. Journal of Accounting and Economics 26, 319–351 (1999)
4. Jensen, M.C., Meckling, W.: Theory of the Firm: Managerial Behavior, Agency Costs, and Ownership Structure. Journal of Financial Economics 3, 305–360 (1976)
5. Smith, C., Stulz, R.: The Determinants of Firms' Hedging Policies. Journal of Financial and Quantitative Analysis 20, 391–405 (1985)
6. Smithson, C., Simkins, B.: Does Risk Management Add Value? A Survey of the Evidence. Journal of Applied Corporate Finance 17, 8–16 (2005)

A Method of the Cloud Computing Security Management Risk Assessment

Hongbing Wang[1], Feng Liu[1], and Heng Liu[2]

[1] National Computer network Emergency Response technical Team/Coordination Center of China ,Beijing 100029
[2] Institute of Computing Technology Chinese Academy Of Sciences, Beijing 100190
whb6606@126.com, justskylife@163.com, liuheng@software.ict.ac.cn

Abstract. In order to effectively carry out cloud computing security risk management, the paper designed a model of overall cloud computing security management risk assessment, and put forward a specific risk assessment methodology. It designed a typical environment of cloud computing, conducted the experiment by the risk assessment methodology, and analyzed the experiment results. The experiment showed that the risk assessment analysis methodology could effectively reveal the vulnerability and risk of security management in a cloud computing environment, which is of great significance on the Cloud Computing Security Management Risks (CCSMR).

Keywords: Cloud computing, security management, risk assessment.

1 Introduction

With more extensive and in-depth application of cloud computing, security of cloud computing is also given special attention to. Since more and more organizations outsource their traditional assets, the security risks are partly transferred to the cloud services providers or depend on them. And these organizations concern more about how to manage the services providers and what security risks comes with the management. Therefore, the traditional technical method of security risk assessment which centers around the assets should give way to the business-focused one in the environment of cloud computing.

European Network and Information Security Agency (ENISA) released Cloud Computing Risk Assessment report, in which ENISA pointed out the advantages and security risks in cloud computing, provided some feasible recommendations and designed a set of assurance criteria to assess the risk of adopting cloud services [1] [2]. Jay Heiser and Mark Nicolett, two analysts at Gartner, co-authored the report of Assessing the Security Risks of Cloud Computing, which listed numerous security risks in cloud computing [3]. Motahari-Nezhad et al. pointed out the potential risks in using outside cloud services [4]. Introducing risk management into cloud computing, J. Oriol Fit´o and Jordi Guitart suggested the SEmi-quantitative BLO(Business Level

D. Zeng (Ed.): Advances in Computer Science and Engineering, AISC 141, pp. 609–618.
springer link.com

Objectives) -driven Cloud Risk Assessment (SEBCRA) as the core sub-process of the cloud risk management approach [5]. All of these researches have laid a solid foundation for cloud computing security management assessment. However, they barely established corresponding or complete models and methods of the security management risk assessment in consideration of the specific characteristics of cloud computing environment. There were neither quantitative descriptions of cloud computing security risks, nor illustrations of the many-to-many relationships of the assets, the vulnerabilities and the risks in the cloud computing environment.

This paper designs a method of the cloud computing security management risk assessment on the basis of the above researches. It mainly commits to assessing Cloud Computing Security Management Risks (CCSMR) to clarify distribution of the risks, impact level, occurrence possibilities, correlation between risks and assets, correlation between risks and vulnerabilities and so on.

This paper firstly establishes the relational model of the cloud computing security management risk assessment. This model is comprised of elements such as asset of the cloud computing organizations and importance of the assets, vulnerabilities of the cloud computing organization and the vulnerability levels, the security management risks of cloud computing, correlation between the risks and the assets, correlation between the risks and the vulnerabilities, the impact levels of the risks, the occurrence likelihood of the risks and the risk valuations. Then, this paper presents a method of cloud computing security management risk assessment, consisting of defining asset, making vulnerabilities, analyzing management risks and having quantitative calculation of risks, such as the total level of the correlated assets, the total level of the correlated vulnerabilities, calculation of the risk occurrence possibilities, possible impacts and valuations of each risk. A comprehensive analysis is carried out based on the risk valuation calculated, and finally this method is had an experimental analysis in a typical cloud computing scenario to reveal the CCSMR.

2 Calculation Model and Assessment Method of the CCSMR

2.1 Calculation Model

The corresponding assessment model of security management risks is designed due to the special characteristics of CCSMR assessment.

Definition 1: The *asset* of the cloud computing organization refers the tangible and intangible assets, the level of whose importance can be represented by A_i [1,5];

Definition 2: The vulnerability of cloud computing organization security management refers to the security management vulnerabilities in business operation and management, whose level can be represented by V_i [1,5] ;

Definition 3: The security management risk of cloud computing organization refers to the risks in business operation and management in the cloud computing environment, which can be represented by R_i ;

Definition 4: The total level of correlated assets refers to the total levels of all assets correlating to the specific risk Ri, which can be represented by $A(R_i)$.

$$A(R_i)=max(A_j)+0.1*(m-1),j=1,...,m$$

A_j represents the level of the *j-th* asset correlating to risk R_i, while m represents the amounts of the correlated asset.

Notes: the total asset level should be based on the highest level of the correlated asset with regard to the amounts of the other correlated assets. Thus, the total asset level will increase 0.1 if the amounts of the other correlated assets rise 1.

Definition 5: the total level of the correlated vulnerabilities refers to the total level of all vulnerabilities correlated to the specific risk R_i, which can be represented by $V(R_i)$.

$$V(R_i)=max(V_j)+0.1*(m-1),j=1,...,m$$

V_j represents j-th level of the vulnerability correlated with risk R_i, while m represents the amounts of the correlated vulnerabilities.

Notes: The total level of vulnerability should be based on the highest level of correlated vulnerabilities with regard to the amounts of the other correlated vulnerabilities. Thus, the total level of vulnerability increases 0.1 if the amounts of the other correlated vulnerabilities rise 1.

Definition 6: the impact level refers to the impact level caused by specific risk $R_{i.}$, which can be represented by $I(R_i)$ [1,5].

Definition 7: the possibility level refers to the likelihood of the specific risk R_i, which can be represented by $P(R_i)$.

Definition 8: the risk valuation refers to the calculated cloud risk value of the specific risk R_i, which can be represented by $C(R_i)$.

$$C(R_i)= A(R_i)* V(R_i)*I (R_i)*P (R_i)$$

2.2 Assessment Method

The CCSMR assessment method is designed as shown in Fig 2, according to the above assessment model. Below is the detailed procedure.

Step 1: define the environment and the object, describe them, evaluate them, identify asset of the cloud computing organization and assign values according to their importance.

Step 2: identity security management vulnerabilities of the cloud computing organization in business operation and management, and assign vulnerability levels.

Step 3: Identify security management risks of the cloud computing organization. That is mainly to identify security management risks in cloud computing operation and to analyze the correlated assets and their vulnerabilities.

Step 4: Calculate the total level of the asset correlated to each cloud computing security management risk and the total correlated vulnerability level.

Step 5: calculate the occurrence possibility $P(Ri)$ of each CCSMR and possible impact $I(Ri)$, as well as the cloud risk valuation $C(Ri)$of each risk.

Step 6: have a comprehensive analysis based on the calculated cloud risk valuation $C(Ri))$.

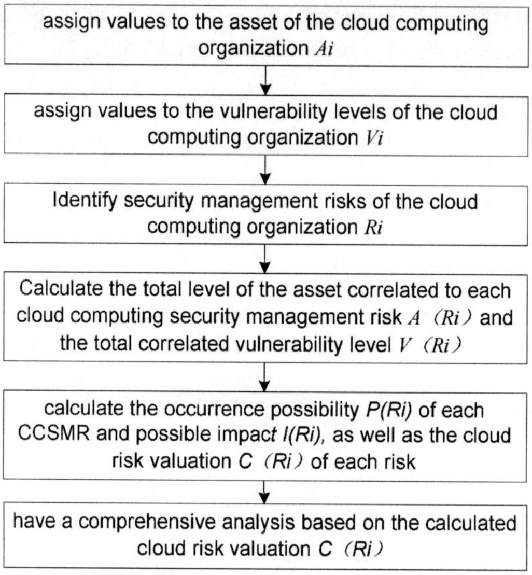

Fig. 1. Assessment method of the CCSMR

3 Experiment Result and Analysis

3.1 Experiment Environment

The corresponding assessment model of security management risks is designed due to the special characteristics of CCSMR assessment.

Suppose Company X is the cloud organization, which provides real time e-business via the Web services interface and the distribution services through the trade portal site. It offers the real time price data and charts, the order history and the report on the inventory control at the real time trade portal site. Considering cost, flexibility, reliability and the other factors, the cloud organization Company X plans to transfer the business to a cloud infrastructure. Fig 2 shows the case of Company X's cloud computing scenario, in which Company X is served by 5 cloud services providers for stronger service competencies and better business development.

- Cloud provider CP-A offers IaaS for information publishing.
- Cloud provider CP-B applies DaaS model (database as a service) to data memory and management.
- Cloud provider CP-C manages customer relationships. When the user needs to get access to things provided by CP-A, CP-C plays the role of federated identity provider for authentication.
- Cloud provider CP-D offers the network infrastructure on which the internal and external business depends.
- Cloud provider CP-E offers network security services and carries out vulnerability assessment, equipment management, security log analysis, patch management, regularity audit and real-time security monitoring.

- CP-D is locally relied on in Company X's business management, such as management of the user-oriented accounting, SLA and cloud services provider. And Company X itself is responsible for the local management of human resources, payrolls, office desktop application and researches, etc..

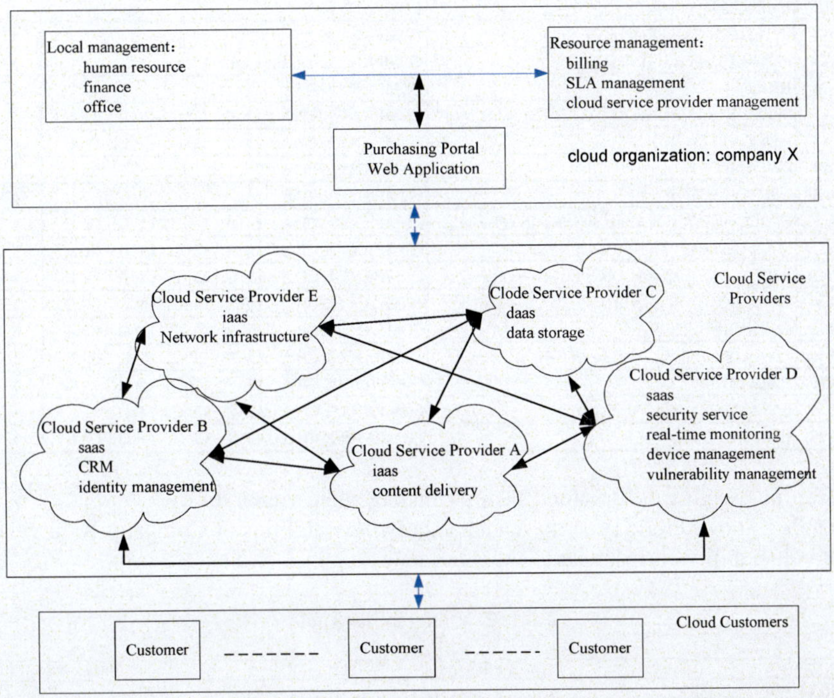

Fig. 2. The cloud computing scenario

As a real-time e-business trade site, it faces with business risks mainly as delayed data preparation which may lead to huger business losses. The other requirements include highly reliable requested execution such as the display of database query results and http requested web service execution, data integrity which may otherwise causes financial losses, confidentiality of data whose disclose means financial losses and integrity and vulnerability of application which may otherwise effect the stable and reliable operation of business systems.

However, are these risks complete? To what extent will these risks impact on? What are the occurrence possibilities? How about the risk management? These questions will be answered with our method of the cloud computing security management risk assessment as follows.

3.2 Experiment Procedure

Step 1: Assign values to the cloud assets.

The asset table (table 1) is drawn up though preliminary analysis and assignment, in which each asset corresponds to one asset level.

Table 1. Cloud computing assets

No.	*ASSET*	*OWNER[ACTORS OR ORGANISATION INVOLVED]*	*Ai*
A1	BUSINESS REPUTATION	COMPANY X	5
A2	CUSTOMER TRUST	COMPANY X	5
A3	EMPLOYEE LOYALTY	COMPANY X	4
A4	INTELLECTUAL PROPERTY	COMPANY X	4
A5	PERSONAL SENSITIVE DATA	COMPANY X/CP-A, CP-B, CP-C	5
A6	HR data	COMPANY X	4
A7	SERVICE DELIVER-REAL TIME SERVICES	COMPANY X/CP-A, CP-B, CP-C, CP-D, CP-E	5
A8	ACCESS CONTROL	COMPANY X/CP-A, CP-B, CP-C, CP-D, CP-E	4
A9	CRESENTIALS	COMPANY X/CP-B	5
A10	USER DIRECTORY	COMPANY X/CP-C	4
A11	CLOUD SERVICE MANAGEMENT INTERFACE	COMPANY X/CP-A, CP-B, CP-C, CP-D, CP-E	5
A12	MANAGEMENT INTERFACE APIs	COMPANY X/CP-A, CP-B, CP-C, CP-D, CP-E	3
A13	INTRA- AND EXTRA- CLOUD NETWORK	COMPANY X/CP-A, CP-B, CP-C, CP-D, CP-E	4
A14	PHYSICAL HARDWARE	COMPANY X/CP-A, CP-B, CP-C, CP-D, CP-E	3
A15	PHYSICAL BUILDINGS	COMPANY X/CP-A, CP-B, CP-C, CP-D, CP-E	4
A16	CLOUD PROVIDER APPLICATION	COMPANY X/CP-A, CP-B, CP-C, CP-D, CP-E	4
A17	OPERATION LOGS	COMPANY X/CP-A, CP-C, CP-D, CP-E	3
A18	BACKUP OR ARCHIVE DATA	COMPANY X/CP-A, CP-B, CP-C, CP-D	3
A19	CERTIFICATION	COMPANY X/CP-A, CP-B, CP-C, CP-D, CP-E	5

Step 2: Assign value to cloud security management vulnerability.

Table 2 shows cloud security management vulnerabilities of Company X, in which each vulnerability corresponds to one vulnerability level.

Table 2. Vulnerability Table

No.	*VULNERABILITIES*	*OWNER*	*Vi*
V1	AAA VULNERABILITIES	CP-A	5
V2	USER PROVISIONING VULNERABILITIES	SaaS CP-A	4
V3	REMOTE ACCESS TO MANAGEMENT INTERFACE	SaaS CP-A	4
V4	HYPERVISOR VULNERABILITIES	CP-E	5
V5	LACK OF RESOURCE ISOLATION	DaaS CP-C	5
V6	LACK OF REPUTATIONAL ISOLATION	CP-A,CP-B,CP-C,CP-D,CP-E	5
V7	COMMUNICATION ENCRYPTION VULNERABILITIES	CP-A,CP-C	4
V8	LACK OF OR WEAK ENCRYPTION OF ARCHIVES AND DATA IN TRANSIT	CP-A,CP-C	4
V9	NO SOURCE ESCROW AGREEMENT	CP-A,CP-E	4
V10	POSSIBILITY THAT INTERNAL (CLOUD) NETWORK PROBING WILL OCCUR	CP-A,CP-E	3
V11	LACK OF FORENSIC READINESS	CP-A,CP-E	4
V12	SYNCHRONIZING RESPONSIBILITIES OR CONTRACTUAL OBLIGATIONS EXTERNAL TO CLOUD	CP-A,CP-C	4
V13	SLA CLAUSES CONTAINING EXCESSIVE BUSINESS RISK	CP-D	5
V14	POOR KEY MANAGEMENT PROCEDURES	CP-B	4
V15	CERTIFICATION SCHEMES NOT ADAPTED TO CLOUD INFRASTRUCTURES	CP-E	3
V16	NO POLICIES FOR RESOURCE CAPPING	CP-D, CP-E	4
V17	INADEQUATE RESOURCE PROVISIONING AND INVESTMENTS IN INFRASTRUCTURE	CP-E	4
V18	LACK OF INFORMATION ON JURISDICTIONS	CP-A,CP-B,CP-C,CP-D,CP-E	3
V19	LACK OF SECURITY AWARENESS	CP-A,CP-B,CP-C,CP-E	4
V20	UNCLEAR ROLES AND RESPONSIBILITIES	CP-A,CP-E	4
V21	NEED-TO-KNOW PRINCIPLE NOT APPLIED	CP-D	4
V22	APPLICATION VULNERABILITIES	CP-A	4
V23	LACK OF, OR A POOR AND UNTESTED, BUSINESS CONTINUITY AND DISASTER RECOVERY PLAN	CP-A,CP-C,CP-E	5
V24	SYSTEM OR OS VULNERABILITIES	CP-C	3
V25	LACK OF STANDARD TECHNOLOGIES AND SOLUTIONS	CP-A,CP-C	5
V26	POOR PATCH MANAGEMENT	CP-E	3
V27	LACK OF SUPPLIER REDUNDANCY	CP-A	3
V28	INADEQUATE OR MISCONFIGURED FILTERING RESOURCES	CP-A	3

Step 3: Identify cloud security management risks.

Table 3 shows results of the correlation analysis of cloud organization's security management risks on the basis of asset and vulnerability analysis, in which each risk can be correlated to several assets or several vulnerabilities.

Table 3. Analysis table of risk correlations

No.	*RISK*	*ASSETS*	*VULNERABILITIES*
R1	LOCK-IN	A1,A5,A7	V25
R2	LOSS OF GOVERNANCE	A1,A2,A3,A5,A7	V20,V12,V25,V16,V18
R3	COMPLIANCE CHALLENGES	A19	V25,V15,V11
R4	LOSS OF BUSINESS REPUTATION DUE TO CO-TENANT ACTIVITIES	A1,A5,A7	V5,V6,V4
R5	CLOUD SERVICE TERMINATION OR FAILURE	A1,A2,A3,A7	V27
R6	RESOURCE EXHAUSTION (UNDER OR OVER PROVISIONING)	A1,A2,A7,A8	V16,V17
R7	ISOLATION FAILURE	A1,A2,A3,A7	V4,V5,V6,V10
R8	CLOUD PROVIDER MALICIOUS INSIDER - ABUSE OF HIGH PRIVILEGE ROLES	A1,A2,A3,A4,A5,A6,A7	V20,V21,V1,V22,V17
R9	MANAGEMENT INTERFACE COMPROMISE	A1,A2,A3,A7,A11	V1,V3,V24
R10	INTERCEPTING DATA IN TRANSIT	A1,A2,A4,A5,A6,A18	V1,V8,V7,V10
R11	DATA LEAKAGE ON UP/DOWNLOAD, INTRA-CLOUD	A1,A2,A3,A4,A5,A6,A10,A11	V1,V8,V7,V11
R12	DISTRIBUTED DENIAL OF SERVICE (DDOS)	A1,A2,A7,A11,A13	V24,V28
R13	UNDERTAKING MALICIOUS PROBES OR SCANS	A1,A2,A7	V17,V22,V25
R14	CONFLICTS BETWEEN CUSTOMER HARDENING PROCEDURES AND CLOUD ENVIRONMENT	A4,A5,A6	V19,V20,V12
R15	RISK FROM CHANGES OF JURISDICTION	A1,A2,A4,A7	V11,V18
R16	DATA PROTECTION RISKS	A1,A2,A5,A7	V5,V18,V10
R17	NETWORK BREAKS	A7	V28,V24,V5,
R18	PRIVILEGE ESCALATION	A5,A6,A10,A8	V1,V4,V2,V20,V21,V28
R19	SOCIAL ENGINEERING ATTACKS	A1,A2,A3,A4,A5,A6,A8	V19,V2,V5,V7
R20	NATURAL DISASTERS	A1,A2,A5,A6,A7,A18	V23

Step 4: Calculate the total level of all risk correlated to assets and that correlated to vulnerabilities. The calculation is based on definition 5.

E.g. $A(R_1)=Max(A_1,A_5,A_7)+0.1*(3-1)$

$=5+0.2$

$=5.2$

$V(R_1)=V_{25}=5$

Table 4 shows the values of the other items calculated in the same way.

Table 4. Calculation of correlation levels

No.	*RISK*	*A (Ri)*	*V (Ri)*
R1	LOCK-IN	5.2	5
R2	LOSS OF GOVERNANCE	5.4	5.4
R3	COMPLIANCE CHALLENGES	5	5.2
R4	LOSS OF BUSINESS REPUTATION DUE TO CO-TENANT ACTIVITIES	5.2	5.2
R5	CLOUD SERVICE TERMINATION OR FAILURE	5.3	3
R6	RESOURCE EXHAUSTION (UNDER OR OVER PROVISIONING)	5.3	4.1
R7	ISOLATION FAILURE	5.3	5.3
R8	CLOUD PROVIDER MALICIOUS INSIDER - ABUSE OF HIGH PRIVILEGE ROLES	5.6	5.4
R9	MANAGEMENT INTERFACE COMPROMISE	5.4	5.2
R10	INTERCEPTING DATA IN TRANSIT	5.5	5.3
R11	DATA LEAKAGE ON UP/DOWNLOAD, INTRA-CLOUD	5.7	5.3
R12	DISTRIBUTED DENIAL OF SERVICE (DDOS)	5.4	3.1
R13	UNDERTAKING MALICIOUS PROBES OR SCANS	5.2	5.2
R14	CONFLICTS BETWEEN CUSTOMER HARDENING PROCEDURES AND CLOUD ENVIRONMENT	5.2	4.2
R15	RISK FROM CHANGES OF JURISDICTION	5.3	4.1
R16	DATA PROTECTION RISKS	5.3	5.2
R17	NETWORK BREAKS	5	5.2
R18	PRIVILEGE ESCALATION	5.3	5.5
R19	SOCIAL ENGINEERING ATTACKS	5.6	5.3
R20	NATURAL DISASTERS	5.6	5

Step 5: Calculate the occurrence possibility $P(R_i)$ of all risks and their possible impacts $I(R_i)$ as well as cloud risk valuation $C(R_i)$. *Calculate* $C(R_i)$, *according to Definition 8, e.g.* $C(R_1)=5.2*5*0.75*3=58.5$.

Table 5 shows valuation calculated in the same way.

Table 5. Cloud risk valuation

No.	*RISK*	*A (Ri)*	*V (Ri)*	*P (Ri)*	*I (Ri)*	*C(Ri)*
R1	LOCK-IN	5.2	5	0.75	3	58.50
R2	LOSS OF GOVERNANCE	5.4	5.4	1.00	5	145.80
R3	COMPLIANCE CHALLENGES	5	5.2	1.00	4	104.00
R4	LOSS OF BUSINESS REPUTATION DUE TO CO-TENANT ACTIVITIES	5.2	5.2	0.25	4	27.04
R5	CLOUD SERVICE TERMINATION OR FAILURE	5.3	3	0.25	5	19.88
R6	RESOURCE EXHAUSTION (UNDER OR OVER PROVISIONING)	5.3	4.1	0.50	4	43.46
R7	ISOLATION FAILURE	5.3	5.3	0.50	5	70.23
R8	CLOUD PROVIDER MALICIOUS INSIDER - ABUSE OF HIGH PRIVILEGE ROLES	5.6	5.4	0.50	5	75.60
R9	MANAGEMENT INTERFACE COMPROMISE	5.4	5.2	0.50	5	70.20
R10	INTERCEPTING DATA IN TRANSIT	5.5	5.3	0.50	4	58.30
R11	DATA LEAKAGE ON UP/DOWNLOAD, INTRA-CLOUD	5.7	5.3	0.50	4	60.42
R12	DISTRIBUTED DENIAL OF SERVICE (DDOS)	5.4	3.1	0.50	4	33.48
R13	UNDERTAKING MALICIOUS PROBES OR SCANS	5.2	5.2	0.50	3	40.56
R14	CONFLICTS BETWEEN CUSTOMER HARDENING PROCEDURES AND CLOUD ENVIRONMENT	5.2	4.2	0.25	3	16.38
R15	RISK FROM CHANGES OF JURISDICTION	5.3	4.1	0.75	3	48.89
R16	DATA PROTECTION RISKS	5.3	5.2	0.75	4	82.68
R17	NETWORK BREAKS	5	5.2	0.25	4	26.00
R18	PRIVILEGE ESCALATION	5.3	5.5	0.25	4	29.15
R19	SOCIAL ENGINEERING ATTACKS	5.6	5.3	0.50	4	59.36
R20	NATURAL DISASTERS	5.6	5	0.25	4	28.00

Different from the traditional risk levels, the risk values in Table 5 indicate scope of the assets affected by risks, amounts of the vulnerabilities involved, occurrence possibilities and possible impacts.

3.3 Experimental Results

1) Distribution of cloud risk valuation $C(R_i)$.

Fig 3 suggests R_2-R_3-R_{16} should be the priority, R_8-R_9-R_7-R_{19}-R_{11}-R_1-R_{10}-R_{15}-R_6-R_{13} be the second one and the R_{12}-R_{18}-R_{20}-R_4-R_{17}-R_5-R_{14} the last one in risk management.

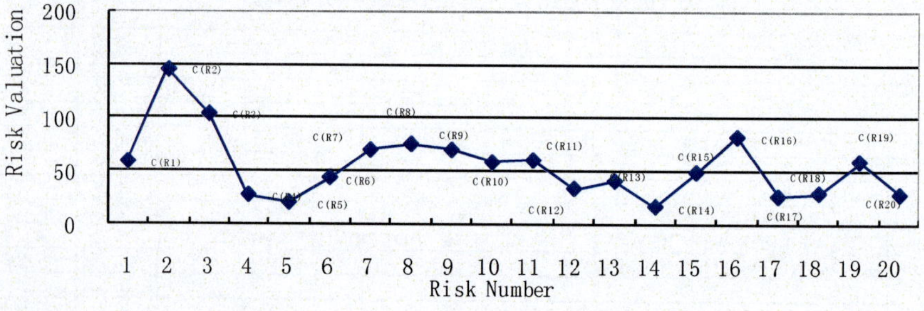

Fig. 3. Diagram of the risk valuation

2) Distribution of risk valuation and impact level.

From the aspect of controlling impacts, Fig 4 proves R_2-R_5-R_7-R_8-R_9 should be concerned first, R_3-R_4-R_6-R_{10}-R_{11}-R_{12}-R_{16}-R_{17}-R_{18}-R_{19}-R_{20} the second and R_1-R_{13}-R_{14}-R_{15} the last in risk management.

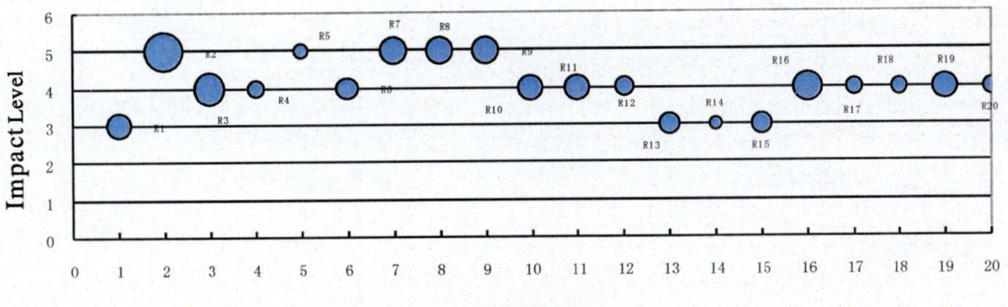

Fig. 4. Distribution diagram of risk valuations and impact levels

3) Distribution of risk valuations and occurrence possibilities

From the aspect of occurrence possibility, Fig 5 indicates R_2-R_5-R_{15}-R_{16}-R_1 should be concerned first, R_6-R_7-R_8-R_9-R_{10}-R_{11}-R_{12}-R_{13}-R_{19} the second and R_4-R_5-R_{14}-R_{17}-R_{18}-R_{20} the last in risk management.

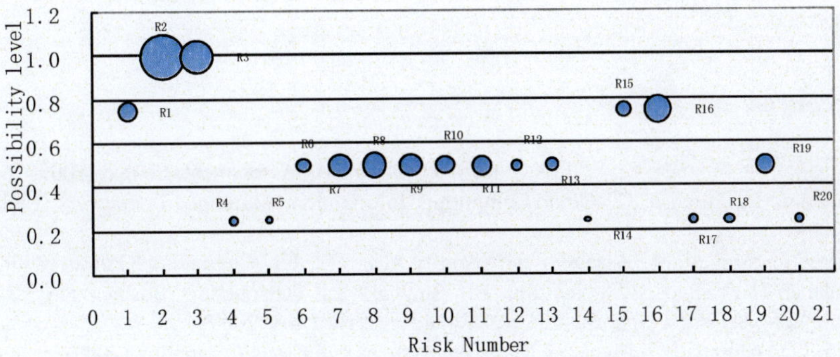

Fig. 5. Distribution diagram of risk valuations and occurrence possibilities

4) Analysis of the risk & asset and the risk & vulnerability correlation

Based on the above statistics analysis, the correlation of the risk and the asset can be calculated. So does the correlation of the risk and the vulnerability. Fig 6 indicates the relationship among assets , vulnerabilities and risks. Thus, related assets and relationships of vulnerabilities and risks can be dealt with more completely in the process of managing and controlling risks.

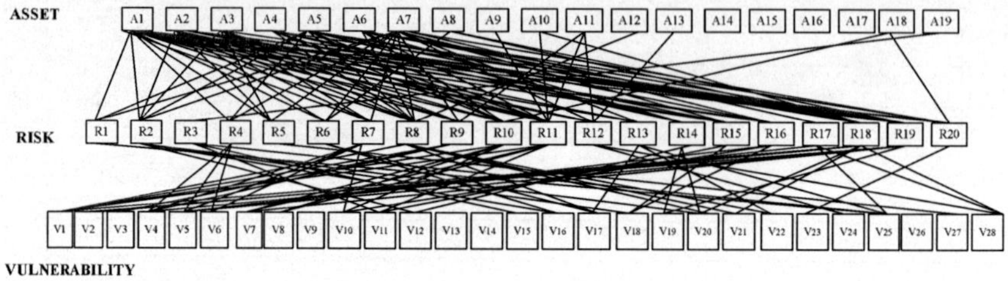

Fig. 6. Distribution Diagram of the risk & asset and the risk & vulnerability correlation

4 Conclusions

This paper puts forward the assessment method of the CCSMR. It contributes to better revelation of the special security management risk in the environment of cloud computing through experimental analysis. It also more vividly presents distributions of cloud security management risks as well as correlations of risk valuation with impact level, occurrence possibilities, asset correlations and vulnerability correlations. Therefore, it is of certain reference value to the study of cloud computing security risk management.

References

1. Catteddu, D., Hogben, G.: ENISA Cloud Computing Risk Assessment. ENISA (2009)
2. Catteddu, D., Hogben, G.: Cloud Computing Information Assurance Framework. ENISA (2009)
3. Heiser, J., Nicolett, M.: Assessing the Security Risks of Cloud Computing. Gartner (2008)
4. Motahari-Nezhad, H., Stephenson, B., Singhal, S.: Outsourcing Business to Cloud Computing Services: Opportunities and Challenges. HP Labs (2009)
5. Fitó, J.O., Guitart, J.: Introducing Risk Management into Cloud Computing (2009)

Research on Logistics Outsourcing of Retail Chain Enterprise Based on Fit between Supply and Demand

Yuan Xu and Qiong Yao

Management School, Jinan University, Guangzhou, China
{enclos,yaoqiong}@126.com

Abstract. In view of the current situation of retail chain logistics outsourcing and the third-party logistics company in China, combining with the significance of modern retail chain logistics outsourcing, This paper proposes that the logistics capability of both don't match currently. Based on previous studies, it seeks to find the retail chain logistics outsourcing strategy that match the ability between logistics outsourcing demander and provider.

Keywords: Retail Chain Enterprise, Logistics Outsourcing, Third-party Logistics, Decision Analysis.

1 Introduction

Since the 1990 s, with the development of information technology and the global economy, enterprise outsourcing becomes popular. As an important part of outsourcing, logistics outsourcing has grown rapidly under the promotion of international division of labor and professional cooperation. To concentrate on enhance core competitiveness, the production enterprise and retail enterprise entrust their logistics business to professional logistics company in the form of a contract, referred to as logistics outsourcing. The widespread notion in the past is that, only big businesses, especially multinational company is the subject of logistics outsourcing, for large scale in the division of labor and the professional expertise provide them with more outsourcing logistics conditions. However, with the development of technology innovation and lowering threshold, small and medium-sized enterprises began to be active in the field of logistics outsourcing. Especially with overall open of Chinese retail market in December 11, 2004, many foreign retailers, such as Wal-Mart, Carrefour poured into China. With their solid operational strength and perfect logistics distribution system, they have brought a huge to Chinese retail chain enterprises management. In response to such challenge, China's retail chain enterprises have looked for support and cooperation from the external environment, and outsource their logistics, which acts as an effective way to participate in international chain retail competition.

2 The Status in Quo of Retail Chain Logistics Outsourcing

Since the first chain enterprise founded in Guangdong in 1990, chain operation has witness a rapid development. However, relative to the booming retail business, the

D. Zeng (Ed.): Advances in Computer Science and Engineering, AISC 141, pp. 619–625.
springer link.com

development of logistics outsourcing of retail chain of slightly lags, which restricts the development of retail chain enterprises in China. According to statistics from the China association of chain operation, most of Chinese retail chain enterprises have their own logistics distribution center, less dependent on the third-party logistics for outsourcing. These distribution centers vary in size and quality, and their management is comparatively in confusion. Another logistics supply and demand survey from Chinese storage association shows that 74% of Chinese retail enterprise logistics is implemented by the supplier, compared to 13% by the third-party logistics. This is far behind logistics outsourcing in developed countries. Most of China's retail chain enterprises have not yet accepted logistics outsourcing, and still bear the ideas of "big and complete, small and complete". Limited by scale a lot of medium-sized and small retail chain enterprisesannot smoothly implement outsourcing, moreover, the lack in services and profits decrease the inclination for the third-party logistics enterprise to provide service. In addition, the complexity of logistics outsourcing itself renders logistics outsourcing to the third-party logistics company more difficult.

In addition to the characteristics of the retail chain enterprises, the third-party logistics has a role in causing the status quo. In abroad, the third-party logistics has become the dominant force of modern logistics industry. In the United States, Japan and other developed countries, the third-party logistics occupied more than 30% of the market share, while in our country the proportion is only 10%. The development of the third-party logistics in our country started rather late, and compared to the third-party logistics system in foreign countries, there is great gap in the professional service, individuation, information technology, systematic management.

Professionalization is reflected in whether the third-party logistics company has the professional skills from the whole process from logistics solution design to the implementation and control.

Individuation is reflected in whether the outsourcing party can provide personalized demand of logistics solutions. Information technology is reflected in logistics information process of commercialization, and data so on; Systematic management embodies in whether the third-party logistics company has a set of perfect management system to support the development of the business and execution of the company. In these areas, the third-party logistics companies in China do not implement well.

3 The Significance of Retail Chain Logistics Outsourcing

Logistics outsourcing is also known as "third-party logistics". The earliest description of third-party logistics is based on the U.S. Council of Logistics Management, a customer service survey of consumers called it "logistics service provider" in 1988.At present for third-party logistics is not a uniform definition. A more authoritative in our definition is: "a business model that other than by the supply-side and demand-side outside of the logistics enterprises to provide logistics services". Popularization and application of information technology makes the traditional flow of commerce and logistics market can't meet the requirements of competition and enterprise development, with the continuous change of commerce flow, a new division of labor organizations - to undertake logistics outsourcing services to third-party logistics company came into being. Companies can focus their limited resources on key advantage business, be easy to construct their own core competencies, reduces the cost of business operations, improve the professional level of logistics services via logistics outsourcing. Many domestic and foreign scholars try to find the drives of modern retail chain logistics outsourcing, the examination of motives for the logistics outsourcing can provide more information[1]. Here we will set out some representative points of view that summarized as follows:

3.1 Reduce Costs

A lot of research studies have shown that companies choose outsource because of reducing costs. From the perspective of the theory of transaction cost, scholars built a model and found that reducing transaction costs is the key drivers of corporate outsourcing; other scholars reached a similar conclusion from the perspective of the theory of economies of scale. Yang Yixing studied retail enterprises logistics outsourcing from the perspective of supply chain, he felt that scholars should focus on the core value chain competitiveness and cost while examining the corporate boundaries[2].Due to the advantage in third-party professional division logistics in the professional division and labor efficiency advantage and efficiency, the third-party logistics help companies reduce the cost on the related equipment, technical inputs so that reduce of costsof the supply chain, such as transportation costs, inventory costs.

3.2 Enhance the Core Competitiveness

Some scholars, from the perspective of the product value chain theory, analyzed the reasons for the development of outsourcing business. They believed that outsourcing is a new form of international division, the international division of labor is no longer confined to the production stage, and the entire value chains in the product are involved. In addition to the value chain as already mentioned, another factor is the core competitiveness. Yang Yixing divide retail value chain into the following sections: procurement, inventory, distribution, transportation and sale. The decision-making part includes procurement; distribution, sales, and the operating part include inventory and transport. The core competitiveness of retail enterprises should lie in the decision-making parts which highly focus on Information and technology such as distribution and sales. In the other hand, inventory and transportation should therefore be excluded from the firm boundaries and be implemented by logistics outsourcing. He believes that retail logistics outsourcing should focus on inventory and shipping, in fact, this conclusion is so limited, not only the inventory and shipping but also loading and unloading, packaging and processing can be included because logistics outsourcing is a comprehensive business activity,. Generally speaking, retail enterprises should constantly cultivate and strengthen its core competitiveness by study the value chain, excluding the link that does not have the advantage, looking for external cooperation, and complementing each other, and constantly improving the value chain.

Some scholars explained the motivation for outsourcing from a comprehensive perspective.Zhang Fen-xia, Liu Jing-jiang found offshore outsourcing is one consequence of the development of economy, policy and technology[3]. LI Fei explained the motives for outsourcing from the external environment and internal driving force thought the research of United States outsourcing [4].Yuan Sheng-nan [5], Chen Ye-ling, Yu Xiao-lei and other scholars found that, apart from reducing costs and improving core competitiveness, the retail chain's logistics outsourcing can reduce risk, increase flexibility, make use of the resources that enterprises do not have [6].

4 Decision Analysis of Retail Chain Enterprises Logistics Outsourcing

Two essential parts are contained in logistics outsourcing, demand and supply sides, which are both irreplaceable. Whenever talking about logistics outsourcing, the decision analysis of outsourcing can never be avoided. For the demand side, three problems should be focused: which supply type they should choose, what to outsource and how much they should outsource. The reason why a plenty amount of chain retailing cannot provide a satisfied logistics outsourcing is that they have not give themselves a distinct recognize about the logistics system of their own and the supply style of the other side. This failure makes them not able to match well, which lead to the operation be lack of coupling. Therefore, in the circumstance when the two side abilities cannot be changed, it is a perfect way to improve the matching quality between them. Such improvement can not only solve the bottleneck effect of the retail chain enterprises but also promote the development of the third-part logistics of our country. In this section, we firstly divide the supply side ——retail chain enterprises—— in our country. Then we will pay attention to who to outsource, what to outsource and how much to outsource. Finally a strategic decision of matching the retail chain enterprises and third-part logistics can be established.

4.1 Forms of Retail Chain Enterprises

When the logistics outsourcing of the enterprise is ongoing, it is necessary to be aware of the personal actual situation and the external environment (mainly focus on third-part logistics). In the view of the company, the logistics outsourcing forms vary from individual to individual. The classification of retail chain enterprises is multiple according to different standard. In accordance with the type of business, it can be classified as supermarket retailing, department store retailing and specialty store. On the basis of scale, it can be sorted into large-scale and small-and-medium scale. In this article, we use the latter.

4.2 Outsourcing Provider

Enterprises must consider how to select the outsourcing provider at the moment of making logistics outsourcing decisions. The provider does not have to choose the best, but should be the most appropriate, with maximum efficiency and lowest cost solution to help solve the problem of enterprises logistics. Generally, retail chain enterprises in the choice of third-party logistics, according to the following three criteria to examine:

4.2.1 Professional Technology

As already mentioned, addition to the general characteristics of the logistics the retail chain showing more features such as variety, small batch, high delivery rates, more distribution points, quick delivery requirements and advanced information systems.

This requires the strong third-party logistics companies must have the expertise including inventory management, distribution processing, and logistics technology and transportation operations management.

4.2.2 Service Level

The third-party logistics company not only need to have professional technology, also should have strong service consciousness and service level. Logistics services level is a general description of logistics services. Generally, we measure the overall level of service through the evaluation of service quality. The definition about logistics service quality come from the study of Tennessee University in USA. From the view of customer, they summed up the communication quality, order release quantity, quality of information, process orders, precision products, goods intact degrees, quality of goods, error handling and timing and nine indicators to measure the level of logistics services. The nine evaluations are not entirely suited to the different demander and different sectors are different in the degrees of importance. For example, retail chain enterprises emphasize information quality, precision and timing and other goods in high demand.

4.2.3 Informatization Level

Along with the development of modern logistics industry, the application of information technology in logistics industry is more and more widely. Using information technology can improves the efficiency and cultivate the core competence, which is the inevitable means in the in the market competition. Scholars points out that the general structure of the third-party logistics enterprise information system including: interface system--include ERP interface module, order system interface module and operating system interface module[7];Operation management system--include warehouse management system, transportation management system, distribution management system; Order processing system; support& decision-making system. Operation management system is the basic logistics information system, support&decision-making system is the highest levels of the logistics information system, The two systems in the entire logistics enterprise information system is more important than the others, this article evaluate the level of information management systems rely mainly on the operational system and support&decision-making system.

4.3 Content of Outsourcing

After determining the outsourcing provider, the next question is what the retail chain enterprises should be outsourced to the third-party logistics companies. Logistics is an integrated service, includes transportation, distribution, handling, storage, packaging, distribution processing, information transmission and etc. Retail chain enterprises need to stand on their own point of view to analyze which aspects of the business need outsourcing. According to previous research, this paper divides the logistics of retail chain enterprises into the aspects of procurement, transportation, inventory, distribution, logistics processing, after-sales service. From every aspects of logistics, transportation, inventory and distribution need to invest large fixed costs in the initial, with retail sales has the seasonal variation characteristics, the requirements of the inventory and distribution is great contrast in the peak season and off-season, Consider the financial strength and the risk aversion, for the small and medium retail chain enterprises should outsource the transport, inventory and distribution to third-party logistics companies. Large retail chain enterprises have advantages in the capital and scale, have the ability to establish their own logistics and distribution centers, and self-logistics has high flexibility to meet the chain's logistics requirements and customer needs in time. More and more large retail chain enterprises began to focus on building its own logistics and distribution center. Even so, large retail chain enterprises also need logistics outsourcing, because any company resources are limited, it is difficult to develop the logistics business to be all-inclusive, such as large retailers like Wal-Mart, its online shopping delivery and distribution services are also outsource to the third-party logistics companies. With the development of China's third-party logistics companies, some large retail chain enterprises have begun to try logistics outsourcing. Large retail chain enterprisesan flexibilitily choose specific outsourcing areas according to their own development strategies and operational superiority companies.

4.4 Degree of Outsourcing

One last question, we need to define the boundaries of the retail chain enterprises, that is, we need to define the degree of logistics outsourcing. According to the scope of the outsourcing we divided the logistics outsourcing into following three models:

4.4.1 Part Business Outsourcing

It is also called special business outsourcing. It means parts of one complete logistics work are outsourced, and the other parts are operated by the enterprise itself. It's good for enterprise to adopt suitable outsourcing model according to own advantages and disadvantages in the logistics functions.

4.4.2 Overall Outsourcing

Overall outsourcing is also called one-stop outsourcing service. It means the complete logistics work is outsourced, and enterprise itself is out of the work. Advantage of this way is that it can break the enterprise's original internal management pattern, eliminate the influence of the artificial factor, and improve the efficiency of the logistics. The weakness is that there is high risk, it needs good external environment and the third-party logistics company with strong ability to be guarantees.

4.4.3 Comprehensive Outsourcing

Comprehensive outsourcing means several logistics functions are outsourcing at the same time. Such outsourcing can choose a third-party logistics company or several companies. It could be an overall logistics functions` outsourcing, or a part of some functions` outsourcing. So, this way of logistics outsourcing puts forward a higher request to third-party logistics and the whole social logistics service system due to flexibility and variability.

Generally speaking, large retail chain enterprises have wide distributions throughout many provinces. As a result, most of them have built their own distribution center. Large stores cooperate with the third-party logistics company to improve operational efficiency as much as possible. At present, most large retail chain enterprises use part outsourcing to outsource some of their own weak parts to relative more professional third-party logistics. It is benefit to form complementary advantages. With the development of logistics environment and third-party logistics, more and more large retail chain enterprises choose to outsource the whole logistics to third-party logistics companies. While medium and small retail chain enterprises don't have enough money, it doesn't pay to build own distribution center, and retail chain enterprises also can't generate economies of scale by outsourcing because of scale limitations. So medium and small retail chain enterprises are apt to use joint distribution, which means these enterprises outsource logistics to the same third-party logistics due to their common or similar logistics demands. Although this way ensures economies of scale just like large enterprises, it also increases risk and uncertainty. Joint distribution binds these enterprises together. Once one company`s problem arises, others will be influenced by chain reaction. In order to decrease this uncertainty, enterprises should increase investment in logistics, bring third-party logistics into own supply chain and build alliances with them. Overall outsourcing and comprehensive outsourcing will be a better choice for medium and small retail chain enterprises under this way.

5 Conclusion

With the continuous inrush of the transnational retail Enterprises giant magnate, the local enterprises are currently facing the unprecedented pressure and challenge. The bottleneck effect of logistics management has seriously restricted the development of our national retailing enterprises and the ability to participate in international competition. The logistics outsourcing can not only lower the cost and enhance the core competitiveness but also reply the potential tension in the future. The utilization of logistics outsourcing in the retail chain enterprises will be more and more wide-ranging. When the capacity of retail chain enterprises and the third-part logistics is vested, the improvement of their matching quality can not only solve the bottleneck effect of the retail chain enterprises but also promote the development of the third-part logistics of our country.

References

1. Huang, Y.-H.: Analysis on the Relationship between Logistics Service Cost and Logistics Service Level. Logistics Technology 2, 22–25 (2007)
2. Yang, Y.-X.: Study on Logistics Outsourcing of Retail Chain Enterprise from the perspective of the supply chain. Modern Business 2, 11 (2008)
3. Yao, J.-M.: Empirical Research on Decision Agent and Relation of Enterprises Logistics Outsourcing Strategy in China. Commercial Research 5, 1–8 (2010)
4. Li, F.: On the Impact of Logistics Model on Retailers' Competitive Advantage. Journal of Business Economics 1, 14–21 (2009)
5. Yuan, S.-N.: Analysis on the Status of Retail chain logistics outsourcing. Manager Journal 7, 75 (2010)
6. Chen, Y.-L.: Analysis on the Retail chain logistics outsourcing. Science & Technology Economy Market 7, 88–89 (2009)
7. Yang, D.-H.: Motives, Determinants and Impacts of Outsourcing: A Summary of Research. Economic Management 2, 51–56 (2008)

The Public Crisis Management in Micro-bloging Environment: Take the Case of Dealing with Governmental Affairs via Micro-blogs in China*

Zhengyu Xia[1], Qian Yu[2], Li Wang[2], and Zhan Xu[2]

[1] Institute of Media Science, Communication University of China, Chaoyang, Beijing, China
[2] Institute of Communication Studies, Communication University of China, Chaoyang, Beijing, China
xia_zhengyu@cuc.edu.cn

Abstract. Micro-blog, an applied model as the booming development of new media technology, has already transformed to a public opinion supervision stage in the social public affairs from a purely internet social interaction tool. This change has brought a paramount big chanllege to the governmental organization in information collecting, releasing and disseminating openly. This essay focuses on the specific situation of dealing with governmental affairs via micro-blog in today's China, its purpose is to make research about the roles which micro-blog has played in the public crisis management, and also find the deficiency in it, with suggestions given to the governmental organization's micro-blogs in dealing with the public crisis management.

Keywords: Micro-blog, public crisis management, governmental micro-blog.

1 Introduction

Micro-blog, a applied model of the internet, with its characteristics such as timely convenience, openess, highly interaction, has been broadly accepted by the netizens since 2010. As an internet development research report carried by CNNIC shows, the amount of the micro-bloging user has arrived 19.5 million until the end of June in 2011. As a tool from which people can obtain information, share and exchange multiple information with others, micro-blog is also gradually becoming an important channel for people to assemble public opionion and to show public feelings at the same time. Facing the fact that micro-blog has already transformed to a public opinion supervision stage in the social public affairs from a purely internet social interaction tool, the public crisis management via micro-blog has become a urgent agenda, in which need carry research. Since the first half of the 2011, there are a paramount large number of governmental microblog booming in Chinese society and these governmental micro-blogs have different classes, such as the organizations', the officials' or the polices'. Therefore, the new phenomenon with micro-blog's application in public crisis management, has proved effectively that dealing with governmental events via micro-blog has been a promising new model.

* This paper is funded by the project "Digital New Media Content Production, Integration, Operation and Regulation (2009)" of Beijing Municipal Special Fund for Cultural and Creative Industries.

D. Zeng (Ed.): Advances in Computer Science and Engineering, AISC 141, pp. 627–633.
springerlink.com

2 The Roles Which Micro-blog Play in Public Crisis Management

Public crisis management often be carried in some certain media environment, it is inevitable to be influenced and constrained by this media internally or externally. The new characters that micro-blog differentiate with other media forms, therefore, these predetermine that the public crisis management will has many changes in the micro-blogging environment. The analysis of the roles which micro-blog has played in the public crisis management is helpful in giving effective strategies about the new changes that will happen.

Role No.1: the source of crisis, the possibility of micro-blog to be the direct source of public crisis is strengthening.

Micro-blog's communication characteristics such as timely convenience, openess, highly interaction has made every user to be the information senders, therefore, the gate-keeping power carried by the traditional media authorities has been passed on to the vast users of micro-blog. The messages sended by netizens via micro-blog mostly have not been selected or filtered, so to a certain extent they can express their ideas and air their opinions freely and personally via micro-blogs. By the way, the information is transmitted horizontally and flowing freely in a relatively transparent information enviroment. As the figures showed by a research carried by Beijing internet media association, in the first half of 2011, micro-blog has surpass the BBS in public opinion, also has been second public sentiment source among all media. And micro-blog has occupied 18.8% hot pot incidents' sources from the internet. Before 2011, the top three main internet public opinion carriers are BBS, blog and the news follow-up on internet, however, in 2011, micro-blog has become the first choice for people to disclose the explosive and unusual information as its booming rapidly. As the figures showed by a public opinion monitor, there are 15 affairs which first happen in micro-blog among the top 80 internet public opinion in the first half of 2011, for instance, *GuoMeimei's gossip, micblog broadcasting a hygiene official's gossip in SuYang JiangSu, the explosion affair in FuZhou JiangXi, the burglary in Forbidden City, the helping young beggers campaign, the elopement gossip of the investor WangGongquan* and so on. All the above affairs are so called public crisis affairs. Therefore, it can be seen that micro-blogs have played the role of negative incentive in the emerging and explosive phase of the public crisis affairs. What's more, with the viral fission track disseminative characteristic, micro-blogs will accelerate the public crisis affairs to erupt rapidly.

Role No.2: the catalyst of crisis, micro-blog has played a driving role in dissemination of the public crisis affairs.

Compared with BBS, blog or other new media patterns, micro-blog has the overwhelmingly rapid speed in both message releasing and message disseminating phases, which has make a breakthrough in the time and space dimension limitation of the information disseminating between the traditional media and new media. According to the internet public opinion analysis report carried by Beijing internet media association, in 2011, almost the hot pot affairs happening in internet in China have been disseminated effectively by micro-blogs. The public opinions usually will be varied and enlarged by micro-blogs, and the leadership given by micro-blog in public opinion is becoming gradually obvious.

Take the case of the 'Forbidden City Scandal'. Since the May 9, in 2011, the 'Forbidden City' has released the burglary message in its official micro-blog, then the users of micro-blogs have participated in this affairs in the very first time. The netizens paid greatly durative attention to the every investigative phases carried by the Police, and the administrators of the Forbidden City Museum has made a very unwise move in dealing with the public opinion in this burglary affair, therefore this purely antique burglary has caused a series chain reaction from the public, developed into many interlinking public crisis affairs such as the Forbidden City's Typo scandal, Forbidden City's Clubhouse scandal, Forbidden City's dismissal scandal and so on.

The viral fission track disseminative characteristic of micro-blog has enlarged the disseminative range and the influence degree of public crisis affairs. Micro-blog is playing the role of catalyst in this phase, it stimulate the disseminative process of crisis affairs, strengthen the attention rate that the public paying on these affairs, also play a driving role in dissemination of the public crisis affairs.

Role No.3: the tool of crisis management, micro-blog has become an important tool in the management of public crisis communication.

In the new media environment, because of the instantnaeity and highly interaction that micro-blog has, governmental organizations can state their standpoints clearly and take a stand via micro-blog very instantly after the public crisis affairs happen. Besides, with micro-blog government organizations can pay attention to the feedback from the netizens and give corresponding answers to their questions at the very first time, which make effective two-way interactive communication with the public. These above make micro-blog has become one of the important information communication tools used by governmental organizations in the management of public crisis communication at the present day society.

With the purpose to explore the new interactive communication tool, the Governmental Information Office in Yunnan Province opened the first Chinese governmental micro-blog called 'Yunnan Micro-blog' in the Luoshiwang, Kunming traders assembling affairs. By using this micro-blog, the Governmental Information Office gave the summary explanation to the public about this affairs and calm down the query wind and wave immediately, occupying the superiority in messages sending and making the facts in front of the gossips. Besides, in the Shanxi earthquake gossip in February 21, in 2010, some netizens appealed to Seismological Bureau of Shanxi to open a specific micro-blog to send messages about the earthquake at the very first time, which will be helpful in refuting a rumor so as to maintain the social security environment. The appeal to the authorities and offices to open their official micro-blogs shows this new media form can produce positive influence for governmental organizations in the management of public crisis communication. What's more, in the '7.23' Yongwen railway crash outsized accident, micro-blog had demonstrated its value and positive influence fully in the public crisis affairs. Soon after the accident happened, the all levels governmental organization micro-blogs in Zhejiang Province had send the messages about the recue situation in accident scene, the source of damage and the tempo of investigation in accident. Except for been the first information source and the stage for seeking help, micro-blog also had made the most of its advantage in refuting rumors about the accident, warning the public about the deceptive messages, recommending unpaid blood donation. The huge communication influence that micro-blog had played in the accident rescue, which also obtained public acceptance among the query and discontented mood from the netizens.

3 The Deficiency of Micro-blogs in Public Crisis Management

In the public crisis management, micro-blog played an important role in the aspect of crisis source, catalyzator and solution. However, for some technological restriction, there also exists certain deficiency when using micro-blog to manage the public crisis.

The first one should be the fragmentation of information in micro-blog, which has contributed to the insufficiency of government's information management. The micro-blog is characterized as being published within 140 words, and it makes the content become less logic. In 1971, Herbert A. Simon, the famous economist, has pointed out that the receivers' attention is what the information consumes. This claim is also suitable for the communication of micro-blog. The fragmentation of information has put the government into a dilemma when encountering on public crisis and trying to establish the information response mechanism. During the whole process of crisis management, the fragmentation makes it more difficult for the government to find out related information. In consequence, the government finds it hard to disclose the truth in time and realize the public right to know.

Besides, the fragmentation of information has impeded the process of publishing important messages on crisis for the government. In micro-blog, the information can be mass, unstable and diversified, so that all the useful, meaningless and even dangerous messages would be poured into the information collection system. As a result, in such a fragmental communicating environment, the premise for the government to effectively manage public crisis is to intergrate the information of value and expose it to the media and public in a distinct and logic way.

In addition, the anonymity of internet user can lead to the lack of responsibility and erratic emotion, which makes the network discourse space filled with false and artifactitious information. During the whole process of crisis outbreak, all internet users can release information, true or false, by using micro-blog. Without efficient management, the false information would turn into internet rumour and become bigger obstacle for the crisis management.

The last but not the least, most micro-blogs run by government agencies have not achieved the object of managing public crisis as expected. In recent days, micro-blogs like this in China are acting as information publishing corner in one piece, while take less steps in interacting with other netizens or solving problems. Take the micro-blogs of officials as an example, it is a hard question to define its role: whether it should behave like a space to talk about work or private life. The officials have to be confronted with the conversion between their public role and personal identity, the difference between ordinary people and themselves in discourse representation space and freedom, which contributed to the improper use of micro-blogs and wasted its advantages for public crisis management.

Current incidents like the Guo Meimei and Red Cross gossip and the the burglary in Forbidden City have outbroken in micro-blog and aroused the netizens' highly query. Even though the related agencies have opened their micro-blogs, they did not give response to those queries actively in time, and the micro-blogs run by the government agencies are being suspicious as launching projects designed to building their own images. At the same time, Governmental micro-blogs research report of China published by Fudan University has revealed that till Mar. 2011, in the micro-blog of Renming, Sina and Tencent, the number of governmental micro-blogs which have a certain amount of fans and high publishing frequency comes to more than 400, and that of the government officials is more than 200. It can be found that the governmental micro-blogs have not influenced a lot in managing public crisis in China.

4 Coping Strategies in Public Crisis Management via Governmental Micro-blogs

As a new communication form, micro-blogs play more crucial role in public crisis management on one side, while on the other side, there also exists some shortages in it. In such a network environment, several strategies should be taken to keep the crisis information open to the public and maintain the social security in managing crisis via governmental micro-blogs.

First and foremost, the government should seize the initiative and timeliness in information disclosure. For the fragmentation of information in public crisis communication via micro-blogs, the government should have the initiative in hands to disclose related information, which includes the flitter of the false and selection of the true, and the release of the latter one to maintain the social stability. Specific operation, the government officials or related people should be recommended to use their own micro-blogs to establish an information disclosure system and deliver the messages to a maximum range. Moreover, since the client of micro-blogs has been upgraded from the traditional network to 3G mobile internet, the mobile phone can also be used as a decent platform to disclose information. With that device, the management of public crisis should be expedited.

In the end of June, 2011, Nanjing released Suggestions on Strengthening Governmental Micro-blogs Establishment, which claimed immediate micro-blogs publishment about the related information in the outbreak of emergency incidents and catastrophic events. Besides, the governmental micro-blogs should give quick and proper response to the topical news, report the progress of incidents in time and take coping strategies. In usual days, the government should provide service for the convenience of the public via micro-blogs. In consideration of the increasing influence of internet opinion leader, the government should pay more attention to them through the way of point-to-point or point-to-plane, combination of individual and gathering, online and offline, in order to collect special opinion and invite them to the hearing of livelihood issues.

In the micro-blog communication environment, the fragmentation of information have brought several negative impacts for the government to manage public crisis, such as the unclear sources, unsmooth communication, false information, narrow channel and defective rules and regulations. In the face of such negative issues, what the government agencies can do is to take advantages of the convenient, fast and interactive characteristics of micro-blogs, disclose important information actively and promptly in public crisis coping so that it can decrease the risk and regain the public trust to achieve the social stability and security.

In the second place, the government should ensure the accuracy in disclosing information and set up its authority. In the background of social democratization processing, the technological advantages of micro-blogs have made more people directly involved in the public decision through them, then to influence the public opinion. However, the messages transmitted by the common netizen are not powerful enough or even have serious hidden danger of false information which would arise internet rumor. If the government cannot master the voice in those areas and establish its own authority, then it will fail to grasp the real feedback and information of value, which could lead to the lost of initiative in managing public crisis events.

As the traditional and new media merging with each other together, the social information system are full of various opinion, most of which are noises. The claim of government is absolutely the most authoritative, especially in the public crisis management. Messages delivered by the government can prevent the deterioration of crisis and maintain the social stability. In the risk incubation and outburst period, the government agencies should provide effective information via micro-blogs, interact with public immediately and integrate the information accurately and comprehensively. As a result, the government can make full use of micro-blogs as a significant tool to supervise the public opinion, communicate with public, disclose important and precise information so that it can establish its authority and credibility.

By the end, a supervision mechanism could be established and perfect to enforce a benign interaction between the government and netizen through micro-blogs. Most emergency incidents are occurred in the social sensitive areas, if not being correctly coped with, they would easily attract public attention and even arouse additive effects in internet public opinion. What's more, the virus spreading of micro-blogs would enlarge the transmission range and sphere of influence of crisis events.

Consequently, take the new changes in the communication of public crisis into account, the government agencies should have an objective system to evaluate the public opinion and determine the situation. On this premise, it can take effective, scientific and institutionalized coping strategies and establish relevant management model. In the early period of crisis events, the parties to the case should be aware of people's ability to collect information and have a right evaluation of its own situation and the pros and cons. Avoiding passive, negative and arbitrary response, the government need to have insight of the real public opinion and take active, positive way to interact with public. In addition to this, the government agencies can also launch a internet press conference via micro-blogs, which enable them to answer questions from the netizen and interact with them as soon as possible.

5 Conclusion

As the new media technology are upgrading continuously, micro-blog has become one important of communication channels at the present society. The timely convenience, highly interaction and the viral fission track disseminative characteristic of micro-blog have made it play a more vital role in in the management of public crisis communication for governmental organizations. However, the management of public crisis communication in the micro-blogging environment also are facing with the problems leaded by the micro-blog message fragmentation which are paradoxically with governmental organizations dealing with public crisis affairs.

Nevertheless, in the trend that micro-blog has played a significant role in the management of public crisis communication, the governmental organizations should have the awareness to wipe off the fragments in releasing and dealing with the public crisis affairs via micro-blogs and make the most use of micro-blogs to select and manage information in order to implement the government affairs transparent, build a more soundly information communication system for governmental dealing with public crisis affairs ,with maintaining the society security.

References

1. Wu, Y.: Crisis Management. The Suzhou University Press (2005)
2. Du, G., Shao, H., Lu, Y.: The Management of Corporation Crisis Communication in New Media Environment. The Modern Communication Journal (2009)
3. Wang, J.: The Strategies for Public Crisis in Self-media Environment. The News Journalism Journal (2011)
4. Fu, P.: The Fragment Dilemma And Strategies for Government Information Openness. The Southeastern Communication Journal (2011)
5. The report of the micro-blog's characteristics and its users using. Beijing Internet Media Association, http://www.dratio.com/2010/0816/103613.html
6. Berrisford, S.: How will you respond to the information crisis? Strategic Communication Management (2005)
7. James, D.: When your company goes code blue. Marketing News (2000)

Long-Term Memory Properties in Oil Future Market and Its Fluctuation

Lu Deng

Central University of Financial and Economics, Beijing, 100081
denglu521@sina.com

Abstract. This article aims to find the long-memory characteristics in oil future market. We have adopted semi-parameter methods, ARFIMA model and FIGARCH model to make empirical analysis on the daily price, daily return rate and its absolute value series. The result shows that long-term memory property do exists in the oil future market especially its fluctuation. In particular, the FIGARCH model shows that the response of the investment risk to the shock is a long-term memory process and its influence may hold in a very long time and its impulse response function decline slowly with a hyperbolic rate.

Keywords: Oil Future Market, Fluctuation, Long Memory.

1 Introduction

Although trading in fuel oil futures has existed in China, trading covering petrol, diesel oil, crude oil futures is still a vacancy needs to be filled, which means China is in a passive and disadvantageous position over the issue of international oil price. Moreover, recent wild fluctuations of international crude oil prices bring great influence to our national economy. Therefore, a deep analysis of international oil futures pricing and predictability of fluctuation is beneficial to reduce bad impacts of our economy as well as improve the pricing mechanism of product oil.

In 1960s-1970s, efficient market hypothesis put forward by Fama become the main theory of Financial Economics. However, this random walk model of asset prices was criticized by both the theoretical and the practical work. In the earlier studies, Mittni and Rachev [1] analyzed that the series of return on assets share the character of "shape changes at the beginning and slow changes in the end". Recently, research on the long memory character of financial time series has been in the spotlight [2][3][4]. Mandela Block put forward the "Joseph Effect" of share prices which enlightens the development of two time series models, namely, ARFIMA model and FIGARCH model. It is these two models that change the research method of theory and practice horizon.

Based on a commentary and analysis of the documents concerned with this area [5][6], we found that long memory model has seldom been used in the research on oil futures prices, yield rate and its fluctuation. Therefore, we analyze the long-term memory of oil futures market by using GPH and RGSE parametric approach and

D. Zeng (Ed.): Advances in Computer Science and Engineering, AISC 141, pp. 635–640.
springer link.com

building ARFIMA model and FIGARCH model so as to testify "Joseph Effect "do exist in oil futures market.

2 Long Memory Modeling Method

The RGSE estimate is a Quasi-Maximum Likelihood Estimator (QMLE), and essentially a local Whittle estimator in frequency. Consider a covariance stationary process x_t, t=1,2,...,n. its spectral density $f(\omega)$ satisfies $f(\omega) \sim G\omega^{1-2H}$ when $\omega \to 0$, where $G \in (0,\infty), H \in (0,1)$, parameter H is called self-similarity parameter, and its relation with long memory parameter d is $\hat{H} \approx \hat{d} + 0.5$. The estimator of H maximizes the following objective function $R(H)$:

$$R(H) = \log \hat{G}(H) - (2H-1)\frac{1}{m}\sum_{1}^{m}\log \omega_j, \hat{G}(H) = \frac{1}{m}\sum_{1}^{m}\omega_j^{2H-1}I(\omega_j), j = 1...m\,. \tag{1}$$

Bandwidth m is no more than 1/2 of the samples, and under conditions somewhat stronger, we can derive $\sqrt{m}(\hat{H} - H) \to_d N(0,1/4)$.

ARFIMA model [7][8] is as follows, where ε_t is white noise, and all the roots of $\varphi(L) = 1 + \varphi_1 L + \cdots \varphi_p L^p$ and $\theta(L) = 1 + \theta_1 L + \cdots \theta_q L^q$ are outside the unit circle, $(1-L)^d$ is the fractional difference operator, defined as the integer power of L binomial expansion to infinite order.

$$\varphi(L)(1-L)^d (y_t - \mu) = \theta(L)\varepsilon_t\,. \tag{2}$$

Currently, for equation (2) of the ARFIMA model, the main estimation method is maximum likelihood estimation. The Exact Maximum Likelihood Estimator (EMLE) was used to simultaneously estimate all parameters of ARFIMA model.

Baillie, Bollerslev, Mikkelson [2] consider a long memory conditional variance process, that FIGARCH(p, d, q)process, the expression is as follows:

$$\{1-\beta(L)\}\sigma_t^2 = \omega + \{1-\beta(L) - \varphi(L)(1-L)^d\}\varepsilon_t^2 \tag{3}$$

where $\varphi(L)$ and $\beta(L)$ are polynomials on L,

$\beta(L) = \beta_1 L + \beta_2 L^2 + \cdots + \beta_p L^p, \varphi(L) = 1 - \varphi_1 L - \varphi_2 L^2 - \cdots - \varphi_q L^q$, satisfy $0 < d < 1$, the roots of $\varphi(L)$ and $[1-\beta(L)]$ are outside the unit circle.

In this paper, the quasi-maximum likelihood estimator QMLE is used to improve the effectiveness of estimators.

3 Data and Long Memory Analysis

This thesis selects daily oil futures price (from Monday to Friday) p_t from New York Stock Exchange with the time span ranging from Nov. 23, 1987 to Aug. 31, 2007.

The total sample size is 5160. We take the logarithm of p_t first, and then take its difference which is the daily return rate of oil futures r_t, use ar_t to represent the absolute series of this return rate.

3.1 Long-Term Memory Properties with GPH and RGSE Method

We use GPH and RGSE methods to estimate d. As to three series p_t, r_t , ar_t, Table 1 lists the GPH estimator of d, asymptotic standard deviation, OLS estimated standard deviation, and t statistic under different bandwidth $m=T^{power}$.

As can be seen from the table 1, estimates of d with three different bandwidths in p_t pass the significant test. The estimated values, similar to one and another, are very close to 1. Therefore, oil futures prices are close to random walk processes. RGSE to estimate d is 0.966 and very similar with the GPH estimator.

About r_t, GPH estimator is 10% significant with power = 0.5, and 5% significant with power = 0.55 and 0.525. Three Estimated value are all very close to -0.15. The return rate r_t can be seen as a moderate memory processes. As to the RGSE method, the estimator of d is -0.074, and the standard deviation is 0.016. Similar to the GPH estimator, it also shows moderate memory properties.

As to ar_t , viewed from the GPH estimator ,the three estimates of d in different bandwidths was all significant and very close to 0.5, Showing a strict long memory process, the series is stationary and mean reverted. RGSE method shows the same. It means that the fluctuation of daily return rate or the investment risks demonstrating significant long memory characteristics, which is predictable.

Table 1. The GPH estimator under three different bandwidths

power		0.5500	0.5250	0.5000
Estimated value	p_t	0.9753	0.9318	0.8882
	r_t	-0.1448	-0.1648	-0.1727
	ar_t	0.46	0.54	0.63
OLS standard deviation	p_t	0.0473	0.0502	0.0541
	r_t	0.0661	0.0805	0.0942
	ar_t	0.069	0.079	0.087
t statistic	p_t	20.62	18.56	16.42
	r_t	-2.1920	-2.0480	-1.8333
	ar_t	6.67	6.84	7.24

3.2 Establishment of Long Memory Models

From the above analysis, the oil daily return series has long memory feature. However, the general ARMA time series modeling can only reflect the short-term characteristics of series. Therefore, we consider applications of ARFIMA model in equation (2) to modeling r_t.

Table 2. The ARFIMA model of oil futures daily return rate

Variable	Parameter estimator	Standard deviation	t statistic	Q^2 (20)
d	-0.0668	0.0132	-5.06	
AR(1)	0.3168	0.0039	81.67	667.2
MA(1)	-0.3324	0.0033	-99.27	

Note: AR (1) represents the first order autoregressive term, MA (1) represents the first order moving average term, $Q^2(20)$ is the Q statistic of squares residuals with lag order 20.

Here, we focus on the fractional parameter and their corresponding short-term parameters. First, we removed the conditional mean which can be predicted with the market information and name it r_{1t} to represent the non-expected return rate. Meanwhile, it also avoids the trouble of estimating intercept. Using Rats6.2, the effective observation is 5159, 21 iterations converges, the likelihood function value is -1813.67. Estimates show (Table 2), AR(1) and MA(1) term are significant, respectively, as 0.3168 and -0.3324, the estimator of fractional parameter *d* is -0.067, so you can say that the series is stationary, and the estimated Value and GPH, RGSE are only little difference. In summary, the return rate series is a stationary, moderate memory processes with fractional parameter less than zero, so the response of the shock of innovation is hyperbolically decay to 0.

Furthermore, the $Q^2(20)$ statistical value is 667.2, which is significant at the 5% level. This indicates that there is still some autocorrelation in squared residuals and should be to further describe. Here, we choose to use FIGARCH model.

Table 3. FIGARCH model of daily return rate series

(*p*,*d*,*q*)	μ	ω	β_1	φ_1	β_2	φ_2	*d*	*Q*(20)	$Q^2(20)$
(1,0,1)	0.02	0.04	0.9322	0.063	—	—		19.7	37.67
	(0.0266)	(0.0148)	(0.0118)	(0.011)				(0.48)	(0.01)
(2,0,2)	0.029	0.046	0.13	0.097	0.76	0.0145		20.45	34.06
	(0.0277)	(0.017)	(0.039)	(0.02)	(0.04)	(0.017)		(0.43)	(0.025)
(1,1,0)	—	0.028	0.9339	—	—	—	1	19.34	38.6
		(0.0028)	(0.0027)				(—)	(0.5)	(0.007)
(1,*d*,0)	—	0.2	0.4224	—	—	—	0.488	20.2	30.3
		(0.06)	(0.063)				(0.059)	(0.45)	(0.065)

Note: The mean equation $r_t = \mu + \varepsilon_t$; $\varepsilon_t \sigma_t^{-1}$ satisfy the i.i.d. N(0, 1) and the number in parentheses is the significance of the test, same as follows.

When $d = 0$, the model becomes GARCH (*p*, *q*), When $d = 1$, the model becomes IGARCH (*p*, 1, *q*). In order to compare FIGARCH model and the traditional GARCH model, the text using QMLE to estimate four forms of model: ①FIGARCH (1, 0, 1) model, also GARCH(1,1) model; ②FIGARCH(2, 0, 2)model, also GARCH (2, 2) model; ③ FIGARCH(1, 1, 0)model, also IGARCH model; ④FIGARCH (1, *d*, 0) model. All the results are in Table 3, robust standard deviation estimator are given in parentheses. $Q(20)$and $Q^2(20)$ represent the standardized residuals and squared residuals in Q statistics with lag of 20, *signif* shows the probability greater than the critical value.

First of all, except the constant μ, the other parameters are all significant at the 5% level (This is because we have removed the mean of the series).Secondly, the $Q(20)$ in the four models are all greater than the critical value with *signif* greater than 5%, indicating that the model does not exist autocorrelation in residual.

Thirdly, by comparison of four models, the sum of GARCH (1,1) model coefficients is 1, indicating that the process shows strong IGARCH characteristics. Therefore, we establish IGARCH(1,1,0) model, and find that the two model are very similar. Further, we establish FIGARCH model. Obviously, FIGARCH (1, d, 0) model is better fitting the original data. It because only $Q^2(20)$ in FIGARCH model accept the null hypothesis, which means only FIGARCH (1, d, 0) model shows no conditional heteroskedasticity. It can be concluded that the response of volatility in return rate to the shock of innovation is a long memory process, the fractional parameter is 0.488.

Finally, as to ar_t series, we also established different forms of FIGARCH model to describe the variance of its conditional heteroscedasticity.

Table 4. The absolute value of oil futures daily return series FIGARCH model

(p,d,q)	ω	β_1	φ_1	β_2	φ_2	d	$Q(20)$	$Q^2(20)$
(1,0,1)	0.069 (0.003)	0.924 (0.0039)	0.07 (0.032)	—	—		36.9 (0.01)	36.36 (0.014)
(2,0,2)	0.085 (0.012)	0.122 (0.064)	0.096 (0.007)	0.75 (0.06)	0.019 (0.01)		39.14 (0.006)	32.05 (0.043)
(1,1,0)	0.029 (0.0028)	0.9331 (0.0027)	—	—	—	1.00 (−)	37.7 (0.009)	38.5 (0.008)
(1,d,0)	0.25 (0.047)	0.4122 (0.06)	—	—	—	0.479 (0.057)	40.2 (0.004)	29.78 (0.074)

Note: The mean equation $ar_t = \varepsilon_t$; constant term is not significant to be removed.

Table 4 shows the FIGARCH (p, d, q) model of daily oil futures prices estimated by QMLE. Robust standard deviations are given in parentheses. All parameters are significant at the 5% level. Besides, since the mean of the differential return rate is zero, it is no longer included in the estimated equation. $Q(20)$, Q^2 (20) and *signif* have the same meaning as above.

It can be seen, similar to that r_t series, the sum of coefficients in GARCH (1, 1) model is also very close to 1, the establishment of IGARCH (1, 1, 0) model estimation results also show that the parameters of the two models is very close. In addition, GARCH (2, 2) model is better compared with the above two, for its $Q^2(20)$ statistic is smaller which means it is better to describe the conditional heteroskedasticity. However, its probability to accept the null hypothesis is still smaller than 0.05.

Finally, we establish FIGARCH model. Estimation results show that the standardized residual of FIGARCH (1, d, 0) model is more close to white noise, and compared with other models, the residuals of the model no longer exists conditional heteroskedasticity, which can be demonstrated by $Q^2(20)$ value, which shows a greater probability than 0.05. So that FIGARCH model is better fitting the original data. The fractional estimator in FIGARCH model parameter is 0.479, indicating that similar to r_t series, the variance of ar_t series also has a long memory response to the shock of innovation. But the shock is not permanent, it is slowly decaying.

4 Conclusion

To sum up, this thesis gives a deep analysis of the Long memory property in the international oil futures market. We modeling and estimate the fractional parameter in oil futures' daily price, return rate and its absolute value series by adopting GPH and RGSE parametric approach and ARFIMA and FIGARCH model.

According to the estimated results, we found that daily oil futures price is very close to a random walk process, while its return rate series shows the characteristics of a moderate memory process. Meanwhile, the series of its fluctuation and its absolute value's return rate displays obvious long memory characteristics. It indicates that the response of the investment risks in oil futures market to the shock of innovation is a long memory process and its impulse response function decrease hyperbolically. Besides, research also finds that FIGARCH model is useful in capturing the fluctuation or long-term characteristics of investment risk.

From the above analysis, we come to the conclusion that return rate series of international oil futures has obvious long memory features, so traditional random-walk model is no long useful in describing future oil prices. In practice, we can adopt long memory model like FIGARCH to predict future oil price and its fluctuation so as to take proper measures to keep macro-economy development steady and sound.

References

1. Mittnik, S., Rachev, S.T.: Modeling asset Returns with alternative stable distribution. Econometric Review 12, 261–330 (1993)
2. Baillie, R.T., Bollerslev, T., Mikkelsen, H.O.: Fractionally integrated generalized autoregressive conditional heteroskedasticity. Journal of Econometrics 74, 3–30 (1996)
3. Ding, Z., Granger, C.W.J.: Modelling volatility persistence of speculative returns: A new approach. Journal of Econometrics 73, 185–215 (1996)
4. Andersen, T.G., Bollerslev, T., Diebold, F.X., Labys, P.: Modelling and forecasting realized volatility. Econometrica 71, 579–625 (2003)
5. Floros, C.: Modelling Volatility using GARCH Models: Evidence from Egypt and Israel. Middle Eastern Finance and Economics (2) (2008)
6. Kang, S.H.: Asymmetric Long Memory Feature in the Volatility of Asian Stock Markets. Asia-Pacific Journal of Financial Studies 35(5), 175–198 (2006)
7. Granger, C.W.J., Joyeux, R.: An introduction to long memory time series models and fractional differencing. Journal of Time Series Analysis I, 15–39 (1980)
8. Hosking, J.R.M.: Fractional differencing. Biometrika 68, 165–176 (1981)

The Development and Countermeasures of Circular Economy in Gansu Province

Wang Si-yuan[1,2] and Dong Yuan[2]

[1] College of Geography and Environmental Science, Northwest Normal University, Lanzhou, China, 730070

[2] Lanzhou University of Finance and Ecnomics, Lanzhou, China, 730020

wsy0846@sina.com

Abstract. With poor natural conditions, harsh ecological environment and weak economic foundation, those made very negative impact to the sustainable development of Gansu Province. In recent years, with economic development and population increase, the shortages of water, land, mineral and other resources would be a urgent problem. As a new eco-economy, circular economy can provide the best way to the sustainable development of the economic, society, resources and environments in Gansu. In the process of developing circular economy, the countermeasures to enhance sustainable development capacity gradually would be found.

Keywords: Circular economy, Gansu province, countermeasures.

1 Introduction

"Circular economy", it is made by the U.S. economist, K · Paulding in 1960s. In the mid 1990s, it emerged in China. All the academic communities defined the concept from different angles such as comprehensive utilization, environmental protection, technological paradigm, economic growth patterns and the perspective of the broad and narrow, etc. The fundamental purpose of circular economy is to require in the economic process to minimize the input of resources, and to avoid and reduce waste. Waste recycling only reduce the amount of final disposal of waste. Circular economy is follow the basic principles of the "3R" principle, namely "reduction, reuse, recycle." Reduction refers to reduce the quantity of material into the processes of production and consumption. Reuse refers to extend the time for Products and services. Recycle refers to try to reuse the waste in order to reduce the amount of final disposal. Priorities for waste disposal are: avoiding----- recycling --- --- final disposal.

The first to give full consideration to conserve resources, improve the unit production products to resource utilization, prevent and reduce the generation of waste; then to recycle the pollutants and wastes to put them back to the economic cycle; only when the avoidance and recycling can not be realized ,it would finally allow to be done waste treatment. Circular economy requires in accordance with laws of ecology to organize the process throughout the production, consumption and waste treatment, its nature is a kind of ecological economic.

D. Zeng (Ed.): Advances in Computer Science and Engineering, AISC 141, pp. 641–646.
springer link.com

2 The Plight of Sustainable Development in Gansu

First, in recent years, Gansu Province has made great development on the socio-economic and other undertakings, but due to most areas in ecologically fragile areas, poor anti-disturbance ability, low productivity, serious soil erosion, landslides, mud-rock flow, desertification, dust storms, together with economic development and population increase, the lack of resources such as water, land, mineral and other constraints on the economic and social development more and more seriously, the capacity for sustainable development is worrisome.

Second, the economic developments only depend on energy and resources, but actual situation are inefficient and extensive use of resources, per unit of output with large energy consumption. Processing and utilization of resources is still a "resources - products - waste" in a one-way extensive use patterns. As resource-based cities, Jinchang, Baiyin, Jiayuguan, Yumen, Yaojie are facing the situation of resource depletion. In Gansu province, the proven storage of coal are 8.6 billion tons, but the per capita is only 330 tons, that is 55% of the national per capita level. The proven storage of oil are 360 million tons, that is only 1.2% of the country's proven storage. the per capita water resources is 1150 cubic meters, only the half national average level. The 45 major mineral resources, except for some nonferrous metal resources of which the storage are relatively abundant, the iron, lead, zinc and other major mineral resources are seriously deficient. The industrial structure is mostly in energy and heavy, which means needing more industrial raw materials.

Third, on the whole, the weak capacity for sustainable development in Gansu is rooted in ally less developed economy, mainly as follows: the total economic output is small, the per capita level is low ,the self-development capacity is weak, the science and technology capability of independent innovation is not strong, economic growth is slow, the use of re-utilization is low, the environmental pollution is serious.

3 Developing Recycling Economy Is the only Way for Sustainable Development in Gansu

Based on the above analysis, to improve and enhance capacity for sustainable development in Gansu province must develop circular economy from the following aspects:

3.1 Promoting the Transformation of Economic Growth Mode

In Gansu, industrial system based on resource development that leading to high resource and energy consumption, it also brought the "bottlenecks “to economic and social development. Therefore, according to the "reduction, reuse, resource-based" principle, the way of lower input, high output, low consumption, less emissions, recycling, and sustainable is the best choice. Pushing forward the building of a resource-saving society and implementing energy, water, land and materials and other resource-conserving system, reducing the resources and energy consumption, improving resource utilization efficiency. Actively developing wind, solar and other clean energy and renewable energy, with special highlighting water conservation, accelerating the construction of water-saving society.

3.2 Changing the Concept of Development and Enhance the Innovation Capability of Technology

In Gansu province, increased investment is main power of economic growth, but the capacity of the independent scientific and technological innovation are insufficient. Consequently, we must continue to implement the " science and technology" strategy to further strengthen the strong economic and technological cooperation with universities, research institutes and enterprises in our country, and accelerate the transformation and application of advanced technological achievements, establish the system of enterprises as the mainstay, including industry, academia, research institutes. And also actively support a number of key enterprises to establish R & D center to develop products with independent intellectual property rights.

3.3 Increasing the Pilot and Typical Enterprises Model to Expand Their Influence

In the key industries and areas, industrial parks and cities , promoting circular economy pilot projects. Through these pilot projects, putting forward the circular economy model, major technology areas and major project areas, a comprehensive evaluation index system of circular economy, the policies and measures which improving the renewable resource recycling network and promoting the recycling of renewable resources. mading plan on circular economy model of constructing transforming the industrial parks and cities, fostering advanced models, in order to accelerate the development of circular economy to provide an example and reference.

4 The Status of Development of Circular Economy in Gansu Province

4.1 Circular Economy Pilot Projects Carried Out Extensively

In recent years, Gansu province has increased the intensity of the work from three levels to promote circular economy: First, to promote circular economy on the regional economy ,Jiayuguan City, Wuwei City, Baiyin City, Pingliang City and Xigu District of Lanzhou City as the pilot cities of promoting circular economy; Second, developing circular economy among symbiotic enterprises and inter-industry, organizing upstream and downstream enterprises to develop cascade recycling of industrial solid waste; Third, developing circular economy among the key enterprises, Jinchuan Company, Baiyin Company, Jiuquan Steel Group have embarked on the resource-based enterprises transforming to the integrated resources processing enterprises as the circular economy pilot models.

Case 1, Jiuquan Steel Company use the regenerative heating technology for comprehensively utilization of blast furnace gas, coke oven gas can be uesd to expand the urban civil supplies; through the production of recycled water systems, saving water nearly 35 million cubic meters each year.

Case 2, Wuwei Paper Company use the waste water recycling technology in 100,000 tons paper-making project construction, adopt the advanced multi-disc vacuum filter recycling processing technologies to realize the cycle of White Water 100% utilization in paper-making; use biological aerated filter technology to deal with gray water, the annual volume of recycled gray water are 5.29 million cubic meters, recycle rate reached 88%; use alkali recovery process technology to deal with black liquor, the annual volume of recycled alkali are more than 8900 tons.

Case 3, Baiyin Mineral Company generated lead and zinc mine tailings in the process of smelting, in addition to being used as raw materials for cement companies, also being used in other areas, such as TianYuan company use lead cast zinc dross recycling of tailings oxidation zinc to produce ultra-fine highly active zinc oxide, annual output achieved 3,000 tons, the sales income reached 45 million yuan; Tianxiang Chemical Building Materials company use mineral tailings to produce new wall materials, annual production amount nearly achieved 10 million tons and sales income reached 500 million.

4.2 The Difficulties Encountered

Gansu Province carried out a number of attempts in building conservation-minded society and developing recycling economy, but there are still many problems and difficulties as follows:

First, in the building of a conservation-oriented society, the understanding of the importance of energy resources to support the sustainable development strategy and its basic role is insufficient.

Second, there's a lack of relevant laws and supporting policies and measures. The country's circular economy development laws have not been issued.

Third, there's a paucity of incentives and funds of support. In energy-saving aspects the country has not yet projected relevant preferential policies. Although some preferential policies for comprehensive utilization of resources have been introduced, but mainly concentrated in cement, building materials, wall materials and a few other areas, and the preferential policies that can actually be enjoyed are minimal.

Fourth, the majority of financing projects face more difficulties in implement. Comprehensive utilization of resources, especially energy-saving projects, compared with the general common projects, social and environmental benefits are obvious, but the economic returns are relatively poor. In the current mechanism of market economy and business conditions, the enthusiasm for the majority of enterprises to invest in comprehensive utilization of energy resources is not very high. Combined with the limited financing channels and difficult bank loans and many other factors, make the project implement difficult.

Fifth, there is a need to further standardize the market order. At present, due to many factors, "low-cost illegal, but high-cost compliance" phenomenon to some extent still exists and there is an urgent need to further strengthen the supervision to effectively regulate them.

5 Development of Circular Economy Policy Ideas in Gansu

5.1 Speed Up the Construction of Policies and Study of Regulations and Formulate the Development of Circular Economy Planning

Accelerate the "Gansu Province to implement Cleaner Production Promotion Law," the "comprehensive utilization of resources in Gansu Province Regulations", "Gansu Province Waste Materials Recovery and Utilization Management Regulations" and other local laws and process of the legislative regulations. Under the guidance of scientific concept of development, speed up formulation of plans in energy efficiency, water conservation, comprehensive utilization and recycling of renewable resources, in some special key areas, to guide the work of the province's development of circular economy.

5.2 Establish a Strict Evaluation Index System

Researches are needed to develop the province's key industries and the main raw material consumption standards of per product unit, and standards for power consumption, water consumption, the rate of final disposal of waste, water cycle, utilization, management indicators such as integrated resource recovery system, and strict the management of market access. Prohibit construction and use of the eliminated technology, skills and equipment, strictly limit or prohibit the high-energy, high water consumption, high pollution and waste of resources in industrial development.

5.3 Promote the Technology Development, Popularization and Application in Key Areas

Focus on key industries, key enterprises, key city (region), industrial solid waste as well as new energy, renewable energy, resource conservation and alternative energy cascade utilization; extend the related industrial chain links, "zero Emission "technology, recycling technology, green re-manufacturing technology, cost reduction and other key technology. Organize research and development; introduce, digest and absorb a number of domestic and foreign advanced technologies, make breakthroughs in key technological bottlenecks.

5.4 Implement a Number of Key Projects of Recycling Economy

Focus on some key enterprises and high-energy-consuming enterprises such as Jinchuan Company, Jiuquan Steel Group, Silver Company, Golden-based group, Lanzhou petrochemical, and other 33 key enterprises in the work of energy-saving and comprehensive utilization of resources; focus on major cities such as Lanzhou, Jiayuguan, Tianshui, Jinchang in the use of wastewater recycling, garbage recycling and safe disposal of items; Focus on a number of new energy and renewable energy development and utilization projects, including the Jiuquan wind resources such as rich wind farm construction project area , solar water heaters and other solar energy equipment promotion items, nomadic areas and areas without electricity; promote the use of solar photovoltaic power projects, livestock manure, crop straw to produce biogas and other rural energy construction projects.

5.5 Make Full and Better Utilize the Country with the Support of the Preferential Policies for Environmental Protection Industry

Concerning cement, new wall materials enterprises that use fly ash, ore slag, slag, etc. more than the proportion of 30% and the product quality fits national standards, subtract the value-added tax; the costs for the enterprise's research and development of new environmental products, new technologies, new processes should be included in management fees. Expedite the implementation of the new energy contract management mechanisms to preferential policies to encourage other provinces, foreign and private capital to enter the new environment-protection industries and other fields.

6 Conclusion

Gansu is one of the old industrial bases in the country; mainly due to energy and raw material based heavy-duty industrial structure, the economic development depend a lot on energy and resources. In recent years, with the rapid economic and social development and accelerating process of industrialization, the province's facing increasingly prominent resource and environmental problems that caused a certain impact on the economic and social development. Faced with the grim situation of resources and the environment, to take the road of new industrialization, speed up the economic growth mode transformation, accelerate the construction of resource-saving and environment-friendly society and develop a recycling economy is the strategic choice of sustainable development of Gansu Province.

References

1. Yan, X.-B.: Discussion on the Development of Countermeasures for the Circular Economy. Sci-Tech Infromation Development & Economy (13), 85–87 (2009)
2. Gansu Statistics Bureau: Gansu Statistical Yearbook 2005. pp. 75–127. China Statistics Press (2005).
3. Qian, H.-T.: The only way for sustainable development to develop circular economy in the western rural regions. Theory (1), 84–85 (2009)
4. Shi, P.-J., Yang, Z.-B.: Evaluation on the Sustainable Development Capacity in Gansu Province. System Sciences and Comprehensive Studies in Agriculture (8), 301–306 (2007)
5. Song, L.: Sustainable development of circular economy and building a harmonious society. Journal of Beijing Normal University (Social Science Edition) (3) (2008)
6. Miao, H.: Analysis and reference of circular economy development pattern in developed countries. Special Zone Economy (7), 84–85 (2009)
7. Information on, http://finance.sina.com.cn

Using the Contingent Valuation Method (CVM) to Analysis the Tourist Value of Zhangye Danxia Landform

Wang Si-yuan[1,2] and Xu Zhong min[3]

[1] College of Geography and Environment Sciences, Northwest Normal University, Lanzhou 730070, Gansu, China
[2] Lanzhou University of Finance and Economics, Lanzhou Gansu, 730020
[3] Laboratory of Watershed Hydrology and Ecology, CAREERI, CAS, Lanzhou Gansu 730000, China

Abstract. According to the results, the respondents' WTP medium value of the non-use value to the Zhangye Danxia landform is 10 yuan / person, and the average is 12.52 yuan / person. The WTP by using the number of maximum likelihood estimates is 15.82 yuan / person. As the respondents who are from the major state-owned units, the non-use value of Zhangye Danxia landform is between 70.73-111.90 million yuan. Among the social factors such as income, gender, education and age, they are lack of the correlation to WTP. But such as occupations, the awareness of scenic spots, the preference of scenic spots, the concern about the conditions of the environment and the support for environmental restoration or protection are closely interrelated with WTA. There is a very significant impact on the value of WTP by occupation, income and the awareness of scenic spots. But the influence of age, gender, education, the awareness of scenic spots on the value of WTP is not remarkable. And the support for environmental restoration or protection are not relevant to the value of WTP. In addition, the regression analysis results showed that such as age, education, occupation, the awareness of scenic spots and environmental changes , these factors have a negative correlation with the value of WTP, and the preference is positively associated with WTP. This is identical to the results of the contingency table analysis.

Keywords: CVM (Contingent Valuation Method), Danxia tourism resources, contingency table analysis, WLS (Weighted Least Square Method).

1 Introduction

The tourism value of scenic resources is the economic value within of the currency of the performance. Scientific assessment the economic value of scenic resources and integrated into tourism development cost - benefit analysis into the system, It is well be beneficial promoting the scenic resources of scientific management and protecting the scenic tour, promoting sustained and healthy development. Scenic resources are a special kind of public resources, and therefore can not assess the value of assets using the general method. Currently the most widely used in practice was the travel cost method and the conditions of valuation method.

With the expanding of the conditions of the valuation method, the scope of the object from the beginning of the environmental items of leisure entertainment value extended

D. Zeng (Ed.): Advances in Computer Science and Engineering, AISC 141, pp. 647–654.
springer link.com

to a widely used assessment of the effectiveness of environmental improvement and environmental damage losses. This valuation method is applied to the Zhangye landform area of non-use value assessment, the use of the method to estimate the area of non-use values, and analysis in the valuation process of the social factors impact on willingness to pay.

2 Contingent Valuation Method and Research Progress

Contingent valuation method is one of the most effective way to assessment of ecological and environmental which non-market public goods. CVM is the subjective satisfaction of departure, and using utility maximization principle, so that respondents in the hypothetical market and the environment in response to an article on the maximum willingness to pay (maximum Willingness to Pay, WTP), or minimum willingness to accept compensation (minimum Willingness to Accept, WTA), and the use of econometric methods to calculate the value.

2.1 CVM Mainly Research Progress

CVM idea originally put forward by the Ciriacy-Wantrup[1], they realized that the soil erosion control measures will have a public goods nature of the "positive external benefits (extra market benefits)", but this benefit can not be directly measured by survey people's willingness to pay these benefits to evaluate these benefits. Davis in 1963, the first formal study of Maine forest land will be used CVM camping, hunting, entertainment value [2]. In the 1960s, people gradually realize that the two main non-use value, option value and existence value is, the total economic value of environmental resources is an important part, as was only a way to assess non-use values of the method, CVM quickly obtain a wide range of applications. After forty years of development, in the West, CVM is applied to various fields, in addition to recreational and aesthetic value, but also involves water[3], biodiversity[4], ecosystem restoration[5], culture and artand many other fields' valuation[6].

The application of CVM in developing countries is later than in Western countries, and initially mainly used in water supply and public health, then gradually applied to the tourism, recreation and national parks, and now extended to the surface water quality, health and biodiversity protection, etc.[7]. China in the last century, the introduction of CVM research methods, applied CVM also carried out some research, the first is Xue Yuan Changbai Mountain biodiversity conservation value of the research[8], the use of payment cards in the study guide technology. Xu Zhongmin, etc. using the CVM for Ejinaqi value of ecosystem restoration study, not only takes into account all possible deviations, and fractions of open and two questionnaires were compared [9,10],. Since the introduction of a relatively short time, our application of CVM in the environmental field study carried out in a preliminary stage.

2.2 The Economic Basis of CVM

CVM theoretical assumptions personal utility by market goods x, q, and environmental goods or services, the impact of s personal preferences, assuming that s the same personal preferences, the individual utility function can be expressed as u (x, q).

Income market in commodity prices, p y and the constraints of personal consumer utility (u) can be maximized by solving the equation (1) to achieve [11]:

$$\text{Max u (x, q)} \tag{2-1}$$

$s.t: \sum p_i x_i \leq y$ (i=1, 2, 3..., n is the type of goods for the market)

Using equation (2-1) can be obtained by a group demand function: $x_i = h_i(p, q, y)$ defined on the basis of the indirect utility function. Suppose p unchanged, environmental goods and services into q from q0 q1, if this change is an environmental improvement, the individual's utility will increase, that is $u_1 = v(p, q_1, y) \geq u_0 = v(p, q_0, y)$ Indirect utility function can be used to represent the utility to improve:

$$v(p, q_1, y - C) = v(p, q_0, y) \tag{2-2}$$

Where the change of equation (2-2) in compensation income of C means that when environmental improvements, from changes to q0 to q1, individuals wish to maintain the utility to pay the same amount of money. CVM approach is guided through the questionnaire survey respondents out of the compensation changes in C. Assumed to improve the environment, C, which means that the individual WTP. If environmental degradation, in order to effect changed, individuals need to increase the income of some C, which is a personal WTA.

3 Application Evaluation Criteria Valuation Method Scenic Tourist Area of Zhangye Danxia

3.1 Introduction to the Study Area[12]

Zhangye landform includes two groups that is north and south, the basin is located in Zhangye oasis in the middle, which the North group about 25 km from Zhangye city, north-south width of 1-5 km from east to west about 20 km. Group located in the Qilian Mountains in the south to Sunan silver Township Center, altitude 2000 meters north-south width of 5-10 km from east to west about 40 km. Zhangye landform wonders formed 600 million years ago, is located in Zhangye City Linze, Sunan County, an area of about 510 square kilometers. Here is the only hilly landscape and color landform complex area, is "China Geography" magazine China one of the most beautiful of the seven landform, is the great development potential of natural and cultural tourism resources. The Former Head of Geography Zhongshan University, China landform Tourism Development Research Association Honorary President Mr. Huang Jin visit many times , Then he arrived the conclusion that" China's first landform is Zhangye landform, Zhangye colored hills is China's first".

3.2 Analysis of Samples

The survey used face to face interview method, the sample includes Zhangye City issued six counties under the jurisdiction of the city, issued a total of 260 samples. The survey questionnaire response rate is high, 258 samples of feedback, the feedback was 99.2%, excluding missed and do not meet the standard responses after the 247 valid questionnaires, the effective rate of 95.5%. 247 respondents in the socio-economic characteristics are summarized below:

(1) Sex: Male 160; female 87. Man accounting for 65% of the total number.
(2) Age: 30 years of age 69, accounting for 28%; 31-50 years old and 151 people, 61%; 51 people over the age of 27, accounting for 11%. That means the sample structure is reasonable.
(3) Education: University degree or above 8, 3%; college degree 142, accounting for 57%; high school diploma and 73, accounting for 30%; 19 junior high school education, accounting for 8%; primary education, 5, accounting for 2 %. That means the sample groups, a higher cultural level.
(4) The level of understanding on landform: quite familiar with 55, accounting for 22%; have some knowledge of the 131 people, 53%; not many know there are 60 people, 25% do not know 1. That means most people to Landform have some knowledge.
(5) Area of the current situation of environmental degradation: the deterioration that has 63 people, 26%, that there is no deterioration of the 93 people, 37% do not know the situation of 91, accounting for 37%. Sample groups identify most people are more optimistic about the area's environmental conditions.
(6) Whether the implementation of agreed landform scenic ecological restoration and protection plan: the consent of the 207 people, accounting for 84%; do not agree with 40, accounting for 16%. That the vast majority of the sample group of people have environmentally conscious.

3.3 Analysis of Frequency of the Willingness to Pay and WTP Value Statistics

The recycling of 247 respondents willingness to pay and WTP values, statistics show that willingness to ensure the existence of individual landform sustainability have to pay 101 people, 41%. Table 1 according to the cumulative distribution, cumulative frequency analysis of the distribution closest to the median value is $ 10.

Table 1. Frequency of willingness to pay

WTP value	Number of people	Frequency	Cumulative frequency
1	6	0.06	0.06
2	6	0.06	0.12
3	4	0.04	0.16
5	13	0.13	0.29
6	1	0.01	0.30
8	4	0.04	0.34
9	1	0.01	0.35
10	29	0.29	0.64
20	8	0.08	0.72
30	2	0.02	0.74
50	12	0.12	0.86
100	12	0.12	0.98
400	2	0.02	1

From the Table 1, we can find most respondents (86%) focused on willingness to pay 50 yuan, the mean willingness to pay can be calculated as:

$$E(WTP) = \sum b_i p_i = 12.52$$

Willingness to pay distribution is more symmetrical, the middle of willingness to pay is 10.00 yuan. Therefore, the average willingness to pay (12.52 yuan) and the intermediate value (10.00 million) is very close.

3.4 Analysis of Socio-economic Factors Impact on the WTP Values

Table 2 is the use of contingency table methods, analysis of the impact of social factors on aggregate WTP values. Generally speaking, occupation, income level, to understand the extent of the WTP values are very significant impact; and age, sex, educational level, on the scenic state of understanding of other factors on WTP values the value of the impact is not significant; while the area to restore and protect whether to support, and WTP of pay no correlation with the value.

Table 2. The impact of the social factors on WTP values

Factors	X2	Freedom	Sig. level	correlation	WTP charact.description
AGE	14.42	8	0.07	conspicuous correlations	the following 30 years old is low, between the ages of 30 to 50 comparatively high
SEX	2.28	4	0.07	correlations	men pay more than the women
PRO	38.31	16	0.00	conspicuous correlations	Government workers and business unit worker pay higher
EDU	23.47	20	0.10	correlations	University degree pay comparatively high
Income level	29.5	16	0.02	conspicuous correlations	High earners pay more than the value of low-income
FAR	19.14	8	0.01	conspicuous correlations	more scenic know, the higher pay
PRE	20.39	16	0.20	correlations	people have been to scenic spot pay high
Scenic spot status	13.46	8	0.10	correlations	Understand the current environment deterioration of the scenic spot to pay a higher value
payment	5.74	12	0.93	Non-correlations	In cash donations to a conservation organization
Pay reason	17.11	8	0.03	conspicuous correlations	Ensure the permanent existence of more nature reserve more

3.5 Social Factors and the Regression Analysis for WTP

In order to determine the direction and the influence of the social factors in WTP, it is necessary to use regression analysis for relative factors . In considering the results of the contingency table above, regarding (PAY) as the dependent variable, SEX, AGE,

EDU, PRO,FAR, PRE, ENV, whether to support and protect (YES). Through the Breusch-pagan heteroscedastic inspection, found several data in different variance, must use weighted least squares regression to eliminate the influence of different variance. In considering heteroscedastic, which is weighted least squares in the questionnaire data regression, get the following regression results. In return, the WTP get the logarithmic processing. The results of the regression analysis show in table 3. At the same time, we can get regression equations for:

$$\begin{aligned} LnPAY &= 140.30 - 0.49AGE - 11.11EDU - 3.12PRO - 11.03FAR \\ &\quad + 5.35PRE - 18.51YES \end{aligned}$$

from the combination of regression analysis , the coefficient of factors such as , age, education level, profession, the extent of knowing, the environment changes are negative, that suggest these factors for willingness to pay are negative correlation; but the coefficient of their preference degree are positive ,which indicates it is positive correlation with the willingness to pay . The analysis is basically the same with of the chi-square results before .

Table 3. Regression analysis of impact of social factors on WTP

variable	coefficient
Constant	140.30(9.44)*
SEX	-5.59(-2.15)
AGE	-.49(-3.16)*
EDU	-11.11(-6.76)*
PRO	-3.12(-3.57)*
FAR	-11.03(-3.92)*
PRE	5.35(3.01)*
ENV	-3.70(-1.72)
YES	-18.51(-6.06)*
R2	0.90
R	0.89
Durbin-Watson	1.60
LnPAY	15.82

3.6 The Estimation of Danxia Landform

As a common human natural heritage, hang ye danxia landform is important danxia landform research base and scenic spot, it's related to the interests of local and national people even the world. So the scope of the total WTP value assessed is important . First , this research omitted foreign respondents . At the same time, the research of the respondents mainly concentrated in Zhangye.In this article, we use the population size of Zhangye to evaluate WTP gross. However, how to determine the scope of the estimation after a suitable samples of the population. Apparently, using the total population in the region are not reasonable enough, because part of the total population such as rural farmers, children, the old man, they obviously are not likely to pay, so, the total population of sample shall be based on the urban workers. This investigation focused on enterprise or business unit, they are the most representative respondents.

Considering the problems, this paper chooses three population sample in Zhangye to calculate WTP gross:(1) the state-owned economic unit worker (2) all the town worker, including urban collective institution, other economic units; (3) all the employees, including urban and rural town individual worker, rural collective and individual worker. According to the sampling survey of CVM , combined with samples of population, the gross of WTP show in table 4.

Table 4. The estimated value of WTP of Zhangye Danxia landform

ITEMS	Total number (people)	Median pay (ten thousand)	The mean willingness to pay($ten thousand)	The mean maximum likelihood(ten thousand)
State-owned economic units	70732	70.73	88.70	111.90
the town worker	21003	21.00	26.34	33.23
Urban-rural employment	662265	662.27	830.48	1047.70

4 The Main Conclusion

(1) For the recovery of 247 questionnaires willingness to pay and WTP value statistics show that,101 people willing to ensure protection of sustainable existence Danxia landform and individual pay expenses person ,41% . The most close to accumulative total of frequency distribution of a value analysis is 10 yuan. From the willingness to pay frequency distribution list, we can see most of the respondents (86%) have willingness to pay 50 yuan, the mean of willingness to pay 12.52 yuan. Using maximum likelihood estimation get willingness to pay 15.82 yuan. The approximate evaluations of willingness to pay are similar, which shows that the willingness to pay for the survey results are reliable.

(2) in general, profession, income level, and the extent of knowing are the very significantly influence in the value of WTP; And the age, sex, education level, the status of scenic spot are not significant effect on WTP value ; And whether to support to restore and protect the scenic spot has no correlation on WTP value.

(3) from the combination of regression analysis , the coefficient of factors such as ,age,education level, profession,the extent of knowing, the environment changes are negative,that suggest these factors for willingness to pay are negative correlation; but the coefficient of their preference degree are positive ,which indicates it is positive correlation with the willingness to pay . Theanalysis is basically the same with of the chi-square results before .

(4) due to the respondents mainly comes from the state-owned economic unit worker, and non-use values of danxia landform can be considered are between 70.73-1.119 million yuan.

5 Policy Suggestions

From the result, in order to better protect danxia landform and promote the further development of the social economy, the construction of infrastructure for scenic spot should be strengthen, the accessibility of the scenic spot should be increased. At the same time, we should fully develop unique scenic spot, with the local natural resources unique cultural resources, and create the first-class tourism products. In addition, we still should strengthen the publicity of tourism scenic spots, expand the scenic spot of social well-known, develop the more the tourism market.

References

1. Venkatachalam, L.: The contingent valuation method: a review. Environment Impact Assessment Review 24, 89–124 (2004)
2. Davis, R.K.: The Value of Big Game Hunting in a Private Forest. In: Truncations of the 29th North American Wildlife and Natural Resources Conference. Wildlife Management Institute, Washington, DC (1964)
3. Mitchell, R.C., Carson, R.T.: An Experiment in Measuring Willingness to Pay for Intrinsic Water Pollution Control Benefits. US Environmental Protection Agency Report (1981)
4. Nunes, P.A.L.D., van den Bergh, J.C.J.M.: Economic valuation of biodiversity: sense or nonsense? Ecological Economics 39, 203–222 (2001)
5. Loomis, J.B., Kent, P., Strange, L., et al.: Measuring the total economic value of restoring ecosystem services in an impaired river basin: Results from a contingent valuation survey. Ecological Economics 33(1), 103–107 (2000)
6. Thompson, E., Berger, M., Blomquist, G., Allen, S.: Valuing the arts: a contingent valuation approach. Journal of Cultural Economics 26, 87–113 (2002)
7. Whittington: Administering contingent valuation surveys in developing countries. World Development 26, 21–30 (1998)
8. Xue, D.: Non-use values assessment with Changbai Mountain Nature Reserve biodiversity. China Environmental Science 20(2), 141–145 (2000)
9. Xu, Z., Zhang, Z., Cheng, G.: The total of ecosystem restoration valuation with Ejinaqi. Geographical Science 57(1), 107–116 (2002)
10. Xu, Z., Zhang, Z., et al.: Open and closed conditions comparative study of valuation methods - in order to restore Ejin total economic value of ecosystem assessment. Ecology 23(9), 1841–1850 (2003)
11. Hanemann, W.M., Kanninen, B.: The statistical analysis of discrete-response CV dates. Department of Agriculture and Resources and Policy Division of Agriculture and Natural Resources. University of California at Berkeley. Working paper No.798 (1998)
12. China · Zhangye, http://www.zhangye.gov.cn
13. Li, Z., Pan, W.: Econometrics, 2nd edn. Higher Education Press, Beijing (2005)
14. Zhang, Z., Xu, Z., Long, A., Gongzeng, T.: Study to restore the value of ecosystem services assessment with Zhangye City in Heihe River Basin. Journal of Natural Resources 19(2), 885–893 (2004)

The Research in the Path on Cultivating the Cultural Identity of Hong Kong and Macau Students: According to the Analysis of Jinan University's Individual Case Survey

Ying Hong*, Yuexiong Huang, and Xiujun Wu

Jinan University, Guangzhou, Guangdong, 510632
thong168@qq.com

Abstract. This paper discusses the road of cultivating the cultural identity of the students who come from Hong Kong and Macau. As the modern society supporting the cultural globalization and diversification, the cultural identity seems to be a symbolic value which has a power to unite people. And on the realistic level it becomes a foundation to encourage the civilization share, value consensus and spirit co-build. China is an ancient civilized country and the culture has extremely complexity and layering in both sides of diachronic and synchronic, which raises the depth requirement of cultural identity. Through the multi-theme and large scale cases research within the Hong Kong and Macau students, we summarized some key factors of cultural identity in such group. Base on the research, we designed some focused methods for cultural identity cultivating, which can supply several useful, idiographic suggestion of educational countermeasure on the macro level cultural identity.

Keywords: cultural identity, Hong Kong and Macau students, cultivating path.

By fostering the cultural identity of Hong Kong and Macau Students to maintain social harmony and development, cultural integration, acceptance and implementation of policy, economic development of mutual support, we hope can cultivating their ability for mutual understanding and recognition of different cultures re-examine and the ability to learn the essence of the traditional nation in the history of the political separation of the environment. Jinan University strives to cultivate individuals with a spirit of patriotism and multicultural heritage of high-quality.

Effective analysis and research of cultural identity of Hong Kong and Macau students based on the documents' analysis, the use of questionnaires and interview methods, and SPSS software are used to analyze the status of the cultural identity

---

* Ying Hong (1977-), research assistant in Jinan University, Research: university education informatization.

This article is a special task humanities and social science research projects of the Ministry of Education in 2010: Research of cultural identity of Hong Kong and Macao, phase of research results).

D. Zeng (Ed.): Advances in Computer Science and Engineering, AISC 141, pp. 655–663.
springer link.com

within Hong Kong and Macau students. For the status quo, workable and effective nurturing ways are proposed based on the educational experience of many-years. We should Cherish our own history, because protecting literature and national language which carry the traditional thoughts and culture, inheriting the tradition of national festivals and promoting its cultural implication, preserving the material and non-material cultural relics as well as the historical heritage which reflects the national culture are containing the most basic values of our nation. As these are all the most meaningful things for the Chinese nation's cultural identity. At the same time we should strengthen the culture of education (especially history, language and national conditions), regard the Chinese cultural heritage and identity as an important goal.

1 Add Conceptual Knowledge According to the Person's Cognitive Order

With China's broad cultural knowledge and awareness of the initial completed, we teach students to appreciate China's development and civilization on the perspective view. By arranging the general education curriculums, the students will acquire conceptual and historical knowledge. If there is any confusion on the concept and concept of "covering some truth", the students will be unaware of who to follow and what to do, what's more they would lost the confidence of Chinese culture. When the students stand on our territory, we have the opportunity to tell them how his country is like!

The 16th National Congress of the CPC report have defined the national spirit of patriotism, unity and peace-loving, industrious and brave, self-improvement and hardworking , which reveals that the national spirit with patriotism as the core contains three dimensions of a national consciousness, cultural identity and civic character. Based on the consensus that cultural identity is the foundation to achieve national recognition, and the way to enhance the national cohesion, school build a patriotic education model which set the culture identity as the base and national education as means.With the creation of "Introduction to Chinese traditional culture", "Outline of Modern China", "Introduction to Contemporary China", "Life cultivation and legal basis "(including Hong Kong and Macau basic Law) such four rational structure, functional complementation, a relatively stable national education curriculums, the model make "Diversification and Unity; Harmonization and Difference" ideas as the starting point, and the national awareness, cultural identity and civic character as the content, to help students achieve a consensus on the intellectual, emotional resonance, a common pursuit of the will of the cultural consciousness, as well as the national spirit and patriotism[1] Meanwhile, university also follows four modules, which is "multi-cultural module", "literary and artistic module", "Philosophy of History module", "Technology and social, political module" to set up a large number of bond elective courses. For instance, multi-cultural segment includes the "Christian and Western cultures," "World Heritage", "American language and culture", etc. Through these courses, the school aims to enable Hong Kong and

[1] Shuzhuo Jiang, The conditions of cultural identity-based education model to explore [N]. Guangming Daily, November 22, 2008.

Macau students with profound Chinese culture, a broad round of global vision, a noble aesthetics and critical thinking, of which the knowledge-based cognitive to understand and grasp is the main point.

History is the nation's memory, is the cohesion core of the national spirit. In Jinan University, Hong Kong, Macau, Taiwan students more, Macau, Hong Kong, respectively, have long been occupied by Portugal and the United Kingdom, Taiwan has 50 years of Japanese occupation, with both sides across the Straits in the state of hostility and isolation. Therefore, to the non-mainland students of which Hong Kong, Macau, Taiwan students are the majority, the modern history of China and Hong Kong, Macau, and Taiwan's modern history education is particularly important.The school will guide them through the history of education to fully understand the true face and evil intentions of the specific groups' "destroy the family, we must first delete its history" (the so-called "desinification") , to enhance their motherland's, national identity and emotion, to support reunification of the motherland, to make the determination of becoming the patriotic people who contribute to promote Hong Kong and Macau's prosperity and stability, as opposed to Taiwan independence, as well as the reunification of the motherland. In school, a specific content is the study of "one country two systems" national policy, and the Basic Law of Hong Kong and Macau Special Administrative Region in knowledge learning. Before the return of Hong Kong and Macau, legal system in mainland China and Hong Kong and Macau legal system does not have organic, interactive relationship; after the Return, in the level of jurisprudence, Hong Kong and Macau have been transferred to the legal validity of the Chinese Constitution as the final normalization.[2] Therefore, we should systematically and carefully guide and direct them to learn and understand the Basic Law, so they can really understand that the Hong Kong and Macau Basic Law, taking into account the interests of all sectors of society, not only retains the tested and effective components which the existing political system in Hong Kong and Macau have, but also fully embodies the requirements of a democratic system which obey the development of gradual and orderly manner in the region. All is conducive to maintaining Hong Kong and Macau social stability and economic prosperity.[3] With this understanding among these Hong Kong and Macao students, they will consciously or unconsciously guide and influence their parents, family and friends, so that letting more people understand and observe the law. At last it will make a contribution that cannot be underestimated for the prosperity and stability of Hong Kong and Macao.

Non-mainland students' cultural formation is easy to lose focus. With continued acceptance of the local political and cultural propaganda, they live in their own residence for a long time and the main idea has formed the set. For example, students in Hong Kong and Macao live permanently to Chinese traditional culture as the female parent culture, but also compatible with Western culture environment. As a

---

[2] Hongyi Chen. Law to reflect the return of Hong Kong in 1997 [A]. The rule of law, the spirit of the Enlightenment and the modern law [C]. Beijing: China University of Political Science Press, 1998.

[3] Bangguo Wu. In the ninth meeting of the Tenth National People's Congress speech [R]. 2004-04-26,2010-10-03 http://www.locpg.gov.cn/big5/gjldrnxg/wubangguo/200702/t20070225_1899.asp

consequence, cultural diversity promotes the diversity of their ideas, but it obscures the nature of their understanding of the female parent culture, resulting to loss the focus of the orientation in the choice of traditional Chinese culture. Our approach is that emerge the development of Chinese history into the curriculums, rather than dating only from the point of view to explain the cultural achievements of the era like other similar programs; the main purpose of the courses is to enable students to master the method of analysis and integration. Therefore, the courses make the traditional Chinese way of thinking and cultural context as the main spiritual structure, with inserting or enumerating a large number of historical facts and cultural phenomenon, so as to allow students to understand the main spirit of traditional Chinese culture and practical significance in just semester's system learning. When students acquire a lot of fidelity information which owns to female cultures from the educational mainstream channels, their understanding of the parent culture will become clearer.

2 Cultural Expansion According to the Individual Living Environment

"The Founding of a Republic", "Bodyguards and Assassins" , these films is such the model which meet the psychological characteristics of young people, and even a tendency to exaggerate and entertainment, through the interactive mode, to tell young people the history and to achieve the cultural communication –setting the lifeblood of historical development as a fundamental, by means of the way that young people loved, i.e. star idol groups, business operations, so that young people are willing to go to the cinema, willing to participate in commentaries, willing to doubt, willing to find a point of view from the book, which as the fact that complete our guide, and there is no fixed pattern of education, education is only the final result.

We can use such a number of media in the university to carry out some way to let the students enjoy movies on weekends and make them gain something useful from the film and television works in which we choose specifically. In fact, with the revival of many cultural folk villages and the re-ignition of the national traditions and culture, the holiday of Chinese New Year and a variety of traditional festivals make a lot of traditional culture and folklore in the recovery.

With the broad and deep impact of customs, The ways of people's thinking, values, and habits are all subject to its constraints. If the school can provide the individual with timely holiday experience through which students can feel the production of cultural festival atmosphere, understand the external environment in our region and their ancestor's simple emotions. If the school experience can provide the individual with timely holiday experience through which students can feel the production of cultural festival atmosphere, understand the external environment in our region and their ancestor's simple emotions. For example, Tomb Sweeping and Hakka's "An funeral" – "two burial" are involved in Chinese culture with rich sense of family and family ties which is the difference in emotional characteristics between China and the other countries, especially Western countries.

Patriotic, protecting the country are very profound truth but hard to be presented. However, through the united spirit and mutual help competition, the saying of no sweat no sweet and the culture of integrity to be respected is easy to be accepted and understood when eating dumplings or playing games, which also depicts a solid concept.

Well-known Chinese musicians and songwriter Chen Xiaoqi said, 'Guangdong, Hong Kong and Macau culture are important parts of Lingnan culture, which has a high degree of geographical and cultural identity. Therefore, the three places should have more exchanges in popular culture.[4] Jinan University's Motion Picture and Hip-hop Associations are such good culture media groups, with the share in the popular music and popular film and television, virtually discussion becomes inevitable, and mutual understanding of films' background of live life and culture becomes natural. Here, the advertently cultural assimilation and collision, with the exchange of mutual recognition and acceptance in the study. Jinan University promotes the international folk dance to students.

The international folk dance started in the early nineties is the mass of the original activities targeted at a certain organization, making participation become stronger with more colorful forms of content. Since then, Jinan University's International Folk Dance attracts various attentions, getting the recognition and love by other colleges and universities around the country.

3 Promotion of Cultural Fusion According to the Person's Emotion Need

Experiencing the inspiration under the same stars creates a sense of belonging to the motherland, enhancing our national pride, cohesion and solidarity as well. In fact, everyone will concern about the success of their friends or relatives as well as the people and events happened in their place of residence, and the continuing people and things at the place of their residence.

After the 5.12 earthquake, Jinan University's teachers and students held a fund-raising and memorial ceremony, where the atmosphere is quite sad. We can feel the suffering of our fellow citizens as well as the love of national government.

Afterward, a lot of Hong Kong and Macao students felt that they saw a different China. In Hong Kong, they had previously always see the fund-raising which they regard a good job. However, at Chinese mainland, the majority they hear from the domestic newspapers is that raising funds did not earmark. And now they realized the truth. Because they saw the classmates were anxiously waiting for news about Sichuan and pay attention to people suffered form the tragedy seems they are from one family. Some students even organized the volunteer groups to the disaster areas to help the victims during the summer holiday. Since some departments have students in Sichuan, their roommates accompanied them, waiting for the news, helping to find ways to solve the problems. All these phenomenon let them recognized and redefine mainland China, which is different from what they heard. They would believe in such a China.

[4] Zhe Wu. First "Pearl Forum" held in Guangdong, Hong Kong non-governmental forces to encourage direct dialogue [N]. 2009-09-11.2010-10-02.
http://www.pprd.org.cn/news/dongtai/200909/t20090911_66554.htm

In fact, cultural identity is a process, during which the people from a unified area create the common life, common experience, also the common culture, values and values of orientation. Without such process in life, paying no attention in nurturing of the emotional needs and feelings condensation in the process, we will lose a lot opportunities of education and sharing. We cannot and will not create a disaster, but we can make advantage of the disaster to unify our thinking, to generate the strength of cohesion and highlights our ability and building.

4 The Improvement of Responsibility According to People's Sense of Society

Non-mainland students have a good habit of complying with the rules. For the sake of helping students maintain good quality and good habits, we should provide them with a positive, supportive and healthy environment of the rules. The first step is to find common words to explain rules, in order to shorten the cultural distance, and pave roads for the cross-cultural education in the process of rule creation and implementation. When the rules of teaching education has become more structured, more in line with the law and more clear, when students accept within this rule consciously, during the process of expecting their own learning effect and value, they can determine their own learning and life goals, and enhance motivation in the accumulation of intrinsic rewards.

With the rapid development of the network, both Hong Kong and Macao students and mainland students will not help to tie them down in the virtual world. In the online world, the values, ethics, social values and historical values, which many people take for granted, have been made a mockery of ridicule style: the plot can be tampered, history can be juggled, lofty images can be derieded. The young immersed in the such a meaningless but easy exciting entertainment to comfort, or anesthesia the mind under the pressure of fatigue in the real life.

In February 2006, a newspaper in Hainan did a special investigation of on the "The most concerning things the college students care", with the "economic development" accounted for 85% of those surveyed, the "employment problem" followed with the account for 80% , "student loans for poor students" and "economic development prospects" tied for third, accounting for 65%, meanwhile the list ranked from No.4 to No.12 were "education fees", "City traffic problem", "medical reform", "urban environment", "cultural facilities", " social security problems", "health food" and "social assistance issues". This reflects the fact that contemporary college students not only concern with the reality and current world, but also the development of today's society and their own future with a relatively clear understanding, which is a certain typical.[5]

Any society group is bound by the law. As a social phenomenon, although education has a variety of forms and means, cultivating students with law education to prepare for participation in social life is an important way to achieve the social

[5] Hainan News Network. "Hainan's four universities' students most concerned topic "special findings [EB / OL]. 2006-3-1. 2010-10-3.
http://www.hainnu.edu.cn/yy/Article_Show3.asp?ArticleID=186

function of education. Hong Kong and Macao students accept better civic education and have relatively clear civic responsibility. Special law enforcement and government regulations are often used the method of execution, which will cause their attention and become the focus of their concern and the standard of behavior.

To help students maintain good quality and good habits, we should provide them with a positive, supportive and healthy environment of the rules, and the first step is to find words to explain the common rules, in order to shorten the cultural distance, and pave roads for the cross-cultural education in the process of rule creation and implementation. When the rules of teaching education has become more structured, more in line with the law and more clear, when students accept this rule consciously, during the process of expecting their own learning effect and value, they can determine their own learning and life goals, and enhance motivation in the accumulation of intrinsic rewards.

With the student scholarships election criteria-Comprehensive discussion and evaluation program and the requirement of students reading and the procedure of signing the Student Handbook, Jinan University strengthen students' sense of responsibility for their own behavior, to understand the criterion and to protect their rights and regulate their own behaviors better.

5 Start to Cultivate the People According to Their Own Development Needs

Reasonable idea of cultural identity should take personal development and improvement as the kernel value[6], regard the improvement of the ability and innovation as means and achieve the mutual shaping by the mix of an individual's unique identity and social identity (or collective identity) as a perfect way. The three arguments: Firstly, from the point of view at which it ought to be, the cultural identity's ultimate value is the liberty of individuals'. Human's and personal integrity is th Secondly, capacity is an important component of personality. The development of capabilities and innovation are embodiment of perfect personality and also the prerequisite of real freedom of the humanity. Thirdly, the formation of cultural identity depends mainly on the mutual shape between the individual identity and social identity (or collective identity).

Because the cultural identity not only has individual property, as Friedman said, cultural identity is carried in the blood of individuals, but also has a collective nature. At the same time, individual identity and social identity or collective identity is itself the sense of social interaction shape.[7] With individual development and capacity building in the sense of honor, students will develop individual human qualities, which in return make the love of the school building complex. The specific method includes ways to guide students to apply for research project and access to the opportunity of cultural tours, and so on.

---

[6] Zhiwen Deng of personality based on the [B]. Social psychology of development and innovation [C]. Wuhan: Central China Normal University Press, 2003:61-66.

[7] Chengbing Wang Contemporary interpretation of identity crisis of human science [M]. Beijing: China Social Sciences Publishing House, 2004:169.

With establishment of a learning organization (student group or workshop), we guide Hong Kong and Macao students' development: on the focus of academic science and technology culture, the front of the annual academic science and technology festival and the opportunity of the participation in the national and Guangdong Province "Challenge Cup" , College Students' Academic Work Competition, Business Plan Competitions, the school guide, support and organize students to carry out multi-channel academic science and technology activities to stimulate student's interest and love to learn, to improve the ability to for students to engage in academic training and exercise science and technology activities, resulting to the ability to to improve their science and technology awareness, adjusting knowledge structure, improving the quality of theoretical, ideological and moral cultivation.

The school also strongly support the academic science and technology community, making it as the extension and branches of classroom teaching, expanding the space of academic science and technology activities. Take Mr. Lee, who graduated from Information Science 06, for example, his basic knowledge, is more solid among the students in Hong Kong and Macao, with a strong desire for knowledge as well as strong hands. A lot of mainland students regard him as a role model. Based on his situation, he set up the Network Volunteer Teams of Jinan University and served as captain, which had effectively in attracting Hong Kong and Macao students and mainland students to learn through their own professional and academic discussion, receiving social recognition and personal advancement while providing service to others. In a series of activities, the professional salon, the Work Competition, mutual teaching of the technology and volunteer activities into the community, Mr. Lee fully demonstrated his own ability, while cultivating the leadership and communication skills except the professional ones, which made him and his classmates enjoyed this organization, and benefit from it. Their organization's team participated in the National Electronic Design Contest, and won the first prize in Guangdong Province. Because his outstanding skills and academic performance, finally he obtained the quota of graduate admission without test in Jinan University, maintaining good human relations with full gratitude to school's cultivation.

It is the best and fastest way to help the students to be the talented ones that discovering the advantages of their own strengths to help them achieve their desire to study and be strong. A good seed will quickly grow if there is the timely provision of the suitable environment, Jinan University attaches great importance to encourage and guide students to participate in after-school research projects sponsored by the school experts, enabling them develop new capabilities, practical work, rigorous style of study, hard work and careful attitude to enhance mutual collaborative team spirit and social responsibility in the direct research.

Hong Kong and Macau is China's a good window, and Hong Kong and Macao students' national identity affects the people of the world's understanding and recognition of China; Hong Kong's and Macao's policies also have radiation effects on Taiwan; the performance of Hong Kong and Macao students' policy thinking on the state will affect the future development of Hong Kong and Macao, as well as the understanding of Taiwan students, which is so very important.

As the so-called quasi-group of intellectuals which is known as a special spirit groups, they take the important task of the door's that opening of the public' ignorance on their shoulders. They possess advanced knowledge of culture with a keen social insight to gain world recognition by highlighting them in the public domain with the ideological power. Any student, when he becomes a real intellectual, he is definitely not just for themselves analyzing the situation, but for the situation of social groups - questioning the status quo, giving the critical voice to effectively guide the public's grasp of social reality. Also their profound criticism of torture will lead the public to the issue in depth and open the truth of the question matter.

Therefore, the college students should establish a benign concept of cultural identity. Not only for their own development needs, but also the call of social development and progress in reality.e truth expression of freedom.

Study on Chinese Culture Recognition of Students from Hong Kong, Macau– Analysis Based on Case Investigation in Jinan University*

Ying Hong[1], Xuefeng Wang[2], and Yuhuan Sun[1]

[1] Jinan University, Guangzhou, 510632
[2] South China Normal University, Guangzhou, 510631
thong168@qq.com

Abstract. Cultural globalization brings effect on modern society in two aspects: firstly, it brings the culture absolutism, a cultural pattern mainly focusing on capitalistic value and overriding other cultures in the name of "general generation"; secondly, culture relativism. Under this condition, every culture begins their retrospection on culture self-awareness so as to communicate with the world culture with independent national culture. However, two seemingly opposite role in a regional pattern of civilization, it is undoubtedly tangled and integration, creating a more obvious cultural theme, culture identity under the condition of multi-culture. This paper, taking students from Hong Kong and Macau in Jinan University, makes a comprehensive analysis on their cultural identity, idea and structure. According to systematical analysis, we get intuitive operational parameter to reason the reasonable appearance and further depth in structure of culture recognition. The conclusion we made can serve as the basic altitude pattern in exploring culture transformation.

Keywords: culture recognition, culture identity, Hong Kong, Macau, multi-culture.

This Culture identification is the reflect of national culture attitude, which formed and developed in a certain regional nation.[1] Culture recognition within a region indicates that individual and the whole group conform to some tradition (value, lifestyle esp.) and people take it as the revaluation and behavior standard. Not only is it the reflection of economic condition, but also reflects the relationship formed in the long-term contacting practice between various nation with the same region .More than the premise and base for intercourse with others ,culture recognition is the key for a harmony world. So we conclude that only by strengthen culture recognition and improve cooperation and communicating between different cultures, can glorious human culture come into being.

* This article is the Ministry of Education in 2010 a special task humanities and social science research projects: "cultural identity of Hong Kong and Macao health research" phase of research results.

[1] Ministry of Education, Humanities and Social Science Fund Project (Grant No.: 10JDSZ 3041).

D. Zeng (Ed.): Advances in Computer Science and Engineering, AISC 141, pp. 665–673.
springer link.com

Hong Kong and Macau are inalienable of the PRC, and people there share the same blood, same culture, same language, and same living habit with those on mainland. As Professor Zheng Xiaoyun puts it, they are inborn with the basic factor inculture recognition —recognition to culture noumenon, and recognition to concept such as family and clan, while recognition to other factors and to other culture may particularize itself due to the living conditions and historical vicissitude.

Students from Hong Kong and Macau, for being brought up in a society of capitalism and cultivated by western ideas, bear duplicity in their ideology, which means that while their thoughts are based on traditional Chinese culture, we still can tell that they are influenced by western thinking. Typically, they admire the idea of" benevolence, righteousness, courtesy, wisdom, loyalty" while still hold the value for individualism, the belief that money is supreme and concern for material gain.

This paper investigates common condition of culture recognition. To insure the reliability and random, we take means of questionnaire and case interview covering students from Hong Kong and Macau from all departments by random sampling. 800 questionnaires are issued covering students from Hong Kong and Macau in all majors, the majority of which is management and journalism according to the distribution of students. The survey is conducted under the instruction and bears no record of name of interviewee. 789 questionnaires making up98% of the total are recollected, and amount of male students is approximately equal to that of female students. So we believe the conclusion drawn from the survey can serve as a reflection of real thinking and sensation on culture recognition, culture recognition of students from Hong Kong and Macau, who are learning in Jinan University.

1 Familiarity with Chinese History and Culture

For most students involved in the survey, they are willing to accept traditional knowledge and get more information about their mother land, and are always doing so. So they have a good understanding of the doctrine of Confucius and Mencius, the Four Books, and tradition knowledge likewise.

From the questionnaires we can see that students do well in the doctrine of Confucius and Mencius (Q8), "Benevolence, righteousness, courtesy, wisdom, loyalty"(Q9), and the Four Books(Q10). And students not only comprehend the evolution of ancient culture but also express their own original ideas.

The result shows that students should be applauded in their perception of Chinese culture. There are two main reasons leading to this. Firstly, students who come back to mainland to continue their education usually receive more traditional cultural edification from family. Secondly, series courses on traditional Chinese culture lead them to a better understanding of this culture. According to the survey, senior occupy a 12% beyond those in junior years in accurate traditional cultural cognition.

Being proud to be a Chinese people, they are inclined to learn, to understand motherland, and to absorb the quintessence of traditional culture into their own need for improvement and richness, taking active part in group activities and culture experience from family, school, and society. Some students are especially good at traditional art and entertainments such as Chinese painting, dancing, kongfu, and 76% of whose without a skill also have the wish to learn(Q23). We can conclude that as long as there is a media to spread traditional culture, most of the students will join in. In addition, students show great interest in folk craft, food, literature, and conventions for they are a vivid description of traditional culture.

2 They All Concern about National Development

Culture selection is the select and adoption of culture itself in the period of turbulence. Any culture is a value system consisting of value significance which is the core of the culture and a form of structure. In any culture mode, there must exist its own production mode, political system, and the culture recognition system which can internalize into national psychology as well.

Value is the standard established to shape people's exercise, guide behavior, and modify the way people do things as well indicate the confirmation of taking people as subjectivity. As different culture orientation embody different culture mode, the understanding of the essence of culture determines the selection and reconstruction of culture. Culture selection, as a reflection of relationship between culture and subject needs , is also the guarantee as well as synthesis of subjective cognition and rational judgment, emotional experience and will.

It's hard work for all citizens to build their value system and cultural spirit depriving from traditional culture and even students on mainland find it confusing to judge and make a choice in the question of culture. Culture's significant influence on individual and nation is that when the nation is faced with a tough time, people will conduct consistent behavior in the belief of same cognition. The great rejoicing we feel after bidding successfully for the Olympic Games, the helping hands from all over the country after "5.12 earthquake" and rollsteinfluten in Zhouqu, are all vivid examples of our strong power of will and high spirit of solidarity and sense of community

When can culture serve the same role in politics? It's urgent to find out the developing rhythm so as to come into an agreement on country development.

In the questionnaires, we ask them to explain some items concerning traditional culture in the field of law, economy, academics, geology and technology in order to know how is their recognition of culture. The ranking comes into four grades, 10 point represents a well comprehension, and 6 knows it but not specific, 3 only heards of but not know, and 1 point means never know it completely. The statistical result is as following: (the bigger the number, the better is the recognition. Here the comparison of recognition id represented by comparison of data).

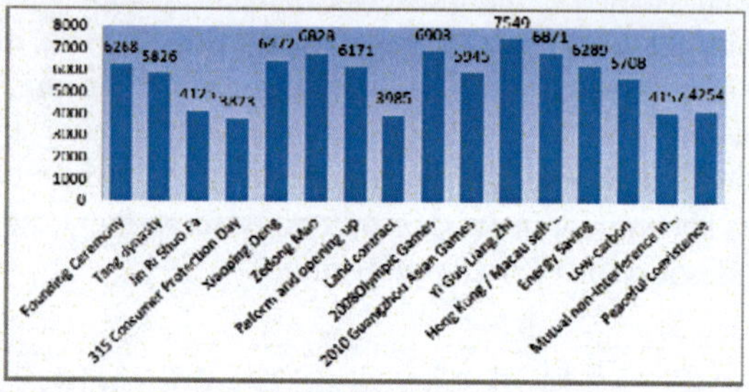

Fig. 1. Your familiarity with the content of the following

As is shown, the item "one policy, two systems" ranking the highest, and then comes the Olympic games. So we can say that they have clearly understanding of political system, and know much about the items indicating national strength. Many students serve as volunteer in Asian Games, which is a direct representation of sense of social responsibility and deep love for beloved country. At the same time, grades for "Hong Kong and Macau autonomy" is high mainly due to affection for homeland, strong awareness of autonomy and their concern about relationship between mainland and particular regions.

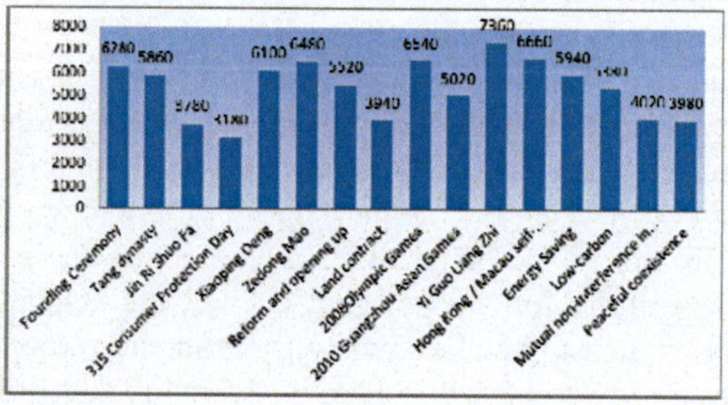

Fig. 2. Your familiarity with the content of the following

As is shown, recognition level of low grade students differ from the general condition.(seen from fig.1 and fig.2): there exists a subtle increase, which means school course helps strength understanding of traditional culture, especially in law. The concern for autonomy is the second while the average is the third, we can tell from this that experience on mainland strength their emotional connection with homeland and identity recognition as well.

When being asked "do you think it is responsibility for university students to protect traditional culture", 64% say yes. But when it comes to the term "traditional culture is now under impact, do you think it necessary to prevent that", 43% think it depends. Traditional cultural will co-exist with the new culture, but which will take the leading position will depend on which is more adaptable to the long term development for our nation. Compared with students in mainland, they are obviously more magnanimous and more frank for new cultures.

People influenced by capitalism ideology see China still lag behind and be exclusive in the seventy to eighty of 20th century. Only by communication can this misunderstanding be solved and judgment of culture be improved. People in Hong Kong and Macau still feel confused about many policies in China and they need more time to understand the changes taking place through China's development. Although at some important time, they take side of China due to their recognitions to identity as Chinese, they lack sufficient knowledge of Chinese political environment.

That vividly describes the fact that they have not completely habituated themselves to the school activities (especially those whose meaning is too externalized). Sometimes they cannot get desire to explore themselves or cannot find a way to what they are interested in. so we suggest that study on approachability and importance should attract more attention.

3 A Willingness to Take Part in Traditional Activities

Democratic education instill the students the idea that citizens should have a more active role and have a say in government's administration. Some takes no pride when being called as Chinese, for Chinese sometimes means littering, jumping the queue and break orders which is the fact to some degree that we cannot deny. So we have no reason to criticize them for not feeling proud for being as Chinese, but that we have to show them although there exist some wrongdoings, we are making progress and china is becoming more and more powerful in economy and society system and people are living a better life now.

Most activities students from Hong Kong and Macau take part in are superficial, not into real understanding of life in mainland. They are mostly active among students coming from same place and what they come across is knowledge on book and common daily life, lacking in the way or intuition to get a clearer understanding of mainland students and their life. For in their mind china is lagging out and exclusive so it's no doubt that they cannot feel the pride to study in mainland. A deeper communication and exchange of idea is thus significant to expand their mind and view. And it's also a problem that some students are reluctant to make efforts in team activities as they accentuating individualism and concentrate on themselves rather than the group as a whole. 29% of students show their appreciation and willingness to join while 64% say that although they like to but whether will take action will depends when asked about their attitude for activities meant to protect or spread the traditional culture.

That vividly describes the fact that they have not completely habituated themselves to the school activities (especially those whose meaning is too externalized). Sometimes they cannot get desire to explore themselves or cannot find a way to what they are interested in. so we suggest that study on approachability and importance should attract more attention.

4 Stronger Legal Thought

Grown up in Hong Kong and Macau, they have had their own peculiar thinking and value judging model. For example, it is commonly accepted that public constrains or social rules, as external formula, may finally turn out to be code of conduct guiding behaviors as the internal formula. The external formula influences the public through internalization, indicating that the social rules play an important role in restricting the individuals. That is why students from Hong Kong have a more mature idea of law and are more inclined to obey social rules.

Most students seem to be familiar with terms in Q.27 such as "legal report","3.15 customer right day", "Hong Kong and Macau autonomy", "one country two political systems" ,through which they keep pace with what is happening in china in both economic and legal sphere. All students interviewed held that they will turn to legal process for help when their rights are violated. "Legal report", the so-called law lunch for Chinese people, analyzing different kinds of cases and legal principles. Watching such programs on one hand shows students' concern for present situation of china, on the other hand suggests that they have a strong legal consciousness. In their opinion, mainland maintains a different legal system which needs improvement and people on the mainland need more modification on their behavior. As to "3.15 customer right day", they think too much attention is paid to the publicity but it just did not hit the mark. With personal experience, their understanding of the "one country, two political systems" is much more profound than mainland students just for that only when they are personally in the scene and combine idea with practice can man internalize it into personal need or principle.

5 More Civilized

We cite from the article of Xu Yiqi "comparison on relationship between traditional thinking mode and that in Hong Kong" that Hong Kong citizen have easier access to taking part in community affairs. Although Hong Kong is densely populated, citizens there don't have difficulty in taking part in community affairs mostly due to the responsibility for community. Compared with people in other cities, we can see that there is an obvious inclination for citizens of Hong Kong to care about political activity, to execute their right to vote, to call the police when the right is violated, or provide clues even they are not involved in, and to arrest a crime when necessary. We can conclude that citizens in Hong Kong have the consciousness to fulfill the responsibility for society and do some contribution as they can. When meeting someone in trouble such as an injured tourist, a traveler, and people who suffers from drugs or mantle disorder. Also they have the sense to make convenience for others in public, crowed streets or elevators and also the willingness to lend a helping hand to the majority. It is to some point influenced by the thinking "social responsibility" from west they take it granted to repay the society and are disposed to contribute time, money to charity, and denote blood and organs.

Growing-up experience matters in shaping the sense of identity in terms with what and to what degree students' identities go. More profound Influence by western culture explicitly explains why Hong Kong students attach more importance to democracy and standard and pay more attention to execution of rules in china than students from Macau. Example of good behavior and spread of uncivilized phenomenon make them halt between two opinions when it comes to their cognition of china.

Students' imitate behavior is an essential process of propagation and prevalence of sub-culture model. Some of the college students' cultural pattern, whether in basic necessities of life, such as food, clothing shelter and transportation ,first popular in the part of students, and then gradually expanding end in fashion and go deep into the students' identity view of the daily life and society. So we can tell that the teachers' and classmates' example plays an important role to the students.

Huang Qianyu comes from Hong Kong. He graduated from Jinan University in 2005 majoring in enterprise management. When he was in school, he served as Chair of the Students' Union, vice-president of Mission ACSF of Guangdong province, vice-president of NUS(National Union of Students. On July21th, 2005, he and his girlfriend who also studied in Jinan University began the one-year volunteer service in the slope town of baise region of Guangxi. They are the first Hong Kong students recruited in the volunteers service planned to western project in three years. Huang Qianyu was impressed by what a former volunteer said that "if you devote to the west for one year, you will harvest the whole life touched." When he talked about his choice, he always said: "everyone devotes to the west for one or two years, we can change the destiny of the west people." To see whether it differs from Hong Kong students to Macao students, we make a discussion among 100 students chosen randomly from different departments. The result is as following:

Table 1. Student type and their reaction

Descriptive Statistics

	Mean	Std. Deviation	N
Student Type	1.46	.501	100
Reaction	1.25	.435	100

We can see from the Table 1 that there are not significant discrepancy between both, the average is 1.46, if the figure is 1.5, that means the number is almost the same. The result indicates that students from both Hong Kong and Macau are easily impacted by companions' behavior.

Table 2 indicates that correlation coefficient is 0.07 when the probability of occurrence is 49.2% , whether the students are influenced has no inevitable relationship between the students' category. One can easily get affected by peers and model effect is efficient among peers. The behavior of their peers influence each other, the example role in between peers is effective.

Table 2. Correlation between student type and reaction

Correlations

		Student Type	Effect
Student Type	Pearson orrelation	1	.070
	Sig. (2-tailed)		.492
	N	100	100
Reaction	Pearson orrelation	.070	1
	Sig. (2-tailed)	.492	
	N	100	100

6 Strong Patriotic

According to the survey, 81% students feel not uncomfortable when been regarded as Chinese, which is quite different from before they must iterate they were from Hong Kong or Macau. Rather than the prosperity in economy and society, 33% of the interviewed students take pride as a Chinese for the booming in culture. Absolutely, students evaluate highly the value and affirmation of culture. and their demand for culture stronger. implying that students form Hong Kong or Macau are on the sameculture recognition base, they develop their own understanding and thinking about culture.

It may help to increase their sense of national pride if we conduct them to visit some well-known domestic enterprises, now more about with the development of high-tech industries, and participate in research of social problems. 80% students deem that it is progress made in legal system and talents cultivation that impress them most, and most of them now hold a rational attitude to China's development and change, can accept China's current shortage.

7 Higher Life Satisfaction

As to satisfaction survey, we adopt the "the questionnaire about research on universities all over the world" edited by Diener whose questionnaire is reliable and efficient approved by cross-culture research. Scholars in china, Zheng Xue and the fellows revised the subjective happiness index table according to the practical situation of Chinese students. The questionnaire consists of happiness index table, life satisfaction index table, positive emotion index table, and negative emotion index table. Subjective happiness index ranges from 1(very unhappy) to 9(very happy), covering 5 items. Likewise, life satisfaction index ranges from 1 (strongly object) to 7(approve) covering 5 items. Positive emotion index ranges from 1(never within a week) to 9(happens in the whole week) covering 6 items, and negative emotion index covers 8 items. In this research, the coefficient of consistency of the questionnaire was 0.715, the coefficient of consistency the Subjective happiness index is 0.762, while the life satisfaction index is 0.762, the Positive emotion index and negative emotion index is 0.810 and 0.803.

We calculate the average $\overline{X}$ and standard deviation SD of four dimensions of subjective happiness index. After comparison with norm, we get the result as showing below:

Table 3. Various dimensions of subjective satisfaction scale score

	Average	Standard deviation
Overall subjective satisfaction	28.67	5.65
Life satisfaction	21.28	6.47
Positive emotion scale	30.88	8.53
Negative emotion scale	29.58	10.04

The result indicates that the former three dimensions of students from Hong Kong and Macau is ranking higher than the average and the negative emotion index is below the average (36 score). We draw inclusion that students from Hong Kong and Macau are relatively satisfied with their life, holding a positive attitude, and are capable to get negative emotion in control and treat all sorts of problems in more tolerant attitude.

Grounded on survey data and literature, this paper draws a comprehensive analysis on reasons for discrepancy in students' culture recognition who comes from Hong Kong and Macau as well as a reflection on particularity on cultivating.

References

1. Zheng, X.: Cultural identity on the. China Social Sciences Publishing House, Beijing (2008)
2. Wang, Y., Wan, M.: Steel national identity and its Implications for China's national education. Comparative Education Research (8) (2004)
3. Zheng, F., Gao, X.: Economic Globalization and the cultivation of national identity. Educational Research and Experiment (3) (2005)

Brain Drains or Brain Gains in Physicians Emigration? Evidence from MENA and East European Countries*

Ahmed Driouchi[1], Cristina Boboc[2], Molk Kadiri[1], Emilia Ţiţan[2], and Nada Zouag[1]

[1] Institute of Economic Analysis & Prospective Studies (IEAPS)
Al Akhawayn University, Ifrane, Morocco
{A.Driouchi,N.Zouag,M.Kadiri}@aui.ma
[2] Bucharest Academy of Economic Studies, Department of Statistics and Econometrics,
15-17 Calea Dorobantilor, 010552, Bucharest, Romania
{cristina.boboc,emilia.titan}@csie.ase.Ro

Abstract. This is a contribution to the new economics of skilled labor emigration that focuses on the mobility of physicians inside European Union from sending MENA and East European countries. Economic models under risk neutrality and aversion are used. The findings show that the education could change significantly the results on the emigration benefices. The empirical results based on the available data on emigration by Docquier and Marfouk (2006, 2008) and Bhargava, Docquier and Moullan (2010) allowed further use of the model to understand the current trends in the emigration of physicians and the border between brain gain and brain drain. The countries included in the study are all exhibiting brain gain under 1991-2004 emigration data. Each country is encouraged to anticipate the likely effects of this emigration on the economy with the increase of health demand, the domestic wages and the increase in education capacity for medical doctors.

Keywords: Physicians' emigration, wages, human capital, risks.

1 Introduction

Physician migration is a complex and multifaceted phenomenon. Imbalances in the production of physicians lead to workforce shortages and surpluses. Many authors analyzed these imbalances. The movement of high skilled human capital from developing to developed countries can have many positive effects. Brain drain improves human capital through ex-ante motivations to be highly educated, creates positive externality on total factor productivity by helping technology diffusion from the receiving countries, decreases information risks and triggers more foreign direct investment inflows (Marchiori et al., 2010).

However, the empirical findings of Beine, Docquier and Özden (2009) suggest that education-based selection rules are likely to have moderate impact. Bhargava, Docquier and Moullan (2010) quantified the effects of physician emigration on human

* This paper is based on a report supported and financed by FEMISE Association.

D. Zeng (Ed.): Advances in Computer Science and Engineering, AISC 141, pp. 675–682.
springer link.com

development indicators in developing countries. The model used suggests a positive effect of migration prospects on medical training but the magnitude of this effect is too small to generate a net "brain gain" in the medical sector. These authors underline also that stopping physician brain drain has a small impact on human development. De la Croix and Docquier (2010) explore the complementarities between highly skilled emigration and poverty in developing countries through a model with human-capital accumulation, highly skilled migration and productivity. Their results show that two countries sharing the same characteristics can exhibit different impacts on poverty.

However, the analysis of the behavior of skilled labor denotes some degree of aversion towards risk that is not really taken into consideration by the literature on skilled labor migration. So, the analysis of labor decisions under risk is important in the process of identifying the optimal human capital and the optimal emigration rates for skilled labor (Schechter, 2005; Schechter, 2006). Other authors emphasized the relationships between the levels of initial wealth, income and levels of risk aversion (Rabin & Thaler, 2001; Chetty, 2003).

In the new member states of EU, Blanchflower et al. (2007) show that the propensity to migrate is correlated with income per capita, unemployment rates, and life satisfaction. Unhappiness with their lives, dissatisfaction with their salaries and working conditions, small number of good jobs and employment insecurity are the key reasons for Eastern Europeans to migrate in the Western European countries (Blanchflower and Lawton, 2008).

Studies by Nassar (2010) insist on the role of education received by the skilled migrants in relation to the transfers in the case of Egypt. Migration benefits the country in terms of receiving education and medical services that result from migrants' remittances. Sika (2010) finds that highly skilled emigration patterns from Egypt, to the OECD and the Gulf, contribute positively to the development process of Egypt. But, Ghoneim (2010) views that the deteriorating Egyptian education system producing less qualified labor lead to increasing emigration as a result of excess labor supply.

This paper deals with the migration of physicians into European Union from MENA and CEE countries. Economic models under risk neutrality and aversion are used. The model used in this paper is not different from the one developed in Driouchi et all (2010). After the underlying assumptions, the cases of risk neutrality and aversion are introduced with their related comparative statistics. The empirical results based on the available data on emigration by Bhargava, Docquier and Moullan (2010) allowed further use of the model to understand the current trends in the emigration of physicians and the border between brain gain and brain drain.

2 The Economic Model

The model used in this paper is not different from the one developed in Driouchi et al. (2010). The basic features of this model are from Stark et al., (2005). After the underlying assumptions, the cases of risk neutrality and aversion are introduced with their related comparative statics.

Labor productivity in a given economy is represented by β. It is equivalent to private returns to labor, as in Stark et al (2005). In the context of this model, β takes values β_S in the source and β_D in the destination countries. It is assumed here that emigration decisions are uniquely based on the levels of β that can be either β_D or β_S with respective probabilities *m* and *(1-m)*. In this model, each emigrant (given the static nature of the model) seeks a level of education *h* (considered as an individual investment in human capital) under the linear cost function *ch* with *c* being the unit cost of education. Furthermore, the level of education *h* is valued through a production function $g(h)=ah^{\gamma}$ (the output of human capital) where $0<\gamma<1$, $g'(h)>0$, $g''(h)<0$ and *a* is the talent of individuals. Each agent is consequently assumed to get (as a student) or to have the level of education *h* (after graduation) based on the maximization of an objective function $V(h)=\beta_S g(h)-ch$ in the absence of emigration (closed economy) and his expected utility in case of emigration (open economy). This latter option is the one considered in this paper where the model is accounting for risk neutrality and risk aversion.

The aggregate stock of human capital remaining in the country (non emigrant) is:

$$H_N=(1-m)N\left[\frac{c}{\gamma a\left[m(\beta_D-\beta_S)+\beta_S\right]}\right]^{\frac{1}{\gamma-1}}$$

Under absence of emigration (m=0), the stock of human capital in the country of origin is:

$$H_{N0}=N\left[\frac{c}{\gamma a\left[\beta_S\right]}\right]^{\frac{1}{\gamma-1}}$$

The relative domestic human capital remaining in the source country is:

$$\left(\frac{H_N}{H_{N0}}\right)=(1-m)\left[m\left({\beta_D}/{\beta_S}-1\right)+1\right]^{\frac{1}{1-\gamma}}$$

Depending on the level of *m*, relative wages and gamma, the changes in the relative human capital relative to gamma can be positive or negative.

In Driouchi et all (2010) it is proved that the maximal emigration rate among other results is given by: $m_N^*=\frac{\left[\beta_D-(2-\gamma)\beta_S\right]}{(\beta_D-\beta_S)(2-\gamma)}=\frac{\left[(\beta_D/\beta_S)-(2-\gamma)\right]}{((\beta_D/\beta_S)-1)(2-\gamma)}$

Again, m_N^* is a function of relative wages and of γ. Increases (respectively decreases) in γ leads to increases (decreases) in the relative net human capital gains. The optimal emigration rate increases (decreases respectively) with increases (decreases) of the relative wages (wage in destination relative to that at the origin). The optimal rate of emigration changes also in the same direction with changes in γ.

The level of m that equates H_N with H_0 is achieved with $m=0$, with $H_N=H_0$ and with $m=m^*$ because of the concavity of H_N.

This implies that m^{**} is given by the second zero of the following equation:

$$(1/(1-m)^{(1-\gamma)})=m\left({\beta_D}/{\beta_S}-1\right)+1$$

The values of m^{**} indicate how the economy enters the net brain drain phase. The higher is m^{**}, the better off is the economy as the brain drain occurs at higher probabilities of emigration. Lower m^{**} is an indication of higher brain drain and then the sensitiveness of the economy to the loss.

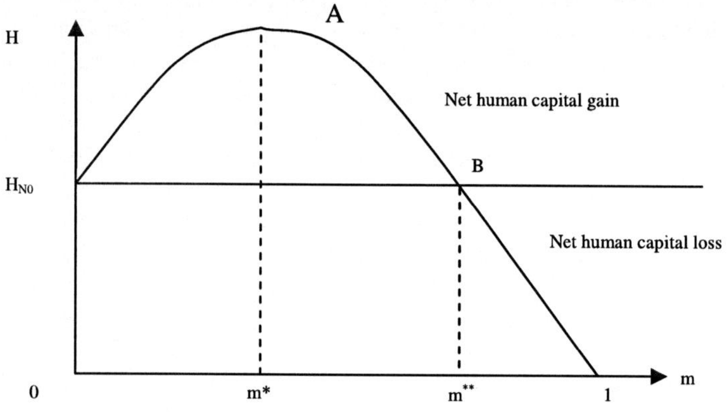

Fig. 1. The shape of H as a function of m

In case of risk aversion, a constant relative risk aversion (CRRA) function is used (Harrison & al, 2005) as $U(x)=\frac{x^{1-r}}{1-r}$ or $U(x)=\frac{x^{\alpha}}{\alpha}$, ($\alpha \in]0,1]$), where $\alpha = 1-r$ and r is the CRRA coefficient. The optimal value for the emigration rate is given by:

$$m^* = \frac{\left[\beta_D^{\alpha} - (2-\gamma\alpha)\beta_S^{\alpha}\right]}{(\beta_D^{\alpha} - \beta_S^{\alpha})(2-\gamma\alpha)}$$

The optimal emigration rate that can be obtained for the maximization of *H* appears to be directly related to most of the parameters of the problem. It has to be noted though that the numerator should be positive in order to meet the conditions imposed on *m*. This leads to the following restriction: $1 \le (2-\gamma\alpha) \le \frac{\beta_D^{\alpha}}{\beta_S^{\alpha}}$.

This condition implies that *(2-γα)* is the minimal value for the relative productivity or relative wage below which migration is not optimal.

The above results are shown in Fig. 1 where point A refers to the maximum of *H* attained at m^*. Point B corresponds to m^{**}[1] where H_R starts getting lower than H_{R0}[2].

Proposition 1: A net human capital gain (brain gain) results when the value of human capital, under different values of emigration rate, is superior to the value of the initial human capital under the absence of emigration. The human capital gain can reach a maximal value at m^* and returns to its initial value at m^{**}, while brain drain starts when human capital is lower than H_{R0}.

---

[1] m^{**} is the solution of the following equation: $(1-m)\left[m(\beta_D^{\alpha}-\beta_S^{\alpha})+\beta_S^{\alpha}\right]^{\frac{1}{1-\gamma\alpha}} = \left(\beta_S^{\alpha}\right)^{\frac{1}{1-\gamma\alpha}}$

[2] H_{R0} is the value of H_R attained at $m=0$ with $H_0 = N\left[\frac{c}{\gamma a^{\alpha}\beta_S^{\alpha}}\right]^{\frac{1}{\gamma\alpha-1}}$

Proposition 2: H/H_0 under relative risk aversion is lower than the level occurring under risk neutrality. This says that higher attainment in relative human capital is achieved under neutrality to risk.

Proposition 3: The optimal emigration rate (m^*) under relative risk aversion is lower than the level occurring under risk neutrality. This says that higher attainment in optimal emigration is reached under neutrality to risk.

3 Empirical Investigations

3.1 Descriptive Analysis

This analysis is based on the data on the emigration of physicians provided by A.Bhargava, F.Docquier and Y.Moullan (2010). Prior versions of this database are by F.Docquier and A.Marfouk (2006 and 2008). This is a new panel data on physicians' emigration rates (1991-2004). The main countries of destination are UK, USA, France, Canada, Germany, Belgium, Australia, Italy, Sweden, Switzerland and Austria.

The current empirical investigation focuses on some MENA countries: Morocco, Algeria, Tunisia, Libya, Egypt, Jordan, Syria, Turkey and Yemen and CEE countries: Bulgaria, Czech Republic, Estonia, Hungary, Latvia, Lithuania, Poland, Romania, Slovakia and Slovenia.

The total emigration rate related to all destinations ranges from 0.1% to 15%. Intermediate levels are recognized for the remaining countries with values between two and four percent.

While the trends expressed over the period 1991-2004 are constant or decreasing for most of the countries in the MENA region. In case of CEE countries, apart from Slovakia and Czech Republic we observe during the analysed period an increasing trend in emigration rate with 1% in average.

The emigration rate needs to be correlated with the domestic availability of medical doctors. When this latter variable is measured by the number of physicians per 1,000 people for each country, large variations appear between countries. Egypt attains a level above two doctors per 1,000 people. The other MENA countries are largely below two doctors per 1,000 people with most of them being between one and 1.5. Romania, Poland and Slovenia attained levels below three doctors per 1,000 people. The other CEE countries have up to three doctors per 1,000 people, Latvia and Lithuania had attaining levels up to four.

3.2 Testing for the Validity of the Theoretical Model

This conducted in different steps with discussion on different parameters where the first one is related to education, the second one to relative wages while the last one introduces the behavioral parameter related to risk aversion. The assessment of these three sets of parameters will allow for the calculation of the theoretical relative domestic capital in each economy. These values are then compared with the observed relative capital as it is shown the published data on emigration of medical doctors. The theoretical and observed means of the values relative to each country and over 1991-2004 are then compared.

The parameters γ and α are estimated from regression models where the dependent variable is the yearly number of graduates from medical schools (GMS) and the independent variable is the number of registered students per year in medical schools (RSPYMS) for the same countries and the same period of time. Lower γ is equivalent to lower productivity of medical education while higher γ is equivalent to higher productivity of medical schooling. By using these parameter estimates, country specific effects are discussed in the next subchapters.

In order to compute the wage in destination country compared to the wage in the source country, we have chosen the average annual gross earnings in health and social care in each source country compared to the average annual gross earnings in health and social care for developed countries in EU, the old members (EU15). In case of MENA countries we have considered a weighted relative wage of physicians in source compared to European and American destinations. Different National and International sources are used in order to find the medical doctors wages. Table 4 summarize the relative wages to be used in the calculations.

Estimates of the constant relative risk aversion (CRRA) coefficient appear to be varying throughout the economic literature but all estimations tend to be around 1. Chetty (2003) found that positive uncompensated wage elasticity can result in a CRRA coefficient below 1.25, while the labor supply literature indicates that CRRA coefficient is close to 1. Szpiro (1986) found that the degree of relative risk aversion (the inverse of the CRRA coefficient) is approximately 2 (meaning a CRRA of 0.5).

Halek and Eisenhower (2001) address the issue distinguishing between pure and speculative risks in order to understand risk aversion. Among their findings, they established that under both pure and speculative risks, individuals who already proved to be risk-takers by migrating across national borders are less risk averse compared with the native population. Also, unemployed people are more disposed to risk their current income for the possibility to double it (Haled and Eisenhower, 2001). Harrison, Lau and Rutström (2005) found that the Danish population exhibited constant risk aversion attitudes with coefficients around 0.45, 0.68 and 0.97. These attitudes are found to vary that with different socio-economic and demographic factors.

As assumed above, medical doctors study medicine in their countries of origin but have to make decisions to emigrate by the end of their studies or later.

The theoretical aggregate level of human capital that stays in the country is derived from the model with the introduction of the values of the parameter related to education γ and the values of β_D and β_S with the observed emigration rate m for each country. These parameters have been discussed in the previous two sections. Different values of the relative risk aversion coefficient are used including the one related to risk neutrality. The calculations provide the theoretical values of the relative net human capital that are given by the theoretical model (the theoretical H/H_0).

The observed relative human capital is obtained from the database considering the lowest value of m as corresponding to H_0 with H as related to the other values of the emigration rates (human capital is considered to be the variable Total number of physicians). This allows for the computation of the observed H/H_0. Then we have performed of a *t-test* to compare the means of the theoretical and observed relative human capital in each country. If there is no evidence for rejecting the hypothesis of equality of means for theoretical and observed values then the model can be validated. By comparing observed and theoretical relative human capital for each country, we have computed the theoretical relative human capital under risk neutrality and by using the estimated gamma and a gamma equal with 0.25, a fixed value of productivity of education. These values are compared with the relative human capital observed.

In most of CEE countries and Libya, the null hypothesis is rejected with the level of significance of 0.05, which means that the model could not be validated under risk neutrality (Bulgaria, Czech Republic, Lithuania, Hungary, Poland, Romania, Libya). In Slovakia and all the other MENA countries we have obtained that there is no evidence for rejecting the null hypothesis with a significance level of 0.05. Therefore, the models could be validated under risk neutrality.

Afterward we have computed the theoretical relative human capital under three risk aversion hypotheses, with alpha 0.33, 0.5 and 0.75 and by using the gamma estimated for each country. These values are compared with the relative human capital observed. In Bulgaria, Hungary, Poland, Romania and Slovakia, the model could be validated under risk aversion with alpha equal with 0.5. In Czech Republic, Lithuania and Slovakia there is no evidence for rejecting the null hypothesis with a significance level of 0.05 under risk aversion with alpha equal to 0.75. Therefore in Czech Republic, Lithuania and Slovakia the model could be validated under risk aversion with alpha equal to 0.75. We observed that the model for Slovakia could be validated under all hypotheses: risk neutrality, risk aversion with alpha equal to 0.5 and risk aversion with alpha equal to 0.75. Given that the highest p-value is obtained for the risk aversion model with alpha equal to 0.5, we will consider that Slovakia is a risk adverse country with alpha equal to 0.5.

In case of MENA countries only for Libya the test is validated under risk aversion with the constant risk aversion coefficient of 0.75.

Given these results, it can be inferred that the theoretical model used in this research does reproduce the data observed about the annual rates of emigration of medical doctors in all studied countries as they are given by the database used. This is in favor of using the theoretical model selected to discuss the cases of each country.

4 Discussion and Findings

Based on the results attained and given the level of emigration rate observed over the period 1991-2004, all countries included in this analysis appear to be benefiting from brain gains in the area of medical doctors. This means that domestic medical education is operating such that economies can still support the emigration of medical doctors at the current number of physicians in each country. Different results might be attained if the current staffing rates are further increased above the current observed trends. Under the current domestic educational system and with the current conditions of staffing, the countries under study appear to be enjoying brain gains. But, there are major variations expressed by the countries. The following set of graphs are the representations for each country, of the function $(1-m)\left[m\left(\left(\frac{\beta_D}{\beta_S}\right)^{\alpha}-1\right)+1\right]^{\frac{1}{1-\alpha\gamma}}-1$ in three cases of productivity of Health education (the estimated productivity for each country, gamma equals 0.1 and gamma equals 0.5).

The countries with higher number of physicians per 1000 people enter later (for higher values of emigration) in brain drain then the countries with lower number of physicians per 1000 people. This could be easily explained by the fact that the countries which dispose by a sufficient number of physicians, could "export" physicians to other countries which need this human resource.

This means that domestic medical education is operating such that economies can still support the emigration of medical doctors. Different results might be attained if the current staffing rates are further increased above the current observed trends.

References

1. Beine, M., Docquier, F., Rapoport, H.: Measuring International Skilled Migration: New Estimates Controlling for Age of Entry. The World Bank Economic Review 21(2), 249–254 (2007)
2. Bhargava, A., Docquier, F., Moullan, Y.: Modeling the Effect of Physician Emigration on Human Development, http://ssrn.com/abstract=1555775 (retrieved 2010)
3. Blanchflower, D.G., Saleheen, J., Shadforth, C.: The Impact of the Recent Migration from Eastern Europe on the UK Economy. IZA Discussion Paper No. 2615 (2007)
4. Chetty, R.: A New Method of Estimating Risk Aversion. National Bureau of Economic Research, Working paper 9988 (2003)
5. De la Croix, D., Docquier, F.: Do Brain Drain and Poverty Result from Coordination Failures? CReAM Discussion Paper Series 1009. Centre for Research and Analysis of Migration (CReAM). Department of Economics, University College London (2010)
6. Docquier, F., Marfouk, F.: International Migration by Education Attainment. In: Ozden, C., Schiff, M. (eds.) International Migration, Remittances & the Brain Drain. World Bank and Palgrave Macmillan, Washington (2006)
7. Docquier, F., Lohest, O., Marfouk, A.: Brain Drain in Developing Countries. The World Bank Economic Review, 1–26 (2007)
8. Docquier, F., Rapoport, H.: Documenting the Brain Drain of «la Crème de la Crème»: Three Case-Studies on International Migration at the Upper Tail of the Education Distribution. Discussion Papers - IRES (2009)
9. Driouchi, A., Kadiri, M.: Emigration of Skilled Labor under Risk Aversion: The Case of Medical Doctors from Middle Eastern and North African Economies. MPRA Paper 22810. University Library of Munich, Germany (2010)
10. Halek, M., Eisenhauer, J.G.: Demography of Risk Aversion. The Journal of Risk and Insurance 68(1), 1–24 (2001)
11. Harrison, G.W., Lau, M.I., Rutström, E.E.: Estimating Risk Attitudes in Denmark: A Field Experiment. Scandinavian Journal of Economics 109(2), 341–368 (2007)
12. Marchiori, L., Shen, I.-L., Docquier, F.: Brain Drain in Globalization: A General Equilibrium Analysis from the Sending Countries' Perspective. Work in Progress, Reprint Submitted to the Economic Journal (January 2010)
13. Nguyen, D.T.: Heterogenous Talent and Optimal Emigration. Paper Presented at the Conference on Growth, Development, and Macroeconomic Policy, December 17-18, Delhi, India (2004)
14. Rabin, M., Thaler, R.H.: Anomalies: Risk aversion. Journal of Economic Perspectives 15(1), 219–232 (2001)
15. Schechter, L.: Traditional Trust Measurement and the Risk Confound: An Experiment in Rural Paraguay. Journal of Economic Behavior and Organization 62(2), 72–292 (2007)
16. Stark, O., Casarico, A., Devillanova, C., Uebelmesser, S.: The New Economics of the Brain Drain: Mapping the Gains (2005)
17. Szipiro, G.: Measuring Risk Aversion: An alternative Approach. Review of Economics and Statistics 68(1), 156–159 (1986)

A New Probe into the Essence of Teaching

Zhiyuan Lü[1] and Zhiping Huang[2]

[1] Department of Physics, School of Science, Inner Mongolia University of Technology, Hohhot, Inner Mongolia 010051, China
[2] Hohhot Vocational College, Hohhot, Inner Mongolia 010050, China
zyuanlv@163.com

Abstract. The argument on essentialism and anti-essentialism in philosophy leads to a dilemma on the study of teaching essence. In order to get out of the dilemma, we may ignore the argument and just consider "essence" as epistemological faith and strategy. As teaching is a kind of purposeful social practice, the author summarized the essence of teaching from the aspects of practice and purpose, which is the cognition and cultivation for both teacher and students. This viewpoint of teaching essence could offer us new perspectives to many teaching problems.

Keywords: Essence, Teaching essence, Cognition, Cultivation.

1 Essence of Teaching

For several decades, scholars have been argued about the essence of teaching, but we have not got a satisfactory result today. Facing this situation, someone lost patience and think it may be a unanswered questions, while someone is trying to change the problem to approach the essence of teaching , that is: don't ask "what is teaching", instead we should ask "what is not teaching"; furthermore, someone denied the existence of "essence "by the reason of anti-essentialism in philosophy. They insist that thing's essence is multidimensional, thus the problem of teaching essence is cleared up. Although we haven't got the commonly-accepted answer, objectively speaking, by exploring the essence of teaching for years, we really deepen our understanding about teaching, and promote the development of the teaching reform. In this case, we should not stop the exploration of the essence of teaching. This means that we should admit the concept of "essence" and the existence of the teaching essence. But in order to avoid debate of essentialism and anti-essentialism in philosophy ,the author would like to regard the " essence " as an epistemological beliefs which is needed in theoretical construction -----making believe the " essence "is indeed, and explore it sincerely ,reflect a kind of human's pursuit of comprehensive and profound understanding things; On the other hand ," essence " provides us a strategy that find knowledge and organize knowledge, that is, through "one" to deduce "many", and organize "many".

So-called teaching essence is the inherent regularity that makes teaching become its own and distinguish with others. Teaching is a kind of objective practice, therefore, the

D. Zeng (Ed.): Advances in Computer Science and Engineering, AISC 141, pp. 683–687.
springer link.com

essence of teaching should be summarized both in practice level and objective level to show its uniqueness. The author insists that the essence of teaching is the cognition and cultivation for both teacher and students. Here cognitive and cultivation is both verb and noun. As verb, cognitive is the performance of study that get knowledge, and cultivation refers to one's change in body and mind, specifically, it means the individual growth in aspects of intelligence, ability, physical, emotion, attitudes, values , etc. In verb meaning, cognitive and cultivation reveal the essence of teaching in practice level. As noun, cognition and cultivation which means literacy and knowledge are the result that students and teachers get from teaching practice. They are used here to reveal the essence of teaching in purpose level.

Teaching means that teacher teach the student knowledge and student learn knowledge. For the student, learning knowledge is a cognitive process; for the teacher, teaching is also a cognitive process. The teachers' cognition refers to knowledge accumulation and preparing lessons before the class (preparing knowledge, preparing students, preparing teaching methods), it also refers to judge the class condition during the teaching process, and reflect teaching after class. In the visible cognitive process, invisible cultivation process is on the way quietly. Engels had expressed the idea like this, to understand the reason of one's current status, you'll have to go to his past life to find the answers. In fact, environment always play an important role in one's life, and its influence is objective and inevitable. Different from other practice, the cultivation of teaching is not only a natural result but also a conscious intervention result, either it is an individual hard work result. After a long period of study, these contents, such as learning content, learning method, communication between teachers and students, campus cultural and so on, will be reflected in the students' intelligence, non-intelligence and body. On the other hand, after long-term teaching, teacher would consciously or unconsciously get certain literacy, certain professional attitude, certain teaching style and special personality charm.

To sum up, with the cognitive behavior and desire, teacher and students are gradually changed in aspect of intelligence, non-intelligence and body, these change what we hope are spontaneous, self-conscious, inevitable. The cognition and cultivation of teacher and students reflect the uniqueness of teaching both in practice level and objective level. Different answer for the following question will be the dividing line between various schools: Which is more important? cognition or cultivation? What kind of knowledge should be taught prior in school? What kind of qualities should be cultivated prior in school? How to better achieve the cognition and cultivation?

2 Interpretation of the New View of Teaching Essence

Cognition and cultivation are both behavior and purpose. From the point of view of personal development, cognition and cultivation form a category. They promote each other and restrict mutually. Cognition develops a means for cultivation, also cognition process is a cultivation process, the content and depth of the cognition have influences on the cultivation content and process; conversely, cultivation provides support and power for cognition, also it influences the cognitive efficiency and cognitive contents choice. If one doesn't value the cultivation, cognition won't develop further and sustainable; while cultivation will be impossible, if one doesn't value the cognition.

Education is to train a person's career, cultivate people not only refers to teach knowledge and skill, but also to make student develop a certain character and abilities that educators expect, so that when external education stops, people will still develop actively and effectively towards the direction that educators previously expected. Teaching is the most important way of education, therefore, it ought to realize the dual task that imparts knowledge and skills and develops character and abilities. It is not a complete education, if we implement teaching only in the cognitive level and, not pay attention to students' body and mind, because the knowledge and skills cannot be automatically transformed into qualities and abilities that educators hope. Lack of the due qualities and abilities, education will depart from its original social mission.

In the teaching activities, the students' cognition mainly means learning knowledge and skills, and a research into study method. For the better cognition, research into study method which is often neglected in teaching is necessary. The cultivation of students mainly refers to a stable development that educators expect in the following aspect of intelligence, abilities, qualities, attitude, behavior and body. Certainly, cultivation is both a subtle influence from environment and an active effort from the student. Relatively speaking, cultivation is a more hidden feature of the teaching, so it is neglected some degree in current teaching, the student lack of innovative capacity, low humanities, social adaptability is not strong and other common issues are related to this. "Teaching" and "learning" is a pair of contradictions, "teaching" restricts the "learning", "teaching" is the principle contradiction. The new view of teaching essence defines the teaching essence as cognition and cultivation for both teacher and student, rather than the student one side, thus develop the full attention to teacher's role in the teaching.

Teachers' cognition mainly includes two aspects, one is to research knowledge deeply, so that the teacher can understand the knowledge comprehensive. The other is teaching research and exploring teaching laws, so that the teacher can find suitable teaching methods to improve the teaching qualities. The cultivation of teachers is the inevitable result of the long-term teaching, also is the result of active pursuit. The teachers' cultivation mainly lies in professional knowledge, professional attitude, teaching style, personality, etc. At present, the education's influence on students is significantly decline compared to the family and the society, one of the reasons is the teacher can't influence the students by their own learning literacy, personality charm. Obviously, the teachers' cognition and cultivation restricts the student's cognition and cultivation.

3 Theoretical Significance of the New View of Teaching Essence

First of all, the new view of teaching essence has a significance of methodology. At present , there are two completely opposite views, one is that things have its " essence " and that teaching has only essence; Another is that things may not have its only "essence", and that thing's essence is multidimensional, accordingly , the essence of teaching is multidimensional. The above difference roots in the essentialism and anti-essentialism in philosophy. To avoid this debate, the author use "essence" on the view of ontological significance, but not on the view of epistemology. In my opinion, the "essence" reflects a kind of human's pursuit of understanding things comprehensively. By believing the existence of "essence", we could get endless power for exploring the world. On the other hand, due to the "essence" is the thing's basic attribute, it decides other attributes of things, therefore, we can deduce a series of knowledge from "essence". It is a wise strategy that using "essence" to organize knowledge, and to spread knowledge. In conclusion, the author regards the "essence" as an epistemological beliefs and strategy, opens a new door for the teaching essence research. Also, as teaching is a purposeful practice activity, the author put forward that the essence of teaching should be generalized from two aspects of practice and purpose, and don't be in favor of going from all directions.

Secondly, the new view of teaching essence has an important significance for the construction of "teaching theory". "Teaching essence is the core problem of the teaching theory and practice, and is of value to teaching theory and practice. In terms of teaching practice, teaching essence restricts the teaching purpose and the teaching task, and influences the teaching design, implementation and evaluation, eventually, determines the quality of talents. In terms of theory of teaching, teaching essence restricts construction of teaching theory and the establishment of the subject of teaching theory "[1]. Not hard to see, to define the essence of teaching as "cognition and cultivation of both teachers and students", works well to radiate above mentioned teaching theory and practice issues, and provide a new theoretical perspective for these issues. There is no doubt, that the new view of teaching essence will have a positive impact on the construction of teaching theory subject , I think even like this, the necessity of studying teaching essence just lies in the construction of teaching theory subject. It is a pity that, so far, none of works of teaching theory deduces its content system from the perspective of teaching essence. In addition, the new view of teaching essence will lead to new teaching research, and will push the original edge problems into the mainstreams, thus, it would broaden teaching theory field of study. Over the years, affected by influence of special recognition, both teacher and student always focus their attention on cognition, also, teaching research and teaching evaluation is guided by cognition, the study on cultivation seriously lags behind. The new view of teaching essence would make the following question become urgent subjects: cultivation of intelligence and character of students, students' assessment, and the cultivation and evaluation of teacher, etc.

All along, people compare the teacher to "candles", which dedicates itself to light others. Teachers' own life value is distorted, and become vassals of the student. The new view of teaching essence defines the teachers' cognition and cultivation as the integral part of the teaching essence, it injects new vitality into teachers' life value, so the teachers' personality will become plumper and healthier. To know oneself, know others and social, understanding the mysteries nature, perfect self are the eternal spiritual pursuit of human. Confucius, the tutor of Chinese, had said: "in years of 15, I decided to study; in years of 30, I established my life principle; in years of 40, I had no doubts about everything, in years of 50, I knew my fate; in years of 60, I understood what I heard; in years of 70, I follow my heart to do things rightly". He also said: "I have no pity to die in the night, as long as I understand the great Tao in the morning". Confucius had been seeking on the road of cognition and cultivation all his life. Cognition and cultivation itself is a lofty pursuit for human life, and this pursuit is consistent with teachers' professional mission!

References

1. He, X.: What on Earth is Teaching Essence. J. Ningxia Univ. 1, 18 (2001) (in Chinese)
2. Nan, J.: Rational Reflect the Essence of The Teaching. J. Edu. Stud. Shanghai 2 (2007) (in Chinese)
3. Shi, Z.: Essentialism, Anti-essentialism and Pedagogic Study in China. J. Edu. Res. 1 (2004) (in Chinese)
4. Xue, J., Zhao, C.: Perplexity and Question of Study of Teaching Essence. J. Theor. Pract. Edu. 11 (2002) (in Chinese)

Empirical Study on thc Long-Memory Components in Asset Returns

Yang Chun-xia, Zhang Ying-chao, and Wu Hong-fa

School of Information and Control,
Nanjing University of Information Science and Technology
Nanjing, 210044, People's Republic of China
yangchunxia@ustc.edu

Abstract. In this work, the presence of long-memory components in asset returns is investigated. For a stock market, it surely possesses various periodic components. Furthermore, for various markets, the relationship between power and period of cycle components follows same power law with an exponent close to -0.92, spanning in a wide scale range from several days to nearly four years. This empirical result indicates that such power relationship may be universal and will enlighten modeling of market volatility.

Keywords: Econophysics, financial markets, long-memory components.

1 Introduction

The presence of long-memory components in asset returns has important implications that optimal consumption or savings and portfolio decisions may become extremely sensitive to the investment horizon. Since 1971, when Mandelbrot first considered the possibility and implications of persistent statistical dependence in asset returns[1], this topic has attracted widespread attention. Several empirical studies have lent further support to Mandelbrot's findings [2,3,4,5,6]. Recent investigations have also uncovered anomalous behavior in long-horizon stock returns[7,8]. In the past, based on the rationality of the investors and the theory of random walk[9,10], the Efficient Market Hypothesis dominated the finance literature. Today, it is facing impugnation and challenge from the above studies, which indicate that traditional tests of the capital asset pricing model and the arbitrage pricing theory are no longer valid and the usual forms of statistical inference do not apply to such time series. There need a new appropriate alternative to explain why persistence occurs. Several playful models have been introduced like the stochastic multiplicative point process[11,12], parametric statistical model also called randomly modulated periodicity (RMP)[13], a mixed process with random noise and deterministic patterns[14], and so on. Some of these models contend that return series can essentially be viewed as that composed of trend and cycle components, as well as additive noise. Among these dominant cycles oscillating at different frequencies, there would possibly be a cycle expected to be the business one. This recognition of persistent cycles presents a new perspective on the reasons underlying market volatility. Some empirical studies indeed have identified a

D. Zeng (Ed.): Advances in Computer Science and Engineering, AISC 141, pp. 689–696.
springer link.com

few cycles, such as 4-year cycle induced by Presidential Elections that occur every 4 years, 28 days periodic variation caused by lunar cycle and so on[15,16]. But how many cycle components are there existing in the market? What is the relationship between them? Who governs cycles oscillating? To construct an appropriate model to explain market volatility, these questions must be answered. Besides this hypothesis of the trade and business cycles, the $1/f$ noise in economic data is usually studied as long range dependence[17]. It has been shown repeatedly that the volatility autocorrelation functions of economic time series, such as series of stock prices over days, weeks or months, do not decay exponentially but reach a non-zero asymptote and remain there, albeit often at a low value, indicating that economic events some distance in the past continue to have an influence on current prices, which may be explained by the multiplicative point process model of trading activity proposed by Gontis, Kaulakys et al.[11,12]. But the situation for $1/f$ noise is that simply taking the fast Fourier transform(FT) of the autocovariance function may give an unreliable estimate of the power spectrum. Here the present study just focuses on the above questions. After obtaining various rhythmic modes of a given market, we will use a cyclical approach to describe the movement of returns over time and investigate the relationship between power and period for various cycles.

2 Data

The empirical data used in this paper includes the closed prices of Standard and Poor's 500 Index (SP500) over the period from October 22 1982 to April 12 2004, Hang Seng Index (HS) from December 31 1986 to October 9 2007, Dow Jones Industrial Average Index (DJIA) from May 26 1896 to May 31 2000, NASDAQ Composite Index (NASDAQ) from November 12 1984 to April 12 2004, ShangHai Stock Market Index (SHH) from December 19 1990 to September 28 2007, NIKKEI 225 Index (N225) from January 4 1984 to October 9 2007, TSY YLD NDX (TYN) from February 2 1962 to April 12 2004, Singapore Straits Times Index(SSTI) from December 28 1987 to October 9 2007. The original data of the time series was separately recorded. After

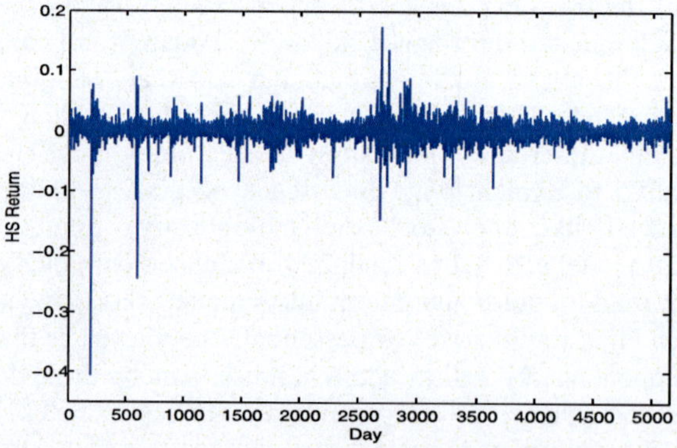

Fig. 1. HS return time series

deleting dates with records, then aligning the closed prices by date for each contract, a set of time series was obtained. Here $p(t)$ denotes the price of an index at time t, the return $r(t)$ at time t simply is the relative variation of the price from t to $t+1$: $r(t)=(p(t+1)-p(t))/p(t)\simeq\ln p(t+1)-\ln p(t)$. Fig.1 shows HS return time series over the period from December 31 1986 to October 9 2007.

3 Methodology

A cycle is a term to indicate repeating patterns, specific to recurrent events, such as seasons, elections, etc. For a business cycle, Sargent offers two closely related definitions by considering a single series governed by a stochastic difference equation[18]. By the first definition, a variable is said to possess a cycle of a given frequency when its covariogram exhibits damped oscillations of that frequency. By the second definition, a given time series has a cycle if a peak occurs in its spectral density function. According to these definitions, the estimate of periodicity for various cycle components can be achieved. By the latter definition, the spectral density function is the magnitude of the discrete Fourier transform of the time series. So, a spectral peak observed at a specific frequency f implies the existence of a sinusoidal component in the time series oscillating at that particular frequency, i.e., implies the existence of a cycle with a period $T=1/f$. Other than analysis in the spectral (or frequency) domain, the same results can also be produced by the analysis in the time domain. Furthermore transformation of the signal representation from one domain to another is carried out by computing the Fourier transform (FT) and the inverse Fourier transform (IFT). Historically, Fourier spectral analysis has ever dominated the data analysis and almost been applied to all kinds of data. Although it is valid under extremely general conditions, there are some crucial restrictions that the system must be linear and the data must be strictly periodic or stationary. Otherwise, the resulting spectrum will make little physical sense. Unfortunately, the price time series is typical of nonlinear and non-stationary signal. Therefore Fourier spectral analysis can't be directly used to process it. Fortunately, the Empirical Mode Decomposition (EMD) introduced by Huang et al. especially well fits analyzing such time-series representing non-stationary and nonlinear processes[19,20,21]. Its essence is to identify the intrinsic oscillatory modes by their characteristic time scales in the data empirically and then decompose the data accordingly. So, the decomposition is based on the direct extraction of energy associated with each time scale. After the EMD is performed, the generated Intrinsic Mode Functions (IMFs) will take on intrinsic oscillation patterns corresponding to their specifical scales, which can be estimated by the above second definition. In EMD, there are two steps to create IMFs. Step-1: First, all the local maxima in the return time series $r(t)$ are identified and connected by a cubic spline line $U(t)$ as the upper envelope. Then the procedure is repeated for the local minima to produce the lower envelope. Both envelopes will cover all of the original time series. Their mean is designated as $m_1(t)$, and the difference between the return data and $m_1(t)$ is the first component $h_1(t)$.

$$m_1(t) = (U(t) + L(t)) / 2 \tag{1}$$

$$h_1(t) = r(t) - m_1(t) \tag{2}$$

If $h_1(t)$ satisfies the following conditions, it is an IMF : (i) $h_1(t)$ is free of riding waves; (ii) It takes on symmetry of the upper and the lower envelopes with respect to zero; (iii) The numbers of zero crossings and extremes are the same, or only differ by unity. If $h_1(t)$ is not an IMF, the sifting process has to be repeated more times until to reduce the extracted signal to an IMF. In the next sifting process, $h_1(t)$ is treated as the original datum.

$$h_{11}(t) = h_1(t) - m_{11}(t) \tag{3}$$

Again, if the function $h_{11}(t)$ does not yet satisfy criterion (i)-(iii), this sifting procedure continues up to k times until $h_{1k}(t)$ is an IMF:

$$h_{1k}(t) = h_{1(k-1)}(t) - m_{1k}(t) \tag{4}$$

Step-2: The first IMF is designated as $imf_1 = h_{1k}(t)$. Then imf_1 can be separated from the rest of the data by

$$d_1(t) = r(t) - imf_1(t) \tag{5}$$

where the difference $d_1(t)$ is the first residue. Since $d_1(t)$ still contains information of longer period components, it is treated as the new data and subjected to the sifting process Step-1. Following the above procedures of Step-1 and Step-2, more intrinsic modes imf_i can be found until the last one. The final residue will be a constant or a monotonic function which represents the general trend of the time series. Finally, the following results are obtained.

$$r(t) = \sum_{i=1}^{n} imf_i(t) + d_n(t) \tag{6}$$

$$d_i(t) = d_{i-1}(t) - imf_i(t) \tag{7}$$

where $d_i(t)$ is a residue.

4 Empirical Results

When EMD is performed on the return time series of HS, N225 and TYN, the results of periodicity estimate for various cycles can be achieved, the results of other stock indexes can be referenced in Ref.22.

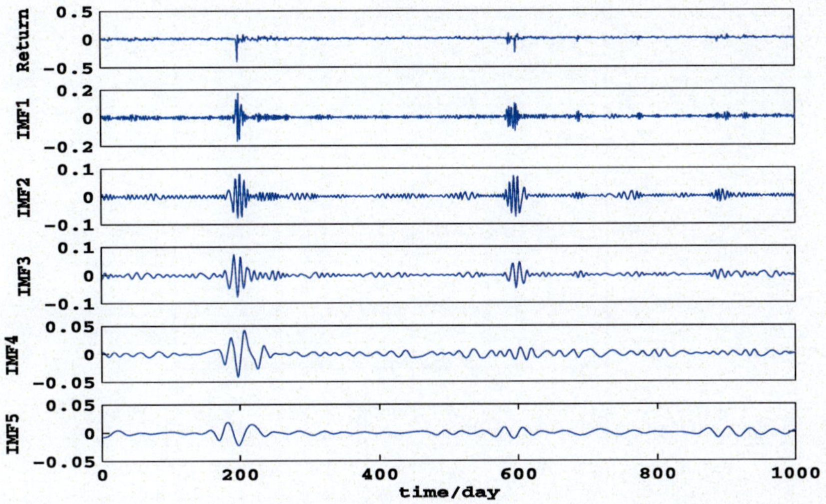

Fig. 2. Return and the first 5 IMFs obtained by EMD method from HS, only the initial 1000 points are shown

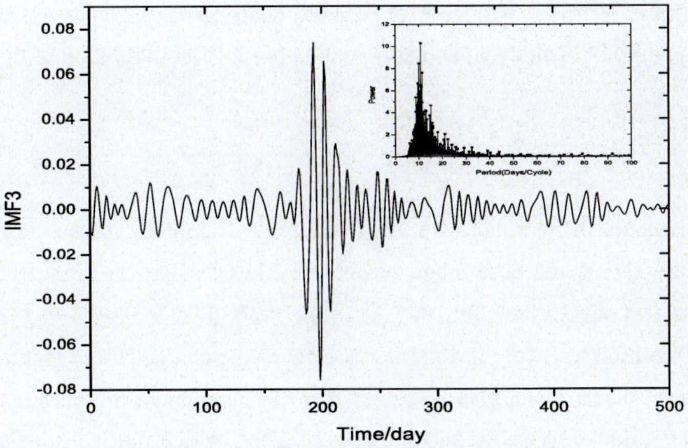

Fig. 3. IMF3 of HS, the insert is its periodogram achieved by the second definition

For HS return time series, there exist 11 cycle elements, Fig.2 only shows its first 5 components. Fig.3 shows the IMF3, which possesses a cycle of 10.79 days. As the reference Ref.22 indicates, return time series surely is composed of a trend and cycle components oscillating at different frequencies. Furthermore, for various markets, there exist the same periodic oscillation patterns in accord with real phenomena. Among these cycles, the long-term one sets trends that often last for a long time, while the other short ones posit the intermediate-term trends.

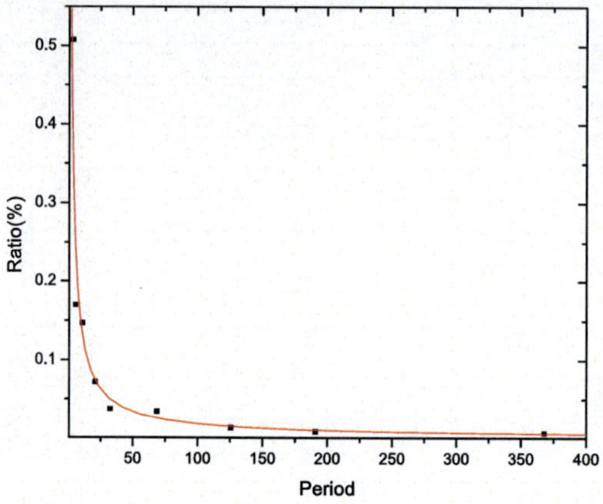

Fig. 4. Percentage of every cycle components at different time scales

It is their interactions that induce prices moving in a manner that may initially appear to be random, but with study show an underlying order. The fact that the same periodic oscillation patterns exist in various markets is the very reflection of such underlying order. Nextly the question which cycle is the most dominant one is considered. In order to investigate every cycle's contribution to $r(t)$, one can define an average energy of each cycle and compute its proportion in relation to the whole.

$$E = \sum_{t=1}^{n} x_t^2 / n \tag{8}$$

where x_t is denoted the value of a IMF at time t and E is its average energy. For HS return series, the percentage of every cycle is shown in Fig.4. It is obvious that the first one IMF_1 has not only the highest frequency shown in Fig.2 but the biggest contribution to $r(t)$. It dominantly catches the characteristic feature of return $r(t)$. Besides, the fact can also be found that the percentage of every cycle decays slowly, reach a non-zero asymptote and remain there, which motivates us to consider the relationship between power and period of every cycle. Fig.5 shows that it follows power law spanning in a wide time scale range from several days to nearly four years and the exponent is -0.90 ± 0.06. For other indexes, it is the same case, their exponent values can also be obtained, seen in the table 1, which indicates that this characteristic may be universal for different markets.

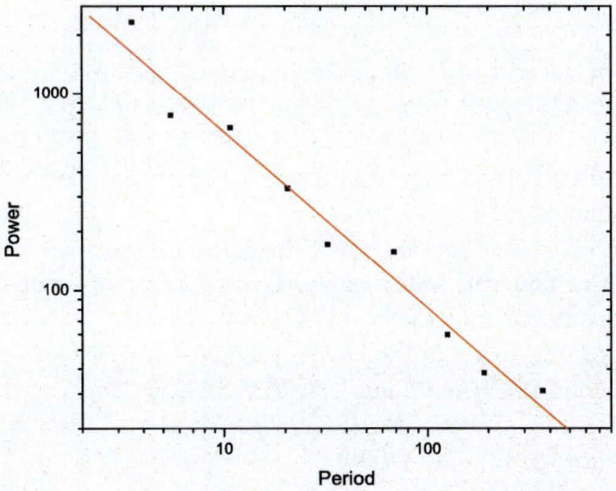

Fig. 5. Relationship between power and period for HS

Table 1. Power law exponents for various indexes

Index	exponent	Index	exponent
HS	0.90±0.06	SP500	0. 90±0.04
DJIA	0.81±0.04	NASDAQ	0.99±0.05
SHH	0.98±0.04	N225	0.92±0.03
TYN	0.92±0.06	SSTI	0.96±0.05

6 Conclusion

In this work, the presence of long-memory components in asset returns is investigated. For a stock market, it surely possesses various periodic components. Furthermore, for various markets, there not only exist same periodic oscillation patterns, but the relationship between power and period of cycles follows same power law with an exponent close to -0.92, spanning in a wide scale range from several days to nearly four years. This empirical result indicates that such power relationship between power and period may be universal, and will enlighten the modeling of market volatility.

Acknowledgments. This work is supported by the National Natural Science Foundation of China (Grant No 60874111), by the College Science Foundation of Jiangsu Province (Grant No 07KJD120128), by Qing Lan Project of Jiangsu Province, by the College Students Practice Program of Jiangsu Province (08CX0010) and by the Program of Social Science Institute of Jiangsu Province(B0808).

References

1. Mandelbrot, B.: When can price be arbitraged efficiently? A limit to the validity of the random walk and martingale models. Review of Economics and Statistics 53, 225–236 (1971)
2. Greene, M.T., Fielitz, B.D.: Long-Term Dependence in Common Stock Returns. Journal of Financial Economics 4, 339–349 (1977)
3. Thaler, R.H.: The January effect. Journal of Economic Perspectives, 169–177 (1987)
4. Harris, L.: A transaction data study of weekly and intradaily patterns in stock returns. Journal of Financial Economics 16, 99–117 (1986)
5. Peters, E.E.: Chaos and order in the capital markets—a new view of circles, price, and market volatility. John Wiley and Sons, New York (1991)
6. Barkoulas, J., Baum, C., Travlos, N.: Long memory in the Greek stock market. Applied Financial Economics 10, 177–184 (2000)
7. Cajueiro, D.O., Tabak, B.M.: Testing for time-varying long-range dependence in volatility for emerging markets. Physica A 346, 577–588 (2005)
8. Yang, C.-X., Zhang, Y.-C., Wu, H.-F., Zhou, P.-L.: The Origin of Volatility Cascade of the Financial Market. In: Shi, Y., van Albada, G.D., Dongarra, J., Sloot, P.M.A. (eds.) ICCS 2007. LNCS, vol. 4490, pp. 114–120. Springer, Heidelberg (2007)
9. Bachelier, L.: Ann. Sci. Ecole Norm. Sup. 3, 21–86 (1900)
10. Fama, E.: Efficient Capital Markets: A Review of Theory and Empirical Work. Journal of Finance 25, 383–417 (1970)
11. Gontis, V., Kaulakys, B.: Multiplicative point process as a model of trading activity. Physica A 343, 505–514 (2004)
12. Gontis, V., Kaulakys, B.: Modeling long-range memory trading activity by stochastic differential equations. Physica A 382, 114–120 (2007)
13. Hinich, M.J., Serletis, A.: Randomly modulated periodicity in the US stock market. Chaos, Solitons and Fractals 36, 654–659 (2008)
14. Chen, P.: Trends, Shocks, Persistent Cycles in Evolving Economy: Business-Cycle Measurement in Time-Frequency Representation. In: Barnett, W.A., Kirman, A.P., Salmon, M. (eds.) Nonlinear Dynamics and Economics, ch. 13, pp. 307–331. Cambridge University Press, Cambridge (1996)
15. Booth, J.R., Booth, L.C.: Is presidential cycle in security returns merely a reflection of business conditions? Review of Financial Economics 12, 131–159 (2003)
16. Dichev, I.D., Janes, T.D.: Working Paper. University of Michigan (2001)
17. Bak, P., Tang, C., Wiesenfeld, K.: Self- organized criticality. Physical Review A 38, 364–374 (1988)
18. Sargent, T.: Macroeconomic Theory. Academic Press, New York (1979)
19. Huang, N.E., et al.: The Empirical Mode Decomposition and the Hilbert Spectrum for Nonlinear and Non-stationary Time Series Analysis. Proc. R. Soc. Lond. A 454, 903–995 (1998)
20. Rilling, G., Flandrin, P., Goncalv, P.: IEEE-EURASIP Workshop on Nonlinear Signal and Image Processing. Grado (1) (2003)
21. Wu, M.-C.: Journal of the Korean Physical Society 50, 304–312 (2007)
22. Yang, C.-X., Wu, H.-F., Zhang, Y.-C.: Periodic components and characteristic time scales in the financial market. Modern Physics Letters B 22, 2571–2578 (2008)

Strategies of Small Apparel Suppliers' Upgrading in China: Analysis Based on Global Commodity Chain

Pan Mingtao

School of Economics, Huazhong University of Science and Technology,
No. 1037, Luoyu Road, Wuhan, 430074, China
bettapisces@yahoo.cn

Abstract. In recently China, many middle and small private firms closed down in Shenzhen, Wenzhou. Many of them are labor-intensive industry, such as apparel, footwear, toy and household electrics. The observed reasons are obstacles of accessing credit caused by financial policies; wage increasing and inputs price inflation, etc. The impacts include short-turn or long-turn factors, but it reveals the symptoms in low-tech manufacturing that depending on low-cost competitive advantages is unsustainable. This paper based on Global Commodity Chain（GVC） analysis, and focus on apparel industry. The main conclusion is in the processes of apparel GVC upgrading, the assemble or manufacture stages must migrate, and the apparel firms should change their roles and functions in commodity chain. It is long-term tendency.

Keywords: Global Commodity Chain, GVC, apparel industry, upgrading; role changing.

1 Introduction

Many Chinese small and middle firms are facing survival dilemma, in coastal areas, where were the origin of traditional commodities exporting, such as apparent,

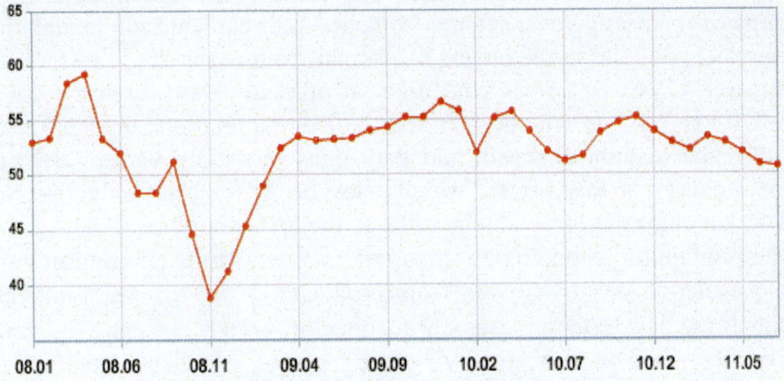

Fig. 1. PMI of Manufacturing,2008.1-2011.7
Data source: http://data.eastmoney.com/cjsj/pmi.html

D. Zeng (Ed.): Advances in Computer Science and Engineering, AISC 141, pp. 697–701.
springer link.com

footwear, toys, electronics components, etc. The PMI of Chinese manufacturing in July 2010, published by CFLP (Chinese Federation of Logistics & Purchasing), is 50.7%, bottoms at the timeliness form January 2008 to May 2011, after the crisis in 2009(See Figure 1). However, the HSBC's PMI, especially focus on small and middle firms, is 49.3%. It's a signal that manufacturing is shrinking, or stagnation. And the contemporaneous media noticed the business failure of labor intensive firms in Wenzhou, Dongguan and Shenzhen.

We all know while the economic growth goes on, low wages and material prices in unsustainable. In China, when the labor force transforming form rural to urban exhausted, the cost advantages couldn't exist anymore. Not only the wages, but also material price, transforming cost and environment pressures is up going. The underlying issue is that as margin industry depicted by Kojima(1984), the labor intensive, low-skilled manufacturing, such as appeal industry, would be eliminated or shifted out gradually in a developing country like China.

This paper illustrated the changing of global apparel industry, based on the GVC(Global Value Chain) approaches(Gereffi,1999). And focus on the strategies of apparel firms in China, especially small firms as local supplier.

2 The Changing Roles and Strategies in Global Apparel Industry

The global apparel industry, form materials & inputs sourcing, design, garment trim and marketing, is a value chain driven by buyer (Gereffi and Frederick, 2010; Kaplinsky, 2004; Gereffi et al.,2005). In this buyer-driven value chain, the leadership is the global brand like Liz Claiborne, and retailers like JC Penny, Wal-Mart, etc. The essential character is the allocation of production and trade in global market is levered by leader firms: retailers, brand marketers and manufacturers. The most valuable activities are brand, marketing, design, but not production.

There are three dimensions in GVC analyzing: role, function and relation. As big buyers, the leader firms' leverage on local suppliers is conducting by the global souring strategies. The local spoilers, competing with each others for orders must enhance functions, except for qualified garments, they also provides logistic services, such as delivery and inputs souring, even designs. Furthermore, there are many intermediaries firms as souring agent between buyers and supplies for the consideration of transacting cost saving. Compare to direct-sourcing model, the agent-sourcing model is popular in the buyers need small volumes.

The apparel GVC is some kind like a pressure transmission system. The competition in the final commodity market requires the retailers, branders responding quickly when the fashion changed, and they must providing various designs every season continuously at low price , which relay on a flexibly production and trade system. This becomes the force of upgrading to the local suppliers around the world.

With the production technologies spreading and new firms entering in developing countries, the market power had shit from producers to sellers. Many global brands increase investment in intangible assets to develop brands, enhance or consolidate competitive advantage in marketing. And the tangible, product related activates are outsourced to the ODM or OEM firms. For example, the production of United States' companies such as Levi's, has been migrated to Mexico and Central American and Caribbean countries in production-sharing arrangements. On the other hand, Fashion-oriented companies like Liz Claiborne or most big retailers, buy from a large number of small contractors, with most of these factories located in Asia.

The processing of production shift form America to East Asia is repeating right now, form East Asia, like Hong Kong, Taiwan, Korea to emerging market in Vietnam, Cambodia, Indonesia, etc. And the companies in pro-exporting area has established a sourcing network connect the apparel exporting countries and the apparel consuming countries. Like, Li & Fung the former OEM firm based in Hong Kong has become an OBM firm coordinating and managing the apparel production and trade network.

The developing of global sourcing network in apparel industry provides new chance for small firms to compete with large companies in niche-market. The fabless pattern is a new business model, allow small firms like design studio focus on only one stage in entire apparel value chain and outsource others at low cost. It provides much more differentiation and variation in final products markets.

At the same time, leader firms adjusting their sourcing strategies to realization supply chain respectively. Most apparel merchants pick capable, multifunctional companies to be their long-term strategic partner, and reduce reliance on one supplier by diversifying their supplier reverses.

3 The Upgrading Pressures and Obstacles in China

The most urgent issue in front of Chinese apparel suppliers is the rising manufacturing cost. Recently years, labor shortage after Spring Festival had been more and more common. Apparel factories have to strike for workers with other higher wages industries, such electronic processing. And in Vietnam, India or other countries in Southeast Asia, the average wages is 1/4 to 1/3 in China (Neidik and Gereffi,2006). With other cost factors, such as inflation, material and power prices, environmental pressure, the cost advantage of "made in China" is fading away(Bair and Gereffi,2003).

The pressures not only come from market, but also from downstream buyers. For cost consideration and focus on core activities, the buyers, from retailers, mass merchants, sourcing-agent to production contractors, all want to give up non-core activities and shift their responsibilities to next tier subcontractors. This means the local supplies need promote their functions such as acknowledge about international standards, production expertise increasing, modular production, flexible production, fast delivery, etc.

However, there are also obstacles in the upgrading, especially for the marginal firms. The strategies of leader firms emphasizing long-term stability of supply chain created an exclusive commodity chain, in which most small firms at a disadvantage state. Apparel retailers gain the most added values by control brands, market and outlets. The small firm usually lack capital, market knowledge, technical skills and social capital for upgrading.

4 Trajectories and Strategies of Upgrading

At industry or country level, upgrading is the processing of labor intensive production become capital intensive and technology intensive production(Porter,1998). At firm's level, it's associated with moving beyond assembly and full-package supply to original design manufacturing (ODM) and original brand manufacturing (OBM).

The upgrading in global apparel value chain has its own continuous processing or course of evolution. Table 1 summarized the regional trends of industry upgrading in global value chain in Asia, the countries joined global value chain early has moved its apparel industries to upstream and provide capital intensive or technology intensive products such as fibers, fabrics and machineries to other labor intensive, assemble oriented countries.

Table 1. Regional Trends and Stages in Asian Apparel Industry

Country/Region	**Stage & Function**			
	Garments Production	Textile	Spinning Fiber	Machinery
Japan	1950s to early 1960s	Mid- 1960s		Post-1990s
Hong Kong Korea TaiWan	Latterly 1960s to early1980s	Post-1980s		
Mainland China Indonesia Thailand India Pakistan	Post-1980s	1990s		
Bangladesh Cambodia Vietnam	Mid-1990s to latterly 2000s			

Source: Gereffi et al.,2010.

For Chinese firms nowadays, the first consideration of upgrading maybe remove their factories in cost raising costal area to the countries or regions price level, especially wages, is relative lower . The central and western provinces in China or emerging market countries in Southeast Asia, even Africa Country are optional.

About two decades acceleration on international business knowledge, productions skills and funds could be the supporting for the organizational and functional promoting of Chinese firms to set and manage a trade and production network. However, additional investment still needed to build a more progressive information and logistics system for the management and coordination. A competitive supply chain must be flexible, quick response and carry out strict quality standard more efficiently. The investment in intangible aspect is equally important, firms need education and training their employees or workers to accumulate the human capital stocks. Technical skills in production as well as skill in management, product development, design and market search all important for the function improving.

Beyond that, three are organizational issues for apparel firm upgrading. The essential matter is the role changing form 'supplier' to 'buyer' in apparel value chain.

One of the choices is like talked above, apparel firms can develop private brand and compete with big brand in the niche market. For Chinese firms, there are a huge domestic consuming market can provide conditions for the national brand development. Compared to abroad company, local firms have geographic and cultural advantage because closed to the market.

In addition, China has built a relatively complete vertical industry chain form material to consumer services in the past two decades. This can be the foundation of the companies want to create organizations to market and network the country/region apparel industry, coordinate the activities among diversified suppliers and buyers.

5 Conclusions

The traditional exporting industrial in China, like apparel, toy, furniture, consumer electronics, have been beset by the current economic recession after crisis, which has lower demand for apparel exporters. At the same time, those firms have to facing the intensified competition from poor and small emerging market countries. However, put aside the short-run impacts, we must understand that the low cost advantages is unsustainable, which also low profits. In any case, China has to upgrading its assemble-oriented industries.

For apparel firms, the general strategies has interpreted above, include (1) investing in intangible aspects, like brand development, marketing, or network building, (2) improving functional to acquire more added value, and (3) changing role in apparel value chain, become seller or organizer.

At the country or industry level, it's necessary to take various policies to accelerate firm transforming, like tax and financial support for technical upgrading and brand promotion; promote FDI or joint ventures to develop domestic vertical capabilities.

In all those policies and strategies, the underlying purpose is to encourage firm to join in the global apparel value chain, acquiring the necessary knowledge about market and production skills through international businesses.

References

1. Bair, J., Gereffi, G.: Upgrading, uneven development, and jobs in the North American apparel industry. Global Networks 3(2), 143–169 (2003)
2. Gereffi, G.: A commodity chains framework for analysing global industries. Institute of Development Studies (1999)
3. Gereffi, G., Humphrey, J., Sturgeon, T.: The governance of global value chains. Review of International Political Economy 12(1) (2005)
4. Gereffi, G., Frederick, S.: The Global Apparel Chain, trade and the Crisis: Challenges and Opportunities for Developing Countries. Global Value Chains in a Postcrisis World: A Development Perspective 157 (2010)
5. Kaplinsky, R.: Spreading the gains from globalization: What can be learned from value-chain analysis?. Problems of Economic Transition 47(2), 74–115 (2004)
6. Kojima, K.: Japanese and American direct investment in Asia: a comparative analysis. Hitotsubashi Journal of Economics, 251–235 (1984)
7. Neidik, B., Gereffi, G.: Explaining Turkey's emergence and sustained competitiveness as a full-package supplier of apparel. Environment and Planning A 38(12), 2285 (2006)
8. Porter, M.E.: Competitive advantage: creating and sustaining superior performance: with a new introduction. Free Pr. (1998)

Evolution of Shanghai as an International Financial Centre with Non-equilibrium Statistical Mechanics

Huibin Du, Qiongqiong Xia, and Linxue Wei

Institute of Systems Engineering, Tianjin University,
Tianjin 300072, China
duhuibin@tju.edu.cn, xia_1989_xue@126.com, wlxly.4@163.com

Abstract. This paper, from the viewpoint of complex system, applies non-equilibrium statistical mechanics to analyze the development of Shanghai as an international finacial centre. And then the evolutionary patterns of Shanghai as an international finacial centre are simulated by SOM. The simulation results show that the achievement of financial industry relies heavily on Shanghai's overall economy; however, financial markets, such as securities market, gold market, and futures market, actually don't match such a rapid development of Shanghai's overall economy.

Keywords: non-equilibrium statistical mechanics, international finacial centre, SOM.

1 Introduction

In March 2009, the State Council of the People's Republic of China announced an ambitious plan of making Shanghai become a major international financial centre by 2020, on par with Hong Kong, Singapore, New York and London. This plan attracts more attention of scholars both at home and abroad.

Kindleberger (1974) stated that 'banking and finacial centres performed a medium of exchange function and an inter-spatial store of value function.[1] Jao (1997) classified financial centre into national or domestic centre, regional centre, and global centre from a geographical perspective; functional centre and booking centre from teleological perspective; traditional centre, financial entrepot and offshore banking centre from historical perspective.[2] Goldberg, Helsley and Levi (1988) suggested that the major factors influencing financial centre development are the level of overall economic development, international trade, financial intermediation, and the stringency of financial regulations. [3] Reed (1981) argued that a financial centre must first be a national financial centre before it became an international financial centre.[4] Poon (2003) studied the hierarchy of financial centre through cluster analysis.[5] However, few researches focused on the dynamic evolutionary mechanism of international financial centre.

This paper firstly builds up a model from a new perspective of non-equilibrium statistical mechanics that can explain the self-organized evolutionary features of a complex system. The rest of this paper is organized as follows: section 2 proposed a

D. Zeng (Ed.): Advances in Computer Science and Engineering, AISC 141, pp. 703–710.
Springerlink.com

theoretical framework; numerical simulations and results were obtained in section 3; and some conclusions were given in section 4.

2 Statistical Dynamics Analysis on a Complex System

The self-organized feature of a complex system is influenced by many factors interacting with each other. However it is very difficult to explain or express this kind of pattern, structure or process using available theories, because of uncertainty of the environment and the complicate interaction of various impacting factors. So a new theoretical framework [6] is introduced and applied to indicate the self-organized feature of a complex system.

The driving forces in a complex system can to be expressed as $x_1, x_2, \ldots, x_n$. Similar to classical statistical theories, all possible microstates compose a continuous range in Γ space. We set $\mathrm{dx} = \mathrm{d}x_1 \mathrm{d}x_2 \cdots \mathrm{d}x_n$ be a volumetric unit in Γ space and define $\rho(\mathbf{x}, t)$ as a distribution function of ensemble at time t. We assume that the flux J including energy, information and people can be taken as coupled interactions among subsystems or factors. Thus, J can be written as: [6, 7, 8]

$$J = \eta + \sum_i \gamma_i x_i + \sum_{ij} \gamma_{ij} x_i x_j + \sum_{ijk} \gamma_{ijk} x_i x_j x_k + \sum_{ijkl} \gamma_{ijkl} x_i x_j x_k x_l + \cdots \quad (1)$$

Let us maximize Eq. (1) under the following constraints,

$$\langle x_i \rangle = f_1, \langle x_i x_j \rangle = f_2, \langle x_i x_j x_k \rangle = f_3, \langle x_i x_j x_k x_l \rangle = f_4, \cdots \quad (2)$$

where $f_1, f_2, f_3, f_4, \ldots$ represent interacting rules among subsystems or factors.

By using Lagrange multiplier, we can obtain that[7, 9]

$$\rho = e^{\mu + \sum_i \sigma_i x_i + \sum_{ij} \sigma_{ij} x_i x_j + \sum_{ijk} \sigma_{ijk} x_i x_j x_k + \sum_{ijkl} \sigma_{ijkl} x_i x_j x_k x_l + \cdots} \quad (3)$$

Eq. (3) as a potential function is defined as

$$\Phi(\sigma, x) = \mu + \sum_i \sigma_i x_i + \sum_{ij} \sigma_{ij} x_i x_j + \sum_{ijk} \sigma_{ijk} x_i x_j x_k + \sum_{ijkl} \sigma_{ijkl} x_i x_j x_k x_l + \cdots \quad (4)$$

Accordingly, by transformation of

$$\xi_k = \sum_{i=1}^{n} a_{ki} x_i \quad (5)$$

Then the potential function can be described as

$$\overline{\Phi}(\lambda, \xi) = \xi + \sum_k \lambda_k \xi_k^2 + \cdots \quad (6)$$

where we actually take ξ_k as structural parameters or evolutionary patterns of complex systems, because ξ_k is a linear combination of all kinds of factors or subsystems expressed by $x_1, x_2, \ldots, x_n$.[10, 11]

According to the connection of potential functions and dynamic equations, we deduce the stochastic evolutionary equations from Eq. (5)

$$\dot{\xi}_k = \lambda_k \xi_k + S_k(\xi_1, \xi_2, \cdots, \xi_n) + F_k(t), k = 1, 2, \cdots, n \tag{7}$$

Inactive patterns $(\lambda_k < 0)$ represent the eliminated patterns that cannot survive to form, while the active patterns $(\lambda_k > 0)$ stand for survival patterns that will win in the competition.[12, 13] In other words, patterns are largely dependent on parameter λ.

3 Numerical Simulations

The theoretical analysis is given above. Here, we will use SOM simulating technique to analyze the patterns of Shanghai as a financial centre.

3.1 Self-Organized Map Network

A self-organized Map (SOM) network is a kind of artificial neural network (ANN), which is trained using unsupervised learning to produce a low-dimensional representation of the input space of the training samples, called a map.[14] The SOM, different from other clustering techniques, applies a neighborhood function to preserve a topological ordering of the classes. The principle of SOM is consistent with above-mentioned theoretical framework of non-equilibrium statistical dynamics. The connected values a_{ki} and the value of order parameter ξ can be obtained by SOM algorithm.

3.2 Financial Sector's Role on Overall Economy of Shanghai

The economic development of Shanghai is influenced by different sectors interacting with each other. This paper analyzes the development of Shanghai from eight sectors, in which Primary Industry, Industry, Construction, Transport, Storage, and Post, Wholesale and Retail Trades, Financial Intermediation, Real Estate are expressed as $x_1, x_2, \ldots, x_7$, respectively; and total imports & exports is taken as x_8.

We choose the period between the first five-year plan and the eleventh five-year plan to analyze the development of Shanghai as a financial centre. "Five-Year Plan" is a part of China's national economic plan. Except three years of adjustment for the national economy from1963 to 1965, China carried out the first Five-Year Plan in 1953, and now is implementing the 12th Five-Year Plan. Data for this study is obtained from "Glorious 60 Years: Historical Statistics of Shanghai", "Shanghai Statistical Yearbook", and "Shanghai Finance Yearbook". Through above theoretical framework and simulation technique, we can obtain ξ_i $(i = 1, 2, \ldots, 12)$ values in Fig. 1 that can describe the comprehensive patterns of Shanghai's overall economy in all periods.

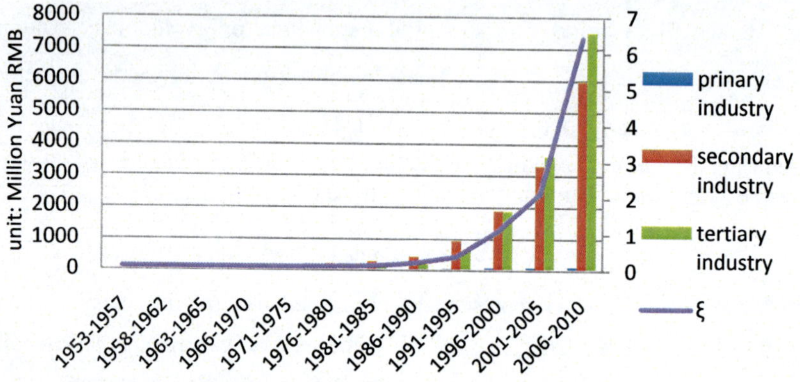

Fig. 1. Development of Shanghai's overall economy during 1953-2010

Fig. 1 shows that there is a turning point of Shanghai's economy from the period of the 9th Five-Year plan. Since 1996, the gross products of tertiary industry has been increased greatly, which has exceeded that of secondary industry. ξ_i in Fig. 1 shows the same trend that the overall economic development of Shanghai has experienced a rapid rise since 1996 as well. In order to find out the driving forces of those changes occurred in Shanghai's economy, the contributions of all indices to ξ_i , i.e. the contributions of all sectors to the development of Shanghai's economy, are shown in Fig. 2. In this paper, ξ_{ij} $(i = 1, 2, \dots, 12;\ j = 1, 2, \dots, 8)$ represent the contribution of x_j to ξ_i; and ξ_i $(i = 1, 2, \dots, 12)$ represent the development patterns of Shanghai's overall economy at the i th period.

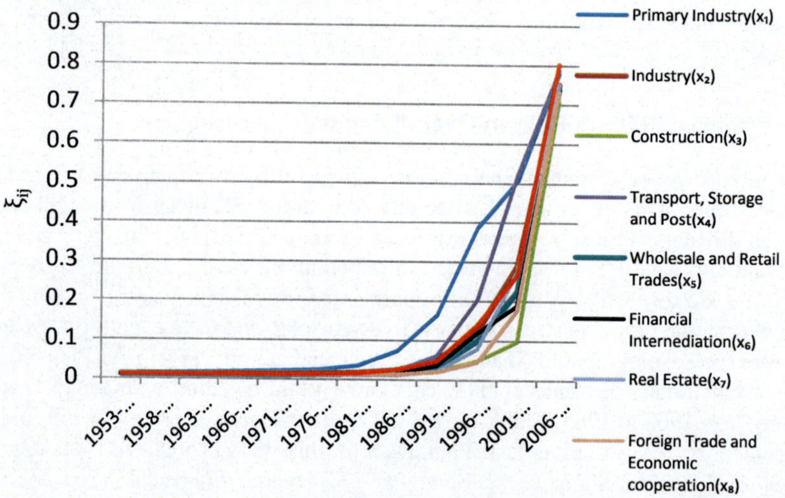

Fig. 2. Contributions of eight sectors to Shanghai's overall economy

"$\xi_i/8$" in Fig. 2 can be explained as average contribution. The contributions to ξ_i of primary industry, wholesale and retail trade are obviously higher than average contribution; while the contributions of other sectors are close to average one. Financial sector didn't play a leading role to accelerate such rapid economic development of Shanghai as we expected.

3.3 Evolution of Shanghai as a Financial Centre

The evolution of Shanghai as an international financial centre is driven by a series of factors. Here, we select 19 impacting factors that could be divided into three categories: the subject of financial centre（x_1: employees in financial sector; x_2: saving deposit balance of financial institutions; x_3: saving deposit balance of foreign-funded financial institutions）, the object of financial center（x_4: turnover of securities; x_5: RMB transactions in inter-bank market; x_6: turnover of future; x_7: turnover of the Shanghai gold exchange; x_8: cash income of banks; x_9: premiums income）, and the environment of financial centre（x_{10}: resident population; x_{11}: gross domestic product (GDP); x_{12}: value-added of the tertiary industry; x_{13}: total value of imports and exports; x_{14}: actually utilized foreign capital; x_{15}: infrastructure investment; x_{16}: total expenditure on science and technology; x_{17}: average compensation of employees; x_{18}: volume of freight handled in ports; x_{19}: legal professional qualification obtained by law firms）. [15] The interaction of these 19 factors will spontaneously induce a certain pattern or structure for Shanghai's financial development.

Through above-mentioned theoretical framework and simulation technique of SOM, the values of ξ_k $(k = 1, 2, \ldots, 10)$ representing the evolutionary patterns of Shanghai as a financial centre from 2000 to 2009 are obtained in Table 1 and Fig. 3.

Table 1. Distribution of ξ_k

Year	2000	2001	2002	2003	2004	2005	2006	2007	2008	2009
k	1	2	3	4	5	6	7	8	9	10
ξ_k	0. 28	0. 44	0. 71	1. 48	2. 36	3. 51	5. 10	8. 51	11. 0	13. 2

The values of ξ_k $(k = 1, 2, \ldots, 10)$ were increased fast from 2000 to 2009. ξ_k represented the comprehensive patterns of Shanghai's financial industry, which were formed by the interaction between these 19 impacting factors. ξ_k from 2000 to 2003 shown as ξ_1, ξ_2, ξ_3, and ξ_4 in Fig. 3 gradually converged to steady attractors after 100-200 steps' simulation, illustrating the self-organized patterns and evolutionary process of Shanghai's financial development are relatively stable. The simulation results of ξ_k from 2004 to 2009 have finally arrived at stable values,

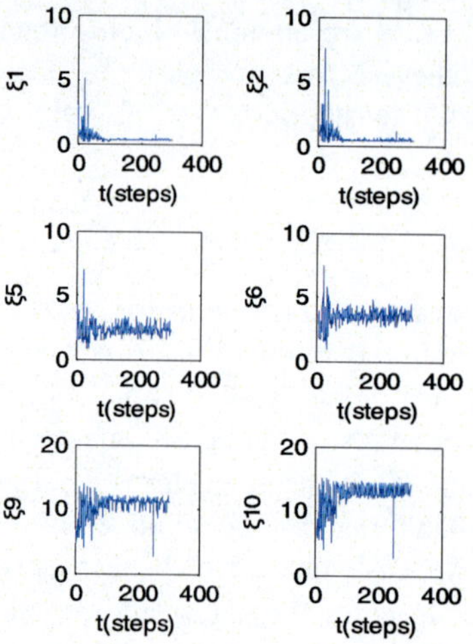

Fig. 3. Convergence of ξ_k

around which there are bigger fluctuation. This can be explained that the patterns of Shanghai's development as a financial centre during 2004-2009 were still active and continuously changing.

In order to understand the contribution of 19 factors to the evolutionary process of Shanghai as a financial centre, we calculate the values of ξ_{kl} ($k = 1, 2, ..., 10$; $l = 1, 2, ..., 19$) with the same simulation technique. In Fig. 4(a) and Fig. 4(b), the contributions of saving deposit balance of financial institutions (x_2), saving deposit balance of foreign-funded financial institutions (x_3), premiums income (x_9), GDP (x_{11}), value-added of the tertiary industry (x_{12}), actually utilized foreign capital (x_{14}), infrastructure investment (x_{15}), total expenditure on science and technology (x_{16}), average compensation of employees (x_{17}), and legal professional qualification obtained by law firms (x_{19}) are close to the average level. These five factors: employees in financial sector (x_1), cash income of banks (x_8), resident population (x_{10}), total value of imports and exports (x_{13}), volume of freight handled in ports (x_{18}) contribute more than the average level in Shanghai's development as a financial centre. In contrast, we can find in Fig. 4(d) that turnover of securities (x_4), RMB transactions in inter-bank market (x_5), turnover of future (x_6), and turnover of Shanghai gold exchange (x_7) contribute less than the average level. These results indicate that financial markets, such as securities market, futures market, and gold market, actually don't match such rapid growth of Shanghai's overall economy.

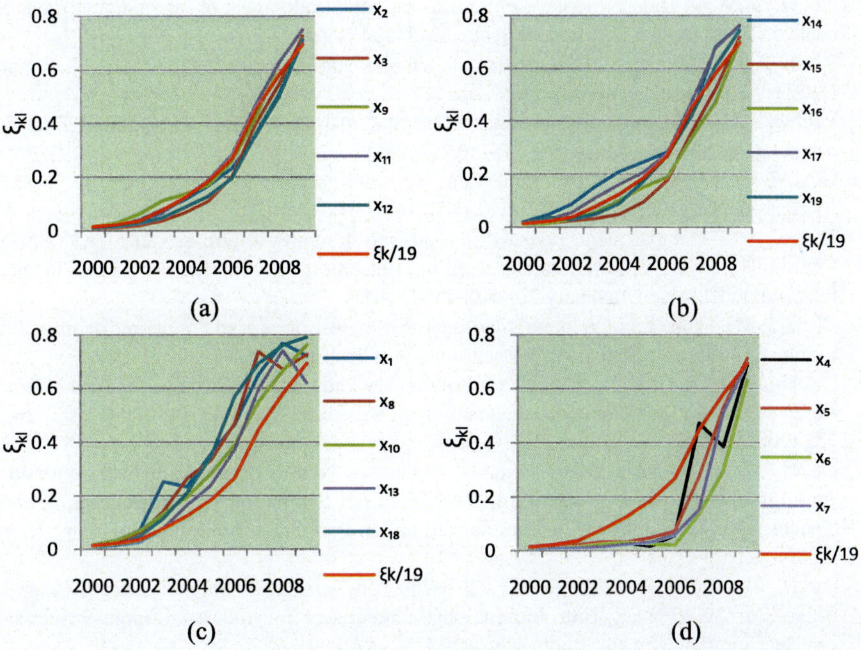

Fig. 4. Contributions of 19 factors to Shanghai's financial industry

4 Conclusion

Due to the strategy of making Shanghai become a global financial center from the State Council of China, Shanghai can avoid the competition with other large cities in China, such as Beijing and Tianjin, so that the resources, especially financial resources in China could be allocated more reasonable. Shanghai should also take advantages of current ports and trades to become an international shipping, trade and financial center at the same time. In order to match the overall economic development of Shanghai, financial market, especially securities market, goal market and futures market, which are major obstacles for making Shanghai become a global financial centre, should be further developed.

Acknowledgements. This research has been supported by the National Natural Sciences Foundation of China (Grant no.70801042). Thanks to Dr. Lihe Chai of Tianjin University's School of Environmental Science and Engineering for research guidance.

References

1. Kindleberger, C.P.: The Formation of Financial Centres: A Study of Comparative Economic History. Princeton University Press, Princeton (1974)
2. Jao, Y.C.: Hong Kong as an International Financial Centre: evolution, Prospects and Policies. City University of Hong Kong Press, Hong Kong (1997)

3. Goldberg, M.A., Helsley, R.W., Levi, M.D.: On the development of international financial centers. The Annals of Regional Science 22, 81–94 (1988)
4. Reed: The preeminence of international financial centers. Praeger Publishers, New York (1981)
5. Poon, J.P.H.: Hierarchical tendencies of capital markets among international financial centers. Growth and Change. 34, 135–156 (2003)
6. Chai, L.H., Shoji, M.: Self-organization and self-similarity in boiling systems. ASME Journal of Heat Transfer 124, 505–517 (2002)
7. Krause, A.E.: Compartments revealed in food-web structure. Nature 426, 282–285 (2003)
8. Chai, L.H., Wen, D.S.: Hierarchical self-organization of complex systems. Chemical Research in Chinese University 20, 440–445 (2004)
9. Chen, L.M., Chai, L.H.: A theoretical analysis on self-organized formation of microbial biofilms. Physica A: Statistical Physics and it Application 370, 793–807 (2006)
10. Du, H., Mao, G., Chai, L., Zheng, J.: Evolutionary Patterns of a City as a Financial Center Based on Non-equilibrium Statistics. International Journal of Arts and Sciences 3, 148–153 (2009)
11. Du, H., Xu, H., Shi, Z., et al.: Evaluation on competitiveness of service industry with non-equilibrium statistical mechanics. In: 2009 IEEE 6th International Conference on Service Systems and Service Management, Xiamen, Fujian Province, China, June 8-10, pp. 76–78 (2009)
12. Du, H., Shi, Z., Jia, Y.: Change of Total Factor Efficiency of Financial Service Industry in China. In: 2009 IEEE 16th International Conference on Industrial Engineering and Engineering Management, Beijing, pp. 1339–1343 (2009)
13. Zhang, X., Chai, L.-H.: A new statistical dynamic analysis on ecosystem patterns. Environmental Modeling & Asssessment 15, 519–529 (2010)
14. Kaski, S., Kangas, J., Kohonen, T.: Bibliography of self-organizing map (SOM) papers: 1981-1997. Neural Computing Surveys 1, 102–350 (1998)
15. Gao, S., Du, H.: Construction and application of financial center competitiveness evaluation index system. Commercial Times 20 (2009)

Chinese Agricultural Growth in Post-reform Era

Mengquan Xu

School of Economics, Huazhong University of Science and Technology,
No. 1037, Luoyu Road, Wuhan, 430074, China
xumengquan@foxmail.com

Abstract. Household responsibility system has brought a period of sound development for Chinese agriculture. However, the positive effect of the system has been released completely in the past 30 years. For China, it is vital to keep the sustainable growth of agriculture. The paper attempts to explain the growth though the empirical analysis of factorinputs, and provides the suggestion for the transformation of the Chinese agriculture growth.

Keywords: Household Responsibility System, Post-Reform Era, Factorinputs, Intensive Growth Form.

1 Introduction

According to Xiaoping Deng's viewpoints, there are two essential transformation for the development of Chinese agriculture in the long run. The first transformation is to use the household responsibility system, instead of People's Communes, we call it reform. The second transformation is to adapt the scientific farming and the needs of the production socialization development, moderate scale management and developing the collective economy, that is to realize intensive growth pattern. Most of the researches attribute the growth of Chinese agriculture to factorinputs, technological progress and system reform. Yifu Lin especially stated that the household contract responsibility system was the main drive for the implementation of the promoting function of agricultural growth (Yifu Lin, 1992). These studies demonstrate that Chinese agriculture has successfully finished the first transformation. Nevertheless, it has been 30 years after the implementation of the household responsibility system, so, we can say that Chinese agriculture has entered into Post-Reform Era. Has the promoting effect of reform been completely released? It is an issue which is worth to think about. Since has implemented reform in 1978, the form of development of Chinese agriculture is extensive growth pattern, and the growth of agriculture consumed a lot of resources and energy. But now the speed of growth is slowing. It indicates that the positivity effect, which comes from the system reform, has been blunted. In this case, how can the Chinese agriculture keep growing? Someone will certainly think that the second transformation will be the new driving force for growth. But, we must realize that it is a long-term and hard process to finish the transformation, even China has entered the late-mid stage of industrialization and has the basic condition in which the extensive form of growth can transform into intensive form. This essay will use data from 1988 - 2008 to conduct empirical research and tries to answer the following questions: in the Post-Reform Era, what

D. Zeng (Ed.): Advances in Computer Science and Engineering, AISC 141, pp. 711–715.
Springerlink.com

China can rely on to achieve sustainable agricultural growth? And how to adjust the development pattern?

This paper is structured as follows: Section 2 analyzes the role of factorinputs on the growth of agriculture; Section 3 selects the appropriate model for empirical analysis; Section 4 Concludes.

2 The Roles of Factorinputs on the Growth of Chinese Agriculture

In general, elements are divided into land, labor, fertilizer, capital, power and irrigation. Statistics show that China commissariat sows a peak of about 113,787,000 hectares in 1998, then decreased and stabilized in 2006-2008. Coincidentally, China's grain output is exactly reached the historical peak of 512 million tons in 1998, then continued to decline, began to rise until 2006 to reach 529 million tons in 2008. Figure 1 shows that there is a clear positive relationship between China's grain output and Arable land after the 1990s. It shows that the contribution of land for agricultural growth is very obvious.

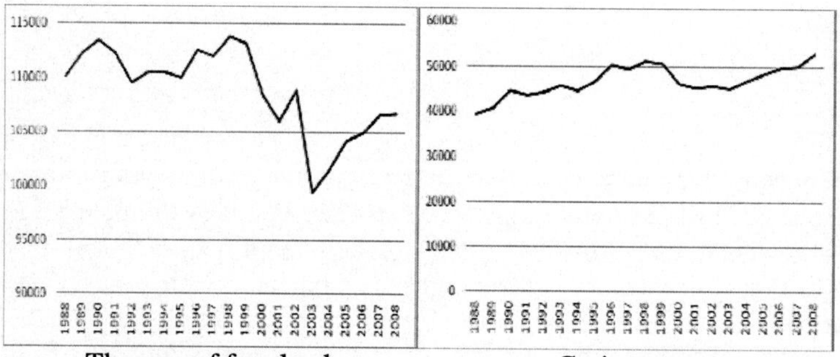

The area of farmland (Unit: One thousand hectares)

Grain output (Unit: Ten thousand tons)

Fig. 1. The area of farmland and the grain output of China

Data source: Compilation of Chinese Agricultural Statistics 1949-2008

As the labor in China is relatively surplus resource, the effect of labor on production is not very obvious. The rural population engaged in agriculture reached a peak of the 330 million in 1999, the rural population continued to decline to 283 million in 2008. Compared with the rural population, food production increased. It shows the contribution of labor to agricultural growth is becoming limited. Although Chinese agriculture is still in the traditional stage, pesticides, inorganic fertilizers and other inputs have reached a high level. The use of chemical fertilizer is 1% more than the world average level.

Yifu Lin and Shenggen Fan (Yifu Lin, 1992 and Shenggen Fan, 1997) believe that the contribution of fertilizers on agricultural growth is obvious. But now, the situation has changed dramatically. Although used largerly, the utilization of chemical fertilizers and pesticides is very low. It is calculated that food production can be averagely increased 11.56kg every year by using 1kg fertilizer during 1980-1985. The amount was 8.24kg during 1986-1990, and dropped sharply to 2.03kg during 1991-1995! Another comparison might be more directly, the national used 40.83 million tons of fertilizer and produced 512 million tons of food in 1998. The use of chemical fertilizers in 2008 reached a total of 52.39 million tons, but the output of food is 529 million tons. The use of fertilizer increased by 28.3% while food production only increased by 3.3%. With the increased use of fertilizer, agricultural production in China shows the characteristics of diminishing marginal returns.

Chinese agriculture consumes huge amount of water, which is accounted for 70% of the total water consumption society. Shenggen Fan's (1997) research indicated that the elasticity of agricultural output for irrigation was the highest among other elements and reached 0.311. As for rural power, China has builted 432 rural electrification counties hydropower in 2006-2010.

3 Data, Model and Empirical Result

3.1 Data

The data are obtained from the "Compilation of Chinese Agricultural Statistics 1949-2008". The time span of the data is from 1988 to 2008. The data categories include agricultural output, land, labor, fertilizer, capital, irrigation and electricity. Among them, the agricultural output is the dependent variable, is measured as the total agricultural output. The Land is measured as crop planting area; Labor refers to all persons engaged in agriculture, forestry, animal husbandry and fishery; Fertilizers measured as the total amount of the using of fertilizers and pesticides; Capital is measured as the total power of various agricultural machinery; Irrigation is measured as one thousand hectares of effective irrigation area; Power in is measured as power consumption in rural areas. All variables are converted to index form (1988 is the base period).

3.2 The Model

The Model is similar with the model which Yifu Lin used (Yifu Lin, 1992), but it has some changes. The model includes agricultural output, land, labor, fertilizer, capital, irrigation and power. Assuming the effectiveness of institutional factors has been completely released, so weed it out of model. The form of the model is following:

$$Y=\alpha+c1Land_t+c2Labor_t+c3Capital_t+c4Irr_t+c5Fer_t+c6Pow_t+\varepsilon_t \quad (1)$$

Where:

Y=log index of total agricultural output

Land=log index of crop planting area

Labor=log index of amount of the farmer in China

Capital=log index of the total power of various agricultural machinery

Irr=log index of one thousand hectares of effective irrigation area

Fer= log index of the total amount of the using of fertilizers and pesticides

Pow= log index of power consumption in rural areas

C is the coefficient to be estimated,a coefficient of say, c=3 means that a 1 percent point increase in the dependency rate is associated with a 3 percent increase in the agricultural output.

3.3 Result

Using OLS to estimate the expression to be the following equation:

$$\begin{aligned} Y=&0.01+0.336^{***}Land_t-0.357^{***}Labor_t+-0.0043Capital_t+0.975^{***}Irr_t \\ &[0.75]\quad [3.632]\qquad [-3.89]\qquad [-0.041]\qquad [5.140] \\ &-0.008Fer_t+0.156^{**}Pow_t+\varepsilon_t \\ &[-0.109]\quad [2.424] \end{aligned} \tag{2}$$

Adjusted R2=0.808

Notes: t-values in parentheses.***Significant at 1% level;**significant at the 5% level.

The result shows that the coefficient of land, labor, irrigation and power is significant. One percent point increase in land is associated with a 0.336 percent increase in the agricultural output. It is consistent with many previous studies.

One percent point increase in labor is associated with a 0.357 percent decrease in the agricultural output. This may be because of diminishing marginal utility of labor produces. As we know, the area of land is limited, in 2006, China's per capita arable land area is 1.40 acre, while the world's is 3.45 acre. It must be noted that 60% of China's population are farmers. Much more labor reduces the productivity. In other words, China's rural labor force is surplus. But the situation is gradually improving with many rural people coming to urban for work.

It is surprising that the coefficient of irrigation is 0.975. The number is even higher than the coefficient of land. It indicates that water shortage is a serious issue for Chinese agriculture. Since 1950, due to drought, China's direct average annual reduction of food is more than 10 billion kilograms, accounting for 60% of the loss of food which caused by the natural disasters. In China, more than half of farmland do not have water conservancy facilities. 80% of the existing irrigation facilities have built before the 1970s, and many of them are age-old, inefficient, without maintenance and ancillary facilities. So, there will be a significant reduction in agricultural output when drought comes, and there is about 326 million acre of farmland affected by drought every year.

It should also be noted that the coefficient of power is 0.156, while the coefficient of capital is not significant. There are certain relations between power and capital. The capital invested in agriculture, then usually transformed as agricultural machinery. So far, in China, lacking of living power for half of the farmers is still a serious issue. There is not adequate power for agricultural machinery to work properly.

Although the coefficient of fer is not significant, the negative relationship between fer and output may have reference value. It indicates that the effect on output through using fertilizers and pesticides has been getting smaller and smaller. This is because that the abuse of chemical fertilizers has reduced soil fertility.

4 Conclusion

The growth of Chinese agriculture is depending on the inputting of land, labor, irrigation and power in post-reform era. In other words, the growth still comes from factorinputs in the long term in which the growth mode will transform from extensive to intensive. Although the increase in cultivated area can promote agricultural growth, a substantial increase in cultivated land is impossible. After all, land is a rare resource in China. But we should not be pessimistic as not all the land has been fully utilized. For the various reasons, there are many low-yield farmlands in China. According to statistics, China's low-yielding fields account for about 70%. So improving the efficiency of land use is necessary. The contribution of irrigation to agricultural growth is the biggest, so China should intensify the construction of water conservancy facilities in rural areas. Power is complementary with capital. Lacking of enough machines or mechanical power is less likely to provide effectively support to the agriculture growth. Oppositely, the agriculture growth could be accelerated through coordinating the relationship between capital and energy.

In summary, in post-reform era, without considering the technical and institutional factors, Chinese agriculture can keep sustainable growth by increasing the input of land, water, and power. But in the end, it is still extensive mode of growth. To achieve intensive growing pattern, China need to pay more attention to four following aspects: improving the efficiency of land use; adjusting the agricultural labor force; increasing the efficiency of irrigation and achieving a reasonable collocation of power and mechanical.

Due to limitations, the essay do have some shortcomings. First, as the lack of data, it is less likely to take account into technological progress ; Second, some northern rural areas in China have actually achieved intensive production. It is very meaningful that those northern areas could be compared with other areas. But there is not much data available to conducting the research.

References

1. Lin, Y.: Rural Reforms and Agricultural Growth in China. American Economic Review 82, 34–51 (1992)
2. Fan, S., Pardey, P.G.: Research, productivity and output growth in Chinese agriculture. Journal of Development Economics 53, 115–137 (1997)
3. Schultz, T.W.: The Economic of Agriculture. McGraw—Hill (1953)
4. Feng, H.: Sustainable Agricultural Development Theory and Practice. Xinhua Publishing House, Beijing (2006)
5. The Rural Social Economic Investigation Department: Agricultural statistics compiled 1949-2008, State Statistics Bureau (2009)
6. Cheng, X., Guo, Z.: Science and education - the only way to revitalize agriculture. People's Education Press, Beijing (2003)

Research on Course Quality Evaluation Method of University Bilingual Education

Min Qi, Helin Cui, Yangyu Fan, and Guiqing He

School of Electronics and Information, Northwestern Polytechnical University, 710072 Xi'an, China
drqimin@nwpu.edu.cn

Abstract. Course quality evaluation is a basic issue and main section of bilingual education. This paper researches the course quality evaluation theory which is of characters of university bilingual education. It first discusses the current study situation in this field. And then, proposes the detail operation methods of bilingual course quality evaluation from the teaching materials, special knowledge learning, English ability promotion and experimental skill. The research is of benefit to the development of bilingual education.

Keywords: Bilingual education, course quality, evaluation.

1 Introduction

Course evaluation is an unignorable research field in the university education practice and innovation. Since R. W. Tyler published the "Basic Principles of Curriculum and Instruction" in 1949, it became an important component of the activities in curriculum development. The content of the course evaluation enriches more and more after the middle of the 1960's, and gradually expands to the course plan, course content and course target etc [1].

As the fast development of the information popularization, economic globalization and education internationalization, bilingual education becomes an unavoidable mission to Chinese universities [2]. Chinese bilingual education is now in the primary stage of exploration period. The teaching level in different universities and areas distinguishes a lot, and the condition of comprehensive course evaluation to bilingual education is not mature yet. Whereas one aspect of course evaluation, course quality evaluation, is the most urgent problem needed to be solved which is related to the improvement of students' learning effect, teaching material, education environment and teaching method. With the help of course quality evaluation, people can find the weakness and shortcoming during the bilingual course performing, and optimize the teaching level, teaching materials, and students' quality, etc. The course quality evaluation has a practical feasibility and is significant to bring the benefits into full play. In addition, it also sets a foundation to further bilingual course evaluation.

2 Analysis of Current Course Quality Evaluation Research

In recent years, experts have researched on course evaluation model deeply and published many researching articles. Some representative kinds of course evaluation

D. Zeng (Ed.): Advances in Computer Science and Engineering, AISC 141, pp. 717–721.
Springerlink.com

models such as traditional expert model, Tyler's objective model, Sufflebeam's CIPP model, Stake's Countenance model, and Ethnographic Evaluation, etc. These evaluation models have different features in the value orientation, method and participant. Many of which are proposed to focus on the weakness of traditional evaluation method.

Modern education theory considers that knowledge is a process and knowledge system always sets in variant condition. In order to evaluate students rationally, people must pay attention to the detail performance of students in study process and then carry out procedural evaluation. Also people should evaluate a course from different level and aspect to reach the goal of comprehensive course evaluation.

There are three evaluation models in Chinese current higher education course evaluation: expert evaluation, peer evaluation and leader evaluation. Expert evaluation has limitation to accurately reflect each teacher's teaching situation and each course's quality because the limited amount of supervisors and attending lectures. Peer evaluation takes the form of attending lectures each other among colleagues. As an independent evaluation model, it's hard to persist effectively. Leader evaluation is much influenced by the evaluators' administration authority. Besides, direct evaluation is also a common method which takes by the students' direct participation of secret ballot.

Recently, China has got much progress in course innovation and made certain researches on course evaluation [3-6]. Although there are still many problems needed to be solved in detail operation, the notion to course evaluation has been changed. Public opinions definitely consider setting up an evaluation system, which with multicomponent evaluation items, different kinds of evaluation methods and focus not only on results, but also process. The evaluation's functions of improving teaching practice, promoting the development of teachers and students are stood out. The phenomena of excessive laterality to knowledge remembrance, written examination and electing are changing. That is to say, it has been the consensus in bilingual education to innovate in current bilingual course evaluation methods and form a better mechanism of evaluation and guidance [7-8].

3 Bilingual Course Quality Evaluation Method

Bilingual education is a newborn object in Chinese education innovation. The evaluation mechanism is still in exploring step. By researching some problems that occur in bilingual education quality evaluation, the corresponding evaluation methods are proposed as follows.

3.1 Similarity Evaluation between Original Textbook and Chinese Syllabus

Material is the incarnation of teaching theory and teaching method. In China, the main teaching material is the English original textbook and adapted teaching material. The majority of adapted teaching material also comes from the original textbook, so the selection of original textbook is the most important step. It is undoubted that using English original textbook can get much benefit for student to learn language, be familiar with western thought patterns and expand the blending advantages of Chinese and Western thinking habits. Simultaneously, we should consider the requirement of Chinese syllabus to subject course. Some English original textbook is very different from the Chinese course requirement, because of different social and life environment. The textbook content is too low or too partial, and can not adapt to Chinese situation. Therefore, how to select an appropriate material should have a criterion. The steps to evaluate similarity between English original textbook and Chinese course syllabus are as follows.

First, referencing Chinese course syllabus and according to the formula (1), calculate the content score of the English original textbook. That is:

$$SCORE = \sum_{i=1}^{n} \lambda_1 a_i + \lambda_2 b_i + \lambda_3 c_i \tag{1}$$

Where, n is the amount of chapters. When calculating, compare with Chinese course syllabus. a_i is the number of terms of optional contents in chapter i. b_i is the number of terms of general compulsory contents in chapter i. c_i is the number of terms of important compulsory contents in chapter i. λ_1, λ_2, λ_3 are weights of each terms respectively. *SCORE* is the total score of the English original textbook. The expert can contrast the English original textbook to Chinese course syllabus, and get the score value denoted by $SCORE_1$. For instance, we can assign λ_1, λ_2, λ_3 as 1,3,10 values to stress proportion of important contents.

Because the *SCORE* is absolute value, and is lack of the objectivity for comparison, we use a relative variable *SIMILARITY* to compare. The formula of calculating *SIMILARITY* is as follows.

$$SIMILARITY = \frac{SCORE_1}{SCORE_2} \times 100\% \tag{2}$$

Where, $SCORE_2$ is the total score of Chinese course syllabus, and also be calculated by formula (1) similar with $SCORE_1$.

SIMILARITY represents the similar degree between English original textbook and Chinese syllabus. The English textbook with a higher *SIMILARITY* is more suitable as a bilingual teaching material for Chinese university.

3.2 Professional Knowledge Studying Effect Evaluation of Bilingual Students

Due to using English original material and part contents of a lesson is taught in non-heritage language, the lagging problem in language will make students set back to some extent when they are thinking. According to the teachers, they usually need to explain the contents repeatedly because of the language limitation existing in both teachers and students. The case causes that teachers can't accomplish lesson in time or can't interpret knowledge sufficiently, and leads to compress the teaching contents.

Therefore, the professional knowledge study effect of students should be evaluated in time. Check regularly whether the bilingual students can comprehend the content that Chinese course syllabus prescribes. Observe and contrast the variation of the academic achievements between bilingual students and non bilingual students who have the same course. If there is problem, solve it in time to avoid the decline of students' studying quality.

3.3 English Ability Increase Evaluation of Bilingual Students

Bilingual education' goal is that the student can read English academic articles, express their academic idea freely, and can summarize the knowledge they have learnt in English. There are some methods that can check and evaluate the increase of students' English ability:

(1) Method that combines the quiz and regular exam. This is a common checking method, but the style of quiz is different. First let students record English videos before and after attending the bilingual course. To increase the objectivity, we can adopt questions from teacher and student performances an impromptu speech. By contrasting these two videos, the teacher judges the increase of student's English ability.

(2)Using non-mutual effect method of the Ethnographic Evaluation to collect information, observe and make out the variation of students' English ability in natural situation during the class teaching, for example, topic discussion, answering question and other interactive activities.

(3)The students can be divided into several groups, and catch two parts video with a few minutes by mobile phone, computer, camera or other devices. One is to give an academic content discussion. The other can discuss the western culture. Demand each person must have performance in the video and take it as a part of total score. Excellent work can be appreciated by the class and discuss freely to the interesting part. This method can arouse students' studying interest. At the same time, the video preparation and appreciation are also a vivid learning process. In the whole process, it is beneficial to have a synthesize evaluation to studying factors in learning process, such as the students' emotion, attitude and etc.

3.4 Experiment Ability Evolution of Bilingual Students

Because the experiment course focuses on the specification of student operation ability and creation, we observe student from three aspects, which are knowledge and skill, process and method, emotion and attitude. In knowledge and skill, we focus on the understanding ability in experiment theory and method, the normalization in using experiment equipment, experiment data analysis and processing ability and report writing. In process and method, we focus on the experiment design ability, the reasonability in experiment steps, experiment method consideration and modification. In emotion and attitude, we focus on whether the student has a good cooperation spirit and communication attitude with other students. The quantization can adapt formula (3):

$$EXPERIMENT = \sum \delta_1 a_1 + \delta_2 a_2 + \delta_3 a_3 + \delta_4 a_4 \tag{3}$$

EXPERIMENT is the experiment ability score. Where, a_1 is the score that reflects the student operation steps normalization during the experiment; a_2 shows the experiment data score; a_3 is the experiment report score; a_4 gives rationality and efficiency score of the experiment steps; δ_1, δ_2, δ_3, δ_4 are the weights of each parameter; a_1, a_2, a_3, a_4 can be changed in different university. The student's score is denoted by $EXPERIMENT_1$. When a_1, a_2, a_3, a_4 are maximum value, the experiment ability full score is got and denoted by $EXPERIMENT_2$.

ERATE which can reflect student experiment ability is defined as formulas (4). Lager *ERATE* value shows stronger ability.

$$ERATE = \frac{EXPERIMENT_1}{EXPERIMENT_2} \times 100\% \tag{4}$$

4 Conclusion

It is the aim of bilingual education to bring up comprehensive talents of international consciousness, improve professional's English ability, increase Chinese international competitiveness and enhance the economy communication and cooperation. Course quality evaluation, which is a weakness in bilingual education researches, is the basic issue and kernel step in bilingual courses development. Its research is benefit for promising bilingual education. This paper is to give a research on course quality evaluation theory which is of characters of university bilingual education. It is a basis to an overall study of course evaluation system of bilingual education.

References

1. Cao, D.: The research on obstacles and strategies of bilingual education in Chinese university from perspective of ISD. Jiangxi Normal University, Jiangxi (2005)
2. Zeng, M.: Thinking of the improvement to bilingual education in university. Education Exploration, 48–49 (2009)
3. Yuan, X., Shi, Y.: Research and practice of bilingual teaching for electrical and electronic technology. Journal of Electrical & Electronic Education, 66–68 (2010)
4. Wang, S.: The primary idea of construction of the western university bilingual education quality evaluation and supervision system. Journal of Changchun University of Science and Technology, 149–150 (2010)
5. National top-quality course resource network, `http://www.jingpinke.com/index`
6. Zuo, G., Gao, H., Duan, L.: The measure model and its accomplishment of course evaluation. Higher Education of Science, 27–31 (2007)
7. Hao, L.: The research of theory and practice about curriculum assessment and curriculum evaluation of colleges and universities. Lanzhou University, Lanzhou (2007)
8. Hu, X., Tan, M.: Reform and practice of bilingual teaching in major course of information specialty. Higher Education Forum 61, 36–38 (2009)

Discussion on the Bilingual Teaching Methods of Chinese University

Min Qi, Jiajun Zhou, Helin Cui, Yangyu Fan, and Guiqing He

School of Electronics and Information, Northwestern Polytechnical University, 710072 Xi'an, China
drqimin@nwpu.edu.cn

Abstract. As Chinese economy is brought into the development trend of globalization and the communication with other countries is becoming more and more frequent, cultivating of bilingual personnel has become one of important goals of Chinese education. This paper first studies the bilingual education modes of some representative countries which have abundant experience in these aspects, then discusses some issues deeply about bilingual education in Chinese university concerned with its present situation. The issues are about Chinese bilingual education mode, transitional mode and most concerned problems, as well as the bilingual teaching methods. It is helpful to improvement of Chinese university bilingual education.

Keywords: Bilingual education, education mode, teaching method.

1 Introduction

In the late 1980s, under the influence of economic globalization, bilingual education was paid more and more attention in many countries. In order to enhance the ability of international communication and cooperation and cultivate high-quality talents with international competitiveness, bilingual education of specialized courses was implemented and expanded in Chinese university since 2001. Bilingual education is benefit to students to learn and thought in two languages, and makes the course learning and language learning supplement each other. Also, it makes the students understand different cultures easier and is helpful for improving their intercultural communicative competence.

Many experts in the field of higher education and bilingual education have carried out a great deal of research works in bilingual education. Many universities combine with their actual situation to formulate management regulations. All those works have greatly promoted the development of bilingual education. But in Chinese universities, the problems of lacking bilingual teachers, lacking teaching materials and lacking teaching skills are still common phenomenon and doubtful opinions has existed. In this case, how to improve the bilingual teaching' effect becomes an increasingly protruding problem. This paper researches the bilingual teaching methods deeply, and hopes to be helpful for bilingual teaching improvement of specialized courses in Chinese university.

D. Zeng (Ed.): Advances in Computer Science and Engineering, AISC 141, pp. 723–729.
Springerlink.com

2 Bilingual Education of Representative Countries

2.1 Canada

In 1960's, in order to eliminate national resentfulness and public pressure, Canada announced that both English and French are official language. Then, the French Immersion bilingual education comes into being [1]. French Immersion education can be divided into early mode, middle mode and late mode. The early mode gives children the French immersion education when they are in kindergarten or grade one, the middle mode begins from grade four or five, and the late mode begins from grade six, seven or eight.

As the students grow into university and the social needs increase, universities start bilingual education, and the most successful experience is from Ottawa University. Ottawa University succeeds in bringing immersion bilingual education from primary and middle school to university of its specialized course teaching, and establishes "protective specialized course teaching" mode. The second language teacher is appointed to specialized course teacher, and the students are given a relaxed "protective environment" to learn knowledge. In the mid of 1980's, the university begin "auxiliary course" that the native students and second language students learn in the same class. At the same time, there is a 90 minutes language lesson a week which was added as a supplement to meet the need of language study.

2.2 America

America adopts submersion bilingual education and focus on minority and immigrant students. The purpose of the special education mode is to help these students to study English and assimilated into mainstream society [2]. The minority and immigrant students with primary English are in the same class with the students whose native language is English, and a competitive study is formed. No extra English lessons to be given, it is a kind of swimming or sinking education mode, and the students struggle and grope by themselves. The purpose is to make sure that English limited students learn English as soon as possible. Submersion bilingual education includes the forms of structured immersion bilingual education, submersion with sheltered English, and submersion with withdraw classes.

American bilingual education is a kind of transitional bilingual education. The final goal is to make English as a dominant language.

2.3 Singapore

Early, the aim of bilingual education in Singapore is to improve all of the students' bilingual ability and cultivate their intercultural communicative competence. Later it gradually changes to improve those of the students who have certain language learning ability. American transitional bilingual education mode is referenced at the beginning, and then Canadian immersion bilingual education mode is adopted in 1970's. After the mid 1980's, there appears three-direction distributed mode with Singapore characteristics as a main bilingual education form. This education mode divides students into three different kinds of classes respectively according to their academic achievements [3].

2.4 Australia

Australian bilingual education started from 1985, Benowa state High School is one of the earliest bilingual education schools in Australia. The goal is to improve students' English ability and cultivate their positive attitude when they face to other language and culture. It is a part French immersion bilingual education mode, and the rate taking French as teaching language in bilingual class increases from 54% to 63% year by year as grade grows. The teacher encourages students to comprehend new words and concepts by inferring and guessing, and provides students with variant chances to use French, such as answering questions, conversation show, playing a role as teacher, group and class discussion, and writing in French, etc. The school also arranges bilingual students to have 3-4 weeks communication activities to French speaking area every year.

Countries above have different political purpose and microcosmic factors in bilingual education, but their ample bilingual education experiences are useful references for Chinese bilingual education.

3 Bilingual Education of Chinese University

3.1 Bilingual Education Mode in China

In China, English is just a foreign language, and Chinese is absolutely dominant language. For students, native language is the most frequently used language and they take English as the second language to learn. They cannot completely separate themselves from the native thinking method. As bilingual education is not language teaching, academic knowledge should be first and its teaching quality permitted no decreasing. Under this situation that there are no enough ideal environment and condition for too much English using in class, an incompletely immerse teaching mode is a better choice for specialized course learning. It is appropriate to the primary period of Chinese bilingual education, and the students can have academic thinking completely during their English ability is progressing [4]. According to other countries' successful experience of bilingual education, 50 percent is the best rate for the second language using in class.

Chinese current bilingual education can be divided into primary, transitional and advanced mode, according to the English usage rate in class.

(1) Primary mode. Use English original teaching material and English writing, but the teacher gives lesson in Chinese, students' homework and examination are in Chinese too. This mode is appropriate to students who are weak at English ability.

(2) Transitional mode. Use English original teaching material and English writing, and the teacher alternately speaks in English and Chinese. Homework and examination are given in English, while the answers are partly in English. In this mode, students are easy to obtain professional knowledge and the requirements to teacher are lower. This is especially suitable for the initial bilingual education.

(3) Advanced mode. Use English original teaching material, English writing, the homework and examination are given in English. Both teacher and students in class speak almost in English, and the homework and examination are finished also almost in English.

The requirement is higher to teacher and students in this mode, and it is fit for teachers who have abroad experience and for higher grade students with better English.

3.2 Transitional Mode and Most Concerned Problems

Teaching Methods of Transitional Mode. Though English is used in different ways and different rate in teaching, almost all successful bilingual education experiences in China belong to the transitional mode at present. The teaching methods of transitional mode are as follows:

(1) Bilingual auxiliary teaching. Simple contents are taught in English and hard parts are taught in the form of English courseware with Chinese explanation. For example, for the review at the beginning and summarizing at the end of a class, the oral English presentation is over 30 percent. For the harder part, the oral English presentation is at about 20 percent. This mode emphasizes that the oral English should be used as much as possible in class, and all homework and examination are arranged and finished in English.

(2) Progressive teaching. At first, Chinese is the dominant language in class. As students become more and more familiar with specialized terms, English is gradually increased in class.

(3) Module teaching. English is as the dominant language in some chapters during each semester.

The most concerned problems are related to teachers and students.

Problems That Teachers Should Pay Attention. English-Chinese translation is a common phenomenon in bilingual education which teachers should specially pay more attention to and avoid it as much as possible. Although Chinese students actually are good at English reading and comprehension, they usually understand the sentences only after translating English into Chinese unconsciously, because understanding and remembering in mother language is more fluent. In this status, it is much necessary for teacher to lead students to cultivate the habit of thinking in English step by step.

Some teachers spent a lot of time on vocabulary translation and explanation overly in detailed, or on the selection of so-called "appropriate" Chinese vocabulary. This approach interrupts students' thinking fluency in English, and actually strengthens Chinese. It is no beneficial for developing the habit of English thinking. In addition, home works similar to English-Chinese translation will greatly limit students' English writing speed and ability, also the sentences has Chinese feature obviously.

The focus of bilingual education is not for the language learning but the actual content that the language to express, so training the language application ability of students is more important [5]. In class, the teacher should take good care of his role, and help students learning professional knowledge in English atmosphere. By understanding foreign culture, the students can immerse deeply to understand the professional issues, in stead of dwelling on a vocabulary or a grammar whether it is proper [6].

Problems Easily Occur to Students. A common problem is that oral expression is not based on English expression, but on the English syntax in translation approach. After students say one or half sentence, syntax errors were found and immediately corrected, which causes the sentence not fluent. According to experts' advice, the teacher can guide students to cultivate and strengthen the habit: if an error is found, as long as it does not affect the understanding, just remember it and correct next time. After a period of training, the oral English will be better.

In order to avoid translation habits, when doing their homework, the students can read several examples first and then give the answers by imitating English expression. By this means, the English writing ability will be improved along with the learning of professional knowledge.

It has been proved by the experiences that transitional mode is comparatively most effective at present. As long as teachers and students pay attention to overcome their own weaknesses and constantly improve the ability of bilingual teaching and learning, the bilingual education of Chinese university will transit to the advanced mode gradually.

3.3 Bilingual Teaching Methods

Spiral Teaching Method. In order to make the students from communicating by simple English to think in English during the professional knowledge learning, the spiral teaching method can be adopted, according to the acceptance ability, level of expertise and general English level of students. It is a cycle-repeated and spiral way of teaching. There are three steps: First, let students learn English vocabulary. Second, guide students to express professional knowledge in English. Last, help students to apply English and thinking in English. These steps are mutual penetrated, continuously repeated and spiral so as to promote the students' professional knowledge and English level ceaselessly [7]. It is mainly adopt pre-question to extend teaching and learning process to outside class.

The pre-question teaching method is to allow students to review in the final 10 minutes of each class. The teacher gives some questions according to the class that learned and to be learned. Students are required to prepare these questions to answer or discuss in the next class. The purpose is to make students not only study professional knowledge, but also pay more attention to English understanding and expressing ability, also not only study hard in class, but also make deep preview after class. This can improve the efficiency of class teaching.

Case Teaching Method and Situational Teaching Method. The teaching activities are performed by case, task and competition. Students' learning enthusiasm can be greatly aroused through participation in discussions and role-playing. The knowledge accumulation is increased by collecting professional materials. The team cooperation consciousness and skills are enhanced by outside class training. At the same time, the English expression ability and logical ability in professional thinking are also improved by class presentation [4]. In all of these processes, students can experience and understand the language impressively, and the professional knowledge is improved significantly.

Interactive Teaching Method. In the interactive teaching method, both teacher's leading role and students' subjective initiative role are given to full play. It is an important method to improve teaching effect, and should take full advantages in bilingual education. There are several concrete forms:

Use heuristic teaching approach. Lead the students to research and find problems, and stimulate their studying and learning interest.

Encourage students to ask questions in simple English expressions, in order to help them to overcome the English speaking problem.

Provide discussion time in class. This can help to create interactive class atmosphere. The current experiences of Chinese bilingual education have shown that once the students start to speak English, their bilingual learning enthusiasm is also improved.

Add interactive parts in class. Wonderful and appropriate interaction can relieve fatigue, make students concentrate and understand knowledge much easier, and test students' listening effect and so on.

In addition, the student listening comprehension is generally in lag condition. With multimedia aids, the students' attention and understanding ability can be greatly improved by vivid visual objects, such as pictures, simulated graphics, flash animation, video, and program display [8]. It is good for teacher to spend more time on the key point explanation.

4 Conclusion

As Chinese economy is brought into the development trend of globalization and the communication with other countries in politics, economics, and culture is becoming more and more frequent, cultivating of bilingual personnel has become one of important goals of Chinese education development. The fundamental purpose of bilingual education is to understand other countries' specialized education system, get a mutual promotion at professional knowledge learning and English learning, and enlarge communication and cooperation with other countries.

There's a famous saying "There is a method in teaching but there is no method of settlement in teaching". Teachers should use different teaching methods flexibly according to the students' level and course characteristics in bilingual education. At present, the mode of bilingual education needs not to be restricted by the form. Teachers' quality, teaching materials, students' English level and other factors must to be considered. With continuous exploration and gradually improvement, the bilingual education in Chinese university will evolve into advanced mode finally.

References

1. Cao, D.: The Research on Obstacles and Strategies of Bilingual Education in Chinese University from the Perspective of ISD. Jiangxi Normal University, Jiangxi (2005)
2. Yuan, P., Yu, L.: A Comparative Study: Immersion Bilingual Education in Canada and Submersion Bilingual Education in the U.S. Comparative Education Review 76, 86–90 (2005)
3. Long, Q.: A Study on Status Quo of Scientific Bilingual Instruction and Related Theoretical Issues. Nanjing Normal University, Nanjing (2006)
4. Zhang, W.: On the Application of Case Teaching Method in Tourism Management Teaching. Liaoning Normal University, Liaoning (2010)
5. Feng, C., Li, G.: The Research on Obstacles and Strategies of Bilingual Education in Chinese University. Education Exploration, 67–68 (2008)
6. Xiang, Z., Zhou, C.: College bilingual teaching method and the teaching means practice and thinking. Journal of Peking University (Philosophy and Social Sciences), 252–253 (2007)
7. Li, M., Yuan, C., Li, Y.: A Study on the Characteristic and Method of Bilingual Teaching in Science. Journal of Harbin University, 132–135 (2010)
8. Ye, X., Wang, X., Wang, M.: Research on the teaching reform of the bilingual course "Design and Implementation of TCP/ IP Networks". Journal of Xi'an University of Post and Telecommunications, 135–138 (2009)

A Study on Intellectual Capital-Driven Mechanism of Growth in Technology-Based New Ventures*

Yiyang Fan and Minghui Xue

Business School, University of Shanghai for Science and Technology,
Shanghai 20003, China
fyyqq@usst.edu.cn

Abstract. Among the factors that promote technology-based new ventures (TBNVs) to grow, the role of intellectual capital has drawn people's attention. Based on the definition of the technological entrepreneurship, intellectual capital and their relationship, this paper builds up the intellectual capital-driven growth model of technical entrepreneurial enterprise and analyzes the interactive mechanism between intellectual capital and technology entrepreneurs, in order to seek the source that makes business grow fast, thus offers the theoretical foundation for Enterprise to manage its intellectual capital effectively.

Keywords: Intellectual Capital, TBNVs, Growth Mechanism, Competitive Advantage.

1 Introduction

In 1996, Organization for Economic Cooperation and Development (OECD) used the word "knowledge economy" in the documents of international organizations for the first time, and pointed out: the modern economy is the "knowledge-led economy". For the knowledge-based enterprises, intellectual capital is no longer the supporting role of physical capital and financial capital, but the main factor that decides the productivity and competitiveness. At the late 90s in 20th century, the ratio of U.S. companies of market value and book value is about 2:1, the equivalent 2 times of the same target of the United States average between 1945-1990. Why is there such a big gap? Undoubtedly that is the work of intangible capital, it is intellectual capital. In the context of modern knowledge economy, the growth of high-tech business mainly depends on the acquisition of intellectual capital, the accumulation of value-added and realization of value, and its essence is the operation of intellectual capital. So this

* Sponsoring Information: Shanghai Municipal Research Project of Education and Science 2011 (B11041); Curriculum Development Project by Shanghai Municipal Education Commission (2010[67]).

D. Zeng (Ed.): Advances in Computer Science and Engineering, AISC 141, pp. 731–736.
Springerlink.com

article's study how does the intellectual capital drive growth of TBNVs is of important practical significance.

2 Paper Preparation

2.1 Technological Entrepreneurship

All the time, entrepreneurship is considered an important source of national innovation and economic growth. Technology-based new venture is a special form of entrepreneurship; it is often seen as the regional economic transformation of the propeller in China and emerging economic development countries. But the academic community did not give a unified definition for technological entrepreneurship. From previous research, the difference between technological entrepreneurship and general entrepreneurship is whether entrepreneurship is related to technological innovation and belongs to high-tech-intensive entrepreneurship. There are two view points for entrepreneurship: organization creation point and opportunity point, the definition of TBNVs also has the two perspectives that considers innovative and dynamic are the core of TBNVs [1].

Similar to the division of the entrepreneurship: personal and business, TBNVs can be divided into independent TBNVs and company TBNVs. Independent TBNVs is a single individual entrepreneurship, with emphasis on new enterprises; the company TBNVs is the company's individuals or groups' entrepreneurial, emphasizing on the company internal entrepreneurship and innovation related to technology[2]. Based on the above theory, this paper defines TBNVs as the dynamic process that provides customers with better products or services using the technology innovation or technological achievements.

2.2 Intellectual Capital

Foreign scholars have recognized the intellectual capital is becoming the core of human resources to create economic wealth long ago. As early as 1836, Senior put the "intellectual capital" as a thesaurus of human capital, he believes intellectual capital is the knowledge and skills that people possess. In 1969, American economist John Kenneth Galbraith first proposed the concept of intellectual capital, in his view, intellectual capital is a kind of intellectual activity, a dynamic capital rather than fixed capital form [3]. The editor of U.S. "Fortune" Thomas A. Stewart (1991) first systematically defined the meaning of intellectual capital in his classic paper, he pointed out that intellectual capital has become the most important asset of the United States [4].

Based on the specific business environment in China, the intellectual capital is defined as " the unique knowledge which is in the form of knowledge and dynamic, can create value and keep competitive advantage to enterprises" in this paper, and according to Table 1, the intellectual capital structure of TBNVs is divided into four dimensions: human capital, structural capital, relational capital and innovation capital.

Table 1. Different structure models of intellectual capital

Structural Category	Structure of Intellectual Capital	References
Two-dimensional	Human capital and structure capital	Edvinsson and Malone(1997)
	Stakeholder resources and structural resources	Marr and Schiuma(2001)
Three-dimensional	Human capital, structural capital and customer capital	Petrash (1996), Ross(1997) and Stewart(1997)
	Employee competence, internal structure and external structure	Sveiby(1997)
	Human capital, organizational capital and relational capital	Bontis(2000),Straman(2000) and Seeman(2000)
Four-dimensional	Human capital, market capital, infrastructure assets and intellectual property assets	Brooking(1996)
	Human capital, structural capital, customer capital and innovation capital	Fornell et al. (1990)

3 Intellectual Capital-Driven Growth of TBNVs

3.1 Model Construction

In the knowledge economy times, technology is progress rapidly, TBNVs needs to grow under the guidance of theory, learn and improve them continuously to gain competitive advantage, so as to keep pace with times. Penrose (1914-1996) is one of the most influential scholars about the growth theory of modern enterprise, she revealed the importance of business growth in the process of resources, and made the simple analysis framework: "Enterprise Resource → business growth". Kazanjian divided the growth of TBNVs into four stages: the phase of conception and development of the research, commercial application stage, the entrepreneurship growth stage and the entrepreneurship maturity stage [5].

Organizational learning theory considers that the growth of TBNVs is the gradual process of learning [8]. In fact, the enterprise will face many problems at every stage of the growth, and the leading issues decide the growth stage where the enterprise is. Only that TBNV develops the knowledge and ability continuously, and successfully respond and resolve these leading issues, can drive the continuous development of TBNVs [6], the company's intellectual capital is knowledge and ability to solve the leading problem. As there are different issues at each stage, the knowledge capital that needs are different, so the structure of intellectual capital in each stage of the TBNVs is dynamic. Figure 1 shows the structure of intellectual capital in different stages.

The growth of TBNVs is a dynamic evolutionary process. Within the enterprise, the intellectual capital (the entrepreneur's experience, ability and employee's technical skills) has important practical significance to attract the risk investment and promote its listing, but also can effectively promote enterprise's rapid growth; Baum's research results indicate that the entrepreneurship's environment (dynamic, threatening and complexity) has a significant indirect influence to the growth of entrepreneurship, and

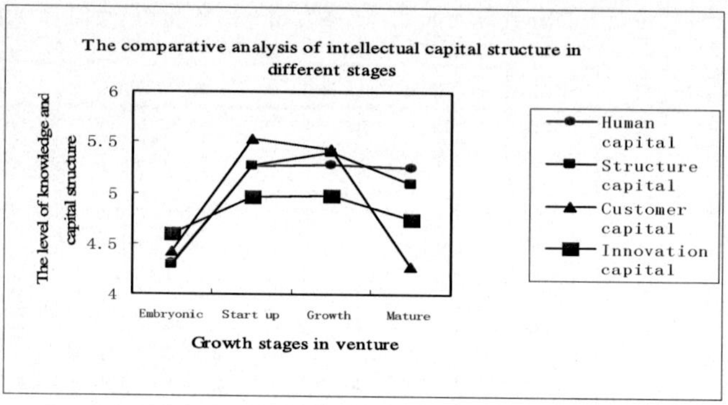

Fig. 1. Characteristic curve of the intellectual capital structure in different growth stages of ventures

its external competitive environment, institutional environment and the choice of corporate strategy will influence the growth of entrepreneurship; based on the internal and external environmental changes and strategic adjustments, entrepreneurship should re-match and accumulate the intellectual capital, in order to promote the sustained high growth performance of TBNVs.

Based on the above ideas, this paper points out the dynamic model of the growth of TBNVs driven by intellectual capital (Figure 2), to describe a more systematic enterprise resources, external environment and internal growth of TBNVs.

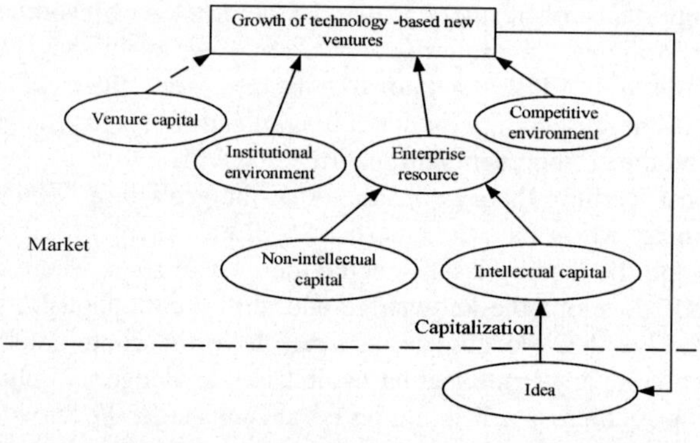

Fig. 2. Dynamic model of TBNVs driven by Intellectual capital

3.2 The Intellectual Capital-Driven Growth of TBNVs Mechanism

As the resource's view that business is the organic combination of tangible resources and intangible resources, the realization of its competitive advantage is to take advantage of these unique resources continuously[8], while, Sanchez and others' studies suggest that intellectual capital is the key resource to obtain competitive advantage [9]. Therefore, this paper puts the enterprise's resources - intellectual capital as the basis of the analysis, builds the dynamic model of intellectual capital-driven growth of technology (Figure2), to clarify the TBNVs can possess competitive advantage and keep up with the times only by increasing their capital stocks and flows of knowledge continuously.

There are many problems (the same or similar problems, different problems) that the technology entrepreneurs would encounter at every stage of the process. However, there are some dominant problems at each stage, companies only respond to these issues and successfully resolve them before entering the next phase to continue to progress. Problems encountered in the current stage may happen again at the next stage, for the same or similar problems, companies should institutionalize or routinize the solution, which can increase the stock of their knowledge, improve problem-solving abilities and promote enterprises grow rapidly. Robert termed that the process as learning cycle of organization (Figure3). Different problems will be solved by entrepreneurs' ability and the knowledge and innovation of technicians.

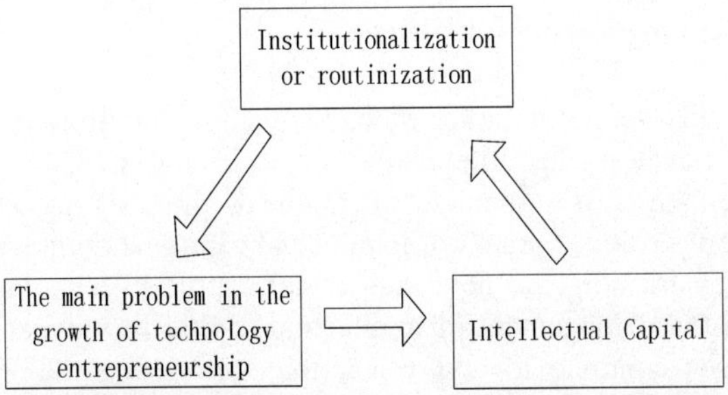

Fig. 3. Learning circle of problem-solving in the stages of TBNVs

Driven by the intellectual capital, TBNVs solves problems continuously and promotes companies' growth, makes their overall ability to accumulation, use and sublimation continuously, so that enterprises can gain competitive advantage constantly. But the TBNVs must carry on technological innovation continuously, and maintain core competencies, so that enterprises have the sustainable competitive advantage in a dynamic environment.

4 Intellectual Capital-Driven the Cornerstone of the Growth of TBNVs

In today's volatile market environment, TBNVs should have the dynamic capabilities to adapt to the new change constantly, it is built up through the interaction of intellectual capital's structural elements , may be the ability of individuals or groups of enterprises, but more importantly, the assets owned by the enterprise. Whether TBNVs has a dynamic competitive advantages, depends on the special resources which is value, rarity, and impossibility to be replaced and imitated, intellectual capital is with above a few features, so that enterprises gets special resources more easily than other companies in a dynamic environment. At the technical level, the dynamic adaptability of TBNVs requires that the R & D activities are innovative, and its creativity, technology and inventions transfer into intellectual capital as soon as possible to promote the value of company at the same time, so that enterprises can combined with external environment and resources, grow rapidly ahead of other companies in a dynamic environment.

Based on the above discussion, the growth of technology start-ups is essentially a spiral growth process of gaining a competitive advantage constantly by intellectual capital. In other words, companies institutionalize or routinize the solution of the same or similar problems to form company's existing intellectual capital, at the same time, companies innovate and form the new business capabilities continuously by interaction with various elements of intellectual capital, to form the competitive advantage and promote business growth. Thus, intellectual capital is the core resources to gain a competitive edge in a dynamic environment for the TBNVs, is the cornerstone of driving business's growth.

5 Conclusions and Implications

From the above, in the modern knowledge economy environment, TBNVs only accumulate and develop the intellectual capital continuously to maintain its competitive advantage, institutionalize or routinize the existing knowledge at the same time, carry on the development of intellectual capital and innovation, to promote business growth by forming the new intellectual capital. Technology entrepreneurs should strengthen the management of intellectual capital in a dynamic environment, to make enterprises continue to add value, thus promoting the virtuous circle of maintaining resources and the competitive advantage dynamically.

References

1. Shane, S., Venkataraman, S.: Gusest editors' introduction to the special issue on technology entrepreneurship. Research Policy 32(2), 181–184 (2003)
2. Tushman, M.L., Anderson, P.: Technological discontinuities and organizational environments. Administrative Science Quarterly 31(3), 439–465 (1986)
3. Edvinsson, L., Sullivan, P.: Developing a Model for Management Intellectual Capital. Europen Management Journal 14(4), 358–364 (1996)
4. Brainpower, S.T.A.: How intellectual capital is becoming America's most valuable Asset. Fortune, 40–56 (1991)
5. Kazanjian, R.K.: Relation of dominant problems to stages of growth in technology-based new ventures. Academy of Management Journal 31(2), 257–279 (1988)
6. Andries, P., Debackere, K.: Adaptation in new technology-based ventures: Insights at the company level. International Journal of Management Reviews 8(2), 91–112 (2006)
7. Macpherson, A., Holt, R.: Knowledge, learning and small firm growth: A systematic review of the evidence. Research Policy 36(2), 172–192 (2007)
8. Leonard-Barton, D.: Well springs of knowledge:Building and sustaining the sources of innovation. Harvard Business School Press (1998)
9. Sanchez, P., Chaminade, C., Olea, M.: Management of intangible:an attempt to build a theory. Journal of Intellectual Capital 1(4), 201–220 (2000)

Study on Professional Ability Improvement for Non-professional Librarians in University Library

Tianfeng Zhao[1] and Jianbo Cao[2]

[1] Library, Zhejiang Normal University, 321004 Jinhua, China
[2] College of Engineering, Zhejiang Normal University, 321004 Jinhua, China
ztf@zjnu.cn, caojianbo2008@163.com

Abstract. Non-professional librarians occupy a large proportion in university library staff. Their professional ability directly affects the library service and development. It is worthy of our exploration and contemplation how to improve the professional ability of non-professional librarians. Through investigation and analysis, some suggestions were proposed in the paper. Posts should be allocated reasonably. The confidence of non-professional librarians should be enhanced. Effective training, coaching guiding mechanism, and humanistic management should be established.

Keywords: University library, Non-professional, Professional ability.

1 Introduction

As the social life rhythm accelerates ceaselessly, people pursue high efficient, bright and smooth way of life and work more and more digitally. Networked information has become a fashion and essential. Digitization and network have entered deep into each field of the society rapidly. Opportunity is brought, and challenge is also brought [1]. As a service department for the readers, university library must keep pace with the times, not only meet the needs of readers, but also lead and inspire people to catch up with and surpass the pace of the times in knowledge. Such a difficult and urgent task needs considerable business personnel, and needs to constantly optimize the human resource. However, non-professional librarian in libraries of domestic universities generally has very large proportion, which impacts the development of the library and can not be ignored. Most of them are not library professional or no professional staff. They do not have the library professional background, but play their role in their respective positions on the positive efforts. Therefore, for smooth functioning of the university library, their strength is indispensable in the past and now, even the future. How to improve their business ability, increase their self-confidence and a sense of belonging, to better promote the development of library, and meet the requirements of the times, is worthy of our deep thinking and exploration.

2 The Current Situation of Non-professional Librarians in University Library

Through consulting relevant literature, it is found that there are a large number of non-professional librarians in many college and university libraries, for an example,

D. Zeng (Ed.): Advances in Computer Science and Engineering, AISC 141, pp. 737–740.
Springerlink.com

in Tianjin Foreign Studies University library non-professional librarians share about 70% in 2009 [2]. In the library of Zhejiang Normal University where the author of this paper works, non-professional librarian also occupies a large proportion.

Non-professional librarians' major before differs very far from professional library mostly, even no relationship between. There are a lot of people having engaged on non-professional library work for many years, and professional and working experience differ very far from the work what they engaged in the library now. Therefore, they aren't confident or not unadapted for the current situation so that their business level is relatively low and they have the negative attitude to improve their business ability.

3 The Suggestions Strengthening Non-professional Librarian Professional Ability

3.1 Allocate Posts Reasonably and Enhance Confidence of Non-professional Librarians

Non-professional librarians have their own strengths, and the different position of library needs different skill and quality. In addition each position is an important part of the whole library, so it is necessary to make every librarian have positions of honor, love their jobs, and be full of confidence. Only if work with the self-confidence and sense of honor, they take a more positive attitude to their own development and the development of library, and expand their service level actively.

Non-professional librarians should be appointed to appropriate post basing on their professional director and characteristics in the library work, striving to give full scope to them to do their best. The development of the library has undergone changes with each passing day. No longer it only contains the narrow borrowing and returning books in the traditional sense, and it has increased in the number of new functions and services. In addition, the targeted subject service has been put on the agenda [3]. Because modern computer operation has grown in popularity, not only the library science professionals are needed, but also other professional talents could play their strengths and embody the advantage [4]. Many professionals, such as in mathematics, physics, chemistry and so on, are available to display their talent and ability in the library subject service. Foreign language is also used more and more widely, for example, many foreign readers and visitors often need to be received and helped in the libraries of college and university. Computer application is the same, and computer professionals are also needed to repair computer and carry out statistical data. Related leadership should consider the expertise of non-professional librarian in the organization of work, and make them adapt to their work at the beginning in the library. As a result, their self-confidence would be increased, so that they can work actively in the library, and then their business could be expanded in the next work, so they can be more and more qualified for other work of library [5].

3.2 Practical and Effective Training and Coaching

Most of the non-professional librarians demand on themselves in a strict way, read work-related books, consult modestly, and try to learn the professional knowledge through their own efforts to improve the level of their business level, but self-endeavor in individual as the unit lacks of direction, not only wastes a lot of time and energy, often gets half the result with twice the effort, but also sometimes gives up halfway. University library can have a definite object in view of elaborate organization to carry on the overall training and guidance, which may be more effective to foster non-professional librarians' professional ability.

Many university libraries have organized a number of professional training, have also taken a variety of forms of non-professional librarian professional ability training and guidance, such as organizing seminars given by professional , having colloquium between professional and non-professional librarians, attending massive training taken the lead by outstanding professional librarians with plentiful experience to group, which have little effect. The reason is that the training and guidance often separate from a professional librarian's practical level and starting point is too high which baffle them. Therefore, it not only is no helpful to raise the level of business, but also will hit the non-professional librarians' aspirant confidence [6]. Hence, it is necessary to understand what is required by the non-professional librarians and provide effective scientific training and guidance.

3.3 Establish Guiding Mechanism

In university library, non-professional librarians are easy to have a sense of inferiority. Most of their work is incongruent with their own studied professional knowledge, even seems irrelevant. However, an important index of professional title evaluation of university rests with the papers and the projects published which must be about operating post. This is too difficult to address this problem, so the fear psychology and the negative attitude are produced easily in improving business ability, even it came to their minds that they want to give up. Often there is the case that a non-professional librarian who works diligently and conscientiously doesn't pursue progress but work completed. The business level including learning library professional knowledge is negatively treated. Non-professional librarians' title requirements are very low, even they give up acquiring professional library series title and maintain the present status.

If guiding mechanism is established to guide and help the non-professional librarians to enhance business level, learn library professional knowledge actively, pay attention to the development trend of library, find the problems of library development and explore the strategy, and approach for the development of library, they will involve into the library enterprise actively, and also improve their business level actively.

In guiding mechanism, on the one hand, non-professional librarians can learn from the experienced professional librarians by the "impart, help, lead" way, whose professional level is relatively high. They could help the non-professional librarians whose business capacity is weak in fixed or indefinite way. On the other hand, library could set up some research groups, and non-professional librarians are assigned to each team, participating in the project and other activities, so that their professional level and the ability of scientific research are improved under the leadership of the groups.

3.4 Humanistic Management

Non-professional librarians need to be valued, but need to be encouraged, recognised and cared more. Library directors should pay more attention to non-professional librarians' growth, difficulties and requirements, adopt flexible and effective measures, solve their problem, trust them and try to allocate some challenging work to them to encourage their ambition. The defects of their business should be criticized and corrected sincerely. Their success and progress were praised appropriately and affirmed timely. Their difficulties in the work should be given help and inspiration. Let them feel that the university library is the cradle for their growth and there is rich nutrition and broad soil for their progress and growth, also there is business facilitation ladder for their business level. As a result, they can be very confident, optimistic to overcome professional obstacles so as to improve their operational capacity, enhance oneself, at the same time, also work better for the library's development.

4 Conclusion

The improvement of non-professional librarians' professional ability has a very important influence to development of the university library. The realistic problems should be faced actively, so this paper embarks from the reality and has putted forward four suggestions, hoping to offer some help for non-professional librarians and the development of library.

References

1. Gao, S., Wu, X.L., Zhao, Y.M.: Research on Library Development in Recent Years. Journal of State Library 67, 23–29 (2009)
2. Cao, Y.: Accelerate College Library Professional Librarians' Pace. Journal of Tianjin Occupation College 12, 145–147 (2010)
3. Zhang, C.H.: The Library Service for the Teaching and Scientific Research Practice Exploration. Journal of Weifang Education Institute 3, 79 (2009)
4. Li, J.J., Wang, Z.Z.: Higher Vocational Library Resources Service for Teaching and Research. Net Wealth 4, 115 (2010)
5. Dai, Y.D.: On Non Library Science Professionals Thinking on Human Resources Development. Library and Information 2, 62–63 (2004)
6. Li, X.F.: How University Library to Strengthen the Young Librarians Talents Team Construction. China Electric Power Education 133, 184–185 (2009)

Research and Analysis of the Forming Mechanism of the Financing Difficulty of Wenzhou's Medium-Sized and Small Enterprises

Xuwei Chen

School of Management, Zhejiang University Hangzhou 310029
Carter Building, Huajiachi, Zhejiang University
268 Kaixuan Road, Hangzhou, Zhejiang 310029
jeky_chen@hotmail.com

Abstract. Wenzhou is an outstanding representative in terms of Chinese private economy and medium-sized and small enterprises (MSEs). Besides market factors, the foremost factor perplexing and restricting the existence and development of MSEs is financing difficulty, so it's of universal, practical significance for solving the financing difficulty of MSEs to analyze effectively the cause and mechanism of forming such difficulty of Wenzhou's MSEs. By analyzing their characteristics and development dilemma and deeply researching MSEs' own factors, financial institutions' business trend and social environment, this article demonstrates the forming mechanism and reason of their financing difficulty and puts forward business thought and suggestions.

Keywords: MSEs, financing difficulty, forming cause, analysis.

1 Financing Status of Wenzhou's MSEs

1.1 Present Development Status of Wenzhou's MSEs

Wenzhou's regional economic is obvious, characterized by developed light industry. It possesses a large market share throughout China in terms of clothing, footwear, printing, plastic woven bags, pens, locks, lighters and shavers, and there spring up a number of excellent MSEs to the extent that they have become a "locomotive" of Wenzhou's economy, exerting huge influence on its whole economy and giving rise to the "Wenzhou Mode" which is famous all over China and even throughout the world. The healthy development of MSEs relates to the future of its economy as an important guarantee for the sustained growth of such mode.

1.2 Characteristics and Development Dilemma of Wenzhou's MSEs

The "Wenzhou Mode" features labor-intensiveness. Based on specialized division of labor and socialized cooperation, Wenzhou has made the following achievements: small commodities supported by large markets, small parts by large complete sets. With the change in domestic and international economic situation in the past years,

D. Zeng (Ed.): Advances in Computer Science and Engineering, AISC 141, pp. 741–745.
Springerlink.com

Wenzhou's MSEs have been faced with numerous disadvantages, including large appreciation of RMB, price rising pressure of raw materials arising from inflation, impact caused by labor deficiency and salary increase, and electricity limiting and energy conservation, all of which constitute major reasons for their dilemma. In addition, the State takes a moderately tight monetary policy so that it becomes more difficult for them to raise funds. When Wenzhou's economy shows a sign of marked slow growth, how MSEs overcome the dilemma as its economic mainstay is vital for the sustained, healthy and rapid development of its economy.

2 Status of Financing Difficulty of Wenzhou's MSEs

2.1 The Question of Their Financing Difficulty from the Perspective of Financing Structure

Over the last 30 years since China's reform and opening-up, the financing of Wenzhou's MSEs has the following unique features: they generally seek internal financing first and then external financing. In the 1990s, Wenzhou's private enterprises' own capital, bank loans and private capital form a proportion of 60:24:16 in their total capital. With the growth of MSEs, the scale of social investment has expanded so that their financing manner has changed increasingly. As a general trend, their own capital still makes the largest proportion, private loans are relatively reduced, bank loans increase by a big margin and regular banking services occupy a more and more important position in the financing structure of non-public enterprises.

First, internal financing is a basic capital source for the establishment and development of MSEs. According to the information from the People's Bank of China, Wenzhou Central Sub-branch, small enterprises with less than 50 employees obtain over 90% of their total capital from their owners, founding members and families and in Wenzhou's private enterprises self-raised funds account for over 95% of their seed capital, so internal financing is their most basic financing manner and a basis for enterprise development, especially when MSEs are just founded so that they are week in profitability and has no credit record, for which they are hard to finance themselves externally. Therefore, it's inevitable for them to encounter financing difficulty.

Second, regular financial capital with loans from banks and credit cooperatives as the core is a major source for MSEs' external financing. According to the information from the People's Bank of China, Wenzhou Central Sub-branch, their loans account for 61% of their total liabilities on the average, and when MSEs are faced with capital difficulty, 78.9% of them will consider borrowing loans from banks and credit cooperatives at first, 17% of them will treat private lending as their first choice and 4.1% of them will prefer to default in paying for their goods. Moreover, the loan requirements of 60.5% of the enterprises are basically met, but 29.1% of them are not satisfied owing to a large loan gap, and 10.4% of them fail to get loans on the whole. When different enterprises face banks as the most important external financing source, their "treatment" is rather different.

Table 1. Indicates the situation of banks' refusing enterprise application for loans and the percentage of China's bank loans in enterprises' capital source

Enterprisescale(persons)	*Less than 51*	*51-100*	*101—500*	*More than 500*
Percentage of refusal times (%)	78.92	57.87	44.18	24.34
Percentage of bank loans (%)	54	20.6	22.6	47.1
Yearsof enterprise (years)	< 2	2—4	4—5	> 5
Percentage of refusal times (%)	72.44	49.07	46.97	45.79
Percentage of bank loans (%)	12	5.2	17.6	25.3

Source: Yang Enqun, *Financing of Medium-sized and Small Enterprises*, Edition 1, Democracy & Construction Press, 2001

As can be seen from the above table, enterprises with different scales face rather different difficulties in obtaining bank loans, and banks refuse MSEs that desire most to get bank loans, but compete for large private enterprises that hardly desire loans.

Table 2. Situation of Financing Difficulty of Wenzhou's Private Enterprises with Various Scales

Enterprisescale (person)	*Less than 51*	*51-100*	*101—500*	*More than 500*
Percentage of enterprises considering financing difficult (%)	93	83	65	54
Registered capital ('0,000)	< 100	100—1000	1000-5000	> 5000
Percentage of enterprises considering financing difficult (%)	82	77	59	36

Source: The People's Bank of China, Wenzhou Central Sub-branch

According to the investigation by the China Academy of Social Sciences, to the question "why are enterprises difficult to obtain loans from regular financial institutions", 38.7% of the enterprises respond by saying that is mainly due to the small scale of MSEs, inadequacy of their mortgaged assets and difficulty in finding guarantors, 25% of them reply that it's ascribable to the low attraction of investment projects, 20.2 % think national policies are partial to large and medium-sized enterprises, and 8.9% regard the discrimination of financial institutions as the reason. Furthermore, 48% and 24% respectively hold financing difficulty is mainly because of the low level of business administration or of the distortion of financial statements and the low transparency of information.

Third, private lending serves as an important supplement for external financing of MSEs. The source of such external financing is diversified, because besides bank loans, other liabilities primarily include private loans, internal financing and enterprises' trade credit. According to the investigation, private lending has occurred to 92.3% of enterprises. The scope of this type of financing activity generally features the descending relations to enterprise owners, so the first choice is borrowing from relatives and friends; the second internal financing; the third borrowing from ordinary social relations; and the last usury from professional lenders. However, private lending features interest rates higher than those of bank loans and a shorter term, hence it suits temporary capital turnover and cannot become a routine channel for enterprise financing.

2.2 Analysis of the Reasons for Financing Difficulty of Wenzhou's MSEs

2.2.1 MSEs' Own Factors Constitute a Basic Reason for Financing Difficulty

MSEs have inherent financing obstacles. MSEs are hard to raise funds by issuing stocks or bonds through direct financing, Wenzhou's MSEs are engaged in labor-intensive industries with low technological content and there exist a few problems such as undue competition of the same trade, generally low profitability and small self-accumulation. Consequently, MSEs can do nothing but borrow loans from banks to meet their capital demands. In the eyes of commercial banks, MSEs have such demerits as irregular internal management, low social credit, low finance transparency and high trading cost, in particular their general difficulties in furnishing enough mortgaged assets and finding third party guarantors, so that the loans are highly risky. Therefore, considering the safety, fluidity and profitability, these banks disagree to grant such loans. Of the difficulties, guarantee difficulty is a major problem for the loans of MSEs.

2.2.2 Financial Institutions' Business Preference Is an External Reason for Financing Difficulty

2.2.2.1 Market Preference of Granting Loans to Large Organizations Instead of Small Ones

In macroeconomic control, all commercial banks concentrate on granting loans to large industries and large customers, embodying the attitude of not supporting MSEs in customer choice. From the angle of practical operation, banks operating in a traditional manner cannot focus on MSEs because a customer manager maintains at most 30 credit customers and there are 300,000 MSEs in the whole city. To avoid excessive investment in human resources and material resources, they need to balance their investment and output in their business development.

2.2.2.2 The Higher Authorities Tend to Regain Part of Loan Powers to an Excessive Extent

In process of adjusting the credit structure and controlling the total amount, some commercial banks raise the extent of centralized management and their higher authorities regain part or all of the powers to examine and approve loans, so that misplacement in time and space occurs to the examination and approval of loans to MSEs and the credit demands of some MSEs fail to be met in time, for which the financing difficulty of MSEs becomes evident.

2.2.3 The Weakness of the Social Environment Is a Bottleneck Resulting in Financing Difficulty

2.2.3.1 Narrow Direct Financing Channels

For MSEs in their growth period, internal financing seldom meets their capital demands while in external financing, direct financing channels are very narrow and both securities markets and stock markets set a threshold hard to cross; the domestic growth enterprise market is at its initial stage, so the capacity of the capital market is limited.

2.2.3.2 Backward Social Credit Development

The imperfectness of the social credit system is always a major barrier restricting the development of China's financial sector. The country's enterprise credit system is yet to be built completely and the information asymmetry between its MSEs and its financial institutions is extremely serious, so banks are hard to have full access to the complete information on enterprises, cannot understand their actual financial condition and have larger difficulty in making clear their operation risks and proceeds. For this reason, financial institutions are especially cautious of granting loans to MSEs, making them encounter financing difficulty owing to information asymmetry.

To sum up, financing difficulty is a major existing problem harassing the development of Wenzhou's MSEs, and for commercial banks, information asymmetry and guarantee difficulty are major barriers of loans to MSEs. Consequently, to appease financing difficulty, it's imperative to solve these two problems. The traditional banking mode hardly helps to solve them, so breakthroughs and innovations must be made to overcome the barrier affecting the financing between MSEs and financial institutions, so as to create a favorable condition for the healthy growth of Wenzhou's MSEs.

References

1. Yang, F.: Research into the Regional Financial and Ecological Environment Dominated by Private Economy – Case Analysis of Wenzhou. Socioeconomic System Comparison 3 (2008)
2. Gong, R., Wang, F.: Research into the Present Development Status of Wenzhou's Private Finance and Its Countermeasures. Reform and Development 1 (2009)
3. Li, Y., Yang, S.: Financing of Medium-sized and Small Enterprises and Banks. Shanghai University of Finance & Economics Press, Shanghai (2001)
4. Cai, L.: Analysis of Quality of Wenzhou's Banking Credit Assets. Zhejiang Finance 10 (2008)

The Relationship between Product Quality and Transfer Price in the Downstream of Pig Supply Chain*

Kai Hu and Xiao-qing Gan

Business School of Jiujiang University, Jiujiang, P.R. China
cnkaiman@163.com

Abstract. In the relationship between pig-raising enterprise and slaughtering enterprise, there exist a principal-agent problem with the quality as core, namely the pig-raising enterprise having information superiority, and slaughter and processing enterprise is in the relative asymmetry position. Slaughtering and processing enterprise is the principal, and pig-raising enterprise is agent. The contract is a typical Stackelberg game model, namely it's the Leader-Follower of the bi-level programming problem. This paper explores in the decentralized decision-making condition, what the relationship is between hog price and quality, and profit of supply chain. Study shows that: In the wholesale price contract, the pig-breeding enterprise decide the relationship between pig price and quantity, and the slaughtering enterprise determines the pig prices and slaughter rate relations, that is to say the slaughtering enterprise can make encouragement and restriction for quality of pigs through the pig price; and In the wholesale price contract, quality incentive to porkers enterprise from slaughtering enterprises is effective. Pig-raising firm undertake responsible for pig quality, and bear the risks and benefits for the quality. Its profit is increasing function for slaughter rate. And slaughtering enterprises are not responsible for slaughter rate.

Keywords: Product quality, pig supply chain, transfer pricing.

1 Introduction

In the relationship between pig-raising enterprise and slaughtering enterprise, there exist a principal-agent problem with the quality as core, namely the pig-raising enterprise having information superiority, and slaughter and processing enterprise is in the relative asymmetry position. Slaughtering and processing enterprise is the principal, and pig-raising enterprise is agent.

Under the condition of the market, interests of pig-raising enterprise are unstable. As for the slaughtering enterprises, the fixed investment is very large. According to the survey, 70 millions RMB is needed in fixed asset investment to a slaughter with 1.2 million pigs in factory. And added with investment in equipment and working capital, the totaling expenditure will be at 1.5 billion RMB. Therefore, there must to improve the equipment utilization to ensure they are not losses. In this case, in the

* Project 70962002&71162023 supported by National Natural Science Foundation of China.

D. Zeng (Ed.): Advances in Computer Science and Engineering, AISC 141, pp. 747–754.
Springerlink.com

beginning of investment, the slaughter firm would sign a number of pig acquire contracts with pig breeding business enterprises.

The contract is a typical Stackelberg game model, namely it's the Leader-Follower of the bi-level programming problem. What the pig-raising firm will do is to decide the pig price based on the object of its profit maximum. And what the slaughter will do is to decide the quantity. Slaughtering enterprises take market risks, and access to market profit. And the pig-raising enterprise takes responsible for product quality risk, and gets the fixed-profit related to quality. This paper explores in the decentralized decision-making condition, what the relationship is between hog price and quality, and profit of supply chain.

2 Symbol Set

M ——slaughtering enterprises, p ——pork price, λ ——Slaughter rate / net meat rate (pork weight / pig weight) , $\lambda \in (0,1)$, b ——Slaughtering byproducts income (such as visceral, head, hoof, etc.), c_{mf} ——Slaughter enterprise daily fixed costs (including fixed asset allocation, Labors, etc.), S ——pig-raising enterprise, p_s —— pig price, n ——the quantity of pigs, w ——the weight of each pig, p_f ——the price of feed, k ——Feed return rate, p_b ——Piglet price, c_{sf} ——fixed costs of hog firm (including fixed asset allocation, artificially, etc.), q ——The pork processing quantity, and the pig quantity is $q/w\lambda$, υ ——The final residual product salvage value, $(0 \le \upsilon < \lambda p_s + \dfrac{c_{sf}}{\lambda q w})$, h ——unit losses caused by shortage of stock.

c_e ——Slaughter enterprise unit inventory cost $(0 \le c_e < \upsilon)$, For porkers' enterprise, its cost including purchasing piglet cost, cost of feed pigs and fixed cost, and the cost function is $C_s = n \cdot (wkp_f + p_b) + c_{sf}$.

Porker-raising firm's income mainly from selling pigs to slaughter enterprise, its function is, $R_s = nwp_s$

The profit function of pig-raising firm is

$$\pi_s = R_s - C_s = n(wp_s - wkp_f - p_b) - c_{sf} \tag{1}$$

For pig-raising firm, their earnings and cost changes everyday. From its profit function, the feed return rate and the pig price p_s is the key factor to its profit with other inputs keep unchanged.

The pork demand $D \ge 0$ is stochastic, and its density function is $f(x)$, and distribution function is $F(x)$. $F(x)$ is Continuously differentiable and incremental, and $F(0)=0$, $\overline{F(x)}=1-F(x)$, $\mu=E(x)=\int_0^{\infty} xf(x)dx$ is the expect value of demand D.

3 Centralized Decision-Making Model

The sales quantity of slaughter enterprise is

$$S(q)=\min[q,D]=\int_0^q xf(x)dx+\int_q^{\infty} xf(x)dx$$
$$=q-\int_0^q F(x)dx \tag{2}$$

$\Rightarrow$

$$S'(q)=1-F(q)$$

The final residual pork quantity is

$$I(q)=(q-D)^{+}=q-S(q) \tag{3}$$

$$I'(q)=1-[1-F(q)]=F(q)$$

And the sales that Unable to meet is

$$L(q)=(D-q)^{+}=\mu-S(q) \tag{4}$$

$$L'(q)=F(q)-1$$

The expect profit of slaughter enterprise is

$$\pi_m=\frac{bq}{w\lambda}+pS(q)+\upsilon I(q)-c_e I(q)-hL(q)-\frac{p_s q}{\lambda}-c_{mf} \tag{5}$$

The expect profit of pig-raising enterprise is

$$\pi_s=\frac{q}{w\lambda}(wp_s-wkp_f-p_b)-c_{sf} \tag{6}$$

And the profit of supply chain is

$$\Pi_{SC}=\pi_s+\pi_m$$

$\Rightarrow$

$$\Pi_{SC}=pS(q)+\upsilon I(q)-c_e I(q)-hL(q)$$
$$+\frac{b-wkp_f-p_b}{w\lambda}q-c_{mf}-c_{sf}$$

⇒

$$\frac{\partial^2 \Pi_{SC}}{\partial q^2} = -(p+h+c_e-\upsilon)f(q) \prec 0$$

Hence, Π_{SC} is concave function, there have

$$\frac{\partial \Pi_{SC}}{\partial q} = p+h+\frac{b-wkp_f-p_b}{w\lambda} - (p+h+c_e-\upsilon)F(q)=0 \quad (7)$$

The function of the supply chain for balance slaughter capacity is
⇒

$$q_{sc}^* = F^{-1}\left(\frac{p+\frac{1}{w\lambda}b+h-\frac{1}{w\lambda}p_b-\frac{kp_f}{\lambda}}{p+c_e+h-\upsilon}\right) \quad (8)$$

Where, $p+\frac{1}{w\lambda}b$ is the Unit weight sales price of slaughtered pigs, and $\frac{1}{w\lambda}p_b+\frac{kp_f}{\lambda}$ is Unit production cost, In the whole supply chain profit maximization conditions, the pig breeding capacity of the enterprise is $\frac{q_{sc}^*}{w\lambda}$.

4 Decentralized Decision-Making Model

In the wholesale price contract, the pig-raising firm problem is:

$$\max_{p_s} \pi_s = \frac{q}{w\lambda}(wp_s - wkp_f - p_b) - c_{sf} \quad (9)$$

s.t.

$$\begin{cases} \max_{q} \pi_m = \frac{bq}{w\lambda} + pS(q) + \upsilon I(q) - c_e I(q) - hL(q) - \frac{p_s q}{\lambda} - c_{mf} & (10) \\ \frac{bq}{w\lambda} + pS(q) + \upsilon I(q) - c_e I(q) - hL(q) - \frac{p_s q}{\lambda} - c_{mf} \succ 0 & (11) \end{cases}$$

Where, the formula 9 is incentive constraint, and formula 10 is participation constraint. Solve above problem by using backward induction.

For

$$\frac{\partial^2 \pi_m}{\partial q^2} = -(p+h+c_e-\upsilon)f(q) \prec 0 \tag{12}$$

So, π_m is concave function, q is decision variable, and let the first derivative of formula 10 equals to 0, so there has:

$$\frac{\partial \pi_m}{\partial q} = \frac{b}{w\lambda} + pS'(q) + \upsilon I'(q) - c_e I'(q) - hL'(q) - \frac{p_s}{\lambda} = 0 \tag{13}$$

$\Rightarrow$

$$(p+c_e+h-\upsilon)F(q) - (p+h+\frac{1}{w\lambda}b - p_s\frac{1}{\lambda}) = 0 \tag{14}$$

Hence, the slaughter enterprise optimal order quantity of pigs is

$$F(q) = \frac{p+h+\frac{1}{w\lambda}b - p_s\frac{1}{\lambda}}{p+c_e+h-\upsilon} \tag{15}$$

$\Rightarrow$

$$q_m^* = F^{-1}\left(\frac{p+h+\frac{1}{w\lambda}b - p_s\frac{1}{\lambda}}{p+c_e+h-\upsilon}\right) \tag{16}$$

Comparative formula 16 and formula10, there has:

$$\frac{p_s}{\lambda} = \frac{wp_s}{w\lambda} \tag{17}$$

And

$$\frac{1}{w\lambda}p_b + \frac{kp_f}{\lambda} = \frac{p_b + wkp_f}{w\lambda} \tag{18}$$

Where, wp_s is the pork's Sales Revenue of each pig, and $p_b + wkp_f$ is the cost of piglets and breeding of each pig, obviously, there has

$$wp_s \succ p_b + wkp_f \tag{19}$$

$\Rightarrow$

$$q_m^* \prec q_{sc}^* \tag{20}$$

Namely, the order quantity of pigs in wholesale price contract is less than the optimal quantity in supply chain coordination.

From formula 10, the pig price p_s with wholesale price contract is:

$$p_s = \lambda[(p+h)-(p+c_e+h-\upsilon)F(q)]+\frac{b}{w} \tag{21}$$

There has

$$\frac{\partial p_s}{\partial \lambda} = (p+h)-(p+c_e+h-\upsilon)F(q) \tag{22}$$

$\Rightarrow$

$$\frac{\partial p_s}{\partial \lambda} = (p+h+c_e-\upsilon)[1-F(q)]+(\upsilon-c_e) \tag{23}$$

For $p \succ \upsilon$, $\upsilon \succ c_e$, so there has $\frac{\partial p_s}{\partial \lambda} \succ 0$, p_s is the increasing function for λ, and is linear function.

And there has

$$\frac{\partial p_s}{\partial q} = -\lambda(p+c_e+h-\upsilon)f(q) \tag{24}$$

For $p \succ \upsilon$, $\lambda \in (0,1)$, so there has $\frac{\partial p_s}{\partial q} \prec 0$, p_s is the decreasing function for q.

Take formula 21 into formula 6, can get:

$$\pi_s = [bq-q(wkp_f+p_b)]\frac{1}{w\lambda}+q[(p+h)-(p+c_e+h-\upsilon)F(q)]-c_{sf} \tag{25}$$

$$\frac{\partial \pi_s}{\partial \lambda} = \frac{q(wkp_f+p_b-b)}{w\lambda^2} \tag{26}$$

Where, wkp_f is the breeding cost of each pig, p_b is the piglet cost of each pig, b is the sell price of byproduct for each pig after slaughtering. So, there has $wkp_f+p_b-b \succ 0$, so $\frac{\partial \pi_s}{\partial \lambda} \succ 0$ is always been right, and that means the pig-raising enterprise profit function is the increasing function of λ, that is to say the slaughter enterprise's price incentive for λ is effective.

Take formula 21 into formula 5, there has

$$\pi_m = -q[(p+h)-(p+c_e+h-v)F(q)]+pS(q)+vI(q)-c_eI(q)-hL(q)-c_{mf} \quad (27)$$

So, can get $\frac{\partial \pi_m}{\partial \lambda}=0$. That means the slaughtering enterprise profits have nothing with slaughter rate λ

That means pig slaughtering enterprises do not undertake the risk of the quality, and the risk of the quality under taken by the pig breeding business.

5 Numerical Example

Take a pig supply chain consist of slaughter enterprise with one million annual slaughtering capacity and pig-raising enterprise. The various parameters are shown as table 1.

Table 1. Parameters of the case

Parameter	Value	Parameter	Value
p	10.5	λ	0.7
b	80	w	100
p_f	1.9	k	2.57
p_b	210	v	6
h	0	c_e	0
c_{mf}	35000	c_{sf}	50000

$$F(q)=(\frac{q}{z})^r \quad (z \succ q, r \succ 0) \Rightarrow f(q)=F'(q)=\frac{1}{z}rx^{r-1}$$

$$S(q)=S(xz)=z-\int_0^x z(\frac{q}{z})^r d\frac{q}{z}=z[x-\int_0^x y^k dy] \quad (28)$$

Suppose the demand distribution function of pork is $F(x)=x^r \quad (x\in[0,1])$,

let $x=\frac{q}{z}$, there has $F(q)=(\frac{q}{z})^r \quad (z \succ q, r \succ 0)$

$$\Rightarrow f(q)=F'(q)=\frac{1}{z}rx^{r-1}$$

$$S(q) = S(xz) = z - \int_0^x z(\frac{q}{z})^r d\frac{q}{z} = z[x - \int_0^x y^k dy] \quad (29)$$

Let $z = 280000, r = 6$, then the result shown in table 2.

Table 2. The results of numerical example

Parameter	Value	Parameter	Value
q_m^*	201050	p_s	7.718
π_m	71300	π_s	161200
Π_{sc1}	232500	π_m / Π_{sc1}	30.67%
π_s / Π_{sc1}	69.33%	Π_{sc2} / Π_{sc1}	109.29%

6 Conclusion

The contract is a typical Stackelberg game model, namely it's the Leader-Follower of the bi-level programming problem. What the pig-raising firm will do is to decide the pig price based on the object of its profit maximum. And what the slaughter will do is to decide the quantity. Slaughtering enterprises take market risks, and access to market profit. And the pig-raising enterprise takes responsible for product quality risk, and gets the fixed-profit related to quality.

References

1. Wu, X.-M.: Pork Quality and Safety Management System in China - An Empirical Analysis Based on consumers and producers in Sichuan Province PH.D Thesis. Zhejiang University (2006)
2. Sun, S.-M.: Study on high quality pork interior negotiators price. System Sciences and Comprehensive Studies in Agriculture 19, 109–111 (2003)
3. Cachon, G.P.: Supply chain coordination with contracts. In: Graves, S., de Kolc, T. (eds.) Handbooks in Opertions Research and Management Science: Supply Chain Management, North Holland (2003)
4. Wang, Y., Gerchak, Y.: Supply chain coordination when demand is shelf-space dependent. Manufacturing and Service Operations Management 3(1), 82–87 (2001)
5. Lariviere, M., Porteus, E.: Selling to the newsvendor: an analysis of price-only contracts. Manufacturing and Service Operations Management 3(4), 293–305 (2001)
6. Parlar, M.: Game theoretic analysis of the substitutable product inventory problem with random demands. Naval Research Logistics Quarterly 35, 397–409 (1988)
7. Hu, K.: The Pricing and Profit Distribution in Downstream of Pig Supply Chain with the Revenue Sharing Contract. In: CCCM 2010, vol. 3, pp. 575–580 (2010)

Empirical Study on China's Agricultural Productivity under the Binding of Environment

Yi Chen* and Huijun Liu

Rongzhi College of Chongqing Technology and Business University, Jingkou Industrial Park, Shapingba District, Chongqing Municipality, China, 400033
chenyi_rzxy@sina.com

Abstract. Agricultural pollution has become the important restraining factor of sustainable agricultural development in China. Taking into account the environmental factor, the China's agricultural environmental technology efficiency in 30 provinces over the period 1999 to 2008 has been estimated using the directional distance function. And the growth of China's agricultural TFP has been measured using the Malmquist-Luenberger productivity index. The main conclusions are as follows: There is no distinct relationship between the change of ML productivity index and the gross output value of farming, forestry, animal husbandry and fishery. And average annual TFP growth measured by Malmquist index is much higher than by ML index; the value is 3.1% and 1.8% respectively, which indicates that agricultural pollution has given a great influence on the measure.

Keywords: Agricultural, TFP, Environmental factor, Malmquist-Luenberger index.

1 Introduction

From the input-output point of view, the factors of economic growth can be divided into production inputs growth and total factor productivity growth. It is generally acknowledged that transformation of economic growth is relying on the total factor productivity (TFP) improved, not capital or material resources inputs. Therefore, we usually study efficiency of economic growth with total factor productivity. It is the same with the study on agricultural growth.

China is big country with heavy agricultural proportion. According to the fifth national population census, China's rural population accounts for 63.8% of the total population. And agricultural productivity growth is the core of the national wealth growth in developing countries [1], such as China. But with the increase of agricultural production inputs, the agricultural pollution is more and more serious. The existing research results indicate that China's agriculture pollution has become a great threat to drinking water source and water environment [2, 3]. And agricultural pollution has become the important restraining factor of sustainable economic development in China. Therefore, considering the environmental factors of agricultural pollution, re-estimated agricultural total factor productivity growth of China's provinces has important practical significance.

---

* Corresponding author.

D. Zeng (Ed.): Advances in Computer Science and Engineering, AISC 141, pp. 755–760.
Springerlink.com

The status of agricultural productivity in China has been the concern of many economists. In recent years, the main researches on China's agricultural total factor productivity are as follows: Zeng Xianfeng[4] used DEA method to estimate the agricultural TFP in China's 28 provinces between 1980 and 2005. It was pointed out that the average growth rate is 2.2%. Zhou Duanming[5] calculated China's average annual growth rate of agricultural TFP from 1978 to 2005 was higher, at 3.3%. Li Gucheng[6] empirically analyzed China's agricultural TFP from 1988 to 2006 in considering the human capital elements, and he pointed out that the annual growth rate is 3.49%.However, with SFA-Malmquist productivity index model, Quan Jiongzhen[7] suggested that China's agricultural TFP growth speed was slow. The annual growth rate was only 0.7% between 1978 and 2007. Although the growth rates of each study are inconsistent, most researchers deem that China's agriculture TFP are regional imbalance. And since 1990, China's agricultural TFP growth mainly depends on agricultural technological progress, but not technical efficiency improvement. Recently, a few researchers begin to study China's agricultural growth under the binding of environment. Xue Jianliang[8] found that during 1990-2008 the environmentally-adjusted agricultural TFP growth shows a decreasing trend and varies in different periods within the range of 0.09—0.6%.

To sum up, estimating China's agricultural TFP under the binding of environment, can investigate China's agricultural development condition more comprehensively, and be helpful in promoting agricultural sustainable development.

2 Research Methods

2.1 The Directional Distance Function

Chung[9] constructed directional distance function to quantification different output:

$$\vec{D}_0^t(x^t, y^t, b^t; g_y, -g_b) = \sup\,[\beta:(y^t+\beta g_y, b^t-\beta g_b) \in P^t(x^t)]\,. \tag{1}$$

For a given input x and environmental technology structure, "good" output and "bad" output expand and contract in the same proportion, and β is the largest possible quantity of output(y) growth and pollutants(b) decrease.

2.2 Environmental Technology Efficiency

Environmental technology efficiency is defined as the ratio of actual "good" output to frontier output in environmental technology structure, and can be expressed as:

$$ETE = \frac{1}{1+\vec{D}_0^t(x^{t,k'}, y^{t,k'}, b^{t,k'}; y^{t,k'}, -b^{t,k'})}\,. \tag{2}$$

Environmental technology efficiency is an index between 0~1. When observation points in the environmental production frontier, the directional distance function value is 0, and the environmental technology efficiency is 1.

2.3 Malmquist-Luenberger Productivity Index

Consideration of environmental factors, total factor productivity index can be represent by Malmquist-Luenberger productivity index[10] (ML index):

$$ML_t^{t+1}=\left[\frac{1+\vec{D}_i^t(x^t,y^t,b^t;g^t)}{1+\vec{D}_i^t(x^{t+1},y^{t+1},b^{t+1};g^{t+1})}\times\frac{1+\vec{D}_i^{t+1}(x^t,y^t,b^t;g^t)}{1+\vec{D}_i^{t+1}(x^{t+1},y^{t+1},b^{t+1};g^{t+1})}\right]^{\frac{1}{2}}$$
$$=\left\{\frac{1+\vec{D}_i^{t+1}(x^t,y^t,b^t;g^t)}{1+\vec{D}_i^t(x^t,y^t,b^t;g^t)}\times\frac{1+\vec{D}_i^{t+1}(x^{t+1},y^{t+1},b^{t+1};g^{t+1})}{1+\vec{D}_i^t(x^{t+1},y^{t+1},b^{t+1};g^{t+1})}\right\}^{\frac{1}{2}}\times\left\{\frac{1+\vec{D}_i^t(x^t,y^t,b^t;g^t)}{1+\vec{D}_i^{t+1}(x^{t+1},y^{t+1},b^{t+1};g^{t+1})}\right\}$$
$$=MLTECH_t^{t+1}\times MLEFFCH_t^{t+1} \quad (3)$$

ML productivity index can be divided into technological progress index (MLTECH tt+1) and technical efficiency index (MLEFFCH tt+1) to analyze their contribution to the growth rate of TFP.

3 Data and Variables

The data used is agricultural input and output data of China's 30 provinces, autonomous regions and municipalities during 1999 ~ 2008 (not including Tibet). All of them are from years of the China Statistical Yearbook.

1. "Good" output: "good" output is measured by gross output value of farming, forestry, animal husbandry and fishery (1999 constant prices).

2. "Bad" output: "Bad" output is refers to the environmental pollution caused by agricultural production. It is depicted by the standard emissions intensity (EI) of total nitrogen (TN), total phosphorus (TP) and chemical oxygen demand (CODCr).

Firstly, this paper measures emissions of agricultural pollutants (Ej) by unit investigation method [11]. Pollution unit is showed in Tab.1.

Secondly, compute the standard emissions (E) of TN、TP and CODCr according to surface water quality standards Ⅲ level.

Lastly, add cultivated land area and grassland area as a land area (AL) of each province, then compute the standard emissions intensity (EI) of TN、TP and CODCr. (EI＝E /AL)

3. Input: this article selects the first industry professional numbers and total power of agricultural machinery as input index.

Table 1. Agriculture non-point source pollution unit

Sort	Unit	Index	Unit
chemical fertilizer	nitrogen fertilizer phosphate fertilizer Compound fertilizer	dosage	Ten thousand tons
livestock	Large livestock, hog, sheep	annual inventory	Ten thousand

4 Empirical Study

4.1 The Comparisons between ML Productivity Index and the Environmental Technology Efficiency in Agriculture of Each Province

The average annual values from 1999-2008, including ML productivity index and the environmental technology efficiency of different provinces have been calculated (see Table 2) . Both of them took the binding of environmental factor into account, but they had the different intrinsic meanings. Also see in Table 2, the top five provinces in the ML productivity index rank, is not the top five in the environmental technology rank. For example, ML productivity index of Sichuan province is in the low-ranking, although its environmental technology efficiency is in the high-ranking, and Chongqing Municipality has the opposite condition.

On the left side of Table2, the ML productivity index is ranking first fourteen, whose environmental technology efficiency is above 0.7 (excluding Chongqing). On the right side of Table2, the ML productivity index is in low-ranking, whose environmental technology efficiency is below 0.7 (excluding three provinces). This indicates that, the meanings of ML productivity index and environmental technology efficiency are different, but the reflections of state in agricultural economy development of different provinces are similar.

Table 2. ML productivity index and environmental technology efficiency of each province

rank	Province	ML index	ETE	rank	province	ML index	ETE
1	Jiangsu	1.056	0.991	15	Guangxi	1.015	0.696
2	Shandong	1.054	0.746	16	Yunnan	1.014	0.637
3	Hebei	1.052	0.757	17	Jingxia	1.011	0.528
4	Heilongjiang	1.033	0.756	18	Shanxi	1.010	0.663
5	Hainan	1.031	0.996	19	Sichuan	1.008	1.000
6	Xinjiang	1.029	1.000	20	Henan	1.007	0.612
7	Liaoning	1.028	0.966	21	Gansu	1.007	0.646
8	Zhejiang	1.028	0.772	22	Shanxi	1.005	0.543
9	Jilin	1.026	0.884	23	Hunan	1.003	0.655
10	Fujian	1.026	0.999	24	Qinghai	1.003	0.568
11	Chongqing	1.026	0.559	25	Hubei	1.002	0.817
12	Guangdong	1.022	1.000	26	Anhui	1.001	0.631
13	Tianjin	1.021	0.849	27	Jiangxi	0.982	0.719
14	InnerMongolia	1.019	1.000	28	Guizhou	0.979	0.645

4.2 The Comparisons between ML Productivity Index and the "Good" Output in Agriculture

The agricultural TFP growth was fluctuating between 2000 and 2008. The agricultural TFP growth speeded up between 2000 and 2002, and descended in 2003 and 2004, and the increase and decrease of TFP growth took place by turn repeatedly between 2005 and 2008. The comparison between the agricultural TFP and agricultural "good" outputs (gross output value of farming, forestry, animal husbandry and fishery) indicated that there was no distinct relationship between the change of ML productivity index and the "good" outputs. The details between 2003 and 2004 showed that the agricultural TFP growth was only 0.3% which was 1.3% lower than last year, while the growth rate of "good" outputs was 7.5% which was 3.6% higher than 2003. This may be caused by the sharply increasing agricultural pollution in 2004, and the ML productivity index decreased after equalizing the "good" and "bad" outputs. It follows that the ML productivity index can reflect the influence of environment pollution on agriculture productivity.

Table 3. The agricultural ML index and the agricultural TOV index

year	ML index	"good" output	TFP growth rate	"good" output rate
2000	1.004	103.6	0.4%	3.6%
2001	1.014	104.2	1.4%	4.2%
2002	1.019	104.9	1.9%	4.9%
2003	1.016	103.9	1.6%	3.9%
2004	1.003	107.5	0.3%	7.5%
2005	1.038	105.7	3.8%	5.7%
2006	1.021	105.4	2.1%	5.4%
2007	1.034	103.9	3.4%	3.9%
2008	1.011	105.7	1.1%	5.7%
Average	1.018	105.0	1.8%	5.0%

4.3 The Comparisons between ML Productivity Index and Malmquist Index

In order to estimate the effects of environment factors on the agricultural TFP, this paper adopted two kinds of index to calculate the agricultural TFP (for the length of the article, table 4 omitted). The Malmquist index excluding the effects of environment pollutions showed the average annual increase of agricultural TFP was 3.1% between 1999 and 2008, among them the average increase of technical progress rate was 5.6% and the average decrease of technical efficiency was 2.4%. The ML index which including the effects of environment pollutions indicated that the growth of agricultural TFP was only 1.8%, and the technical progress rate and the technical efficiency were 2.1% and -0.3% respectively. The calculated results showed that the agricultural TFP will decrease distinctly while considering the environment factors.

5 Conclusion

This paper estimated the China's agricultural environmental technology efficiency and the growth of China's agricultural TFP by ML productivity index. The main conclusions are as follows:

(1) ML productivity index and environmental technology efficiency have different connotations, but they reflect each province's agricultural growth similarly, because both of them take into account environmental factors.

(2) There is no obvious relationship between ML productivity index change and gross output value of agricultural growth rate changes. This may be caused by that ML index considers the environmental factors while the other one does not consider environmental factors.

(3) Comparing ML productivity index with Malmquist productivity index, we found that agricultural pollution has given a great influence on the agricultural TFP.

Due to the obvious influence on agricultural productivity measure, agricultural pollution ought to be considered in measuring the agricultural economic development.

References

1. Johnson, D.G., Richard, T.: Ely lecture: Agriculture and the Wealth of Nations. American Economic Review 87, 1–12 (1997)
2. He, C., Fu, B., Chen, L.: Non-point Source Pollution Control and Management. Chinese Journal of Enviromental Science 19, 87–91 (1998)
3. Quan, W., Yan, L.: Effects of Agricultural Non-Point Source Pollution on Eutrophication of Water Body and Its Control Measure. Acta Ecologica Sinica 22, 291–299 (2002)
4. Zeng, X., Li, G.: Estimate the Agricultural Production Efficiencies and Analysis It's Convergence: 1980~2005. Quantitative & Technical Economics (5), 81–92 (2008)
5. Zhou, D.: Technical Progress, Technical Efficiency, and Productivity Growth of China's Agriculture. Quantitative & Technical Economics (12), 70–82 (2009)
6. Li, G.: Human Capital and TFP Growth of Regional Agriculture in China. Journal of Finance and Economics 35, 115–128 (2009)
7. Quan, J.: Evidence of China's Rural TFP Growth: 1978~2007. Chinese Rural Economy (9), 36–47 (2009)
8. Xue, J., Li, B.: Environmentlly-Adjusted Measurement of China's Agricultural Total Factor Productivity. China Population, Resources and Environment 21, 113–118 (2011)
9. Chung, Y.H., Fare, R., Grosskopf, S.: Productivity and Undesirable Outputs: A Directional Distance Function Approach. Journal of Environmental Management 51, 229–240 (1997)
10. Fare, R., Grosskopf, S., Pasurka, C.: Accounting for Air Pollution Emissions in Measuring State Manufacturing Productivity Growth. Journal of Regional Science 41, 381–409 (2001)
11. Nai, S., Du, P., Chen, J.: Evaluation on Non-point Source Pollution based on Unit Analysis. J. Tsinghua Univ (Sci. & Tech.) 44, 1184–1187 (2004)

The Study on the Follow-Up Audit in China

Che Gui-juan

School of Economic & Management, Wuhan University, P.R. China, 430072

Abstract. With the increase of national macro-control efforts and the frequent occurrence of major emergencies, Chinese government has been investing more and more funds and other resources for the public expenditure. In order to supervise these resources and enhance the use efficiency, China has carried out follow-up auditing on some significant projects. Follow-up auditing has become an important means of safeguarding national economic security and improving efficiency in the use of public resources. But the study on the basic theories of the follow-up audit is very poor. This paper tries to study these basic theories.

Keywords: Follow-up audit, concept objectives, theory basis attribution.

1 Introduction

President HU Jin-Tao had made important instructions on the deepening the reform of fiscal and taxation system, improving the public finance system and improving financial management performance when he presided over the Politburo collective study of 2009. He also required that the audit work should make sure the funds and projects of people's livelihood can be comprehensively implemented, the government major investment projects are implemented and the government public power can be exercised correctly.

When the 2009 National People's Congress answered the reporters, Premier WEN Jia-Bao said : "The public finance funds where the audit to follow up where". He said again after December 25, 2009 heard the work reports of the leadership of the China National Audit Office said: "The audit maintain national economic security, the need for capital investment in major projects to implement "follow-up auditing." The Director of Audit, LIU Jia-Yi at the end of 2009, the national audit working conference said: "We have carried on the follow-up audit in the whole processes for the state large investment projects in 2009." " and we will enlarge the follow strength ". On the national audit working conference on December 28-29, 2009, Liu Jia-Yi put forward the audit work task of 2010, he emphasized that the auditors need to strive to complete the "eight tasks" and the "three requirements ", and he did more detailed discussions to enlarge the follow strength for audit.

Obviously, follow-up audit has become an important means of national audit, which safeguard national economic security played an important role, then, theoretically know how to track the audit, for example, what is the follow-up audit? What is it based on? What kind of the audit does it belong to? What is the main effect as the follow-up audit? So, this paper tries to give a preliminary analysis of these issues.

D. Zeng (Ed.): Advances in Computer Science and Engineering, AISC 141, pp. 761–768.
springerlink.com

2 General Outline

The follow-up audit has already become an important part of the national government audit in many countries. But the development and application of follow-up audit is late in our country. Although from the 90's in last century there have been some cases of the follow-up audit, but until 2005, the China audit academic association definitely affirm that the follow-up audit is one kind of four performance audit modes. In July, 2008 the National Audit Office put forward "The National Audit Office Work Development Program for 2008-2012". And this program submits a positive exploration of the whole process of the implementation of follow-up auditing on the large investment projects and the major public emergencies. Since 2008, the National Audit Office and some local audit institutions had some breakthrough in follow-up audit, and have made certain achievements. The last two years, audit institutions has carried out the follow-up auditing mainly concentrated on the large investment projects of livelihood, the special resources development, the environmental protection projects, the major public emergencies and the implementation of significant national policies. Such as the financial income and expenses of the OCOG(Organizing Committee of Olympic Games) and the engineering of the Olympic Games, the Beijing-Shanghai high-speed railway construction project audit, the Second West-East (West) audit, the project fund audit on the water pollution control engineering in Taihu Lake, Jiangsu Province, the Wenchuan earthquake relief and post-disaster recovery and reconstruction audit, new special funds by investment tracking and auditing.

In the earthquake relief funds and materials tracking audit, the audit authority make recommendations at any time in accordance with the auditing requirement. Including identifying problems at any time and urging the rectification at any time. The audit recommendations had been made in many ways more than 3640 recommendations, and more than 2940 among them were adopted. At the same time, 570 regulations and systems were made according to these audit recommendations which ensured that the relief funds and materials could be used scientific, rationally and effectively. At the end of October 2009, there were 72 important items as the follow-up audit items on the disaster recovery and reconstruction, and 65 items had been started and 11items had been completed, accounting for the number of audit items for 90% and 15% respectively; there were 753 schools should be follow-up audited, 718 had started, 480 had been completed, accounting for the number of the investigation for 95% of and 64% respectively. The National Audit Office had carried out the follow-up audit on the financial income and expenses of the Beijing OCOG(Organizing Committee of Olympic Games) and the engineering of the Olympic Games since the September, 2005. These audit work was carried out twice a year and effectively control the costs, then improve the capital efficiency. As a result the end balances was more than 10 billion Yuan, exceeding the original budget of 410 million Yuan, so the budget scale was better controlled. The National Audit carried on follow-up audit on 93 items in 102 engineering items of Olympic Games, the total investment of the audit items was 19.455 billion Yuan, accounting for 99.82% of the total investment. But the audit project settlement amount approved 15.253 billion Yuan, compared to 16.67 billion Yuan of the reported amount of project settlement, there was reduction of 1.417 billion Yuan and there was reduction rate of 8.5%.

The follow-up audit projects have been carried out successfully and have achieved remarkable results and effectiveness. These successful cases give us confidence in the future follow-up audit. After the earthquake inYushu, the national Audit office in less than a week has put forward the plan of the follow-up audit on its relief funds and materials. The 2010 Shanghai World Expo is the second major government investment projects after the Beijing Olympic Games recently. From March 2005, the Shanghai Audit Bureau adopted a multi-linkage audit model, which was based on national audit, organized by the Shanghai local Audit Bureau and cooperate with part of the county audit institutions and social professional intermediaries, according to the progress of projects, to track the financial audit of special funds on the organization operating funds for the Expo, Park venue construction projects and associated with the Expo organizers. Xu Aisheng, the Fixed asset investment audit Division chief of the national Audit office in the interview with Xinhua News Agency said that in 2011 the national Audit office of China will announce funds audit result of Shanghai World Expo.

In the theoretical studies on the follow-up audit, there are many articles on the specific operation, but the basic theory of the audit is uncovered.

3 Concept, Procedures and Target of the Follow-Up Audit

3.1 The Concept of the Follow-Up Audit

There are several definitions of the follow-up audit.

As a method of audit, the follow-up audit refers to the whole process of audit supervision on the items or the fund when the items are being implemented or the funds are collected, distributed, managed and used.

The follow-up audit refers to the method of the dynamical audit that the audit authorities, according to the needs of users of the audit product, supervise the whole processes of the capital flows and the orbit of power, including before, during, and afterwards the items.

The follow-up audit refers to the audit institutions and auditors, according to state laws and regulations and system requirements, make the whole process supervision, including before and during the items, on the fiscal year funding for key projects appropriated by the relevant departments of the Government's and the critical funds (funds) to ensure that these funds are used truly, legitimately and effectively,

These three concepts clear the following points: Firstly, the follow-up audit is a way of the audit; secondly, the follow-up audit is the comprehensive supervision on the whole process of audit matters; thirdly, the follow-up audit is a dynamic process. But they failed to account for the core and vital of the follow-up audit.

Explained by Chinese Dictionary, follow-up closely behind them is to catch up or surveillance, apparently the follow-up audit refers to tracking the audited object, monitoring the authenticity, legitimacy and effectiveness of the financial accounts of the audited object.

The crucial point of follow-up audit is "tracking", that is audit authority finding problem at any time can make recommendations simultaneously, instead of auditing the completion of the engineering items.

The definition of the follow-up audit is defined by the author as: the follow-up audit is a dynamic and comprehensive audit, it is necessary to the object being audited on the authenticity and rationality of the financial accounts, but also really concerning about the effective use of funds by the audited object. The audit found problems, propose solutions and measures and establish and improve relevant systems.

3.2 The Steps of the Follow-Up Audit (Program)

Plan: the audit authority will normally schedule the follow-up audit into the audit project plan of first year. And those follow-up audit items need to be multi-year audited will be included in next year the audit plan as the "continued examination" projects

Setting up the Project: the choice of audit project is known as the setting up of audit Project, the audit organization determines that the audit units and audit matters of process. The current main areas of the follow-up audit involved in the government financial allocation, capital of construction project set up by the Development and Reform Commission, relief funds and donor funds managed by relevant government departments, the government procurement funds operated by the government financial sector, and so on.

Implementation: the audit institutions usually formulate a unified implementation of the audit program and the audit can be segmented, but they should make the arrangements of the whole process for the follow-up audit. It doesn't need to make another plan of implementation for the multi-year follow-up audit.

Report: the audit authority need to prepare a complete audit report. For the Multi-year audit items, the audit team should submit annual audit reports at the end of the year. The annual audit report generally includes the basic information the follow-up audit, the audit found problems, recommendations and rectification of the situation of the audited entity.

Archives work: the audit authority may require the audit team to develop hand for filing archive and one follow-up audit item should be archived into one file. For the multi-year follow-up audit items, the audit team should develop staff for the annual collection of files, organize and keep them until the end of the audit. The archive work of all of the audit filing will be unified. Particularly, the follow-up audit notice, the audit plans, programs, audit evidence, the content of exchange of views and the audit reports must be placed on file to ensure integrity of the folloe-up audit files.

3.3 Objectives

Someone identified the objectives of the follow-up audit as "no problem, small problem or no big problem". We think that it is questionable.

Because the items of follow-up audit include the large investment projects, the sudden public financial provision (such as earthquakes, floods and other catastrophic events), it is too general to set the audit objectives at "no problem, little or no major problems issue ". " no problem, no big problems or little problems "is only from the point of view of financial waste, fraud, but does not think over the return of the investments. So we think the objectives of the follow-up audit can be defined as: the follow-up project to be audited timely and supervised fully to ensure the authenticity and rationality of the financial accounts of the audited units. At the same time, the follow-up audit can make the promotion of the use efficiency of public resources and social contributions, enable the investigation of problems, the promotion of reform, the improvement of the system, the strengthening of management and the improvement of economic efficiency together.

4 The Theoretical Basis of the Follow-Up Audit

4.1 Responsible Government Theory

The responsible government is the core meaning of modern government. A responsible government must do its utmost to meet the needs of society and the masses, undertake the community responsibility because of the consequences of their actions. Creating a responsible government is the central requirement, an important task of building a harmonious society and the general trend of the international public administration, more is the need to achieve sustainable economic and social development.

To create a truly responsible government, it is inevitable to supervise the government's public policy-making and the use of public resources. The Supervision of public decision is an important protection to make sure the public decision scientific and democratic. The resources available to government are public resources and the effective use and the conditions of management for the public resources are the process and results of the exercising of the government rights. Government audits, through auditing the use and management of public resources, oversee the authority of government, thus avoiding the occurrence of corruption, to ensure decision-making scheme is reasonable and lawful, to identify the gap between the target of public decision and the actual performance, identify the problems and find out the solution to the problems.

With the development of society, meaning of the government's public accountability and the public focus are also changing. In the early, the public accountability of government concerned by the public focused on the financial aspects. That is requiring the activities in the budget expenditures to comply with all laws, rules and regulations, and all the expenses to be true and legitimate. With the increasing sound democratic system, especially with the rise of new public management, public accountability requirements of government are increasing. The public require that government not only ensure the revenue and expenditure legally, manage resources according to the law, but also use resources effectively.

The follow-up audit is a pre-audit supervision, monitoring and supervision afterwards or something in the "trinity" of unity. This kind of audit strengthen the supervision on power, which can expose, stop, correct and combat the corruption and the rent-seeking behavior in economic affairs in time.

4.2 Management Control Theory

Control is imposing some on the object in order to make the process of object behavior or change in accordance with the standard to achieve the purpose of the act or process.

In 1948, the famous American mathematician N. Wiener (Norbert Wiener) published his first monograph of the control theory, "Cybernetics or Control and Communication in the Animal and the Machine". The monograph laid the foundation of a unified point of view and explaining control and communication problems of various systems, which marked the birth of cybernetics. Since then, ideas and methods of control theory had penetrated into almost any field of natural science and social science.

A basic feature of cybernetics is to test system in dynamic (movement and change) process and keep the system in equilibrium or steady state. This fundamentally changes the method of researching systems.

The related disciplines which are inextricably linked with the control theory include information theory and systems theory. These three are together called "Three Theories". "Three Theories" for the modernization of management provides an effective method, which look the management activities as a dynamic system, the information analysis system as the basis of internal and external links, the control as a means of system optimization.

In control theory, the "control" is defined as: you need to obtain and use information to "improve" one or some of the features or development of the controlled objective, and the effects selected which base on the information and influence the controlled objective is called control. Thus, the control is based on the information, all information transfer is for control, and any control will depend on feedback to achieve. Information feedback is an extremely important concept of the control theory. In layman's terms, feedback refers to information transmission by the control system out, then sending the result again back to its role and impacting the information output once again, thus play a controlling role. The purpose to control the implementation of the plan is that the results were compared with the standard at any time, if the permissible range of deviation is more than the planned range of deviation, the prompt corrective measures will be taken to realize the relative stability of the system's activities. At last, the organization can achieve its goals.

Feedback, information and control are the three elements of control theory. We think of audit item as a system. Such control by means of follow-up audit can achieve the objectives within the shortest possible time, consuming as little as possible of the human, material and financial. Or consuming the same time, resources, conditions, can achieve the best target state for the output of the system. System to achieve their goals, we need the corresponding control activities, so do not deviate from the desired objectives which can be achieved, that is the control system optimization purposes. It's obviously, the follow-up audit is actually the audit functions of executive control, through real-time audit to the audited projects, timely access to information, to identify problems, recommend corrective and timely advice to the relevant departments in order to avoid unnecessary loss.

4.3 The Immune System Theory of Auditing

As a physiological function, immunity is the body recognizes "self" and "non-self" antigens, tolerates the formation of natural immune of self antigens, and excludes the role of non-self antigens. The immune system is able to identify in vivo "non-self material" (usually the outside bacteria), which will be the elimination or exclusion of the overall project collectively. It is the body's defensive structure to protect their own, March 2008, the Auditor General Liu Jia-Yi, proposed the view of audit playing the "immune system" function, this is the biological "immune system" theory extended to the audit area. Since then, the national audit system responded positively, whether theorists or practitioners, on "the audit theory of the immune system" and conducted active exploration. Audit theory of the immune system, "the modern state as a national economic security audit 'immune system', has the responsibility to feel the risks earlier, has the responsibility to identify problems more accurately, have the responsibility to mobilize national resources and the ability to solve problems, resist the 'disease 'The proposal has the responsibility to never stay in against a time, matter, a single' disease 'at the same time, promoting their healthy function, improved mechanisms, build a strong defense. "

The theory of the immune system updates the audit concept. Prior to this, the audit work is mainly post-supervision, the focus of the investigation and prosecution of major cases, although effective, deterrent, but the negative impact has been formed, the economic loss has occurred. The audit is the audit trace theory of the practical application of the immune system, tracking audit as the immune system of organisms, as in the "disease", the time found that "virus" of the invasion, and automatically resist, while increasing the organism's own resistance.

For the dynamic tracking; the follow-up audit is through the whole process of audited items, the development of the audited entity is always paid attention to, the data database of the daily financial and major decision-making of the audited entity is established. Collecting information, maintaining the work links with the corporate board of supervisors and the internal audit department and taking the initiative to find the main problem can discover and expose the problem timely, which the supervision of the Trinity including before, during and after. And so play the role of against and prevention. At the same time, the audit authority can identify problems at any time, propose the revisable opinions at any time and urge the audited units excluded from the institutional barriers, institutional shortcomings and plugging loopholes in management through the follow-up audit. All of these measures can promote establishing institutional mechanisms scientific and promising, so the follow-up audit also play the role of promoting and enhancing.

References

1. Li, X.-W.: What Does The Follow-up Audit Concern. China Audit (12) (2004)
2. Cao, H.-M., et al.: The Follow-up Audit on the Construction Project, pp. 3–12. China Financial and Economic Publishing House, Beijing (2005)
3. The National Audit Office, Audit Results Announcement No. 8 of 2009: The Follow-up Audit Result of Revenues, Expenditures and Construction Project of Beijing Olympic Games. National Audit Office of People's Republic of China (June 19, 2009)
4. Song, Y.-J.: Increase Follow-up Efforts to the "Expo audit" to Play the Role of Auditing Security. China Audit (3) (2010)
5. China audit, Events of National Audit Office in 2009. China audit (4) (2010)
6. Yang, S.-C.: Corrosion Corruption, the Mission of the National Audit. China Audit (1) (2010)

7. The Research Institutes of National Audit Office, Points of Piew on the Follow-up Audit. Audit Research Briefing (10) (2009)
8. Chi, Y.-Z.: Discussion On the Related Issues of the Follow-up Audit. China Audit (3) (2010)
9. Liu, J.-Y.: The National Audit Working Conference Speech. National Audit Office of People's Republic of China (December 28, 2009)
10. Jia-Bao, W.: Audit Work Should Make Greater Contributions to the Economic Development and Social Progress. National Audit Office of People's Republic of China (December 28, 2009)
11. Lan, F.-A.: Full audit the Immune System, Promoting Democracy and the Law, Improve Governance Mechanism. China Audit (5) (2010)
12. Wang, H.-J., Wang, S.-M.: National Audit "Immune System" Building: Orientation and Choice. Audit and Economic Studies (2) (2010)
13. Qin, R.-S.: Public Economic Responsibility Theory and the National Audit Reform. Auditing Research (6) (2004)
14. Audit Events in (2009)
15. The Research Institutes of National Audit Office, The Situation on the Conduct of Research to Track Audit Report (9) (2009)
16. The Research Institutes of National Audit Office, To Carry out the Audit on the Track Reflections (10) (2009)

Discussion about Early Patent Warming Mechanism of Chinese Wind Power Enterprises

Peng Yuanyuan and Nie Ying

College of Literature, law & politics, China University of Mining and Technology

Abstract. The Chinese wind power enterprises have developed in recent years rapidly, however, lacking the core technologies patent have actually limited their further development. Establishing the patent early warning mechanism is an effective measure to improve the wind power enterprise innovation ability and win the market competition. But the insufficiency of the government investment, the imperfectness of laws, the lack of patent awareness and so on are obstructing the steps in building patent early warning mechanism. As the author's view, the establishment of wind power enterprise's patent early warning mechanism should reform from the machine-made interior to exterior, solving the machine-made internal condition problems, and optimizing mechanism external environment, only then it can be realized.

Keywords: Patents, early warning mechanisms, wind power.

1 Introduction

Being affected by the traditional production mode, wind power industry in China is still in the initial stage and has not gotten an impressive development. What's more, because of the great difficulty in technical development of new energy, wind power industry has been seriously restricted by the relative lack of core technology. After the signing of the TRIPS Agreement, patent protection has been widely used in technology protection, research, innovation, market competition and some other fields. So in order to avoid repeating mistakes of a few enterprises in the loss of response to patent disputes, wind power companies need to make a priority to take measures to establish patent early warning mechanism..

2 Value Analysis: Patent Early Warning Mechanism for the Chinese Wind Power Companies

Wind power with its characteristics of clean, renewable and low cost, receives extensive worldwide recognition as a clean energy. China's vast territory and long coastline make inherent advantages of wind power development. Since the "Renewable Energy Law" was promulgated and the "long-term renewable energy development plan" has been introduced, wind power industry has a huge market prospects in China. However, lack of core technology has hindered the development of China's wind power industry, moreover, China has long been subject to the concept

D. Zeng (Ed.): Advances in Computer Science and Engineering, AISC 141, pp. 769–775.
Springerlink.com

of "value technique, contempt patent", after its WTO accession, wind power companies are facing serious patent dispute crisis.

2.1 To Help Raise Awareness of Patent in Wind Power Companies

With the enhancing status of patent in market competition, enterprises which ignore the patents issue will be difficult to live. Wind power industry is relatively a high-tech industry, therefore more emphasis should be put on patent protection and makes it a strategic issues in the way of forming an active patent attitude. Patent early warning requires dynamic monitoring of patent trends and timely analysis of information in market, then by using the early warning indicators, companies can identify knowledge of patent issues and problems advance. Furthermore, according to the characteristics of the patent, enterprises are not only easy to take measures in dealing with patent problems, but also get a good way to find next step of development advice.

2.2 To Better Cope with Patent Layout in Market Competition

Many large multinational corporations have already started patent portfolio in China's wind power industry. In 2004, the Danish company LM began patent layout by applying for pre-bent blade fan technology in China's wind power industry, but China's wind power business perform insensitively, just one year later then they recognize and respond to this patent [1]. Patent early warning mechanism has the ability to detect patent trends of competitors through real-time monitoring of patent information and distinguish other companies' patent layout in complex patent information. It is a fast and effective way to avoid or resolve patent disputes and minimize losses when enterprises predominate actions that the competitor will take.

2.3 A Guide for Enterprises to Have a Independent Research

Wind power technology research has always been the bottleneck of development in China's wind power industry, and many domestic fans perform quite unstable because of lack of the control on core technology, so most fans that china's wind power corporations use have long been the introduction of foreign products, which results a higher costs. If we can digest and exchange the core technology abroad to the use of wind power, thus predominate patented technology by ourselves, it will strongly promote the development of China's wind power industry. Establishing patent early warning mechanism and analyzing patent documents can find many necessary technology, and on the basis of accurate information and data, companies are not only capable to avoid duplication of research and conservation of resources, but also to save costs and reduce study time through deeply study of the patent literature and further technology information.

2.4 An Effective Solution to Patent Dispute

In recent years, with the fast development of wind power industry, patent disputes have become pronounced. The signing of TRIPS agreement provides a legal guarantee for enterprises relying on patent in market competition and consequently forms a new mechanism of competition. Many states increases the intensity of patent protection, and the possibility of patents entanglement goes up. Because patent early warning mechanism can quickly find out the conditions of patent application, approval and invalid, so it's not only effective in the pre-known companies to bypass the mined areas and avoid disputes, but also effective in providing support for patent litigation, patent disputes and reduce damage to business.

3 Problems in the Process of Establishing Patent Early Warning Mechanism in China's Wind Power Companies

3.1 Lack of Executive Investment r

In 2009, China's investment in technological innovation is about 146.1 billion yuan, accounting for about 1.52% of gross national product [2], while in 2005 the GBAORD (government budget allocations and expenditures for research) of United States has beyond 900 billion euro which combined 900 billion yuan[3], there are still many Chinese enterprises are facing the dilemma of R&D funding. In addition, an incomplete input subsidies network and the limitations of large investment targets and conditions are not yet formed a concrete standard that refers to the implementation of investment objects. The extent of patent policy is too narrow so that it obviously can't immensely benefit the Chinese market and commercial patents. Moreover, only some enterprises with a certain size have the qualification to achieve the standard required by the Government and get financial funds, a lot of small businesses are excluded from the preferential policy. Government support for the patent early warning project is still in the testing phase, because of its highly cost and quite complex condition, one separate business operating the mechanism may cost too many and tolerate too much pressure, thus a shortage of motion is inevitably.

3.2 The Aid Policy Isn't Work Well

China now has many patent-related policies, but has not yet formed a more complete system. All areas and industries divide staggered patent policy thus results in implementation difficulties on the lack of environmental policy support. Especially on the issue of combining science and technology policy with national economic, there is no full use of economic support to pave the way for the patent business. On the other hand, ignoring the important economic role and status of patent in the science and technology, and have not take patent into the economic science and technology policy system. China has not been able to form an effective incentive mechanism in technological innovation and patent protection[4]. Domestic policy because of the lack of necessary conditions for the implementation often goes into idle talk, and the policy chain scission, buck-passing between various departments, publicity is not in place career development and other issues in the patent which are also occur frequently make thing worse. Patent early warning mechanism is mentioned as an important part in the "Outline of National Intellectual Property Strategy" and "National patent career development strategy", but specific support policies have not yet been introduced, and ambiguous policy guidance is difficult to achieve the purpose of establishing early warning mechanism.

3.3 The Patent Service Development Is Not Integrated Enough

Although the state has funded the establishment of a public service platform and information retrieval systems, and the phenomenon of universities or parks undertaking national or provincial research project to establish a patent search system is also more common, but China is still a lack of private investment in patent information services. Wind power industry now also has its own proprietary information services, such as the Baoding High-tech Zone, Valley of new energy research center, but still built under a research subject, wind power industry still get no private capital about patent information services. In addition, China's patent agency pattern has taken shape, the patent agencies gradually developed into a complex patent services organization which deal with patent serving, patent counseling, patent Proctoring and others. However, there are still shortages of professional practitioners in the china's patent agencies. Only few employees in ordinary patent offices have their certificate, and if requires more stringent patent analysts, the condition goes more scarce. What's more, the regional economic level of development also determines the size of regional patent agency and its professional degrees, the wind power industry often cite in China northwest, therefore patent agents specializing in the affairs of wind power are scarce.

3.4 The Lack of Enterprises' Own Conditions to Establish Patent Early Warning Mechanism

Influenced by the impact of the traditional large-scale development mode, after a number of preferential policies to develop wind power industry have been proposed, wind power industry in China is showing a bad develop tendencies of "scattered, small, random". On the one hand, wind power companies still lack awareness of patents, R&D cost relatively small proportion and only a small number of patent applications are made by the serving workers which accounts for about 1% and it does not match China's wind power development speed and scale. On the other hand, wind power companies are generally lack of research capabilities and have a greater reliance on imported technology, thus an internal driving force is difficult to form in patent protection. with information retrieval system, information analysis system, the expert team system, management guidance system and others to set an early warning mechanism, companies need restructuring and strategic adjustments which means a huge investment, and that is one factor of the fatigue performance. Finally, lack of internal patent professionals and experts also become an obstruction to establish patent early warning mechanism.

4 Exploration of the Path to Establish Patent Early Warning Mechanism in Wind Power Companies: the Realization of Internal Conditions and the Construction of External Environment

Patent early warning mechanism runs on the focus of information analysis to achieve the alerts purpose. Information analysis relies on professional and material support, so it is necessary to solve these two problems. while the establishment of early warning mechanism is often difficult to assume full responsibility by a single company itself, the favorable external environment is very essential.

4.1 Realization of Internal Conditions

1. Deepening patent awareness of recognition of early warning mechanism
Government can attract the attention of the domestic mass media and then encourage enterprises to increase investment in the patent by the national patent strategy and patent career development strategy, and therefore policy makers must fully consider the policy incentive. Social media is the main propaganda, a broad coalition of television, Internet, newspapers and other media made coverage of patent cases is an effective means of propaganda to achieve purposes. Technology is a key factor of wind power equipment, so introduced the concept of patent into the management level, then corporate R&D, production, sales with patents, and linked patent analysis, market to R&D department closely. At the same time, create incentives to patent and take patent into relevant sector performance evaluation and annual evaluation to implement patent awards system. Moreover, it is a good way to write the patent policy into the articles of association and to disseminate in a staff meeting to instill awareness of the patent [5].

2. Promote the establishment of patent personnel training system to fill the talent gap
Industry authority and large patent agencies can hold training courses and counseling service within the industry to transmit patent professional to wind power company or do on-the-job training. Line can also combine with universities and add professional teaching and training courses that the main content are patent agency, analysis, litigation and judge to train more high-quality patent professionals[6]. On the one hand, hold the patent agent qualification certificates exams on the qualified professionals market to limit the phenomenon of cohabitation in patent talent market, on the other hand, by the course of training to increase the number of trainees and also change status lack of talent about patent. Companies can even introduce training classes to train a number of applications of induction training personnel in a short time.

3. Expand sources of special funds with wind power patents to create a funding mechanism
Enterprises need to change their mode of development and give up extensive expansion enclosure-style to invest a greater proportion of money into technology development and patent protection, while using patent pledge, patents transfer and patents licensing widely for the company to absorb more special funds at the same time. Government should increase investment in the business of patents by tax incentives and found special central bank loans to additional scientific research by reduce patent credit interest rate. In addition, encourage career development of new energy industry by providing special financial support and fiscal incentives; government funding for new energy technologies development should both focused on the best scientific talent to save R&D funds. Line will be used with the industry's appeal to set up a special foundation to attract investment or to accept donations to provide financial assistance for the cash-strapped companies.

4.2 The Construction of the External Environment

1. Improve and implement new energy-related patents regulations to optimize legal environment

Strengthen patent laws on the constraints of patent novelty to prevent invalid patents disturb the market order; emulate the United States to set up patent court to trial patent litigation cases specifically[7]; standardize patent search report to enhance the patent application standards, and strengthen the patent application report review to improve the quality of patent applications; promote in-service applications to support the enterprise enhance creativity; improve patent license transaction system, address the loss of unknown cause of regulations; adjust our system of compulsory licensing for enterprise development and livelihood conditions; improve patent law execution environment by enhance the execution of patent law.

2.Optimizing the market environment by a comprehensive and coordinated develop tendency

Government use financial or tax policy and award incentives to make a great guide to patent information services and market-oriented commercial development, and to collect private capital to patent information services; support agency exchange to the multiple service patent agency doing services advice, information retrieval, patent Proctor and others from a single patent application. Wind power enterprises can combine with the Customs and the State Intellectual Property Office on the establishment of appropriate information systems to connect required patent information by set up their own patent department, team or professional consulting services systems with the commissioning team in the overall manufacturing of wind turbine and so on. Maintain long-term good relations of cooperation with an expert team for R&D direction and advice. For small and medium enterprises which have no strength alone to establish early warning mechanism, the joint establishment of industry can save costs on industry-specific information retrieval system and benefit sharing.

3.Optimize the policy environment by develop and implement a reasonable government support policies

Combine rational patent policy and patent strategy to supported research related policies in China to form a close and effective policy system to create a good atmosphere for its implementation; policy makers need to consider regional differences to provide the conditions for the coordinated regional development; encourage a full development of wind power in areas where are favorable of wind power industry, and develop a viable performance evaluation program to avoid local governments' false performance; executive organization should appropriately reduce the approval process to broaden the conditions for approval of new energy industries, and make a regular visits and monitoring on the new energy projects, prevent the support fund being used for other purposes. Wind power enterprises should make full use of national patent policies and new energy policy support to create the conditions of establish patent early warning mechanism. Carry out policy and advocacy work training within the industry timely so that to guide enterprises to reasonable use of government incentives and also provide a reference to the industry.

5 Summary

To construct a patent early warming mechanism needs to a hard working and a long journey by all wind power enterprises' investment and the national support. China has a huge resource advantage in developing winder power industry, but lack of core technical and recognition of patent, and the short of skilled worker and expert makes situation worse. It is essential to build internal and external conditions to solve these problems and consequently reach the goal. On today's circumstances and on the practice of China, even though the guide and support of government can't be ignored, corporations' change in develop strategy should be highlight, too.

References

1. Xue, C.: Create spear and shield of patent: to form a pant early warming mechanism. Chinese Wind Power (August 6, 2008)
2. Science and technology ministry: Chinese science and technology investment increased by 40 times in 20 years (September 17, 2009), http://shengbuzhang./20090917/103611.shtml
3. Global science and technology investment Highlights. ISTIC.7-8 (April 30, 2008), http://stsurvey.nstl.gov.cn/htm/stview/starcbin.jsp?db=MCN2007
4. Liu, H.: Career development strategy for the national patent (2011-2020) (December 22, 2010), http://www.istis.sh.cn/list/list.asp?id=6929
5. Li, L.: Vimicro patent early warning mechanism. Corporate Magazines (09), 40–44 (2008)
6. Yang, Y.: Guild system in the building industry, the role of intellectual property strategy. Industry and Technology Forum, 167 (July 9, 2008)
7. Patent law Imperfect,200 defendants to PRB a year (September 8, 2009), http://aooknn.com/detail.asp?id=167.

The Legalization of Policy Discretion and Its Realization Path

Liang Zhiming

College of Business, Qinzhou University

Abstract. Policy discretion should give full play to its own through both policy and legal double advantage, between public policy and criminal law's unique location, in order to improve their own of rule of law into concrete ideas, in order to break the traditional criminal policy is the legal policy and internal tensions. The traditional criminal policy already cannot adapt to the requirements of the law, the need of its reconstruction, through the path choice, realize the rule of law as a criminal policy transformation of the criminal policy goals.

Keywords: Legalization, Policy Discretion, Realization Path.

1 Introduction

According to the formulation main body, to set up a program as well as binding and mandatory features other external logic can be differences, policy and law in both obvious separately. From a historical perspective, the criminal policy in a long a period in fact plays a substitute the function of the law, criminal policy is often seen to be legal, as comrade dong bi-wu, says: "we usually said the policy of the party and the country according to law, and the policy referred to here also plays the function of the law." [1] but as long as the policy is a minimum level of general, open, stability, and other basic legal characteristics, policy will obtain the form in the relatively independent power special attributes, on the exterior is made similar law itself, policy and law between both tension is difficult to fundamentally avoid. The criminal policy of policy and legal internal tension exists because there are basic reason is that the criminal policy itself functions and structure, thus the inherent tension is difficult to eliminate. But in our country the law of grand words, through our criminal policy shall rule of successive ascensions to elements of the criminal policy of gradually realize the rule of law.

2 The Connotation of the Legalization of Policy Discretion and Its Implementation

The government by law principle is the basic principle of the criminal policy, the so-called law principle is to point to the criminal policy formulation, implementation, and feedback and perfect the law must be in the basic principle of within the framework. The principles of the rule of law and is essentially deal with the

D. Zeng (Ed.): Advances in Computer Science and Engineering, AISC 141, pp. 777–782.
Springerlink.com

relationship between the criminal policy. China's traditional ideas that criminal policy is the soul of the criminal law and the core, the criminal law is the embodiment of the criminal policy of criminal policy, with the criminal law is formulated and had direct significance. This kind of knows only that one of the aspects of the things, but neglects the criminal law and the criminal policy between the adverse effect of more complex relationship. "Criminal policy is the soul of criminal law and criminal law is the core, the criminal policy of culture and the standard" article, [2] here should be regarded as the criminal policy of criminal policy ideas. Criminal policy thought is the soul of the criminal policy system, mature of the criminal policy of thought in the shall, when necessary, into criminal law, the criminal policy legalization. In the criminal policy in the process of legalization, criminal policy is undoubtedly the criminal law soul and core. The criminal policy of criminal policy thought that legalization value choice and a basic policy into criminal law, which makes the criminal policy won authoritative and stability. In other words, the criminal policy of criminal policy legalization is the legalization of the most important and the most reliable forms, but not all of the criminal policy of thoughts to into law, only those who mature, stable criminal policy ideas in legislation when necessary to into the criminal law.

On the other hand, in criminal law making, after a possible criminal law is a matter of policy. Criminal law is a kind of abstraction, and the system of rules of real life is complicated, and the specific form of the case law applicable may not directly, the one-to-one in criminal law to find the answers, this needs in the implementation of the criminal law in the policy process law of processing, namely the criminal law policy, or in the execution of the criminal law in criminal law policy process, the processing. The criminal justice which requires combine principle with flexibility, but the policy of the criminal law of criminal law must be in the framework of criminal law, and cannot transcend the principles of criminal law must be respected and abide by, even in the criminal law under the condition of the bear, also can't through the criminal policy means to supplement the criminal law, this is the basic requirement of the country ruled by law.

Therefore, from the criminal policy and criminal law on the movement to see, both is two-way movement, two-way constraints of the relationship, namely the criminal policy need to legalization, the criminal law also requires policy change, criminal policy thought is the core and soul of criminal law and criminal law is the criminal policy of specific basis and the boundary. But, no matter be the criminal policy of legalization or criminal law policy change, all should follow the principles of the rule. Criminal policy is a pursuit of maximizing the efficiency of the crime prevention and control the internal tension, criminal policy involving constraint of the criminal suspect or criminal body free occasions many, therefore the criminal policy formulation and implementation of the impact of the easy to cause the law and damage, it is necessary in criminal policy emphasis on the principles of the rule of law, adhere to the basic principle of the criminal policy of restraint.

The rule of law is not only the criminal policy of criminal policy level by level to the law, the process of change is also the criminal law will continue to put the criminal policy of internalization of the process. The specific terms, the rule of law of criminal policy mainly includes the criminal policy of legislation and the criminal policy judicial two is a process in the legislative or judicial process, will be shown the criminal policy of value in the legislative or judicial process, is also reflected in the principles of the rule of the security of the criminal policy of legalization and criminal law policy change two interactive process. With a view of criminal policy legalization is the end of the criminal policy; we think this is mainly aimed at the criminal policy in the field of the rule of law legislation performance. China's criminal law need to reflect the value orientation of the criminal policy in the establishment, but after had his independent character and value, but at this time of legal status can be moved more can't be replaced by criminal policy. Law can never and criminal policy to follow suit. So, the criminal policy of the rule of law in legislation form of expression should also be different, not to the content of the criminal policy according to law as the intact, but as long as the law reflects the criminal policy goals and value line. For those short-term or for some aspects of the criminal policy, such as "strike hard" criminal policy is not to need to pass legislation in the form of expression comes out, For long-term criminal policy, such as "combined with wide concerning", "tempering justice with mercy" and so on the criminal policy can be expressed through legislation.

3 The Confirmation and Consensus of Legalization of Policy Discretion

Criminal policy is not without foundation of people, it was sometimes to create, for a potential, trying to find the prevention and control of crimes law, is not a kind of people to be unique. To the harmonious society under the background of criminal policy legalization construction, need the people's joint effort. Our country for a long time, on the one hand, is the master of the policy, the universal use, but on the other hand is policy of ruling by law vacant. Modern society as a diversified development of society, the society people, thinking concepts, interest groups are appear differentiation, so the pattern of diversified criminal policy of ruling by law must also confirmed and consensus is more, the group level.

The criminal policy of ruling by law and consensus has confirmed level, mainly include the following aspects:

1. Political leaders the confirmation and consensus. Criminal policy and other social rules on all belong to social determinism in the category of superstructure, by social economic foundation and political factor and influence, that in which important political leader is the confirmation and consensus. According to Marxism's point of view, and that people are the main body of creation history and reality, but political leaders is to take people realize create guidance. In the rule of law and realize the criminal policy of constructing socialist harmonious society, the process of political leaders on the criminal policy function, status and perfect the confirmation and consensus has decisive role. Criminal policy is a legal and policy of the double properties, in modern society above the dual nature of all cannot leave political elements of the influence and support and independent existence. China is the people's democratic dictatorship of socialist countries, promoting the rule of law in criminal policy in the course of political leaders, the political wisdom, authority influence, overall planning for the development of the rule of law of criminal policy provides

guidance. Criminal policy as a public, the political leaders of the major political responsibility is to according to the collective consciousness of the populist sex, irrationality, and emotional, conservative, the local sex even the dialectical and the justice of scientific deconstruction and analysis, use political wisdom and authority art, rationally guide ordinary people to power the heyday of the mature and rational national mentality, scientific understanding crime law and criminal cause of crime, makes rational and peaceful social psychological reaction.

2. The confirmation and the social public consensus. Criminal policy for promoting the rule of law is ultimately the criminal policy of the rule of law, the scientific quality service for the society, the final is the interest of the public services, and will be able to get the public's approval and follow. But this goal but may be long and hard, and in this goal before the rule of law of criminal policy need to get the confirmation and the social public consensus. Criminal field face "crime" phenomenon in general social the public understand often is in negative, harmful, except to the spirit, such as general in the spoken language expression of the "evil" and "bad", "justice" and so on, the public not to the understanding of the criminal policy is also often out of utilitarian standpoint, such as cracking down on crimes and punish offenders, often can't rational treatment crime and criminal. The criminal policy of ruling by law to get the social public confirmation and consensus, itself is itself a good approach to improve the quality of the rule by law. The social public confirmation and consensus criminal policy for the rule of law are important, but should pay attention to the social public also with a strong emotional reward, the collective unconscious, such as emotional factors, and negative in the criminal policy in the process of rule by law to achieve social public confirmation and consensus, also need to think about how to prevent the problems such as most people irrational.

3. Social elite the confirmation and consensus. Modern society a remarkable phenomenon is what we call social elite, generally believe that the rise of social elite refers to the society has the outstanding talented person, and takes it as a standard will social group is divided into elite and the public class or the public class, and called it the elite and the grassroots. The social elite put forward main is political science for interpretation on the democratic system development and use of the concept, but in the political elite socialist is most view negative thought tidal current. The rule of law of criminal policy to social elite the confirmation and consensus, not equal to said don't need the social public confirmation and consensus on social elite, but the confirmation and consensus for the criminal policy of rule by law to advance and develop an irreplaceable role. Social elite generally mastery of the political, economic and cultural resources beyond the average level of the society, in social political power system, in the economic resources possession, the cultural resources to have an advantage on. Social elite itself is always there, forever in the society, but it has only a few in knowledge discourse, interpersonal network, the wisdom of inheritance and resource allocation has the most impact. Social elite especially intellectual elite criminal policy in the process of the rule of law will take the lead in a confirmation, consensus and follow role.

4. The judicial authority the confirmation and consensus. The criminal policy of the final link to fully implement the rule by law in the judicial process, especially in the process of the practice of criminal justice, and judicial organs can confirm that the criminal policy of rule of law of criminal policy connotation, the necessity and

feasibility of rule by law, active push reached consensus on the criminal policy of ruling by law, it is a criminal policy of ruling by law decisive link. No matter the criminal policy of theoretical analysis, how to comprehensive and scientific value advocate, no criminal judicial practice with the self-consciousness of the can be empty. The judicial organ of the criminal policy in the confirmation and the rule of law, mainly in consensus of criminal case investigation, prosecution, trial process, also reflected in the offenders and the victim's rights, the more reflected in the social public law propaganda and implants. The judicial organ of the criminal policy of ruling by law shall be confirmed and consensus is conscious, from the mission of the judicial organ function, and should not be mechanization, localized, passive behavior, the judicial organs shall be of the existence and development and it's connected.

5. The social public opinion the confirmation and consensus. In the modern society of mass media and information channels have developed to the diversification of unprecedented degree, the role and influence beyond the ordinary person often imagine, "the reporter is nonetheless an cloud" really don't exaggerated. In the areas of the law and criminal field, public opinion, the media, news programs, the film and television works etc information communication forms of play widely information feedback, mood, understanding the role of the rally, but not to be overlooked is the negative effect is more and more strong. There was a time, criminal cases of the legal system of programs, the film and television works a lot of reports, some for better ratings, in which a detailed description of a crime, the offender's attitude change means, blood field, etc, and even describe the offenders reconnaissance means and the judicial organs of the special investigation methods, to a considerable degree, in the opposite role play. Promote the process of rule by law of criminal policy can't be in the vacuum under the environment of, many measures, work link, the case needs to be in a certain public situation, so social public opinion to the criminal policy of confirmation and consensus is essential, it is also the public's right to know, is the responsibility of the news public opinion. But as the above points out, the social public opinion, the media, news programs, the film and television works etc information communication forms because also has utilitarianism, artistic exaggeration sex, emotional features, such as the need to resolve and the negative effect; carry forward its positive role.

4 The Planning of Legalization of Policy Discretion

The criminal policy of ruling by law, namely the planning and formulate the criminal policy plan and specific measures suggestion, design and final determination process for specific implementation is the criminal policy of ruling by law and puts forward the perceiver and choose the process of policy options.

Criminal policy belongs to the relationship between the people's basic rights and freedoms, national power and public affairs boundary of important issues, so it is very important for the legalization of main body, the criminal policy formulation main body is to ensure that the legalization of the criminal policy of ruling by law the organizational safeguard. Criminal policy formulation main body called for the legalization of criminal policy subject has social public authority, and criminal policy formulation main body conform to China's political system and principles of the rule.

According to this request, in our country present stage the legalization of criminal policy formulation main body should be ruling party and the country's central authorities, including the ruling party and the nation's collective leadership of the party's political leaders, the central agency, and the national legislature, etc.; May rule out the main body of the state administrative organs have the ruling party and country, local agencies at all levels. One should be aware of is the place to have criminal policy formulated the legalization of reason; our country is a significant difference history. Chinese, local country, giving local certain discretion and special policy is deal with the relationship between the central and local important ways, but we think that the criminal policy formulation should not give place, giving local criminal policy formulation rights will lead to countries of the rule of law in chaos and criminal field of rule of law missing.

Acknowledgement. This project is funded by Guangxi Department of Education(Name of the project: The Study of Competition and Cooperation Among Local Governments in Guangxi Beibu Gulf Economic Zone, number of the project: 200911MS240, The Government legal coordination and innovation in Regional Economic Cooperation, number of the project: 201106LX539). It is also the science project of Qinzhou University (Name of the project: The Study of Administrative Agreement of Regional Economic Cooperation in Beibu Gulf Economic Zone, number of the project: 2010XJKY-32C).

References

1. Benkler, Y.: The Wealth of Networks: How Social Production Transforms Markets and Freedom. Yale University Press, New Haven (2006)
2. Brogan, B., Vreugdenburg, S.: You've Got Mail': Accountability and End User Attitudes to Email Management. In: Proceedings of the 4th International Conference on e-Government (ICEG), October 23-24. RMIT University, Melbourne (2008)
3. Bruns, A.: Life Beyond the Public Sphere: Towards a Networked Model for Political Deliberation. Information Polity 13, 71–85 (2008)
4. Chawner, B.: Spectators, not Players: Information Managers' Use of Web 2.0 in New Zealand. The Electronic Library 26(5), 630–649 (2008)
5. Dearstyne, B.W.: Blogs, Mashups, and Wikis: Oh, my! Information Management Journal 41(4), 7–11 (2007)

The Sustainable Development of the Sports Industry in China

Liu Na and Liu Kun

JiangXi University of Traditional Chinese Medicine, China
liuk0791@yahoo.com.cn

Abstract. The sports industry in today is known as "the most profitable industry", it to blow the new industry of the whirlwind spread all over the world. Sports is the highlight of the topic, the modern media is sports attracted much attention in the various projects to bring into the country. Today's sports industry has become a huge industry system, and satisfy the public spirit entertainment need and, at the same time, create great wealth. Combining with the development of the sports industry in China, the basic content of sports industry and the development trend in detail, and the sports industry in China, the problem in the process of development put forward countermeasures.

Keywords: Sports industry, present status, countermeasure.

1 Introduction

The sports industry is and sports all about the production and business operation activities. In socialist market economy, since the development of sports market in China very quickly. Along with the development of market economy in China, the sports industry will become a pillar industry of the national economy. Implementation of sports industrialization and related reform meet people's fundamental interests, accord with the requirement of market economy in China, also the needs of the development of national economy.

2 The Current Situation and Existing Problems of the Sports Industry in China

2.1 The Current Situation of the Sports Industry in China

Although the sports industry in China started late, but after nearly 20 years of development, the sports industry is taking shape; Although the whole scale of sports industry and other industries also is not very big, but compared in the socialist market economy development process, it has become a unique characteristic of the industry, is becoming a new growth of the national economy of a window. Overall, our country initially formed various kinds of sports industry and, through the DuoZhong ownership coexist and common development of the industry development in the new pattern, mainly in the following aspects:

D. Zeng (Ed.): Advances in Computer Science and Engineering, AISC 141, pp. 783–787.
Springerlink.com

2.1.1 The Sports Industry in China Is Narrowing the Gap with Developed Countries

1990 years in the sports industry in the second development, with the peak period of about 140 years of our country sports industry development process, from the middle of the twentieth century, starting when bud formation, management experience expanded scale, overall development, and business areas and so on four development period, to 21 century sports industry market mature, about 50 DuoNian experience of time, that is, our country sports development in the world economy than the peak of the developed countries has about 90 DuoNian shorten the time.

2.1.2 Sports Market Gradually Development, Market System Have Been Formed the Basic Framework

According to the theory of market economy, the market is connected production, circulation, distribution and consumption center link. The sports industry though is the production spirit and service product mainly industry, but also want to follow the rules. The sports industry products directly into the market, direct in the market, so the output to exchange the development of the sports industry in the development of sports market as the prerequisite. In the very great degree of sports market development and prosperity of the decision of sports industry development and prosperity.

2.1.3 The Sports Industry Development in Areas of Expanding, and Industry Development Gradually Improve Quality and Efficiency

The development of the sports industry in addition to the basic form of the market has been outside the territory, because the market in the economic development of the resource allocation role, many sports assets and resources are shown tremendous development potential and the potential market advantage. Development business mainly from two aspects: one is the tangible assets management, and the second is the development and utilization of the intangible assets.

2.1.4 The Social Investment Do Industry Development Trend, the Emergence of a Large Number of Soon Conforms to Modern Enterprise System, DuoZhong Ownership Sports Club, Sports Enterprise and Enterprise Group

With the high-speed development of national economy and the people's standard of living rise, social sports consumption demand rapid expansion. Many fitness entertainment is stronger loved by the public sports become the hot spot of the social investment. In recent years, a large number of different system of ownership, different sizes of sports business enterprise such as jiang surges, except for some large famous brand enterprise to run football, basketball, volleyball outside, still have a lot of small and medium enterprises engaged in investment hot spots the consumption market, such as fitness and strong and handsome, bowling, billiards, tennis etc.

2.1.5 Initially Formed a Guarantee of the Development of Sports Industry, Multi-level and Multi-channel Forms of Financing Mechanism

The socialist market economic system established to make market under the state's macro-regulation, for resources together.In the allocation of resources, the government is no longer a single configuration main body, the government has a mainly the macro economic and leading correction function.

2.2 Chinese Sports Industry Development Problems

2.2.1 Industry Scale Is Not Large, and Unbalanced Development, Obvious Regional Differences

Compared with the developed countries sports industry, the sports industry in China is only at the preliminary stage, and urban and rural residents in China as a family unit for spots consumption in 100 yuan the following account for 58.3% of the total. 100-200 or 27.8% of the total, 201 yuan of above accounted for only 13.9% of the total. The sports industry in the developed countries the proportion of economic structure, and the rise in China's economic structure, the sports industry only accounts for about 0.5% of gross domestic left. [1]

2.2.2 The Sports Industry Structure Defects

Industry remains to be further improve the quality of sports industry unreasonable internal structure is mainly manifested in the ontology industry development were not big sports industry, the core subject position did not reflect, ontology market development is not enough form pillar advantage. Sports fitness entertainment industry as the main body of the essential function embodied sports industry, the development is still in the primary stage, its own potential and advantage is still not fully play: as a sports related industry, such as sports advertising, sporting goods industry is developing a rapidly.

2.2.3 Sports Intangible Assets Had Loss

The planned economy system, the training of competition system China has made a batch of famous in the world of sports advantages, such as gymnastics project, table tennis, badminton, swimming, diving, women weightlifting, women's middle, etc. Although they are in the world sports is among the best, but failed to develop its economic value in the world, make sports within the scope of the brilliant achievements in economic field have become meaningless. Instead, the United States made the Nike basketball, Italy's football made adidas, British gymnastics became famous, rocket company.

2.2.4 Restrict the Development of the Sports Industry in China Is Factors

First of all, the economic development of our country, a low level on the whole of sport consumption of main factors. In 2002, China's per capita GDP of nearly $1000, this is only just get out of the least-developed countries level. Needless to say with the developed countries more than $per capita GDP2 compared to one, and the world is the per capita nearly $5000, is also very obvious gap compared.

Second, the lack of sports industry development business operation mechanism and lack of financial support. Lack of business operation system and lack of enough money to support make business sports organizations lack of the earnings for the purpose of consciousness, can't make sports business institutions to establish modern enterprise system with the operation mechanism, not the profits of the business as its purpose, unable to inspire investors of sport industry interests and desires. On the one hand, sports department in the development of industry purpose is to make up for the lack of sports operating funds, original little, not may get more money in the development of industry, and through the industry development revenues from most used for sports, career development, and was rarely used for capital accumulation, enlarged reproduction. On the other hand the sports industry for more than individual manage, its funds originally not abundant, after the first, the lack of liquidity in, the ability to resist risk is very poor. To avoid this kind of situation we will use of the capital market in the developed countries for reference to promote the development of sports industry experience, walk the sports industry and financial effectively combine the notion, to solve the sports industry in China is in the process of the development of the problems of insufficient funds.

Third, the urbanization level is low, limiting the sports consumption fast development, the gap between urban and rural areas and the development of the sports consumption imbalance state. China is the world's urban population level is low, one of the countries accounted for 30% of the urban population, much lower than the 50% of the world's level. Because China's urbanization rate lag in industrialized speed, make whole national consumption ability is not restricted the development of the sports direct speed. In addition, the gap between eastern and western economic development to increase, also make the sports consumption appear in large imbalances.

Fourth, the lack of government's input, and guide the construction of legal system lags behind, the poor, is restricted by the sports consumption development important factors. Because our country at all levels of government spending pressure big, overstaffed organs, burden, and obviously deficiencies in the sports commitment. In the social commitment, because the sports consumption, investment risk expected efficiency is not stable, and economic growth slowed in recent years and the social security system reform has not arrived, the social investment proportion lesser also.

3 Our Country Sports Industry Development Strategy

3.1 Strengthen the Control of the Government

Deepen the reform of the management system of sports, transformation of government functions to meet the challenges of the WTO, we must deepen the reform hinder the development of the sports industry in China is relevant management system, to change the policy functions, but is not the government refused to macroeconomic control. In today's world, no one of the market economic countries and regions, a completely free of the market economy, don't let the government's intervention. But first, the government function, it must be transformed by micro management into a macro management, by direct management into indirect management, and concrete ways in the following aspects:

First, for the development of the sports industry to create a good policy environment, the development of the sports industry into a new round of the general development of planning in.

Second, make perfect strategic development planning. Third, the transformation government function, reduce planning, increase the marketability, for the development of the sports industry create a favorable external environment. Fourth, we should build a socialist sports. market system.

3.2 Further Exploration and Development of the Sports Industry Market

Sports has become a big industry, the world's current sports industry worth about 400 billion dollars, and increasing at the rate of 20% per year. Every year in all countries on the market only football turnover reached $250 billion

4 Summary

The sports industry is a new industry, it's healthy and rapid development of physical education in our country, for stimulating consumption expanding domestic demand and promote the national economy progress and socialist spiritual civilization construction, plays a very important role. In the market economy to the early stage of building the sports industry in China, although the tide of the economy by many unprecedented new problem, but the trouble to create new life. The new century as long as we can seize the opportunity, make in time of the development of the sports industry and promote the sports strategic planning of the market mechanism and the competition mechanism, the sports industry reforms will succeed to the remarkable success and economic benefits.

References

1. Gao, Y.: Theory of industry evolution and urbanization development. Modern Management Science 1, 54–57 (2002)
2. Xu, B.-L.: The sports industry in China to the theoretical study of the problems in the survey and research. Beijing Sport University Journal 25(2), 148–151, 172 (2002)
3. Ni, G.: Sports industry financing structure and way of research. Guangzhou Sports Institute Journal 23(1), 15–18 (2005)
4. Li, H.: WTO on the impact of sports industry research. Journal of Shenyang Sports College (2), 25–27 (2006)
5. Zhao, G.-J., et al.: Enterprise development strategy choice. Tianjin university press, Tianjin (2000)

On the Innovation in the Scientific Research Administration and Evaluation under the Modern University System

Su Junhong, Li Gaohong, and Yu Xiaoning

Department of Science & Technology, Xi'an Technological University, Xi'an, 710032

Abstract. The university system is established on the basis of the academic nature to guide the administration and operation of a university and to determine its survival and development. This paper expounds the modern university system and the connotation of its innovation, analyzes five main problems arising in the scientific research administration and evaluation under the modern university system, and by absorbing foreign advanced experience in innovative university system, gives a solution that only strengthening the importance, level and efficiency of administration by professors can improve the modern university system with Chinese characteristics.

Keywords: Academics, University system, Innovation, Administration.

1 The Problems Arising in the Current Scientific Research Administration and Evaluation

The reform and innovation of the scientific research administration and evaluation of a university is one of the important contents of modern university system. At present, there do exist some problems with the current scientific research administration and evaluation of the universities in China.

1.1 The System of Scientific Research Administration Is Not Conducive to Innovation

First, independent and spontaneous research activities don't adapt to the development of modern science and technology with the characteristics of the multidisciplinary research, the complex study objects, and the group work of research. Second, the current assessment system can hardly produce high level research groups and significant research results. Third, the research staff follow the model of working separately, and team work is not valued. The cooperation and communication between departments, laboratories and researchers don't suffice. As a result, the advantage in group work can't be embodied. Teachers, since the beginning of their career, have to choose projects and write papers only for their professional title promotion. So they don't have a stable research direction. Consequently, it turns very hard to make the achievements of scientific and social value.

D. Zeng (Ed.): Advances in Computer Science and Engineering, AISC 141, pp. 789–794.
Springerlink.com

1.2 The Concept of Scientific Research Administration Is behind the Times

Both the scientific research administrators and the research staff only focus their attentions on applying for projects and awards, and transforming the achievements, with a keen interest in scientific research funds. Too strong an interest in the number of awards, papers and approved projects has triggered the violation of the law of the development of science and technology and that of education. Under this circumstance is innovation greatly devalued and comprehensive mind lacked in the scientific research administration.

1.3 The Administration Admits Too Much Interference in the Scientific Research Field

The ranking of universities in China has driven some schools to place too much emphasis on the fame as one of universities of Scheme 985 and 211. The school authority, under the influence of input-output theory, requires the teachers to produce research results of great value in a relatively short period of time after the fund is allocated. In such a system, the teachers are treated as machines, and the intellectual achievements as products. But the high level research results involve a number of factors including increased investments. The administrative mode doesn't adapt to the development of modern university scientific research management and can hardly create an environment favorable for innovation.

1.4 The System of Scientific Research Evaluation Is behind the Times

At present, the university scientific research evaluation system focuses on the results and data of researches, while scant attention is paid to the process and the ideas of them. The number of the papers and books published and the amount of research fund have become the measurement of the research level of teachers, and even the priority for teachers' professional title promotion, their annual assessment and competition for posts.

1.5 Scant Attention Is Paid to Intellectual Property Rights and Research Results Transformation

The current university scientific research evaluation system focuses on academic value instead of practical benefits. Little attention is paid to intellectual property rights and research results transformation. Many teachers view patent application as a task. When the application is approved, they would turn a blind eye to them. As a result, the application for patent is out of proportion to its implementation.

2 The Solutions to the Problems

2.1 Emancipating the Mind to Initiate the System of Scientific Research Management and Evaluation

(1) Promoting multidisciplinary research

The combination of the humanities and the natural sciences can lead to theoretical innovation. The multidisciplinary research may, to a great degree, broaden and deepen a single subject research.

(2) Funding research results of great significance

Projects are to be implemented to finance basic researches, such as offering self-chosen subjects, encouraging free academic exploration, establishing funds for publication of academic achievements and that of academic books, and for keeping academic journals. Projects are also to be implemented to transform research results. The measures take the form of establishing special funds for policy consultancy to encourage the production of the policy consultation reports which are of great value and which can meet the great demand of the government, and for promoting the cooperation between the research institutions and the governments of different levels The increased proportion of supportive funds is to be given to the research projects of great significance and of an amount of funds.

(3) Encouraging service-based scientific researches

It is required that the results of research be applied to the places like factories, villages, schools and communities. The study on the big social problems and practical ones are encouraged so that the ability to solve the problems can be improved.

(4) Reforming the current scientific research evaluation system

The current scientific research evaluation system is supposed to achieve the transformation from static assessment to dynamic assessment, from single evaluation to comprehensive evaluation so as to form a reasonable scientific research evaluation system. The current scientific research evaluation system which focuses itself on quantity rather than quality, results than process must be changed. It is very necessary to establish a fair, open, just evaluation mechanism, to form an effectively competitive, incentive and restraint system, to encourage teachers to make the creative research results of great value, and to explore a new evaluation means which combine team evaluation and individual evaluation together, spur researchers to do initiative work for the achievements of self-creation.

(5) Achieving internationalization

More attention should be paid to international academic exchange and cooperation. It is necessary to establish international academic exchange funding to finance the joint projects between our teachers and foreign universities and research institutions, to finance the international publication of the papers or books written in foreign languages by our teachers.

(6) Improving the current model of scientific administration and organization

The universities should strengthen the authority of professorship in scientific administration and organization so that the teachers can be fully engaged in research work. They should also formulate the strategic plan for scientific research development, create an academic atmosphere of aggressiveness, competition, exploration and tolerance to failure, and serve the research staff wholeheartedly so as to draw, make, empower them and to keep their minds on work. They should offer devolution to their subordinates for more flexibility too.

2.2 Promoting the Independent Innovation and Social Service Capability of University Scientific Research Institutions

It is required that the concept, system of university scientific research management be innovated. Universities should fully play its role as the pioneer in the national independent innovation system, integrate scientific study and personnel training, strengthen the construction of innovation groups, academic atmosphere and scientific base and explore a scientific evaluation mechanism. The cooperation between universities and scientific research institutions and enterprises should be greatly encouraged so as to combine production, teaching and research .

2.2.1 Innovating the Concept of Scientific Research Management

① The innovation of the function orientation of university scientific research management

Universities have to change the traditional model in which the functions of different departments are fragmented and each department develops itself separately, to integrate various goals and resources, to integrate teaching, scientific research, management, service into a system. University scientific research management is supposed to support personnel training, to promote research results' production and transformation, and to involve in marketing so that technology and economy can be well combined. The function of university scientific research management has to be changed from administration to service, from being economic-benefits-oriented to being comprehensive-benefits-oriented, from technological result assessment to comprehensive indexes assessment. Establishing key index system, empowering the research staffs, perfecting service system, establishing innovation group--all these methods can be adopted to promote innovation.

② The innovation of the concept of university scientific research management

The university scientific research management must establish the concept of man-orientation, resources integration and market involvement. Teachers are the body of the scientific research, and their independence and originality must be respected. Discussion, supervision, and open evaluation will help to redefine the schools' obligation of formulating the goals, supervising the results, and motivating the researchers. In addition, the school authority should optimize resources allocation according to the market demands, the law of scientific study instead of the leaders' arbitrary will or order. Last, it is required that the scientific research meet the great demands of socialist market economic environment.

③ The innovation of the system of university scientific research management

The reform aims for an efficiency-based university scientific research management system. All the barriers, during the reform, should be got rid of. Ensure that the research staffs devote their energy and enthusiasm to innovation by means of simplifying management procedures and improving management efficiency.

2.2.2 Optimizing Assessment and Motivation Mechanism

First, it is necessary to improve the scientific research assessment system that is suitable to various types of universities. Under the direction of the scientific development concept, we must understand that at present there do exist various types of universities, each having some specific subjects. Today's Chinese universities can be classified into research type, teaching type, research-teaching type and teaching-research type. Each type of universities should be evaluated according to their own conditions. According to the results of evaluation can the effective incentive system be implemented to achieve rapid development of science and technology? Besides, the scientific discipline of systematic evaluation should be established. To ensure the implementation of assessment, the assessment index system should also be segmented and simplified. The index system can not reflect all the contents of value disciplines, but has to fit in with them basically. Scientific research assessment must, systematically, define some basic elements and other interrelated factors, and dynamically, assess some elements which might be constantly changing. Information collected must be objective and reliable, and the choice of the elements must be carefully considered to ensure the feasibility of the assessment. Next, pre-assessment research and post-assessment supervision must be well combined to improve assessment, and in turn, to promote the development of science and technology.

2.2.3 Establishing Service-Based Scientific Research Management System

First, it is very necessary to establish service-based scientific research management system to ensure sustainable development of scientific research. A transition from organization to service should be achieved. The research staffs are supposed to be in charge of exploration, application for research projects and independent operation. Besides, service-based scientific research management means support rather than guide, evaluation rather than determination. Second, it is advisable that we establish an open, internationalized service model to promote the transformation of research achievements. The awareness of openness must be intensified so that we can capitalize on both domestic and overseas resources. International co-operation is greatly encouraged to achieve academic construction, personnel training and adequate support to scientific research. The goals and practical measures must be decided to promote the establishment of a service-based scientific research management system.

2.3 Combining Production, Teaching and Research Well to Promote the Transformation of Research Results

Various forms of co-operation between universities and enterprises and scientific research institutions including co-operative research, personnel exchange, joint innovation institutions, staff training to form an effective mechanism in which production, teaching and research are well combined. A model of "Three-part-cooperation", namely the co-operation between universities, communities and scientific parks, should be introduced into the society. And in this way will some regional comprehensive and open innovation leagues be formed so that some quite valuable research results can be implemented and the role of universities as the think tank can be exploited.

2.3.1 Promoting the Combination of Production, Teaching and Research to Achieve the Goals of the Industry or That of the Government

With great desire for development and innovation, small and medium-sized enterprises demand the support from the government to promote innovation for lack of funds and experience. Therefore, by means of rational science and technology development projects, small and medium-sized enterprises innovation funding, local science and technology plan, the list of much-in-demand technologies and products, innovation will be dramatically promoted. In fact, the development of small and medium-sized enterprises is indispensible to the establishment of "Production-teaching-research" strategic league.

2.3.2 Establishing Incentive System to Promote Innovation

Independent innovation includes originality, integration of technologies, and re-innovation on the basis of absorption of overseas advanced technologies. The problem of backward incentive system has not been solved so far. The current title determination system in universities and scientific research institutions focuses on academic papers rather than research results transformation. The current scientific research funds management rules require that the research staffs working in the units financially supported by the government are not allowed to allocated the research funds freely. If the rules are strictly enforced, the research staffs are, in fact, doing research work voluntarily. The fact is that most of the researchers work in some universities and scientific research institutions. As a result, the initiative of the research staff is greatly influenced. To establish scientific personnel evaluation system and flexible distribution system will become important incentives for innovation.

Acknowledgements. This work was supported by Social science fund project of The Ministry of Education of China under grant No. 08JA630065.

References

1. Zhou, X.: A New Theory of the System Innovation and the Scientific Development of Universities. China Higher Education Study (4) (2009)
2. Liu, E.: The Problems with University Scientific Research Evaluation and the Solutions. Higher Engineering Education Study (1), 39–42 (2004)
3. Feng, H.: The Exploration of University Scientific Research Evaluation System Reform. The Technologies and Its Industrialization of Chinese Universities (12), 56–57 (2006)
4. Fuming, Weihua, Zhangqun: Reflect on the University Science and Technology Evaluation System. Science and Technology Management Study (11), 118–120 (2006)
5. Chen, Y.: University Scientific Research Management: An Analysis on the Basis of National Innovation System Theory. Ningbo University Journal (science and education page) (2), 56–58 (2009)
6. Huang, Y.: Some Problems with University Science and Technology Management System and Policy Innovation. Scientific Study (3), 96–98 (2000)
7. Liu, Z., Chen, X., Yang, J.: University Scientific Research Management under New Circumstances. Technology and Innovation Management (3), 15–18 (2005)

Numerical Simulation on Distribution Characteristics of Particle Distribution Uniformity in a Radial Style Diesel Particulate Filter

Gong Jinke[1,2], Tian Chan[2], and Wu Gang[2]

[1] State Key Laboratory of Advanced Design and Manufacturing for Vehicle Body
Changsha, China

[2] College of Mechanical and Vehicle Engineering, Hunan University
Changsha, China

{tianchan789,wug1999}@163.com

Abstract. In this paper three-dimensional mathematical model of gas-solid two-phase in a radial style diesel particulate filter (DPF) is established and flow computation software was used to simulating distribution characteristics of gas-solid two-phase and particle distribution uniformity. Several influence parameters such as flow velocity, divergence angle and particle diameter was investigated. Through the verification analysis, the simulation calculation results can reflect the laws of gas-solid two-phase flow inside this DPF.The results indicate that distribution uniformity of particle could be improved by reducing inlet velocity, divergence angle and particle diameter. The study is useful for structural parameters optimization designing and controlling particle regeneration of radial style diesel particulate filter.

Keywords: diesel particulate filter, gas-solid two-phase flow, three-dimensional numerical simulation.

1 Introduction

At present, the diameter of traditional axial diesel particulate filter is relatively so large that during the regeneration , it is esay to be cracked for uneven heating ,and high power microwave source is also required. In view of these problems, Gong Jinke, Wang Shuhui etc invented a radial style diesel particulate filter in which rotating filter block was used [1].It could realize continuous regeneration with many advantages such as lower power microwave source and more evenly heating[2]. Particle distribution uniformity has important significance for the optimization design of DPF structure parameters and the further study of DPF regeneration characteristics.

In this paper, gas-solid two-phase flow simulation analysis is made in radial style diesel particulate filter ,using computational fluid dynamics software.Through calculation and analysis, we can get the characteristics of particle distribution uniformity in a Radial Style Diesel Particulate Filter with the changes of structure parameter.

D. Zeng (Ed.): Advances in Computer Science and Engineering, AISC 141, pp. 795–805.
Springerlink.com

2 Basic Mathematical Equations

2.1 Equation of Continuity

equation of particle phase

$$\frac{\partial(\delta\alpha_s\rho_s u_s)}{\partial x}+\frac{\partial(\delta\alpha\rho_s v_s)}{\partial y}+\frac{\partial(\delta\alpha\rho_s w_s)}{\partial z}=\sum\dot{m}_{gs} \tag{1}$$

equation of gas phase

$$\begin{aligned}&\frac{\partial[(1-\alpha_s)\delta\rho_g u_g]}{\partial x}+\frac{\partial[(1-\alpha_s)\delta\rho_g v_g]}{\partial y}+\\&\frac{\partial[(1-\alpha_s)\delta\rho_g w_g]}{\partial z}=\sum\dot{m}_{sg}\end{aligned} \tag{2}$$

porous medium

$$\frac{\partial(\delta\rho_f)}{\partial t}+\frac{\partial(\delta\rho_f u)}{\partial x}+\frac{\partial(\delta\rho_f v)}{\partial y}+\frac{\partial(\delta\rho_f w)}{\partial z}=0 \tag{3}$$

2.2 Momentum Conservation Equation

equation of particle phase

$$\begin{aligned}&\frac{\partial(\delta\alpha_s\rho_s v_s u_s)}{\partial x}+\frac{\partial(\delta\alpha_s\rho_s v_s v_s)}{\partial y}+\frac{\partial(\delta\alpha_s\rho_s v_s w_s)}{\partial z}=\\&-\alpha_s\frac{\partial(\delta p)}{\partial i}+\nabla\cdot(\delta\vec{\tau_s})+\\&\sum(R_{gsi}+\delta\dot{m}_{gs}v_{gs})+\alpha_s\rho_s(F_{si}+F_{lift,si}+F_{Vm,si})+\alpha_s\frac{\mu}{\kappa}V_s\end{aligned} \tag{4}$$

porous medium

$$\frac{\partial\left(\delta\rho_f u_j\right)}{\partial t}+\frac{\partial\left(\delta\rho_f \vec{u}u_j\right)}{\partial x_j}=\rho_f f+\frac{\partial}{\partial x_i}\left(\mu\frac{\partial\delta u_j}{\partial x_i}\right)-\frac{\partial(\delta p)}{\partial x_j}+\frac{\mu}{\kappa}u_j \tag{5}$$

equation of gas phase

$$\frac{\partial[\delta(1-\alpha_s)\rho_g u_g u_g]}{\partial x}+\frac{\partial[\delta(1-\alpha_s)\rho_g u_g v_g]}{\partial y}+\frac{\partial[\delta(1-\alpha_s)\rho_g u_g w_g]}{\partial z}= -(1-\alpha_s)\frac{\partial(\delta p)}{\partial i}+\nabla\cdot(\delta\vec{\tau_g})+ \sum(R_{sgi}+\delta\dot{m}_{sg}u_{sg})+\alpha_g\rho_g(F_{gi}+F_{lift,gi}+F_{Vm,gi})+(1-\alpha_s)\frac{\mu}{\kappa}V_g \tag{6}$$

In all the equations above, $i = x$、y、z is direction x、y、z ,V= u、v、w is velocity in direction x、y、z , μ 、 ρ_f the viscosity and density of fluids respectively, κ is permeability.

2.3 Turbulence Equations in Standard k Epsilon Model

k equation

$$\frac{\partial(\rho_m u_m k)}{\partial x}+\frac{\partial(\rho_m v_m k)}{\partial y}+\frac{\partial(\rho_m w_m k)}{\partial z}= \frac{\partial}{\partial x}\left(\frac{\mu_{t,m}}{\sigma_k}\cdot\frac{\partial k}{\partial x}\right)+\frac{\partial}{\partial y}\left(\frac{\mu_{t,m}}{\sigma_k}\cdot\frac{\partial k}{\partial y}\right)+\frac{\partial}{\partial z}\left(\frac{\mu_{t,m}}{\sigma_k}\cdot\frac{\partial k}{\partial z}\right) +G_{k,m}-\rho_m\varepsilon \tag{7}$$

In this equation ,k is Turbulent Kinetic Energy,Turbulent Dissipation Rate, ρ_m is mixture density of Gas and particle phase, $\mu_{t,m}$ is Turbulent viscosity.

ε equation

$$\frac{\partial(\rho_m u_m \varepsilon)}{\partial x}+\frac{\partial(\rho_m v_m \varepsilon)}{\partial y}+\frac{\partial(\rho_m w_m \varepsilon)}{\partial z}= \frac{\partial}{\partial x}\left(\frac{\mu_{t,m}}{\sigma_\varepsilon}\cdot\frac{\partial\varepsilon}{\partial x}\right)+\frac{\partial}{\partial y}\left(\frac{\mu_{t,m}}{\sigma_\varepsilon}\cdot\frac{\partial\varepsilon}{\partial y}\right)+ \frac{\partial}{\partial z}\left(\frac{\mu_{t,m}}{\sigma_\varepsilon}\cdot\frac{\partial\varepsilon}{\partial z}\right)+\frac{\varepsilon}{k}(C_{1\varepsilon}G_{k,m}-C_{2\varepsilon}\rho_m\varepsilon) \tag{8}$$

In equation (8) , σ_ε is the number of Prandtl corresponding to turbulent dissipation rate , $C_{1\varepsilon}$ 、 $C_{2\varepsilon}$ 、 σ_ε are constants of the model ,According to the values recommended by Launder and later experimental verification , $C_{1\varepsilon}=1.44$, $C_{2\varepsilon}=1.92$, $\sigma_\varepsilon=1.3$.

3 Geometric Model and Boundary Conditions

3.1 Geometric Modeling Grid

Using Gambit software , the mesh of this Diesel Particulate Filter is generated. The grid is hexahedron -Wedge Hybrid and the mesh size is seted as small as possible which is conducive to get the stability numerical solution.The model grid is shown in Figure1 (section a and b are x = 35mm, x = 130mm)

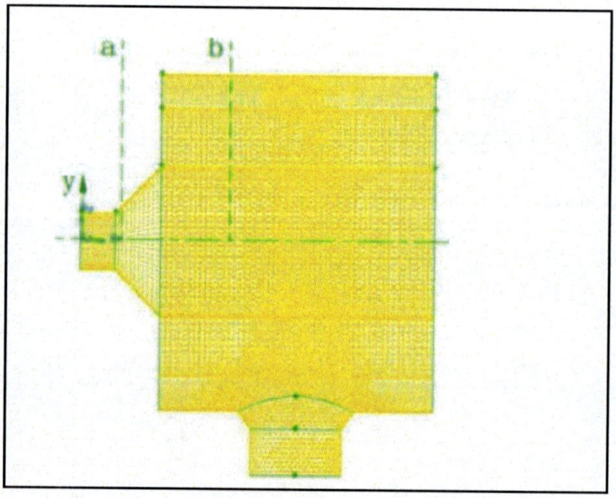

Fig. 1. Grid of a radial style diesel particulate filter

3.2 Calculation of Boundary Condition Setting

The flow rate in the DPF is not high, so the gas can be considered as incompressible. The velocity and pressure is used as boundary condition of entrance and exit. Pressure outlet boundary condition is set to 0Pa, considering only the entrance and exit pressure relative value, ignoring the filter body resistance to airflow and all surface conditions are set to the no-slip boundary condition.Gas phase material density and dynamic viscosity at 300 degrees C : ρ =0.3827kg/m3, μ =2.946Pa s , Thc volume fraction of particle phase is 0.05 [3]. Using the turbulence intensity and hydraulic diameter as Turbulent input conditions ,turbulence intensity can be get according to equation (9) calculation:

$$I = 0.16\,\mathrm{Re}^{-1/8} \tag{9}$$

$$\mathrm{Re} = UD_H / \nu \tag{10}$$

In these equations , I-turbulence intensity, D_H- entrance pipe diameter, m ; ν - kinematic viscosity, $m^2\,s^{-1}$.

4 Results of Numerical Calculation and Analysis

4.1 Influencing Factor of Particulate Concentration Distribution

1)The effect of exhaust velocity

In this case , parameter is seted as Table 1. Through calculation , the particulate concentration distributions of different exhaust velocity in the expansion tube section (x=35mm) and the filtering body section (x=130mm)are shown as figure 2 and figure 3.

It indicate that : In the two section, the particle concentration showed a parabola distribution. The exhaust velocity does not change the structure of particle concentration field.

In calculation section x=35mm , under these three exhaust velocity (20 m / s, 40 m / s, 80 m / s) particle concentration minimum values were 0.04997, 0.01757, 0.01716. The minimum velocity, the smaller particle concentration minimum values. As the exhaust velocity decreases , the expansion tube within the region of the vortex strength is weakened, the role of particles separation around the eddy center area is also corresponding weakened, so the minimum value become larger.

In the filter section (x=130mm), Particle concentration maximum value and the minimum value difference showed a trend of decrease. When the exhaust flow rate is 20m / s, particle concentration curve almost parallel to X axis and the particle concentration distribution uniformity is best..

Table 1. Differcnt exhaust velocity

Fixed value	Dilation angle		Filtering body length
	90°		240 mm
Change value (exhaust velocity)	20 m/s	40 m/s	80 m/s

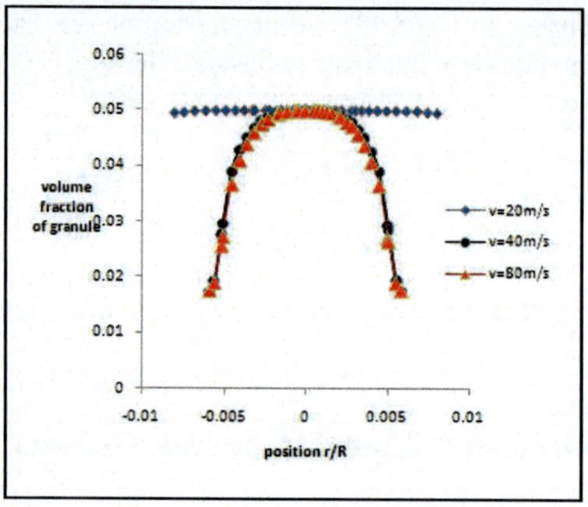

Fig. 2. Curve: granule concentration of different velocity in x=35mm

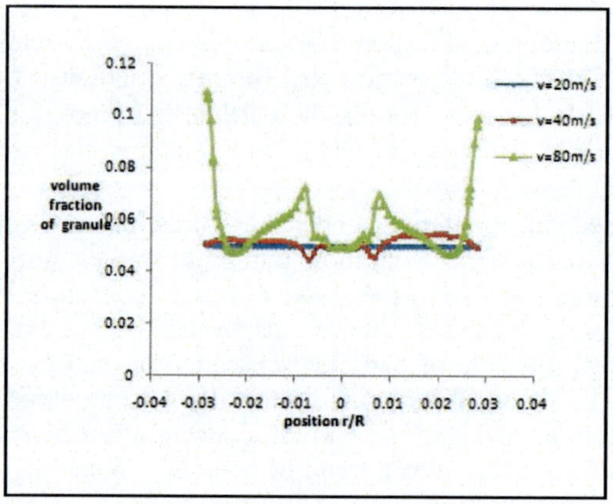

Fig. 3. Curve: granule concentration of different velocity in x=130mm

Comprehensive analysis shows that : with the entrance exhaust velocity decreases , in the expansion tube area and a filtering body segment, particle concentration distribution uniformity becomes better, that is , particulate trap particle has better concentration distributy uniformity.

2) The effect of particle diameter

In this case , parameter is seted as Table 2. Through calculation , the particulate concentration distributions of different exhaust velocity in the expansion tube section (x=35mm) and the filtering body section (x=130mm)are shown as figure 4 and figure 5.

Table 2. Different particle diameter

Fixed value	Dilation angle	Filter body length	exhaust velocity
	90°	240 mm	40 m/s
Change value (particle diameter)	1μm	0.5μm	0.1μm

It indicate that : In the two section, the particle concentration showed a parabola distribution. The exhaust velocity does not change the structure of particle concentration field. As the particle diameter decreases, particulate concentration distribution in the expansion tube regional curves more quietly.

In figure 4: In the filter section, when the particle diameter is 0.1μm ,there is small fluctuations of Particle concentration curve (remained in the 0.05). This is because inertia is small, the small diameter particle has good flow property.

Analysing and comparing Figures 4 and 5 , it is easy to find that in the expansion tube area, near the centerline axis , particle concentration reaches its maximum value , but in the filter section shaft near the center , the particle concentration is minimal. This is because backflow effect in the cross section of x = 35mm is more obvious and particles are aggregated in the vicinity of the axis under the action of the air flow. In a word , with good flow property, concentration distribution curves of small diameter particles particle are smoother.

3) The effect of dilation angle

In this case , parameter is seted as Table 3. Through calculation , the particulate concentration distributions of different exhaust velocity in the expansion tube section (x=35mm) and the filtering body section (x=130mm)are shown as figure 6 and figure 7.

In figure 6 ,It is indicated that , in the region of expansion tube , particle concentration distribution is axial symmetry and slope of particle concentration curves increces with the expansion of dilation angle.When the dilation angle increases to 120°, curve inclination change from acute angle to obtuse angle suddenly and distribution uniformity becomes bad. This is because as the expansion angle increases , in the expansion tube , eddy current generated gradually and increases slowly.

In figure 7 , in the filter section (x=130mm) , When dilation angle is the 60° and 90° , the particle concentration did not change much and the slope of the particle concentration distribution curve is nearly 0 , though When dilation angle was increased to 120° , concentration curve begins to appear fluctuation. This is because compared with other expansion angle , in the filter body section of the particulate trap of 120° , the vortex and circumfluence phenomenon is very apparent, particles affected by the air-flow relatively large.

In general, the expansion angle has great influence on particle concentration distribution , DPF with small expansion angle on the whole particle distribution is uniform than the big one.

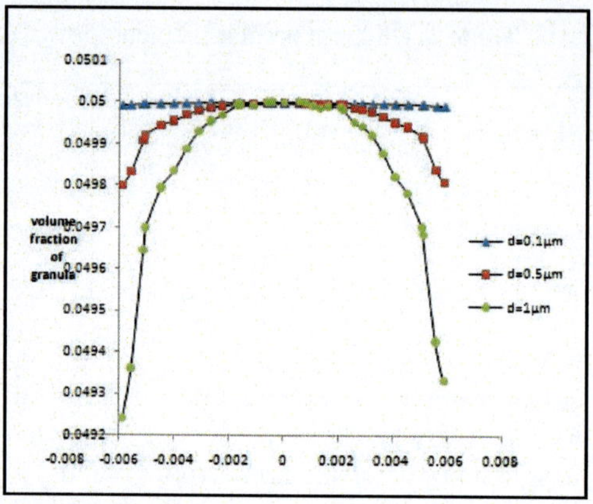

Fig. 4. Curve: granule concentration of different particle diameter in x=35mm

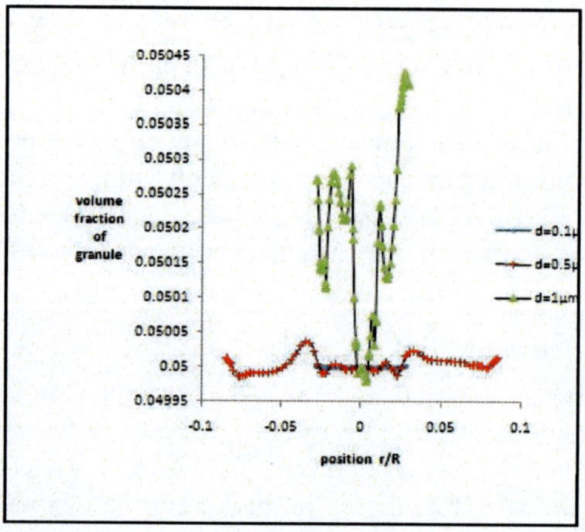

Fig. 5. Curve: granule concentration of different particle diameter in x=130mm

Table 3. Different dilation angle

Fixed value	**exhaust velocity**		**particle diameter**
	40 m/s		1µm
Change value (dilation angle)	60°	90°	120°

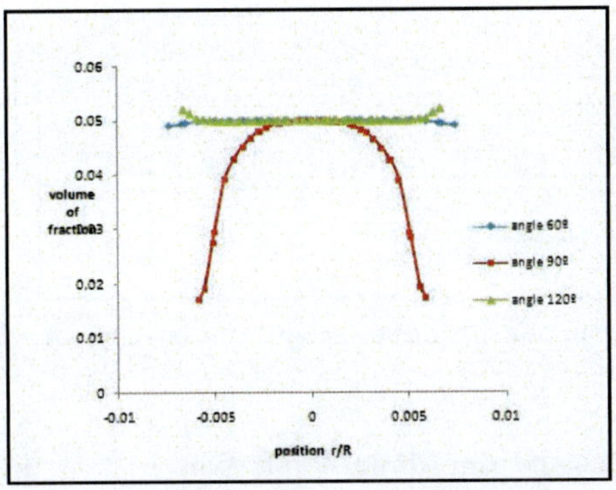

Fig. 6. Curve: granule concentration of different angle in x=35mm

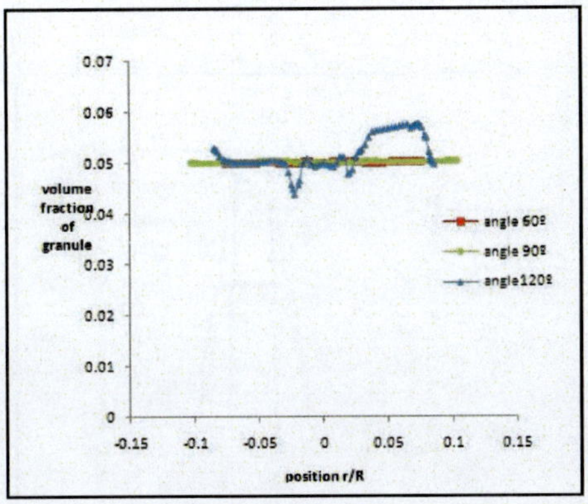

Fig. 7. Curve: granule concentration of different angle in x=130mm

5 Experimental Verification

Considering great influence which the distribution characteristics of velocity has on the distribution characteristics of particle distribution uniformity in a Radial Style Diesel Particulate Filter and the difficulties to measure the particle velocity separately , in this paper only gas flow velocity were measured and comparative analysis was made with the result of numerical simulation to verify if the gas-solid two-phase flow model is correct.

5.1 Experiment Method

Isothermal, stable test devices are established on stable test table to measure the distribution characteristics of velocity.The structure parameters of DPF which is used in the experment is in the following : Dilation angle is 90° and Filtering body length is 240mm.Schematic diagram of experimental setup is shown in figure 8.

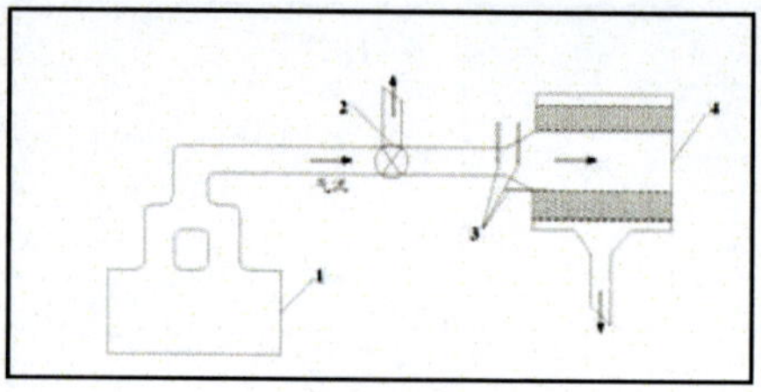

Fig. 8. Schematic diagram of experimental setup 1- air compressor 2 -valve 3 - flow meter 4 -particulate trap

5.2 Experement Results and Model Validation

When the exhaust velocity is 20m / s, the velocity values of gas in x = 35mm, x = 65mm and x = 120mm, x = 190mm, x = 250mm these five section were measured and compared with numerical simulation values and the results of comparison is shown in Figure 9 and Figure 10.

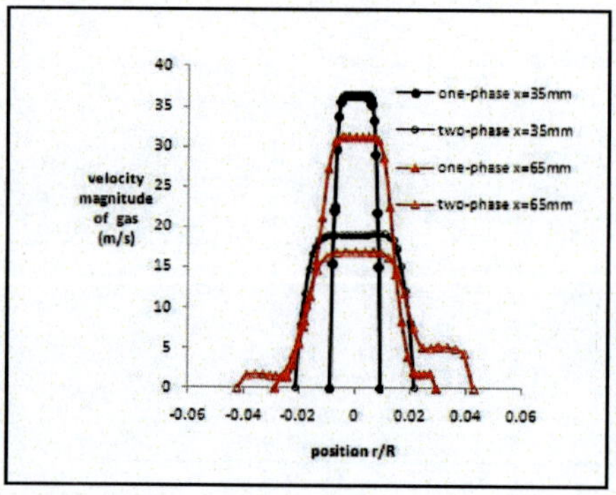

Fig. 9. Curve: comparing result of gas velocity in expansion pipe area

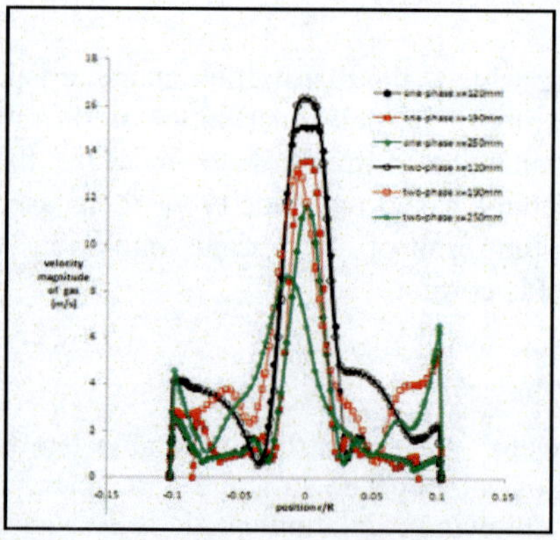

Fig. 10. Curve: comparing result of gas velocity in filter area

6 Conclusion

- The gas solid two phase flow model established in this paper is correct. In the error range allows, it is in line with the actual situation, the analysis results have a certain reference.
- The distribution of particles in the Y axis plus or minus two direction is basically symmetrical. The smaller entrance exhaust velocity, the smaller effect particle by eddy current, so the particle distribution uniformity is better.
- In a radial style Diesel Particulate Filter, particle diameter is smaller, the effect of reflux is smaller, so the particle distribution was more uniform.
- Expansion angle is bigger, more prone to eddy current. Reducing the expansion angle is beneficial to improving the particle distribution uniformity.

Acknowledgmants

This paper is supported by the projects following ,great thanks to them.
The natural science foundation of Hunan Province (10JJ6080).
The National Natural Science Fund Projects (50876027).
The national "863" project project (2008AA11A116).

References

1. Gong, J., Wang, S., Wu, G., et al.: A diesel particulate matter reduce emissions of methods and devices. China's patent, ZL200710035168.6 (September 17, 2008)
2. Gong, J., Wu, G., Wang, S., et al.: Numerical analysis on distribution characteristics of flow velocity in a radial style diesel particulate filter. Journal of Agricultural Engineering 26(4), 119–126 (2010)
3. Liu, Y.: Numerical of gas grain two-phase flow characteristics inside a diesel particulate filter (Master bachelor's thesis). Hunan University, Hunan (2007)
4. Zhou, L.: Numerical simulation the turbulence two-phase flow and the burning. Tsinghua University Press, Beijing (1991)

Edge Detection from Remote Sensing Images Based on Canny Operator and Hough Transform

Jing Xi[1] and Ji-Zhong Zhang[2]

[1] Faculty of Surveying and Mapping
Xi'an University of Science and Technology
Xi'an, China
[2] Geographic Information Co., Ltd.
Aerial Photogrammetry and Remote Sensing Bureau
Xi'an, China
xijingjing@gmail.com, zjz9108651@sohu.com

Abstract. At present, the widespread edge detection algorithms, such as Roberts edge detection, Prewitt edge detection, Sobel edge detection and Marr edge detection, detect the edge using the variation of one-order or two-order directional derivative near to the edge usually. As a result, the noises can be detected in the image regarding as the edge points, and the real edge maybe missed due to noise interference. Because the remote sensing images have a disadvantage of containing noises, the better detecting results are difficult to obtain using the above edge detection algorithms. Combined with the characteristics of remote sensing images, including rich edge information and a lot of noise mixed, this paper present a edge detection algorithm for remote sensing based on the Canny operator and Hough Transform. From the experiment introduced in the paper, the proposed edge detection algorithm has some noise immunity, and can explore the edge information accurately in the remote sensing image.

Keywords: Edge Detection, Canny Operator, Hough Transform, Noise, Remote Sensing Images.

1 Introduction

The information obtained from the images is one of the important sources to acquire knowledge for humans. In the information time, it is an important issue for how to auto-process, understand and explain the massive image data during the whole process of society informatization, therefore, it has become a key issue to extract the image features from the remote sensing images. The feature extraction consists of three parts, including areal feature extraction, linear feature extraction and point-like feature extraction. The basic characteristic of the areal feature extraction and linear feature extraction is expressed by the edge of straight line. The edge detection technique of remote sensing image has play a important role in the fields of aviation, aerospace, military reconnaissance, disaster forecasting and other fields, and the key issue is to solve the conciliation between the accuracy of edge detection and the performance of noise immunity.

D. Zeng (Ed.): Advances in Computer Science and Engineering, AISC 141, pp. 807–814.
Springerlink.com

At present, the widespread edge detection algorithms, such as Roberts edge detection, Prewitt edge detection, Sobel edge detection and Marr edge detection, detect the edge using the variation of one-order or two-order directional derivative near to the edge usually. As a result, the noises can be detected in the image regarding as the edge points, and the real edge maybe missed due to noise interference.

The Robert gradient operator is just to calculate the gray difference of adjacent pixels with the pixel range of 2×2, however, it is sensitive for noise, and it cannot suppress the effect of noise. In order to reduce the effect of noise during the edge detection, Prewitt operator enlarges the size of template of the detection operator from 2×2 to 3×3 to calculate the difference operator. Based on the Prewitt operator, Sobel adopts the method with weight to calculate the difference for 4 – neighborhood which not only can detect the edge point, but also can further suppress the effect of noise. Marr-Hildreth operator is completed based on the Laplace operator which has an important biological and physiological significance. Because of the sensitivity for noise of Laplace operator, in order to reduce the effect of noise, firstly, the image is smoothed with normal Gaussian distribution function, and then the Laplace operator is adopted to detect the edge. Because the remote sensing images have the disadvantage of high noise, thus the above algorithms are difficult to obtain the better results.

2 Edge Detection If Rs Images Based on Canny Operator and Hough Transform

2.1 Canny Operator

Recently, canny edge detection operator is widely used in the field of digital image process, which has the good detecting performance, good positioning performance and the uniqueness of edge locating, but it is also easy to detect the meaningless edge. Edge strength and edge gradient direction of the digital image can be calculated with the canny operator which can service for the later edge points, and it can be described as Equation (1).

$$G_n = \nabla G; \qquad n = \frac{\nabla G * f(x,y)}{|\nabla G * f(x,y)|}; \quad \nabla G = \begin{bmatrix} \frac{\partial G}{\partial x} \\ \frac{\partial G}{\partial y} \end{bmatrix} \tag{1}$$

n is the edge direction, ∇G is gradient vector, and ∇G*f(x,y) is edge strength.

Utilizing the separability of Gaussian functions, two convolution filter templates of ∇G are decomposed as two one-dimensional rank filter.

$$\frac{\partial G}{\partial y} = kx\exp\left(\frac{-y^2}{2\sigma^2}\right)\exp\left(-\frac{x^2}{2\sigma^2}\right);$$

$$\frac{\partial G}{\partial x} = kx\exp\left(\frac{-x^2}{2\sigma^2}\right)\exp\left(-\frac{y^2}{2\sigma^2}\right) \tag{2}$$

Convoluted Equation (2) with the image $f(x, y)$, we can get:

$$E_x = \frac{\partial G}{\partial x} * f(x, y); \qquad E_y = \frac{\partial G}{\partial y} * f(x, y) \tag{3}$$

$$A(i, j) = \sqrt{E_x^2(i, j) + E_y^2(i, j)}; a(i, j) = \arctan\left[\frac{E_y(i, j)}{E_x(i, j)}\right] \tag{4}$$

$A(i, j)$ reflects the edge intensity of the point (i, j) on the image, $a(i, j)$ is the normal vector of the point (i, j) on the image. The pixel can be regarded as the edge points of the image satisfying the following three conditions: (1) the edge intensity of the point is bigger than the edge intensity of the two adjacent pixel points along the gradient direction, (2) the angle difference between the two points is smaller than 45°, and (3) the maximum edge intensity of the 3 x 3 in the field is less than a threshold value.

2.2 Hough Transform

The Hough transform is a technique which can be used to isolate features of a particular shape within an image. Because it requires that the desired features be specified in some parametric form, the classical Hough transform is most commonly used for the detection of regular curves such as lines, circles, ellipses, etc. The purpose of the technique is to find imperfect instances of objects within a certain class of shapes by a voting procedure. This voting procedure is carried out in a parameter space, from which object candidates are obtained as local maxima in a so-called accumulator space that is explicitly constructed by the algorithm for computing the Hough transform. The typical transform is the polar coordinate equation using straight line shown as follows:

$$\rho = x \cos\theta + y \sin\theta \tag{5}$$

According to the above method, the point on the two value image corresponds to a sine curve in the new parameter space, namely a sinusoid duality. Let many points on the planar, each point on the straight line is respectively corresponding to the polar coordinates of a sine curve. If the sine curves have the same intersection point (ρ', θ'), then, the corresponding line equation can be described as follows:

$$\rho' = x \cos\theta' + y \sin\theta' \tag{6}$$

Let the width of the image is w and the height is h, the value range of ρ is $[-\sqrt{w^2 + h^2}, \sqrt{w^2 + h^2}]$ in the polar coordinates space, the value range of θ is $[-90^\circ, 90^\circ]$.

2.3 The Method Integrated the Canny Operator and Hough Transform

Canny operator and Hough transform can be said to be have the different advantages, and if we combined the two methods, we can not only anti noise but also can detect some discontinuous edge which can be describes as three steps: (1) extract the edges of the image, hidden the non main information of the image and keep the edge information of the representative target feature. The procedure is mainly based on the remote sensing image imaging principle, gray distribution features, combining the Canny operator, and to acquire the straight line on the processed image, (2) according the characteristics of the objects, select the number of the target and position accuracy using the Hough transform, and further to detect the straight line based on the edge image from the Canny operator, and (3) let the endpoint of the detected straight line by Hough transform as a seed growing point, and further to make the connected processing and labeling for the edge image by Canny operator.

3 Experiment Results and Analysis

In the paper, the IKONOS high resolution image data of the southern suburbs of Xi'an province is selected as the experimental image data as shown in Figure 1(a), the size is 286 pixels x 347 pixels and the resolution is 4 m. the purpose of the experiment is to extract the edge information of the road from the image. The detected result by Roberts operator is shown in Figure 1(b) which contains much more noise, and the edges have obvious discontinuity, not continuous. The detected result by Sobel operator is shown in Figure 1(c) which also contains much more noise, but the edge discontinuity has improves a lot, continuity of slightly. The detected result by Marr operator is shown in Figure 1(d) which contains little noise, and the edge discontinuity also has improves a lot, continuity of slightly, but some false edge becomes more obvious. Figure 1(f) shows the detected straight line by Hough transform which detect the all major roads and not for the false edge. Figure 1(f) shows the detected edge image using the proposed method in the paper which almost does not contain noise and has the better continuity. Figure 1(h) shows overlay image by the detected edge and original image,

(a) Original image

Fig. 1. Comparison among the different edge detection operator results images

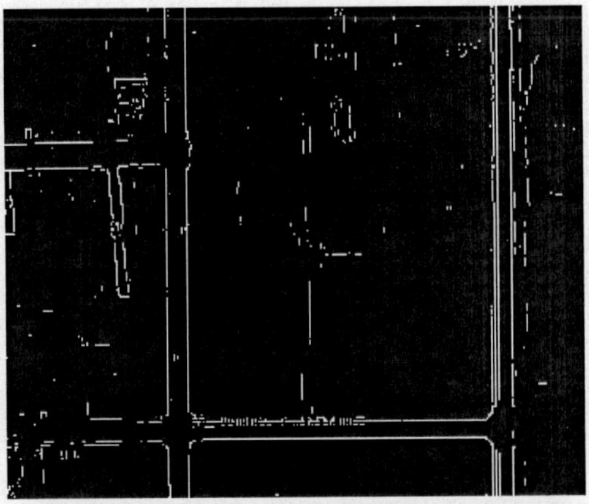

(b) Detected edge image by Roberts operator\

(c) Detected edge image by Sobel operator

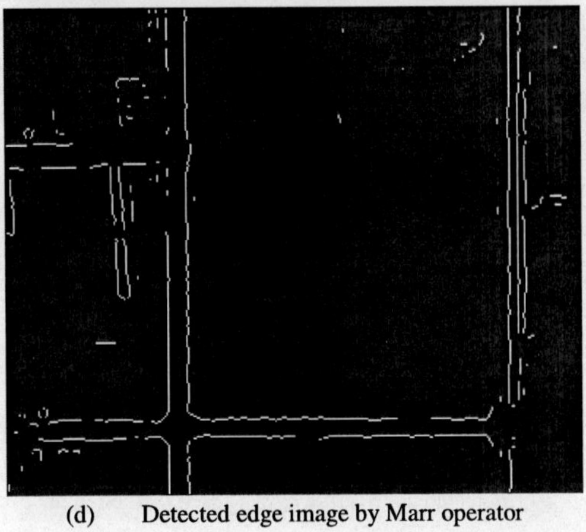

(d) Detected edge image by Marr operator

Fig. 1. *(continued)*

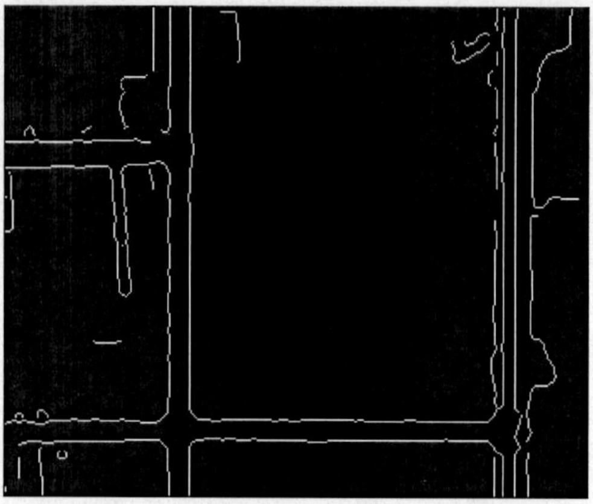

(e) Detected edge image by Canny operator

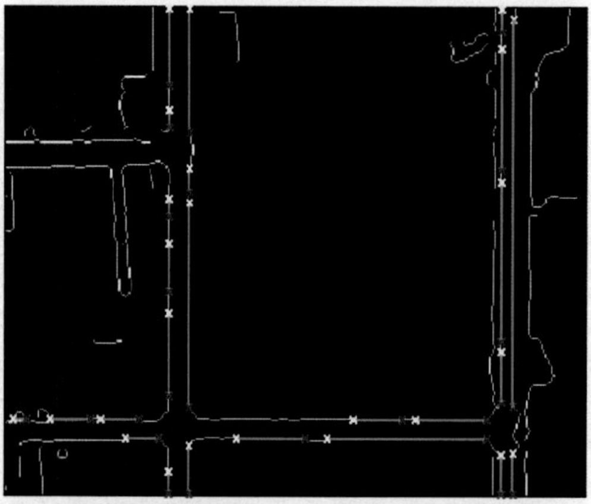

(f) Detected edge image by Hough transform

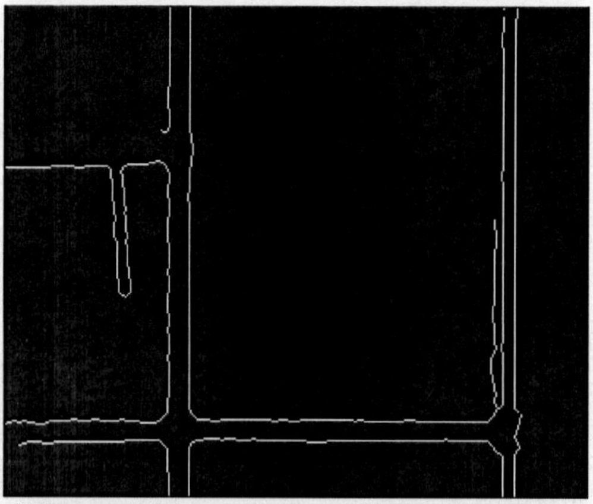

(g) Detected edge image by proposed method in the paper

Fig. 1. *(continued)*

(h) Overlay image

Fig. 1. *(continued)*

4 Conclusions

This paper proposes a new edge detection method for remote sensing image based on the Canny operator and Hough transform according to the characteristics of remote sensing image like rich edge information and mixed with a lot of noise. From the experimental results, we can make the conclusion that the proposed method to detect the edge information the remote sensing image has a certain noise resistance, and can inhibit the false edge effect, and can acquire the edge information accurately

The edge detection accuracy and the anti-noise performance is a pair of contradictions, the proposed method combined the Canny operator and Hough transform not only utilize the higher position ability of Canny operator, but also use the anti-noise characteristic of Hough transform. The propose method is validated from the experiment for the IKONOS high resolution image and comparison among the different edge detection operator results images.

References

1. Shi, W.Z., Zhu, C.Q., Wang, Y.: Road Feature Extraction from Remotely Sensed Image: Review and Prospects. Acta Geodaetica Et Cartographica Sinica 30(3), 257–262 (2001)
2. Shen, Y.: Research on edge detection technique based on morphological. University of Electronic Science and Technology (master thesis), Chengdu (2008)
3. Yang, S.B.: Introduction of image edge detection technology. Journal of Wuhan Institute of Chemical Technology 22(1), 73–76 (2003)
4. Zhou, S.G., Xu, Y.: To Extract Roads with No Clear and Continuous Boundaries in RS Images. Acta Geodaetica Et Cartographica Sinica 37(3), 301–307 (2008)
5. Jia, Y.H.: Digital image process. Wuhan University Press, Wuhan (2004)

6. Wang, Z.S., Li, C.H.: Improved Classical Hough Transform Applied to Building Detection and Recognition. Journal of Image and Graphics 10(4), 463–467 (2005)
7. Wu, Z.C., Wan, Q.T., Liang, J., Zhou, Z.: Line Detection in Remote Sensing Images Using Hough Transform Based on Granular Computing. Geomatics and Information Science of Wuhan University 32(10), 860–863 (2007)
8. Thomas, M.L., Ralph, W.K.: Remote Sensing and Image Interpretation. Publishing House of Electronics Industry, Beijing (2003)

Author Index